Pierre Dustin

Microtubules

With a Foreword by K. R. Porter

With 177 Figures

Springer-Verlag
Berlin Heidelberg New York 1978

Prof. Dr. Pierre Dustin

Laboratoires d'Anatomie Pathologique et de Microscopie Electronique, Université Libre de Bruxelles, rue aux Laines 97, B-1000 Bruxelles

ISBN 3-540-08622-6 Springer-Verlag Berlin Heidelberg New York
ISBN 0-387-08622-6 Springer-Verlag New York Heidelberg Berlin

Library of Congress Cataloging in Publication Data. Dustin, Pierre. Microtubules. Includes bibliographical references and index. 1. Microtubules. I. Title. QH603.M44D87 574.8'74 77-27968.

Printed in Germany.

Reproduction of the figures: Gustav Dreher GmbH, Stuttgart

Typesetting, printing and bookbinding: Konrad Triltsch, Graphischer Betrieb, D-8700 Würzburg

2131/3130-543210

Foreword

The author of this remarkably comprehensive review, PIERRE DUSTIN, has performed an invaluable service in bringing together in one volume the observations and theory on microtubules that have accumulated over the last fifteen years. He has understood the magnitude of the task from the beginning and has met it thoroughly and, I must say, courageously. From here on, and for many years to come, young investigators, and some not so young, will have a ready reference as they seek to discover what has been done and remains to be done in achieving a better understanding of these important cell components.

Since the early 1960's when it became clear that the filaments (microtubules) making up the 9+2 complex of cilia and flagella were to be found very widely in cells, microtubules have attracted an ever-increasing amount of attention. Now it is known, as reviewed in this volume, that they influence the morphogenesis of anisometry in the shapes of cells and cell extensions; that they function as frames for the intracellular movement of granules and chromosomes; that they are subject to control in their assembly and disassembly by externally applied substances such as colchicine and cyclic AMP; and that they display abnormalities in their numbers and orientations in transformed (malignant) cells, to mention only a few of their several functions. Much less is known about the normal control over their disposition in cells and how they engage and interact with the cytoplasmic matrix to bring about its various translocations. This much is certain, the study of these and other questions of microtubule science will be greatly facilitated by the existence of this volume. It is to be hoped that Professor DUSTIN will find the time and enthusiasm to produce successive editions.

KEITH R. PORTER

Contents

Chapter 3 General Physiology of Tubulins and Microtubules

Chapter 4 Microtubule Structures: Centrioles, Basal Bodies, Cilia, Axonemes

Chapter 5 Microtubule Poisons

Chapter 6 Cell Shape

Chapter 10 The Role of MT in Mitosis

Chapter 11 Pathology and Medicine

Chapter 12 Outlook

Introduction

Where is the wisdom we have lost in knowledge?
Where is the knowledge we have lost in information?

T. S. Eliot
The Rock.

"There are no authorities left, since only those who have no use for their knowledge can acquire it: doubtless a general predicament in all natural sciences these days."

F. Chargaff
The Sciences 15 : 21 – 26, 1975

It is indispensable to define the purpose and limitations of writing a monograph on a subject as vast as that of microtubule research. The name "microtubule" was coined by Slautterback some 14 years ago [23], and already in 1966 the ubiquity of these structures was mentioned in a fundamental article written by Porter [21]. Microtubules (MT) are present in all eukaryotic cells with the single exception of the anucleated red blood cells of most mammals. They appeared about one billion years ago, at the same time as the nucleus, marking the turning point form pro- to eukaryotes. They have maintained since that period of time a remarkable constancy of structure and of chemical composition. Their discovery was a result of electron microscopy and of the advent of better fixation and embedding procedures. They had however been suspected by microscopists since the end of last century, in the form of the marginal bundle of erythrocytes, the neurofibrils, the mitotic spindle fibers, and the complex structures of cilia and flagella.

The study of MT is closely linked with that of a few drugs which are known to combine specifically to some sites of their constitutive molecules—the tubulins. The first of these is colchicine, known since 1889 as a poison of mitosis [20], rediscovered as such in 1934, and which was the subject of a monograph in collaboration with O. J. Eigsti, published in 1955 [11]. At that time colchicine, apart from its very ancient use as a treatment in acute crises of gout, was mainly studied for its destructive and specific action on the mitotic spindle, leading to the arrest of cell divisions at metaphase—the so-called "stathmokinetic" effect [7]. This had numerous applications in various biological fields, the most notable being the use of colchicine for the production of artificial polyploid or amphidiploid species of plants (such as *Triticale*), and for the amplification of the mitotic index in the germinative zones of animals—the "colchicine method", which has been supplanted by the use of tritiated thymidine. Another important consequence of these studies was the demonstration in 1956 [31] of the exact number of chromosomes in man, and the subsequent use of colchicine in most studies of modern cytogenetics. In 1955, only a small number of actions of colchicine on non-dividing cells were mentioned, and mainly considered as non-specific toxic side-effects.

The high degree of specificity of colchicine was emphasized, as no other chemical appeared to be able to destroy the spindle filaments at such low concentrations. It was mentioned that "... any work which helps to solve the problem of spindle inactivation by this complex molecule may throw more light on the physiology of the peculiar fibrous protein which constitutes the spindle". We know now that this protein, called *tubulin* since 1968 [1], is present in all cells, and that its participation in the movements of chromosomes at mitosis is only one of the large number of its activities, linked with the determinism of cell shape, cell movements, secretion, and growth.

Other powerful poisons of tubulin were discovered in the early sixties, and the alkaloids extracted from *Catharanthus roseus* (*Vinca rosea*) [29] have demonstrated an equally specific action on tubulins, and moreover have proved, contrary to colchicine, to be most useful in the treatment of cancer (vinblastine, vincristine). The greatest change in perspective since twenty years ago has however been the demonstration that the tubulin poisons not only affect mitosis, but are remarkable tools for the analysis of many cellular activities. This has been the consequence of the discovery of MT, and this discovery is closely linked with colchicine, as the use of tritiated colchicine, in the hands of Taylor and his collaborators, between 1965 and 1968, was to provide the tool necessary for the isolation and purification of the receptor protein, *tubulin* [1, 4, 28, 29].

The growth of research in this field has been and remains momentous, and papers by the hundred have been published each year in the last decade. It may appear unsound to attempt to write a monograph on the subject of MT, especially if it is planned to cover most aspects of these structures. The problem of dealing with the flow of information is considerable, but some kind of synthesis should be attempted.

Having covered twenty years ago the literature on colchicine, which was at that time already extensive, and having written several reviews on this subject and on MT [8, 9, 10], I felt that a synthesis of the works scattered in many periodicals was possible. It was also a help to have lived all the modern history of the subject, from the early work of A. P. Dustin and F. Lits, to the recent progress of MT chemistry and physiology. All these developments have shown that MT are fundamental organelles and that their importance for cell biology is at least as great as that of structures such as mitochondria or lysosomes. Their multitudinous functions show, in all eukaryotes, some constant features which become more apparent when all aspects of their activity are compared. This study must comprize the complex structures made of assembled MT—with other specific proteins—such as the centrioles, the cilia and flagella, the axonemes and other specialized organelles.

The ubiquity of MT demands that cells of all types be considered. The limitations of space, and the trends of recent work, will put more stress on animal than on plant cells. Unicellulars will be studied in several chapters, for they provide a wealth of information on the possibilities of MT assemblies, which were already suspected by protistologists long before the term microtubule was coined [12].

The history of MT research began with medicine, as *Colchicum* had been known since the end of the 18th century—and probably already in antiquity—as a cure for articular pain and in particular for gout, a position it still holds today. The discovery of the properties of the alkaloids of *Catharanthus* has brought to medicine at least two remarkable drugs, vinblastine and vincristine. Their action is closely related to their fixation on specific receptor sites of the tubulin molecule. More recently, many studies on the action of colchicine and on the medical uses of the *Vinca* alkaloids and some other MT poisons in gout and also in other inflammatory diseases, have been published, and Chapter 11 will be devoted to some medical problems related to MT functions.

In the last few years, several conferences and symposia have been devoted to MT and problems related with MT activity: the most important are the conference organized by the New York Academy of Sciences in 1974 [25] and the symposium which took place in Belgium in 1975 [3]. This was followed by a most important conference, held at Cold Spring Harbor in September 1975, and which covered in detail all the aspects of cell motility, comprising the role of MT in cell structure and movement [13]. This contribution completes the book edited in 1975 by Inoué and Stephens, and devoted mainly to cell motility [16]. Several other reviews of problems related to MT have been published since 1970, and should be mentioned here [2, 3, 5, 6, 14, 15, 17, 18, 19, 22, 23, 24, 26, 27, 30, 32].

Considering this wealth of information, one may wonder whether one more book was necessary. However, few reviewers have attempted to cover the whole field of MT research, from fundamental data to medical applications. The New York meeting neglected cilia and flagella, while Cold Spring Harbor was not concerned with the role of MT either in secretion or in neuroplasmic flow. These subjects may seem far apart, but the MT provide the link which leads to a synthetic understanding of all these problems. It is hoped that this may indicate which functions of MT are constant and which are not, and lead to a clear definition of the role, in all cells, of these organelles.

The literature covered, which numbers several thousand references, is mainly that of the last ten years. It extends to the end of 1976, with a few exceptions for 1977.

Although the tempo of new publications on this subject remains high, many important problems related to tubulins, their assembly, their control, their action, still await a solution. It is thought however that the time is ripe for an overall review. This will aim, as far as possible, to emphasize the unity of life through the variety of the structures made of tubulins and of cell functions which depend on their integrity.

The decision to write this book alone was reached with the purpose of giving the greatest unity to the text, and with the hope of completing a manuscript in the shortest time necessary, when the subject is moving fast and new papers are published each day. The author is well aware of the great dangers of this decision, and takes full responsibility for the errors which may slip into a book covering so many aspects of cell biology. He hopes that through descriptions of complicated facts and events, some idea of the marvellous adaptations of the living cell may become apparent.

References

1. Adelman, M. R., Borisy, G. G., Shelanski, M. L., Weisenberg, R. C., Taylor, E. W.: Cytoplasmic filaments and tubules. Fed. Proc. **27:** 1186 – 1193 (1968)
2. Bardele, S. F.: Struktur, Biochemie und Funktion der Mikrotubuli. Cytobiologie **7:** 442 – 487 (1973)
3. Borgers, M., De Brabander, M. (eds.): Microtubules and Microtubule Inhibitors. Amsterdam-Oxford: North-Holland Publ. Co. 1975, New York: American Elsevier Publ. Co., Inc. 1975
4. Borisy, G. G., Taylor, E. W.: The mechanism of action of colchicine. Binding of colchicine-^{3}H to cellular protein. J. Cell Biol. **34:** 525 – 534 (1967)

5. Bryan, J.: Microtubules. Bioscience **24:** 701 – 711 (1974)
6. Burnside, B.: The form and arrangement of microtubules: an historical, primarily morphological review. Ann. N.Y. Acad. Sci. **253:** 14 – 26 (1975)
7. Dustin, A. P.: L'action des arsenicaux et de la colchicine sur la mitose. La stathmocinèse. C. R. Assoc. Anat. **33:** 204 – 212 (1938)
8. Dustin, P. Jr.: New aspects of the pharmacology of antimitotic agents. Pharmacol. Rev. **15:** 449 – 480 (1963)
9. Dustin, P. Jr.: Microtubules et microfilaments: leur rôle dans la dynamique cellulaire. Arch. Biol. **83:** 419 – 480 (1972)
10. Dustin, P.: Some recent advances in the study of microtubules and microtubule poisons. Arch. Biol. **85:** 263 – 287 (1974)
11. Eigsti, O. J., Dustin, P. Jr.: Colchicine, in Agriculture, Medicine, Biology, Chemistry. Ames, Iowa: The Iowa State College Press 1955
12. Fauré-Fremiet, F.: Microtubules et mécanismes morphopoiétiques. Année Biol. **9:** 1–61 (1970)
13. Goldman, R., Pollard, T., Rosenbaum, J. (eds.): Cell Motility. Cold Spring Harbor Laboratory 1976
14. Hepler, P. K., Palewitz, B. A.: Microtubules and microfilaments. Ann. Rev. Plant Physiol. **25:** 309 – 362 (1974)
15. Hepler, P. K.: Plant microtubules. In: Plant Biochemistry (3rd ed.) (eds.: J. Bonner, J. F. Varner). New York: Academic Press 1976
16. Inoué, S., Stephens, R. E. (eds.): Molecules and Cell Movement. Soc. of Gen. Physiologists Series, Vol. 30. New York: Raven Press; Amsterdam: North-Holland Publ. Co. 1975
17. Margulis, L.: Colchicine-sensitive microtubules. Intern. Rev. Cytol. **34:** 333 – 361 (1973)
18. O'Brien, T. P.: The cytology of cell-wall formation in some eukaryotic cells. Bot. Rev. **38:** 87 – 118 (1972)
19. Olmsted, J. B., Borisy, G. G.: Microtubules. Ann. Rev. Biochem. **42:** 507 – 540 (1973)
20. Pernice, B.: Sulla cariocinesi delle cellule epiteliali e dell'endotelio dei vasi della mucosa dello stomaco e dell'intestino, nelle studio della gastroenterite sperimentale (nell'avvelenamento per colchico). Sicilia Med. **1:** 265 – 279 (1889)
21. Porter, K. R.: Cytoplasmic microtubules and their functions. In: Ciba Foundation Symposium on Principles of Biomolecular Organization. London: J. and A. Churchill 1966
22. Roberts, K.: Cytoplasmic microtubules and their functions. Progr. Bioph. Mol. Biol. **28:** 373 – 420 (1974)
23. Slautterback, D. B.: Cytoplasmic microtubules. I. Hydra. J. Cell Biol. **18:** 367 – 388 (1963)
24. Snyder, J. A., McIntosh, J. R.: Biochemistry and physiology of microtubules. Ann. Rev. Biochem. **45:** 699 – 720 (1976)
25. Soifer, D. (ed.): The Biology of Cytoplasmic Microtubules. Ann. N. Y. Acad. Sci., Vol. 253. New York 1975
26. Stephens, R. E.: Microtubules. In: Biological Macromolecules (eds.: S. N. Timasheff, G. D. Fasman), Vol. 4, pp. 355 – 391. New York: Dekker 1971
27. Stephens, R. E., Edds, K. T.: Microtubules: Structure, chemistry, and function. Physiol. Rev. **56:** 709 – 777 (1976)
28. Taylor, E. W.: The mechanism of colchicine inhibition of mitosis. I. Kinetics of inhibition and the binding of H^3-colchicine. J. Cell Biol. **25:** 145 – 160 (1965)
29. Taylor, W. I., Fransworth, N. R.: The Catharanthus Alkaloids. New York: Dekker 1975
30. Tilney, L. G.: Origin and continuity of microtubules. In: Origin and Continuity of Cell Organelles (eds.: J. Reinert, H. Ursprung). Heidelberg: Springer 1971
31. Tjio, J. H., Levan, A.: The chromosome number of man. Hereditas (Lund) **42:** 1 – 6 (1956)
32. Wilson, L., Bryan, J.: Biochemical and pharmacological properties of microtubules. Adv. Cell Mol. Biol. **3:** 21 – 72 (1974)

Acknowledgments

The task of preparing, in a relatively short time, the manuscript and the figures of a book covering so many fields of biology would have been impossible without the facilities available at the University of Brussels. My warmest thanks go to my secretary, Mrs. D. Libert, who not only typed the final manuscript, but also checked my text for inaccuracies or repetitions, and helped considerably in the preparation of the lists of references. Throughout two years, while performing all the menial jobs of secretary of the Laboratory of Pathology, she remained helpful and smiling, although the number of folders was increasing, and many last-minute changes were necessary. Without her activity, it would have been impossible to meet the deadline imposed by the publisher.

The preparation of the illustrations is another important work for a book like this. Mr. R. Fauconnier, Head of the Iconography Department of the Medical School, made the drawings, and prepared the photographic illustrations. His help has made it possible to present the multiple aspects of mictotubule morphology with the greatest clarity for the reader.

Several microtubule specialists gave me their advice during the years of writing. I should like to single out J. R. McIntosh, who took the time to read over the chapter on mitosis, and suggested many interesting changes. Many authors helped me also with important suggestions during the preparation of the figures.

I should thank also all the collaborators of this laboratory, who enabled me to spend so much time in front of my typewriter. I am grateful also to those who provided me with the opportunity to attend, during the preparation of this book, several international meetings where I met many of the research workers whose names are repeatedly mentioned in the lists of references: the New York Academy of Sciences meeting in 1974; the symposium organized at Beerse in Belgium by Janssen Pharmaceutica in 1975; the Cold Spring Harbor Conference on Cell Motility in the fall of 1976; the First International Congress on Cell Biology in Boston in 1976; the workshop on mitosis at the Deutsches Krebsforschungszentrum in Heidelberg in 1977.

I should also like to mention here the help provided by the Institute of Scientific Information (Philadelphia), through *Current Contents, Life Sciences,* which has become an essential tool for research.

Many authors helped me by sending on request some of their best photographic documents, or by authorizing me to have some documents redrawn. Without them, this book could not have been published. Illustrations from work done in this laboratory were provided by my collaborators Mrs. J. Flament-Durand (Figs. 2.1, 4.9, 9.1, 9.2, 9.4, 9.6), Mrs. P. Ketelbant-Balasse (Fig. 8.4), Mrs. D. Derks-Jacobovitz (Fig. 4.8), Dr. J.-P. Hubert (Fig. 5.23) and Mrs. A. Anjo (Fig. 10.17).

During the preparation of this book, I was supported by grants from the Belgian National Fund for Scientific Research (F. N. R. S.), the Belgian National Fund for Medical Research (F. R. S. M.; grants no. 20.472 and 3.4539.76), and by the Rose et Jean Hoguet Foundation for Cancer Research (Brussels).

Last but not least, I wish to thank all those at Springer-Verlag, whose kind help and excellent advice were indispensable for bringing to a favorable end the publication of this book.

Permission for reproduction of figures was obtained from the following periodicals:
Anatomical Record (Fig. 5.8)
Annals of the New York Academy of Sciences (Figs. 2.6, 4.10, 5.4, 8.1, 9.5)
Annotationes Zoologicae Japonenses (Fig. 5.16)
Archiv für Protistenkunde (Figs. 3.9, 3.10, 3.17)
Archives of Internal Medicine (Fig. 11.3)
Arthritis and Rheumatism (Fig. 11.1)
Biochimica et Biophysica Acta (Fig. 2.25)
Brain Research (Fig. 9.9 b)
British Journal of Haematology (Fig. 11.2)
Cancer Bulletin (Fig. 11.4)
Cancer Research (Fig. 5.20)
Cell and Tissue Research (Fig. 4.25)
Chromosoma (Fig. 10.16)
Cytobiologie (Fig. 10.1)
Cytobios (Fig. 10.6)
Developmental Biology (Figs. 3.14, 4.4, 6.8, 6.9, 6.16)
Diabetes (Figs. 8.2, 8.3)
Experimental Cell Research (Figs. 2.3, 2.9, 2.24, 5.15, 7.8)
International Review of Experimental Pathology (Fig. 6.5)
Journal of Cell Biology (Figs. 2.2, 2.5, 2.12, 2.15, 2.27, 3.1, 3.4, 3.5, 3.11, 4.5, 4.6, 4.15, 4.16, 4.18, 4.23, 5.7, 5.19, 5.21, 6.7, 6.13, 7.7, 9.8, 10.3, 10.4, 10.10, 10.15)
Journal of Cell Science (Figs. 3.8, 3.12, 3.13, 4.7, 4.11, 4.12, 4.17, 5.17, 6.2, 6.15, 6.18, 10.2)
Journal of Experimental Medicine (Fig. 7.4)
Journal of Experimental Zoology (Fig. 4.20)
Journal of Molecular Biology (Figs. 5.5, 5.22)
Journal of Morphology (Fig. 10.5)
Journal of Protozoology (Fig. 3.16)
Journal of the Reticuloendothelial Society (Fig. 3.3)
Journal of Supramolecular Structure (Fig. 2.14)
Journal of Ultrastructure Research (Figs. 2.7, 2.26, 3.18, 6.1, 6.10, 6.11, 9.3)
Journal of Virology (Fig. 3.7)
Metabolism (Fig. 5.9)
Nature (Fig. 2.17)
Protoplasma (Figs. 4.2., 10.12)
Tissue and Cell (Fig. 4.19)
Triangle (Sandoz Journal of Medical Sciences) (Fig. 6.3)
Zeitschrift für Zellforschung (Figs. 4.26, 6.6)

Authorization has been granted for the reproduction of the following figures from books:

Borgers, M., De Brabander, M. (eds.): Microtubules and Microtubule Inhibitors. Amsterdam: North-Holland 1975 (Figs. 7.5, 10.14)
Busch, H. (ed.): Methods in Cancer Research. Vol. XI. New York: Academic Press 1975 (Fig. 10.13)
Dyck, P. J., et al. (eds.): Peripheral Neuropathy. Philadelphia: W. B. Saunders 1975 (Fig. 9.7)
Eigsti, O. J., Dustin, P: Colchicine—in Agriculture, Medicine, Biology and Chemistry. Ames, Iowa: The Iowa State College Press 1955 (Fig. 6.17)
Ghadially, F. N.: Ultrastructural Pathology of the Cell. London: Butterworths 1975 (Fig. 4.24)
Goldman, R., et al. (eds.): Cell Motility. Cold Spring Harbor Laboratory 1976 (Figs. 2.8, 2.13, 2.16, 2.18 a, b, c, 2.22, 2.23, 2.28, 5.11)
Grell, K. G.: Protozoology. Berlin, Heidelberg, New York: Springer 1973 (Fig. 7.6)
Inoué, S., Stephens, R. E. (eds.): Molecules and Cell Movement. New York: Raven Press 1975 (Figs. 2.14, 10.7)
Little, M., et al. (eds.): Mitosis: Facts and Questions. Berlin, Heidelberg, New York: Springer 1977 (Fig. 5.14)
Sleigh, M. A. (ed.): Cilia and Flagella. London: Academic Press 1974 (Fig. 7.3)

Chapter 1 Historical Background

1.1 Microtubules (MT)

1.1.1 Definition

Although a correct understanding of the structure and function of MT cannot be reached until several chapters later, it may be useful for the reader to start with the following definition: *"Microtubules are proteinaceous organelles, present in nearly all eucaryotic cells, made of subunits assembled into elongated tubular structures, with an average exterior diameter of 24 nm and an indefinite length, capable of rapid changes of length by assembly or disassembly of their subunit protein molecules or tubulins, sensitive to cold, high hydrostatic pressures and some specific chemicals such as colchicine and vinblastine, building, with other proteins, complex assemblies like mitotic spindle, centrioles, cilia and flagella, axonemes, neurotubules, and intervening in cell shape and cell motility".*

1.1.2 Early Observations

While MT had been observed as filamentous structures by light microscopists a century ago, a proper understanding of their structure had to await the electron microscope and the improvements of fixation procedures (glutaraldehyde) and observation techniques (epoxy embedding, negative staining).

In the nucleated red blood cells of cold-blooded vertebrates and birds, a fibrillar peripheral structure known as the *marginal band* (or *bundle*) had already been observed by Ranvier [71] and well described by Meves [57; Fig. 1.1]. After staining with gentian violet, and better after a "fixation" with nitric acid, this bundle, which was considered an "elastic zone maintaining the cytoplasm under tension", was seen to be formed of minute parallel concentric fibrils [43, 44].

In the anucleate red blood cells of mammals, this differentiation was missing, except in camelidae: Jolly, who described the erythrocytes of the llama [43], suggested that a relationship may exist between the elliptic shape of the cells in this group of mammals and the marginal bundle—linking for the first time the fibrils with the preservation of cell shape (cf. Chap. 6). A similar structure was found later to support the disk shape of mammalian blood platelets [5].

Other structures peculiar to mammalian red blood cells and related to MT are the rings of Cabot, which have long been known in severe anemias. These are filaments, stained violet by the usual panoptic methods, which appear in blood smears as ring-shaped or figure-of-eight inclusions in adult red blood cells. They

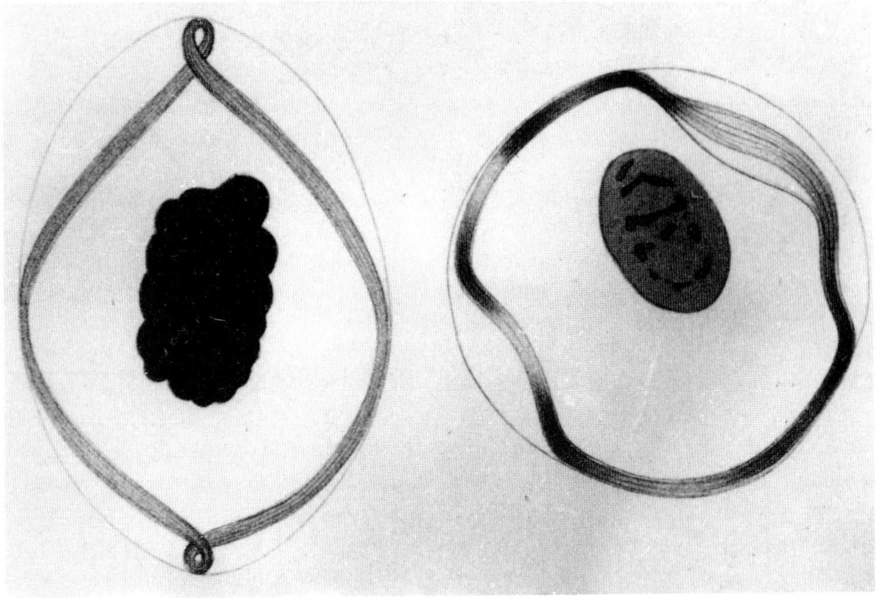

Fig. 1.1. Red blood cells of *Salamandra maculosa.* Gentian violet staining. The cell on the *right* has been treated with a 3% solution of NaCl, and its marginal bundle is twisted. As the hemoglobin is not present, as a consequence of hemolysis, the fibrillar substructure of the marginal bundle is apparent (from Meves [57])

stain the same colour as the fibrils which often link couples of post-telophasic erythroblasts in the bone marrow [8] and which are spindle remnants—the so-called "telophasic body" made of closely packed MT [88; cf. Chap. 10].

The nerve cells contain a variety of fibrils, and these have been known since the early works of Apathy [1] and Bethe [9]; these *neurofibrils,* which were seen in the cytoplasm and the processes of the nerve cells (axons and dendrites), were described as a "plexus of fine filaments". These filaments became fewer and fewer as the dendrites bifurcated. The use of silver nitrate techniques gave pictures of a plexiform network, and the nature of these structures remained mysterious until the advent of electron microscopy.

The *mitotic spindle* was known as a fibrillar structure, while the reality of the fibers was strongly disputed for many years [79]. The spindle birefringence [78], the isolation of the mitotic apparatus [56], the fact that the birefringence disappeared under high hydrostatic pressures [53, 54] while the spindle size decreased, were indications of a dynamic fibrillar substructure.

Complex fibrillar structures associated with *cilia* and *centrioles,* and often linked with the mitotic apparatus, had been studied for years by protistologists (cf. 15). They were seen to play various roles, either maintaining the complex shape of some cells, or intervening in cell motility and in the displacements of chromosomes at mitosis and meiosis. Although *basal bodies* and *centrioles* had been known since the last century, and their relations with cilia and mitotic spindles described, no unifying thread was apparent before ultrastructural observations.

1.1.3 First Ultrastructural Observations

Although the word "microtubule" was proposed for the first time in 1961 [81, 82] and immediately met with great success, several electron microscopical descriptions, before the introduction of glutaraldehyde, epoxy embeddings, and even the staining of sections, had shown the tubular nature of some of the fibrils described by earlier microscopists. In one of the first thorough studies of the structure of cilia and basal bodies, Fawcett and Porter [36] not only clearly demonstrated the 9+2 structure of the cilia, but also illustrated the doublet nature of the peripheral "fibrils". About the structure of these fibrils, they wrote cautiously: "The ... filaments in cross section have the appearance of tiny tubules. This may be their true structure ..." The same year, Porter [69] described the spindle fibrils in an endothelioma of the rat as narrow tubules with a diameter of 25 nm. He noted correctly that these dimensions were similar to those of the subunits of cilia. In the first study of the ultrastructure of centrioles, de Harven and Bernhard [18] reached a similar conclusion: these organelles were found to be built of tubules, the diameter of which (about 20 nm) was comparable to that of the spindle fibers, which these authors clearly illustrated as having a double outline (cf. their Fig. 5). This indicates, contrary to what is sometimes written, that the spindle MT are visible after fixation by osmic acid alone, even at 4 °C. The resemblance between centrioles and basal bodies was pointed out, both organelles being linked with the formation of fibrous proteins.

The marginal bundle of erythrocytes in the toadfish was shown as early as 1959 [35] to consist of a group of tubular structures of about 30 nm in diameter, and this observation was confirmed by several authors in the following years [5].

The MT of nerve cells, often called "neurotubules", were observed by Palay in 1956, although he was first of the opinion that they belonged to the smooth endoplasmic reticulum. His description is however clear: in central nervous system neurons he describes in dendrites "numerous, long, tubular elements ... about 18 nm wide and remarkably straight" [61, 62, 63]. Rosenbluth in 1962 [74] illustrated axons "with cross sections of tubules". Sandborn *et al.* [76], two years later, gave one of the first descriptions of neurotubules in mammalian nerve cells, where they were found in axons and dendrites. In neurons, the presence of a "variable amount of darkly stained content (in their) lumen" is mentioned. In later papers (cf. 69, 70), the identification of neurotubules and MT was confirmed by many authors.

In several insects of the class diptera, the attachments of the muscle cells to the cuticule are made of specialized structures, and many MT are seen on the cytoplasmic side of these: they appear to have mainly a mechanical role (in tension) and were observed and illustrated as early as 1963 by Auber [2], who called them "tonofilaments", with a diameter of 20 – 25 nm: his figures clearly indicate their tubular structure.

Fibrillary structures have long been known to play an important role in the complex internal structures of protozoa. In 1958 Roth [75] indicated that filaments, measuring about 21 nm in diameter, and resembling the central filaments of cilia, were found in various protozoa, in close relation with the flagellae and with the membranes of vacuoles. Similar structures were found in spermatozoa of invertebrates, amphibia, and mammals, and the resemblances with "neurofilaments" indicated. These filaments showed a clear central zone. Similar structures were rapidly described in many species of protozoa (cf. 34): pharyngeal baskets of ciliates, axostyles of flagellates, axopodia of heliozoa, subcortical fibrillary structures of ciliates, cytoplasmic fibrils of trypanosomes. All these structures appeared morphologically identical with the mitotic MT of the same cells.

As MT had been described in plant cells since 1963 [48], it was evident in 1966, as indicated by the important review articles of Porter [70] and Pochon-Masson [67], that a new type of ubiquitous cell organelle had been identified. Many physiological observations had already been made at that date: role in movements of particles in the cytoplasm of protozoa, in the transport of melanin granules in melanophores, in the shaping of cells, in axoplasmic transport. It was also known that MT were destroyed by cold and high hydrostatic pressures, and that they were made of subunits of about 5 nm diameter, probably with a helical disposition and 13 subunits per turn [70]. It was also evident that the assembly and disassembly of MT was very sensitive to changes of cellular activity and that MT appeared often to be assembled at some specialized zones of the cell, such as the pericentriolar dense bodies [7].

1.2 Colchicine: a Specific MT Poison

Although many different chemicals are known to combine specifically with the subunits of MT and prevent their assembly, colchicine (and some of its derivatives) is by far the most powerful and specific. The discovery of its action on the mitotic spindle is closely linked to the discovery of MT, and all recent work has brought conclusive evidence that various actions of colchicine are mediated through MT poisoning.

1.2.1 Discovery of the Cellular Action of Colchicine

This has been told several times, and only the principal landmarks of this work will be outlined here (cf. 32). Colchicine, the active principle of *Colchicum automnale,* also found in various related plants of the liliaceae family, has been known since antiquity as a poison, and its use in medicine is probably very old—as is that of several other drugs used in popular medicine and also acting on MT, such as the resin of *Podophyllum.* The extracts of *Colchicum* were known to allay "rheumatic" joint pains, and in the 18th century a preparation known as the "eau médicinale" was largely used in Europe. Its formula was secret and it was only in 1814 that it was found that colchicum was the active principle [90]. Anton von Storck (1731 – 1803) is credited as having widely prescribed colchicum preparations for the treatment of gout, and to this day colchicine has remained one of the most specific and active drugs for alleviating the excruciating pains of acute gout [73] (cf. Chap. 11). Thomas Syndenham and Alfred Barring Garrod (who wrote in 1876 a classical work on gout) played a great role in bringing the preparations of colchicum in the Pharmacopoeia.

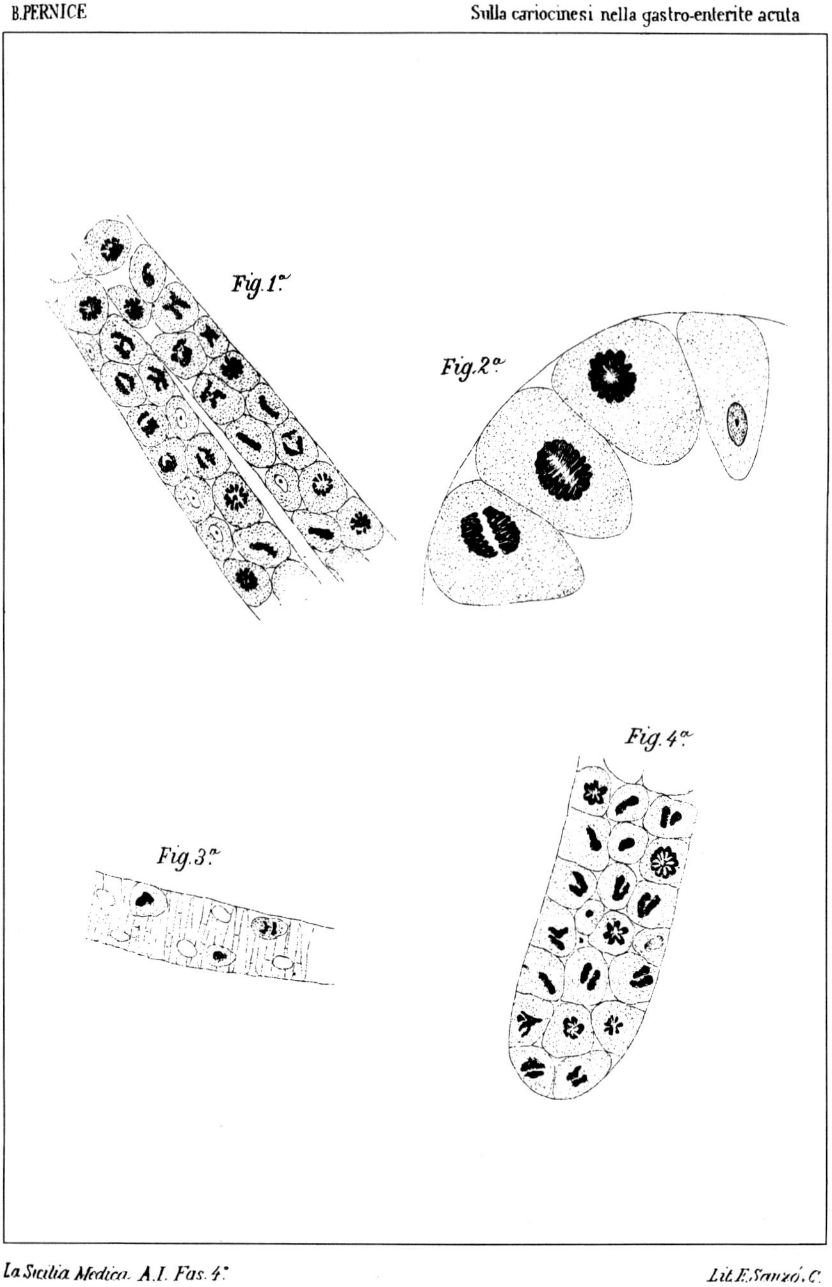

Fig. 1.2. (1), (2), (4): Accumulation of arrested mitoses in the gastric and the Lieberkühn glands of dogs, 24 and 48 h after ingestion of 10 g of tincture of colchicum. (3) shows that endothelial mitoses of the gastric mucosa are also arrested. Many metaphases are in the "star" configuration (reproduction of the plate from the paper by Pernice [65])

In 1883 the active fraction of *Colchicum* had been purified and crystallized. Believed for many years to be a three-ringed aromatic nucleus with a side-chain with a substituted amino group, its true structure, that of a tropolone derivative, with two seven-membered rings, was only elucidated in 1940 [16].

During the 19th century, many cases of colchicine poisoning were reported, and one of the signs of toxicity was diarrhea and intestinal ulcerations. This led in 1889 a Sicilian author, Pernice [65], to study experimentally the action of colchicine in dogs, and to the discovery that the most striking changes were observed in the germinative zones of the small intestine (Fig. 1.2). This contribution was to remain forgotten until 1949 [32].

Pernice studied two dogs who died after the ingestion of 10 g of tincture of colchicum. He noticed that numerous mitoses were visible in the stomach, and that in the glands of Lieberkühn "nearly all the elements are in indirect division". The illustrations of his paper (Fig. 1.2) clearly show the accumulation of arrested "star" metaphases (cf. Chap. 5), a typical effect of colchicine which was only to be rediscovered in 1934.

Although Dixon [19] and Dixon and Malden [20] in pharmacological studies did mention that "a further effect of colchicine is to excite karyokinesis" [19] and that in the bone marrow of injected animals "plentiful mitotic forms can occasionally be observed" [20], this did not attract the notice of other workers. The modern discoveries on colchicine took place in the Laboratory of Pathology of Brussels University, under the direction of Albert P. Dustin, who had been working on the regulation of mitotic activity in the thymus since 1908 [21]. Several studies on the pycnotic destruction of the thymocytes in mice injected with drugs such as diaminoacridine were to lead Selye [80] a few years later to the discovery of the role of corticoid hormones in these thymic changes and to the concept of "stress".

In 1929, A. P. Dustin and Piton [27] had observed that several arsenical derivatives (arsenious oxide, sodium cacodylate) had a similar effect on the thymus, although the destruction of the cells was preceded by a considerable increase in the number of mitoses, which was thought to result from a true stimulation of mitotic growth. Similar changes were described in other tissues, such as the Lieberkühn crypts of the intestine (Fig. 1.3): these changes are identical to those mediated by colchicine, and opened the path for the study of the alkaloid a few years later.

This stemmed from the work of a student in medicine, F. Lits, who, after reading the works of Dixon and Malden [19, 20], suggested to A. P. Dustin a study of the action of colchicine on cell division. It was rapidly found that mice injected with small doses of colchicine demonstrated, in the following hours, a considerable increase of the number of mitotic figures in all germinative regions [50, 51]. For several years, while it was apparent that these mitoses were arrested because of spindle abnormalities, it was thought that colchicine stimulated mitotic activity, and this theme is still to be found in a study by A. P. Dustin of hormone-stimulated growth [25], published after his death in 1942.

However, studies in tissue culture [14, 52] indicated that the increased number of mitoses was explained by their rapid accumulation resulting from their arrest at premetaphase, while the other stages (G1, S, and G2) were not modified. A. P. Dustin, considering the similarity of action of arsenicals and colchicine, proposed

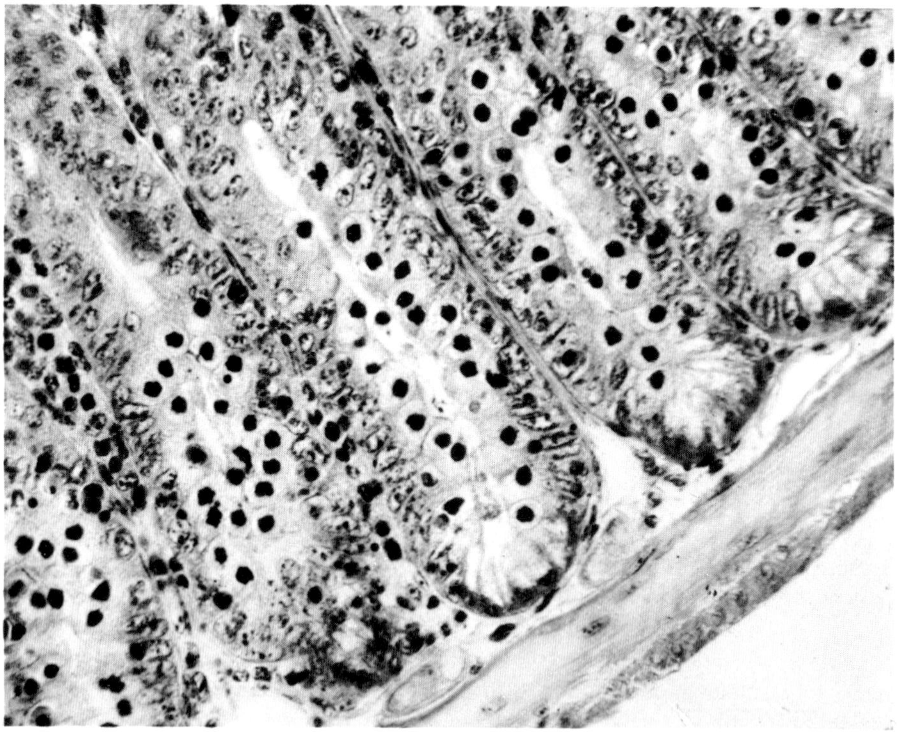

Fig. 1.3. Intestinal glands of a mouse, 9 h after an injection of 2 mg/g of sodium cacodylate. There is a considerable increase of visible mitoses, which are all in a condition of arrest ("ball metaphases") resulting from spindle inactivation (from Piton [66])

in 1936 the term "stathmokinesis" to designate this type of pathological cell division [23].

A few years later a Hungarian botanist, Laslo Havas, came to work at Brussels, and his simple experiments with *Allium* root tips led to the discovery that colchicine modified plant mitoses [26]. These results, demonstrated at the International Cancer Congress in Brussels (1937), led Gavaudan [37] in France to the conclusion that the number of chromosomes was increased in plant cells treated for several days with colchicine, and led Blakeslee, the same year [10], to the demonstration that the alkaloid was an excellent tool for the production of polyploid plants. While animal cells, after colchicine, either degenerate or recover, plant cells may undergo several mitotic cycles without spindles, resulting in a doubling and quadrupling of the number of chromosomes. Colchicine was from that date, and has remained, one of the main tools for the production of polyploid and amphidiploid plants in agriculture, which are today playing a role in the improvement of crops needed for human alimentation. The experimental crossing of *Triticum* and *Secale,* leading to the fertile amphidiploid *Triticale* [40], is one instance of this application of colchicine [92]. In 1942, more than 350 papers on colchicine polyploids had already been published [47].

1.2.2 Colchicine as a Tool

The impressive increase in the number of mitoses found a few hours after an injection of colchicine suggested that the "colchicine method" would be useful for the study of various types of mitotic growth (Fig. 1.4). Quite early it was observed that malignant cells reacted to colchicine (and to cacodylate) by a considerable increase of mitotic figures [22, 24]. Many hormones are known to stimulate mitosis, and the study of their action was greatly facilitated by the use of colchicine. The principles and difficulties of this technique have been discussed [30]. Since the use of tritiated thymidine, colchicine has lost much of its appeal for the study of growth; it remains however a good control method, providing some conditions are met. The low toxicity of the drug is important, but it must not be forgotten that the mitotic increase is only approximately linear, and is followed by the death of many arrested cells. The *Vinca* alkaloids (cf. Chap. 5), which have a more prolonged action, are sometimes preferred, although their effects on the cells are more complex than those of colchicine and consequently less specific.

One of the spectacular developments of the colchicine tool has been cytogenetics. The exact number of chromosomes in man was still unknown before 1956; it was still supposed to be 48, although several authors had failed to confirm this number. Colchicine provided a better way to study chromosomes: it not only increased the numbers of metaphases; by destroying the spindle, it rendered the cells far easier to flatten on a glass slide when placed in an hypertonic solution. It

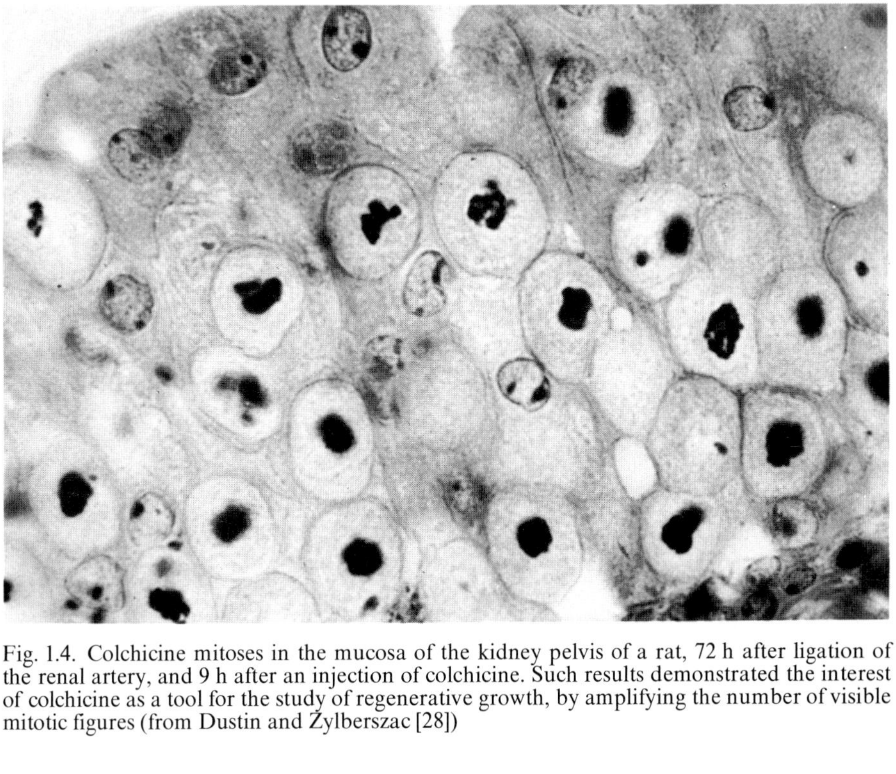

Fig. 1.4. Colchicine mitoses in the mucosa of the kidney pelvis of a rat, 72 h after ligation of the renal artery, and 9 h after an injection of colchicine. Such results demonstrated the interest of colchicine as a tool for the study of regenerative growth, by amplifying the number of visible mitotic figures (from Dustin and Żylberszac [28])

was colchicine metaphases which permitted Tjio and Levan [87], in 1956, to demonstrate without doubt that 46 is the proper diploid number of chromosomes in man. Since that date, cytogenetics has become a major branch of applied cell biology, and colchicine is widely used as a tool for the preparation of slides with many metaphases.

Hence, for many years colchicine was used in applied cytology as a tool to handle chromosomes, and its action on mitosis was believed to be the most important change brought to the cells, intermitotic cells apparently not being affected. In the monograph published in 1955 [31] only a few effects of colchicine which did not appear to be related to mitotic poisoning were mentioned: some changes in the shape of cell walls in plants [38], the great toxicity of the alkaloid for the central nervous system, the increased numbers of leukocytes and platelets following a colchicine injection and, last but not least, its therapeutic action in gout, which remained as mysterious as ever.

1.2.3 Radioactive Colchicine and the Discovery of Tubulin

Radioactive colchicine was first prepared by growing *Colchicum* in an atmosphere containing $^{14}CO_2$ [3, 89]. In mice, this ^{14}C-colchicine was found to disappear rapidly; after four hours no more radioactivity was detected in the brain, the heart, and the blood. The intestine, the kidney and the spleen retained most of the ^{14}C (17.6% of the radioactivity is found in the intestine). This study was not pursued, and only recently a more precise labeling with ^{14}C on the methoxy group of ring C [91] has permitted a study of the distribution of colchicine in man (cf. Chap. 5).

The introduction by Taylor [85] of colchicine labeled with tritium on the methyl group of ring C was to have the most spectacular consequences. ^{3}H-colchicine was found to bind reversibly to human cells in tissue culture. Doses as low as 5×10^{-8} M were found to arrest all mitoses in metaphase without causing other metabolic effects. Taylor concluded that these results could be explained by a fixation of colchicine on a structural protein of the spindle, and suggested that MT might be the site of fixation. Further research [11, 12] indicated that after homogenization the macromolecule binding colchicine appeared in the soluble fraction of the cell, and had a sedimentation constant of 6S. Binding activity was found in a variety of cells, and could be correlated with the presence of MT. No colchicine was fixed by molecules such as actin, myosin, albumin or hemoglobin. HeLa cells, *Arbacia punctulata* eggs, the brain of several mammals, cilia of *Tetrahymena,* sperm of *Arbacia* and the mitotic apparatus of *Arbacia* and *Strongylocentrotus purpuratus* were found to bind relatively large quantities of colchicine, the most active being squid axoplasm.

The biochemical explanation of these results will be discussed in Chapter 2. They mark a turning point in the study of MT and the pharmacology of colchicine, answering for the first time the question asked in 1955 [30] in the following terms: "any work which helps to solve the problem of spindle inactivation (by colchicine) may throw more light . . . on the physiology of the peculiar fibrous protein which constitutes the spindle". Similar results were obtained with another type of tritiated colchicine, with the ^3H on the acyl moeity of ring B [94], in cultured grasshopper embryos. From this date, it was clear that colchicine binding could occur in many cells, and was not limited to mitosis. This was an indirect confirmation of the results of Went [93] on the presence during interphase of a protein with the same immunological properties as the mitotic apparatus. Later research was to demonstrate that this protein was *tubulin*.

1.3 Other MT Poisons

Colchicine is by no means the only chemical which may combine with MT, and many other "spindle poisons" interfere with mitosis in a similar way, although few display the same specificity. The mitotic arresting effect of arsenic has been mentioned; it opened the path to the discovery of that of colchicine. Its study has been neglected for many years, although the tubulin molecule has several –SH groups with which arsenic could readily combine. Apparently, these groups play a role in the assembly of the MT spindle, as suggested long ago by Rapkine [72] and discussed by Mazia [55]. Other –SH reagents, such as heavy metals and an excess of S–S groups, may also interfere with the spindle functions (cf. Chap. 5; 29, 56).

1.3.1 The *Catharanthus (Vinca)* Alkaloids

The history of the discovery of powerful spindle poisons in extracts of the periwinkle *Vinca rosea* (now named *Catharanthus roseus,* 86) has been told several times. Popular medicine attributed to this plant antidiabetic properties; this proved to be inexact but it was discovered that a severe leukopenia appeared in injected animals [17, 42, 58]. This was rapidly demonstrated to be the consequence of a mitotic arrest of the blood-forming cells of the bone marrow, and led to the isolation of new powerful drugs. Among the many indole alkaloids isolated from this plant, two have occupied a central role not only in MT research, but also in cancer chemotherapy: vinblastine (VLB) (formerly named vincaleukoblastine) and vincristine (VCR) (leurocristine). Like colchicine, these alkaloids combine selectively with tubulins and interfere very actively, at low doses, with all functions involving MT assembly. Contrary to colchicine, they have proved to be very helpful in human cancer chemotherapy (cf. Chap. 11). A remarkable effect of the *Vinca* alkaloids is to precipitate intracellular tubulin in the form of crystalline structures [6].

1.3.2 Other Substances of Plant Origin

Several other natural substances are known in popular medicine. Those which have been recommended for the treatment of warts are the most interesting, as they may act by slowing down the mitotic growth of these tumors. Such is the case of the resin of *Podophyllum,* the scientific study of which started in 1947 [84], when it was confirmed to have interesting effects on benign skin tumors. Several mitotic inhibitors have been prepared from this plant, the principal ones being podophyllotoxin and peltatin (cf. 45; cf. Chap. 5). Podophyllotoxin is the active substance of Hecker's liniment, which as early as 1860 was claimed to cure skin tumors. This preparation was made from the dried leaves of *Juniperus sabina,* which contain the active principle [33].

Among the other substances which arrest mitosis at metaphase, extracts of *Chelidonium majus,* which were known in Europe as a cure for warts, have been shown to contain a MT poison, chelidonin [49]. Other active substances are protoanemonin, sanguinarine and cryptopleurine (extracted from *Cryptocaria pleurospora*; cf. 31).

One antibiotic, *griseofulvin,* isolated from *Penicillium griseofulvum* Dierckx in 1939 [59] and used in medical practice as a fungistatic, arrests mitoses in the rat. Plant cell divisions are also affected [60; cf. Chap. 5].

1.4 Action of Physical Agents on MT

The studies of the disruption of spindle fibers under the effects of low temperature and high hydrostatic pressure are of more than historical interest, as they have brought fundamental information about the thermodynamics of MT assembly (cf. Chaps. 2 and 5).

The action of temperature on mitosis has long been known: before the use of colchicine as an agent for the production of polyploid species, heat-shock was the most usual technique [77]. Cold had been known to arrest mitosis since 1890 [39]. In tissue cultures of chick fibroblasts, binucleate cells appear when the anaphase is cooled to a temperature close to 0 °C, as a result of the destruction of the spindle [13]. Further evidence of mitotic changes in cooled cells is to be found in the monograph by Politzer [68]. Later, the effects of cold (3 °C) were found to be quite similar to those of colchicine in the epithelial cells of *Triturus vulgaris* [4]. The cells were arrested in metaphase, with the chromosomes grouped in a "star" configuration or in an "exploded" condition, with chromosomes at the periphery of the cell. The authors quite correctly indicated that in this condition the centriole would occupy a central position (cf. Chap. 10). It was suggested that cold and colchicine would act on the surface of the centromere (kinetochore) and the centrosomes (centrioles): modern work was to show that the assembly of MT takes place at these two sites. Later, it was shown that precise relations existed between the temperature and the birefringence of the spindle, a fact which helped considerably to understand the nature of the mitotic apparatus [41].

The action of high hydrostatic pressures is also important for studies of the mitotic spindle and other structures made of MT. Following the work of Marsland [53, 54] a rapid destruction of the tentacles of *Ephelota coronata* (Str. Wright), a Suctorian Protozoa, was described [46]. Sea-urchin eggs were observed in a special pressure chamber, and above 2000 psi [1], the astral rays were seen to disappear and

[1] Pressure Units. In most biological publications, these are expressed as pounds per square inch (psi). The international unit is the Pascal (Pa) which is worth one Newton per square meter (N/m^2). This is a quite small unit, and the normalized atmosphere (atm) is often preferred. The following figures give the conversion values of these units (cf. 83): One atm = 101,325 Pa = 1.0334 kg/cm^2 = 14.696 psi. One psi = 6894 Pa = 0.068 atm. One Pa = 14.653×10^{-5} psi = 0.98×10^{-5} atm. A detailed conversion table is to be found in the book edited by Zimmerman [95]

the spindle to decrease in size. Above 3000 psi, no spindle was apparent. After the release from pressures as high as 15,000 psi, the astral and the spindle fibers became rapidly visible. The pressures which destroy the spindle also arrest the movements of chromosomes and cytoplasmic cleavage. Recovery cells with many small daughter nuclei (caryomery) are also illustrated in this remarkable paper [64]. These techniques were to be used later and provide important data about the thermodynamics of tubulin assembly (cf. Chap. 2).

1.5 Conclusion

This introduction should help to make clear that although MT were only formally described in the sixties, they had been known under other names since the end of the last century, and had to wait for the improvements of electron microscopy to receive their name, and to be considered ubiquitous organelles. The study of the so-called spindle poisons, whose main effect, at the light microscopical level, was the destruction of the fibrillary structure of the mitotic apparatus, played an important role in the isolation of the specific proteins of MT, the tubulins. From the convergence of these studies the general concept of MT was born.

References

1. Apathy: Das leitende Element des Nervensystems und seine topographischen Beziehungen zu den Zellen. Mitt. Zool. Stat. Neapel. (1897) (cf. also ibidem 1894)
2. Auber, J.: Ultrastructure de la jonction myoépidermique chez les Diptères. J. Microsc. **2,** 325 – 336 (1963)
3. Back, A., Walaszek, E. J., Umeki, E.: Distribution of radioactive colchicine in some organs of normal and tumor-bearing mice. Proc. Soc. Exp. Biol. Med. **77,** 667 – 669 (1951)
4. Barber, H. N., Callan, H. G.: The effects of cold and colchicine on mitosis in the newt. Proc. R. Soc. London B **131,** 258 – 271 (1943)
5. Behnke, O.: Microtubules in disk-shaped blood cells. Intern. Rev. Exp. Path. **9,** 1 – 92 (1970)
6. Bensch, K. G., Malawista, S. E.: Microtubule crystals: a new biophysical phenomenon induced by *Vinca* alkaloids. Nature (London) **218,** 1176 – 1177 (1968)
7. Bessis, M., Breton-Gorius, J., Thiery, J. P.: Centriole, corps de Golgi et aster des leucocytes. Rev. Hématol. **13,** 363 – 386 (1958)
8. Bessis, M.: Living Blood Cells and Their Ultrastructure. Berlin-Heidelberg-New York: Springer 1973
9. Bethe: Über die Primitivfibrillen in den Ganglienzellen von Menschen und anderen Wirbeltieren. Morphol. Arb. **8** (1898)
10. Blakeslee, A.: Dédoublement du nombre de chromosomes chez les plantes par traitement chimique. C. R. Acad. Sci (Paris) **205,** 476 – 479 (1937)
11. Borisy, G. G., Taylor, E. W.: The mechanism of action of colchicine. Colchicine binding to sea-urchin eggs and the mitotic apparatus. J. Cell Biol. **34,** 535 – 548 (1967)
12. Borisy, G. G., Taylor, E. W.: The mechanism of action of colchicine. Binding of colchicine-^{3}H to cellular protein. J. Cell Biol. **34,** 525 – 534 (1967)
13. Bucciante, L.: Influenza di temperature molto basse su mitosis di culture "in vitro". Formazione di cellule binucleate. Protoplasma **5,** 142 – 157 (1929)
14. Bucher, P.: Zur Kenntnis der Mitose. VI. Der Einfluß von Colchicin und Trypaflavin auf den Wachstumsrhythmus und auf die Zellteilung in Fibrocyten-Kulturen. Z. Zellforsch. **29,** 283 – 322 (1939)
15. Cleveland, L. R.: The centrioles of *Pseudotrichonympha* and their role in mitosis. Biol. Bull. **69,** 46 – 51 (1935)

16. Cohen, A., Cook, J., Roe, E.: Colchicine and related compounds. J. Chem. Soc. (London) 194 – 197 (1940)
17. Cutts, J. H., Beer, C. T., Noble, R. L.: Biological properties of Vincaleukoblastine, an alkaloid in *Vinca rosea* Linn. with reference to its antitumor action. Cancer Res. **20,** 1023 to 1031 (1960)
18. De Harven, E., Bernhard, W.: Etude au microscope électronique de l'ultrastructure du centriole chez les Vertébrés. Z. Zellforsch. **45,** 378 – 498 (1956)
19. Dixon, W.: A Manual of Pharmacology. London: Arnold 1906
20. Dixon, W., Malden, W.: Colchicine, with special reference to its mode of action and effect on bone-marrow. J. Physiol. **37,** 50 – 76 (1908)
21. Dustin, A. P.: Recherches d'histologie normale et expérimentale sur le thymus des Amphibiens anoures. Arch. Biol. (Liège) **30,** 601 – 693 (1920)
22. Dustin, A. P.: Contribution à l'étude des poisons caryoclasiques sur les tumeurs animales. II. Action de la colchicine sur le sarcome greffé, type Crocker, de la souris. Bull. Acad. R. Med. Belg. **14,** 487 – 502 (1934)
23. Dustin, A. P.: L'action des arsenicaux et de la colchicine sur la mitose. La stathmocinèse. C. R. Ass. Anat. **33,** 204 – 212 (1938)
24. Dustin, A. P.: A propos des applications des poisons caryoclasiques à l'étude des problèmes de pathologie expérimentale, de cancérologie et d'endocrinologie. Arch. Exp. Zellforsch. **22,** 395 – 406 (1939)
25. Dustin, A. P.: Recherches sur le mode d'action des poisons stathmocinétiques. Action de la colchicine sur l'utérus de Lapine impubère sensibilisé par l'injection préalable d'urine de femme enceinte. Arch. Biol. (Liège) **54,** 111 – 187 (1943)
26. Dustin, A. P., Havas, L., Lits, F.: Action de la colchicine sur les divisions cellulaires chez les Végétaux. C. R. Assoc. Anat. **32,** 170 – 176 (1937)
27. Dustin, A. P., Piton, R.: Etudes sur les poisons caryoclasiques. Les actions cellulaires déclenchées par les composés arsenicaux. Bull. Acad. R. Med. Belg. **9,** 26 – 35 (1929)
28. Dustin, A. P., Zylberszac, S.: Etude de l'hypertrophie compensatrice du rein par la réaction stathmocinétique. Note préliminaire. Bull. Acad. R. Med. Belg. VIe série **4,** 315 to 320 (1939)
29. Dustin, P. Jr.: Mitotic poisoning at metaphase and –SH proteins. Exp. Cell Res. Suppl. **1,** 153 – 155 (1949)
30. Dustin, P.: The quantitative estimation of mitotic growth in the bone-marrow of the rat by the stathmokinetic (colchicinic) method. In: The Kinetics of Cellular Proliferation (ed.: F. Stohlman), pp. 50 – 56. New York: Grune and Stratton 1959
31. Eigsti, O. J., Dustin, P. Jr.: Colchicine, in Agriculture, Medicine, Biology and Chemistry. Ames, Iowa: Iowa State College Press 1955
32. Eigsti, O. J., Dustin, P., Gay-Winn, N.: On the discovery of the action of colchicine on mitosis in 1889. Science **110,** 692 (1949)
33. Eisenmann: Über die locale Wirkung der Sabina. Virchows Arch. **18,** 171 – 172 (1860)
34. Fauré-Fremiet, F.: Microtubules et mécanismes morphopoiétiques. Année Biol. **9,** 1 – 61 (1970)
35. Fawcett, D. W.: Electron microscopic observations on the marginal band of nucleated erythrocytes. Anat. Rec. **133,** 379 (1959)
36. Fawcett, D. W., Porter, K. R.: A study of the fine structure of ciliated epithelia. J. Morphol. **94,** 221 – 282 (1954)
37. Gavaudan, P., Pomriaskinsky-Kobozieff, N.: Sur l'influence de la colchicine sur la caryocinèse dans les méristèmes radiculaires de l'*Allium cepa.* C. R. Soc. Biol. Paris **125,** 705 – 707 (1937)
38. Gorter, C.: De invloed van colchicine of den groei van den celwand van wortelharen. Proc. K. Ned. Akad. Wet. **48,** 3 – 12 (1945)
39. Hertwig, O.: Über pathologische Veränderung des Kernteilungsprozesses infolge experimenteller Eingriffe. Intern. Beitr. Wiss. Med. **1** (1891) (cf. Politzer, 1934)
40. Hulse, J. H., Spurgeon, D.: Triticale. Sci. Am. **231,** 72 – 80 (1974)
41. Inoué, S.: Motility of cilia and the mechanism of mitosis. Rev. Mod. Phys. **31,** 402 – 408 (1959)
42. Johnson, I. S., Wright, H. E., Svoboda, G. H., Vlantis, J.: Antitumor principles derived from *Vinca rosea* Linn. I. Vincaleukoblastine and leurosine. Cancer Res. **20,** 1016 – 1022 (1960)
43. Jolly, J.: Hématies des Tylopodes. C. R. Soc. Biol. **93,** 125 – 127 (1920)
44. Jolly, J.: Traité technique d'hématologie. Paris: Maloine et Fils 1923
45. Kelly, M. G., Hartwell, J. L.: The biological effects and the chemical composition of podophyllin. A review. J. Natl. Cancer Inst. **14,** 967 – 1010 (1954)

46. Kitching, J. A., Pease, D. C.: The liquefaction of the tentacles of suctorian protozoa at high hydrostatic pressure. J. Comp. Physiol. **14,** 410–412 (1939)
47. Krythe, J. M., Wellensiek, S. J.: Five years of colchicine research. Bibliogr. Genet. **14,** 1 – 132 (1942)
48. Ledbetter, M. C., Porter, K. R.: A "microtubule" in plant fine structure. J. Cell Biol. **19,** 239 – 250 (1963)
49. Lettré, H., Lettré, R., Pflanz, C.: Über Synergisten von Mitosegiften. II. Bulbocapnin, Colchicin, N-methylcolchicinamid und ihre Kombinationen. Z. Physiol. Chem. **286,** 138 – 144 (1950)
50. Lits, F.: Contribution à l'étude des réactions cellulaires provoquées par la colchicine. C. R. Soc. Biol. Paris. **115,** 1421 – 1423 (1934)
51. Lits, F.: Recherches sur les réactions et lésions cellulaires provoquées par la colchicine. Arch. Intern. Med. Exp. **11,** 811 – 901 (1936)
52. Ludford, R. J.: The action of toxic substances upon the division of normal and malignant cells in vitro and in vivo. Arch. Exp. Zellforsch. **18,** 411 – 441 (1936)
53. Marsland, D. A.: The effects of high hydrostatic pressure upon cell division in *Arbacia* eggs. J. Cell. Comp. Physiol. **12,** 57 – 70 (1938)
54. Marsland, D. A.: The effects of high hydrostatic pressure upon the mechanism of cell division. Arch. Exp. Zellforsch. **22,** 268 – 269 (1939)
55. Mazia, D.: SH and Growth. Glutathione. New York: Academic Press 1954
56. Mazia, D., Mitchinson, J. M., Medina, H., Harris, P.: The direct isolation of the mitotic apparatus. J. Biochem. Biophys. Cytol. **10,** 467 – 474 (1961)
57. Meves, F.: Gesammelte Studien an den roten Blutkörperchen der Amphibien. Arch. Mikrosc. Anat. **77,** 465 – 540 (1911)
58. Neuss, N., Gorman, M., Hargrove, W., Cone, N. J., Bieman, K., Buchi, G., Manning, R. E.: The structure of oncolytic agents alkaloids vinblastine (VLB) and vincristine (VCR). J. Am. Chem. Soc. **86,** 1440 – 1441 (1964)
59. Oxford, A. E., Raistrick, H., Simonart, P.: Studies on the biochemistry of microorganisms. LX. Griseofulvin $C_{18}H_{17}O_6Cl$, a metabolic product of *Penicillium griseofulvum* Dierckx. Biochem. J. **33,** 240 – 248 (1939)
60. Paget, G. E., Walpole, A. L.: The experimental toxicology of griseofulvin. AMA. Arch. Dermatol. **81,** 750 – 757 (1960)
61. Palay, S. L.: Synapses in the central nervous system. J. Biophys. Biochem. Cytol. Suppl. **1,** 193 – 201 (1956)
62. Palay, S. L.: The morphology of synapses in the central nervous system. Exp. Cell Res. Suppl. **5,** 275 – 293 (1958)
63. Palay, S. L.: The fine structure of secretory neurons in the preoptic nucleus of the goldfish (*Carassius auratus*). Anat. Rec. **138,** 417 – 444 (1960)
64. Pease, D. C.: Hydrostatic pressure effects upon the spindle figure and chromosome movement. I. Experiments on the first mitotic division of *Urechis* eggs. J. Morphol. **69,** 405 – 442 (1941)
65. Pernice, B.: Sulla cariocinesi delle cellule epiteliali e dell' endotelio dei vasi della mucosa dello stomaco et dell' intestino, nelle studio della gastroenterite sperimentale (nell'avvelenamento per colchico). Sicilia Med. **1,** 265 – 279 (1889)
66. Piton, R.: Recherches sur les actions caryoclasiques et caryocinétiques des composés arsenicaux. Arch. Intern. Med. Exp. **5,** 355 – 411 (1929)
67. Pochon-Masson, J.: Structure et fonctions des infrastructures cellulaires dénommées "microtubules" Année Biol. **6,** 361 – 390 (1967)
68. Politzer, G.: Pathologie der Mitose. Protoplasma Monographien n° 7. Berlin: Gebr. Bornträger 1934
69. Porter, K. R.: Changes in cell fine structure accompanying mitosis. In: Fine Structure of Cells. Symposium of the 8th. Congress on Cell Biology. Leiden-Groningen: Voordhoff 1954 (New York: Interscience)
70. Porter, K. R.: Cytoplasmic microtubules and their functions. In: Ciba Foundation Symposium on principles of biomolecular organization. London: Churchill 1966
71. Ranvier, L.: Recherches sur les éléments du sang. Arch. Physiol. **2,** 1 – 15 (1875)
72. Rapkine, L.: Sur les processus chimiques au cours de la division cellulaire. Ann. Physiol. Physicochem. Biol. **7,** 382 – 418 (1931)
73. Rodnan, G. P., Benedek, T. G.: The early history of antirheumatic drugs. Arthritis. Rheum. **13,** 145 – 165 (1970)
74. Rosenbluth, J.: The fine structure of the acoustic ganglia in the rat. J. Cell Biol. **12,** 329 – 359 (1962)
75. Roth, L. E.: A filamentous component of protozoal fibrillar systems. J. Ultrastruct. Res. **1,** 223 – 234 (1958)

76. Sandborn, E., Koen, P. F., McNabb, J. D., Moore, G.: Cytoplasmic microtubules in mammalian cells. J. Ultrastruct. Res. **11,** 123 – 138 (1964)
77. Sax, K.: Effect of variations in temperature on nuclear and cell division in *Tradescentia.* Am. J. Bot. **24,** 218 – 225 (1937)
78. Schmidt, W. J.: Die Doppelbrechung von Karyoplasma, Zytoplasma und Metaplasma. Protoplasma Monographien 11. Berlin: Gebr. Bornträger 1937
79. Schrader, F.: Mitosis. The Movements of Chromosomes in Cell Division (2nd ed.) New York: Columbia University Press 1953
80. Selye, H.: Thymus and adrenals in the response of the organism to injuries and intoxications. Brit. J. Exp. Path. **17,** 234 – 248 (1936)
81. Slautterback, D. B.: A fine tubular component of secretory cells. Am. Soc. Cell Biol., Abstr. 199 (1961), cf. [82]
82. Slautterback, D. B.: Cytoplasmic microtubules. I. Hydra. J. Cell Biol. **18,** 367 – 388 (1963)
83. Sleigh, M. A., MacDonald, A. G. (eds.): The Effects of Pressure on Organisms. Symposia Soc. exp. Biol. 26. Cambridge: Univ. Press 1972
84. Sullivan, M., King, L. S.: Effects of resin of podophyllum on normal skin, condyloma acuminata and verrucae vulgaris. Arch. Dermatol. Syphil. **56,** 30 – 45 (1947)
85. Taylor, E. W.: The mechanism of colchicine inhibition of mitosis. I. Kinetics of inhibition and the binding of H^3-colchicine. J. Cell Biol. **25,** 145 – 160 (1965)
86. Taylor, W. I., Fransworth, N. R.: The *Catharanthus* Alkaloids. New York: Dekker 1975
87. Tjio, J. H., Levan, A.: The chromosome number of man. Hereditas (Lund) **42,** 1 – 6 (1956)
88. Van Oye, E.: L'origine des anneaux de Cabot. Rev. Hématol. **9,** 173 – 179 (1954)
89. Walaszek, E. J., Kelsey, F. E., Geiling, E. M. K.: Biosynthesis and isolation of radioactive colchicine. Science **116,** 225 – 227 (1952)
90. Wallace, S. L.: Colchicum: the panacea. Bull. N. Y. Acad. Med. **49,** 130 – 135 (1973)
91. Wallace, S. L., Ertel, N. H.: Preliminary report: plasma levels of colchicine after oral administration of a single dose. Metabolism **22,** 749 – 754 (1973)
92. Wellensiek, S.: Methods for producing Triticales. J. Hered. **38,** 167 – 173 (1947)
93. Went, H. A.: Some immunochemical studies on the mitotic apparatus of the sea urchin. J. Biophys. Biochem. Cytol. **5,** 353 – 356 (1959)
94. Wilson, L., Friedkin, M.: Synthesis and properties of colchicine labeled with tritium on its acetyl moiety. Biochemistry **5,** 2463 – 2468 (1966)
95. Zimmerman, A. M. (ed.): High Pressure Effects on Cellular Processes. New York-London: Academic Press 1970

Chapter 2 Structure and Chemistry of Microtubules

2.1 Introduction

MT may be described as regular helical assemblies of two slightly elongated protein subunits, tubulins α and β, each of about 55,000 daltons molecular weight. The shape and properties of MT proceed from the linkages of these two closely related molecules. However, other proteins appear to be necessary for tubulin assembly into MT or may become closely linked with MT when these assemble into complex structures such as centrioles, basal bodies, cilia or axonemes (cf. Chap. 4). It is more and more evident that the tubulin subunits may assemble in more than one way, and tubulin "polymorphs" have repeatedly been described, and help to understand the steps and the factors required for assembly. Tubulins are also associated with guanine nucleotides, possibly with enzymes and with non-protein components: the molecular structure of MT may thus be far more complex than a helical association of protein subunits.

This chapter will be mainly concerned with "simple" MT, as found in the cytoplasm, and sometimes the nucleus, of most cells. The specific problems related to more complex structures, such as the doublets of cilia, will be discussed in Chapter 4. The synthesis of tubulin in the cell and its regulation, the relations of MT with cell organelles, and the complex associations of MT in some cellular structures, will be reviewed in Chapter 3.

2.2 Shape and Size of MT

MT were first observed in fixed cells, later as isolated structures. The development of immunohistochemical techniques has provided more recently excellent images of their overall shape and number in whole cells at the light microscope level. The possibility of assembling purified tubulin in vitro has led to a good understanding of its assembly into MT in normal or highly artificial conditions.

In early work, MT were described as *long, slender, rigid, elastic, non-bifurcated* [163]. All immunohistochemical observations have confirmed this picture, although MT appear, when seen in light microscopy, more flexuous than was first imagined [31, 32, 60, 61, 154, 155, 176]. Their diameter is constant and the clear central zone, readily penetrated by uranyl acetate when isolated MT are observed after negative staining, indicates that they are truly tubular. The diameter measures about 25 nm [163], although figures varying between 15 and 35 nm have been mentioned in the

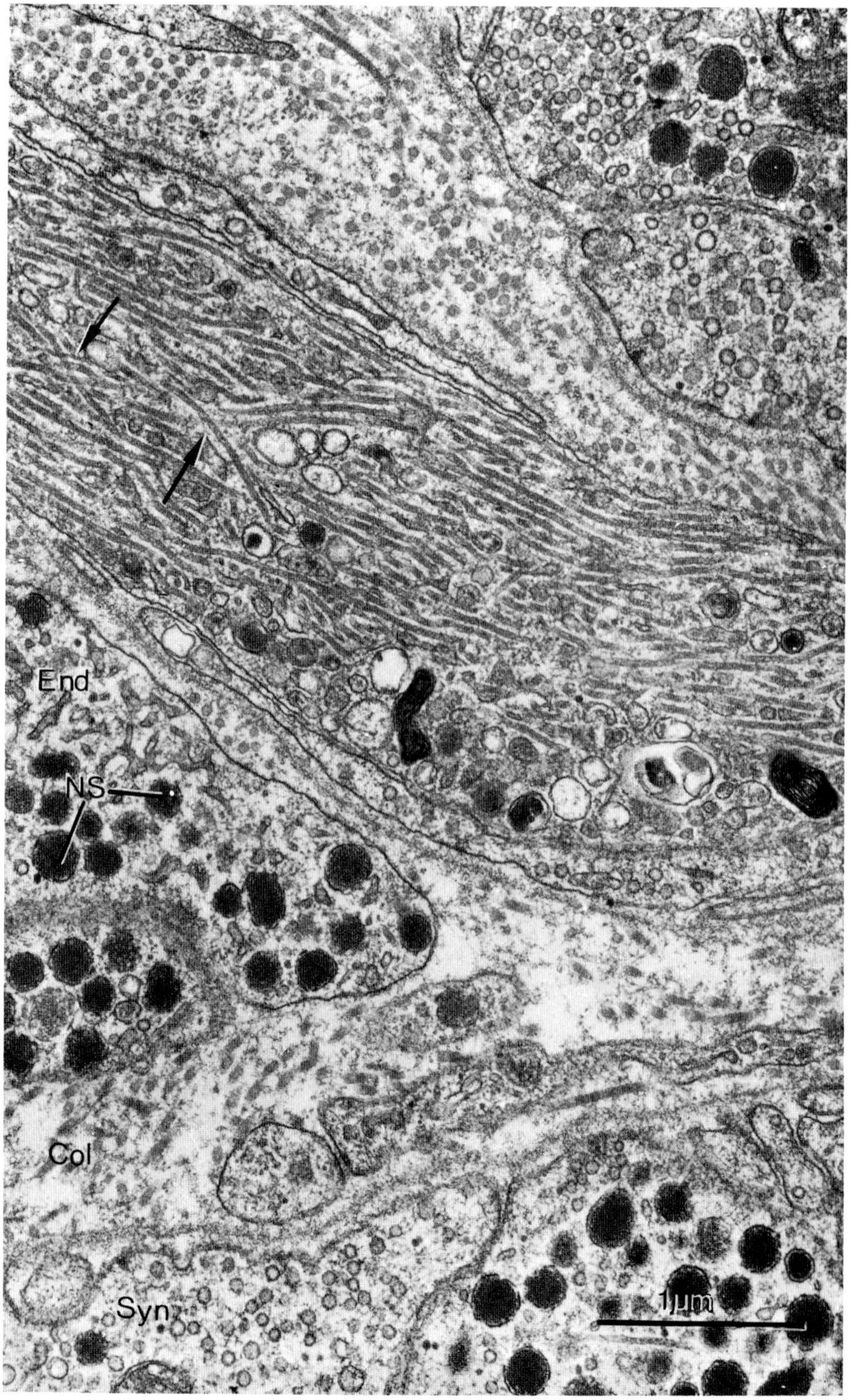
End
NS
Col
Syn
1µm

literature (cf. [162]). A more recent figure is 24 ± 2 nm [36] with a wall about 5 nm thick, and the hollow central core about 15 nm in diameter. These dimensions are directly related to the size and number of the tubulin subunits as described below.

The length of MT varies considerably, it is of a much greater order of magnitude than the diameter, and measured in μm. Variations are related to the linear growth of MT, and to the many roles which they play in differentiated cells. The length may be difficult to measure, for instance in cells where groups of MT form circular bundles at the periphery of the cytoplasm, like in red blood cells and platelets, or helical assemblies around the nucleus, as in some types of spermatogenesis ([133]; cf. Chap. 6). In neurons, there are indications that MT (often called "neurotubules") may be as long as the axons, that is to say thousands of mm (in whales, some nerves exceed 10 m in length). This hypothesis, suggested by Porter [163], has been confirmed by numerations of MT in bifurcated axons, the total number remaining more or less constant [222, 223]. No direct demonstration of the exact length of such MT has been published, however.

MT have been described as "rigid" and "elastic": in fixed cells they appear, at the electron microscope level, nearly straight, and some waviness has often been considered a fixation artefact. However, when observed at the light microscope level, by immunohistochemical techniques, MT are seen to curve and bend. They are known to be strongly bent in cells where they form marginal bundles ([11]; cf. Chap. 6). Isolated MT observed in vitro are straight (contrary to doublet MT, cf. [143]) and may break when bending stresses are too great ([85]; cf. Fig. 2.2).

Incomplete or "C-tubules" have been repeatedly observed, in particular in conditions where MT were undergoing assembly or disassembly: there are reasons to think that this aspect corresponds to one step of MT formation, as will be described below, and could be normally present at the end of normal MT. First observed in blood platelets, during reassembly of MT after destruction by cold [10], C-MT have been described in various mitoses (such as *Haemanthus catherinae,* 122, and *Arbacia punctulata,* 54). Similar open MT have been found in cells treated with poisons such as halothane (cf. Chap. 5). The curved side-arms which unite laterally the MT of the testis of *Gerris remigis* [206] bear some resemblance to these C-MT.

Immunofluorescent techniques [32, 55, 60, 155, 176, 210, 211, 212, 224] have enabled MT to be observed in flattened cells in tissue culture, and give an idea of the real number of these organelles. While in electron microscopy, in most cells, MT do not appear numerous, except in specialized regions—pericentriolar zones, mitotic spindle, marginal bundle of red blood cells and platelets, axons—in immunofluorescence, the whole cytoplasm appears to be supported by a cytoplasmic complex of MT (CMTC) ([32]; cf. Figs. 2.3, 2.4). Theoretically a ±25 nm structure should not be visible in light microscopy, but in fluorescence techniques the limit of resolution may be smaller, as the MT become themselves the light-sources. A

Fig. 2.1. Rat. Posterior lobe of pituitary. Several axons, containing MT, smooth endoplasmic reticulum (*End*), neurosecretory granules (*NS*), and synaptoid vesicles (*Syn*). Collagen fibers (*Col*) between the axons. One large axon shows a great number of MT, which are well preserved after fixation in glutaraldehyde in PIPES buffer (without cacodylate, cf. Chap. 5). The MT are surrounded by a clear ("exclusion") zone. They follow the main direction of the axon, but are not straight, and often their paths are seen to cross one another (*arrows*)

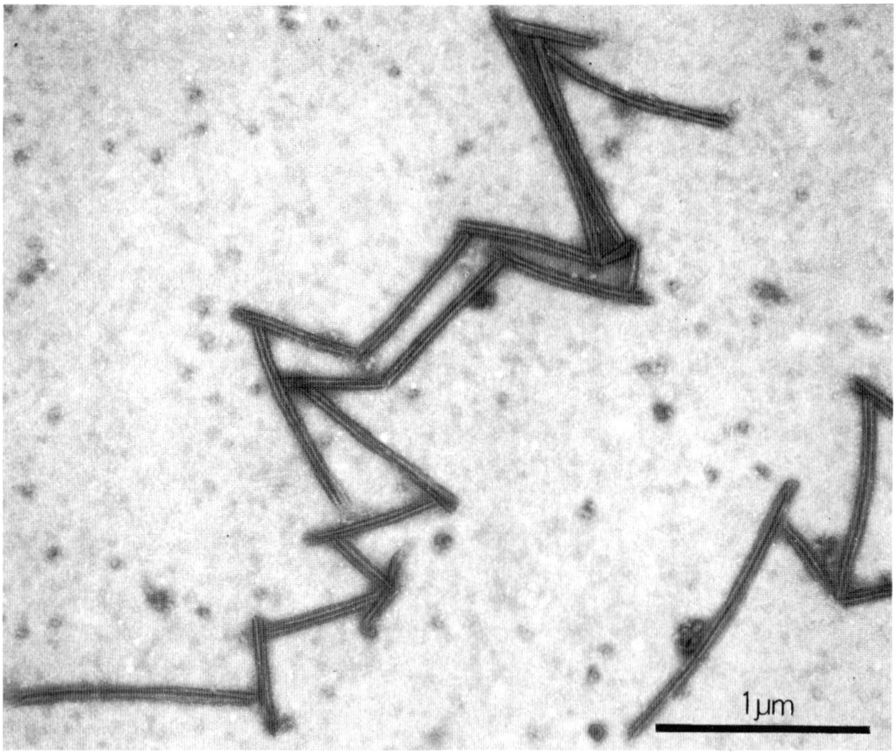

▲ Fig. 2.2 and ▼ Fig. 2.3. (Legends see opposite page)

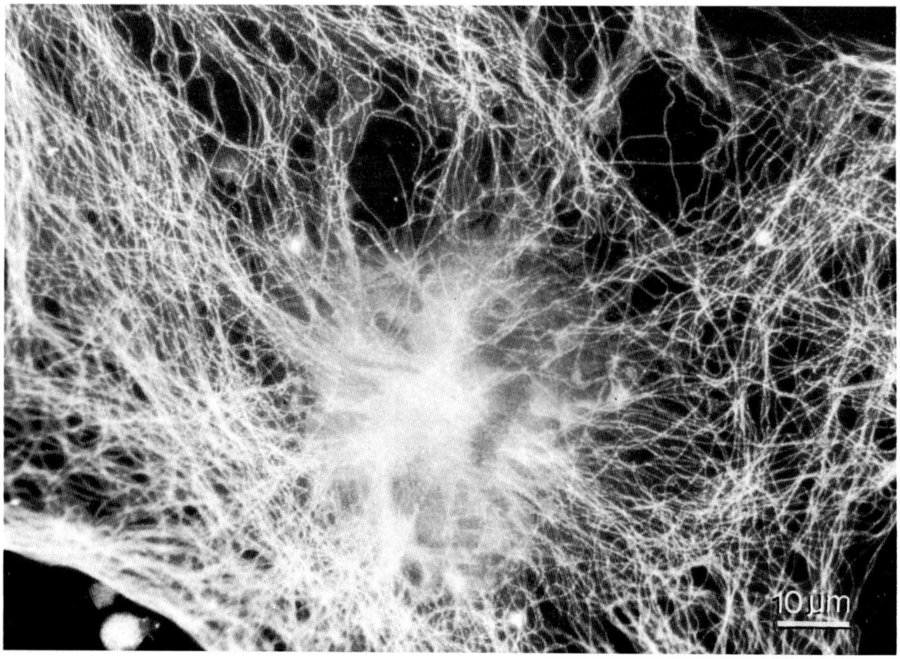

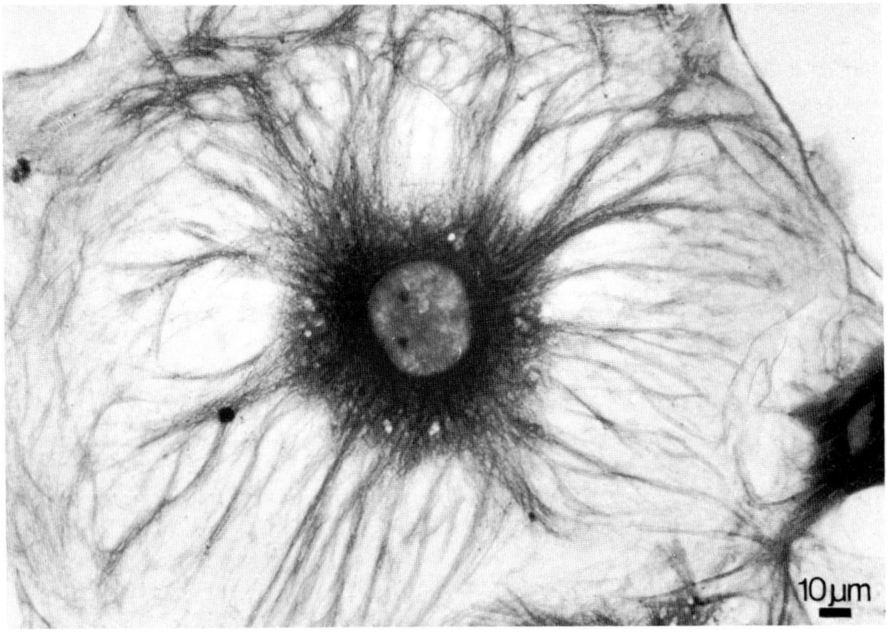

Fig. 2.4. Immunohistochemical staining of MT in an embryonic mouse cell in tissue culture (peroxidase–antiperoxidase method). The MT extend from the nuclear membrane toward the cell periphery where they curve. This method makes possible to check, at the electron microscope level, the MT nature of the stained fibrils (from De Brabander [59])

comparison between high-voltage electron microscopy (HVEM) and fluorescence shows that the images are quite similar [42].

Moreover, the more recent use of immunochemical techniques [60, 61] has enabled a direct observation at the electron microscope level of MT covered with a layer of antibodies which thicken them enough to make them visible in light microscopy: this confirms also that the stained filaments are single MT and not other fibrillar components of the cell.

2.3 Molecular Structure of MT

In the last few years, a considerable amount of information on the structure of MT has been gathered, in particular once it became possible to study the assembly [2] in vitro of tubulin molecules. Progressively, the relations between chemical

[2] The term "assembly" will be preferred to that of "polymerization" used by some authors, as it is a supramolecular phenomenon, involving at least two different proteins, and closer to the formation of a viral capsid than to a chemical polymer [112]

Fig. 2.2. *Triturus* red blood cells. Negative staining in 1% sodium phosphotungstate. Two broken MT showing a strongly contrasted lumen. The cracks and the straight shapes of the fragments suggest that MT are relatively rigid rods (from Gall [85])

Fig. 2.3. Immunofluorescence staining of MT in a glycerinated fibroblast. The curving shape of the MT is noticeable, as is the density of the microtubular cytoskeleton (from Osborn and Weber [155])

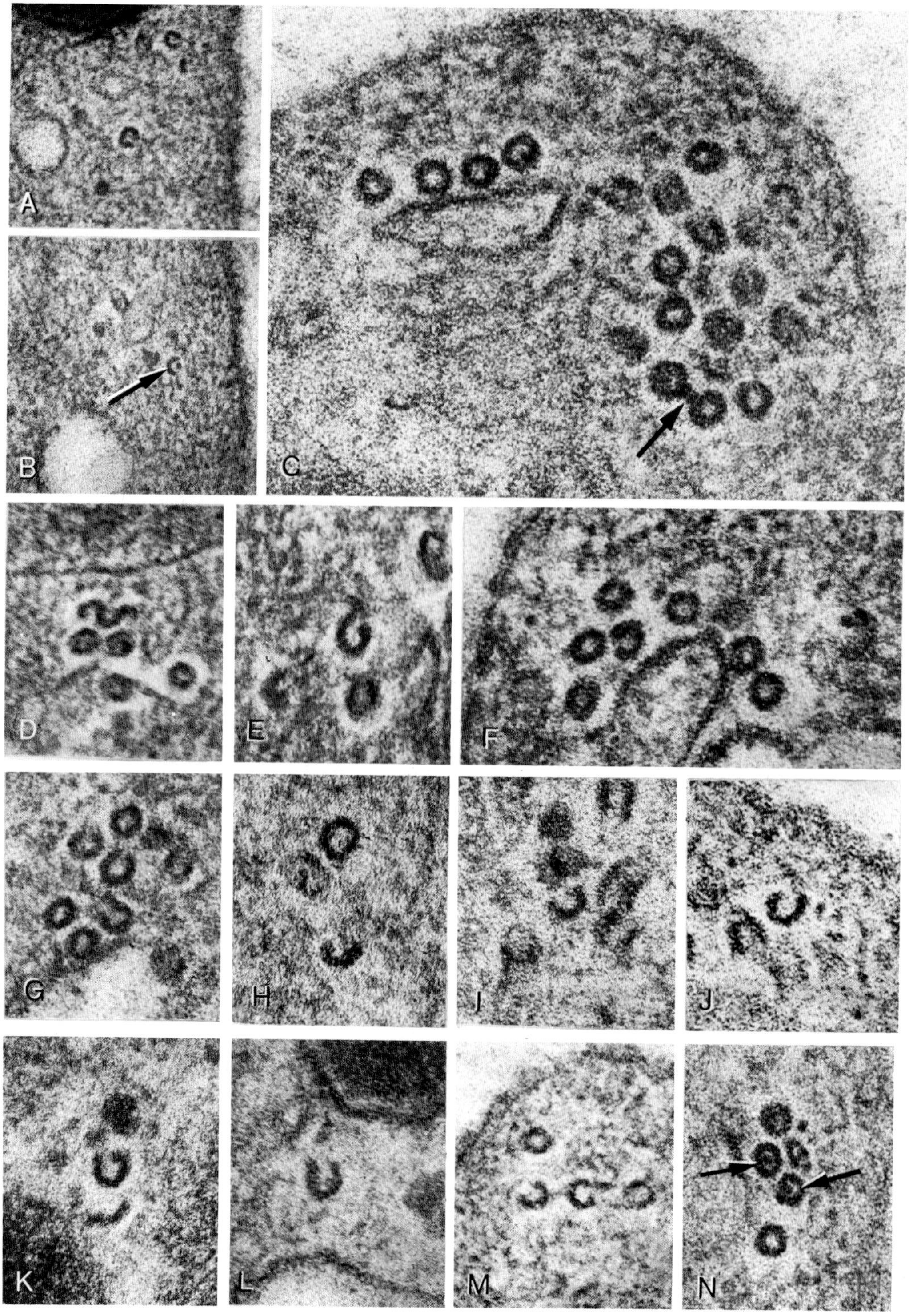
A
B
C
D
E
F
G
H
I
J
K
L
M
N

Fig. 2.5. Human and rat blood platelets, fixed during MT reformation after exposure to 0 °C for 5 min. (A) 10 min reheating at 37 °C. ×95,000. (B) 8 min reheating: incomplete MT with dot (*arrow*). ×95,000. (C) MT connected in pairs (*arrow*). ×224,000. (D) S-shaped MT. ×148,000. (E) Incomplete MT. ×224,000. (F) C-MT. ×224,000. (G) C-MT. ×189,000. (H) Hook-shaped MT after 8 min reheating. ×213,000. (I)–(M) C-MT after 2–12 min reheating. (I)–(L) ×213,000. (M) ×150,000. (N) Two MT with central densities (*arrows*) after 1 h reheating. ×189,000 (from Behnke [10])

structure and supramolecular shape are becoming clearer, while the complex mechanism of growth and its control are better understood.

2.3.1 Methods of Study

Routine electron microscopical techniques have illustrated the shape and the size of MT. While it is often mentioned in the literature that glutaraldehyde fixation played an important role in the discovery of MT [3], it should not be forgotten that those of centrioles, cilia, and mitotic spindle had been clearly observed with routine osmium acid fixation before the advent of glutaraldehyde (cf. Chap. 1). The use of fixatives with tannic acid has provided excellent images of MT in cross-sections, demonstrating the number of subunits, which is usually 13 [204], as already suggested in 1966 [163]. On the other hand, the demonstration of MT after staining by fluorescent antibodies, or by horse-radish peroxydase combined to antitubulins, has brought improved information on the number and location of these organelles in cells cultivated in vitro. These images are comparable to those obtained by the study of relatively thick sections with high-voltage (1 Mev) electron microscopes ([42]; cf. Chap. 10).

Isolated MT, negatively stained, are an excellent starting point for optical diffraction studies which have provided information on the helical assembly of the subunits [2, 3, 4]. X-ray diffraction has also been used in favorable materials, such as the flagellae of spermatozoa [53].

2.3.2 Size of the Subunits

Negative staining of isolated MT demonstrates parallel longitudinal protofilaments, as observed as early as 1963 in human spermatozoa [5]. These protofilaments, which have a beaded structure, are made of subunits with a diameter close to 5 nm [85, 95]. Although the numbers of filaments mentioned in the early literature vary between 10 and 15, MT from most cells have the same size and show a helical arrangement of 13 subunits with a pitch of about 10–25 degrees [44]. From many observations on vertebrates and invertebrates, it is apparent that the basic structure of MT is closely similar in the whole animal kingdom [11, 15, 44, 85, 95, 203].

[3] While divalent glutaraldehyde is often mentioned as the best fixative for MT, it may in some conditions alter the more labile MT in the axonemes of heliozoa and other cells [183]

However, modern studies with the tannic acid method indicate that other numbers of subunits are possible, and in the same animal, the crayfish *Procambarus clarkii*, MT with 13 and 12 subunits may be seen side by side, although in different cells [46, 47]. With the same technique, MT with 15 subunits have been described in the epidermis of the cockroach [147]; this is in agreement with the fact that these MT are larger than usual, their diameter reaching 30 nm. MT with 16 subunits and a dense central core have also been described in the spermatids of the cricket [111]. On the other hand, MT with less than ten subunits have been mentioned [48] in protozoal axopodia (cf. Chap. 3).

In the spermatozoa of *Macrostomum* (Plathelminthes) the cortical MT show discrete rows of 12 protofibrils with a longitudinal periodicity of 8 nm, suggesting the association of tubulin dimers along the fibril length [203]. From all these observations, it is clear that identical or almost identical subunits assemble in rows to form the MT: this has been widely confirmed by two different approaches, the biochemical purification of the subunits, and the mathematical, computer-assisted studies of the diffraction images of micrographs of negatively stained MT.

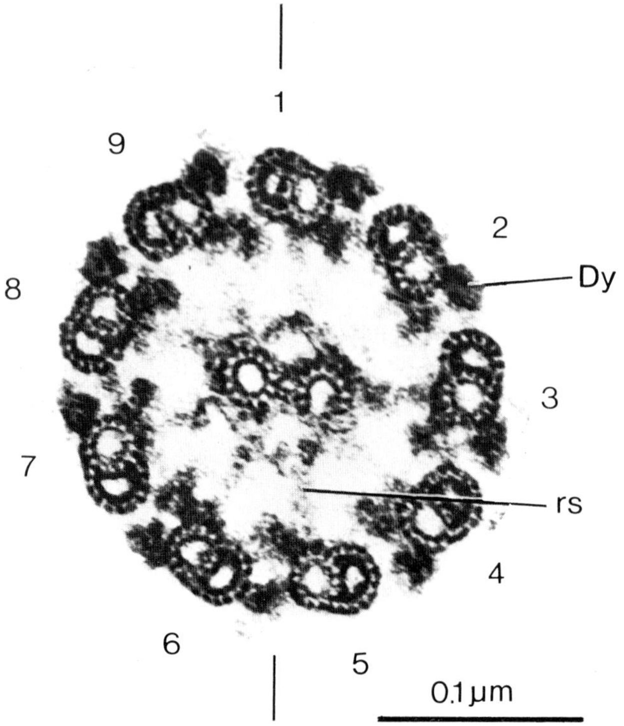

Fig. 2.6. Cross-section of a cilium from *Lytechinus:* the tannic acid fixation stains the periphery of the MT subunits, which appear like tubules with a clear center. Each peripheral doublet of this cilium (cf. Chap. 4) is made of one complete MT with 13 subunits (tubule A) and one incomplete tubule (B), showing about 10 subunits. The two central MT show also 13 subunits. The dense bodies (*Dy*) attached to the peripheral doublets are the two dynein arms (cf. Chap. 4). The radial spokes are poorly visible (*rs*). The plane of bilateral symmetry is indicated by the central pair of MT and by the specialized bridges between doublets 5 and 6 (cf. Chap. 4) (from Fujiwara and Tilney [81])

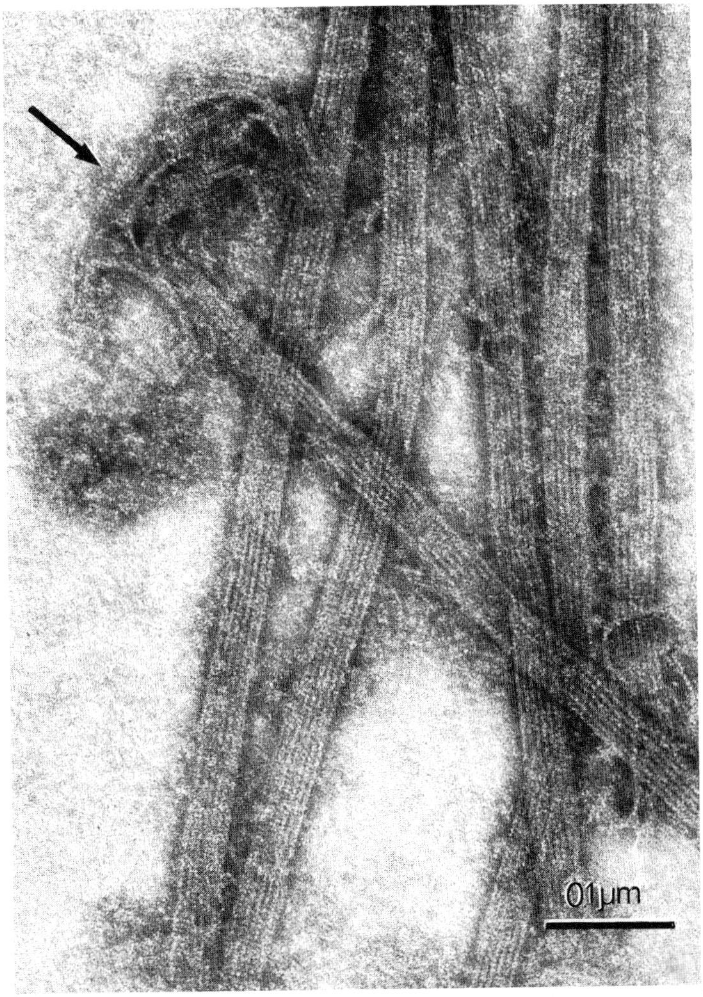

Fig. 2.7. Unfixed MT from rat blood platelet, negatively stained with potassium phosphotungstate at pH 5. The MT show 6 protofilaments, with evidence of beading. One MT is fragmented at one end (*arrow*) (from Behnke and Zelander [15])

The excellent optical definition provided by negatively stained MT, in particular when these are observed flattened and "open", showing all their protofilaments, has been widely used for an analysis of their diffraction patterns. Computer-assisted reconstitutions of the MT lattice from the main diffraction lines compare favorably with the few studies on X-ray diffraction of MT [53]. A first comparison of the diffraction images and computer-drawn aspects from theoretical models led to the conclusion that MT were made of 12 longitudinal filaments, with a triple or quadruple start helix of subunits [103]. Later results confirmed the early suggestion [163] that most MT had 13 protofilaments, for instance in a study of neuronal MT reassembled in vitro (cf. [163]). Diffraction studies indicated that the subunits measured 4 to 5 nm in diameter, which, as was shown previously [182], corresponds to a globular protein of 55,000 daltons molecular weight. The subunits may be slightly elongated and possibly split symmetrically into two lobes [71]. The angle of the helix would be about 10°, which fits in well with a vertical spacing of 4 nm between the subunits. Recent and better X-ray diffraction images of neurotubules are in agreement with these results [52].

Improved studies of MT from *Trichonympha* indicate that the number of protofilaments is 13, and that the tubulin molecules are helically arranged and grouped as heterodimers of α and β tubulin (vide infra), although the exact location of each subunit in the helix is not quite certain. The fact that MT splay out into protofilaments shows that the lateral bonds between tubulin molecules are weaker than the longitudinal ones. The proposed structure (Fig. 2.8 [2, 3]) displays three-, five- or

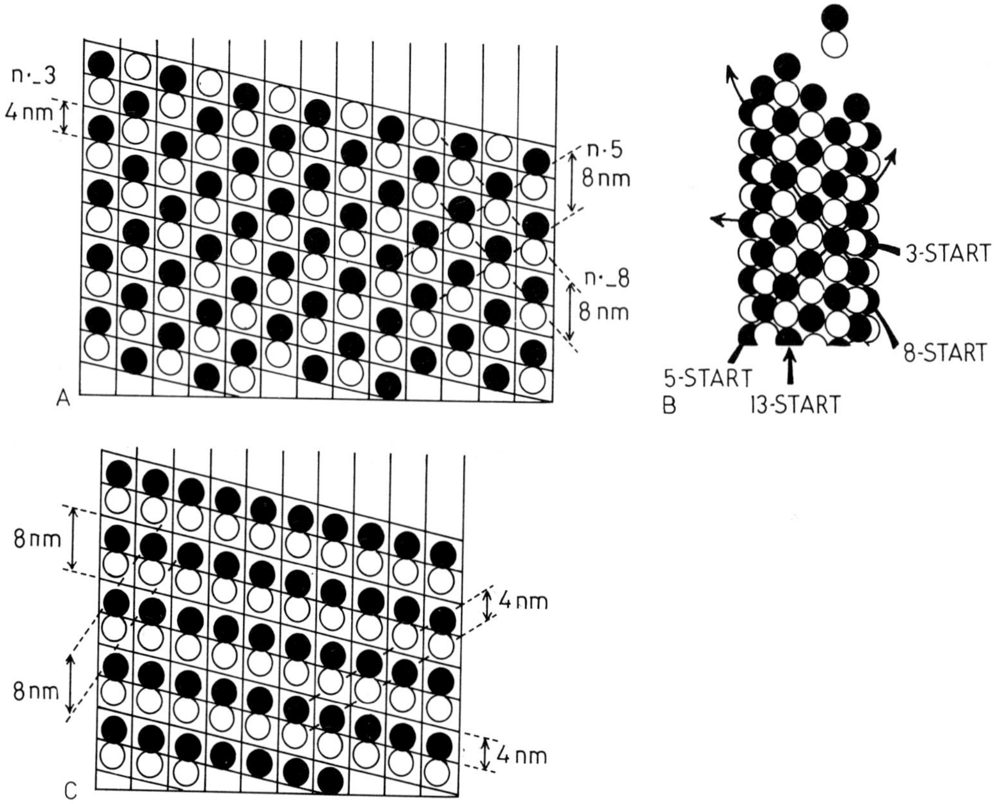

Fig. 2.8 A–C. Substructure of MT. (A) Pattern of subunits in an opened MT with 13 protofilaments. Each of these is made of alternating α and β subunits, represented in black and white, all oriented identically. The spacing between the monomers is about 4 nm; that between the dimers 8 nm, the dimers having a dumb-bell shape. Three possible helices are represented. The three-start helix is made of alternating α and β subunits, contrary to the 5- and 8-start helices. The MT being most probably assembled from $\alpha\beta$ dimers, and growing from one extremity, the growth probably takes place along the 5-start helix: (B) A MT assembled from $\alpha\beta$ dimers. The 3-, 5-, 8-, and 13-start helices are indicated. The 13-start corresponds to the protofibrils. (C) Probable assembly of dimers in the B tubule of a ciliary doublet (cf. Chap. 4). The doublets are not staggered as in other MT (redrawn from Amos et al. [4]) (Cold Spring Harbor Laboratory, Copyright 1976)

eight-start helices. However, if the MT are assembled from heterodimers, as is probable, the basic helices would be the five- or eight-start ones which are formed of identical, similarly oriented dimers [25]. The tubulin helix is left-handed [127]. A three-dimensional model proposed by Amos and Klug [3] shows the possible arrangement of the subunits, depicted as "tilted dumb-bell dimers". The relation of this structure to the lattice of doublet MT will be considered in Chapter 4.

2.3.3 The Central Core

Several workers have described axial densities in MT which have not yet found any proper explanation [96]. These have been found in MT from various animal species and organs. Behnke [11] lists many older references, noticing that these densities do not appear to have been found in mitotic spindle MT.

In the epithelial cells of insects *(Locusta migratoria)*, where bundles of MT link the cells to the cuticle, some MT have a clear central zone, others appear compact [8]. In the neurons of the toad *(Bufo arenarum)*, central nodules of 3 to 5 nm in diameter have been considered as evidence of a migration within the MT [169]; in the lamprey *(Pteromyzon marinus)*, where MT are closely associated with synaptic vesicles, their core is often found to be opaque [188], like in the mechanoreceptors of the cockroach [144]. These uniformly densely stained MT are not necessarily identical to those containing discrete central granules. It appears that MT of nerve cells show such central densities more often than others. Although the idea proposed earlier by Slautterback [184] that a circulation of liquids and of small molecules could take place in the lumen of the MT (cf. [162]) has received no further support, the possibility of molecules migrating in the MT lumen should be kept in mind.

2.3.4 The Clear (Exclusion) Zone

MT are rarely seen to come into contact with one another, even when associated in bundles such as seen in platelets or erythrocytes. In transmission electron microscopy, they are separated by a clear zone, about 10 nm wide [125, 163]. The nature of this zone remains poorly understood and attempts to stain it with various techniques have often failed [193].

It is particularly visible in the cells of insects' ovaries, such as *Notonecta glauca* [134], where large numbers of ribosomes are carried along MT toward the eggs in long cytoplasmic expansions which do not contain any other organelles (cf. Chap. 7). Stains such as ruthenium red coat the surface of the MT, although part of the clear zone remains unstained; Alcian blue, in contrast, stains the MT lumen and all the exclusion zone. With lanthanum hydroxyde, the clear zone is seen to be crossed by narrow unstained extensions of the MT which may be compared to the side-arms described below [193, 194, 195]. The acidic nature of tubulin may explain these staining properties, which could also result from the presence of a mucopolysaccharidic coat on their surface. The clear zone may possibly be explained by an electrostatic repulsion between highly hydrated acidic MT [193].

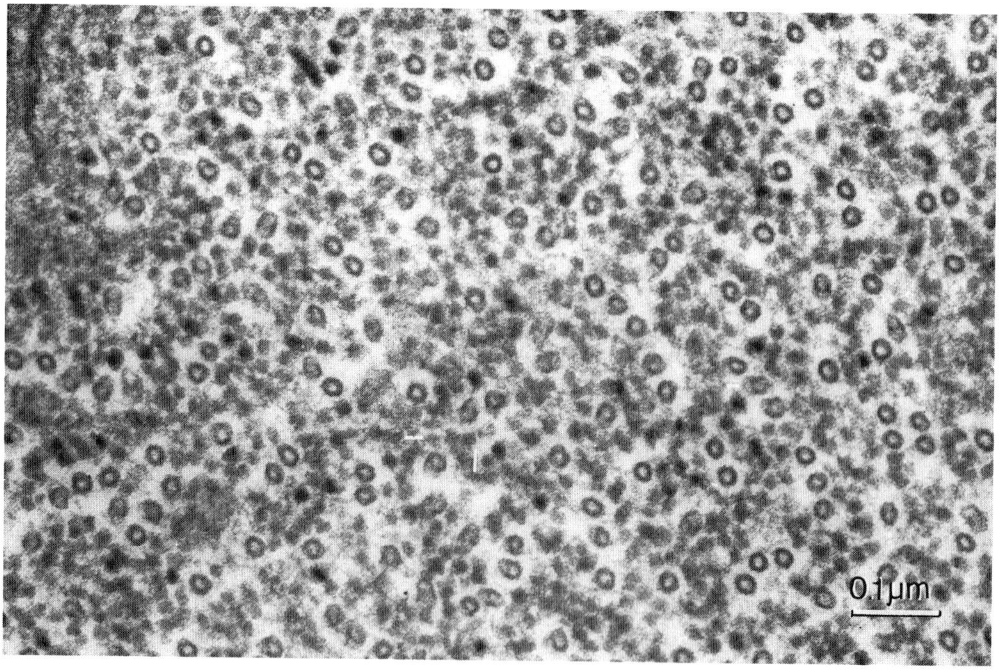

Fig. 2.9. Transverse section through a trophic tube of *Notonecta* ovary nurse cells. Each MT is surrounded by a clear zone that separates it from the surrounding ribosomes, which are carried toward the germ cells (from Stebbings and Bennett [194])

Freeze etching techniques confirm the presence of this empty zone [195]. Negatively stained MT show lateral, "hairy" projections which may connect the MT with the ribosomes. Their length is approximately that of the clear zone. When the MT of the nurse cells are destroyed by large doses of colchicine (1% solution acting for 4 h), the clear zone is no longer visible, the ribosomes becoming homogeneously packed [195]. Similar findings have been reported in the nervous ganglia of *Periplaneta americana*: the exclusion zone and the core of the MT is stained by lanthanum hydroxyde [123]. The axostyles of the flagellates *Saccinobaculus* and *Pyrsonympha* have MT separated by spacings which are crossed by lateral expansions; these empty zones appear larger after freeze-fracturing than after sectioning (dehydration may remove a water-soluble "glycocalyx" of possible polysaccharidic nature; [23]).

The exclusion zone, whatever its chemical nature, indicates that in many conditions MT are kept apart by some structures. This may not necessarily be a continuous sheath, and discrete lateral expansions could maintain the separation between the MT. In dorsal root ganglia of chick embryos, examined in tissue culture, groups of two to six MT in close contact have however been observed [22].

The exclusion zone differs from the "outer component" described by Behnke [12] in MT isolated from brain or blood platelets in a glycerol-dimethylsulfoxide medium, which is a helically wrapped sheet of tubulin around a normal MT; this is one of the many "tubulin polymorphs" which will be described below.

2.3.5 Side-Arms, Lateral Expansions, Links and Bridges

In many cells, MT do not appear "smooth", but covered by delicate wispy filaments, which may or may not connect them, like bridges, to other MT cell structures or organelles. A detailed study of these side-arms has been published by McIntosh [131], who studied the axoneme of *Saccinobaculus* (cf. Chap. 4) and the spermatogenesis of the rooster. The side-arms are regularly spaced, and their periodicity is apparently related to that of the MT helix: distances range from 4 to 48 nm, while a tubulin dimer measures about 8 nm. All side-arms do not form bridges between neighboring MT.

The importance of these side-arms is great: they are related to the proteins which purify with tubulin (vide infra), they may link together MT in stable and complex structures (cf. Chap. 3), and may be active in the MT-associated movements of various particles and organelles (Chap. 7). Moreover, similar lateral bridges are observed in the mitotic spindle MT, and may be related to the movements of chromosomes (cf. Chap. 10). In cilia and flagella, far more complex links join the MT with one another, as described in Chapter 4; the most interesting is the ATPase *dynein* [89, 90].

Some MT side-arms may be destroyed by agents which destroy MT: like MT, they disappear in HeLa cells cooled to 4 °C, and reform after warming, suggesting that they may be made of tubulin or a closely related protein [18].

In the testis of *Gerris remigis,* the MT are linked by curving structures, which are destroyed by colchicine [206]. The links between the helically wound MT of some spermatozoa are also modified by colchicine, growing from 8 to 16 nm, and it has been suggested that they are made of tubulin proteins [135].

In nerve cells, MT may be connected by similar bridges to various organelles (in particular mitochondria, cf. Chap. 3); they stain with lanthanum salts, a fact

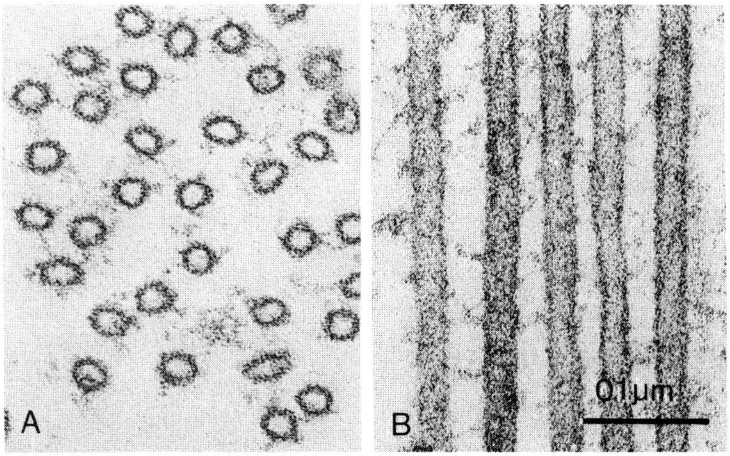

Fig. 2.10 A, B. MT reassembled in vitro with heavy molecular weight proteins: these appear as wispy filaments which on cross-section (*A*) of the MT appear to link these together. The longitudinal section (*B*) shows clearly that these side-arms are regularly spaced along the MT (from Murphy and Borisy [146])

suggesting a possible polysaccharidic nature [45] as observed in the axon of the crayfish *Procambarus clarkii,* where a three-dimensional lattice links the MT with the plasma membrane. The lanthanum staining could be explained by the presence of polyanionic groups, calcium, or mucopolysaccharides. Although similar techniques were used, these results differ from those mentioned above in *Periplaneta,* where lanthanum stained the "exclusion" zone, and not the lateral expansions [123].

The relation of some of these side-arms with the high molecular weight proteins which co-purify with tubulin will be mentioned below.

2.4 Chemical Composition

2.4.1 Technical Aspects

As mentioned in Chapter 1, the purification of tubulin(s) was the consequence of the observation by Taylor [201] of a specific binding of tritiated colchicine to the proteins of the spindle, and the demonstration by Borisy and Taylor [28] that a colchicine-binding protein could be extracted from cells and had a definite sedimentation constant (6S) [181]. The molecular weight of this protein was estimated by electrophoresis to be about 120,000 [178]. More precise methods of electrophoresis indicated that two closely proximate bands were found in tubulin preparations, suggesting that in the native state, the protein was a dimer formed by two nearly identical subunits, the α and β tubulins [181, 182].

The discovery by Weisenberg [221] that these proteins could be found in large quantities in the brain of mammals, and that a purified fraction may reassemble in vitro into MT, provided the proper ionic conditions are met, led to better purification methods, based on repeated cycles of assembly and disassembly [215]. The separation of the different protein components of "microtubule protein", which is a complex mixture (cf. [132]), has benefited from various techniques: electrophoresis, in particular on polyacrylamide gels, molecular sieve chromatography, and chromatography on agarose columns containing colchicine derivatives which combine electively to tubulin [145]. Another method of purifying tubulin relies on the fact that the *Vinca* alkaloids precipitate tubulins in large crystals (cf. Chap. 5) that may be isolated from the cell [33, 34]. However, this procedure may modify the molecule, as tubulin from rat brain is more antigenic in rabbits when isolated by VLB precipitation, and has a greater affinity for colchicine than when prepared by chromatography [207].

2.4.2 The Tubulin Molecule

The early histochemical work of Behnke and Forer [13] on various MT (spermatids of *Nephrotoma suturalis* Loew, spermatozoa of the rat, cilia of the rat's trachea) indicated that papain digestion destroyed the MT, although some were more resistant than others: MT were thus proteinaceous. The experiments on the fixation of tritiated colchicine confirmed that the alcaloid was bound to a substrate which appeared identical with the 6S protein that had been found in the isolated mitotic apparatus (cf. Chap. 10) and was identified in mitotic cells, in spermatozoa, in neurons, in cilia, in summary wherever MT had been described. Adelman [1] suggested the name of *tubulin*, which was rapidly accepted.

Shelanski et al. [182] showed that it was an acidic protein, combined with two molecules of guanine nucleotides (GTP and GDP), with a sedimentation constant of 4.8S at 20 °C, and a molecular weight of about 120,000 daltons. It is a dimer, one unit of which fixes electively the colchicine molecule, each monomer having about the same molecular weight. The protein extracted from the brain of mammals proved to be identical to that from cilia and flagella [221].

A quite different technique—selective extraction of MT protein from the mitotic apparatus by an organic mercurial, sodium meralluride, followed by precipitation by VLB—confirmed the dimeric protein nature of tubulin [19]. Brain tubulin was purified by Kirkpatrick et al. [113]; its molecular weight was 55,000 ±2,000, and the diameter of the monomer was about 4 nm, a figure which was in agreement with the ultrastructural data of negatively stained isolated MT mentioned above.

Bryan [35] has discussed the influence of isolation techniques on the apparent molecular weight. The Stokes radius of the protein is 42–44 nm. The molecule appears to have the shape of a prolate ellipsoid of axial ratio 5 : 7. The tertiary structure of tubulin has been studied by the analysis of its circular dichroism under various temperature, pH and solvent conditions: the dimer appeared to have about 48% random coil, 22% α helix and 30% β structure [208]. These figures are in agreement with the previous finding of 28% helix [168]. They apply to the protein studied at 4 °C—a temperature at which no MT are present, and the colchicine binding is low. At 37 °C, when the binding of colchicine is highest, the protein has no more helical content (half random coil, half β configuration). The action of temperature, nucleotides and MT poisons on the configuration of tubulin, and the complexities of its denaturation process have also been studied by this technique [208]. The aminoacid composition of α and β tubulins is given in Table 2.1.

Table 2.1. Amino acid composition of samples of sea-urchin tubulin (from Luduena and Woodward [129])

Aminoacid	mol/55,000 MW		Aminoacid	mol/55,000 MW	
	α	β		α	β
Lys	24.3	21.1	Gly	40.7	42.4
His	14.2	11.3	Ala	40.5	34.4
Arg	26.1	23.4	Val	33.1	33.1
Cys⁺	9.8	6.9	Met	9.8	16.6
Asp	48.4	54.5	Ile	24.6	18.9
Thr	29.4	31.5	Leu	35.5	36.4
Ser	20.9	24.2	Tyr	17.6	16.8
Glu	69.0	69.2	Phe	21.1	23.6
Pro	25.3	27.7	Trp	ND	ND

⁺ Cysteine determined as carboxymethylcysteine ND: not determined
Figures represent the average of two analyses

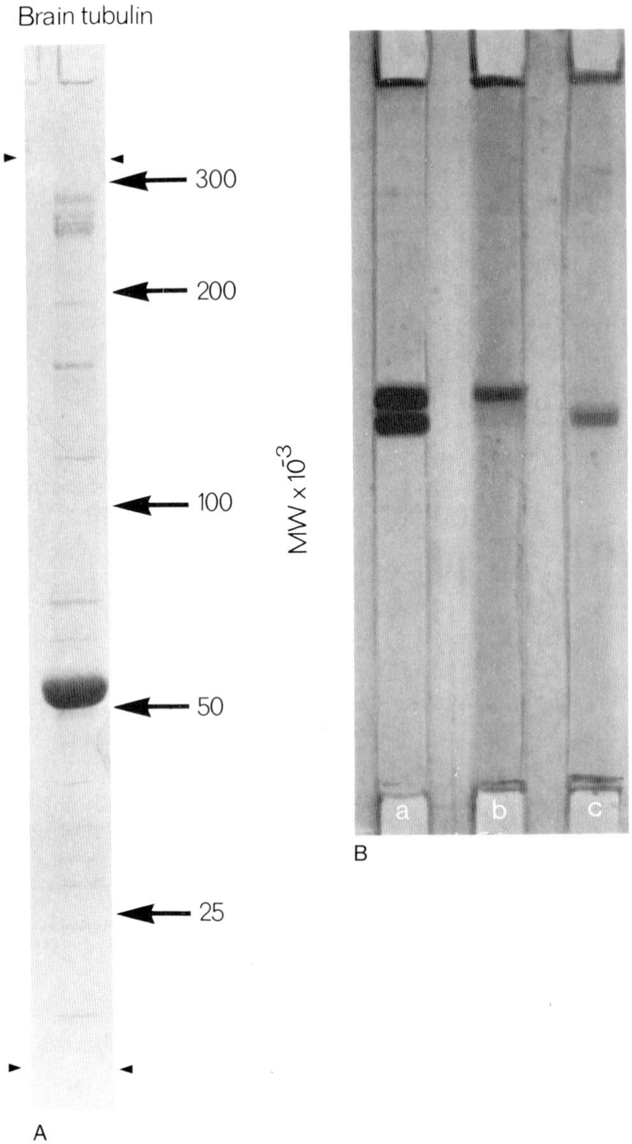

Fig. 2.11 A, B. (A) Electrophoresis of bovine brain tubulin. An homogenate of bovine brain, clarified by centrifugation, was polymerized into MT at 37 °C and the MT collected by centrifugation (50,000 × g for 75 min at 25 °C). The resuspended pellet was disassembled by cold at 0 °C for 30 min, then centrifuged at 100,000 × g for 60 min. The supernatant is stored at −20 °C after addition of 3.3 ml per 10 ml of glycerol. The SDS (sodium dodecyl sulfate) electrophoresis on a 5 – 15% acrylamide slab gel using Laemmli buffers shows several bands of high molecular weight MAPs and the thick band close to 50,000 daltons representing the two tubulins, α and β, which are not separated here (from J. Bryan, personal communication). (B) Separation of α and β tubulins from chick brain. The proteins in the colchicine-binding fraction were concentrated by ultrafiltration, reduced and carboxymethylated, and subjected to electrophoresis in 8 M urea-acrylamide gels. *a*, α and β tubulins. *b* and *c*, purified tubulins separated on 5% acrylamide gels using a 8 M urea system with 25 mM triglycine buffer at pH 8.0 (from Bryan and Wilson [41])

In purified tubulin, the number of sulfhydryl residues, as measured by DTNB and 4,4-dithiopyridine, and expressed per 55,000 molecular weight subunit, is about 7.2 [119, 141]. This value is that of tubulin prepared in the presence of glycerol, while without glycerol only four sulfhydryl residues are detected, indicating probably an oxidation to disulfide bonds [141, 142]. The role of such bonds in MT assembly and the action of –SH reagents on MT will be further discussed in Chapters 5 and 10.

The sulfhydryl reagent, mercury orange, (1,4-chlormercuryphenyl azo-)2-naphtol, strongly stains cytoplasmic and spindle MT in various cells. They then appear dense, without a central clear zone. MT cross-bridges are also stained [98].

2.4.3 α and β Tubulins

The dimeric structure of tubulin was early recognized [180] and amply confirmed by research on the fixation of colchicine and VLB, and the location of the guanine nucleotides. The polyacrylamide gel electrophoresis of tubulin preparations shows two closely located bands, named α and β tubulins, the β subunit having the greater electrophoretic mobility [41]. This separation results from a difference of charge. These two tubulins have been found in nearly all cells studied.

Amino acid sequence analyses [129, 130] have demonstrated that α and β tubulins are closely related (Table 2.2).

Each has been highly stable in the course of evolution, as indicated by the similarities of tubulins from two widely separated species like the chick and the sea-urchin: in α tubulin, no differences were found in the 25 first N-terminal amino acids (Table 2.2). It is likely that α and β tubulins derive from a common ancestor protein. They do differ by the location of their specific binding sites for

Table 2.2. Amino acid sequences of NH_2-terminal regions of α and β tubulins from chick brain and outer doublet microtubules of sea-urchin sperm (from Luduena and Woodward [130])

	1 2 3 4 5 6 7 8 9 10
Chick brain α tubulin	Met-Arg-Glx-Ser?Ile -Ser?Ile -His -Val-Thr-
Sea-urchin α tubulin	Met-Arg-Glu-Ser?Ile -Ser?Ile -His -Val-Thr-
Chick brain β tubulin	Met-Arg-Glu-Ile -Val-His-Ile -Gln-Ala-Thr-
Sea-urchin β tubulin	Met-Arg-Glu-Ile -Val-His-Met-Glx-Ala-Thr-
	11 12 13 14 15 16 17 18 19 20
Chick brain α tubulin	Gln-Ala-Thr-Val-Gln-Ile-Thr-Asx -Ala-Ser?
Sea-urchin α tubulin	Glx-Ala-Thr-Val-Glx-Ile-Thr-Asx -Ala-Ser?
Chick brain β tubulin	Gln-Ser -Thr-Asx -Gln-Ile-Thr-Ala- ? -Phe-
Sea-urchin β tubulin	Glx-Ser -Thr-Asx -Glx-Ile-Thr-Ala- ? -Phe-
	21 22 23 24 25
Chick brain α tubulin	? -Glx-Leu-Try-Ser?
Sea-urchin α tubulin	? -Glx-Leu-Tyr-Ala?
Chick brain β tubulin	Trp?Glx-Val -Ile-Ser?
Sea-urchin β tubulin	? - ? -Val -Ile-Ser?

guanine nucleotides, and the lateral and longitudinal sites necessary for their assembly into tubules. They differ also by the sites of fixation of specific poisons such as colchicine and VLB (cf. Chap. 5). Two definite biochemical properties have been observed: β tubulin is specifically phosphorylated [68] and α tubulin is the substrate of a curious enzyme, tubulin-tyrosine ligase, which, in the presence of ATP, without any specific tRNA, attaches a single tyrosine molecule to the N-terminal group of the molecule. The substrate is the tubulin dimer, and tyrosilation does not appear to interfere with MT assembly [166, 167].

Electrophoretic data indicate that the two tubulins are present in equal quantities in most MT studied. The ultrastructural data suggest that MT are assembled from identical, $\alpha\beta$ dimers, as suggested by Bryan and Wilson [41]. If solubilized tubulin is treated with a cross-linking agent such as dimethyl-3,3′(tetramethylene-dioxy)-dipropioimidate, and studied on an acrylamide gel system capable of discriminating between $\alpha\alpha$, $\alpha\beta$, and $\beta\beta$ dimers, it is found that most tubulin is of $\alpha\beta$ type. Poisons such as colchicine and VLB also stabilize the $\alpha\beta$ configuration [128, 129]. This is also in agreement with the findings on nucleotide relations with tubulins (vide infra). The dimeric nature of tubulin is important to keep in mind when considering the geometry of MT and their mode of assembly.

2.4.4 Other Tubulin Variants

The stability of MT varies considerably, even in the same cell [13] and their reactions to MT poisons such as colchicine, and physical agents such as cold, may also differ. The resistance of MT to osmium acid fixatives is also variable. Variability may result from the association of tubulins with other proteins, as in the doublets of cilia, or from differences of metabolic turnover of tubulins in various MT structures (basal bodies, cilia, and neurotubules being particularly stable). However, the number of tubulins for which an amino acid sequence has been performed is far too small for any generalization to be made at this time, and the possibility that tubulins differ, more or less, in their amino acid sequences cannot be excluded. There are several indications of species differences of tubulins. For instance, the tubulin purified from *Aspergillus* has a lower affinity for colchicine than mammalian tubulin [58] and this seems true for other tubulins of plant origin [97]. That extracted from *Chlamydomonas* cannot assemble in vitro, whatever the concentration, the presence of ions or nucleotides, and the temperature. It cannot co-polymerize with gerbil brain tubulin and inhibits the assembly of this into MT [78, 79]. Antibodies prepared in the rabbit against tubulin from *Chlamydomonas* react only with β tubulin, and it is possible that this is the only form present in this species [161]. However, in a study of tubulin synthetized in vitro from m-RNA prepared from *Chlamydomonas,* undergoing cilia regeneration, α and β tubulins could be detected ([213] cf. Chap. 3).

Other differences between tubulins in complex structures such as the ciliary doublets will be discussed in Chapter 4 (cf. [196, 227, 228]).

From studies on the assembly of MT, Kirschner [114] had suggested that two types (X and Y) of tubulin were present in dog brain extracts. X tubulin has a high colchicine binding capacity, and is unable to form by itself either MT or other structures such as rings or spirals, which are important for the initiation of MT assembly (vide infra). It contains no phosphate. Y tubulin, which migrates faster in agarose columns, has a weak affinity for colchicine, but is capable of forming rings, spirals and MT spontaneously. It would contain 1 mol of phosphate per dimer. The two fractions are incorporated into MT. Apart from the presence of phosphate—it was mentioned above that β tubulin is phosphorylated [68]—these results must be reconsidered in the light of recent findings on MT assembly.

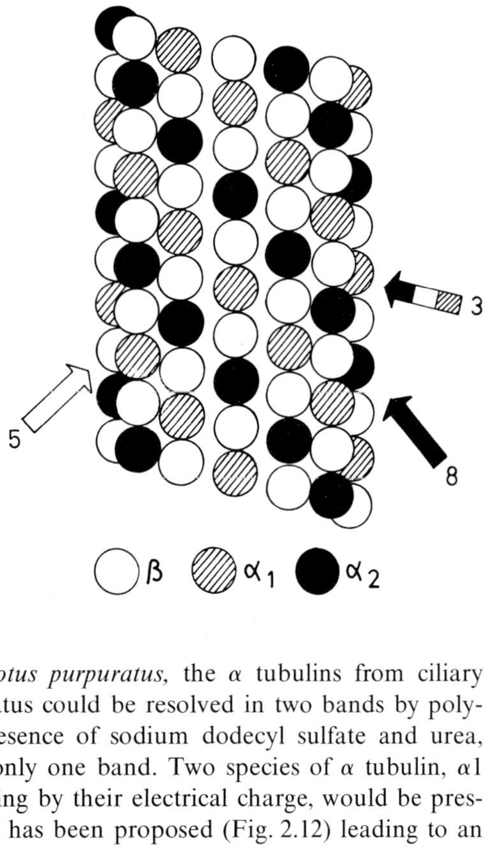

Fig. 2.12. Possible location of α1 and α2 tubulins in MT. Three-start and five-start helices of alternating α1–β and α2–β tubulin dimers. Homogeneous eight-start helices of α1–β and α2–β dimers (redrawn from Bibring et al. [20])

In the sea-urchin, *Strongylocentrotus purpuratus*, the α tubulins from ciliary doublets and from the mitotic apparatus could be resolved in two bands by polyacrylamide electrophoresis in the presence of sodium dodecyl sulfate and urea, while the flagellar doublets showed only one band. Two species of α tubulin, α1 and α2, in equal quantities but differing by their electrical charge, would be present, and a model of the tubulin helix has been proposed (Fig. 2.12) leading to an eight-start succession of dimers and a 13 protofilament MT. This may explain the 16 nm spacing observed in these MT [20].

2.4.5 Antigenicity of Tubulin

As indicated by the amino acid sequence analysis, tubulins have remained very stable in evolution, histones being apparently the only proteins which have undergone less change since the origin of eucaryotes [57]. MT being ubiquitous, one could have expected the tubulins to be weakly antigenic, and animal species to have a natural tolerance to these antigens. However, antibodies against MT have been prepared, and as immunological methods are often more sensitive than biochemical data, they help to detect differences between tubulins with an identical electrophoretic motility and a similar overall amino acid composition.

The first results appeared to confirm the near identity of all tubulins: antibodies directed against the spermatozoa of *Arbacia punctulata* reacted with MT from cilia, spermatozoa and mitotic spindle [82]. In mammals, antibodies prepared by injecting rabbits with tubulin from mouse ascites tumor, and from brains of rat, pig, and calf, gave an identical reaction in Ouchterlony diffusion tests. Tritiated colchicine was used to mark tubulin and to demonstrate the specificity of the reaction. Hence, if MT are not necessarily identical they do have a common antigenic determinant [65]. Rabbit antibodies against MT proteins of *Tetrahymena* combined specifically with the proteins of cilia; however, immunofluorescence studies showed that the oral apparatus and the macronucleus were also stained [199, 200]. Further observations have confirmed the existence of a common antigenic determinant in MT from man, mammals, birds, reptiles, teleostean fishes, and diptera [57]: this does not imply that all these MT are chemically identical. Antibodies prepared against the outer doublets of sea-urchin sperm were used to demonstrate the various MT of mammalian cells—intermitotic and mitotic—as well as the VLB-precipitated MT crystals [210, 211, 212]. Bovine brain tubulin has also been used in similar observations as an antigen: the fluorescent antibodies stain a dense network of fibrils in intermitotic cells, and the spindle structures in mitosis; these fibrils were destroyed by various poisons such as colcemid and low temperatures (cf. Chap. 5), demonstrating their MT nature [30, 31, 32].

More refined studies are bringing further information about the possible multiplicity of tubulins, even in the same species and within the same cell. The ameboflagellate, *Naegleria gruberi* (cf. Chap. 4), has two different sets of tubulin: one is involved in the formation of basal bodies and cilia, which takes place when this unicellular assumes a motile form. None of the preexisting large cytoplasmic pool of tubulin is used in flagellar differentiation, and flagellar tubulin has different antigenic determinants. Most of it is synthesized de novo during the change from an ameba to a flagellate, and may be coded by different genes [83, 84]. This confirms that more than two types of tubulins may exist, a fact which may explain differences of behavior towards various agents of different MT (resistance to fixatives, fragility towards MT poisons) [84]. These facts show how cautious one should remain in view of all-embracing theories of MT functions and structure.

2.4.6 Carbohydrates and Lipids in MT

So far, MT have been considered proteins without any other constituents, except the nucleotides which will be studied below. Conflicting evidence about the presence of carbohydrates (aminosugars in particular) and lipids has often resulted from difficulties of purification of MT. Several cycles of in vitro assembly and disassembly of tubulins have led to a better purification.

Following an early affirmation that tubulin is a glycoprotein [77], porcine brain prepared by the method of Weisenberg et al. [217] was found to contain carbohydrate, consisting of glucosamine, galactosamine, galactose, mannose, fucose and sialic acid. This would imply that two types of oligosaccharide are present, each mole of MT protein dimer containing seven monosaccharide residues [136]. This appears however to result from contaminants, and an improved method of purification has led to a tubulin with no detectable amino sugars, and less than 1.2 mol of neutral sugar per dimer [68].

The arguments for the presence of lipids (probably phospholipids) in MT are at this moment only indirect ones. When MT protein from chick embryonic brain and muscles or from HeLa cells is phosphorylated in vitro, 80% of the trichloracetic acid insoluble radioactivity is extractable in organic solvents; this fraction is mainly phosphatidic acid. In vivo, the phosphorus combined with tubulin is part of a mixture of phospholipids and lecithin. A diglyceride kinase activity was also identified in MT from various tissues [56]. These results have been partially confirmed. When tubulin is purified on DEAE cellulose, a chloroform-methanol extractable phosphate is isolated [69]. After photoinactivation of colchicine bound to tubulin, the alkaloid becomes incorporated into a PCA-precipitable material presumed to be a lipid or phospholipid [37]. The addition of phospholipase A to a brain extract competent to form MT prevents completely the MT assembly at 37°, leading also to the conclusion that tubulin may be associated with phospholipids [37, 38].

2.4.7 The Tubulin-Associated Nucleotides

The early work on purified tubulin indicated that two molecules of guanosine nucleotides (GDP or GTP) were bound to the dimeric molecule of 120,000 molecular weight [181, 182, 221]. Two sites are present: at one, GTP is rapidly bound to tubulin and exchanges with the medium, whereas at the other site GDP is bound much more firmly; it may be transphosphorylated by incubation with GTP [216, 218, 219, 221] or with ATP [16]. Hence the proposal to name these sites "E" (exchangeable) [106, 108, 109] and "N" (non-exchangeable). The fixation of GTP to the E site proceeds within a few seconds. Other nucleotide triphosphates bind very weakly or not at all at this site (ATP, CTP, UTP). Mg^{2+} and free sulfhydryl groups have been considered necessary for the binding of GTP to tubulin [6, 7]. The turnover of GTP on the N site has been explained by the presence of a low GTPase (or ATPase) activity in tubulin preparations and by the action of a transphosphorylase, which may transport the phosphorus from the N to the E site [106, 108, 109]. As transphosphorylation is faster with ATP than with GTP, two hypotheses have been put forward: either that tubulin has a third nucleoside binding site, with a higher affinity for ATP than GTP, and an intrinsic transphosphorylase activity, or that a separate transphosphorylase is present in tubulin preparations [106]. The role of these nucleotides in MT assembly will be discussed.

2.4.8 Enzymatic Activity of Tubulin?

Enzymes are involved in MT assembly: transphosphorylase (for GTP formation on the N site) [106], GTPase (for hydrolysis of GTP to GDP on the E site during MT assembly), perhaps ATPase.

The problem of the relations of tubulin to a cAMP-activated protein kinase remains a subject of debate. It has been mentioned that β tubulin is phosphorylated [68]. Tubulin extracted from the brain of mammals has been considered to act as a phosphokinase [94]. Tubulin precipitated from porcine brain by VLB maintains this phosphokinase activity, which would be an intrinsic property of the molecule [120, 121, 189, 190, 191] or may belong to a contaminant [164, 165]. However, the fact that the protein kinase activity persists in tubulin purified by assembly – disassembly, coprecipitates with tubulin and VLB, cosediments with reassembled MT, and has some unique properties such as inhibition by 5′-AMP, gives support to the enzymatic nature of tubulin or of a very closely associated protein [189, 190, 191]. The technical difficulties in purifying tubulins should not be overlooked and at this moment, it is not evident that any of the enzymatic activities mentioned belongs to the tubulin molecule itself.

The substrate of this phosphokinase remains unclear: in the early work it was suggested that it was tubulin itself. It would be rather surprizing if tubulin is thought to have the enzymatic activity; more recent work [187] suggests that a high molecular weight protein (about 300,000 by electrophoresis) closely associated with tubulin is the true substrate. This protein has been compared to muscle ATPase and to dynein, and its phosphorylation has been suggested to play some role in motility associated with MT. In contrast, tubulin isolated from the blood platelets of the pig has been confirmed to be associated with a protein kinase, the substrate of which would be the tubulin molecule itself [51]. Rat brain tubulin, purified by affinity chromatography, is associated with a highly specific phosphoprotein phosphatase. This enzyme is cAMP dependent, and phosphorylates some of the high molecular weight MAPs associated with the 36S fraction of tubulin. However, the purification on columns containing colchicine may not be quite specific for tubulin, and may retain other proteins [174].

2.4.9 Microtubule-Associated Proteins (MAPs)

When tubulin is purified by any of the methods listed above it is accompanied, on electrophoresis columns, by several associated proteins of higher molecular weight (350,000 – 300,000). Two or three bands are usually observed [24, 132, 175, 185]. These have been found in tubulin preparations from various sources, and their presence, even after repeated cycles of assembly – disassembly in vitro, often in stoichiometrical relations with tubulin [26], suggests that they play a part in the structure of MT. They are not the only proteins which may be associated with tubulins: the *tau* factor, described below, which is a protein of about 70,000 molecular weight, may also be involved in tubulin assembly [157, 214]. These accessory proteins should not be confused with the many (more than 30) proteins which may be extracted at the same time as tubulin from complex structures such as cilia [146], and which are constituents of the complex structures of these organelles (cf. Chap. 4), nor with the heavy assemblies of tubulin molecules (rings, helices) which are intermediate steps in MT assembly.

The possible role of these MAPs will be discussed when considering the problems of tubulin assembly into MT. An important observation is that tubulin may

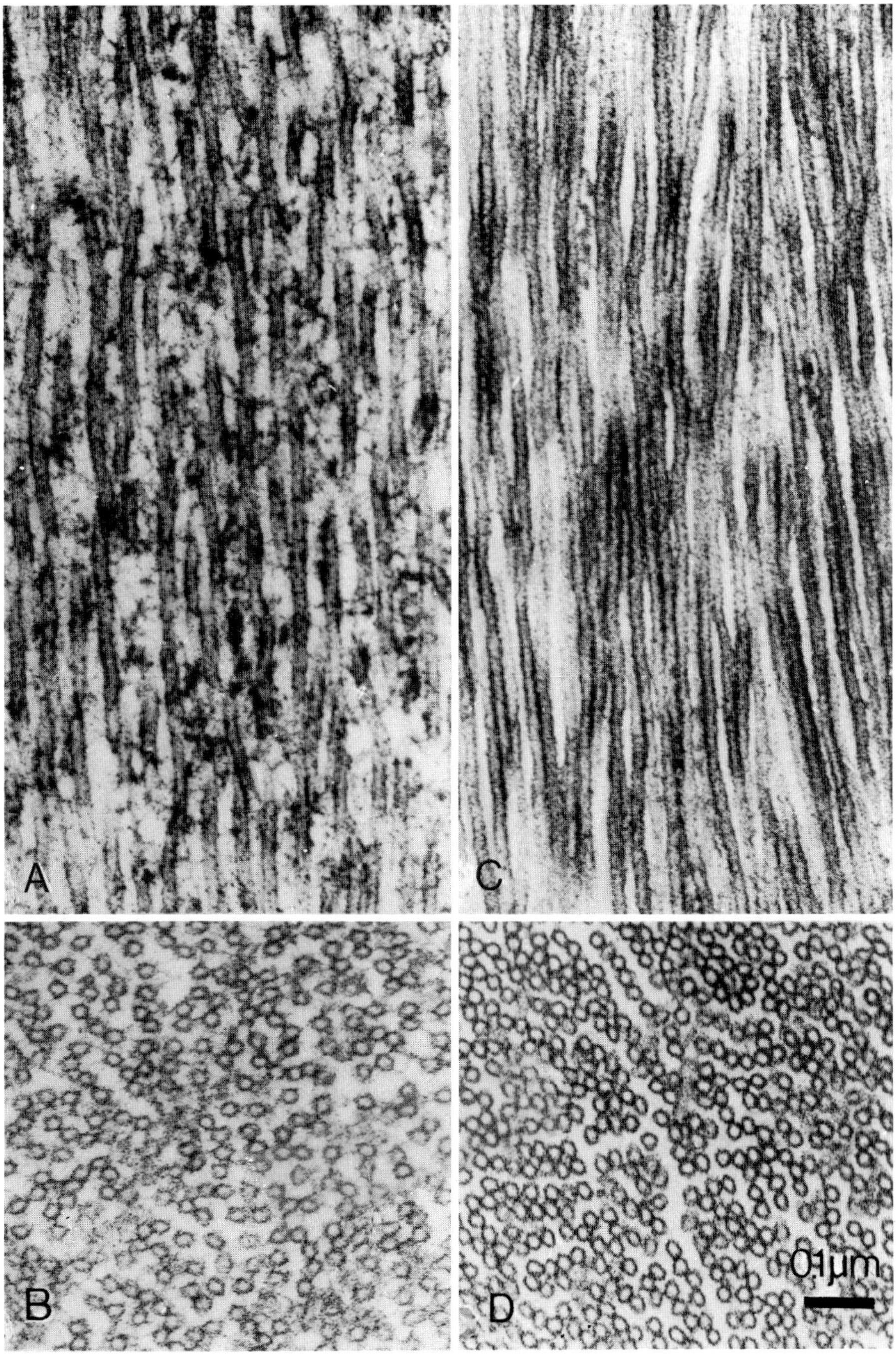

Fig. 2.13 A–D. Thin sections of MT assembled in the presence of high molecular weight proteins (peak 1 MAP fraction): (A) and (B); and in the absence of this fraction: (C) and (D). The MT without MAPs do not show any lateral wispy filaments as in (A) and (B), and pack more tightly (from Sloboda et al. [185]) (Cold Spring Harbor Laboratory, Copyright 1976)

form MT in the absence of these proteins: these tubules then show a smooth surface, and are devoid of the wispy filaments which are related to the side-arms, lateral expansions, and bridges described above [62, 185, 186].

2.5 Assembly and Disassembly of MT

2.5.1 Introduction

The formation of long tubular structures from dimeric tubulin has been the subject of many works in the last few years, and the various steps of assembly, its thermodynamics, the factors controlling the formation of MT in vivo and in vitro are becoming better understood. The problem—which may be compared to the assembly of structures such as viral capsids [112] or other elongated polymers like actin—is complex because of the various possible assemblies in vitro, and the poor understanding of the regulatory factors in vivo. The fact that cells contain a large pool of unassembled tubulin molecules, and that the formation of MT only takes place in relation to changes of cell shape and movement, or in preparation for mitosis, indicates that regulatory mechanisms must exist, if only to prevent all the cytoplasmic tubulin from assembling into MT.

The formation of MT is controlled, in the cell, by several organelles or centers, grouped under the name of "microtubule organizing centers" (MTOC) [160]. These, the role of which is clearly visible after staining with immunofluorescent antitubulins, are the centrioles, the polar regions of the mitotic apparatus, the basal bodies, the kinetochores, the pores of the nuclear membrane.

The discovery that tubulin solutions from the brain could assemble in vitro provided several conditions of concentration, ionic strength, and temperature were fulfilled [215], has provided a tool for the study of the factors necessary for the formation of MT or other associations of tubulin molecules. In the following pages, the problems of assembly will be studied in three different conditions: in tubulin solutions (with or without MAPs), in similar solutions in the presence of nucleating centers—what may be called "site-initiated" assembly [25]—and last, in the living cell where the conditions are far more complex and more difficult to analyze. The assembly of tubulins into structures involving several MT and other proteins—cilia, centrioles, axonemes—will be studied in Chapters 3 and 4. Further aspects of assembly in vivo will be considered in Chapters 6 to 10.

The in vitro assembly of tubulin from other organs than the brain has been described (rat neoplastic glial cells [225], human blood platelets [104]), but it is far from certain that all tubulins may form MT in vitro (cf. [79]).

2.5.2 MT Assembly In Vitro

In 1968, Stephens [196] solubilized the protein from the flagellar outer fibers (doublets) of *Arbacia punctulata* spermatozoa, and found that this solution, when diluted in cold CO_2-free water or in tris buffer at pH 8.0, formed fibrous aggregates which were demonstrated by electron microscopy to consist of ribbons of linearly disposed subunits. Nucleation centers were however indispensable, and no assembly was observed when all MT fragments had been eliminated by centrifugation.

Weisenberg [215] was the first to assemble MT from tubulin extracted from the brain of rats, and purified by high speed centrifugation. The resulting protein solution (not necessarily made of tubulin only) assembled to form normal MT provided several conditions were met: (1) the concentration had to be sufficiently high; (2) guanosine nucleotides were required; (3) calcium ions must be removed, preferably with the strong chelator EGTA[4] which does not remove Mg^{2+} ; (4) the solution should be warmed to 37 °C. The tubules formed rapidly, as could be assessed by turbidity measurements, which increase in proportion to the amounts of MT, independent of MT length [86]. The assembly can be objectivated by electron microscopy, and by measurements of viscosity and sedimentation rates.

Since this observation, much research has been devoted to the various aspects of MT assembly in vitro, and this has been compared to that taking place in the cell. Many conditions have to be satisfied for MT assembly, and this having been studied in various conditions, apparently contradictory results have been published. An attempt will be made here to list thc principal factors which are necessary for assembly in a homogeneous solution of tubulin protein, with or without nucleation centers. Most of the results have been obtained with tubulin extracted from mammalian brain.

2.5.2.1 Biochemistry of Assembly. First, a minimal concentration is required: this is around 1 mg/ml of tubulin, although it may vary according to the techniques used [152]. The optimal pH is 6.9. Strict ionic requirements are necessary, and this may be important for a proper understanding of assembly in vivo. As assembly was first observed in the presence of a chelator for calcium, this ion was thought to play a regulatory role in vivo; however, the inhibitory concentrations required are larger than millimolar, that is to say far greater than those which may be expected to be found in the cell. *Magnesium* is required, as demonstrated by the inhibition by EDTA, which chelates at the same time Ca^{2+} and Mg^{2+}. Large concentrations of Mg^{2+} (more than 20 mM) are however inhibitory. The possible roles of these ions in assembly have been discussed by Olmsted et al. [150, 151, 152].

Low concentrations of Ca^{2+} (10×10^{-6} M) modify the MT of heliozoa, leading to the formation of many incomplete, C-MT [177]. At physiological concentrations of Mg^{2+}, micromolar concentrations of Ca^{2+} inhibit the assembly of brain tubulin, to which Ca^{2+} binds stoichiometrically [99, 171]. Lithium ions ($0.2 - 1.0 \times 10^{-3}$ M) promote rat brain tubulin assembly if small concentrations of Mg^{2+} are present and, like Mg^{2+}, Li^{+} may antagonize the inhibitory action of colchicine or VLB on assembly [17, 192].

The assembly reaction is inhibited by *cold,* and has been considered endothermic [150], a fact apparently in agreement with previous data on the kinetics of assembly in vivo. The enthalpy and entropy increases related to assembly were considered to be the consequence of the volume increase due to the liberation of

[4] EGTA: ethylene glycol bis (β-aminoethylether)N,N,N′,N′-tetraacetic acid

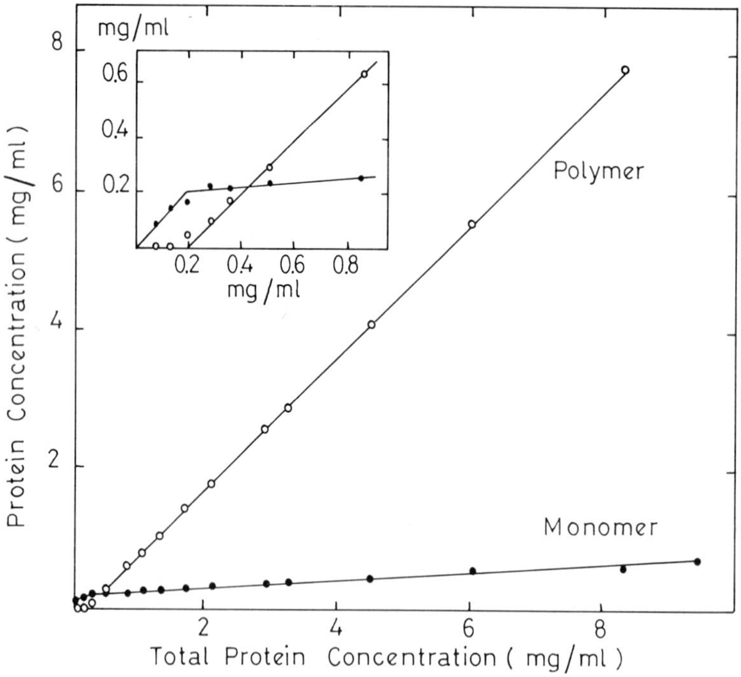

Fig. 2.14. Sedimentation analysis of the dependence of porcine brain tubulin assembly on protein concentration. Purified MT protein was prepared in the presence of 0.5 mM GTP, 0.1 mM $MgCl_2$, and diluted immediately, before incubation at 37 °C. The amount of assembled tubulin was measured after centrifugation. The graph indicates that assembly is proportional to the tubulin concentration. The *inset*—data for the lowest concentrations on an expanded scale—shows a lag indicating the need of critical concentration, probably necessary for the initiation step (redrawn from Olmsted et al. [152] and Johnson and Borisy [110])

bound water [29]. However, this conclusion cannot be accepted without caution today. The observations in vivo (vide infra) are not necessarily related to the assembly of MT alone. Moreover, recent data indicate that the assembly may be essentially athermic, as shown by calorimetry measurements of the enthalpy change of assembling brain tubulin at 17 °C and 25 °C, which is less than 1 kcal per mol of 6S dimer of tubulin [198]. The results previously published, indicating a strong positive enthalpy, may be explained by the nucleation process taking place at the initiation of the assembly, independent of the elongation into MT.

Gaskin [87, 88], who studied the assembly of brain tubulin, concluded from its temperature dependence below 23 °C that the reaction enthalpy was +21 kcal/mol. In contrast disassembly, observed after adding Ca^{2+} or cooling the preparation, had an activation enthalpy of −28 kcal/mol, with first-order kinetics. Disassembly by Ca^{2+} appears however more complex than after cooling, and shows a slow exothermic activation. These figures may be compared to those for actin, which has a positive enthalpy for assembly but, contrary to tubulin, also a positive enthalpy for disassembly (+10 kcal/mol; [87, 88]). Some contradictions of these studies may be explained by the fact that the equilibrium constants decrease at higher temperatures, thus affecting enthalpy and entropy changes [70].

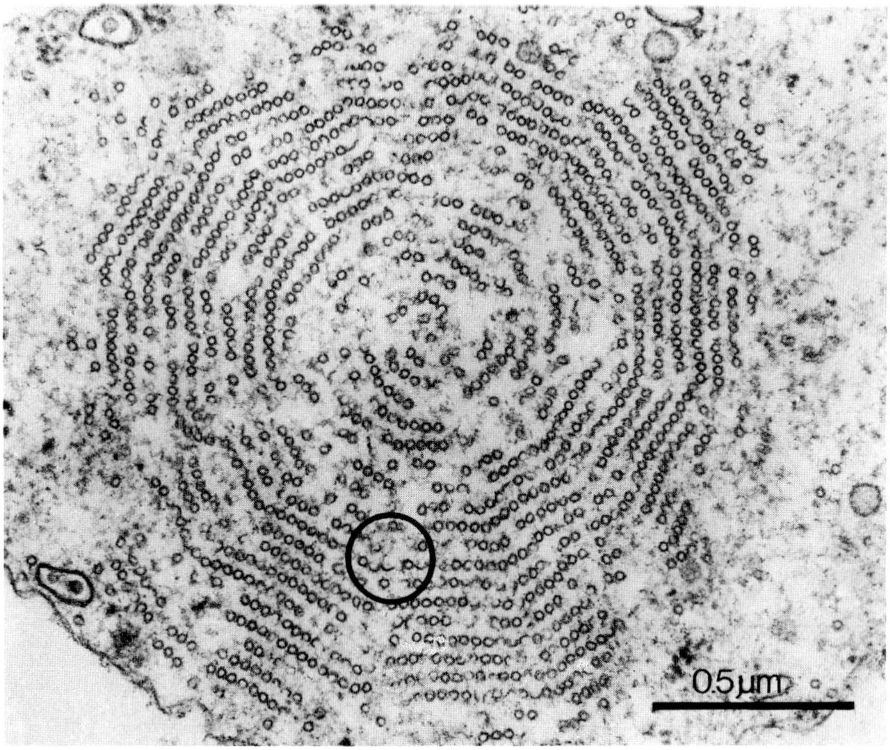

Fig. 2.15. *Actinosphaerium eichorni*. Axoneme treated for 10 min in 0.5 mM Ca^{2+} + 0.5 μg/ml A23187. There are many missing MT (compare with Fig. 3.16). A group of incomplete C-MT is encircled (from Schliwa [177])

These results are important to keep in mind when considering the rapid cycles of assembly and disassembly of MT in living cells.

Many other factors influence MT assembly, even in the simplest conditions. While in artificial conditions purified tubulin may assemble without any other protein, in more physiological conditions *MAPs* appear indispensable. These belong to two groups: the heavy molecular fractions, and the so-called "tau" protein. The heavy molecular weight MAPs are closely associated with tubulin, as evidenced by the fact that after several cycles of assembly – disassembly, they remain present, and appear to be in a stoichiometric relation to tubulin—about 1 MAP molecule per tubulin dimer. The two principal MAPs have a molecular weight of 350,000 and 300,000 daltons. They compose about 6% and 20% of the total MT protein fraction [185, 186]. They may be separated from tubulin by molecular sieve chromatography: the electron microscope observations indicate that the rings of tubulin molecules which are apparent during the first steps of assembly are no longer visible in the absence of MAPs. These proteins appear, in some conditions, indispensable for tubulin assembly in vitro: the initial rate of assembly and the total amount of MT formed are proportional to the amount of MAPs present [185, 186]. As the number of rings is also proportional to the MAPs, these may be considered true initiation factors [24]. The relations between the rings and the MT will be described below. Another important finding is that the amount of wispy filaments associated with the surface of MT appears to be proportional to the amount of MAPs, a fact which has been confirmed by studies on MT assembled in their absence [24, 101, 146].

A high molecular weight protein has been extracted from the flagella of *Arbacia punctulata* and *Chlamydomonas* by 0.5 M KCl and found to promote the assembly

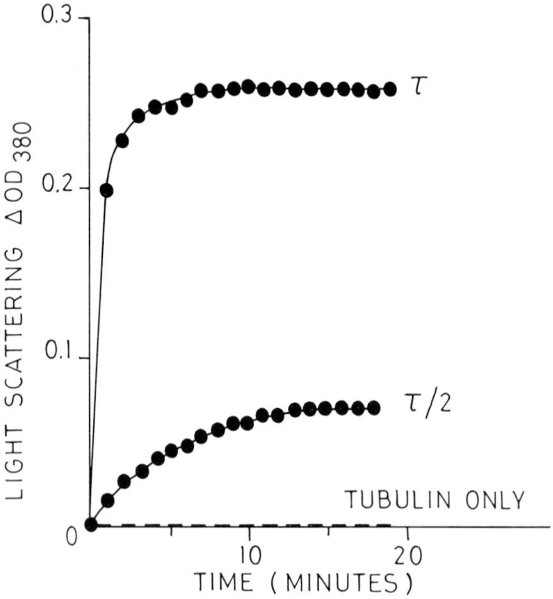

Fig. 2.16. Turbidometric assay of MT assembly in vitro: 0.3 ml of a 4 mg/ml solution of tubulin in MES-EDTA buffer with 3 mM GTP were warmed at 37 °C in the presence of a 1 mg/ml solution of *tau*, or with a half dilution of this solution. No assembly takes place in the absence of *tau*, and this factor acts stoichiometrically (redrawn from Penningroth et al. [157]) (Cold Spring Harbor Laboratory, Copyright 1976)

of calf and chick brain tubulin. Contrary to other MAPs, this fraction leads to the assembly of smooth MT, without any lateral expansions. It contains the flagellar ATPase *dynein* (cf. Chap. 4; [24]).

Another protein involved in assembly has been named *tau* by Kirschner [114, 115, 116] and Weingarten et al. [214]. It is an extremely heat-stable protein (resistant to 100 °C for 5 min), insensible to sulfhydryl reagents. It copurifies with tubulin through several cycles of assembly – disassembly. It may explain the differences between the X and Y tubulins, being mainly associated with the second. Its molecular weight has been estimated to be about 70,000 daltons. It is a very elongated molecule with a length-to-width ratio of 1 : 20 [157]. Further purification by hydroxyapatite chromatography has led to the separation of two fractions, *tau* I (molecular weight: 60,000 – 70,000) and *tau* II, which comprizes heavier components. For Penningroth et al. [157], both the rate of assembly and the yield of MT vary directly with the concentration of *tau* (Fig. 2.16) which would act stoichiometrically and not catalytically. It would mainly affect the longitudinal binding of tubulin dimers, forming first rings and later protofilaments [157]. More recent studies on the relations between accessory factors such as *tau* and tubulin do not however confirm a stoichiometric relation (Borisy, 1977, personal communication). Another suggestion is that *tau* may be a breakdown product of MAPs [132, 185, 186].

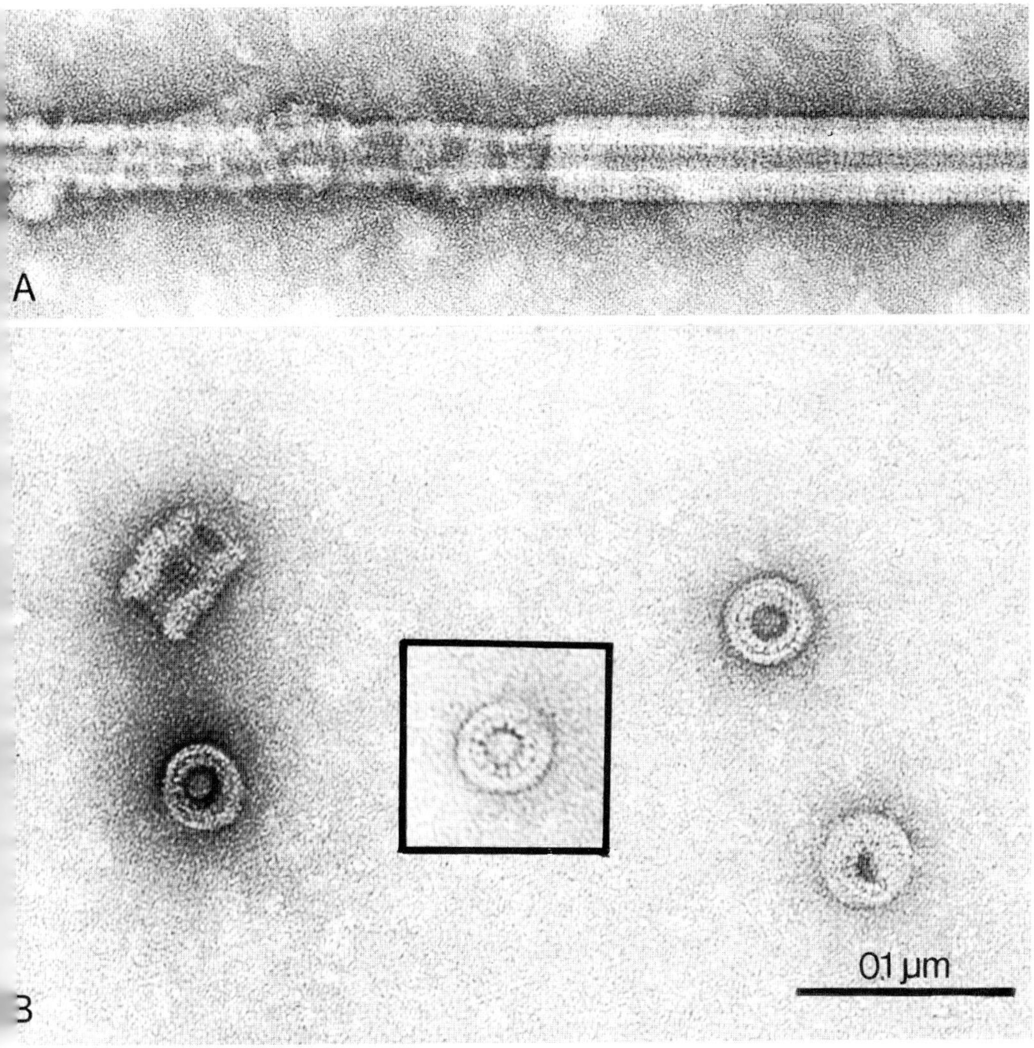

Fig. 2.17 A, B. Decorated MT. (A) Duplex MT induced by low levels of protamine. Uranyl acetate negative stain. 10×10^{-6} M of purified tubulin was assembled in assembly buffer (100 mM MES, 1 mM EGTA, 0.5 mM Mg^{2+}, 1 mM GTP, pH 6.8) with 5×10^{-6} M protamine. (B) Double rings and short segments of duplex MT seen when protamine is added to tubulin at 4 °C. The *inset* shows a ring with 13 inner subunits. Uranyl acetate negative staining (from Jacobs et al. [107])

The ionic requirements for assembly have already been mentioned; further research has indicated that *polycations* help assembly [73, 75], while *polyanions* are inhibitory. The observation that RNA inhibits the assembly of tubulin from sea-urchin extracts [39, 40] led to studies of other RNAs, which all show the same effect. Further experiments indicated that poly(L)glutamic acid had a similar action, which was also found with poly(A)-agarose and various cation exchangers,

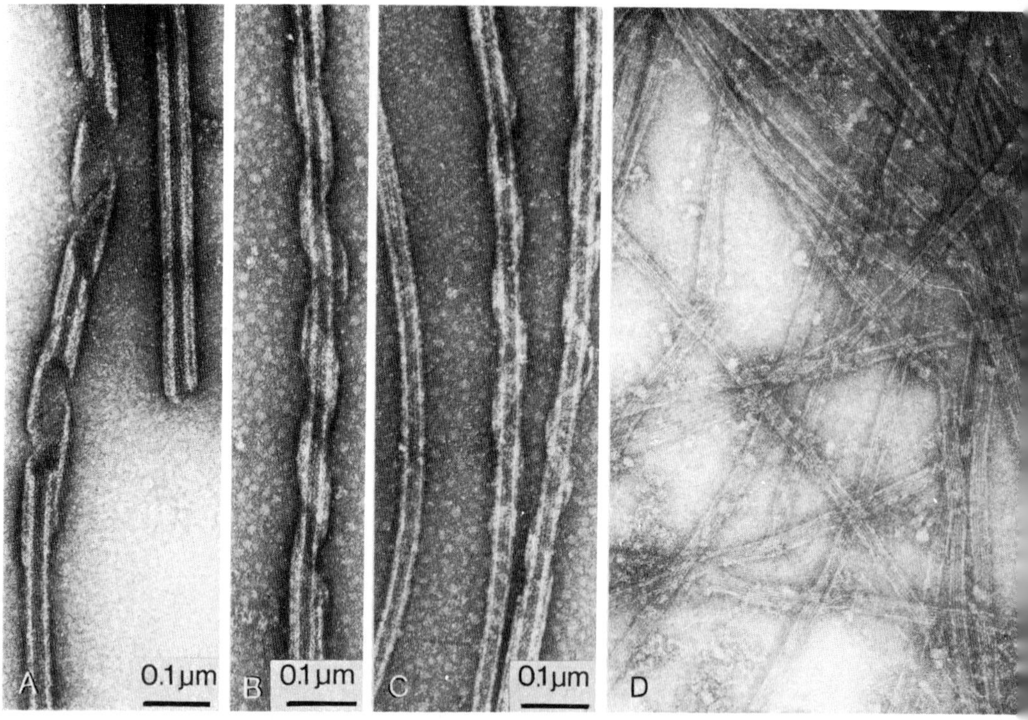

Fig. 2.18 A–D. Normal and double-walled MT assembled from a mixture of phosphocellulose purified tubulin (0.13 mg/ml) and RNase (1.2 mg/ml). (A) *Right:* a typical double-walled MT; *left:* spiral sheet of tubulin partially wrapped around a normal MT. (B) and (C) MT with a helically wrapped sheet of protofilaments, at an angle of about 45° (from Erickson [73]). (D) Normal MT assembled from phosphocellulose purified tubulin plus DEAE dextran of low charge density (0.4 mg DEAE/g of dextran) (original document, by courtesy of H. P. Erickson)

such as phosphocellulose and carboxymethyl cellulose. These substances appeared to act by removing from the tubulin solution a fraction required for assembly, which would be a heat-stable cationic protein [39, 40]. This protein may possibly be identical with the tau factor.

In contrast, several *basic proteins* have been demonstrated to facilitate assembly: RNase A, when added in the cold to a purified tubulin solution, increases the turbidity. Electron microscope studies show the presence of long MT and also of spiral sheets of tubulin molecules [73, 74]. Assembly is possible with RNase at quite low concentrations of tubulin. Another stimulator of assembly is the non-protein polycation, DEAE dextran (diethylaminoethyl dextran). This often leads to the formation of double-walled MT, which at one time were thought to be MT "decorated" with DEAE [170]. Similar double-walled MT are often seen when tubulin assembles in the presence of glycerol [12]. Other complex assemblies such as helical sheets may also be found with DEAE dextran [73, 74] although quite normal MT may be assembled with DEAE dextran in slightly modified conditions (Erickson, personal communication). It has been suggested that the MAPs may act, like the polycations, by electrostatic interactions [74].

Another agent which has often been used in the study of tubulin assembly is *glycerol.* Some apparent contradictions in the literature arise from the fact that glycerol itself promotes assembly, while binding to tubulin in the absence of added nucleotides [180]: about 5 M of glycerol for 1 M of tubulin dimer, and 22 M of glycerol per dimer of MT [64]. Glycerol favors the formation of double rings as intermediates in assembly [64] and enhances also the formation of sheets of tubulin molecules [140].

Evidence that MAPs are not indispensable for tubulin assembly is provided by experiments with *dimethylsulfoxide* (DMSO): in a study of assembly of tubulin separated from MAPs by phosphocellulose chromatography, while the 6S component alone would not assemble into MT, the addition of 10% DMSO leads to the formation of long MT, which disassemble again if DMSO is removed by dialysis. These MT are without any fuzzy surface material, further evidence that this is made of MAPs. Tubules shaped like Cs and Ss are observed: these are incompletely closed sheets of protofilaments. This type of assembly is prevented by cold (0 °C) and 0.1×10^{-3} M colchicine [101].

The role of *guanine nucleotides* in in vitro assembly has been studied with contradictory results. In the model proposed by Jacob [106, 107, 109] both E and N sites are occupied by GDP when MT are formed, phosphate being lost during assembly according to the equation

$$\mathrm{nGTP + n_ET_NGDP \rightarrow nGTP_ET_NGDP \rightarrow (GDP_ET_NGDP)n + nP.}$$

In assembled MT, GDP is not affected by GTP from the medium, suggesting either that its sites are that of tubulin to tubulin binding, or that the E site has an increased affinity for GDP in MT [106, 117]. When MT are disassembled, GTP can again become bound to the E site. The assembly of brain tubulin in vitro is activated by ATP [149, 150] and by GTP [218]. Studies with non-hydrolyzable analogs of guanine nucleotides brought contradictory results. Borisy [26] found no assembly with compounds such as adenylyl- or guanylyl-imidodiphosphate and β,γ-methylene-adenosine- or -guanosine triphosphates, while other authors [7] assembled MT in the presence of 5′-guanylyl-imidophosphate. These MT were found to be more resistant to depolymerization by Ca^{2+} (Table 2.3), while their structure was quite normal. They are also more resistant to dilution of the tubulin solution than normal MT [218]: hydrolysis of the bound GTP may be indispensable for disassembly. ATP promotes the assembly of MT by phosphorylating one GDP unit, later converted again to GDP [108, 117, 118]. However, in the presence of 4 M glycerol, tubulin from brain may assemble directly from $\mathrm{GDP_ET_NGTP}$ without any addition of ATP [118].

Table 2.3. Ca^{2+} sensitivity of MT formed in the presence of GTP and Gpp(NH)p (from Arai and Kaziro [7])

	pmol of tubulin in MT	
	GTP	Gpp(NH)p
Control	67.7	47.1
2.5 mM $CaCl_2$	14.3	44.0

Glycerol may favor assembly of tubulin assembly solutions in the absence of nucleotides [180], perhaps by altering the nucleotide binding site. The non-hydrolyzable analogs favor assembly when glycerol is present [149].

Another factor which is important for assembly is the balance between *S–S and –SH groups.* Indirect evidence that –SH bonds play a role in MT structure is provided by the fact that the colchicine binding of tubulin, which decreases rapidly as a function of time (tubulin "breakdown"), involves the cleavage of one disulfide bond [91] (Table 2.4). The stability of the tubulin dimer may be maintained by an S–S bond. Further data on the role of sulfhydryls will be considered in Chapters 5 and 10.

Table 2.4. Effects of reduced and oxidized glutathione and dithiothreitol on tubulin "breakdown" (from Gillespie [91])

Addition (1 mM)	Percentage of breakdown in 90 min (average ± SD)
None	30 ± 3
GSH	51 ± 3
GSSG	30 ± 4
GHS + GSSG	63 ± 3
DTT	23 ± 3

Assembly in vitro is prevented by most of the MT poisons such as colchicine, the *Vinca (Catharanthus)* alkaloids, benzimidazole derivatives; this will be treated in Chapter 5. It provides evidence that some of the links between tubulin molecules, which are indispensable for assembly, can be specifically inactivated or blocked by relatively small and highly specific molecules.

2.5.2.2 Site-Initiated Assembly In Vitro. So far, the mode of formation of a tubular structure from monomers or dimers has not been considered. It is preferable to study this problem, which is of the utmost importance for understanding the changes of MT in living cells, in relation to the assembly of MT, in vitro, on specialized nucleation sites.

The study of the kinetics of MT assembly in vitro show that when the proper conditions are fulfilled, there is always a short lag before the turbidity increase indicates the formation of MT (cf. Fig. 2.14); this has been interpreted as indicating the necessity of a first step, initiation, preceding that of elongation (cf. [152]). Many tubulins do not form MT except when they are seeded either by small aggregates of tubulin or by fragments of MT. For instance, extracts of sea-urchin eggs, which contain large amounts of tubulin, will not assemble into MT unless

Fig. 2.19. Rat pituitary, posterior lobe, incubated in vitro for 2 h at 37 °C in 10^{-5} VCR in Locke's solution. The central axon shows, besides neurosecretory granules (*NS*) and smooth endoplasmic reticulum (*End*), normal MT which in two places are linked with larger structures displaying oblique striations: the orientation of these depends on the plane of focusing of the microscope, indicating that they result from the wrapping, probably around an intact MT, of a helical thread made of protofilaments, so-called "decorated MT" (cf. Fig. 2.17)

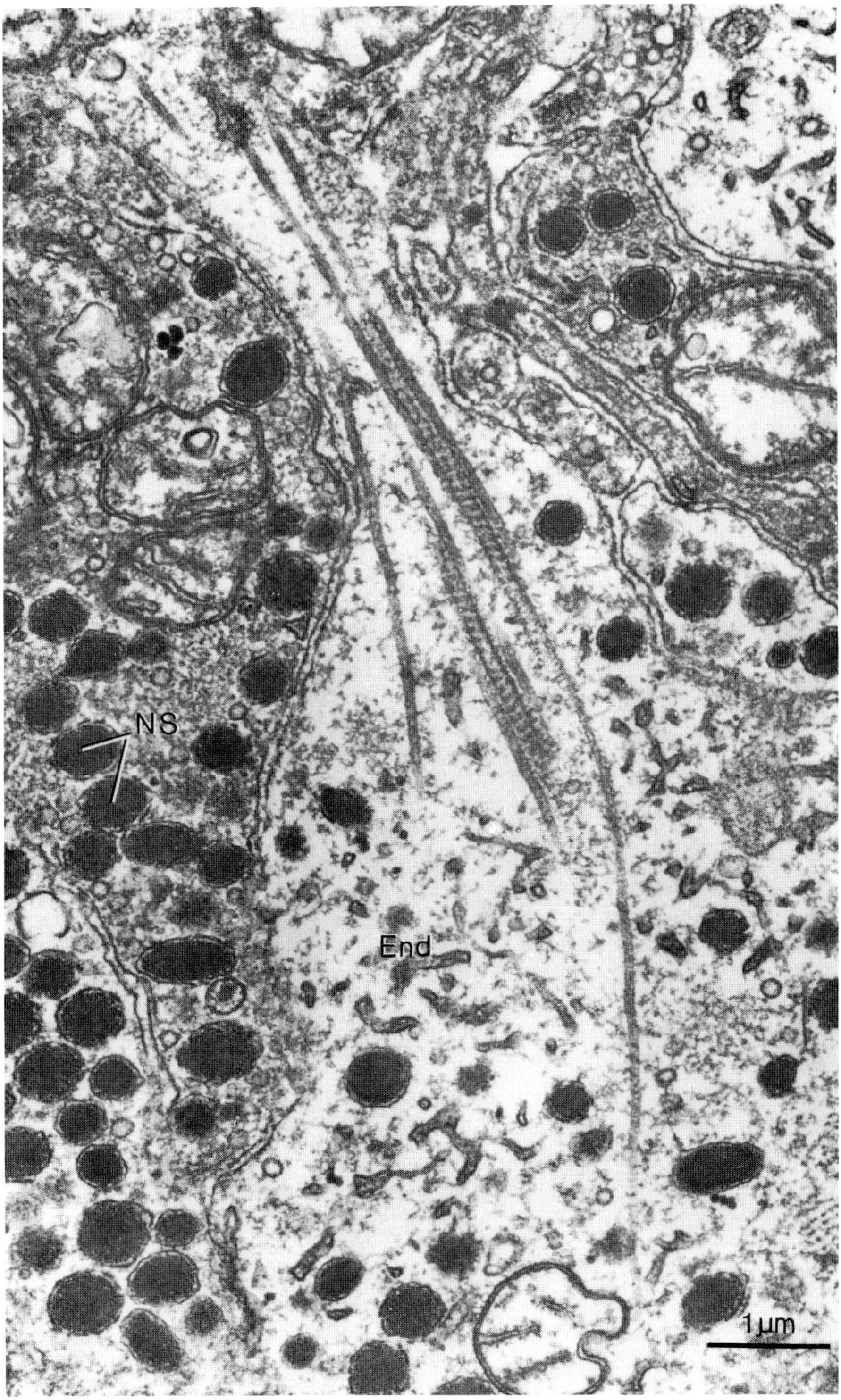
NS
End
1 µm

short fragments of MT from mammalian brain are added as "seeds" [43]. This demonstrates the possibility of heterologous formation of MT.

Porcine brain extracts capable of forming MT in vitro contain disk- or ring-shaped assemblies of tubulin molecules, which appear indispensable for assembly [27]. Depolymerized extracts of porcine brain prepared by the method of Shelanski et al. [179, 180] contained, apart from the 6S protein representing the tubulin dimers, a 36S fraction formed of ring-like assemblies, sometimes spirally wound, which looked like initiation centers for MT formation [114, 115, 116]. Double rings and other complex small assemblies were described in these preparations: they will be further described below.

In most living systems, MT grow in definite directions, from initiation centers such as centrioles or basal bodies.

The assembly of MT from such centers can be studied in vitro, and brings information about the directionality and the type of growth of MT. For instance, fragments of ^{3}H-leucine labeled chick brain MT were incubated in the presence of unlabeled tubulin from the same species in conditions favorable for assembly [63]: autoradiography of the MT showed consistently that the radioactive zone was located closer to one end of the newly formed MT, indicating that growth was mainly in one direction. Similar results were obtained by growing brain MT, in

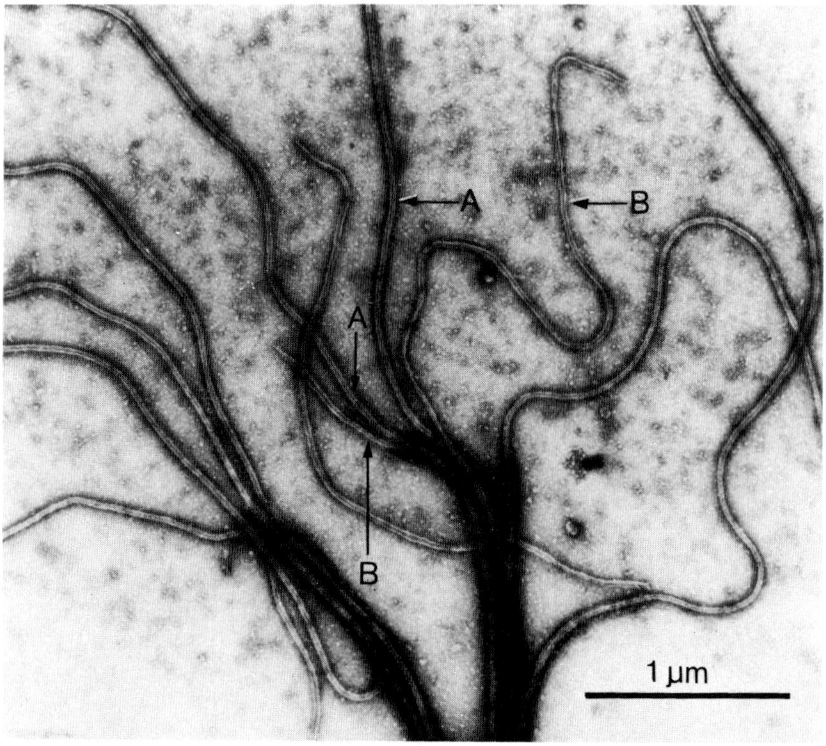

Fig. 2.20. Assembly of chick brain tubulin into a flagellum of *Chlamydomonas*. Although no doublet MT are formed, new MT are assembled in continuity with the A and B MT of the flagellar doublet (from Binder et al. [21])

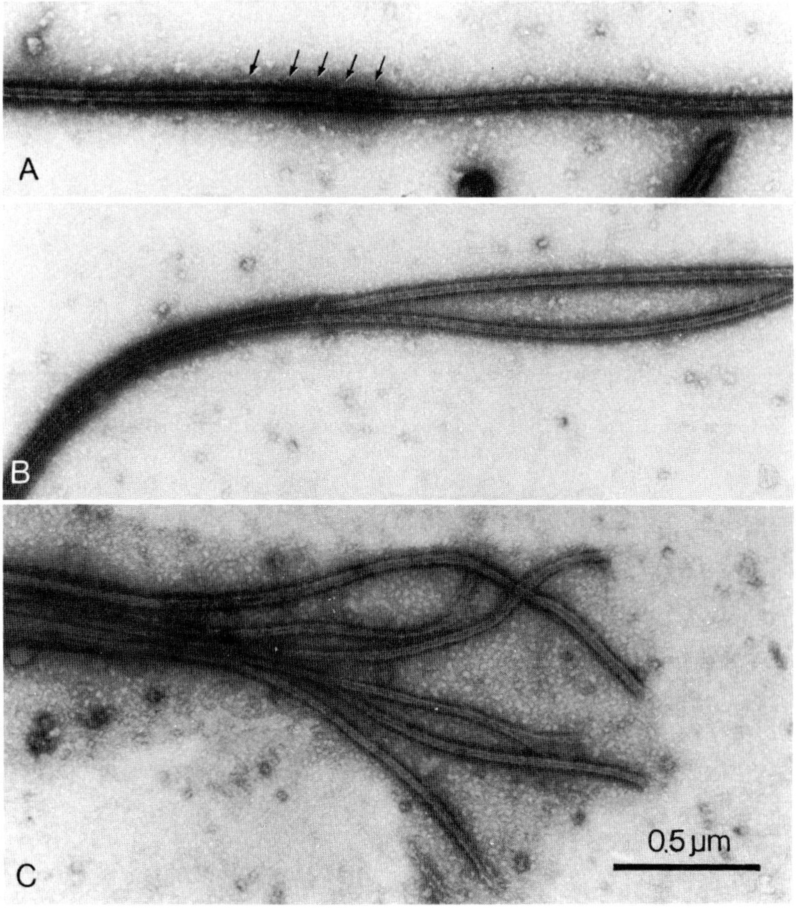

Fig. 2.21 A–C. Assembly of brain tubulin on *Chlamydomonas* doublets. (A) The newly formed tubule is in continuity with the A tubule of the doublet. The *arrows* indicate radial links of the doublet. (B) Formation of MT from brain tubulin on the central pair of MT. (C) Formation of MT on the proximal end of a flagellar axoneme. Uranyl acetate negative stain (from Binder et al. [21])

vitro, on heterologous structures such as *Chlamydomonas* flagellae. However, at higher concentrations of tubulin, some growth at the proximal and distal extremities of the flagellae was observed [170].

Site-initiated assembly has been most useful for the study of the dynamics of MT formation. The results obtained suggest strongly that the MT grow by an assembly of dimers on one end of the helical structure, and that the reaction obeys first-order kinetics, the dimers being added one by one. It is reversible and can be summarized by the equation $M_n + S \underset{-k_1}{\overset{k_2}{\rightleftharpoons}} M_{n+1}$, where M_n is some length of MT, S a subunit (dimer), k_1 and k_2 the constants of disassembly and of assembly. In these conditions, elongation (as opposed to initiation) does not appear to require MAPs [25, 110], which would be mainly related to the formation of the MT side-arms.

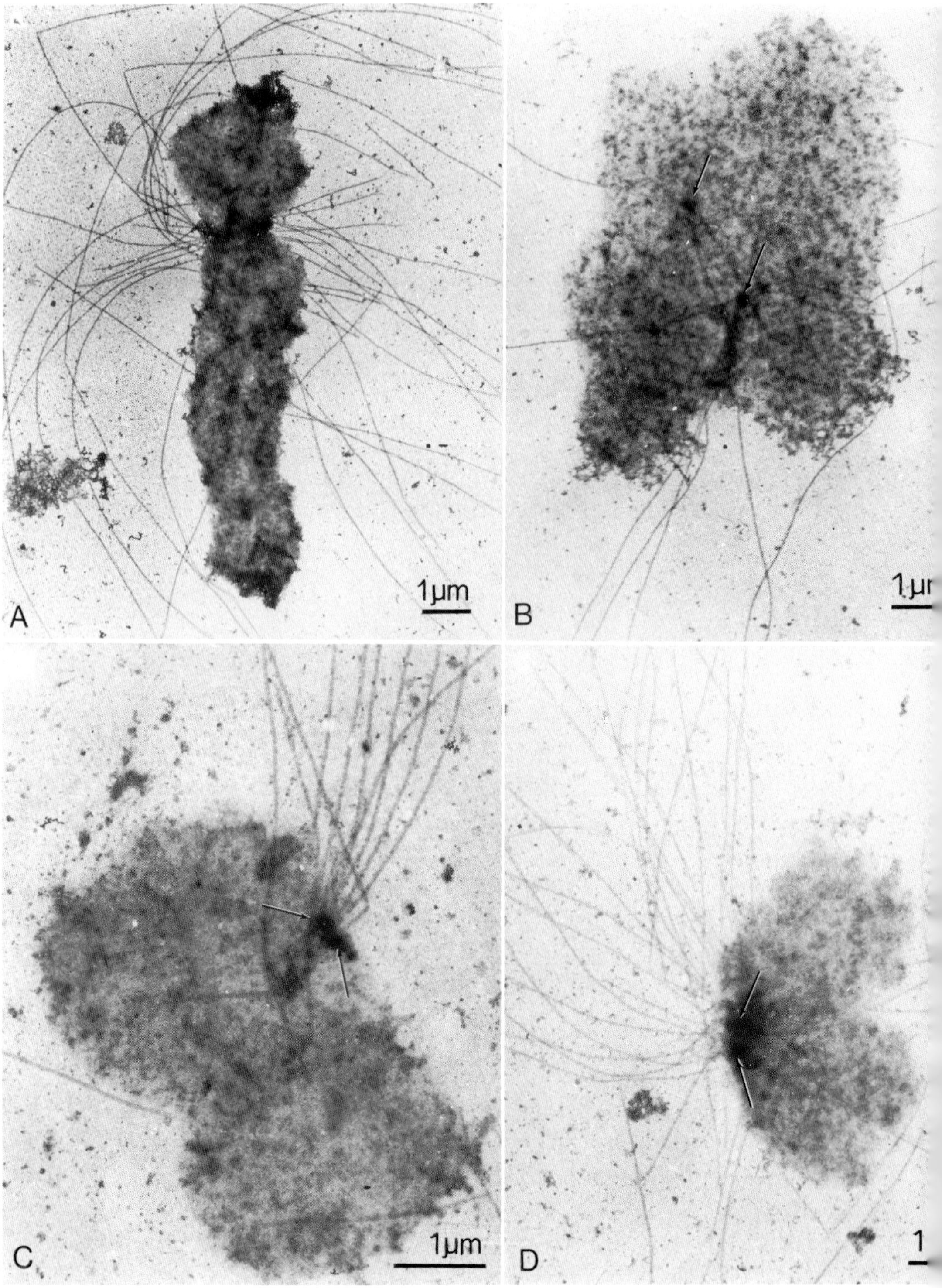
A
1μm
B
C
1μm
D

This mode of growth of MT explains well the observation that assembly may be arrested by quite small quantitites of poisons which attach at the growing end of the tubule [226].

The sites which may promote the assembly of MT in vitro are not species-specific, and fragments of MT may provide the support for the growth of tubulin from unrelated species; the same is true for assembly on more complex structures, such as spindle poles, centrioles, kinetochores ([202]; cf. Chap. 10). Here are some instances of heterologous assembly of MT: unfertilized sea-urchin egg cytoplasm + rat brain tubulin [43]; *Chlamydomonas* basal bodies and axonemes + chick brain tubulin; axonemes of *Arbacia punctulata* spermatozoa + chick tubulin [21, 170]. These observations confirm the close similarity of tubulins from most cells.

When MT grow on sites acting as MTOC, there is evidence that the tubulin molecules are added at the distal end of the growing MT, but some authors admit that growth could proceed from the proximal extremity. Another question which remains without proper answer is whether the MT helix is always oriented in the same direction in relation to the MTOC.

2.5.2.3 Tubulin Polymorphs and Assembly. Experiments with tubulin solutions have in recent years brought much information about the intermediate forms between isolated tubulin dimers and completed MT, and are progressively throwing more light on the mechanisms of growth of the MT helix.

In tubulin solutions prepared from brain, ultracentrifugation indicates that at least two components are present, apart from the MAPs: one sedimenting at 6S, and representing the $\alpha\beta$ dimer, another sedimenting at 36S, and made of rings or spirals [114, 115, 116]. Another 20S fraction, only present at a definite pH and made of rings, would not be an intermediate in assembly but a storage form [66].

The various forms of assembly, which follow more or less the rules defined by Erickson [76] for the formation of tubules from spherical subunits, have been obtained in various conditions of assembly and disassembly, in the presence of various ions and at different pH. Their polymorphism reflects the possibilities of assembling tubulins longitudinally and laterally, and should help to explain why in nature the five-start left-hand helix of 13 protofilaments is preferred. The great plasticity of tubulin assemblies is becoming more and more evident, and the tentative list given here is probably by no means complete.

a) *Rings* and short *spirals*—it may be quite difficult to be sure of the difference, as spiral segments may flatten when stained negatively—are found in solutions which are capable of assembly, and appear to be the initiation sites for further MT growth. They are the main component of the 36S fraction [27, 116] and are probably identical with the "disks" which have been described in a 30S purified fraction of brain tubulin [149, 152]. Relations between rings and MAPs have been mentioned, as the number of rings appears to be proportional to the amount of MAPs present [185, 186]. However, as MT may in some conditions assemble

Fig. 2.22 A–D. Isolated HeLa chromosomes, immobilized on grids, incubated in tubulin in conditions favorable for assembly. Formaldehyde fixation; alcoholic phosphotungstic hematoxylin stain. Assembly of MT on the kinetochores (*arrows*) (from Sloboda et al. [185]) (Cold Spring Harbor Laboratory, Copyright 1976)

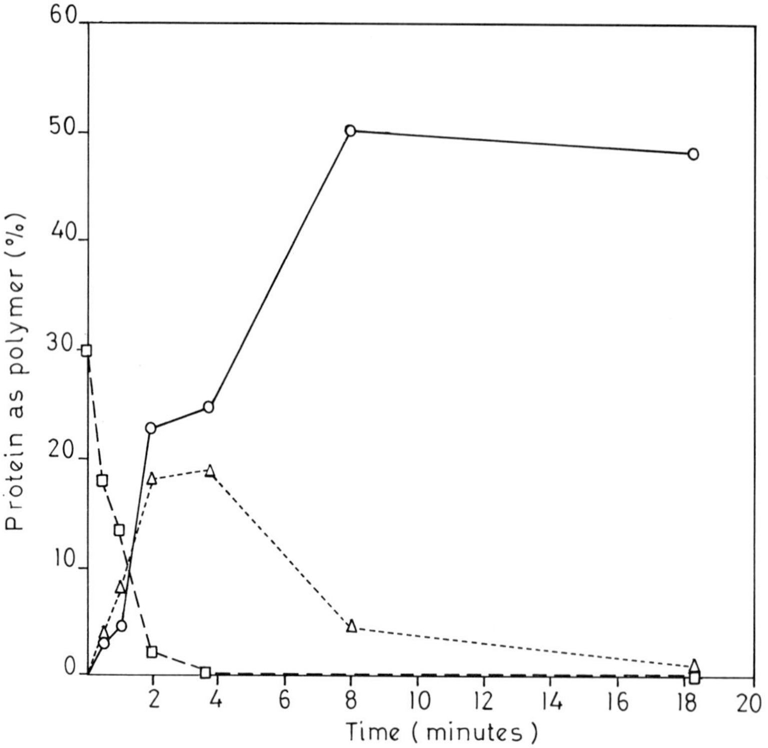

Fig. 2.23. Percent protein as rings (□---□), twisted ribbons (△---△), and MT (○——○) during assembly. Protein in assembly buffer (50 mM ammonium acetate pH 6.4, 1 mM EGTA, 0.5 mM $MgCl_2$) was warmed to 37 °C and GTP was added to 1 mM (redrawn from Penningroth et al. [157]) (Cold Spring Harbor Laboratory, Copyright 1976)

without MAPs, it is not certain whether rings are obligatory intermediates in assembly [39, 148]. While the rings may be a step in the formation of a helical tubule, the disks have been found to assemble also into *stacks* without helical substructure [26, 149, 152].

b) *Double rings,* made of an inner circle of 18 tubulin subunits and an outer one of 24, with a total molecular weight of 2.3×10^6 daltons, have been found in disassembly products of porcine brain [114, 115] and during Mg^{2+}-induced assembly [80]. Double, triple, and four-layered rings have been described in assembling MT in vitro [220]. Partial rings, and coils have also been found in similar conditions [72].

c) Isolated *protofilaments* are rarely observed; they have been mentioned in solutions of assembled calf brain tubulin and are destroyed by cooling to 20 °C, while double rings remain intact [126]. Such filaments—apparently made of end-to-end assemblies of tubulin molecules or dimers—are often seen to curl at the end. Rings have also been observed to uncurl into protofilaments [185].

d) The double rings may assemble into *double-walled MT:* these are found mainly when assembly takes place in the presence of DEAE dextran [73, 75] or

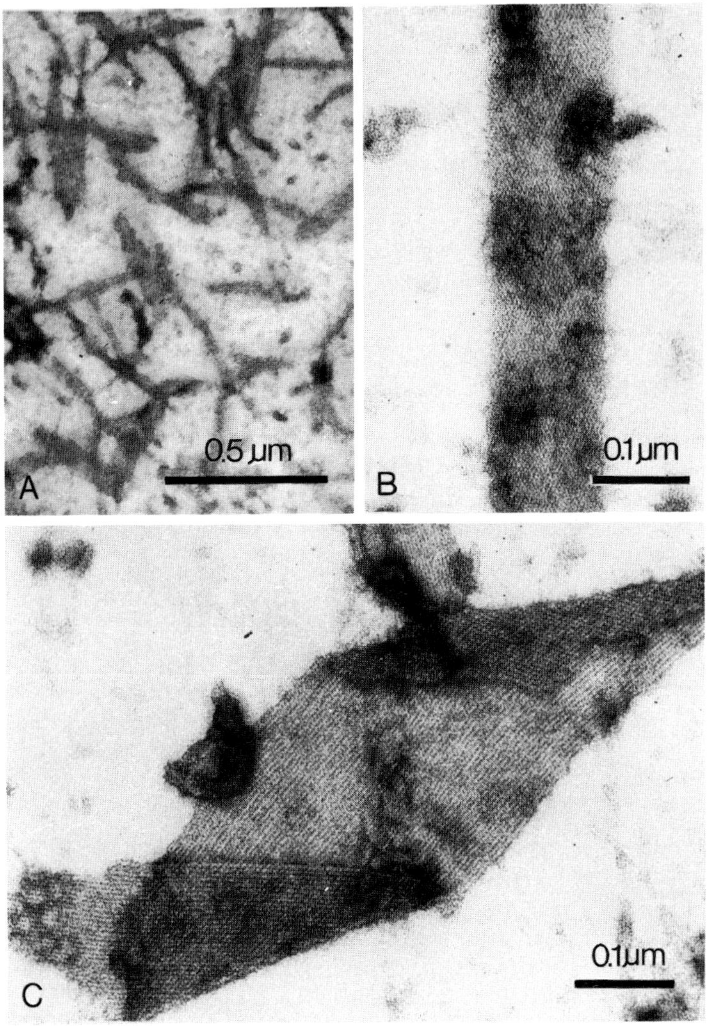

Fig. 2.24 A–C. (A) Several 0.2 – 0.5 µm long sheets of tubulin assembled in the presence of 2.5×10^{-4} M Zn^{2+}. (B) "Giant tube" from the same preparation as in (C), with a diameter of around 130 nm. Both sides of the tube are visible, which makes it look cross-striated. 1% uranyl acetate staining. (C) Sheet composed of 55 protofilaments induced by the addition of 10^{-4} Zn^{2+} to polymerized MT (from Larsson et al. [124])

polycations [74] ("decorated" MT). They result from a spiral wrapping of a sheet of protofilaments around a normal or a large MT which may have up to 16 subunits ([73, 75] cf. Fig. 2.19).

e) *Sheets* of tubulin subunits have not often been described, and many observations suggest that such assemblies have a tendency to curl or to wrap around already formed MT as described above. However, when tubulin assembles in the presence of Zn^{2+} ions, large flat sheets, made of parallely arranged protofilaments, have been demonstrated [124]. Such sheets are an indication that a much larger number of protofilaments than the 13 usually found in MT may assemble side by side. Similar associations have been observed in tubulin assembly solutions by varying the amount of EGTA and the pH. Matsumura [140] has shown the possibility of obtaining sheets with from 8 to 22 protofilaments (1 mM EGTA, pH 5.8 – 6.2), which may curl to form either *twisted ribbons,* or *supermacrotubules,* depending on the pH, in the presence of 1 mM $CaCl_2$. Transitions are found be-

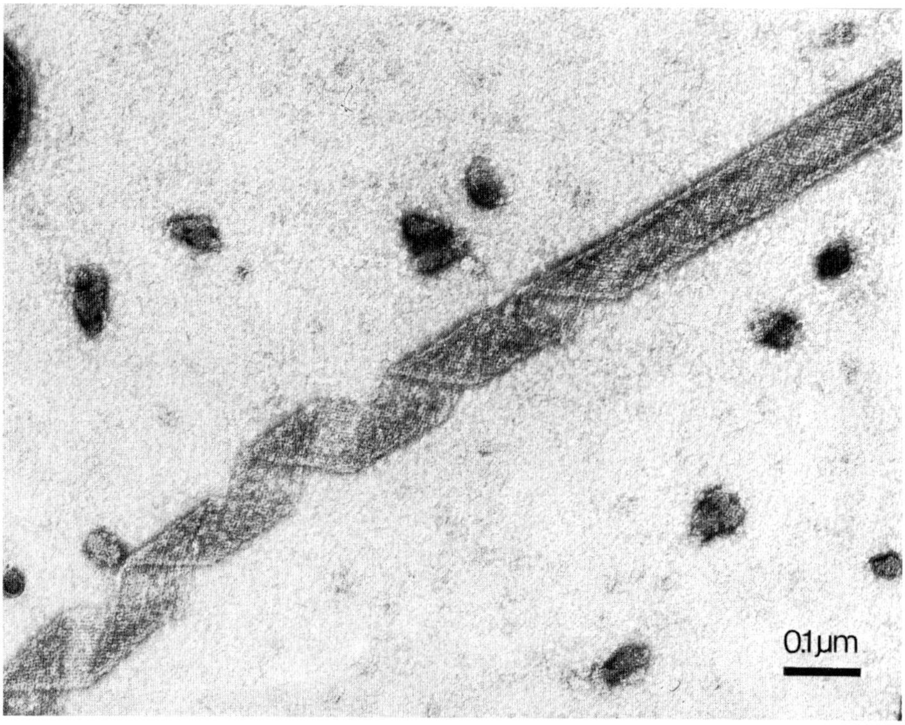

Fig. 2.25. Formation of a supermacrotubule in the presence of 1 mM $CaCl_2$ at pH 6.3 at 37 °C in 0.1 M MES buffer, 0.5 $MgCl_2$ and 1 mM ATP. The tubule is formed by the wrapping of a spiral sheet of tubulin protofilaments (from Matsumura and Hayashi [140])

tween sheets and twisted ribbons, the formation of sheets being favored by glycerol. While the normal MT are considered to be left-handed helices, the twisted ribbons may be right- or left-handed. The importance of calcium lies in the fact that it is bound to tubulin. It is improbable that a flat planar structure is stable: this is in agreement with the skewness of most MT structures such as centrioles, cilia, and axonemes ([38, 140] cf. Chap. 4).

f) These assemblies of curved sheets which close into tubules larger than normal MT are probably related to the so-called *macrotubules* (macroT) which have been mainly observed in vivo, in cells treated with various MT poisons (Chap. 5). These have a diameter of 31 – 52 nm, and were first observed in the

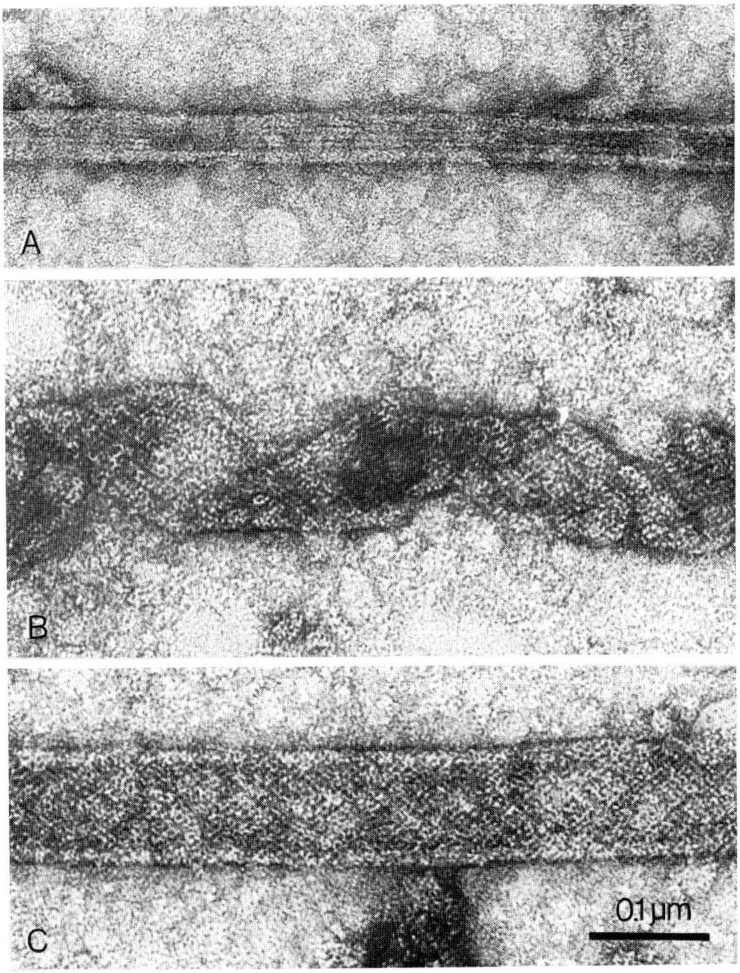

Fig. 2.26 A–C. (A) Control MT from the crayfish. The longitudinal protofilaments are clearly visible. (B) Helically twisted ribbons of tubulin forming macroT after 30 min treatment with 5 mM halothane. (C) MacroT after 30 min exposure to 5 mM halothane demonstrating the oblique assemblies of tubulin subunits (from Hinkley [102])

heliozoan, *Actinosphaerium nucleofilum,* after cooling to 4 °C [205]. This leads to a loss of the rigidity and the birefringence of the axopodia, which have a skeleton made of many MT. Before the MT vanish under the influence of cooling (cf. Chap. 5), macroT appear. They have a helical structure, showing profilaments angled at 45°. MacroT would result from an abnormal assembly of protofilaments, in a twisted structure, resulting from a modification of the lateral bonds between subunits. There are close relations between these macroT and the paracrystalline structures which are formed from tubulin in cells treated by the *Vinca* alkaloids (cf. Chap. 5). MacroT differ from MT by the fact (hitherto unexplained) that they fail to stain by immunohistochemical techniques [59, 61]. A study of the action of

VLB on brain tubulin has led to the conclusion that the macroT would result from the coiling of pairs of MT, after a partial disassembly of MT and the formation of helices of protofilaments: 5 – 9 of these pairs would form the macroT ([209] cf. Chap. 5).

The number of these tubulin polymorphs explains the varied conceptions of the assembly of tubulin. In fact, its observation remains most difficult: in vivo, the

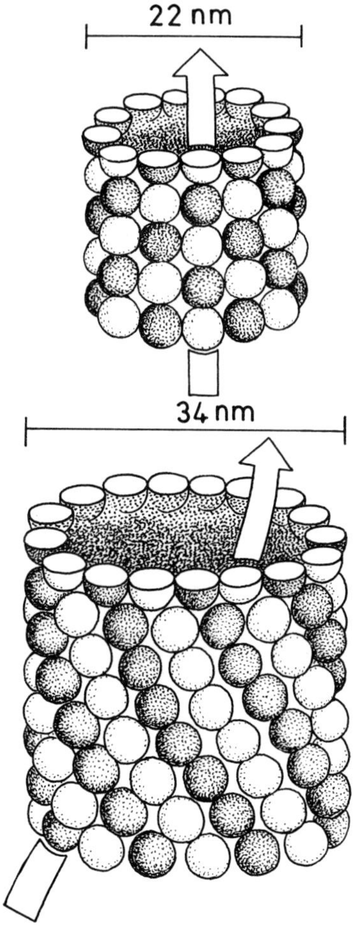

Fig. 2.27. Theoretical formation of a macrotubule from a normal MT. The 34 nm tubule, with 17 protofilaments, would result form the torsion of the protofilaments of a normal MT (13 protofilaments). The *white arrow* indicates the torsion of the protofilaments made of alternating molecules of α and β tubulin. In this model, however, the tubulin molecules being more or less spherical, the increased number of subunits per turn should be apparent on a cross-section, which is not the case, as the macroT do not show definite subunits. This figure is to be compared with the macro- and supermacroT observed in vitro in artificial conditions, resulting from the spiral wrapping of sheets of protofilaments (redrawn from Tilney and Porter [205])

only evidence—although conjectural—of a step in assembly is the presence in cells of C-MT, which may be formed by curled sheets of protofilaments before their closure into MT, and which were mentioned earlier in this chapter. The aspects observed in vitro suggest the following sequence, although more information is clearly needed. The MT—in the site-associated growth which resembles most closely the conditions found in the cell—grow linearly. They require first a step of initiation, and most observations indicate that the formation of rings (or short segments of helices) is this step. From these, protofilaments may grow linearly, although it it not known how many protofilaments may grow at the same time, and how the helical structure appears. These protofilaments, which may result from the uncoiling of the rings, have strong end to end linkages and assemble to form sheets, more or less curved. In normal conditions (perhaps related to this curvature), the number of observed protofilaments does not exceed 11 [39]. The addition of the next subunits leads to the closure of the MT, with an inter-

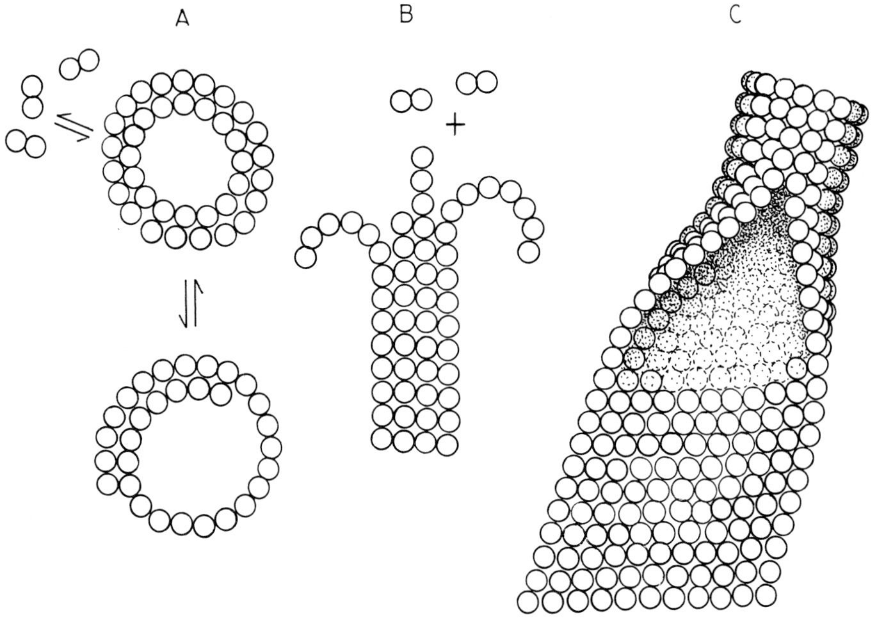

Fig. 2.28 A–C. Possible steps of MT assembly. (A) Equilibrium between tubulin $\alpha\beta$ dimers, rings, and short helices. (B) Assembly of dimers at the end of short protofibrils. The curls may represent unwinded spirals. (C) Formation of a MT from a more or less flat sheet of MT. The sheet is represented here with less than 13 protofilaments, the tubule assembly probably taking place once 12 filaments are assembled (however, the possibility of assembling larger sheets without tubule formation in vitro is illustrated in Figs. 2.25 and 2.26) (redrawn from Penningroth et al. [157], modified) (Cold Spring Harbor Laboratory, Copyright 1976)

mediary step which would be the C-MT. Once the MT starts to assemble, it only grows from one extremity, and this growth results from the fixation of $\alpha\beta$ dimers, all oriented in the same direction, forming the five-start helix demonstrated by diffraction studies and the models presented by Amos et al. [3, 4].

The steps represented in Figure 2.28 summarize the facts described in recent years and the data from the literature [75, 148, 149, 152, 157, 185, 187, 209]. However, most of these are from studies of tubulin assembly in mammals, and mainly from brain, and it is impossible to be sure that all tubulins assemble the same way. As pointed out recently in a study of the tubulin of *Aspergillus,* which has a lower affinity for colchicine than mammalian tubulin, the fact that both may coassemble into MT indicates that throughout evolution the binding sites involved in assembly appear to have been well conserved [58].

2.5.3 Assembly In Vivo

The data obtained in vitro help greatly to understand how MT may assemble in vivo, although they do not explain how this is regulated. The role of "seeds" or initiation sites has been underlined: it is most probable that this plays an important role in vivo, but one cannot affirm that in the proper conditions, MT may not assemble without being attached to any morphologically defined site. There are many observations, in mitotic and intermitotic cells, of MT which do not appear to be linked with any organelle or specialized region of the cell. Anyhow, the pericentriolar regions, the periphery of basal bodies, the kinetochores, the nuclear membrane, and the nuclear pores, are places from which MT can be seen to grow. This is particularly evident with immunochemical methods which allow a visualization of MT at the light microscope level (vide supra).

Apart from the evident role of MTOC in many cells, the thermodynamics of MT assembly in vivo and some aspects of its regulation will be discussed. Many aspects of this problem will be treated in further chapters.

2.5.3.1 Thermodynamic Aspects. Many of the early conclusions about the endothermic assembly of MT resulted from observations on living cells, where changes in the birefringence of spindle MT as a function of temperature and pressure were used to calculate the thermodynamics of MT assembly and disassembly. However, more recent work has shown (1) that the spindle birefringence cannot be entirely explained by the presence of MT and may result from other factors (cf. Chap. 10); (2) that the data on endothermic assembly in vitro are open to serious criticism, as mentioned above [198].

The pioneering works of Marsland [137, 138] and Pease [156] indicated that the fibrillary structure of the mitotic spindle of echinoderm eggs *(Arbacia, Urechis)* was destroyed by high hydrostatic pressures, and that this change was rapidly reversible [229]. Similar experiments on the meiosis of the starfish oocyte *(Asterias forbesis)* led to the conclusion that the assembly of spindle fibers implied the separation from the molecules of tubulin of bound water, with an overall increase in volume. Deuterium oxide had, on the other hand, a stabilizing action on the mitotic apparatus [139].

Between 1952 and 1971, Inoué and several collaborators (Carolan, Sato, and Bryan) studied the thermodynamic implications of these birefringence changes. Inoué [105], from a study of *Chaetopterus pergamentaceus* oocytes treated with heavy water, concluded that the assembly of spindle fibers involved an enthalpy increase of 28 kcal and an entropy increase of 98 eu. Similar results were obtained on other types of mitosis (cf. [197]). In plant cells (pollen of *Lilium longiflorum)* the spindle birefringence disappears at 3 °C. This effect is reversible, and it was concluded that the entropy increased as a linear function of temperature [49, 50].

The action of temperature may be used to calculate the rate of assembly of the mitotic apparatus: in the amoeba *Chaos carolinensis,* the reassembly of the spindle after its destruction at 2 °C takes place at a rate which has been calculated to correspond to 62 subunits per second. As about 10^{15} tubulin molecules are present in the cell, this rate may result from simple diffusion and collisions of the molecules, the MT being self-assembled [92, 93].

A detailed analysis of the thermodynamics of mitotic spindle MT has been published by Stephens [197]; the data are compared with those of other authors using different techniques. They appear to confirm that MT assembly is accompanied by an increase of both enthalpy and entropy ($\Delta H = +54.9$ kcal/mol and $\Delta S = 197$ eu, at 8 °C). The action of cold on the mitotic spindle of *Chaetopterus* has been more recently studied with improved equipment. The optical retardation is considered to be a function of the number of MT (a fact which is not accepted by all authors, cf. Chap. 10). The stability of spindle fibers varies according to their attachement sites (centrioles, kinetochores). The spindle assembly is endothermic ($\Delta H = 28$ kcal/mol; $\Delta S = 101$ eu). The endothermic, "entropy-driven" reaction is comparable to that of other assemblies of protein subunits (tobacco mosaic, actin, myosin, flagellin) [173].

These conclusions remain important, although the spindle is far too complex a system for the analysis of tubulin assembly (cf. Chap. 10); further data obtained in vitro should help to understand the reason of the discrepancies described. A good knowledge of the thermodynamic aspects is most important in any study of the regulation of MT formation in the cell.

2.5.3.2 Regulation of Assembly and Disassembly. One of the most remarkable properties of MT in the living cell is to undergo modifications in number and in location. These are often independent of protein synthesis and result from assembly of MT from a cytoplasmic tubulin pool and their later disassembly.

It is evident that some regulation of this is needed, and one of the problems of MT research is to discover how the cell not only controls the number of MT, but also expresses by some message where and when they assemble.

Many of the mechanisms which appear to be involved are related to the presence of cell structures which play a part in MT formation, as nucleating centers or MTOC: pericentriolar bodies, kinetochores, nuclear pores, basal bodies. At this moment only some general considerations on the known facts about the regulation of assembly in vivo are necessary. The problems of disassembly and of the reutilization of tubulins will also be considered briefly.

From what has been learnt from assembly in vitro, several factors may be supposed to play a role in the cell [38]:

1. Nucleation sites: the MTOC [160] are often dense regions of the cytoplasm from which MT or more complex MT structures, such as centrioles and basal bodies, are seen to arise. Their chemical composition remains obscure and it is not evident whether they are or are not accumulations of unassembled tubulin units.

2. Ions: the role of Ca^{2+} is evident in vitro, but it is uncertain whether the quantities which prevent MT assembly [150, 159] do play a role in the living cell.

3. MAPs: these may have regulatory functions, although normal MT may be assembled in vitro in their absence.

4. Molecules combining specifically with tubulin: colchicine, the *Vinca* alkaloids and the benzimidazoles demonstrate the very powerful and specific action on MT of relatively small molecules (cf. Chap. 5). Many of these (cf. Chap. 5) are synthesized by living matter: their function in Nature is unknown and the suggestion that they—or some similar molecules—may play a regulatory role in physiological conditions should not be dismissed.

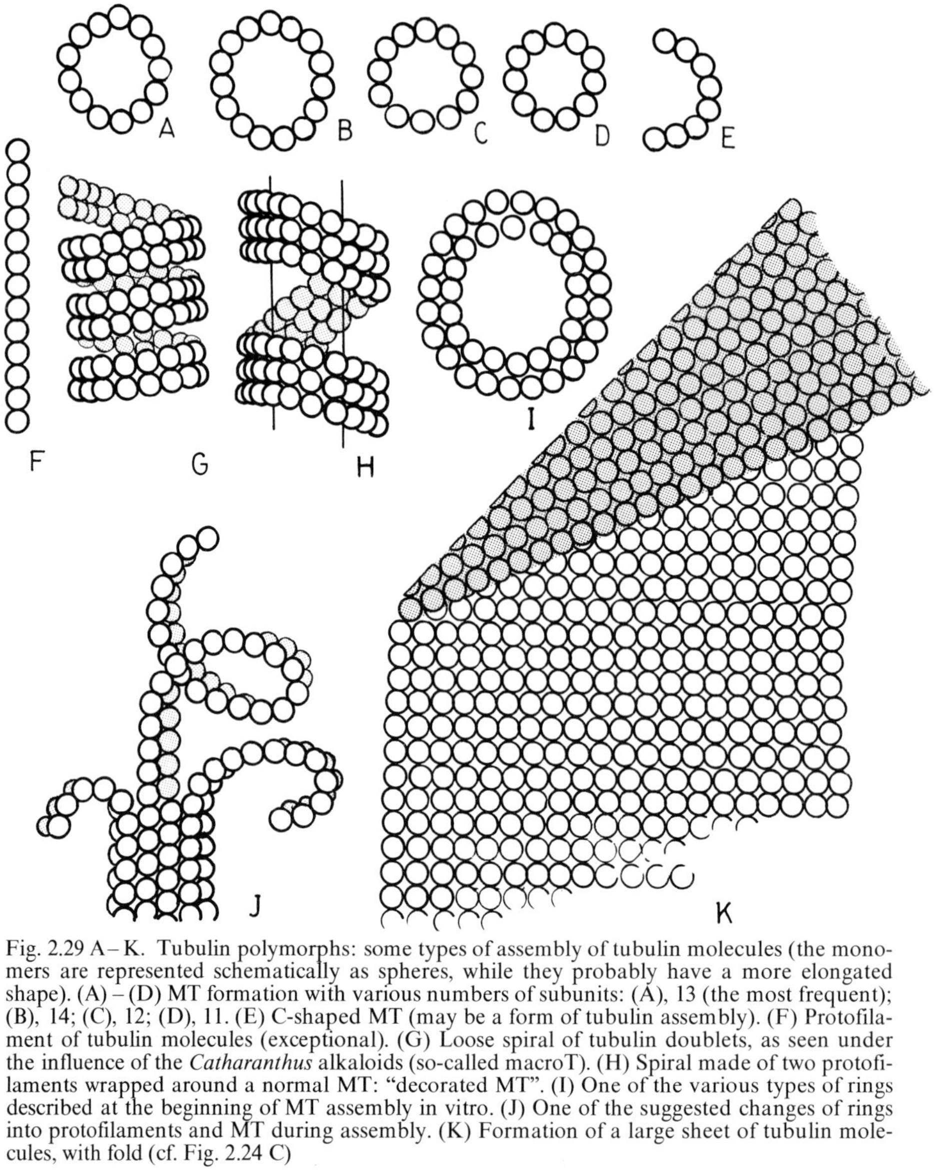

Fig. 2.29 A–K. Tubulin polymorphs: some types of assembly of tubulin molecules (the monomers are represented schematically as spheres, while they probably have a more elongated shape). (A)–(D) MT formation with various numbers of subunits: (A), 13 (the most frequent); (B), 14; (C), 12; (D), 11. (E) C-shaped MT (may be a form of tubulin assembly). (F) Protofilament of tubulin molecules (exceptional). (G) Loose spiral of tubulin doublets, as seen under the influence of the *Catharanthus* alkaloids (so-called macroT). (H) Spiral made of two protofilaments wrapped around a normal MT: "decorated MT". (I) One of the various types of rings described at the beginning of MT assembly in vitro. (J) One of the suggested changes of rings into protofilaments and MT during assembly. (K) Formation of a large sheet of tubulin molecules, with fold (cf. Fig. 2.24 C)

5. Guanine nucleotides appear indispensable for assembly. Phosphorylases or cAMP may also be mentioned here. Tubulin would be in the cell in an inactive state, with GDP on the N site, and GDP or nothing on the E site. Transphosphorylation with formation of GTP on both sites, perhaps affected by the concentration of Ca^{2+}, would be an activation step.

The possible role of cAMP has been studied [187], in relation to the presence of a protein kinase and the phosphorylation of MAPs. Two proteins are phosphorylated in vivo in the chick brain (after intracerebral injection), while little ^{32}P was associated with MT: the significance of this phosphorylation for MT assembly needs further research, and may concern more the structures associated with MT than the MT themselves. In Chinese hamster ovary cells grown in culture, no phosphorylation of MT could be detected [172]. In these cultures, the amount of MT increases when the cells are approaching confluency, and this is paralleled by an increase in cAMP; moreover, dibutyryl-cAMP increased the quantity of polymerized tubulin, although this was not phosphorylated.

When MT or MT structures (cilia, axonemes, mitotic apparatus) disassemble in the cytoplasm, there is evidence that the tubulin subunits remain in the cellular pool and may be reutilized later. Protein synthesis and breakdown play only a minor role in these changes. There are many instances where the destruction by the cell of one MT organelle takes place while others are assembled, as if the controls of assembly and disassembly were independent [23 a]. The problems of resorption of MT by living cells at some moments of their cycle or at some special locations in their cytoplasm will be discussed when the problems of centriole and ciliary growth (Chap. 4), disassembly of MT by various poisons (Chap. 5), growth of MT in nerve cells (Chap. 9), and mitosis (Chap. 10) are considered in more detail.

2.5.3.3 MT Polymorphs. Much less is known about the possibilities of variant assemblies of tubulin molecules in vivo. Mention has already been made of MT with more or less than the usual number of 13 subunits. Helical structures, comparable to the ribbons described in vitro, may be found in cells treated by the *Vinca* alkaloids (cf. Chap. 5). They are closely related to the paracrystalline structures which are formed under the influence of these drugs.

The most frequent change, after various drugs, is the formation of macroT, which have been described above. These may also be related to the *Vinca* crystals and result from the abnormal assembly of helical sheets of tubulin subunits [14, 81].

Another type of assembly, which has not been analyzed here, is that of doublets (as seen in cilia) and triplets (as in centrioles and basal bodies). These assemblies—which will be studied in Chapter 4—are not formed only by tubulin, as MAPs and other accessory proteins are clearly involved in their structure. Contrary to MT, doublets and triplets have not been observed to form in vitro, even when these structures are used as initiating sites: the newly formed MT grow mainly from the MT A of the doublet, which is the only one to have its full complement of 13 protofilaments, contrary to MT B and C.

2.6 Extracellular MT ?

MT are considered by all authors to be mainly intracellular organelles. However, their mechanical resistance is demonstrated in structures where they transmit traction forces in specialized epidermal cells (cf. Chap. 4): while MT are often observed to end in contact with the cell membrane, it is not extravagant to imagine that they could cross it and be found in the extracellular space. It should however be clear that MT are not the only tubular structures made by proteins, and that the few cases of extracellular tubular structures, with similar sizes, are not necessarily identical from a chemical point of view. Some examples may be mentioned here. In *Avena* coleoptiles, numerous extracytoplasmic MT are found between the cytoplasm and the cellulose layer. They have the usual size and are made of 13 subunits. They may play a role in the orientation of the cellulose fibers ([100] cf. Chap. 4). The surface of the spores of *Nosema michaelis* shows extracellular MT made of spherical subunits, from which α and β tubulin have been extracted [67].

The tubular structures, about 14 – 16 nm wide, which cover the microvilli of the gut of aphids, are apparently different from true MT [153]. The same applies to the protein tubules of 27 – 31 nm diameter, made of 10 subunits, and ejaculated with the spermatozoa, as a holocrine secretion, in *Drosophila* [158].

References

1. Adelman, M. R., Borisy, G. G., Shelanski, M. L., Weisenberg, R. C., Taylor, E. W.: Cytoplasmic filaments and tubules. Fed. Proc. **27,** 1186 – 1193 (1968)
2. Amos, L. A.: Substructure and symmetry of microtubules. In: Microtubules and Microtubule Inhibitors (eds.: M. Borgers, M. De Brabander), pp. 21 – 34. Amsterdam-Oxford: North-Holland Publishing Company. New York: Am. Elsevier Publ. Comp. 1975
3. Amos, L. A., Klug, A.: Arrangement of subunits in flagellar microtubules. J. Cell Sci. **14,** 523 – 550 (1974)
4. Amos, L. A., Linck, R. W., Klug, A.: Molecular structure of flagellar microtubules. In: Cell motility (eds.: R. Goldman, T. Pollard, J. Rosenbaum), pp. 847 – 868. Cold Spring Harbor Lab. 1976
5. André, J., Thiery, J.-P.: Mise en évidence d'une sous-structure fibrillaire dans les filaments axonématiques des flagelles. J. Microsc. **2,** 71 – 80 (1963)
6. Arai, T., Ihara, Y., Arai, K., Kaziro, Y.: Purification of tubulin from bovine brain and its interaction with guanine nucleotides. J. Biochem. (Tokyo) **77,** 647 – 658 (1975)
7. Arai, T., Kaziro, Y.: Effect of guanine nucleotides on the assembly of brain microtubules: ability of 5'-guanylyl imidophosphate to replace GTP in promoting the polymerization of microtubules *in vitro*. Biochem. Biophys. Res. Commun. **69,** 369 – 376 (1976)
8. Bassot, J. M., Martoja, R.: Données histologiques et ultrastructurales sur les microtubules cytoplasmiques du canal éjaculateur des Insectes Orthoptères. Z. Zellforsch. **74,** 145 – 181 (1966)
9. Behnke, O.: A preliminary report on "microtubules" in undifferentiated and differentiated vertebrate cells. J. Ultrastruct. Res. **11,** 139 – 146 (1964)
10. Behnke, O.: Incomplete microtubules observed in mammalian blood platelets during microtubule polymerization. J. Cell Biol. **34,** 697 – 701 (1967)
11. Behnke, O.: Microtubules in disk-shaped blood cells. Intern. Rev. Exp. Path. **9,** 1 – 92 (1970)
12. Behnke, O.: Studies on isolated microtubules. Evidence for a clear space component. Cytobiologie **11,** 366 – 382 (1975)
13. Behnke, O., Forer, A.: Evidence for four classes of microtubules in individual cells. J. Cell Sci. **2,** 169 – 192 (1967)
14. Behnke. O., Forer, A.: Vinblastine as a cause of direct transformation of some microtubules into helical structures. Exp. Cell Res. **73,** 506 – 508 (1972)
15. Behnke, O., Zelander, T.: Filamentous substructure of microtubules of the marginal bundle of mammalian blood platelets. J. Ultrastruct. Res. **19,** 147 – 165 (1967)
16. Berry, R. W., Shelanski, M. L.: Interactions of tubulin with vinblastine and guanosine triphosphate. J. Mol. Biol. **71,** 71 – 80 (1972)
17. Bhattacharyya, B., Wolff, J.: Stabilization of microtubules by lithium ion. Biochem. Biophys. Res. Commun. **73,** 383 – 390 (1976)
18. Bhisey, A. N., Freed, J. J.: Cross-bridges on the microtubules of cooled interphase HeLa cells. J. Cell Biol. **50,** 557 – 561 (1971)

19. Bibring, T., Baxandall, J.: Selective extraction of isolated mitotic apparatus. Evidence that typical microtubule protein is extracted by organic mercurial. J. Cell Biol. **48**, 324 – 339 (1971)
20. Bibring, T., Baxandall, J., Denslow, S., Walker, B.: Heterogeneity of the alpha subunit of tubulin and the variability of tubulin within a single organism. J. Cell Biol. **69**, 301 – 312 (1976)
21. Binder, L. I., Dentler, W. L., Rosenbaum, J. L.: Assembly of chick brain tubulin onto flagellar microtubules from Chlamydomonas and sea urchin sperm. Proc. Natl. Acad. Sci. USA **72**, 1122 – 1126 (1975)
22. Bird, M. M., Lieberman, A. R.: Microtubule fascicles in the stem processes of cultured sensory ganglion cells. Cell Tissue Res. **169**, 41 – 48 (1976)
23. Bloodgood, R. A., Miller, K. R.: Freeze-fracture of microtubules and bridges in motile axostyles. J. Cell Biol. **62**, 660 – 671 (1974)
23a. Bloodgood, R. A.: Resorption of organelles containing microtubules. Cytobios **9**, 142 – 161 (1974)
24. Bloodgood, R. A., Rosenbaum, J. L.: Initiation of brain tubulin assembly by a high molecular weight flagellar protein factor. J. Cell Biol. **71**, 322 – 331 (1976)
25. Borisy, G. G., Johnson, K. A., Marcum, J. M.: Self-assembly and site-initiated assembly of microtubules. In: Cell Motility (eds.: R. Goldman, T. Pollard, J. Rosenbaum), pp. 1093 – 1108. Cold Spring Harbor Lab. 1976
26. Borisy, G. G., Marcum, J. M., Olmsted, J. B., Murphy, D. B., Johnson, K. A.: Purification of tubulin and associated high molecular weight proteins from porcine brain and characterization of microtubule assembly in vitro. Ann. N. Y. Acad. Sci. **253**, 107 – 132 (1975)
27. Borisy, G. G., Olmsted, J. B., Klugman, R. A.: *In vitro* aggregation of cytoplasmic microtubule subunits. Proc. Natl. Acad. Sci. USA **69**, 2890 – 2894 (1972)
28. Borisy, G. G., Taylor, E. W.: The mechanism of action of colchicine. Colchicine binding to sea urchin eggs and the mitotic apparatus. J. Cell Biol. **34**, 535 – 548 (1967)
29. Brinkley, B. R., Cartwright, J. Jr.: Cold-labile and cold-stable microtubules in the mitotic spindle of mammalian cells. Ann. N. Y. Acad. Sci. **253**, 428 – 439 (1975)
30. Brinkley, B. R., Fuller, G. M., Highfield, D. P.: Studies of microtubules in dividing and non-dividing mammalian cells using antibody to 6-S bovin brain tubulin. In: Microtubules and Microtubule Inhibitors (eds: M. Borgers, M. De Brabander), pp. 297 – 312. Amsterdam-Oxford: North-Holland Publishing Company. New York: Am. Elsevier Publishing Comp. 1975
31. Brinkley, B. R., Fuller, G. M., Highfield, D. P.: Cytoplasmic microtubules in normal and transformed cells in culture. Analysis by tubulin antibody immunofluorescence. Proc. Natl. Acad. Sci. USA **72**, 4981 – 4985 (1975)
32. Brinkley, B. R., Fuller, G. M., Highfield, D. P.: Tubulin antibodies as probes for microtubules in dividing and nondividing mammalian cells. In: Cell Motility (eds.: R. Goldman, T. Pollard, J. Rosenbaum), pp. 435 – 456. Cold Spring Harbor Lab. 1976
33. Bryan, J.: Definition of three classes of binding sites in isolated microtubule crystals. Biochemistry **11**, 2611 – 2615 (1972)
34. Bryan, J.: Vinblastine and microtubules. II. Characterization of two protein subunits from the isolated crystals. J. Mol. Biol. **66**, 157 – 168 (1972)
35. Bryan, J.: Biochemical properties of microtubules. Fed. Proc. **33**, 152 – 157 (1974)
36. Bryan, J.: Microtubules. Bioscience **24**, 701 – 711 (1974)
37. Bryan, J.: Preliminary studies on affinity labeling of the tubulin-colchicine binding site. Ann. N. Y. Acad. Sci. **253**, 247 – 259 (1975)
38. Bryan, J.: Some factors involved in the control of microtubule assembly in sea urchins. Am. Zool. **15**, 649 – 660 (1975)
39. Bryan, J.: A quantitative analysis of microtubule elongation. J. Cell Biol. **71**, 749 – 767 (1976)
40. Bryan, J., Nagle, B. W., Doenges, K. H.: Inhibition of tubulin assembly by RNA and other polyanions: evidence for a required protein. Proc. Natl. Acad. Sci. USA **72**, 3570 – 3574 (1975)
41. Bryan, J., Wilson, L.: Are cytoplasmic microtubules heteropolymers? Proc. Natl. Acad. Sci. USA **68**, 1762 – 1766 (1971)
42. Burkholder, G. D., Okada, T. A., Comings, D. E.: Whole mount electron microscopy of metaphase. I. Chromosomes and microtubules from mouse oocytes. Exp. Cell Res. **75**, 497 – 511 (1972)
43. Burns, R. G., Starling, D.: The *in vitro* assembly of tubulins from sea-urchin eggs and rat brain: use of heterologous seeds. J. Cell Sci. **14**, 411 – 420 (1974)
44. Burton, P. R.: A comparative electron microscopic study of cytoplasmic microtubules and axial unit tubules in a spermatozoon and a protozoan. J. Morphol. **120**, 397 – 424 (1966)

45. Burton, P. R., Fernandez, H. L.: Delineation by lanthanum staining of filamentous elements associated with the surfaces of axonal microtubules. J. Cell Sci. **12,** 567 – 584 (1973)
46. Burton, P. R., Hinkley, R. E.: Further electron microscopic characterization of axoplasmic microtubules of the ventral nerve cord of the crayfish. J. Submicrosc. Cytol. **6,** 311 – 326 (1974)
47. Burton, P. R., Hinkley, R. E., Pierson, G. B.: Tannic acid-stained microtubules with 12, 13 and 15 protofilaments. J. Cell Biol. **65,** 227 – 232 (1975)
48. Cachon, J., Cachon, M.: Les systèmes axopodiaux. Année Biol. **13,** 523 – 560 (1974)
49. Carolan, R. M., Sato, H., Inoué, S.: A thermodynamic analysis of the effect of D_2O and H_2O on the mitotic spindle. Biol. Bull. **129,** 402 (1965)
50. Carolan, R. M., Sato, H., Inoué, S.: Further observations on the thermodynamics of the living mitotic spindle. Biol. Bull. **131,** 385 (1966)
51. Castle, A. G., Crawford, N.: Phosphorylation of platelet microtubule proteins by an endogenous protein kinase. Biochem. Soc. Trans. **4,** 691 – 693 (1976)
52. Cohen, C., Derosier, D., Harrison, S. C., Stephens, R. E., Thomas, J.: X-ray pattern from microtubules. Ann. N. Y. Acad. Sci. **253,** 53 – 59 (1975)
53. Cohen, C., Harrison, S. C., Stephens, R. E.: X-ray diffraction from microtubules. J. Mol. Biol. **59,** 375 – 380 (1971)
54. Cohen, W. D., Gottlieb, T.: C-microtubules in isolated spindles. J. Cell Sci. **9,** 603 – 620 (1971)
55. Crawford, N., Trenchev, P., Castle, A. G., Holborow, E. J.: Platelet tubulin and brain tubulin antibodies: immunofluorescence of cell microtubules. Cytobios **14,** 121 – 130 (1976)
56. Daleo, G. R., Piras, M. M., Piras, R.: The presence of phospholipids and diglyceride kinase activity in microtubules from different tissues. Biochem. Biophys. Res. Commun. **61,** 1043 – 1050 (1974)
57. Dales, S.: Concerning the universality of a microtubule antigen in animal cells. J. Cell Biol. **52,** 748 – 753 (1972)
58. Davidse, L. C., Flach, W.: Differential binding of methyl benzimidazol-2-yl carbamate to fungal tubulin as a mechanism of resistance to this antimitotic agent in mutant strains of *Aspergillus nidulans*. J. Cell Biol. **72,** 174 – 193 (1977)
59. De Brabander, M.: Onderzoek naar de rol van microtubuli in gekweekte cellen met behulp van een nieuwe synthetische inhibitor van tubulinpolymerisatie. Thesis, Brussels 1977
60. De Brabander, M., De Mey, J., Joniau, M., Geuens, G.: Ultrastructural immunocytochemical distribution of tubulin in cultured cells treated with microtubule inhibitors. Cell Biol. Intern. Rep. **1,** 177 – 183 (1977)
61. De Mey, J., Hoebeke, J., De Brabander, M., Geuens, G., Joniau, M.: Immunoperoxidase visualisation of microtubules and microtubular proteins. Nature (London) **264,** 273 – 275 (1976)
62. Dentler, W. L., Granett, S., Rosenbaum, J. L.: Ultrastructural localization of the high molecular weight proteins associated with *in vitro*-assembled brain microtubules. J. Cell Biol. **65,** 237 – 241 (1975)
63. Dentler, W. L., Granett, S., Witman, G. B., Rosenbaum, J. L.: Directionality of brain microtubule assembly *in vitro*. Proc. Natl. Acad. Sci. USA **71,** 1710 – 1733 (1974)
64. Detrich. H. W., III; Berkowitz, S. A., Kim, H., Williams, R. C., Jr.: Binding of glycerol by microtubule protein. Biochem. Biophys. Res. Commun. **68,** 961 – 968 (1976)
65. Dönges, K. H., Roth, E.: Serological similarity of microtubule proteins. Naturwissenschaften **59,** 372 (1972)
66. Dönges, K. H., Biedert, S., Paweletz, N.: Characterization of a 20S component in tubulin from mammalian brain. Biochemistry **15,** 2995 – 2999 (1976)
67. Dwyer, D. M., Weidner, E.: Microsoporidian extrasporular microtubules. Ultrastructure, isolation and electrophoretic characterization. Z. Zellforsch. **140,** 177 – 186 (1973)
68. Eipper, B. A.: Rat brain microtubule protein: purification and determination of covalenty bound phosphate and carbohydrate. Proc. Natl. Acad. Sci. USA **69,** 2283 – 2287 (1972)
69. Eipper, B. A.: Purification of rat brain tubulin. Ann. N. Y. Acad. Sci. **253,** 239 – 246 (1975)
70. Engelborghs, Y., Heremans, K. A. H., De Maeyer, L. C. M., Hoebeke, J.: Effect of temperature and pressure on polymerization equilibrium of neuronal microtubules. Nature (London) **259,** 686 – 688 (1976)
71. Erickson, H. P.: Microtubule surface lattice and subunit structure and observations on reassembly. J. Cell Biol. **60,** 153 – 167 (1974)
72. Erickson, H. P.: The structure and assembly of microtubules. Ann. N. Y. Acad. Sci. **253,** 60 – 77 (1975)

73. Erickson, H. P.: Facilitation of microtubule assembly by polycations. In: Cell Motility (eds.: R. Goldman, T. Pollard, J. Rosenbaum), pp. 1069 – 1080. Cold Spring Harbor Lab. 1976
74. Erickson, H. P., Scott, B.: Microtubule assembly in DEAE dextran: effect of charge density and MW of the polycation. Biophys. J. **17,** 274 a (1977)
75. Erickson, H. P., Voter, W. A.: Polycation-induced assembly of purified tubulin. Proc. Natl. Acad. Sci. USA **73,** 2813 – 2817 (1976)
76. Erickson, R. O.: Tubular packing of spheres in biological fine structure. Science **181,** 705 – 716 (1973)
77. Falxa, M. L., Gill, T. J. III: Preparation and properties of an alkylated brain protein related to the structural subunit of microtubules. Arch. Biochem. Biophys. **135,** 194 – 200 (1969)
78. Farrell, K. W.: Flagellar regeneration in *Chlamydomonas reinhardtii:* evidence that cycloheximide pulses induce a delay in morphogenesis. J. Cell Sci. **20,** 639 – 654 (1976)
79. Farrell, K. W., Burns, R. G.: Inability to detect *Chlamydomonas* microtubule assembly in vitro: possible implications to the in vivo regulation of microtubule assembly. J. Cell Sci. **17,** 669 – 681 (1975)
80. Frigon, R. P., Valenzuela, M. S., Timasheff, S. N.: Structure of a magnesia-induced polymer of calf brain microtubule protein. Arch. Biochem. Biophys. **165,** 442 – 443 (1974)
81. Fujiwara, K., Tilney, L. G.: Substructural analysis of the microtubule and its polymorphic forms. Ann. N. Y. Acad. Sci. **253,** 27 – 50 (1975)
82. Fulton, C., Kane, R. E., Stephens, R. E.: Serological similarity of flagellar and mitotic microtubules. J. Cell Biol. **50,** 762 – 773 (1971)
83. Fulton, C., Kowit, J. D.: Programmed synthesis of flagellar tubulin during cell differentiation in *Naegleria.* Ann. N. Y. Acad. Sci. **253,** 318 – 332 (1975)
84. Fulton, C., Simpson, P. A.: Selective synthesis and utilization of flagellar tubulin. The multi-tubulin hypothesis. In: Cell Motility (eds.: R. Goldman, T. Pollard, J. Rosenbaum), pp. 987 – 1006. Cold Spring Harbor Lab. 1976
85. Gall, J. G.: Microtubule fine structure. J. Cell Biol. **31,** 639 – 644 (1966)
86. Gaskin, E., Cantor, C. R., Shelanski, M. L.: Turbidimetric studies of the in vitro assembly and disassembly of porcine neurotubules. J. Mol. Biol. **89,** 737 – 738 (1974)
87. Gaskin, F., Cantor, C. R., Shelanski, M. L.: Biochemical studies on the in vitro assembly and disassembly of microtubules. Ann. N. Y. Acad. Sci. **253,** 133 – 146 (1975)
88. Gaskin, F., Gethner, J. S.: Characterization of the in vitro assembly of microtubules. In: Cell Motility (eds.: R. Goldman, T. Pollard, J. Rosenbaum), pp. 1109 – 1123. Cold Spring Harbor Lab. 1976
89. Gibbons, I. R.: Chemical dissection of cilia. Arch. Biol. (Liège) **76,** 317 – 352 (1965)
90. Gibbons, I. R., Rowe, A. J.: Dynein: a protein with adenosine triphosphatase activity from cilia. Science **149,** 424 – 426 (1965)
91. Gillespie, E.: The mechanism of breakdown of tubulin in vitro. FEBS Lett. **58,** 119 – 121 (1975)
92. Goode, M. D.: Kinetics of microtubule assembly after cold disaggregation of the mitotic apparatus. J. Cell Biol. **35,** 47 A – 48 A (1967)
93. Goode, D.: Kinetics of microtubule formation after cold disaggregation of the mitotic apparatus. J. Mol. Biol. **80,** 531 – 538 (1973)
94. Goodman, D. B. P., Rasmussen, H., Dibella, F., Guthrow, C. E., Jr.: Cyclic adenosine 3′-5′-monophosphate-stimulated phosphorylation of isolated neurotubule subunits. Proc. Natl. Acad. Sci. USA **67,** 652 – 659 (1970)
95. Grimstone, A. V., Gibbons, T. R.: The fine structure of centriolar apparatus and associated structures in the complex Flagellates *Trichonympha* and *Pseudotrichonympha.* Phil. Trans. R. Soc. London **250,** 215 – 242 (1966)
96. Gupta, B. L., Berridge, M. J.: Fine structural organization of the rectum in the Blowfly, *Calliphora erythrocephala* (Meig.) with special reference to connective tissue, tracheae and neurosecretory innervation in the rectal papillae. J. Morphol. **120,** 23 – 81 (1966)
97. Hart, J. W., Sabnis, D. D.: Colchicine and plant microtubules: a critical evaluation. Curr. Adv. Plant Sci. **26,** 1095 – 1104 (1976)
98. Hauser, M.: Differentielles Kontrastverhalten verschiedener Mikrotubulinsysteme nach Mercury Orange-Behandlung. Cytobiologie **6,** 367 – 381 (1972)
99. Hayashi, M., Matsumura, F.: Calcium binding to bovine brain tubulin. FEBS Lett. **58,** 222 – 226 (1975)
100. Heyn, A. N. J.: Intra- and extracytoplasmic microtubules in coleoptiles of *Avena.* J. Ultrastruct. Res. **40,** 433 – 457 (1972)

101. Himes, R. H., Burton, P. R., Kersey, R. N., Pierson, G. B.: Brain tubulin polymerization in the absence of "microtubule-associated proteins". Proc. Natl. Acad. Sci. USA **73,** 4397 – 4399 (1976)
102. Hinkley, R. E. Jr.: Microtubule-macrotubule transformations induced by volatile anesthetics. Mechanism of macrotubule assembly. J. Ultrastruct. Res. **57,** 237 – 250 (1976)
103. Hookes, D. E., Randall, J., Hopkins, J. M.: Problems of morphopoiesis and macromolecular structure in cilia. In: Formation and Fate of Cell Organelles (eds.: K. B. Warren), pp. 115 – 173. New York-London: Acad. Press 1967
104. Ikeda, Y., Steiner, M.: Isolation of platelet microtubule protein by an immunosorptive method. J. Biol. Chem. **251,** 6135 – 6141 (1976)
105. Inoué, S.: Motility of cilia and the mechanism of mitosis. Rev. Mod. Phys. **31,** 402 – 408 (1959)
106. Jacobs, M.: Tubulin nucleotide reactions and their role in microtubule assembly and dissociation. Ann. N. Y. Acad. Sci. **253,** 562 – 572 (1975)
107. Jacobs, M., Bennett, P. M., Dickens, M. J.: Duplex microtubule is a new form of tubulin assembly induced by polycations. Nature (London) **257,** 707 – 709 (1975)
108. Jacobs, M., Caplow, M.: Microtubular protein reaction with nucleotides. Biochem. Biophys. Res. Commun. **68,** 127 – 135 (1976)
109. Jacobs, M., Smith, H., Taylor, E. W.: Tubulin: nucleotide binding and enzymic activity. J. Mol. Biol. **89,** 455 – 468 (1974)
110. Johnson, K. A., Borisy, G. G.: The equilibrium assembly of microtubules in vitro. In: Molecules and Cell Movement (eds.: S. Inoué, R. E. Stephens), pp. 119 – 142. New York: Raven Press 1975
111. Kaye, J. S.: The fine structure and arrangement of microcylinders in the lumina of flagellar fibers in cricket spermatids. J. Cell Biol. **45,** 416 – 430 (1970)
112. Kikuchi, Y., King, J.: Assembly of the contractile tail of bacteriophage T4. In: Cell Motility (eds.: R. Goldman, T. Pollard, J. Rosenbaum), pp. 71 – 92. Cold Spring Harbor Lab. 1976
113. Kirkpatrick, J. B., Hyams, L., Thomas, V. L., Howley, P. M.: Purification of intact microtubules from brain. J. Cell Biol. **47,** 384 – 394 (1970)
114. Kirschner, M. W., Honig, L. S., Williams, R. C.: Quantitative electron microscopy of microtubule assembly in vitro. J. Mol. Biol. **99,** 263 – 276 (1975)
115. Kirschner, M. W., Suter, M., Weingarten, M., Littman, D. D.: The role of rings in the assembly of microtubules in vitro. Ann. N. Y. Acad. Sci. **253,** 90 – 106 (1975)
116. Kirschner, M. W., Williams, R. C., Weingarten, M., Gerhart, J. C.: Microtubules from mammalian brain: some properties of their depolymerization products and a proposed mechanism of assembly and disassembly. Proc. Natl. Acad. Sci. USA **71,** 1159 – 1187 (1974)
117. Kobayashi, T.: Dephosphorylation of tubulin-bound guanosine triphosphate during microtubule assembly. J. Biochem. **77,** 1193 – 1198 (1975)
118. Kobayashi, T., Shimizu, T.: Roles of nucleoside triphosphates in microtubule assembly. J. Biochem. **79,** 1357 – 1364 (1976)
119. Kuriyama, R., Sakai, H.: Role of tubulin –SH groups in polymerization to microtubules. Functional –SH groups in tubulin for polymerization. J. Biochem. **76,** 651 – 654 (1974)
120. Lagnado, J. R., Lyons, C. A., Weller, M., Phillipson, O.: The possible significance of adenosine 3′:5′-cyclic monophosphate-stimulated protein kinase activity associated with purified microtubular protein preparations from mammalian brain. Biochem. J. **128,** 95 P (1972)
121. Lagnado, J., Tan, L. P., Reddington, M.: The in situ phosphorylation of microtubular protein in brain cortex slices and related studies on the phosphorylation of isolated brain tubulin preparations. Ann. N. Y. Acad. Sci. **253,** 577 – 597 (1975)
122. Lambert, A.-M., Bajer, A. S.: Dynamics of spindle fibers and microtubules during anaphase and phragmoplast formation. Chromosoma **39,** 101 – 144 (1972)
123. Lane, N. J., Treherne, J. E.: Lanthanum staining of neurotubules in axons from cockroach ganglia. J. Cell Sci. **7,** 217 – 231 (1970)
124. Larsson, H., Wallin, M., Edstrom, A.: Induction of a sheet polymer of tubulin by Zn^{2+}. Exp. Cell Res. **100,** 104 – 110 (1976)
125. Ledbetter, M. C., Porter, K. R.: A "microtubule" in plant fine structure. J. Cell Biol. **19,** 239 – 250 (1963)
126. Lee, J. C., Timasheff, S. N.: The reconstitution of microtubules from purified calf brain tubulin. Biochemistry **14,** 5183 – 5187 (1975)
127. Linck, R. W., Amos, L. A.: The hands of helical lattices in flagellar doublet microtubules. J. Cell Sci. **14,** 551 – 560 (1974)

128. Ludena, R., Wilson, L., Shooter, E. M.: Cross-linked tubulin. In: Microtubules and Microtubule Inhibitors (eds.: M. Borgers, M. De Brabander), pp. 47 – 58. Amsterdam-Oxford: North-Holland Publishing Company. New York: Am. Elsevier Publ. Co. 1975
129. Luduena, R. F., Woodward, D. O.: Isolation and partial characterization of α- and β-tubulin from outer doublets of sea-urchin sperm and microtubules of chick-embryo brain. Proc. Natl. Acad. Sci. USA **70,** 3594 – 3598 (1973)
130. Luduena, R. E., Woodward, D. O.: α- and β-tubulin: separation and partial sequence analysis. Ann. N. Y. Acad. Sci. **253,** 272 – 283 (1975)
131. McIntosh, J. R.: Bridges between microtubules. J. Cell Biol. **61,** 166 – 187 (1974)
132. McIntosh, J. R., Candle, W. Z., Lazarides, E., McDonald, K., Snyder, J. A.: Fibrous elements of the mitotic spindle. In: Cell Motility (eds.: R. Goldman, T. Pollard, J. Rosenbaum), pp. 1261 – 1272. Cold Spring Harbor Lab. 1976
133. McIntosh, J. R., Porter, K. R.: Microtubules in the spermatids of the domestic fowl. J. Cell Biol. **35,** 153 – 173 (1967)
134. MacGregor, H. C., Stebbings, H.: A massive system of microtubules associated with cytoplasmic movement in telotrophic ovarioles. J. Cell Sci. **6,** 431 – 449 (1970)
135. MacKinnon, E. A., Abraham, P. J., Svatek, A.: Long link induction between the microtubules of the manchette in intermediate stages of spermiogenesis. Z. Zellforsch. **136,** 447 – 460 (1973)
136. Margolis, R. K., Margolis, R. U., Shelanski, M. L.: The carbohydrate composition of brain microtubule protein. Biochem. Biophys. Res. Commun. **47,** 432 – 437 (1972)
137. Marsland, D. A.: The effects of high hydrostatic pressure upon the mechanism of cell division. Arch. Exp. Zellforsch. **22,** 268 – 269 (1939)
138. Marsland, D.: Pressure-temperature studies on the mechanism of cell division. In: High Pressure Effects on Cellular Processes (ed.: A. M. Zimmerman), pp. 259 – 312. New York: Acad. Press 1970
139. Marsland, D., Hiramoto, Y.: Cell division: pressure-induced reversal of the antimeiotic effects of heavy water in the oocytes of the starfish, *Asterias forbesi.* J. Cell Physiol. **67,** 13 – 22 (1966)
140. Matsumura, F., Hayashi, M.: Polymorphism of tubulin assembly. In vitro formation of sheet, twisted ribbon and microtubule. Biochim. Biophys. Acta **453,** 162 – 175 (1976)
141. Mellon, M. G., Rebhun, L. I.: Sulfhydryls and in vitro polymerization of tubulin. J. Cell Biol. **70,** 226 – 238 (1976)
142. Mellon, M., Rebhun, L. I.: Studies on the accessible sulfhydryls of polymerizable tubulin. In: Cell Motility (eds.: R. Goldman, T. Pollard, J. Rosenbaum), pp. 1149 – 1164. Cold Spring Harbor Lab. 1976
143. Miki-Noumura, T., Kamiya, R.: Shape of microtubule in solutions. Exp. Cell Res. **97,** 451 – 453 (1976)
144. Moran, D. T., Chapman, K. M., Ellis, R. A.: The fine structure of cockroach campaniform sensilla. J. Cell Biol. **48,** 155 – 173 (1971)
145. Morgan, J. L., Seeds, N. W.: Properties of tubulin prepared by affinity chromatography. Ann. N. Y. Acad. Sci. **253,** 260 – 271 (1975)
146. Murphy, D. B., Borisy, G. G.: Association of high-molecular-weight proteins with microtubules and their role in microtubule assembly in vitro. Proc. Natl. Acad. Sci. USA **72,** 2696 – 2700 (1975)
147. Nagano, T., Suzuki, F.: Microtubules with 15 subunits in cockroach epidermal cells. J. Cell Biol. **64,** 242 – 245 (1975)
148. Nagle, B. W., Ryan, J.: Factors affecting nucleation and elongation of microtubules in vitro. In: Cell Motility (eds.: R. Goldman, T. Pollard, J. Rosenbaum), pp. 1213 – 1232. Cold Spring Harbor Lab. 1976
149. Olmsted, J. B.: The role of divalent cations and nucleotides in microtubule assembly in vitro. In: Cell Motility (eds.: R. Goldman, T. Pollard, J. Rosenbaum), pp. 1081 – 1092. Cold Spring Harbor Lab. 1976
150. Olmsted, J. B., Borisy, G. G.: Characterization of microtubule assembly in porcine brain extracts by viscometry. Biochemistry **12,** 4282 – 4289 (1973)
151. Olmsted, J. B., Borisy, G. G.: Ionic and nucleotide requirements for microtubule polymerization in vitro. Biochemistry **14,** 3996 – 4004 (1975)
152. Olmsted, J. B., Marcum, J. M., Johnson, K. A., Allen, C., Borisy, G. G.: Microtubule assembly: some possible regulatory mechanisms. J. Supramol. Struct. **2,** 429 – 450 (1974)
153. O'Loughlin, G. T., Chambers, T. C.: Extracellular microtubules in the aphid gut. J. Cell Biol. **53,** 575 – 578 (1972)
154. Osborn, M., Weber, K.: Cytoplasmic microtubules in tissue culture cells appear to grow from an organizing structure towards plasma membrane. Proc. Natl. Acad. Sci. USA **73,** 867 – 871 (1976)

155. Osborn, M., Weber, K.: Tubulin-specific antibody and the expression to a substratum. Further evidence for the polar growth of cytoplasmic microtubules in vivo. Exp. Cell Res. **103,** 331 – 340 (1976)
156. Pease, D. C.: Hydrostatic pressure effects upon the spindle figure and chromosome movement. I. Experiments on the first mitotic division of *Urechis* eggs. J. Morphol. **69,** 405 – 442 (1941)
157. Penningroth, S. M., Cleveland, D. W., Kirschner, M. W.: In vitro studies of the regulation of microtubule assembly. In: Cell Motility (eds.: R. Goldman, T. Pollard, J. Rosenbaum), pp. 1233 – 1258. Cold Spring Harbor Lab. 1976
158. Perotti, M. F.: Microtubules as a secretion product. Boll. Zool. **39,** 249 – 254 (1972)
159. Petzelt, C., Ledebur-Villiger, M.: Ca^{++}-stimulated ATPase during the early development of parthenogenetically activated eggs of the sea urchin *Paracentrotus lividus.* Exp. Cell Res. **81,** 87 – 94 (1973)
160. Pickett-Heaps, J. D.: The evolution of the mitotic apparatus: an attempt at comparative cytology in dividing plant cells. Cytobios **1,** 257 – 280 (1969)
161. Piperno, G., Luck, D. J. L.: Microtubular proteins of *Chlamydomonas reinhardtii.* An immunochemical study based on the use of an antibody specific for the β-tubulin subunit. J. Biol. Chem. **252,** 383 – 391 (1977)
162. Pochon-Masson, J.: Structure et fonctions des infrastructures cellulaires dénommées "microtubules". Année Biol. **6** (Series 4), 361 – 390 (1967)
163. Porter, K. R.: Cytoplasmic microtubules and their functions. In: Ciba Foundation Symposium on Principles of Biomolecular Organization, pp. 308 – 345. London: Churchill 1966
164. Rappaport, L., Leterrier, J. F., Nunez, J.: Protein kinase activity, in vitro phosphorylation and polymerization of purified tubulin. Ann. N. Y. Acad. Sci. **253,** 611 – 629 (1975)
165. Rappaport, L., Leterrier, J. F., Virion, A., Nunez, J.: Phosphorylation of microtubule-associated proteins. Eur. J. Biochem. **62,** 539 – 550 (1976)
166. Raybin, D., Flavin, M.: An enzyme tyrosilating α-tubulin and its role in microtubule assembly. Biochem. Biophys. Res. Commun. **65,** 1088 – 1095 (1975)
167. Raybin, D., Flavin, M.: Specific enzymatic tyrosilation of the α chain of tubulin and its possible roles in tubulin assembly and function. In: Cell Motility (eds.: R. Goldman, T. Pollard, J. Rosenbaum), pp. 1133 – 1138. Cold Spring Harbor Lab. 1976
168. Renaud, F. L., Rowe, A. J., Gibbons, I. R.: Some properties of the proteins forming the outer fibers of cilia. J. Cell Biol. **36,** 79 – 90 (1968)
169. Rodriguez Echandia, E. L., Piezzi, R. S., Rodriguez, E. M.: Dense-core microtubules in neurons and gliocytes of the toad *Bufo arenarum* Hensel. Am. J. Anat. **122,** 157 – 168 (1968)
170. Rosenbaum, J. L., Binder, L. I., Granett, S., Dentler, W. L., Snell, W., Sloboda, R., Haimo, L.: Directionality and rate of assembly of chick brain tubulin onto pieces of neurotubules, flagellar axonemes, and basal bodies. Ann. N. Y. Acad. Sci. **253,** 147 – 177 (1975)
171. Rosenfeld, A. C., Zackroff, R. V., Weisenberg, R. C.: Magnesium stimulation of calcium binding to tubulin and calcium induced depolymerization of microtubules. FEBS Lett. **65,** 144 – 147 (1976)
172. Rubin, R. W., Weiss, G. D.: Direct biochemical measurements of microtubule assembly and disasssembly in Chinese hamster ovary cells. The effect of intercellular contact, cold, D_2O, and N^6,$O^{2\prime}$-dibutyryl cyclic adenosine monophosphate. J. Cell Biol. **64,** 42 – 53 (1975)
173. Salmon, E. D.: Spindle microtubules: thermodynamics of in vivo assembly and role in chromosome movement. Ann. N. Y. Acad. Sci. **253,** 383 – 406 (1975)
174. Sandoval, I. V., Cuatrecasas, P.: Opposing effects of cyclic AMP and cyclic GMP on protein phosphorylation in tubulin preparations. Nature (London) **262,** 511 – 513 (1976)
175. Sandoval, I. V., Cuatrecasas, P.: Proteins associated with tubulin. Biochem. Biophys. Res. Commun. **68,** 169 – 177 (1976)
176. Sato, H., Ohnuki, Y., Fujiwara, K.: Immunofluorescent anti-tubulin staining of spindle microtubules and critique for the technique. In: Cell Motility (eds.: R. Goldman, T. Pollard, J. Rosenbaum), pp. 419 – 434. Cold Spring Harbor Lab. 1976
177. Schliwa, M.: The role of divalent cations in the regulation of microtubule assembly. In vivo studies on microtubules of the heliozoan axopodium using the ionophore A23187. J. Cell Biol. **70,** 527 – 540 (1976)
178. Shapiro, A., Vinuela, E., Maizel, J.: Molecular weight estimation of polypeptide chains by electrophoresis in SDS-polyacrylamide gels. Biochem. Biophys. Res. Commun. **28,** 815 – 820 (1967)
179. Shelanski, M. L.: Chemistry of the filaments and tubules of brain. J. Histochem. Cytochem. **21,** 529 – 539 (1973)

180. Shelanski, M. L., Gaskin, F., Cantor, C. R.: Microbutule assembly in the absence of added nucleotides. Proc. Natl. Acad. Sci. USA **70,** 765 – 768 (1973)
181. Shelanski, M. L., Taylor, E. W.: Isolation of a protein sub-unit from microtubules. J. Cell Biol. **34,** 549 – 554 (1967)
182. Shelanski, M. L., Taylor, E. W.: Properties of the protein subunit of central-pair and outer-doublet microtubules of sea urchin flagella. J. Cell Biol. **38,** 304 – 315 (1968)
183. Shigenaka, Y., Tadokoro, Y., Kaneda, M.: Microtubules in Protozoan cells. I. Effects of light metal ions on the Heliozoan microtubules and their kinetic analysis. Annot. Zool. Jap. **48,** 227 – 241 (1975)
184. Slautterback, D. B.: Cytoplasmic microtubules. I. Hydra. J. Cell Biol. **18,** 367 – 388 (1963)
185. Sloboda, R. D., Dentler, W. L., Bloodgood, R. A., Teizer, B. R., Granett, S., Rosenbaum, J. L.: Microtubule-associated proteins (MAPS) and the assembly of microtubules in vitro. In: Cell Motility (eds.: R. Goldman, T. Pollard, J. Rosenbaum), pp. 1171 – 1212. Cold Spring Harbor Lab. 1976
186. Sloboda, R. D., Dentler, W. L., Rosenbaum, J. L.: Microtubule-associated proteins and the stimulation of tubulin assembly in vitro. Biochemistry **15,** 4497 – 4505 (1976)
187. Sloboda, R. D., Rudolph, S. A., Rosenbaum, J. L., Greengard, P.: Cyclic AMP-dependent endogenous phosphorylation of a microtubule-associated protein. Proc. Natl. Acad. Sci. USA **72,** 177 – 181 (1975)
188. Smith, D. S., Jarlfors, U., Beraneck, R.: The organization of synaptic axoplasm in the lamprey *(Petromyzon marinus)* central nervous system. J. Cell Biol. **46,** 199 – 219 (1970)
189. Soifer, D.: Enzymatic activity in tubulin preparations: cyclic-AMP dependent protein kinase activity of brain microtubule protein. J. Neurochem. **24,** 21 – 34 (1975)
190. Soifer, D., Laszlo, A. H., Scotto, J. M.: Enzymatic activity in tubulin preparations. I. Intrinsic protein kinase activity in lyophilized preparations of tubulin from porcine brain. Biochim. Biophys. Acta **271,** 182 – 192 (1972)
191. Soifer, D., Laszlo, A., Mack, K., Scotto, J., Siconolfi, L.: The association of a cyclic AMP-dependent protein kinase activity with microtubule protein. Ann. N. Y. Acad. Sci. **253,** 598 – 610 (1975)
192. Solomon, F.: Characterization of the calcium-binding activity of tubulin. In: Cell Motility (eds.: R. Goldman, T. Pollard, J. Rosenbaum), pp. 1139 – 1148. Cold Spring Harbor Lab. 1976
193. Stebbings, H., Bennett, C. E.: The sleeve element of microtubules. In: Microtubules and Microtubule Inhibitors (eds.: M. Borgers, M. De Brabander), pp. 35 – 45. Amsterdam-Oxford: North-Holland Publishing Company. New York: Am. Elsevier Publ. Co. 1975
194. Stebbings, H., Bennett, C. E.: The effect of colchicine on the sleeve element of microtubules. Exp. Cell Res. **100,** 419 – 423 (1976)
195. Stebbings, H., Willison, J. H. M.: Structure of microtubules: a study of freeze-etched and negatively stained microtubules from the ovaries of *Notonecta.* Z. Zellforsch. **138,** 387 – 396 (1973)
196. Stephens, R. E.: Reassociation of microtubule protein. J. Mol. Biol. **33,** 517 – 519 (1968)
197. Stephens, R. E.: A thermodynamic analysis of mitotic spindle equilibrium at active metaphase. J. Cell Biol. **57,** 133 – 147 (1973)
198. Sutherland, J. W. H., Sturtevant, J. M.: Calorimetric studies of in vitro polymerization of brain tubulin. Proc. Natl. Acad. Sci. USA **73,** 3565 – 3569 (1976)
199. Tamura, S.: Properties of microtubule proteins in different organelles in *Tetrahymena pyriformis.* Exp. Cell Res. **68,** 169 – 179 (1971)
200. Tamura, S.: Synthesis and assembly of microtubule proteins in *Tetrahymena pyriformis.* Exp. Cell Res. **68,** 180 – 185 (1971)
201. Taylor, E. W.: The mechanism of colchicine inhibition of mitosis. I. Kinetics of inhibition and the binding of H^3-colchicine. J. Cell Biol. **25,** 145 – 160 (1965)
202. Telzer, B. R., Moses, M. J., Rosenbaum, J. L.: Assembly of microtubules onto kinetochores of isolated mitotic chromosomes of HeLa cells. Proc. Natl. Acad. Sci. USA **72,** 4023 – 4027 (1975)
203. Thomas, M. B., Henley, C.: Substructure of the cortical singlet microtubules in spermatozoa of *Macrostomum* (Platyhelminthes, Turbellaria) as revealed by negative staining. Biol. Bull. **141,** 593 – 601 (1971)
204. Tilney, L. G., Bryan, J., Bush, D. J., Fujiwara, K., Mooseker, M. S., Murphy, D. B., Snyder, D. H.: Microtubules: evidence for 13 protofilaments. J. Cell Biol. **59,** 267 – 275 (1973)
205. Tilney, L. G., Porter, K. R.: Studies on the microtubules in Heliozoa. II. The effect of low temperature on these structures in the formation and maintenance of the axopodia. J. Cell Biol. **34,** 327 – 343 (1967)

206. Turner, F. R.: Modified microtubules in the testis of the water strider, *Gerris remigis* (Say). J. Cell Biol. **53,** 263 – 270 (1972)
207. Twomey, S. L., Raeburn, S., Baxter, C. F.: Colchicine-binding protein in rat brain: biochemical consistency and immunological discrepancy. Brain Res. **66,** 509 – 518 (1974)
208. Ventilla, M., Cantor, C. R., Shelanski, M.: A circular dichroism study of microtubule protein. Biochemistry **11,** 1554 – 1561 (1972)
209. Warfield, R. K. N., Bouck, G. B.: On macrotubule structure. J. Mol. Biol. **93,** 117 – 120 (1975)
210. Weber, K.: Specific visualization of tubulin containing structures by immunofluorescence microscopy: cytoplasmic microtubules, vinblastine induced paracrystals and mitotic figures. In: Micortubules and Microtubule Inhibitors (eds.: M. Borgers, M. De Brabander), pp. 313 – 325. Amsterdam-Oxford: North-Holland Publishing Company. New York: Am. Elsevier Publ. Co. 1975
211. Weber, K.: Visualization of tubulin-containing structures by immunofluorescence microscopy: cytoplasmic microtubules, mitotic figures and vinblastine-induced paracrystals. In: Cell Motility (eds.: R. Goldmann, T. Pollard, J. Rosenbaum), pp. 403 – 418. Cold Spring Harbor Lab. 1976
212. Weber, K., Pollack, R., Bibring, T.: Antibody against tubulin: the specific visualization of cytoplasmic microtubules in tissue culture cells. Proc. Natl. Acad. Sci. USA **72,** 459 – 463 (1975)
213. Weeks, D. P., Collis, P. S.: Induction of microtubule protein synthesis in *Chlamydomonas reinhardi* during flagellar regeneration. Cell **9,** 15 – 29 (1976)
214. Weingarten, M. D., Lockwood, A. H., Hwo, S.-Y., Kirschner, M. W.: A protein factor essential for microtubule assembly. Proc. Natl. Acad. Sci. USA **72,** 1858 – 1862 (1975)
215. Weisenberg, R. C.: Microtubule formation in vitro in solutions containing low calcium concentrations. Science **177,** 1104 – 1105 (1972)
216. Weisenberg, R.: The role of nucleotides in microtubule assembly. Ann. N. Y. Acad. Sci. **253,** 573 – 576 (1975)
217. Weisenberg, R. C., Borisy, G. G., Taylor, E. W.: The colchicine-binding protein of mammalian brain and its relation to microtubules. Biochemistry **7,** 4466 – 4478 (1968)
218. Weisenberg, R. C., Deery, W. J.: Role of nucleotide hydrolysis in microtubule assembly. Nature (London) **263,** 792 – 793 (1976)
219. Weisenberg, R. C., Deery, W. J., Dickinson, P.: Nucleotide interactions during polymerization and depolymerization of tubulin. In: Cell Motility (eds.: R. Goldman, T. Pollard, J. Rosenbaum), pp. 1123 – 1132. Cold Spring Harbor Lab. 1976
220. Weisenberg, R., Rosenbaum, A.: Role of intermediates in microtubule assembly. Ann. N. Y. Acad. Sci. **253,** 78 – 89 (1975)
221. Weisenberg, R. C., Taylor, E. W.: The binding of guanosine nucleotide to microtubule subunit protein purified from porcine brain. Fed. Proc. **27,** 299 (1968)
222. Weiss, P. A., Mayr, R.: Neuronal organelles in neuroplasmic ("axonal") flow. I. Mitochondria. Acta Neuropath. Suppl. **5,** 187 – 197 (1971)
223. Weiss, P. A., Mayr, R.: Neuronal organelles in neuroplasmic ("axonal") flow. II. Neurotubules. Acta Neuropath. Suppl. **5,** 198 – 206 (1971)
224. Wiche, G., Cole, R. D.: An improved preparation of highly specific tubulin antibodies. Exp. Cell Res. **99,** 15 – 22 (1976)
225. Wiche, G., Cole, R. D.: Reversible in vitro polymerization of tubulin from a cultured cell line (rat glial cell clone C-6). Proc. Natl. Acad. Sci. USA **73,** 1227 – 1231 (1976)
226. Wilson, L.: Microtubules as drug receptors: pharmacological properties of microtubule protein. Ann. N. Y. Acad. Sci. **253,** 213 – 231 (1975)
227. Witman, G. B., Carlson, K., Berliner, J., Rosenbaum, J. L.: *Chlamydomonas* flagella. I. Isolation and electrophoretic analysis of microtubules, matrix, membranes, and mastigonems. J. Cell Biol. **54,** 507 – 539 (1972)
228. Witman, G. B., Carlson, K., Rosenbaum, J. L.: *Chlamydomonas* flagella. II. The distribution of tubulins 1 and 2 in the outer doublet microtubules. J. Cell Biol. **54,** 540 – 555 (1972)
229. Zimmerman, A. M., Marsland, D.: Cell division: effects of pressure on the mitotic mechanisms of marine eggs *(Arbacia punctulata).* Exp. Cell Res. **35,** 293 – 302 (1964)

Chapter 3 General Physiology of Tubulins and Microtubules

3.1 Introduction

At this point, we have reached a good knowledge of the chemical make-up, the structure, and the modes of assembly of MT. Before analyzing the role of tubulin in the formation of far more complex organelles, like centrioles and cilia, and the effects of various drugs on MT, some fundamental data about general aspects of MT physiology will be considered: the synthesis of tubulins in the cell; its regulation; the role of nucleating sites for the assembly of MT; the various associations of MT with cell organelles and with other MT to form complex structures. Like the great polymorphism of tubulin molecules association, MT themselves may become grouped in various numbers to build, with the help of other proteins, complex structures such as the centrioles, the basal bodies, the cilia, the axonemes, and the mitotic spindle: a detailed survey of these will be given in Chapters 4, 6 and 10.

3.2 Synthesis of Tubulin

Before the discovery of the role of MT in the spindle, it had been clearly indicated that most of the proteins of the "mitotic apparatus" (MA) (which, as isolated by Mazia [79, 80], is far from being a pure MT structure and contains other proteins) were already present in the cell before mitosis [143]: in the sea-urchin egg, about 10% of all proteins take part in the formation of the MA [81]. When does their synthesis take place in the eggs, during embryonic growth, and in differentiated cells?

3.2.1 In Embryonic Growth

In the sea-urchin egg, maternal messenger RNA is transcribed during oogenesis, and one of the proteins which is formed in this way is tubulin. The pool of egg tubulin is used for mitosis and in the formation of cilia [5]. The synthesis of tubulin, after fertilization, has been measured either by the incorporation of ^{14}C marked amino-acids or by the fixation of ^{3}H-colchicine [82]: it increases after fertilization, and decreases during mitosis, like other protein syntheses. The proof of the early maternal origin of the mRNA is given by the fact that tubulin synthesis is not modified when eggs are cultivated in actinomycin, which depresses all new mRNA formation [99, 100]. Other evidence is that the synthesis takes place in unfertilized eggs, in anucleate activated eggs, and even in embryos cultivated in the presence of actinomycin plus ethidium bromide [98]. In contrast, tubulin used for the formation of cilia in the growing embryo has been demonstrated to be synthesized early after cleavage in the sea-urchin *Strongylocentrotus droebachiensis* [122]. A quantitative estimation of the amount of tubulin present during the early development of *Spisula* eggs and embryos was made by VLB precipitation in the presence of a known amount of radioactive tubulin: the total is more or less constant from fertilization to the formation of the gastrula. It represents about 3.3% of the total proteins, i.e., about 2.0 – 2.4 10^{-4} μg/egg or embryo [23]. Similar facts have been observed in the echiurid worm *Urechis campo:* an important pool of tubulin is present in the mature eggs; like other proteins, tubulin accumulates continuously during oogenesis [83, 84].

In the soluble fraction of the chick brain, tubulin increases from 20% at 5 – 7 days, to 42% at 13 days of development [6]. The binding of colchicine appeared to decrease with embryonic age. The turnover is rapid, in embryo as well as in adult chicks: four days as found by measuring the amount of VLB precipitable protein [44].

Eggs and embryos of *Drosophila* also contain a pool of tubulin (about 3% of the soluble protein) which is constant during early embryonic growth [98]; this tubulin, which may be prepared in relatively large quantities, is chemically closely similar to that of the sea-urchin. It plays an important part in the organization of the *Drosophila* embryo [47].

3.2.2 In Adult or Differentiated Cells

The synthesis of tubulin has been studied in various conditions where the cell required new MT, such as ciliogenesis or increased mitotic activity. Cycloheximide has been used by several authors as a specific inhibitor of protein synthesis. Most work has been done on protozoa, in relation with the regeneration of cilia. In several flagellates (*Ochromonas, Euglena, Astasia*) cycloheximide was found to inhibit the new growth of axonemes after experimental deciliation [106]. A kind of balance seems to exist between the amount of tubulin assembled in MT and the requirements of the cell, as exemplified by the fact that in *Chlamydomonas,* when one flagellum has been destroyed, the remaining one shortens during the regeneration of the first, as if subunits were transported from one flagellum to another ([105, 107]; Fig. 3.1). During flagellar regeneration in *Chlamydomonas,* tubulin and other flagellar proteins are synthetized: polyribosomes isolated from such cells were associated with tubulin and two proteins comigrating with tubulin, while in control cells no ribosome-associated tubulin was found. The synthesis starts immediately after deflagellation, and between 45 and 90 min after, about 14% of synthetized proteins are tubulins (Table 3.1). When the cells are treated with cycloheximide, flagella grow only to half-size. The polyribosomes were tested on a wheat germ translation system, and α and β tubulins were formed, which could coassemble with chick brain tubulin: this is different from the native tubulin extracted from *Chlamydomonas,* which does not (cf. Chap. 2; [141]).

Cycloheximide may inhibit cilia regeneration by preventing the synthesis of other proteins than tubulin, as suggested by experiments on *Tetrahymena.* Basal body

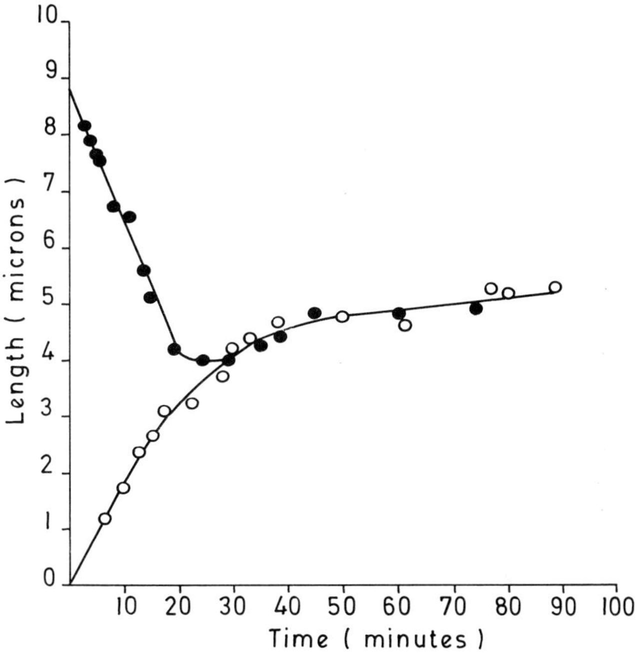

Fig. 3.1. Kinetics of flagellar shortening and elongation in *Chlamydomonas*. Partial shortening of the intact flagellum (*closed circles*) during regeneration of the amputated flagellum (*open circles*) (redrawn from Coyne and Rosenbaum [30])

formation is inhibited in cells recovering from the action of VLB, which does not block the synthesis of tubulin—as measured by the incorporation of ^{3}H-amino acids [16]. In the same unicellular, amino acid deficiency leads to cilia destruction, which takes place without any evidence of autophagy, suggesting that the tubulin returns directly to a cytoplasmic pool [146]. During regeneration, most of the basal bodies and cilia are built from preexisting subunits ("subunit exchange", [145]). Protein synthesis is probably important for other factors which control the assembly of MT [147]. The rapidity of cilia regeneration, which can take place in 90 min after deciliation by the local anesthetic dibucaine [126], also indicates that synthesis plays

Table 3.1. Percentage of protein synthesis devoted to tubulin synthesis in regenerating *Chlamydomonas* after deflagellation (from Weeks and Collis, 141)

Time of polyribosome isolation (min)	Tubulin synthesis as per cent of total protein synthesis
controls	0
15	2.2
30	5.1
40	12.8
60	14.2
90	13.1
120	6.6
180	2.4

only a minor role. In non-growing *Tetrahymena*, little or no synthesis would take place during cilia regeneration [88]: this explains the fact that, after 75% deciliation by $CaCl_2$ and EDTA, when the altered cilia are resorbed, cycloheximide has no inhibitory action, contrary to VLB. On the other hand, if all cilia are destroyed, regeneration cannot take place without protein synthesis, and is prevented by cycloheximide [102]. Protein synthesis is also necessary for ciliary regeneration after ultraviolet irradiation in *Stentor coeruleus* and not after urea deciliation ([22]; cf. Chap. 4). The peculiar problems of tubulin synthesis in the ameboflagellate *Naegleria gruberi*, where apparently two different pools of tubulin are present [48], will be analyzed in Chapter 4.

In the brain of adult mammals, tubulin is actively synthesized and shows a rapid turnover [8]; this is related to the transport of neurotubules to the axonal and dendritic endings and to the "slow" component of neuroplasmic flow (cf. Chap. 9). After intracerebral injection of ^{14}C-leucine, as much as 20% of the microtubular protein is labeled in the 70-day-old mouse. In the nerve endings, MT protein is the only labeled soluble component [8].

In the mouse oviduct, where very active formation of cilia takes place in the last stages of embryonic growth, the fixation of MT protein (as measured by its colchicine-binding properties and by electrophoresis) is high, and declines in the adult animal. This is related to the presence in the epithelial cells of centriole precursors, which appear as dense masses and which are simply "packaged aggregates of MT protein" ([120]; cf. Chap. 4). Further research indicates that in the three-day-old mouse, more than 90% of the tubulin is synthesized de novo, and 75% in the five-day-old mouse. The incorporation of ^{3}H-leucine, detected by electron microscope autoradiography in the centriole precursors, is inhibited by cycloheximide. Colchicine in a concentration of less than 10^{-3} M does not affect the formation of basal bodies and axonemes [39].

A study of the fixation of ^{3}H-colchicine in the regenerating liver of the rat indicates that it increases at 18 h and doubles during the first 36 h. In the following days, tubulin may reach levels as high as five times the normal values. This increased synthesis is parallel with the mitotic activity [21]. The amount of tubulin in HeLa cells, previously synchronized by amethopterin or hydroxyurea treatment, has been measured by the binding of ^{3}H-colchicine during the cell cycle. Synthesis persists during all the cycle, from G1 through S to G2, when it is above average. The quantity measured results from an equilibrium between synthesis and catabolism [73].

3.2.3 Tubulin mRNA

The study of tubulin synthesis and of its structure would considerably benefit from the isolation of the specific mRNA for tubulin. The first results were published in 1973 [64] and indicated the possibility of synthesizing tubulin from a soluble fraction of muscle to which were added polysomes from brain. Further results have been reported, in which the ribosome-containing fraction was provided by a lysate of reticulocytes of anemic rabbits, and the mRNA by chick embryo brain. The mRNA was separated from the total RNA by ultracentrifugation and added to the lysate of reticulocytes in presence of an ATP generating system, chick tRNA, amino acids and ^{35}S-methionine. "Cold" carrier tubulin was added after the reaction had been stopped by RNase. Electrophoresis indicated that the synthesis of both actin and tubulin was enhanced about twelvefold over that of the control. This synthesis took place without the addition of initiation factors [53], but was stimulated, as was the synthesis of actin, by one fraction from embryonic muscle or brain [52]. Similar results have been obtained with cell-free systems from rat brain, containing either membrane-bound or free ribosomes, supplemented with mRNA and initiation factors. The tubulin, when purified, was not contaminated by any other protein [45].

3.3 Role of Hormones

In Chapter 8, many relations between MT and the secretion of hormones will be related. At this moment, we are interested in tubulin synthesis: a few data from the literature indicate that this may be influenced by hormones. (The mitogenic activity of many hormones, involving the assembly of spindle MT, will not be considered here.)

The elongation of the lens epithelium is closely related to the growth of MT (cf. Chap. 6). In the chick, this is stimulated in vitro by small quantities of insulin (1 μg/ml) [94] independent of any stimulation of bulk protein synthesis. The cells treated by insulin double their length; the role of the MT is demonstrated by the fact that colchicine ($20 \times M^{-6}$) prevents this elongation, which is also inhibited by cycloheximide, indicating that protein synthesis (of tubulin?) takes place [94]. Insulin and nerve growth factor (NGF) have been reported to increase the growth of MT (in lens epithelia in tissue culture [93, 94] and in adipocytes [117]); the mechanism of action may be similar, as proinsulin and NGF have a related structure. In Chapter 8, other evidence that insulin secretion may be linked with an increased amount of tubulin synthesis, under the action of glucose or cAMP, will be mentioned [96]. The doses of insulin which are required, and the absence of action of reduced and carboxymethylated A and B chains of insulin suggest that its effect may not be specific [93] but related to the so-called "insulin-like" activity of the serum.

NGF, which is prepared from the salivary glands of snakes and mice (review in [74]), increases considerably the sprouting of axons from sympathetic nerve cells grown in vitro, and the overall growth of the nerve cells. Its action does not result from an increase in protein synthesis, but from a more specific stimulation of the growth cone of the neurons and of the formation of MT [151, 152]. The increased formation of MT (neurotubules) is paralleled by a considerable increase in the number of neurofilaments [3]. This effect is quite specific and is not observed with most other neurons: apart from the sympathetic cells, at any stage of development, only embryonic sensory nerve cells respond to NGF [74]. This last type of cells has been studied with the purpose of assessing the possible role of cAMP in the action of NGF. The inhibitory action of colcemid on neurite elongation in vitro is antagonized both by NGF and dibutyryl-cAMP. The inhibition by cytochalasin is, in contrast, not prevented. It was concluded that increased MT assembly plays a role in elongation, and is stimulated by NGF. cAMP may mediate, like in other models, the hormonal-like action of the growth factor [104].

The growth of an extensive network of neurites by neuroblastoma cells in vitro requires the formation of many MT. However, a careful study of the pools of tubulin, as measured by three different methods, failed to indicate any notable change in the total amount of tubulin in the cells. This stability in cells in which many new MT are formed indicates that a large pool of tubulin must be present, and that MT assembly is regulated by other factors [87].

3.4. Role of Cyclic Nucleotides

The intervention of cAMP and cGMP in tubulin assembly has been mentioned (Chap. 2). There is a growing amount of data indicating that the number of MT may be controlled in some cells by the levels of these nucleotides; however, this concerns the numbers of assembled tubulin molecules, not the synthesis of tubulin. Some of these results will be discussed in relation with phagocytosis (Chaps. 8 and 11). The main results can be summarized here, and have been described under the term of "ying-yang" hypothesis [142, 154] of mediator release: changes related to secretion are inhibited by an increased intracellular content of cAMP, while cGMP has the opposite effect. While agents such as theophylline, colchicine, VLB, inhibit the release of polymorphonuclear granule enzymes, carbamylcholine and cGMP increase the secretory activity and the number of MT. Phorbol myristate also increases the level of cGMP and the numbers of MT in leukocytes [55].

While the number of visible MT appears to be greater after cGMP [89], there is no evidence yet that the synthesis of tubulin is affected by the cyclic nucleotides, which appear mainly to modify the degree of tubulin assembly into MT (cf. however Chaps. 8 and 11).

3.5 Role of Nucleating Sites (MTOC)

The observations in vitro have clearly indicated that the assembly of tubulin into MT can only take place in some definite conditions, and that "accessory" proteins or intermediate forms of assembly, such as rings or stacks, are often indispensable for the initiation of MT growth.

In the living cell, the assembly of MT is poorly understood: while in many cells, MT seem to appear without any preexisting structure, there are indications that differentiated cell regions play a part in MT assembly. It is also evident from many observations that this is polarized, and often, if not always, takes place by an unidirectional addition of subunits (whatever these may be: tubulin molecules or more complex subassemblies) to already formed MT (cf. Chap. 2)

Pickett-Heaps [95] has proposed "microtubule organizing centers" (or MTOC) to name the regions which in the cell appear to initiate the growth of MT. More will be said about these when the complex assemblies of MT, as found in cilia and in the mitotic spindle, are considered (Chaps. 4 and 10). At this moment, some interesting contributions in this field will be summarized.

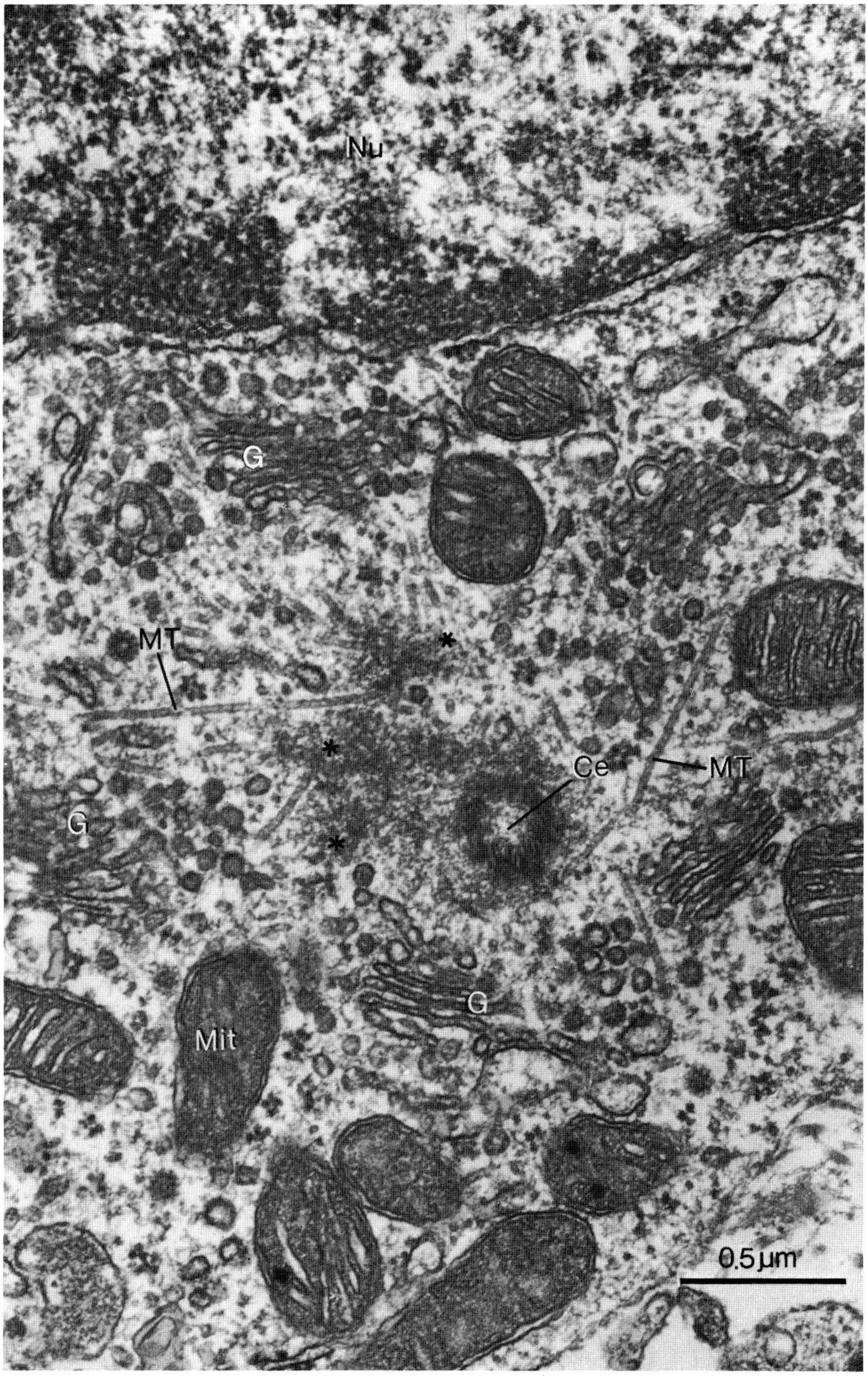

Fig. 3.2. Rat pituitary, posterior lobe. Pituicyte showing MT radiating from a centriole (*Ce*), which is surrounded by irregular masses of granular material (*). Five groups of Golgi vesicles (*G*) are located in the pericentriolar region, and are in close relation with the MT. *Nu:* nucleus; *Mit:* mitochondria

In the early works of Bessis et al. [13] and de Harven and Bernhard [36, 37], the pericentriolar dense bodies had been described as regions from which MT extended into the cytoplasm. These masses of granular electron-dense material play a role in the spindle-fiber attachment in many cells, no MT being directly fixed to the centriolar MT (except when a centriole becomes a basal body). Many mitoses have no centrioles, and in primitive cells polar bodies, more or less related to the kinetochores, play the role of MTOC [123]. In the ectodermal cells of the *Arbacia* blastula, cold destroys the MT without affecting the dense pericentriolar MTOC from which they extend. When the embryos are warmed again, the newly formed MT appear to grow from these bodies [131]. In cells engaged in an active ciliogenesis, the newly formed basal bodies are also born in a dense granular matrix, which may be an aggregate of non-assembled tubulin [38]. In a study of MT formation in *Chlamydomonas,* it has been concluded that the tubulin pool may only assemble to form MT in the presence of "a competent MTOC" which would have a regulatory action [43].

The role of these MTOC has been demonstrated more recently by placing cell organelles—from cells treated by the detergent Triton X-100—in the presence of bovine brain tubulin in a buffer permitting the assembly of the subunits: it was found that in HeLa cells, the MT of which had been destroyed by colcemid treatment, the brain tubulin assembled into MT on kinetochores and centrioles, thus on the two structures to which the spindle MT are attached. In these conditions, some MT may be formed in continuity with the centriole MT (preferably from the distal end) and many MT surround the centrioles, with no definite relation with the pericentriolar bodies which are not apparent by this technique [76, 116]. The implications of these experiments for the mechanics of mitosis will be discussed in Chapter 10. The role of kinetochores as MTOC has been demonstrated similarly by incubating chromosomes from lysed mammalian cells with tubulin solutions ([125]; cf. Chap. 2). Another indication of basal bodies as MTOC is given by experiments where purified basal bodies from protozoa have been injected into unfertilized eggs of *Xenopus:* these induced in a short time asters and cleavage furrows. Surprizingly, other MT structures (flagella, cilia, brain MT) were ineffective. Moreover, fully grown oocytes, although they could be demonstrated to have the same amount of tubulin as unfertilized eggs, did not react to basal bodies nor to other MTOC, suggesting that their tubulin was in a condition preventing assembly [60]. These results indicate that while the MTOC may be important in orienting the MT, other factors necessary for assembly and independent of the MTOC are required (cf. Chap. 2).

Further evidence for the role of MTOC or other more complex structures will be related in Chapter 4. The role of "dense matrices" in the formation of MT and of centrioles is evident, but the biochemical nature of these structures awaits further studies, for instance with fluorescent antitubulin antibodies.

3.6 Disassembly and Reutilization of Tubulin

It has already been mentioned that the equilibrium between MT and tubulin is reversible: in the living cell, tubules are frequently seen to vanish, like the mitotic apparatus at telophase. There are some definite observations which indicate that tubulin is not catabolized and returns to a cytoplasmic pool. This is most evident when complex structures are destroyed within the cell. The shortening of one flagella of *Chlamydomonas,* when the other is amputated, has already been mentioned (3.2.2; [30]). While colchicine may inhibit the flagellar regeneration, cycloheximide is without effect, demonstrating that the assembly proceeds from a preexisting pool. In another protist, *Tetrahymena,* if the cilia are destroyed by high hydrostatic pressures, their basal bodies are resorbed, and their tubulin returns to a common pool [86]. In the regeneration of the oral structures, when the cilia are destroyed by cultivating *Tetrahymena* in a medium deficient in amino acids, less than 6% of the MT proteins are synthesized, suggesting a reutilization of tubulins from the destroyed basal bodies [145, 147].

A detailed study of resorption of MT structures has been published by Bloodgood [18] with detailed tables on the many instances found in protozoa, fungi and algae, and of the various treatments inducing such resorption. Cilia, flagella, axostyles, axopodia, complex MT organs in protozoa, may be resorbed during some steps of the biological cycle or under the influence of various drugs or physical agents. Cilia and flagella may be lost, resorbed, or internalized in various ways before disappearing altogether. There are many examples which confirm that the tubulin returns to a pool from which new cilia or MT structures may be built. Assembly and disassembly seem to be independently controlled. For Bloodgood [18] none of the three following mechanisms explains all the facts: (1) enzymatic breakdown; (2) dynamic equilibrium between tubulin and MT; (3) a combination of these, as resorption of an organelle made of tubulin can take place when a similar one is made in the close vicinity, and the author concludes that "much additional research is necessary before we can ... propose models for the mechanisms of resorption of microtubular organelles and its control."

3.7 Differential Sensitivity of MT

When discussing (Chap. 2) the unicity or multiplicity of tubulins, some instances of variable sensitivity of MT were mentioned, for instance the selective extraction of the two central MT of cilia [51]. While it is probable that most MT are made of the same tubulins (α and β), there are several observations indicating that all MT do not react identically to various physical or chemical agents: this could be the consequence of the presence of other proteins or even of different tubulins, as mentioned in Chapter 2, and introduces in the study of MT a new element of complexity which should not be underestimated.

Behnke and Forer [10] were the first to underline these differences. They showed that in the crane-fly spermatids (*Nephrotoma suturalis* Loew) colchicine destroyed the cytoplasmic MT without affecting the 9+2 tubules of cilia. Treatment by heat (50 °C) destroyed the cytoplasmic MT and, in the cilia, first the central and then the B tubules. A short pepsin treatment, on the other hand, destroyed electively the 9 doublet MT, leaving in their place "holes" larger than the MT, suggesting that some peritubular material may have been digested. Similar findings were reported for rat sperm tails and tracheal cilia, and it was concluded that different classes of MT could be recognized. Many similar reports from the literature, in particular on the differential sensitivity of peripheral and central MT of cilia and flagella, were tabulated.

From these observations stems the idea that at least two types of MT exist: a stable variety, resistant to cold, high hydrostatic pressures, and chemical poisons, and a labile form [130], and that all MT are not structurally and chemically identical [24]. Quite another indication of the complexity of the MT of cilia is given by the study of sterile mutants of *Drosophila,* in which the sperm flagellum is abnormal. A sequence has been described in which the MT disappear in the following order: the internal part of MT B, the external part of MT A, the internal part of A, the external part of B, and last the central and accessory MT [66]. Such differences within MT were already indicated by Behnke and Forer [10], who suggested that the walls of the B tubules were composed of two materials, the portion adjacent to and common with the A tubule, and the remaining part (cf. Chap. 2).

A more remarkable finding is that in the spermatozoa of *Rhynchosicara,* which have flagella made of 400 doublets, with a spiral disposition, colchicine may destroy only one MT of each pair [113, 114].

A detailed study of the resistance of spindle MT of rat kangaroo fibroblasts to cold indicates that a heterogeneity exists, as if cold-resistant and cold-sensitive MT were present [19]. The authors suggest that some may be in a stabilized condition, such as those of the cilia, the centriole, and the telophasic midbody (cf. Chap. 10). They do not think that it is necessary to imagine a physicochemical difference in MT: the environment of the MT (the clear peritubular zone, the dense matrix of the telophasic body) may explain the differences of resistance. This opinion is in agreement with other findings on MT from protozoa [20, 42].

Another fact, the importance of which will become more evident later (Chap. 10), is that differential sensitivities of MT may result in many cases from differences in metabolic turnover. MT which are in a condition of rapid renewal will be more rapidly affected by poisons such as colchicine, which prevent the assembly of their subunits, than more static structures like centrioles and cilia. However, this does not explain all differences of sensitivity, and the role of accessory factors and stabilizing conditions of environment should be kept in mind.

3.8 Relations of MT with Other Cell Organelles and Structures

3.8.1 Introduction

It is exceptional for MT to float freely in the cytoplasm: their close relations with various types of nucleating centers (kinetochores, pericentriolar dense masses, basal bodies) have already been mentioned. MT are closely associated with many cell movements (cf. Chap. 7), and one of the most important and unsolved problems is the relation between the force that moves cytoplasmic particles and MT. Several of these problems will be treated in further chapters: the study of cilia (Chap. 4) will reveal some very specific and close relations between ciliary MT and cell membranes; in Chapter 8, the relations of MT with secretory granules will be considered, and

Chapter 9 will be entirely devoted to the neuroplasmic flow of cytoplasm and organelles and its relation with MT (neurotubules); last, the associations of chromosomes and nuclear membranes in mitosis will be discussed in Chapter 10.

There remains a large number of facts which indicate close links of MT with various organelles, cellular and nuclear membranes, and cytoplasmic inclusions, such as viruses. The relations with the cell membranes will be considered first, for they are particularly important for an understanding of other cell changes, like secretion, phagocytosis, the movements of synaptic vesicles. They may explain how colchicine-binding proteins have been found in membranes [78, 119].

The relations of MT with cell structures may be direct, with a close linkage between both; more frequently, they take place through the lateral expansions or "side-arms" of MT, described in Chapter 2.

3.8.2 Cell Membrane

Two different types of attachment of MT to cell walls may be considered. In several species of invertebrates, there exist strong links between parallel MT and the exoskeleton, the MT acting as mechanical "tendons" between contractile cells and cuticle (cf. Chap. 6). Similar links are also found in specialized cells acting as sensory mechanoreceptors ([127]; Chap. 9).

The fixation of colchicine on cell membranes has been described by several authors: an antibody prepared against rat brain tubulin in rabbits reacts with membranes and with a colchicine-binding protein which may be extracted from those of brain and thyroid cells. It co-migrates with tubulin but its optimum binding temperature for colchicine is higher (54 °C) [14]. With a new method of treatment with albumin before fixation [57], associations of MT with membranes have been demonstrated in the post-synaptic junctions in the rat [144]. Isolated synaptosomal membranes from the brain of man and swine contain a protein of 53,000 MW which has a peptide map and an electrophoretic mobility similar to tubulin monomer [71]. The problem still awaiting a solution is whether MT ends are closely linked to membranes, or if some other related protein, unassembled into MT, is present in this location.

More direct evidence of colchicine receptors in membranes is afforded by the studies, by freeze fracture techniques, of intramembranous particles in embryonic mouse cells. These particles are aggregated by colchicine and VLB at low concentrations (10^{-5} and 10^{-9} M); this action is not mediated by MT, as lumicolchicine (cf. Chap. 5) is also effective. The actions of the drugs are dose- and time-dependent. Cold (4 °C), which destroys MT, does not affect the location of the particles [49].

Membrane-bound tubulin from the brain of guinea-pigs has been isolated and copurified with brain tubulin: the results indicate that the membrane contains a 55,000 dalton protein which seems to be identical to cytoplasmic tubulin, except for a greater thermal stability [15].

There are some indications that a protein similar to tubulin may be present in the membranes of cells which have no MT. In pigeon erythrocytes, colchicine-binding protein is present, as demonstrated either by ^{3}H-colchicine binding to isolated membranes, or by binding of intact erythrocytes to colchicine-Sepharose beads.

This observation, indicating that the tubulin molecules are present on the exterior surface of the membrane, is confirmed by the fact that in the presence of free colchicine (5×10^{-3} M) the beads do not fix any erythrocytes [153]. While avian erythrocytes are still rather complex structures, with a marginal band of MT (cf. Chaps. 2 and 6), human red blood cells may also be affected by colchicine, which stimulates virus-induced fusion. However, the effective doses of the alkaloid are high ($2-10\times10^{-3}$ M), and it is suggested that spectrin may be the colchicine-binding receptor [112].

Several recent studies have indicated that the displacements of membrane receptors for antigens or for lectins such as concanavalin A are influenced by MT poisons, suggesting that MT may control the position of proteins embedeed in the fluid cell membrane. Such gathering together of membrane receptors for "capping" has brought interesting observations, although sometimes contradictory. The changes observed in pathological conditions will be further discussed in Chapter 11.

As it is not evident that these surface changes are identical in all cells, it is preferable to consider separately each type of cell which has been studied and the different substances which have been used to mark the membranes.

a) In *lymphocytes*, capping can be observed in the presence of immunoglobulins. Splenic lymphocytes exposed to antibody directed against immunoglobulins bind these to surface receptors at 4 °C. When these cells are warmed to 37 °C, the receptors for immunoglobulins (which can be observed by immunofluorescence techniques) move toward one pole of the cell. This capping is inhibited by cytochalasin B but not by colchicine [40, 111, 124]. It is also inhibited by iodoacetamide, sodium azide and other metabolic poisons [139, 140], and prevented by the introduction into the cell of Ca^{2+} by an ionophore, suggesting the intervention of MT in the movements of the surface receptors. However, it appears that these results may be "fundamentally different" from those found with the lectins [111].

When lymphocytes are treated by concanavalin A, no caps are formed unless the cells are modified by MT poisons [149, 150] as indicated by Table 3.2. The action of VLB is more readily reversible than that of colchicine, which acts at low concentration (10^{-6} M; [40, 41]); lumicolchicine is ineffective, confirming the role of MT (cf. Chap. 5). Two explanations for these effects have been proposed [149]: either cell-surface receptors interact directly with colchicine-sensitive structures, or MF (actin?; cf. Chap. 7) act as a link between surface receptors and MT.

b) In *fibroblasts*, colchicine has different effects on Con A receptors. In normal cells, it disperses these, while in SV40 transformed cells selected for resistance to

Table 3.2. Effect of MT poisons on Con A capping by lymphocytes (from Yahara and Edelman, 149)

Drug	Concentration (M)	Caps (%) with Con A
—	0	1 ± 1
Colchicine	10^{-4}	24 ± 10
Colcemid	10^{-4}	25
Vinblastine	10^{-4}	52 ± 12
Vincristine	10^{-4}	15
Podophyllotoxin	10^{-4}	10
Lumicolchicine	5×10^{-4}	1
Griseofulvin	10^{-4}	2

Con A, colchicine allows the capping phenomenon to take place [11, 12]. It should be recalled here that transformed cells contain less MT and tubulin than normal ones.

c) In *polymorphonuclears*, capping may also take place and is often closely related to the internalization of surface receptors, which is one of the steps of phagocytosis (cf. Chap. 8). The formation of caps with Con A has led to contradictory results: colchicine was thought by some authors to have no effect [109], while for

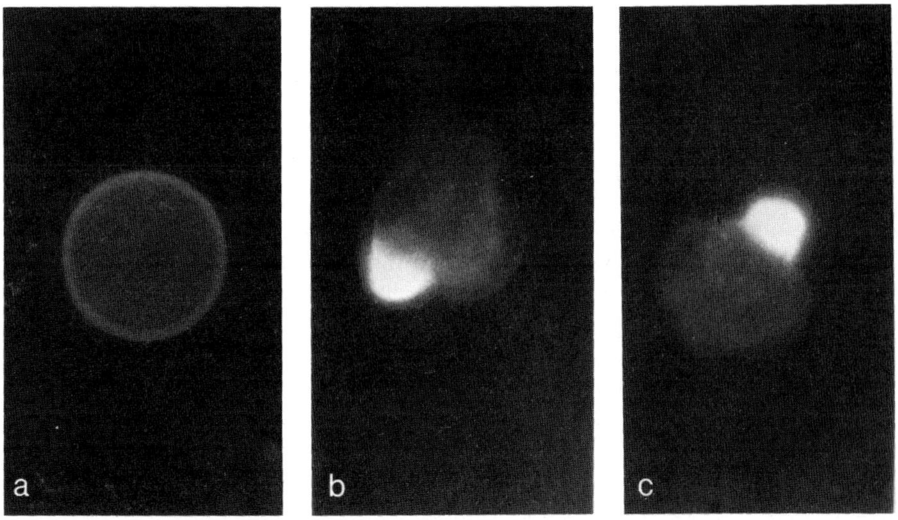

Fig. 3.3. Distribution of fluoresceine-isothiocyanate conjugated concanavalin A on human polymorphonuclear leukocytes preincubated for 30 min: (a) with buffer alone; (b) with buffer containing 10^{-6} M colchicine (capping); (c) with buffer containing 10^{-6} M R17934 (oncodazole). The receptors for Con A, which remain dispersed in normal cells, migrate to one pole (capping) when the MT are destroyed (cf. Chaps. 5 and 11) (from Oliver [90])

others [89] it facilitates capping, as if the mobility of surface receptors was increased by the destruction of MT. This was confirmed by experiments with the MT poison R 17934 (cf. Chap. 5), which also facilitated Con A capping [90, 92]. The cellular action of Con A is complex, and in human polymorphonuclears, it has been shown that within 2 min after Con A treatment, an increased number of MT is visible in the pericentriolar region. Many images of internalized membranes with Con A fixation sites (labeled with ferritin) were also observed [61]. These changes are also related to the discharge of lysosomal enzymes (cf. Chap. 8).

The role of MT in capping is confirmed by the action of the –SH poison, diamide (cf. Chaps. 5 and 10): Con A capping is promoted in human polymorphonuclears and lymphocytes and the MT assembly is inhibited. As similar effects are observed in cells treated with the oxidant, tertiary butylhydroperoxide, a role of reduced glutathione in these changes is probable [91].

Capping can be induced by immune complexes; this is inhibited by relatively large doses of colchicine (10^{-3} M), while lower doses (10^{-5} M) are without action although decreasing the release of lysosomal enzymes [110].

d) Capping has also been studied on cultures of rabbit ovarian granulosa cells, treated by Con A [29]. In the capped region, VLB crystals are more numerous close to the cap, while in normal cells they are dispersed; a study of these cells with fluorescent thiocarbamyl-colchicine indicated also that this accumulated in the capped region [1]. This capping is also prevented by various metabolic inhibitors (sodium fluoride, dinitrophenol, sodium azide). MT poisons do not modify the capping, but prevent the internalization of the modified membrane. In Chinese hamster ovary cells, colchicine has been found, like podophyllotoxin, to inhibit Con A capping; mutant lines of cells, resistant to colchicine, are not affected, a fact which suggests that the drug must penetrate into the cell, probably in order to affect the MT which play a role in the movements of cell membrane receptors [4].

In protozoa, MT may also be necessary for the movement of membrane proteins, as observed in *Tetrahymena,* treated with large (5 mg/ml) doses of colchicine [148]; as the MT are not destroyed, a non-specific action of the drug on the cell membranes is possible.

There are several other indications that colchicine-fixing proteins (tubulins?) may be associated with cell membranes. In the liver of the rat and the mouse, colchicine and colcemid are fixed by the cell membranes [119] which may be closely associated with MT. Recent experiments on fluorescence emission spectra lead also to the conclusion that in rabbit's polymorphonuclear leukocytes, considerable interactions exist between MT and membranes [9].

It is evident that further research is needed in this complex field and that the topographical relations between MT and membrane receptors should be better defined. Similar problems will be met in the study of axoplasmic flow and synaptic endings (cf. Chap. 9). It would be important to know whether the MT themselves are attached to the cell surface, or if this is done via lateral expansions or side-arms or other cytoplasmic fibrils.

Other indirect relations of MT with cell walls, where the MT are at a distance from the membrane but play a role in the shaping of the cell, will be discussed in Chapter 6.

3.8.3 Nuclear Membrane

The position of the nucleus in the cell may depend on the integrity of MT. Relations between the location of chromatin condensations and that of MT will also be mentioned (Chap. 6). An interesting relation is that of MT with nuclear pores: this has been described in neurons, after prefixation treatment with albumin, which stabilizes the MT [57]. It should be compared to the experimental growth of intranuclear MT from the nuclear pores of mastocytes in the rat's intestine [35].

Fig. 3.4 A, B. Linkages between MT and membrane-bound vesicles in *Paramecium caudatum.* (A) Disk-shaped vesicles (*V*) seen in profile lie close to a curved microtubular band (*arrows*). They are located at one side of the MT, from which glycogen particles, ribosomes, and other organelles are separated by a distance of 30 – 40 nm. (B) After collidine-buffered fixation, the cytoplasm is retracted and the vesicles are compressed in a zig-zag pattern, while maintaining their association with the MT (from Allen [2])

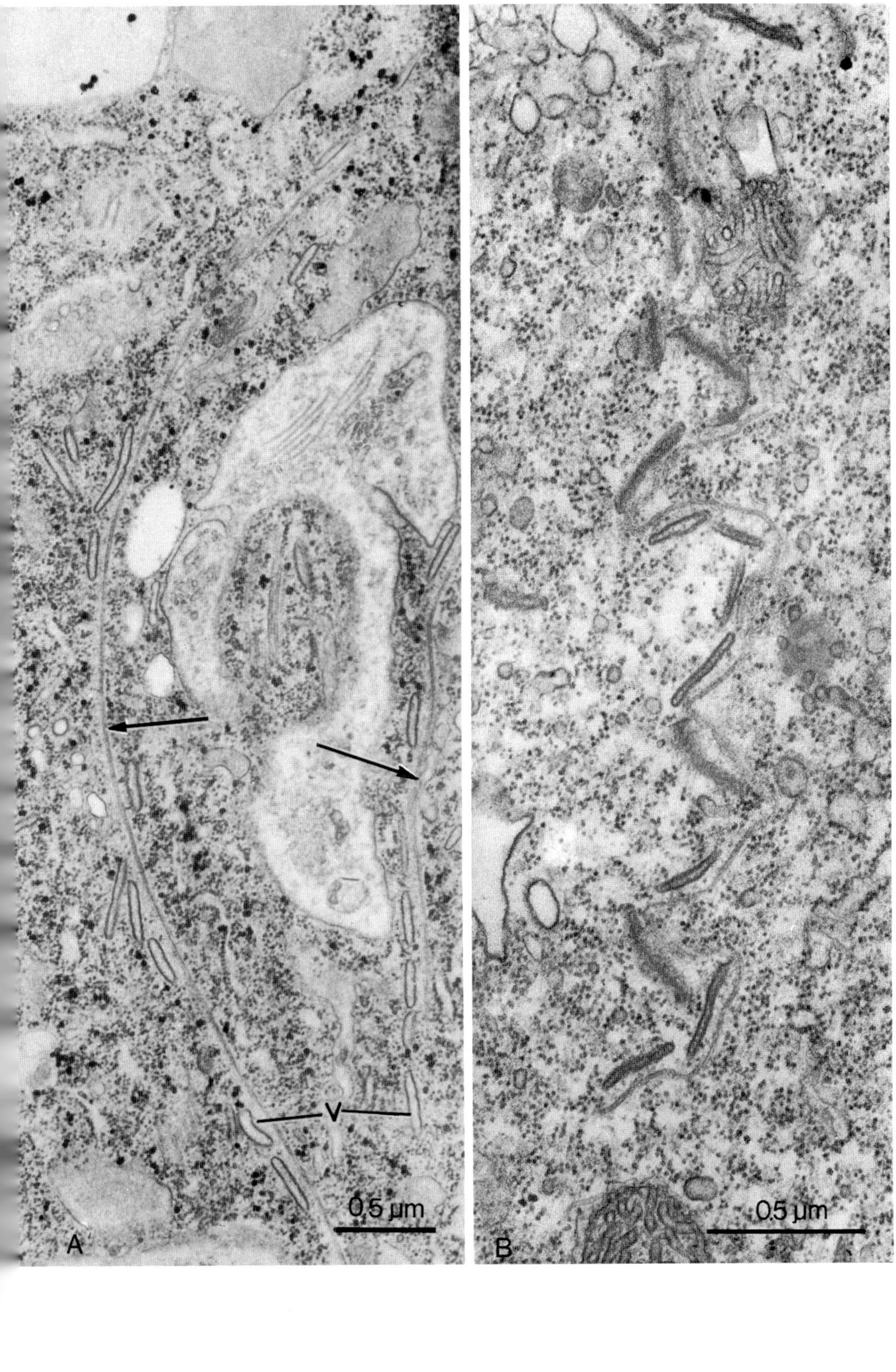
V
0.5 µm
A
0.5 µm
B

3.8.4 Cell Organelles

The movements and the location of various organelles in the cell depend on the integrity of the MT: this is clearly shown in the thyroid cells of the cream hamster, where after destruction of the MT the normal stratification of mitochondria and the Golgi apparatus vanishes, the cytoplasm assuming a more homogeneous aspect [65]. Similar disorganization of intracellular location of organelles has been reported in various cells after destruction of their MT [34].

Associations between mitochondria and MT are found in various cells. They do not involve a continuity between both organelles. The MT play the role of orienting fibers, as in the meiosis of *Nephrotoma* spermatocytes [72], where the mitochondria are parallel to the external MT of the spindle. This may have some importance, and help to provide an equal repartition of mitochondria in the daughter cells. In axons, mitochondria are known to move with the neuroplasmic flow (cf. Chap. 9), and quantitative estimations have shown that they often lie in close association with MT, to which they may be attached by wispy filaments [101].

Other organelles may be transported in relation to MT without any evidence of a firm linkage between both: such is the case of the telotrophic ovarioles of *Notonecta glauca glauca,* where a continuous flow of ribosomes takes place in the nurse cells toward the oocytes within cell extensions containing as many as 30,000 MT

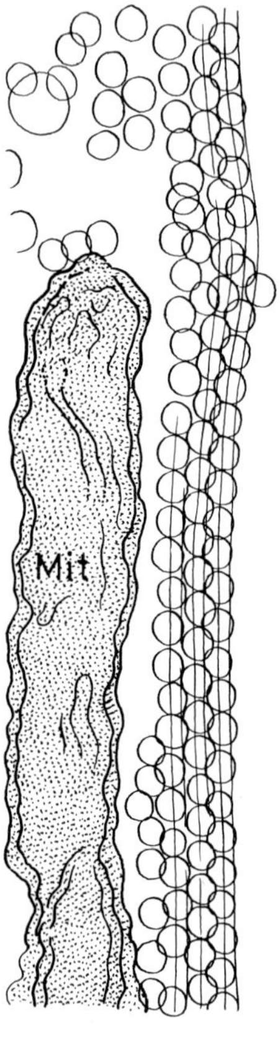

Fig. 3.5. Association of MT and synaptic vesicles in the lamprey (*Petromyzon marinus*). Longitudinal aspect of grouping of synaptic vesicles in *Petromyzon* axons. The vesicles are hexagonally packed around a group of three MT. On the left: mitochondria (*Mit*) (redrawn from Smith et al. [115])

([77, 121]; cf. Chaps. 2 and 7; cf. Fig. 2.9). Other cell organelles may be closely linked with MT, such as the intracytoplasmic vesicles found in *Paramecium caudatum*. These are aligned along MT; when, as a consequence of retraction during fixation, the MT take a zig-zag pattern, the vesicles are deformed accordingly, while keeping their close relation with the MT. This demonstrates some kind of linkage. However, it has not been possible to visualize how the vesicles were attached to the MT: some "bridges" are observed, but the space between vesicles and MT is occupied by a low-contrast material, the nature of which remains unknown ([2]; Fig. 3.4).

The problems of the synaptic vesicles will be discussed in relation with axonal flow (Chap. 9). In the lamprey (*Petromyzon marinus*) the central nervous system shows, in the giant axons, which may contain up to 3,000 MT, a close association of vesicles with MT, about five synaptic vesicles being symmetrically disposed around one MT. About five MT reach each synapse: the vesicles may be carried along them [115] like other particulate bodies (cf. Chap. 9). In the rat spinal cord, cultured in vitro, clusters of synaptic vesicles surround the MT, which often terminate at the synaptic membrane [17].

An association between MT and Golgi vesicles is found in *Euglena gracilis*, where MT are located between the Golgi cisternae: it has been suggested that they may play a role in the transfer of products into or out of the Golgi apparatus [85]. Several observations of changes of the Golgi system when MT are destroyed will be mentioned in Chapter 8. In the oil-body cells of *Marchantia*, microbodies (peroxisomes) are closely associated with surrounding MT during the differentiation of these cells [50].

3.8.5 Viruses

As the components of the envelopes of many viruses derive more or less directly from cell structures and may have close analogies with cell membranes, it is not surprizing to find that some viruses display very intimate relations with MT.

This is especially the case with reoviruses, which are associated with the spindle apparatus [118] and have been the subject of a series of papers [31, 32, 33]. In strain L cells infected by reovirus the MT are seen to be enveloped by a dense material which could be demonstrated, by the use of ferritin labeled antibody, to be antigenically related to the virus [33]. The association of the viruses with MT is an early step in their replication, as ^{3}H-uridine marked viral RNA can be found in the close vicinity of the MT [33]. Already 4 h after infection, the MT are coated with this virus-specific protein. It is interesting to note that the mitotic index of the L strain cells is increased after reovirus infection [33]. These findings have been confirmed [70, 118] and extended to other types of cells, such as the African green monkey kidney and human amniotic cells. The staining of the viral antigens by immunofluorescent techniques has been proposed as a simple method for following the changes in MT. Another interesting finding is that viral coating stabilizes the MT toward colchicine and cold. It could also be demonstrated by this technique that D_2O prevented the reassembly of MT after cooling to 5 °C: this method could permit an accurate study of MT assembly and disassembly [70]. The fixation of reovirus antigens to neurotubules shows that these structures are identical to MT of other cells [56].

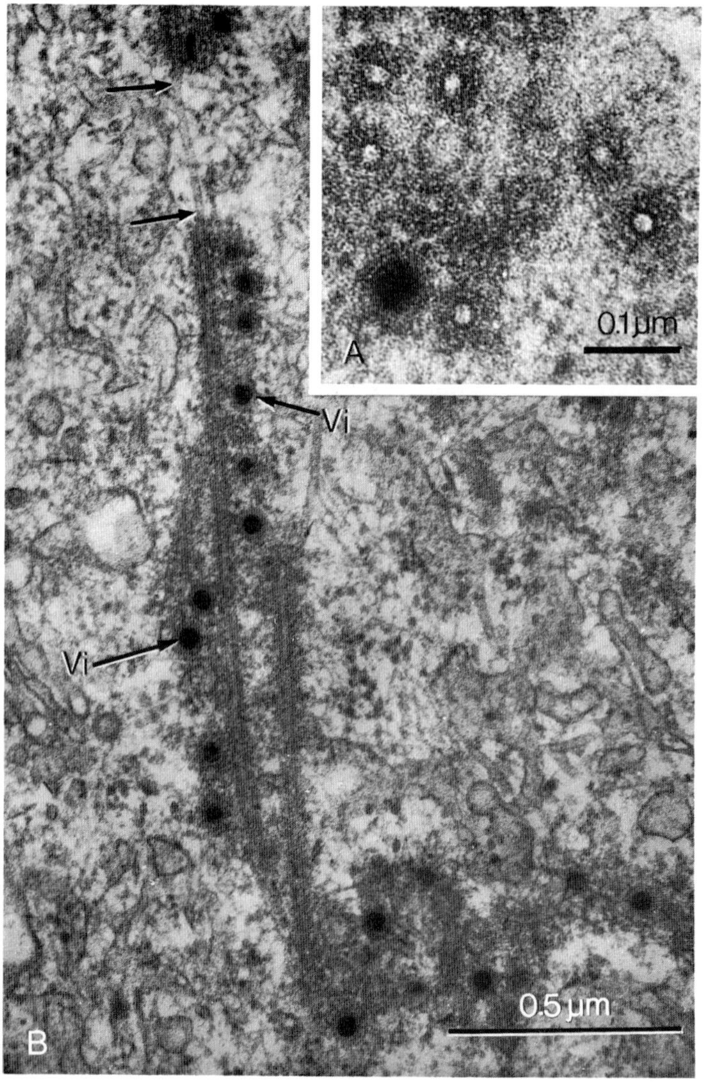

Fig. 3.6. Strain L_2 cells infected by reovirus type 3 (Dearing strain). The MT are coated by a dense material along most of their length (an uncoated segment is indicated by the *arrows*). Several virus particles are embedded in this dense material (*Vi*). The *inset* shows cross-sections of metaphase MT, surrounded by a sheath of dense material. One virus particle, with a dense center, is apparent (from Dales [31])

In HeLa cells, the migration of adenoviruses toward the nucleus takes only a few minutes: it follows the MT, and this association can be demonstrated by treating the cells with VLB: the typical crystalline arrays of tubulin contain viral particles, and the infectious cycle is delayed [32]. The nuclear pore complex is a site of viral fixation; in cells treated with colchicine, the transport of viral particles is not slowed down, although MT are absent, but there is an increase (cf. Chap. 5) of cytoplasmic annulate lamellae, to which the virus becomes attached [32].

The binding of viruses to MT can be studied in vitro. Adenoviruses become attached to the MT, and 90% of the viruses are seen to be associated with the edges of MT, whatever the preparative procedure. This linkage is tight and close and probably involves the hexon capsomeres of the virus, as demonstrated by negative staining. The binding is complete after less than 3 min. It may be that the "edge" binding is an artefact of drying; however, the close linkage between capsomeres and

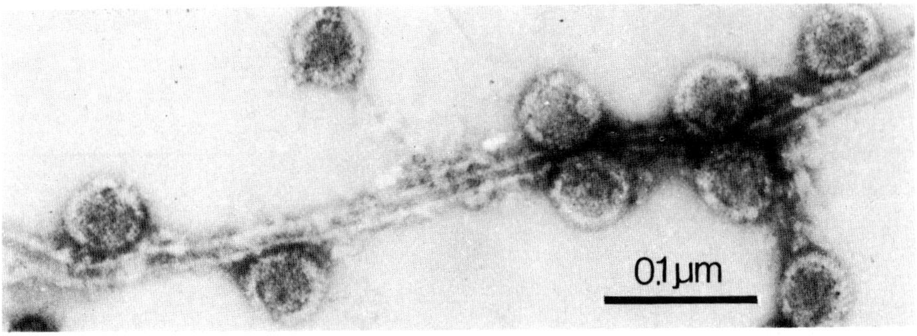

Fig. 3.7. Attachment of adenoviruses to MT. The supporting grid was first covered with adenovirus particles. Rat brain tubulin was then assembled into MT on the same grid. The MT are closely bound to the edge of the adenovirus capsids (from Luftig and Weihing [75])

MT is in agreement with the role played by MT in the intracellular migration of viruses [75]. Further discoveries are to be expected in this field, as it has been shown that viruses are also bound to plant cell MT [67].

3.9 Associations and Patterns of MT

It was shown in Chapter 2 that the association of tubulin molecules may build various polymorph structures. The same holds true on a higher level for associations of MT. Centrioles, cilia, and flagella will be discussed in Chapter 4, and the associations of MT with chromosomes and mitosis in Chapter 10. The structure of centrioles and basal bodies involves the formation of doublets and triplets made of 23 and 33 protofilaments, probably associated with other proteins. At this point, we will be more interested in other types of associations of MT, which display geometrical patterns, the study of which can help us to understand the basic principles of intertubule linkage.

The number of possible associations of MT is limited, but a considerable variety is found in nature. These patterns depend on the functional role of the cellular organs which are made of MT and which, in protozoa, attain an extraordinary complexity, as in the feeding organs of ciliates [97, 134 – 136]. Many of these structures must be understood as groupings of MT linked together by bridges or side-arms, the nature of which remains poorly understood. While in some cases these side-arms are quite conspicuous, in other cells their presence is only indirectly inferred from the position of the MT in relation to one another, and the distances maintained between them. There are also some instances of associations of MT side by side, without any bridging material, like in the cytopharyngeal lamellae of ciliates, such as *Phascolodon vorticella* [137] and *Nassula* ([135, 136]; cf. Fig. 3.13). This however is the exception, and a proper understanding of MT patterns implies a knowledge of the size and orientation of the links between them. One more word about these links: the clearest pictures which are reproduced in all schematic drawings are from sections perpendicular to the long axis of the MT, and there is a tendency to forget that the MT are helical structures and that their assembly takes place in three dimensions (cf. Chap. 2). It is thus important to remember that the links are not necessarily coplanar, and also that the MT often show some degree of skewness which is apparent in the cytopharyngeal baskets of ciliates [137].

Some fundamental problems of MT structure are apparent in studying their associations. Dynein and other molecules are attached to the MT doublets of cilia at fixed locations (cf. Chap. 4): the 13 subunits of MT, which should only differ by the α and β nature of their tubulins, and the fixation sites of guanosine nucleotides and of MT poisons, appear to differ in other ways, as the attachment of dynein always takes place on the same subunit. It may be that the attachment sites of bridges linking MT in complex patterns are related to different properties of the subunits, although this proposal complicates considerably the concept of MT. It is however clear from the various forms of MT associations and patterns that the number of linkages between MT is limited, and is smaller than 13. Six bridges may be observed in dense structures such as the rods encircling the pharyngeal baskets of ciliates [137]. More often, MT are linked by three or four bridges; as mentioned above, it is not evident that all these are in the same plane. The geometry of these links should be related to that of the MT and their subunits. Few attempts have been made to study this difficult problem, for it has only been known for a short time that the usual number of protofilaments of MT is 13—and in Chapter 2, several exceptions to this rule were mentioned. One careful geometric attempt has been made to correlate the number 13 with the presence of six connecting links, and the irregular hexagonal packing which results (see below; [63]).

Many instances of complex association of MT will be described in the next chapters. A tentative classification of the principal patterns will be presented here. Many of these are found in protozoa, which have, during Evolution, made the largest use of MT for the edification of extraordinarily elaborated organs of digestion and of movement.

3.9.1 Lamellae

The simplest type of association is that of several MT side by side, as found in the pharyngeal baskets of ciliates [97, 137] and in the so-called "km" fibers of *Stentor* [62]. In these structures, MT are closely packed side by side and form a sheat, curved at one of its extremities. Their exact mode of linkage, which somewhat resembles that seen in cilia, has not been studied by the new tannic acid staining methods, and it is not clear whether each MT has a complete complement of 13 subunits, or if some subunits are shared by two neighboring tubules.

3.9.2 Three Links

In the axonemes of the acantharia (or heliozoa?) *Gymnosphaera albida*, axonemes are seen to radiate from a central axoplast. On cross-sections, the MT of the axonemes display a pattern of more or less irregular hexagons, each MT being linked by three bridges (perhaps three double links?). A detailed geometric analysis of this pattern shows that it could be related to the 13 subunits of MT, the three bridges forming angles in the ratio 5 : 4 : 4 (two of 110°46′, one of 138°28′). This explains the hexagonal pattern and its irregularity, although strict assembly rules are necessary to group such a large number of MT into hexagons. The angles were measured from electron micrographs of the axoneme and corrected for possible effects of tilting

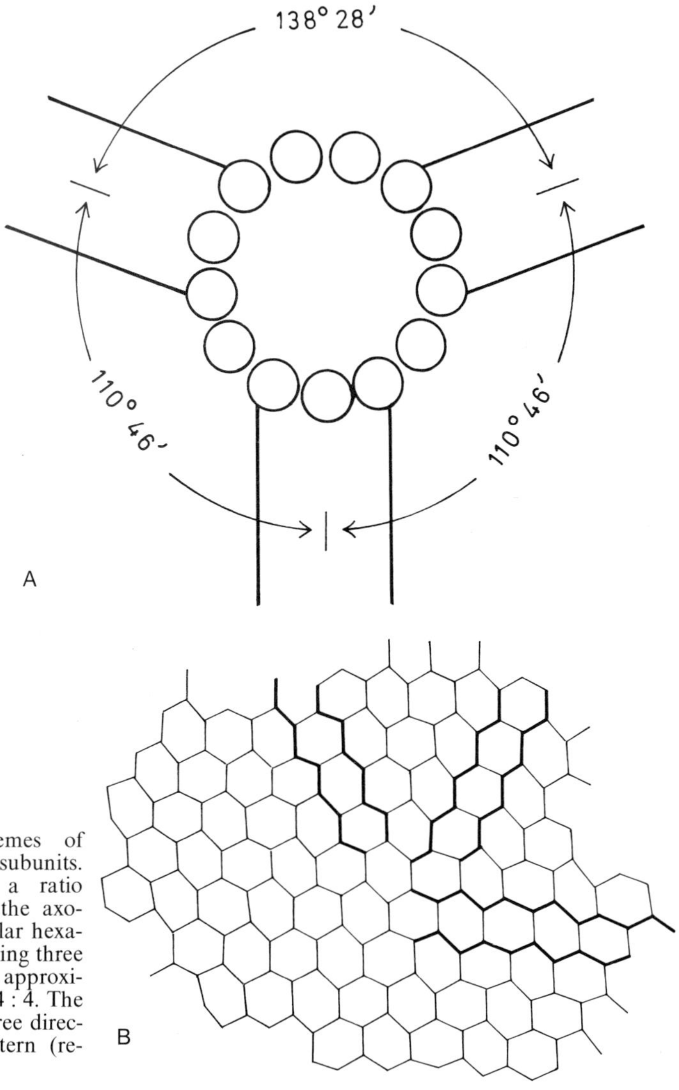

Fig. 3.8 A, B. (A) Axonemes of *Gymnosphaera albida*. 13 subunits. Angles of bridges in a ratio 5 : 4 : 4. (B) Tracing of the axonemal pattern: the irregular hexagone are made of MT having three cross-bridges each, in the approximate angular ratio of 5 : 4 : 4. The *thick lines* indicate the three directions of the zig-zag pattern (redrawn from Jones [63])

or obliquity of the sections. It is from these measurements that the 13 subunit model was constructed as indicated by Figure 3.8 [63].

The axopodia of heliozoa and radiolaria provide some other striking assemblies of MT. In *Nasselaria,* for instance, an apparent hexagonal grouping of MT results from the alternation of MT with two links and MT with three links: these form a duododecagonal lattice, with alternating convex and concave sides (Fig. 3.9; [25]). This

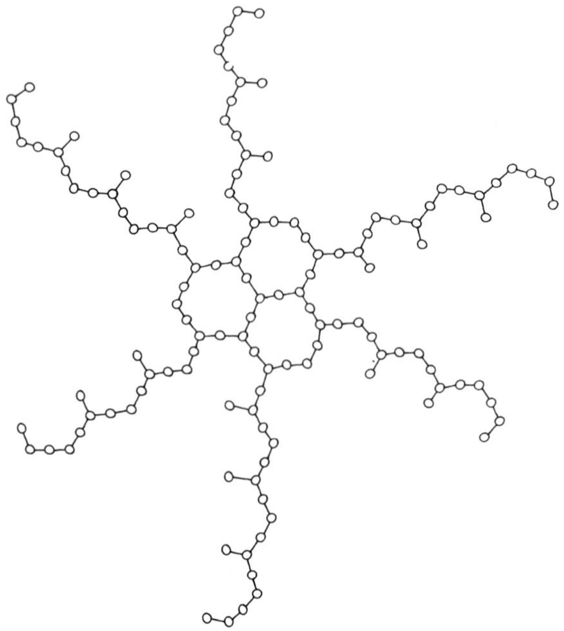

Fig. 3.9. Pattern of MT of the axonemes of *Nasselaria:* starting from a group of irregular duododecagons, alternating links between two and three MT lead to a general hexagonal symmetry (redrawn from Cachon and Čachon [25])

results from the fact, probably related to the geometry of the subunits of MT, that the angles of intertubule bridges are not submultiples of 360°. It is interesting to note that a sixfold symmetry is again found, as in *Actinosphaerium,* although the geometry is apparently quite different [25 – 28]. Each MT would be linked to its neighbor either by four or by six links, grouped two by two, and related to the *nine* subunits of the MT of this species [25]. From this hexagonal lattice, twelve long extensions may radiate into the cytoplasm.

In other species of heliozoa, with a central axoplast, the MT are assembled according to the following rule: starting at the center by two MT with three bridges, sheets of MT are formed by an alternation of two MT with two bridges ("divalents") and one MT with three bridges ("trivalent"). These sheets grow in a skewed way, and only six major rows extend to the periphery, with many lateral rows (Fig. 3.10). This sixfold symmetry, again, does not result directly from the symmetry of the MT themselves, but from the absence of growth of some sheets, related to the packing of the MT [26, 27]. The number of subunits of each MT may vary from one species to another: a most remarkable fact is that starting with MT which do not show 12 subunits, once again the general symmetry is six- or twelvefold.

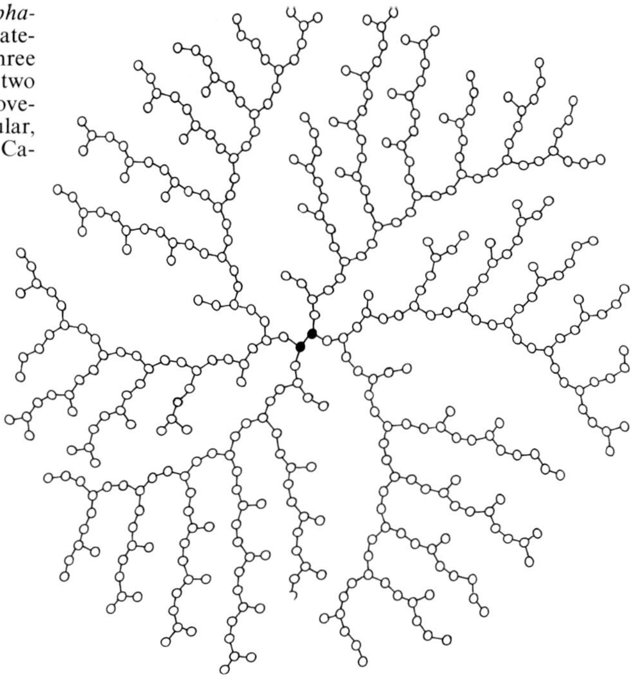

Fig. 3.10. Axoneme of *Sphaerellaria:* the MT are alternately bound to two and to three other MT, starting from two central MT (*in black*). The overall pattern has an irregular, hexagonal symmetry (from Cachon et al. [27])

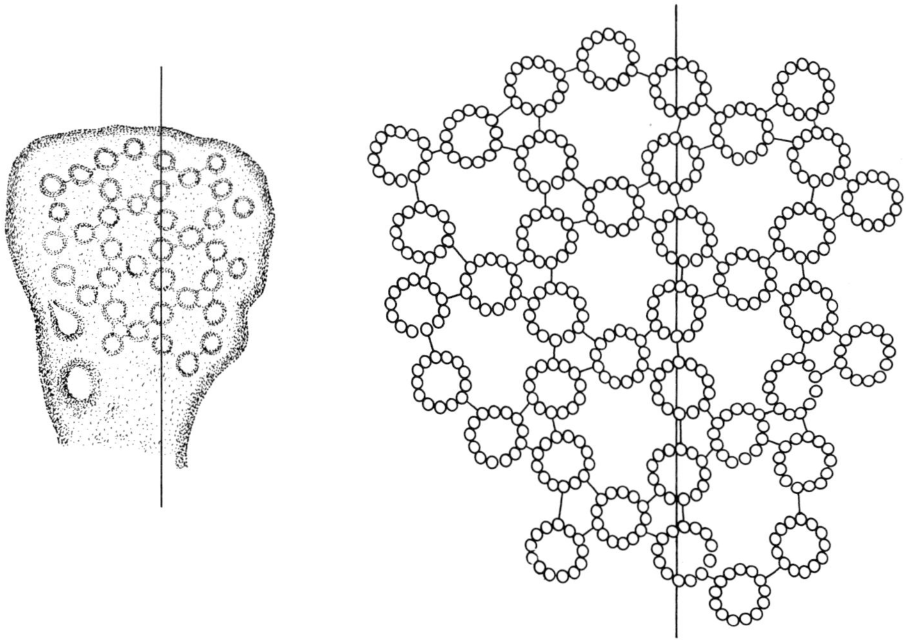

Fig. 3.11. Axoneme of *Raphidiophrys. Left:* transverse section. *Right:* schematical arrangement of bridges if MT have 13 subunits. The MT form irregular hexagons, each MT being linked to four others (redrawn from Tilney [128])

3.9.3 Four Links

Several different patterns may be found with four links uniting each MT to four others. The most complex is that described in the axonemes of the centrohelidian, *Raphidiophrys* [46, 128]. These axonemes radiate from a dense, unstructured centroplast. The packing of the MT delimitates hexagons and small triangles (Fig. 3.11, 3.12). These can be clearly explained by the formation of four small bridges between adjacent MT, and here also it was shown that MT with 13 subunits could assemble more readily and regularly than with 12, two of the links being separated by one subunit, the two most distant ones by four subunits. This pattern is different from that described above in *Gymnosphaera.* A study of the action of cold (0 °C) which destroys the MT, and of the reformation of the pattern after warming, indicates that the centroplast plays no role in the grouping of the MT, which is determined only by the location of the MT side-arms [128].

In the same group of centrohelidian heliozoa, Bardele [7] has published a careful study of the geometrical relations between the MT (Fig. 3.12), and shown that different patterns may arise from a hexagonal grouping of MT (Fig. 3.13). Here also, a sixfold symmetry results from the assembly of MT with 13 subunits, linked three by three.

Other, geometrically less well defined assemblies of MT by four links are known: the links may be nearly orthogonal, leading to the formation of square pat-

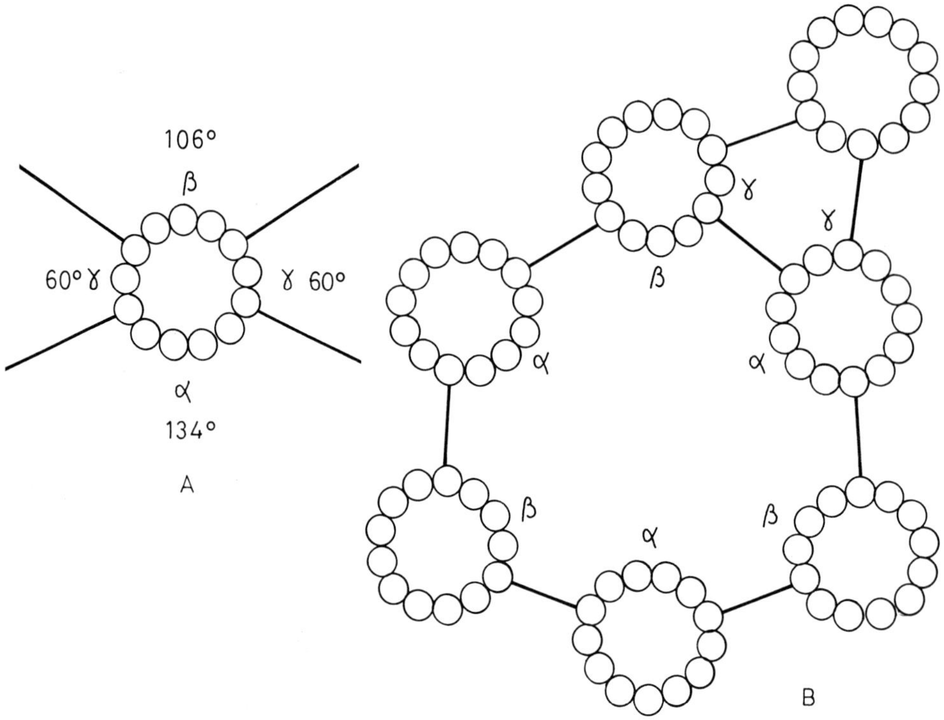

Fig. 3.12 A, B. (A) Suggested links of MT from centrohelidian heliozoa: the four links are separated by angles of 134° (α), 106° (β), and 60° (γ). (B) MT pattern in *Acanthocystis* and *Raphidiophrys* (redrawn from Bardele [7])

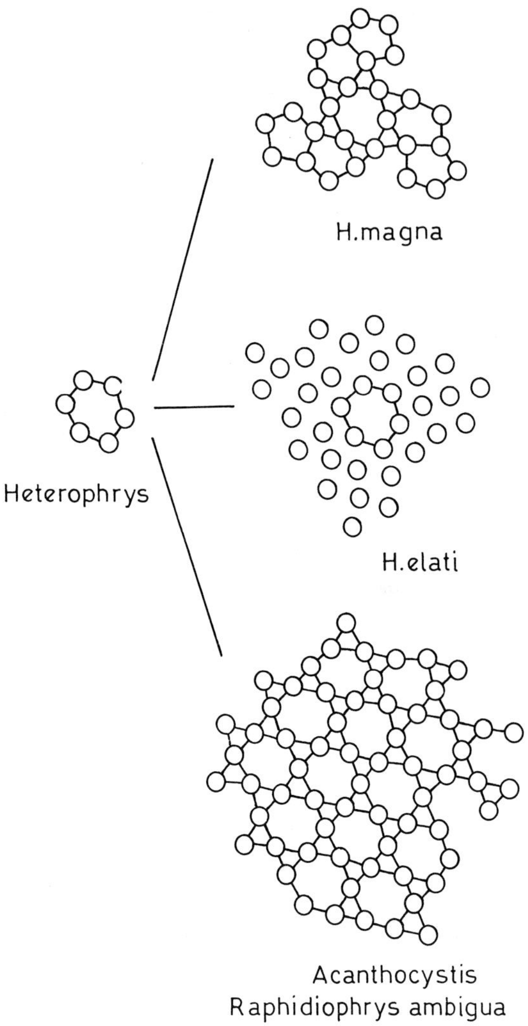

Fig. 3.13. Synopsis of MT patterns in *Heterophrys* and *Acanthocystis* "assembly line." Three patterns are generated from a basal hexagon of MT. In *Heterophrys magna,* pentagons are formed above the basal hexagon (redrawn from Bardele [7])

terns of MT. In elongated structures such as the spermatozoa of coccids (which are formed only of a nucleus and many MT, with no centriole or MTOC), the MT form rings around the central nucleus: discontinuities exist between the rows of MT, similar to dislocations in a crystalline structure. These dislocations may be located at the corners of a hexagon, explaining the hexagonal shape of the spermatozoa when they are tightly pressed one against another. In normal conditions, these dislocations lie on six spiral rows extending outward from the center [59]. In other coccid insects, these filamentous spermatozoa display various spiral arrangements of MT, leading to patterns which display a bilateral asymmetry which is probably linked with the movements of the cells, in the same way as the bending of cilia is linked with the 9+2 asymmetrical structure [103]. The exact nature of the side by side links of these MT is not known.

3.9.4 Six Equal Links

This leads to the formation of the most rigid type of MT assembly. It is not evident that all six links lie in the same plane, and the electron micrographs often show some missing links which may be explained by their spacing along the MT (cf. Chaps. 2 and 4). It is in the pharyngeal baskets of ciliates that rods made of such dense hexagonal packing of MT are found: these structures probably have a mechan-

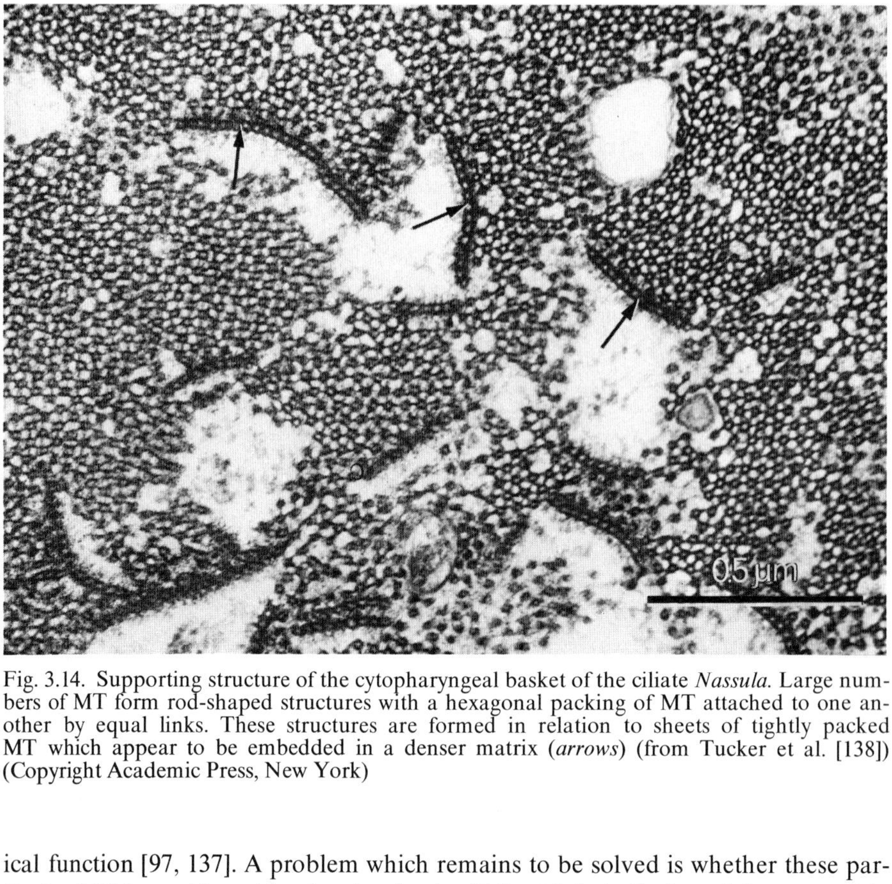

Fig. 3.14. Supporting structure of the cytopharyngeal basket of the ciliate *Nassula*. Large numbers of MT form rod-shaped structures with a hexagonal packing of MT attached to one another by equal links. These structures are formed in relation to sheets of tightly packed MT which appear to be embedded in a denser matrix (*arrows*) (from Tucker et al. [138]) (Copyright Academic Press, New York)

ical function [97, 137]. A problem which remains to be solved is whether these particular MT have 12 or 13 subunits. Again, if the triple-helical model proposed in Chapter 2 is accepted, the pitch of this helix will place the six links in different planes: one could imagine, for instance, that only α or β subunits play a role in connecting the MT.

3.9.5 Unequal Links

The most remarkable associations of MT are found in the axonemes of the heliozoa, which have been extensively studied [68, 69, 129, 132, 133] and which are most favorable material for the study of the action of cold, hydrostatic pressure, and various MT poisons (cf. Chap. 5), and of the relations between MT and intracellular movements (cf. Chap. 7). These axonemes, as observed in *Actinosphaerium* (*Echinosphaerium*) *nucleofilum*, are long and rigid structures made of several hundred parallel MT, laterally linked at regular intervals along their length, and patterned as a double-start spiral (Fig. 3.16) [5]. Each cell has about 100 axopodia which are about 300 μm long, made of a total of $\pm 3 \times 10^9$ molecules of tubulin [108]. The spiral determines twelve sectors by the addition of a regular number of MT at each turn. These MT are connected by long (±31 nm) and short (2 nm) links, the number being eight or more. A single MT, in the central part of the axoneme, may have as many as seven long links (however, this is more theoretical than apparent from the micrographs). The assembly starts from a central parallelogram of four MT.

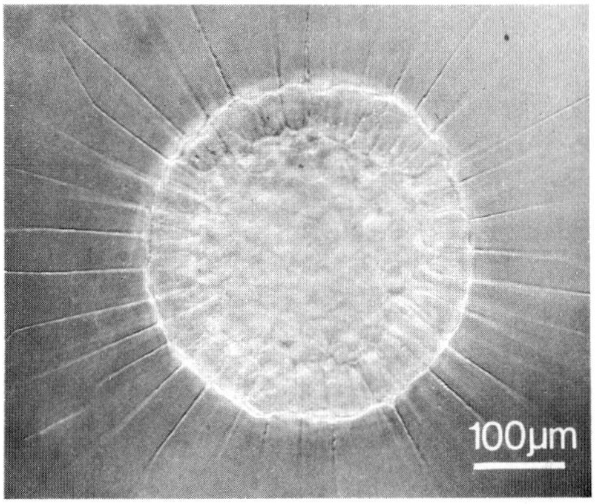

Fig. 3.15. Living *Actinosphaerium* in culture medium. Interference contrast (original document, by courtesy of M. Schliwa)

A quantitative study of the respective number of MT and links shows that the ratios of long links to short ones to MT is approximately 1 : 5 : 30. This clearly indicates that the quantity of linking material is far smaller than that of tubulin, a fact which should be compared with the findings on the MT-associated proteins described in Chapter 2. The geometrical structure of these axonemes implies that the formation of links is controlled: if only short links were to form, a compact assembly of MT (such as seen in other types of protozoa) would be made. No hexagonal packing has been observed in these cells and the numbers and the angles of links must be defined with precision. The number of short links is preferably two, it may be three, while four to six is unacceptable. Roth et al. [108], who have considered the difficult problems involved (problems which are not fundamentally different from those governing all complex patterns), are of the opinion that several different types of tubu-

[5] The geometrical pattern of the MT in the axonemes of *Actinosphaerium* can be defined as a central tesselation of triangles and parallelograms changed into a spiral tesselation by a one-step shift of half of the tesselation, as indicated by Figure 3.17 [54]. It would be interesting to know whether the patterns found in other species of protozoa can be defined geometrically. In *Actinosphaerium*, as pointed out by Cachon [26] and Roth et al. [108], the whole pattern is defined by the central assembly of a parallelogram of MT with long and short links. In other species, such as *Nasselaria*, the position and linkage of the central MT appear also to determine the whole pattern, provided some simple rules of linkage are respected

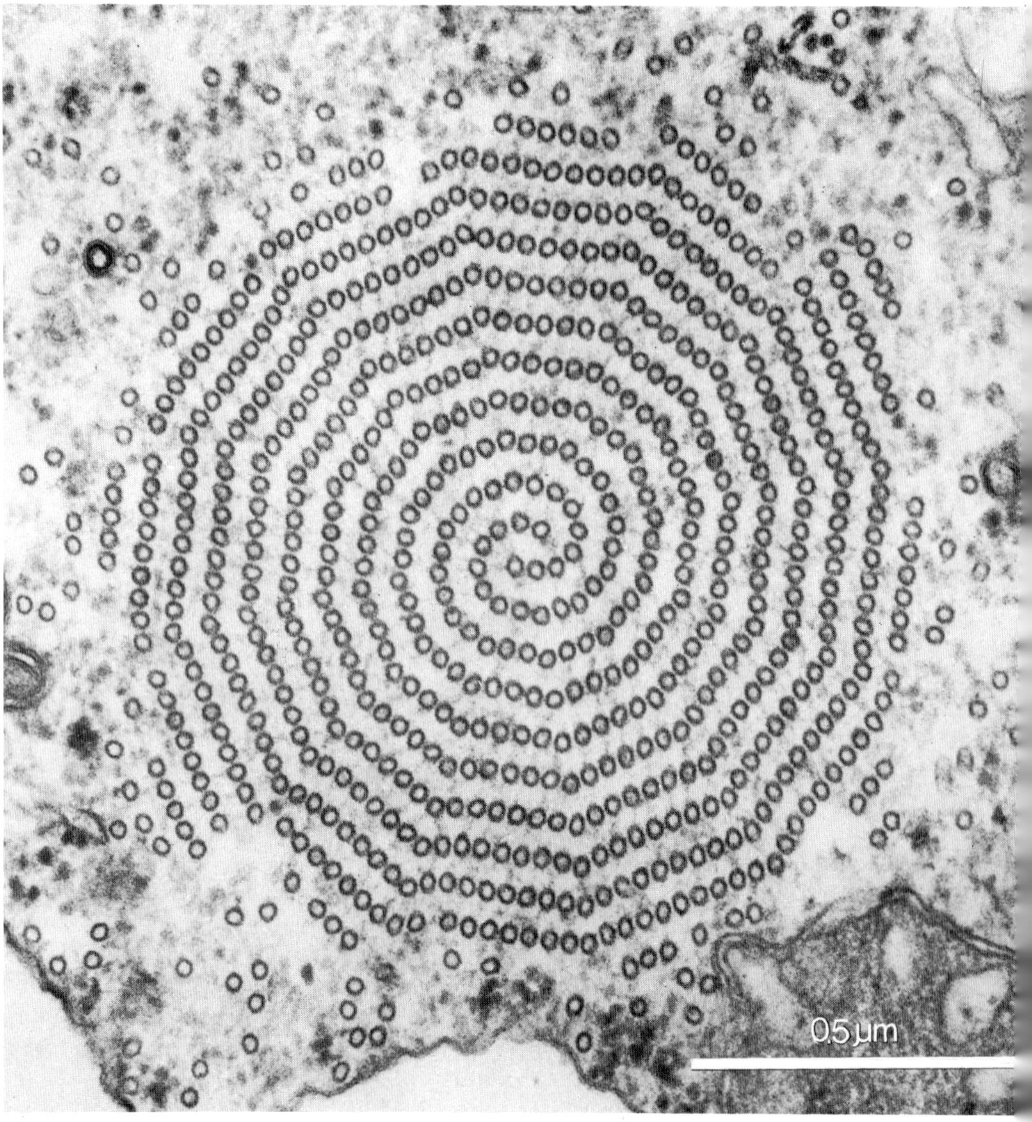

Fig. 3.16. Cross-section of an axopodium of *Actinosphaerium eichorni.* The axoneme is made of two interlocked spirals on MT, building a figure with twelvefold symmetry, the number of MT increasing by one unit in each sector from the center to the periphery. The MT are separated by two definite distances (long and short links). At the periphery, several sectors are incomplete and isolated MT are visible. The probable structure of the links and the central parallelogram are illustrated in Fig. 3.18 (from L. E. Roth)

lins and sites of link attachments have to be admitted. They propose that once a link has been formed, it determines allosteric changes in protein configuration, which may possibly move around the MT helix or along the MT. This transmission of allosteric effects is suggested as the "gradion hypothesis", and the authors think that it may explain some other properties of MT and may be extended to the problems of transport along MT (cf. Chaps. 8, 9, and 10).

The theoretical basis for the formation of these complex axonemes and for their twelvefold symmetry was thoroughly discussed in two recent papers [26, 27]. The exact number of links between the MT is difficult to see, and various structural proposals have been made [cf. 58, 129]. One of the simplest is to imagine that the spiral

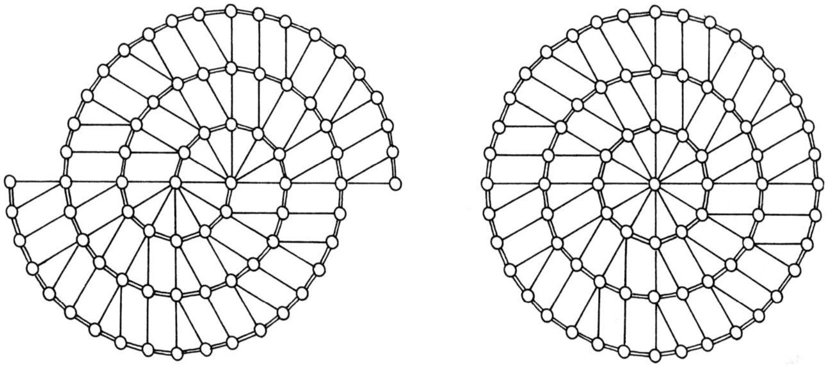

Fig. 3.17. This figure shows how the double spiral arrangement of MT in the axonemes of *Actinosphaerium* is similar to a symmetrical central tesselation of triangles (cf. [54]; drawn here after Cachon et al. [27])

structure starts from two MT separated by a long link. From each of these MT, six other similar links would locate the first MT in a spiral disposition (Fig. 3.18). Each half of this structure would thus have a sixfold symmetry, and following the spiral from one half to another, one more MT would be added in each of the six segments. The MT would have 13 subunits, with the result that the angles could not be exact fractions of 360°, leading to some skewness of the whole structure, evidenced by the slight curvature of each line of MT associated by short links [58, 129]. In this propo-

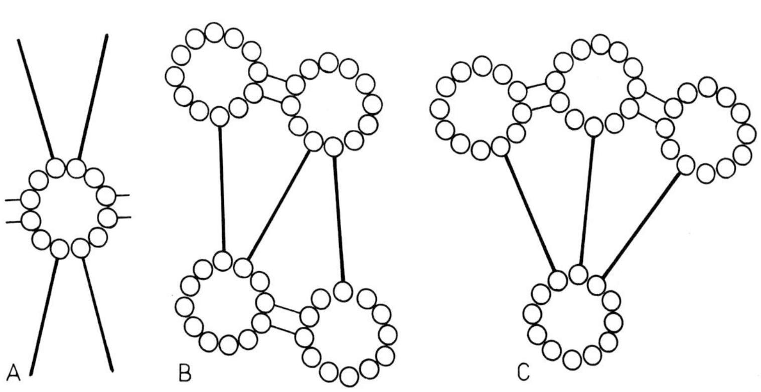

Fig. 3.18 A–C. Linkage patterns in *Echinosphaerium* axonemes. (A) Maximum linkage capability of a single MT. (B) Typical parallelogram arrangement found at the center of the spiral axoneme, and repeated throughout the structure. Maximum linkage is indicated again. (C) MT link pattern at a sector interface. The entire pattern can be described as an assembly of isoceles triangles, with sides about twice the length of the base (redrawn from Roth et al. [108])

sal, which remains partly theoretical, three types of MT would be found: one with two long and four short links, one with three long and four short links, and one with two short and six long links. It is clear that, whatever the model, the size of the MT, and the length of the links, once the central parallelogram is built, it leads to a twelve-fold symmetry, a fact which may be surprizing if the MT have 13 subunits. However, if the number of subunits was 12, one would probably find a more regular hexagonal pattern, as described in other species. One fact which has to be kept in mind is that additional MT may be added at the periphery without necessarily following the spiral, although not contradicting its laws of assembly (cf. Fig. 3.15). It has also been shown that some extra MT may have room between two MT separated by a long link [108].

The importance of these studies for a better understanding of the links between MT must not be neglected: they border on crystallography, and one may imagine that in the near future computers may be taught to draw the remarkable patterns which are described. The most remarkable feature of all is that these structures may be disassembled and reassembled at short notice, indicating that very definite rules of assembly—similar, on a different scale, to chemical bonding—must be obeyed.

References

1. Albertini, D. F., Clark, J. I.: Membrane-microtubule interactions: Concanavalin A capping induced redistribution of cytoplasmic microtubules and colchicine binding proteins. Proc. Natl. Acad. Sci. **72,** 4976 – 4980 (1975)
2. Allen, R. D.: Evidence for firm linkages between microtubules and membrane-bounded vesicles. J. Cell Biol. **64,** 497 – 502 (1975)
3. Angeletti, P. U., Levi-Montalcini, R., Caramia, F.: Ultrastructural changes in sympathetic neurons of newborn and adult mice treated with nerve growth factor. J. Ultrastr. Res. **36,** 24 – 36 (1971)
4. Aubin, J. E., Carlsen, S. A., Ling, V.: Colchicine permeation is required for inhibition of Concanavalin-A capping in Chinese hamster ovary. Proc. Natl. Acad. Sci. USA **72,** 4516 – 4520 (1975)
5. Auclair, W., Siegle, B. W.: Cilia regeneration in the sea urchin embryo: evidence for a pool of ciliary proteins. Science **154,** 913 – 915 (1966)
6. Bamburg, J. R., Shooter, E. M., Wilson, L.: Developmental changes in microtubule protein of chick brain. Biochemistry **12,** 1476 – 1481 (1973)
7. Bardele, C. F.: Comparative study of axopodial microtubule patterns and possible mechanisms of pattern control in the centrohelidian heliozoa *Acanthocystis, Raphidiophrys* and *Heterophrys.* J. Cell Sci. **25,** 205 – 232 (1977)
8. Barondes, S. H., Feit, H.: Metabolism of microtubular protein in mouse brain. In: Alzheimer's Disease and Related Conditions (eds.: G. E. W. Wolstenholme, M. O'Connor), pp. 267 – 276. Ciba Foundation Symposium. London: Churchill 1970
9. Becker, J. S., Oliver, J. M., Berlin, R. D.: Fluorescence techniques for following interactions of microtubule subunits and membranes. Nature (London) **254,** 152 – 153 (1975)
10. Behnke, O., Forer, A.: Evidence for four classes of microtubules in individual cells. J. Cell Sci. **2,** 169 – 192 (1967)
11. Berlin, R. D.: Microtubules and the fluidity of the cell surface. Ann. N. Y. Acad. Sci. **253,** 445 – 454 (1975)
12. Berlin, R. D.: Microtubule membrane interactions: fluorescence techniques. In: Microtubules and Microtubule Inhibitors (eds.: M. Borgers, M. De Brabander), pp. 327 – 339. Amsterdam: North Holland Publ. Co. 1975
13. Bessis, M., Breton-Gorius, J., Thiery, J. P.: Centriole, corps de Golgi et aster des leucocytes. Etude au microscope électronique. Rev. Hématol. **13,** 363 – 386 (1958)
14. Bhattacharyya, B., Wolff, J.: Membrane-bound tubulin in brain and thyroid tissue. J. Biol. Chem. **250,** 7639 – 7647 (1975)

15. Bhattacharyya, B., Wolff, J.: Polymerisation of membrane tubulin. Nature (London) **264,** 576 – 577 (1976)
16. Bieber, R. W., Stone, G. E.: Microtubule protein synthesis in vinblastine synchronized *Tetrahymena.* J. Cell Biol. **55,** 19 a (1972)
17. Bird, M. M., Liberman, A. R.: Microtubule fascicles in the stem processes of cultured sensory ganglion cells. Cell Tissue Res. **169,** 41 – 48 (1976)
18. Bloodgood, R. A.: Resorption of organelles containing microtubules. Cytobios **9,** 142 – 161 (1974)
19. Brinkley, B. R., Fuller, G. M., Highfield, D. P.: Cytoplasmic microtubules in normal and transformed cells in culture. Analysis by tubulin antibody immunofluorescence. Proc. Natl. Acad. Sci. USA **72,** 4981 – 4985 (1975)
20. Brown, D. L., Bouck, G. B.: Microtubule biogenesis and cell shape in *Ochromonas.* II. The role of nucleating sites in shape development. J. Cell Biol. **56,** 360 – 378 (1973)
21. Bucher, N. L. R., Berkley, P.: Synthesis of microtubule protein during regeneration of rat liver. J. Cell Biol. **55,** 31 a (1972)
22. Burchill, B. R.: Effects of ultraviolet light on microtubule formation. J. Cell Biol. **59,** 38 a (1973)
23. Burnside, B., Kozak, C., Kafatos, F. C.: Tubulin determination by an isotope dilution-vinblastine precipitation method. The tubulin content of *Spisula* eggs and embryos. J. Cell Biol. **59,** 755 – 761 (1973)
24. Burton, P. R.: Effects of various treatments on microtubules and axial units of lung-fluke spermatozoa. Z. Zellforsch. **87,** 226 – 248 (1968)
25. Cachon, J., Cachon, M.: Le système axopodial des Radiolaires Nasselaires. Origine, organisation et rapports avec les autres organites cellulaires. Arch. Protistenk. **113,** 80 – 97 (1971)
26. Cachon, J., Cachon, M.: Les systèmes axopodiaux. Année Biol. **13,** 523 – 560 (1974)
27. Cachon, J., Cachon, M., Febvre-Chevalier, C., Febvre, J.: Déterminisme de l'édification des systèmes microtubulaires stéréoplasmiques d'Actinopodes. Arch. Protistenk. **115,** 137 – 153 (1973)
28. Cachon, J., Cachon-Enjumet, L.: Ultrastructure des Amoebophryidae (Péridiniens Duboscquodinida). II. Systèmes atractophoriens et microtubulaires; leur intervention dans la mitose. Protistologia **6,** 57 – 70 (1970)
29. Clark, J. L., Albertini, D. F.: Filaments, microtubules and colchicine receptors in capped ovarian granulosa cells. In: Cell Motility (eds.: R. Goldman, T. Pollard, J. Rosenbaum). pp. 323 – 332. Cold Spring Harbor Lab. 1976
30. Coyne, B., Rosenbaum, J. L.: Flagellar elongation and shortening in *Chlamydomonas.* II. Reutilization of flagellar proteins. J. Cell Biol. **47,** 777 – 781 (1970)
31. Dales, S.: Association between the spindle apparatus and reovirus. Proc. Natl. Acad. Sci. USA **50,** 268 – 275 (1963)
32. Dales, S., Chardonnet, Y.: Early events in the interaction of adenoviruses with HeLa cells. IV. Association with microtubules and the nuclear pore complex during vectorial movement of the inoculum. Virology **56,** 465 – 483 (1973)
33. Dales, S.: Involvement of the microtubule in replication cycles of animal viruses. Ann. N. Y. Acad. Sci. **253,** 440 – 444 (1975)
34. De Brabander, M.: Onderzoek naar de rol van microtubuli in gekweekte cellen met behulp van een nieuwe synthetische inhibitor van tubulin polymerisatie. Thesis, Brussels 1977
35. De Brabander, M., Borgers, M.: Intranuclear microtubules in mast cells. Pathol. Eur. **10,** 17 – 20 (1975)
36. de Harven, E.: The centriole and the mitotic spindle. In: Ultrastructure in Biological Systems (eds.: A. J. Dalton, F. Haguenau). Vol. 3, pp. 197 – 227. New York, London: Academic Press 1968
37. de Harven, E., Bernhard, W.: Etude au microscope électronique de l'ultrastructure du centriole chez les Vertébrés. Z. Zellforsch. **45,** 378 – 498 (1956)
38. Dirksen, E. R., Staprans, I.: Microtubule protein levels during ciliogenesis. J. Cell Biol. **59,** 83 a (1973)
39. Dirksen, E. R., Staprans, I.: Tubulin synthesis during ciliogenesis in the mouse oviduct. Developm. Biol. **46,** 1 – 13 (1975)
40. Edelman, G. M., Wang, J. L., Yahara, I.: Surface-modulating assemblies in mammalian cells. In: Cell Motility (eds.: R. Goldman, T. Pollard, J. Rosenbaum). Cold Spring Harbor Lab. 1976

41. Edelman, G. M., Yahara, I., Wang, J. L.: Receptor mobility and receptor-cytoplasmic interactions in lymphocytes. Proc. Natl. Acad. Sci. USA **70,** 1442 – 1446 (1973)
42. Everhart, L. P. Jr.: Heterogeneity of microtubule proteins from *Tetrahymena* cilia. J. Mol. Biol. **61,** 745 – 748 (1971)
43. Farrell, K. W., Burns, R. G.: Regulation of microtubule assembly. J. Cell Biol. **63,** 98 a (1974)
44. Feit, H., Barondes, S. H.: Colchicine-binding activity in particulate fractions of mouse brain. J. Neurochem. **17,** 1355 – 1364 (1970)
45. Floor, E. R., Gilbert, J. M., Nowak, T. S. Jr.: Evidence for the synthesis of tubulin on membrane-bound and free ribosomes from rat forebrain. Bioch. Biophys. Acta. **442,** 285 – 296 (1976)
46. Fujiwara, K., Tilney, L. G.: Substructural analysis of the microtubule and its polymorphic forms. Ann. N. Y. Acad. Sci. **253,** 27 – 50 (1975)
47. Fullilove, S. L., Jacobson, A. G.: Nuclear elongation and cytokinesis in *Drosophila montana.* Developm. Biol. **26,** 560 – 577 (1971)
48. Fulton, C., Kowit, J. D.: Programmed synthesis of flagellar tubulin during cell differentiation in *Naegleria.* Ann. N. Y. Acad. Sci. **253,** 318 – 332 (1975)
49. Furcht, L. T., Scott, R. E.: Effect of vinblastine sulfate, colchicine and lumicolchicine on membrane organization of normal and transformed cells. Exp. Cell Res. **96,** 271 – 282 (1975)
50. Galatis, B., Apostlakos, P.: Associations between microbodies and a system of cytoplasmic tubules in oil-body cells of *Marchantia.* Planta **131,** 217 – 222 (1976)
51. Gibbons, I. R.: Chemical dissection of cilia. Arch. Biol. (Liège) **76,** 317 – 352 (1965)
52. Gilmore-Hebert, M. A., Heywood, S. M.: Translation of tubulin messenger ribonucleic acid. Bioch. Biophys. Acta **454,** 55 – 66 (1976)
53. Gilmore, M. A., Heywood, S. M.: Tubulin synthesis in a heterologous cell-free system. Ann. N. Y. Acad. Sci. **253,** 348 – 351 (1975)
54. Goldberg, M.: Central tesselations. Scr. Mathem. **21,** 253 – 260 (1955)
55. Goldstein, I. M., Hoffstein, S., Weissmann, G., Chauvet, G., Robineaux, R.: Lysosomal enzyme release and microtubule assembly induced by phorbol myristate acetate. J. Cell Biol. **63,** 113 a (1974)
56. Gonatas, N. K., Margolis, G., Kilham, L.: Reovirus type III encephalitis: observations of virus-cell interactions in neural tissues. II. Electron microscopic studies. Lab. Investig. **24,** 101 – 109 (1971)
57. Gray, E. G., Westrum, L. E.: Microtubules associated with nuclear pore complexes and coated pits in the central nervous system. Cell Tissue Res. **168,** 445 – 454 (1976)
58. Harris, W. F.: The arrangement of the axonemal microtubules and links of *Echinosphaerium nucleofilum.* J. Cell Biol. **46,** 183 – 187 (1970)
59. Harris, W. F., Robison, W. G. Jr.: Dislocations in microtubular bundles within spermatozoa of the coccid insect *Neoseingelia texana* and evidence for slip. Nature (London) **246,** 513 – 514 (1973)
60. Heidemann, S. R., Kirschner, M. W.: Aster formation in eggs of *Xenopus laevis.* Induction by isolated basal bodies. J. Cell Biol. **67,** 105 – 117 (1975)
61. Hoffstein, S., Soberman, R., Goldstein, I., Weissmann, G.: Concanavalin induced microtubule assembly and specific granule discharge in human polymorphonuclear leukocytes. J. Cell Biol. **68,** 781 – 786 (1976)
62. Huang, B., Pitelka, D. R.: The contractile process in the ciliate *Stentor coeruleus.* I. The role of microtubules and filaments. J. Cell Biol. **57,** 704 – 728 (1973)
63. Jones, C. W.: The pattern of microtubules in the axonemes of *Gymnosphaera albida* Sassaki: evidence for 13 protofilaments. J. Cell Sci. **18,** 133 – 156 (1975)
64. Jorgensen, A. O., Heywood, S. M.: Synthesis of embryonic chick microtubule protein (tubulin). J. Cell Biol. **59,** 159 a (1973)
65. Ketelbant-Balasse, P., Nève, P.: New ultrastructural features of the Cream hamster thyroid with special reference to the second kind of follicle. Cell Tissue Res. **166,** 49 – 63 (1976)
66. Kiefer, B. I.: Development, organization and degeneration of the *Drosophila* sperm flagellum. J. Cell Sci. **6,** 177 – 194 (1970)
67. Kim, K. S., Fulton, J. P.: An association of plant cell microtubules and viral particles. Virology **64,** 560 – 565 (1975)
68. Kitching, J. A.: Effects of high hydrostatic pressures on *Actinophrys sol* (Heliozoa). J. Exp. Biol. **34,** 511 – 517 (1957)

69. Kitching, J. A.: The axopods of the sun animalcule *Actinophrys sol* (Heliozoa). In: Primitive Motile Systems in Cell Biology (ed.: R. A. Allen). pp. 445 – 456. New York, London: Acad. Press 1964
70. Kohler, M. R., Spendlove, R. S.: Reovirus infected cells for studying microtubules and spindle poisons. Cytobios **9**, 131 – 142 (1974)
71. Kornguth, S. E., Sunderland, E.: Isolation and partial characterization of a tubulin-like protein from human and swine synaptosomal membranes. Bioch. Biophs. Acta **389**, 100 – 114 (1975)
72. Lafountain, J. R. Jr.: An association between microtubules and aligned mitochondria in *Nephrotoma* spermatocytes. Exp. Cell Res. **71**, 325 – 328 (1972)
73. Lawrence, J. H., Wheatley, D. N.: Synthesis of microtubule protein in HeLa cells approaching division. Cytobios. **13**, 167 – 179 (1975)
74. Levi-Montalcini, R., Angeletti, R. H., Angeletti, P. U.: The nerve growth factor. In: The Structure and Function of Nervous Tissue (ed.: G. H. Bourne). Vol. 5, pp. 1 – 58. New York, London: Acad. Press 1972
75. Luftig, R. B., Weihing, R. R.: Adenovirus binds to rat brain microtubules in vitro. J. Virol. **16**, 696 – 706 (1975)
76. McGill, M., Brinckley, B. R.: Human chromosomes and centrioles as nucleating sites for the in vitro assembly of microtubules from bovine brain tubulin. J. Cell Biol. **67**, 189 – 199 (1975)
77. MacGregor, H. C., Stebbings, H.: A massive system of microtubules associated with cytoplasmic movement in telotrophic ovarioles. J. Cell Sci. **6**, 431 – 449 (1970)
78. Matus, A. L., Walters, B. B., Mughal, S.: Immunohistochemical demonstration of tubulin associated with microtubules and synaptic junctions in mammalian brain. J. Neurocytol. **4**, 733 – 744 (1975)
79. Mazia, D.: The analysis of cell reproduction. Ann. N. Y. Acad. Sci. **90**, 455 – 469 (1960)
80. Mazia, D.: Mitosis and the physiology of cell division. In: The Cell. Biochemistry, Physiology, Morphology (eds.: J. Brachet, A. E. Mirsky). Vol. 3, pp. 77 – 412. New York, London: Acad. Press 1961
81. Mazia, D., Roslansky, J. D.: The quantitative relations between total cell proteins and the proteins of the mitotic apparatus. Protoplasma **46**, 528 – 534 (1956)
82. Meeker, G. L., Iversen, R. M.: Tubulin synthesis in fertilized sea urchin eggs. Exp. Cell Res. **64**, 129 – 132 (1971)
83. Miller, J. H.: An investigation of the microtubule protein in the mature oocytes of *Urechis caupo*. Exp. Cell Res. **81**, 342 – 350 (1973)
84. Miller, J. H., Epel, D.: Studies of oogenesis in *Urechis caupo*. II. Accumulation, during oogenesis, of carbohydrate, RNA, microtubule protein, and soluble mitochondrial, and lysosomal enzymes. Developm. Biol. **32**, 331 – 344 (1973)
85. Mollenhauer, H. H.: Distribution of microtubules in the Golgi apparatus of *Euglena gracilis*. J. Cell Sci. **15**, 89 – 98 (1974)
86. Moore, K. C.: Pressure-induced regression of oral apparatus microtubules in synchronized *Tetrahymena*. J. Ultrastr. Res. **41**, 499 – 518 (1972)
87. Morgan, J. L., Seeds, N. W.: Tubulin constancy during morphological differentiation of mouse neuroblastoma cells, J. Cell Biol. **67**, 136 – 145 (1975)
88. Nelsen, E. M.: Regulation of tubulin during ciliary regeneration in non-growing *Tetrahymena*. Exp. Cell Res. **94**, 152 – 158 (1975)
89. Oliver, J. M.: Microtubules, cyclic GMP and control of cell surface topography. In: Immune Recognition (ed.: A. S. Rosenthal). New York: Acad. Press 1975
90. Oliver, J. M.: Concanavalin A cap formation on human polynuclear leukocytes induced by R 17934. A new antitumor drug that interferes with microtubules assembly. J. Reticuloend. Soc. **19**, 389 – 395 (1976)
91. Oliver, J. M., Albertini, D. F., Berlin, R. D.: Effect of glutathione-oxidizing agents on microtubule assembly and microtubule-dependent surface properties of human neutrophils. J. Cell Biol. **71**, 921 – 932 (1976)
92. Oliver, J. M., Ukena, T. E., Berlin, R. D.: Effects of phagocytosis and colchicine on the distribution of lectin-binding sites on cell surfaces. Proc. Natl. Acad. Sci. USA **71**, 394 – 398 (1974)
93. Piatigorsky, J.: Lens cell elongation in vitro and microtubules. Ann. N. Y. Acad. Sci. **253**, 333 – 347 (1975)
94. Piatigorsky, J., Rothschild, S. S., Wolberg, M.: Stimulation by insulin of cell elongation and microtubule assembly in embryonic chick lens epithelium. Proc. Natl. Acad. Sci. USA **70**, 1195 – 1198 (1973)

95. Pickett-Heaps, J. D.: Aspects of spindle evolution. Ann. N. Y. Acad. Sci. **253**, 352 – 361 (1975)
96. Pipeleers, D. G., Pipeleers-Marichal, M. A., Kipnis, D. M.: Regulation of tubulin synthesis in islets of Langerhans. Proc. Natl. Acad. Sci. USA **73**, 3188 – 3191 (1976)
97. Pyne, C. K., Tuffrau, M.: Structure et ultrastructure de l'appareil cytopharyngien et des tubules complexes en relation avec celui-ci chez le cilié gymnostome *Chilodonella uncinata* Ehrbg. J. Microscopie **9**, 503 – 516 (1970)
98. Raff, R. A., Brandis, J. W., Green, L. H., Kaumeyer, J. F., Raff, E. C.: Microtubule protein pools in early development. Ann. N. Y. Acad. Sci. **253**, 304 – 317 (1975)
99. Raff, R. A., Greenhouse, G., Gross, K. W., Gross, P. R.: Synthesis and storage of microtubule proteins by sea urchin embryos. J. Cell Biol. **50**, 516 – 528 (1971)
100. Raff, R. A., Kaumeyer, J. F.: Soluble microtubule proteins of the sea urchin embryo: partial characterization of the proteins and behavior of the pool in early development. Developm. Biol. **32**, 309 – 320 (1973)
101. Raine, C. S., Ghetti, B., Shelanski, M. L.: On the association between microtubules and mitochondria within axons. Brain Res. **34**, 389 – 393 (1971)
102. Rannestad, J.: The regeneration of cilia in partially deciliated *Tetrahymena*. J. Cell Biol. **63**, 1009 – 1017 (1974)
103. Robison, W. G. Jr.: Microtubular patterns in spermatozoa of coccid insects in relation to bending. J. Cell Biol. **52**, 66 – 83 (1972)
104. Roisen, F. J., Murphy, R. A.: Neurite development in vitro. II. The role of microfilaments and microtubules in dibutyryl adenosine 3′,5′-cyclic monophosphate and nerve growth factor stimulated maturation. J. Neurobiol. **4**, 397 – 412 (1973)
105. Rosenbaum, J., Carlson, A.: Cilia regeneration in *Tetrahymena* and inhibition by colchicine. J. Cell Biol. **40**, 415 – 425 (1969)
106. Rosenbaum, J. L., Child, F. M.: Flagellar regeneration in protozoan flagellates. J. Cell Biol. **34**, 345 – 364 (1967)
107. Rosenbaum, J. L., Moulder, J. E., Ringo, D. L.: Flagellar elongation and shortening in *Chlamydomonas*. The use of cycloheximide and colchicine to study the synthesis and assembly of flagellar proteins. J. Cell Biol. **41**, 600 – 619 (1969)
108. Roth, L. E., Pihlaja, D. J., Shigenaka, Y.: Microtubules in the heliozoan axopodium. I. The gradion hypothesis of allosterism in structural proteins. J. Ultrastr. Res. **30**, 7 – 37 (1970)
109. Ryan, G. B., Borysenko, J. Z., Karnovsky, M. J.: Factors affecting the redistribution of surface bound concanavalin A on human polymorphonuclear leukocytes. J. Cell Biol. **62**, 351 – 365 (1974)
110. Sajnani, A. N., Ranadive, N. S., Movat, H. Z.: Redistribution of immunoglobulin receptors on human neutrophils and its relationship to the release of lysosomal enzymes. Lab. Investig. **35**, 143 – 151 (1976)
111. Schreiner, G. F., Unanue, E. R.: Calcium-sensitive modulation of Ig capping: evidence supporting a cytoplasmic control of ligand receptor complexes. J. Exp. Med. **143**, 15 – 31 (1976)
112. Sekiguchi, K., Asano, A.: Effect of colchicine on virus-induced fusion of human erythrocytes. Life Sci. **18**, 1383 – 1390 (1976)
113. Shay, J. W.: Electron microscopic studies of induced alterations in microtubular elements of *Rhynosciara* spermatozoa. J. Cell Biol. **47**, 188 a (1970)
114. Shay, J. W.: Electron microscope studies of spermatozoa of *Rhynchosciara* sp. I. Disruption of microtubules by various treatments. J. Cell Biol. **54**, 598 – 608 (1972)
115. Smith, D. S., Järlfors, U., Beráneck, R.: The organization of synaptic axoplasm in the lamprey (*Petromyzon marinus*) central nervous system. J. Cell Biol. **46**, 199 – 219 (1970)
116. Snyder, J. A., McIntosh, J. R.: Initiation and growth of microtubules from mitotic centers in lysed mammalian cells. J. Cell Biol. **67**, 409 a (1975)
117. Soifer, D., Braun, T., Hechter, O.: Insulin and microtubules in rat adipocytes. Science **172**, 269 – 270 (1971)
118. Spendlove, R. S., Lennette, E., John, A. C.: The role of the mitotic apparatus in the intracellular location of reovirus antigen. J. Immunol. **90**, 554 – 560 (1974)
119. Stadler, J., Franke, W. W.: Characterization of the colchicine binding of membrane fractions from rat and mouse liver. J. Cell Biol. **60**, 297 – 303 (1974)
120. Staprans, L., Dirksen, E. R.: Microtubule protein during ciliogenesis in the mouse oviduct. J. Cell Biol. **62**, 164 – 174 (1974)
121. Stebbings, H., Bennett, C. E.: The sleeve element of microtubules. In: Microtubules and Microtubule Inhibitors (eds.: M. Borgers, M. De Brabander). pp. 35 – 45. Amsterdam: North-Holland Publ. Co. 1975

122. Stephens, R. E.: Studies on the development of the sea-urchin *Strongylocentrotus droebachiensis.* II. Regulation of mitotic spindle equilibrium by environmental temperature. III. Embryonic synthesis of ciliary proteins. Biol. Bull. **142,** 145 – 159; 489 – 504 (1972)
123. Tanaka, K.: Intranuclear microtubule organizing center in early prophase nuclei of the plasmodium of the slime mold, *Physarum polycephalum.* J. Cell Biol. **57,** 220 – 224 (1973)
124. Taylor, R. B., Duffus, W. P. H., Raff, M. C., de Petris, S.: Redistribution and pinocytosis of lymphocyte surface immunoglobulin molecules induced by anti-immunoglobulin antibody. Nature New Biol. **233,** 225 – 229 (1971)
125. Telzer, B. R., Moses, M. J., Rosenbaum, J. L.: Assembly of microtubules onto kinetochores of isolated mitotic chromosomes of HeLa cells. Proc. Natl. Acad. Sci. USA **72,** 4023 – 4027 (1975)
126. Thompson, G. A. Jr., Baugh, L. C., Walker, L. F.: Nonlethal deciliation of *Tetrahymena* by a local anesthetic and its utility as a tool for studying cilia regeneration. J. Cell Biol. **61,** 253 – 256 (1974)
127. Thurm, U.: An insect mechanoreceptor. I. Fine structure and adequate stimulus. Cold Spring Harbor Symp. Quant. Biol. **30,** 75 – 82 (1965)
128. Tilney, L. G.: How microtubule patterns are generated. The relative importance of nucleation and bridging of microtubules in the formation of the axoneme of Raphidiophrys. J. Cell Biol. **51,** 837 – 854 (1971)
129. Tilney, L. G., Byers, B.: Studies on the microtubules in Heliozoa. V. Factors controlling the organization of microtubules in the axonemal pattern in *Echinosphaerium (Actinosphaerium) nucleofilum.* J. Cell Biol. **43,** 148 – 165 (1969)
130. Tilney, L. G., Gibbins, J. R.: Differential effects of antimitotic agents on the stability and behavior of cytoplasmic and ciliary microtubules. Protoplasma **65,** 167 – 179 (1968)
131. Tilney, L. G., Goddard, J.: Nucleating sites for the assembly of cytoplasmic microtubules in the ectodermal cells of blastulae of *Arbacia punctulata.* J. Cell Biol. **46,** 564 – 575 (1970)
132. Tilney, L. G., Porter, K. R.: Studies on the microtubules in Heliozoa. I. The fine structure of *Actinosphaerium nucleofilum* (Barrett) with particular reference to the axial rod structure. Protoplasma **60,** 317 – 344 (1965)
133. Tilney, L. G., Porter, K. R.: Studies on the microtubules in Heliozoa. II. The effect of low temperature on these structures in the formation and maintenance of the axopodia. J. Cell Biol. **34,** 327 – 343 (1967)
134. Tucker, J. B.: Fine structure and function of the cytopharyngeal basket in the Ciliate *Nassula.* J. Cell Sci. **3,** 493 – 514 (1968)
135. Tucker, J. B.: Morphogenesis of a large microtubular organelle and its association with basal bodies in the Ciliate *Nassula.* J. Cell Sci. **6,** 385 – 429 (1970)
136. Tucker, J. B.: Initiation and differentiation of microtubule patterns in the Ciliate *Nassula.* J. Cell Sci. **7,** 793 – 821 (1970)
137. Tucker, J. B.: Microtubule arms and propulsion of food particles inside large feeding organelle in the Ciliate *Phascolodon vorticella.* J. Cell Sci. **10,** 883 – 903 (1972)
138. Tucker, J. B., Dunn, M., Pattisson, J. B.: Control of microtubule pattern during the development of a large organelle in the ciliate *Nassula.* Developm. Biol. **47,** 439 – 453 (1975)
139. Unanue, E. R.: Cellular events following binding of antigen to lymphocytes. Am. J. Path. **77,** 1 – 20 (1974)
140. Unanue, E. R., Karnovsky, M. J.: Ligand-induced movement of lymphocyte membrane macromolecules. V. Capping, cell movement, and microtubular function in normal and lectin-treated lymphocytes. J. Exp. Med. **140,** 1207 – 1220 (1974)
141. Weeks, D. P., Collis, P. S.: Induction of microtubule protein synthesis in *Chlamydomonas reinhardi* during flagellar regeneration. Cell **9,** 15 – 29 (1976)
142. Weissmann, G., Goldstein, L., Hoffstein, S., Tsung, P. K.: Reciprocal effects of cAMP and cGMP on microtubule-dependent release of lysosomal enzymes. Ann. N. Y. Acad. Sci. **253,** 750 – 762 (1975)
143. Went, H. A.: Some immunochemical studies on the mitotic apparatus of the sea urchin. J. Biophys. Biochem. Cytol. **5,** 353 – 356 (1959)
144. Westrum, L. E., Gray, E. G.: Microtubules and membrane specializations. Brain Res. **105,** 547 – 550 (1976)
145. Williams, N. E.: Regulation of microtubules in *Tetrahymena.* Intern. Rev. Cytol. **41,** 59 – 86 (1975)
146. Williams, N. E., Frankel, J.: Regulation of microtubules in *Tetrahymena.* I. Electron microscopy of oral replacement. J. Cell Biol. **56,** 441 – 457 (1973)
147. Williams, N. E., Nelson, E. M.: Regulation of microtubules in *Tetrahymena.* II. Relation between turnover of microtubule proteins and microtubule dissociation and assembly during oral replacement. J. Cell Biol. **56,** 458 – 465 (1973)

148. Wunderlich, F., Muller, R., Speth, V.: Direct evidence for a colchicine-induced impairment in the mobility of membrane components. Science **182,** 1136 – 1137 (1973)
149. Yahara, I., Edelman, G. M.: Modulation of lymphocyte receptor mobility by concanavalin A and colchicine. Ann. N. Y. Acad. Sci. **253,** 455 – 469 (1975)
150. Yahara, I., Edelman, G. M.: Electron microscopic analysis of the modulation of lymphocyte receptor mobility. Exp. Cell. Res. **91,** 125 – 142 (1975)
151. Yamada, K. M., Spooner, B. S., Wessells, N. K.: Ultrastructure and function of growth cones and axons of cultured nerve cells. J. Cell Biol. **49,** 614 – 635 (1971)
152. Yamada, K. M., Wessells, N. K.: Axon elongation. Effect of nerve growth factor on microtubule protein. Exp. Cell Res. **66,** 346 – 352 (1971)
153. Zenner, H. P., Pfeuffer, T.: Microtubular proteins in pigeon erythrocyte membranes. Eur. J. Biochem. **71,** 177 – 184 (1976)
154. Zurier, R. B., Hoffstein, S., Weissmann, G.: Mechanism of lysosomal enzyme release from human leukocytes. I. Effect of cyclic nucleotides and colchicine. J. Cell Biol. **58,** 27 – 41 (1973)

Chapter 4 Microtubule Structures: Centrioles, Basal Bodies, Cilia, Axonemes

4.1 Introduction

Several cell organelles assembled from MT and other proteins perform a great role in cell biology and display a high degree of complexity. The centrioles, basal bodies, cilia and flagella, have a ninefold symmetry which, with few variants, has been maintained in most species. The association of centrioles with mitosis is found in all primitive plants and in all animal cells; its disappearance in most mitoses of higher plants indicates that centrioles are by no means indispensable for cell division, and that their relation with the spindle MT may be mainly a convenient way for the cell to assure their continuity. Centrioles and basal bodies are closely related and similar in their basic structure: the principal function of centrioles may be to form basal bodies, cilia and flagella. However, several instances of de novo formation of centrioles in cells indicate that these structures, contrary to what was thought, have no genetic continuity. This is in contradiction to the idea [106, 144] that centrioles may have a semi-autonomous existence and could possibly represent the persistence of the symbiotic union of a primitive organism devoid of MT and a spirochete-like unicellular.

Cilia and flagella (these terms are synonymous, the second being mainly used for the longest specialized cilia of motile cells, e.g., gametes) have many functions related to cell growth and motility (cf. Chap. 7). The 9+2 groups of MT found in most cilia are connected by several links, two of which are made of the ATPase protein, *dynein*, which is closely involved in their motility. Cilia play other roles in connecting the organisms with the milieu, most sensory cells having complex appendages which are modified cilia, as in the eye, the ear, the olfactory epithelium and various mechanochemical receptors.

In protozoa, apart from various types of cilia and flagella, the cytoplasmic extensions called axopodia contain most complex assemblies of MT (cf. Chap. 3) in their axonemes, which will be also considered in this chapter.

A detailed study of these MT structures alone would need more than a monograph. Centrioles have been reviewed by Dalcq [34], de Harven [37], Fulton [56, 57], Taylor [172], and Went [189], and Sleigh has edited an excellent book on cilia [155]. The reader is referred to these texts for detailed morphological and physiological studies. The role of centrioles in mitosis will not be studied here (cf. Chap. 10), and only a few documents on the action of MT poisons on centrioles and cilia will be mentioned; these will find their place in Chapter 5.

Our purpose will be mainly to attempt an understanding of the properties of centrioles and cilia in relation to the properties of MT, on which they throw much light. The motility of cilia will be studied with the other movements related to MT (Chap. 7).

Although cilia are at this moment far better known, in particular at the biochemical and ultrastructural levels, centrioles, as relatively simpler structures, will be described first.

4.2 Centrioles

4.2.1 Definition

In light microscopy, centrioles have long been known as dense, minute granules, usually located at the two poles of the mitotic apparatus at metaphase. It was recognized early in the 19th century that these bodies may move toward the cell membrane and become basal bodies from which cilia develop: Henneguy-Lenhossek's theory of the origin of basal bodies has been extensively confirmed [82, 96]. Centrioles appear in the light microscope, under favorable conditions, as elongated, rod-like structures; in some cells, for instance in the neuropteran insects, they may reach at meiosis a length of 8 μm [54] while their ultrastructure remains fundamentally the same as that of smaller and more usual ones. It is however not certain that all large centrioles observed in light microscopy and which have not yet been studied at the ultrastructural level truly correspond to the description of a typical centriole [57].

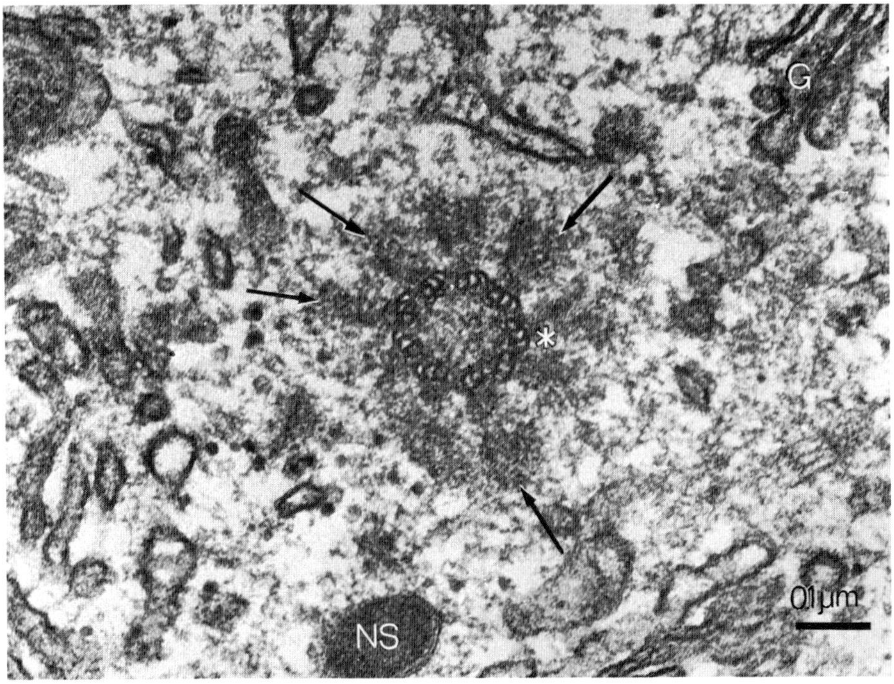

Fig. 4.1. Normal centriole in a pituicyte (glial cell) from the posterior lobe of the pituitary of a rat. The nine groups of triplets (✱) are visible; they are surrounded by an equal number of granular masses (*arrows*). No MT are visible in the cytoplasm. *G* Golgi vacuoles. *NS* neurosecretory granules

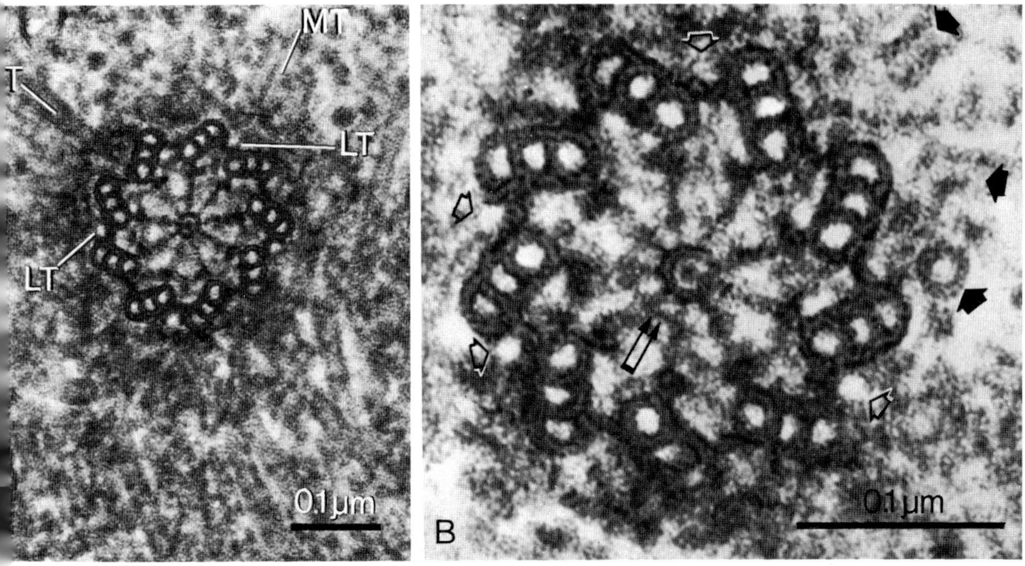

Fig. 4.2 A, B. Centriole structure in the aquatic fungus *Phlyctochytrium irregulaze* Koch. (A) The nine groups of three tubules (triplets) with their approximately 40° angulation show the typical "pinwheel" structure. Many microtubules (*MT*) radiate from the periphery of the centriole. The links between the triplets (*LT*) and the central cart-wheel structure are apparent. (B) At this magnification, the related triplets are seen as made of one complete MT, related to a central hub by a radial link attached to two incomplete MT. The subunits of these MT are apparent. The triplets are linked by bridges between MT A and MT C of the next triplet (*open arrows*). Some substructure of the cytoplasmic MT (*solid arrows*) is visible. The double arrow indicates a fiber connecting the radial spokes (from McNitt [103])

In intermitotic diploid cells, except when centrioles are engaged in the formation of basal bodies, only two or four centrioles are observed. During mitosis, two centrioles are often found at each pole of the mitotic apparatus: one of each pair is smaller and represents a stage in the formation of a new centriole. It ensues that in cells that undergo several mitoses without cytoplasmic cleavage, leading to the formation of multiple or polyploid giant nuclei, like in the bone marrow megakaryocytes or some tumor cells, the number of centrioles may be much larger than four.

4.2.2 Ultrastructure

The modern study of centrioles started with the electron microscopical observations of de Harven and Bernhard [38], which also opened the path for the study of mitotic MT. From the numerous studies which have been published since, the complex architecture of the centriole can be described as follows.

a) *The nine triplets.* The ninefold symmetry is a fundamental property of centrioles, and only few exceptions are known. Throughout evolution this symmetry—which is found also in basal bodies and cilia—has been preserved, although the reason for this remains obscure. Nine groups of MT triplets form an elongated cylinder measuring about 400 nm, clearly showing a lengthwise polarity, with a base recognizable by the presence of denser intracentriolar material, and an end slightly narrower than the base. The triplets are formed of lengthwise associated structures which have the dimensions, and most of the properties, of MT as found in structures such as cilia. However, only one of these MT, the A one, is formed by a regular circular arrangement of subunits. Their exact number remains poorly known, but indirect evidence indicates that this should be, like in ordinary MT, 13 (cf. however Chap. 2). The second tubule, B, is incomplete, and shares some of its subunits with tubule A, like in cilia; if this comparison is correct, it should have only 10 subunits, three being shared by the two MT. The same is probable for tubule C.

These triplets have an unequal length, the C tubules often being slightly shorter; like other MT structures, they may appear straight, but often the centriolar MT and all the centriole appear more or less skewed, with some obliquity of the MT which are slightly twisted in the longitudinal direction [8, 119] and, as mentioned above, inclined toward the center of the structure at its extremity. When observed on transverse sections, the triplets show a most characteristic grouping, which has been compared to a pin-wheel or a turbine, each group of triplets being inclined at an angle of about 40° toward the axis of the centriole

b) *Links between the triplets.* The centriolar MT are nearly always surrounded by a poorly defined, narrow sleeve of electron-dense material, which is mostly seen at the basis of the organelle. On the other hand, links hold together the triplets: from MT C to MT A there is a bridge, which resembles some of the side-arms already described in other MT structures. These bridges are less stained than the tubules. Moreover, at the basis of the triplet cylinder, transversal sections show more or less clearly a series of bridges linking each triplet to a central core of dense material. This has been called the "cart-wheel" structure; its symmetry is enhanced by using

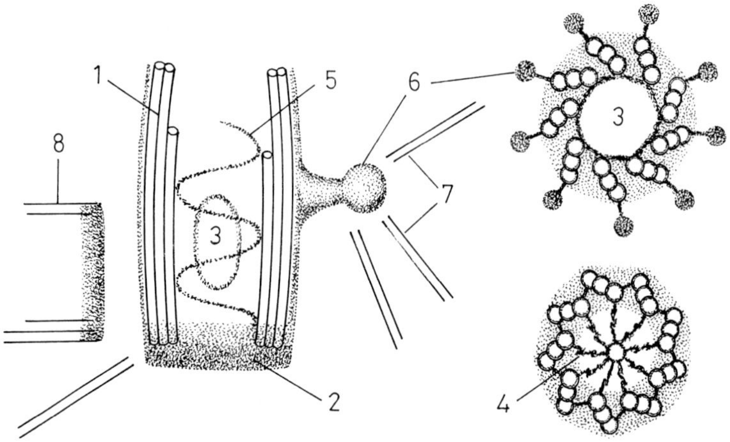

Fig. 4.3. Schematic structure of a centriole pair. The fully grown centriole shows nine groups of triplets, in which the MT C is shorter than the two others. (1) At the base, the tubules are surrounded by a dense matrix. (2) The center of the structure may be occupied by a vacuole. (3) At the base, a typical cart-wheel pattern is visible. (4) The MT A and C of neighboring triplets are attached by a bridge (nexin?). (5) A central strand of helically wound RNA is figured: its exact location remains poorly defined. (6) The centriole is surrounded by nine groups of pericentriolar structures, in the close vicinity of which are attached the cytoplasmic or mitotic MT (7). (8) The daughter centriole, which has not yet reached its adult length, arises perpendicular to the mother centriole, and close to its basal region

the rotatory photographic technique of Markham [37]. This structure is attached to densities of the A tubules which are described as "feet" [57].

c) *Other structures* are more variable. A clear vacuole limited by a thin membrane is sometimes found at the center of the centriole. This region may also be occupied by dense material. More important is the observation that a fine helically wound strand may be located close to the triplets [167]; the possible relation of this with nucleic acids will be discussed below.

d) *Dimensions.* As already mentioned, the size and in particular the length of centrioles may vary considerably. The diameter of the centriole is about 0.15 – 0.25 μm, which explains the fact that it is close to the limit of visibility in the light microscope. The length varies between less than two and several μm. It is also a function of the stage of growth of the organelle.

e) *Pericentriolar structures.* Most centrioles are linked with dense rounded masses located at a short distance from them. These pericentriolar masses ("massules" or centriolar satellites; [21, 39, 97, 168]) have been numbered in favorable cases and may form a circle of nine around the centriole, reflecting its ninefold symmetry. In some conditions, two circles of satellites have been observed [20]. It is from these structures that the cytoplasmic MT are seen to extend. The satellites do not appear to be permanent, and may possibly play a role of nucleating centers in the assembly of MT. They may be compared to similar dense masses which are found in the steps preceding the formation of centrioles (see below).

4.2.3 Biochemistry

The chemical make-up of centrioles is poorly known: as there is no proper method for isolating them in large numbers, no biochemical proof of their microtubular nature has been given. The evidence that they are assemblies of MT is indirect, although quite convincing.

a) The size of the centriolar MT is identical with that of the basal bodies and the MT of cilia, the tubulin nature of which has been clearly demonstrated.

b) Tubulin is found in large quantities in cells which are engaged in the formation of centrioles and cilia, and centriole precursors are related to such colchicine-binding protein [158].

c) Studies with ^{3}H-leucine and radioautography demonstrate that in the mouse oviduct large quantities of newly synthetized tubulin are associated with centrioles and centriole precursors [45].

d) Centrioles, from cells treated by colchicine so as to destroy all MT in the cytoplasm (the centrioles are resistant to colchicine), and extracted with the detergent Triton X, bind newly formed MT when placed in a tubulin solution in conditions permitting MT assembly: the centrioles appear as active MTOC or nucleating centers.

Some of the newly assembled MT are linked with the distal end of the centriole MT [100]. These experiments should be compared to the action of cell extracts, probably containing centrioles, on unfertilized eggs of *Xenopus laevis,* which lead to the formation of asters [81].

It is also evident, from the description of the structure of centrioles, that other proteins must be present: it would be interesting to know whether these are comparable to those found in cilia, such as the linkage protein *nexin* and the ATPase *dynein*. An answer to these questions can only be given from a study of purified isolated centrioles. There are however many reasons to believe that some results obtained from isolated basal bodies apply to centrioles.

The possible presence of nucleic acids in the centrioles has been the subject of much debate, and is linked with the self-replicating property of these structures, which often, but not always, are born in close association with another centriole. This discussion is related to that of the possible symbiotic origin of cilia, as proposed by Sagan [144]. This theory needs no discussion here, as it is evident that centrioles can be formed de novo and cannot be considered self-replicating. Several claims about the presence of DNA in centrioles have been made; they were mainly based on the staining properties of centrioles by acridine orange. However, more precise techniques, such as ^{3}H-thymidine incorporation, have repeatedly failed to show any DNA [76]. On the contrary, studies on the basal bodies of *Tetrahymena,* which can be isolated with the cytoplasmic pellicle to which they are attached after deciliation, indicate that the staining with acridine orange may be explained by the presence of a single-stranded molecule of RNA. The fluorescence is completely abolished after RNase treatment [77]. This RNA would be neither mitochondrial nor transfer nor ribosomal RNA, as indicated by molecular hybridization techniques. Hartman [76, 77] proposed the hypothesis that this RNA plays a similar role to the RNA of ribosomes, that of a nucleation site around which the complex array of MT organizes. This opinion is in agreement with the fact that basal body replication is inhibited by actinomycin D [199], which also prevents the formation of centrioles [36].

A detailed investigation on the effects of RNase and pronase on the basal bodies of *Paramecium* has been published by Dippell [42]. Parts of the axosomes and the lumen complex of these structures are removed by the combined action of the two enzymes. It is suggested that the basal body RNA is single-stranded, but its function remains unknown. These results should be compared to the findings in human lymphocyte cultures treated by the basic and fluorescent dyes ethidium bromide and propidium iodide: various abnormalities of the triplet MT and of the centrally located structures may be the consequence of fixation of the dye on the centriolar RNA [101].

The suggestion of a morphogenetic role, at the molecular level, of a special, centriolar RNA deserves close attention: it may perhaps lead to an explanation of the cause of the ninefold symmetry. Since 1964, when Satir and Satir [149] proposed an explanation based on the possible relation of this periodicity with that of the α helix of protein chains, supposing an axial protein in this configuration with identical amino acids at 40° intervals, no molecular explanation of the ninefold grouping of MT has been proposed. It remains, in the words of Pickett-Heaps [126], “obscure and inexplicable”.

4.2.4 Replication and Growth

During the mitotic cycle, the formation of new centrioles takes place at a fixed moment of the cycle, and leads to the formation of two cells each provided with a normal complement of one or two centrioles. In the formation of basal bodies, new centrioles are formed in the vicinity of other centrioles before migrating toward the cell membrane and leading to cilia growth. While it is not correct to talk about a "reduplication" of the centriole, as a centriole is never seen to grow and then to divide, it is evident that in the majority of cells the birth of this organelle takes place in close relation to a preexisting one. However, as will be described below, a relatively large number of conditions are now known in which centrioles are formed de novo, without any topographical relation with other centrioles, sometimes in cells completely devoid of any such structure. Let us first consider the case where new centrioles grow close to old ones. The problems of atypical centrioles will be considered further.

In mitosis, the centriole cycle has been repeatedly described [18, 37, 57, 62, 127, 138]. The new centrioles, whether formed before prophase or later in the cycle, are

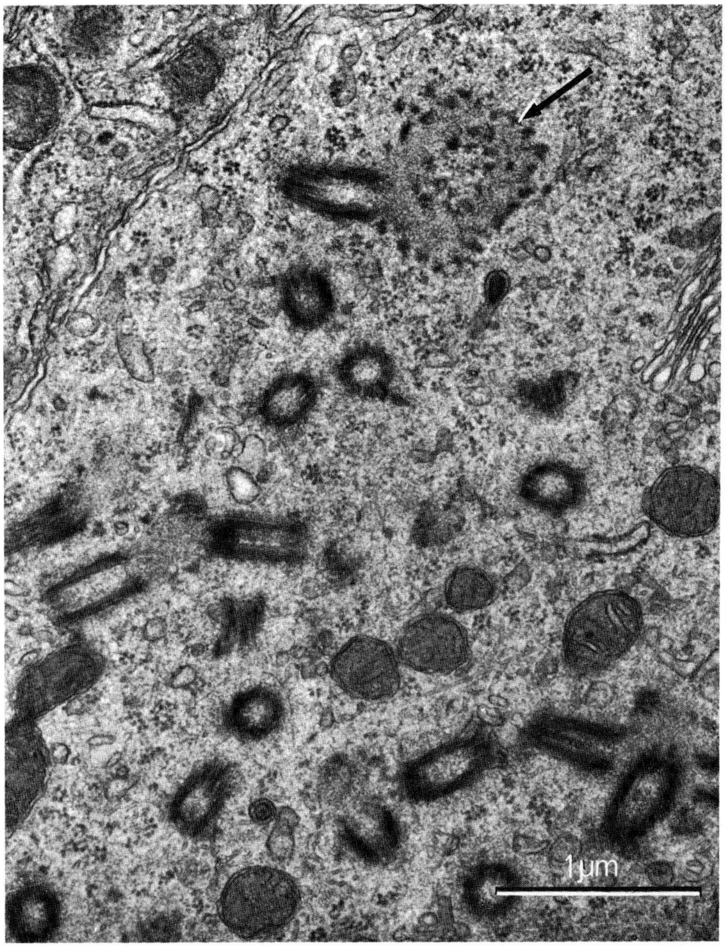

Fig. 4.4. Epithelial cell from the oviduct of an eight-day-old mouse. Numerous centrioles sectioned in various planes. Close to one centriole, a fibrogranular mass, with some scattered densities, is visible (*arrow*) (from Dirksen and Staprans [45]) (copyright Academic Press, New York)

always found at a short distance from the "old" ones, and at right angles to them. They appear to grow progressively to their normal length from a disk of more or less dense material located close to the base of the first centriole. No structure of the old centriole is used in this growth, and the changes which take place suggest that the old centriole acts as an organizing center, at most, for the assembly of the constituents of the new one. This assembly proceeds from base to tip, leading to the formation of a diplosome made by two equal structures. The centriolar triplets do not grow at once, and while the ninefold symmetry is apparent from the start, it is first formed by single MT, which change later into doublets and then into triplets. This has been well demonstrated in the formation of basal bodies and appears to be a general rule. The structure of the triplets, in which only tubule A is complete, is an indirect suggestion that this is the only way they could be assembled [57].

Many problems remain unsolved even in this simple case of mitotic centrioles: from what substance (unpolymerized tubulin?) does the first outline of the new centriole grow, how is its symmetry determined, from what materials and when are the accessory structures (cart-wheel, links, vacuole) built? It would also be important to know how the centriolar replication is linked with mitosis: this is visible even in cells whose cytoplasm does not divide, and indicates some close relation with the genome replication. The formation of procentrioles is inhibited by cycloheximide [125, 135], demonstrating that other proteins than tubulin (which probably comes from a preexisting pool) are involved. DNA reduplication and centriole multiplication are independent, as shown by the fact that arabinosylcytosine prevents the S phase of synchronized L cells without interfering with the formation of the new centrioles. However, the prophasic migration of the new centrioles to the two poles of the nucleus (cf. Chap. 10) does not take place any more [135].

When the replication of centrioles is the first change leading to the formation of basal bodies and then to ciliogenesis, more than one centriole can be seen around the parent one. In the atypical multiflagellate sperm in the snail *Viviparus,* as many as 13 new centrioles grow around a normal centriole [62]. The sequence of formation of basal bodies from centrioles has been studied in cells actively engaged in ciliogenesis, like the mouse oviduct [44], the rat lung [156], the chick tracheal epithelium [91], the rat pituicytes [52], indicating that the formation of new centrioles is a fundamental cellular process necessary for the growth of cilia. However, new centrioles are not necessarily born in association with preexisting ones.

4.2.5 Centriole Neoformation

Since the first descriptions of centrioles, their constant presence in cells of protozoa and metazoa, and their relation to cell division, have led to the opinion that like chromosomes or mitochondria, they are "self-replicating" units. The ultrastructural observations that new centrioles were born in close association with previous ones, and many facts on the centriole and basal body cycle in various types of cells, including protozoa, confirmed this idea. At one time, DNA was thought to be present in centrioles. The symbiotic theory of their origin (and of that of MT) [106] implied that like mitochondria, they had their own genetic information and were self-replicating.

Three groups of facts demonstrate that centrioles may arise de novo, in unstructured cytoplasm, and that they certainly have no continuity in several types of cells. Neoformation of centrioles can be observed during ciliogenesis in metazoans, in the cell cycle of the ameboflagellates, and in the activated egg of several species.

4.2.5.1 In Ciliogenesis. The formation of hundreds of cilia implies the synthesis by ciliated cells of the same number of basal bodies, which can be considered as centrioles which have moved toward the cell membrane. Descriptions of these stages of centriole formation have been given by many authors [9, 44, 91, 108, 156, 159, 165] and will be summarized here. The first indication is the formation in the cell cytoplasm, with no particular relation to other organelles, although often close to the Golgi apparatus, of limited zones of granular cytoplasm. These become irregular, and denser bodies are observed, with a more or less checker-board pattern. These bodies have some resemblance to the pericentriolar satellites described above. These precentriolar condensations can be multiple, and are not spatially related to the normal centrioles of the cell. It is in these zones that the centriolar MT are seen to appear, first as single MT, then as doublets and triplets. The ninefold symmetry may be missing in the first stages, and incomplete centrioles with an open wall and less than nine MT may be found. However, a very precise mechanism is at play, and rapidly the typical pinwheel structure is formed.

In the oviduct of the Rhesus monkey, ciliogenesis can be induced by the administration of estrogens [25]. In a few days, many cilia are formed, and the steps of their assembly have been carefully studied [9]. Apart from a minority of centrioles (about 5%) which arise close to existing ones, most centrioles—which will become basal bodies—are born in the cytoplasm, with no particular connection to any other organelle. The first sign of their differentiation is the formation of a granular mass, more or less spherical, made of fibers embedded in an amorphous material ("fibrous granules"). The first centriolar structures to appear in close relation with these granules are ring-shaped structures. These lead to the differentiation of an axial rod, with radial spokes connected to a series of filaments which form a tubular structure, already showing a ninefold symmetry. It is on the outside of this scaffolding that the first A tubules will start to grow, from the bottom to the top of the centriole. The lower part of this structure is in relation with a spherical mass or "deuterosome" which lies in the cytoplasm in close relation with the fibrous granules. The tubules are formed in the order A, B, and C, and their typical orientation becomes apparent. The cart-wheel is present, in these cells, throughout the length of the centriole; however, once this has moved to the cell surface, and changed into a basal body with its complex associated structures (see below), the cart-wheel disappears. This description, which applies with some minor changes to several other instances of centriole growth, indicates clearly that the ninefold symmetry is not a result of the properties of the MT: on the contrary, the cart-wheel, originating from the fibrous granules as if from an organizing center, imparts its symmetry to the young centriole. The central vesicles sometimes found in the center of centrioles may be remains of cart-wheel structures [9].

A similar mode of formation of basal bodies, independent of the centrioles, from dense fibrous masses and deuterosomes, has been well documented in the oviduct of the quail (*Coturnix coturnix japonica*), where mucous cells can be transformed into ciliated elements by estrogen-progesterone treatment [146, 147].

The three types of centriolar (and basal body) multiplication have been called "kinetosomal mode" when centrioles are formed at right angles to a preexisting identical structure, "deuterosomal mode" when they arise in the cytoplasm close to a dense body such as the deuterosome, and "de novo" mode when centrioles are formed in cells which do not contain any precursor structure, such as the ameboflagellates which will be considered below [128].

The basis of the centriole is often closely related to a dense matrix which, as described above, will persist around the triplets and possibly also as pericentriolar bodies. The other structures will become progressively visible, but may not be apparent before the newly formed centriole has migrated toward the cell membrane and changed into a basal body. In fact, centriole migration is not indispensable: often, the first step of ciliogenesis is the association with the distal part of the centriole, of a cytoplasmic vacuole, deriving from the smooth endoplasmic reticulum, into which the newly formed cilium will grow ("ciliary vacuole") [108]. The fact that similar steps of centriologenesis are found in cells which are normally ciliated and in cells which under normal conditions have no cilia [52] indicates that this sequence must have a fundamental significance. If the dense masses and the checker-board pattern may be imagined as some type of MTOC with unpolymerized tubulin, the problems of the birth of the ninefold symmetry and of the other centriolar structures, and of the regulation of centriole formation and multiplication, remain unsolved. It is apparent from experimental data [25, 166] that changes of cellular metabolism may induce such ciliogenesis. The question becomes then: how does the normal, non-ciliated cell prevent this growth of centrioles? This is in fact closely related to the regulation of tubulin assembly, where it was apparent that the cell is provided with mechanisms which prevent any excessive formation of MT. Here, however, the problem is far more complex.

4.2.5.2 In Ameboflagellates. At least two species of protozoa may change in less than an hour from an ameboid form, devoid of all centriolar structure, to a flagellate, with two or four flagella attached to basal bodies formed from centriolar structures. Since the first description of these events in *Naegleria gruberi* by Schuster [152], this species has been thoroughly studied by several authors [cf. 59], and similar changes found in another protist, *Tetramitus rostratus* [56, 121]. *Naegleria* (of which some strains may be pathogenic to man and cause a mortal meningoencephalitis; [cf. 89]) grows normally as an ameba.

The main results obtained in this most interesting species may be summarized as follows [41, 56, 58, 59, 93]:

a) The change from ameba to a flagellated form is rapid. It takes place when cells are placed in a medium containing a low concentration of cations (distilled water, Tris buffer). The possible role of Ca^{2+} in determining the changes has been suggested by Willmer [193]: it is interesting in the light of the known actions of Ca^{2+} on tubulin assembly (cf. Chap. 2).

b) This change is rapid and dramatic: within one hour in the new medium the cell rounds up and within the next half-hour all cells become flagellated and motile [41]. In this condition, they do not divide: when mitosis is resumed, the cells loose first their two flagellae. A similar change is observed if cysts or amebae are maintained for 45 min at 21 °C under a pressure of 238 atm. (24.1×10^6 Pa): after release of the pressure, 98% of the cells become flagellated within 3½ h [179].

c) The number of flagellae is about two, although cells with a greater number have been observed. If, shortly before the growth of flagellae, the cells are exposed for less than 1 h to a heat-shock (38.2 °C), the number of flagellae increases to an

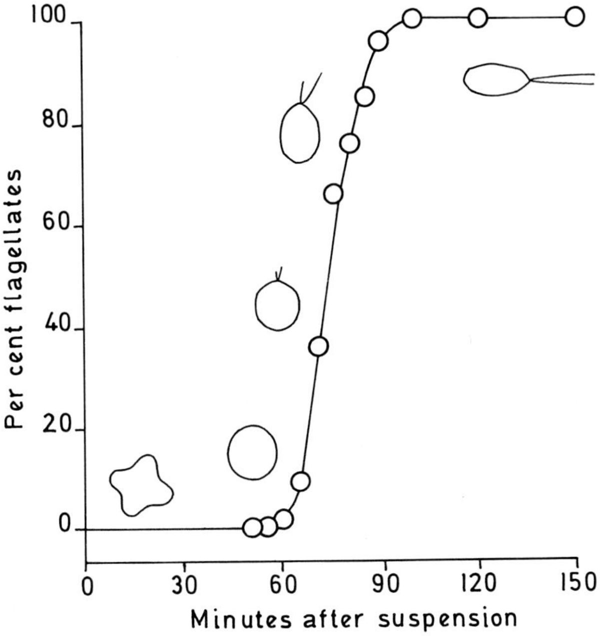

Fig. 4.5. Appearance of flagellae in a transformation population of *Naegleria* in Tris buffer at pH 7.4 (2×10^6 cells/ml) (redrawn from Dingle and Fulton [41])

average of 4.5 cilia per cell, and in some extreme cases as many as 18 cilia are found ("hairy cells"; [40]).

d) Careful studies of the ameba have consistently failed to show any structure comparable to a centriole or even to a procentriole [58]. During mitosis, the spindle is entirely intranuclear [cf. 130, 132, 153]. The MT are straight, and they do not show any special type of attachment to the nuclear poles, which lack any differentiation such as "polar caps" (cf. Chap. 10). The formation of the basal bodies (which have the typical structure of centrioles) and of cilia is not preceded by any "procentriolar" structure [58].

e) The growth of the cilia is very rapid, and calculations indicate that in 30 min about 1.5 million tubulin polypeptides have to be synthesized (at 28 °C). This is probably coded by several identical genes [59].

f) The antigenic features of the flagellar (doublet) MT are different from the cellular tubulin ([59]; cf. Chap. 2) and this property, and also isotopic labeling experiments, show that only a small fraction—1 or 2%—of the large tubulin pool of the cell, which represents about 12% of all proteins, is used in the formation of cilia and basal bodies [60].

g) This may be explained either by the necessity of a special form of tubulin for ciliogenesis or by the segregation of the newly formed tubulin in a special cellular compartment still, at this moment, poorly defined [59]. One explanation could be that the main tubulin pool is located in the nucleus, where it is used for the formation of the spindle in the ameba stage. There are however cytoplasmic MT, separate from the cilia, in the flagellated forms [41], and these are far more numerous than could be accounted for by the intranuclear MT necessary for mitosis (about a fifty-fold excess), the nucleus occupying only $1/20$ of the cell volume (Fulton, 1976, personal communication).

It can be concluded that *Naegleria*—and also the four-flagellated *Tetramitus*, which has been less studied—demonstrates the de novo formation of basal bodies (identical with centrioles) and flagella, in cells where no similar structures are to be found during the ameba stage. These results indicate that the rate of growth of cilia is rapid, and give some hints about the regulation of ciliary growth (role of cations). There is however no evidence of the way this is limited, as a large amount of flagellar tubulin remains in the cells after the two cilia have ceased to grow [59], explaining how it is possible for the same cell to develop a larger number of basal bodies and cilia.

Naegleria is a most interesting model, deserving further study; it is, however, not the only evidence that cilia and all their accessory structures can arise de novo, in cells devoid of any morphological precursors.

Studies on the acrasiales, such as *Physarum polycephalum*, also demonstrate the neoformation of cilia when motile cells are formed during the complex life-cycle of these organisms. The pluricellular stage contains little or no tubulin, although it displays a considerable amount of cytoplasmic motility, related to the presence of actomyosin [88].

4.2.5.3 In Activated Eggs. It had long been known that artificial treatments of activated eggs of various species could result in the formation, even in the absence of a nucleus, of asters, sometimes multiple. In the sea-urchin, *Strongylocentrotus purpuratus*, centrioles could be detected by electron microscopy, and it was concluded that this may mean "that the cytoplasm contains material capable of producing many centrioles" [43]. In a series of experiments, following the observations that heavy water inhibited mitosis ([74]; cf. Chap. 10), it was found that in several species of amphibia (*Xenopus, Pleurodeles, Rana*) a considerable number of cystasters were found after 2 h treatment in 100% or 50% D_2O. Electron microscopy indicated that these were formed around centrioles [183]. In artificial parthenogenesis of *Arbacia punctulata*, centrioles and MT could be seen 60 min after activation, while the inactivated egg did not show any centriole [143]. The experiments of Kato and Sugiyama [92] on *Hemicentrotus pulcherrimus* indicated clearly that activation could have the same result in eggs devoid of any nuclear structure: in non-nucleated half-eggs, typical centrioles were present. All these results could of course be criticized, for it is quite difficult to demonstrate the absence of a small organelle such as the centriole, and it could always be objected that small procentriolar bodies were present but unnoticed. However, it is generally agreed that in unfertilized eggs, centrioles are not to be found. The more recent results of Weisenberg et al. [188] have now, by experiments conducted in vitro, clearly demonstrated that like in other cells, centrioles may arise de novo from unstructured cytoplasm, and result from the assembly of tubulin in the characteristic ninefold symmetrical pattern. In a study of the meiosis of the clam *Spisula solidissima,* MT were found, during interphase, to be associated with dense matrices, perhaps made of unpolymerized tubulin [188]. In a further study on homogenized cytoplasm of activated eggs of this species, MT were observed to polymerize, at pH 6.5 and in the presence of EGTA (cf. Chap. 2), into small "asterlike structures". Electron microscopy demonstrated that these were centered by the typical pinwheel structure of centrioles with nine MT. It is to be noted that pretreatment of the eggs with colchicine did not prevent the formation of these centrioles: this was explained by two possibilities, either that the MTOC was not formed of tubulin, or that tubulin was in a colchicine-resistant state [188]. In these experiments, the centrioles (like in many other mitoses) were in relation with a dense material. This should be compared with the checker-board patterns of granular masses observed in the formation of basal bodies in ciliated cells (cf. 4.2.5.1).

Another experiment indicating the close relations between basal bodies, centrioles and MT in activated eggs may be mentioned here. Purified basal bodies of protozoa (*Chlamydomonas* and *Tetrahymena*) induce the formation of asters and cleavage furrows if injected into unfertilized eggs of *Xenopus laevis,* while flagella, cilia, and brain MT are ineffective. In oocytes, in contrast, none of these materials led to the formation of asters, although the cells are known to contain as much tubulin as eggs, suggesting an inhibitory factor preventing the assembly of MT [81]. These experiments are good evidence of the close relation (or identity) of centrioles and basal bodies. Parthenogenesis has also been realized, in the frog, by injecting a centriole fraction from the testis, or centrioles from sea-urchins [104].

4.2.6 Atypical Centrioles

The centriole, throughout evolution, has been a very stable structure, and few instances of atypical—or pathological—centrioles have been described. While more and more information is gathered about mutants with atypical cilia (cf. Chap. 11) and pathological changes of these centriolar-induced structures are frequent, centrioles, when present, nearly always have the typical pinwheel, 9×3 MT structure. When this is not the case, as in primitive species, either centrioles are replaced by other structures playing a similar role in mitosis, such as polar caps, or they have disappeared almost completely as in higher plants: in gymnosperms they only become visible in the formation of male gametes [182], and in angiosperms they have been completely superseded by pollen-tube growth in fertilization [57].

The "giant" centrioles of neuropteran insects [54] have already been mentioned: it is only their length which is surprizing. In the fern *Marsilea,* basal bodies are normal, but their mode of formation, from a spherical "blepharoplast" on the surface of which small short young basal bodies are packed, is unusual [112]. The blepharoplast appears to be made of densely packed tubules which have about the diameter of the basal bodies.

This dense structure arises twice during the spermatogenesis of *Marsilea vestita,* and divides into numerous centrioles only at the prophase of the ninth division; in other divisions, it occupies the poles of the spindle. It arises each time de novo, and

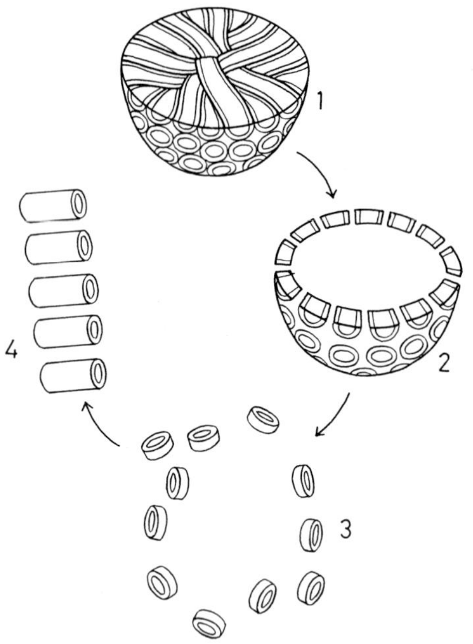

Fig. 4.6. Stages of the transformation of the blepharoplast of the fern *Marsilea* into basal bodies. (1) Solid blepharoplast before last mitosis. (2) Hollow blepharoplast with radially arranged procentrioles, after last mitosis. (3) Break-up of the blepharoplast in the early spermatid. (4) The procentrioles have elongated and align before becoming basal bodies and each growing a cilum (redrawn from Mizukami and Gall [112])

no explanation of its location has been given [83]. Abnormal centrioles are also found in relation with some forms of atypical spermatogenesis. Examples are the flagella made of 70 doublets and 70 singlets in *Sciara coprophila* [123], the spermatozoon of *Littorina* (Gastropoda) with a basal body with only nine doublets [26], and the much more extraordinary multiplication of free kinetosomes, each with the usual nine triplets, in some flagellates living in the intestine of Australian termites. Here, giant "centrioles" grow into 0.5×10^6 disk-like kinetosomes, which lengthen and become associated in chains, all elements of the chain having their triplets in register. Up to 870,000 of these kinetosomes may be found in a single cell [170].

4.2.7 Stability of Centrioles

The possible existence of two types of MT, stable and labile, was mentioned in Chapter 2. Undoubtedly, centriole MT belong to the first category, for once formed they display a considerable resistance to all agents—chemical or physical—which destroy other MT. Basal bodies are identical in this respect, while cilia are more fragile and may be destroyed by various agents. It is indeed frequent that cells may lose their cilia—many techniques of deciliation have been used in protozoa—without altering their basal bodies, which later regenerate new cilia.

The stability of centrioles may result from the fact that their MT are associated in a particular triplet configuration, which perhaps increases their stability, and also from the presence of a dense matrix surrounding these triplets. A similar condition is found in another MT structure, the telophasic bundle of mitosis (cf. Chap. 10) where the spindle remnant, made of MT in a dense matrix, is particularly resistant to poisons which destroy other spindle MT. Centrioles and basal bodies (contrary to other complex structures such as the axonemes of some protozoa) resist high hydrostatic pressures, low temperatures, osmium acid fixation (hence their discovery in electron microscopy before that of the ordinary MT) and MT poisons. These may modify their location in the cell, may in some conditions separate two diplosomes one from another [110], but do not alter the centriolar MT. On the contrary, centriole neoformation may persist in cells treated with MT poisons (cf. Chap. 5).

In an attempt to alter the centrioles, cells were stained with acridine orange or ethidium bromide, which stain the centrioles, and irradiated by an argon laser at a wavelength corresponding to the absorption of the dye. No definite changes were found, while the pericentriolar MT decrease in number, and changes in the pericentriolar region and in the neighboring mitochondria were observed [19].

One other report of centriolar changes in conditions where MT are destroyed should be mentioned. In the aquatic fungus *Saprolegnia ferax* Thuret, very high hydrostatic pressures—14,000 psi—destroyed the MT including those of the centrioles. These however remained visible, for the dense osmiophilic material surrounding the triplets persisted in its normal location, as did also portions of the cart-wheel. This species is characterized by a particular resistance of MT to many agents, such as colchicine, VLB, griseofulvin, low temperatures, and pressure [79]. The molecular weight of *Saprolegnia* tubulin also differs from that of other species [80].

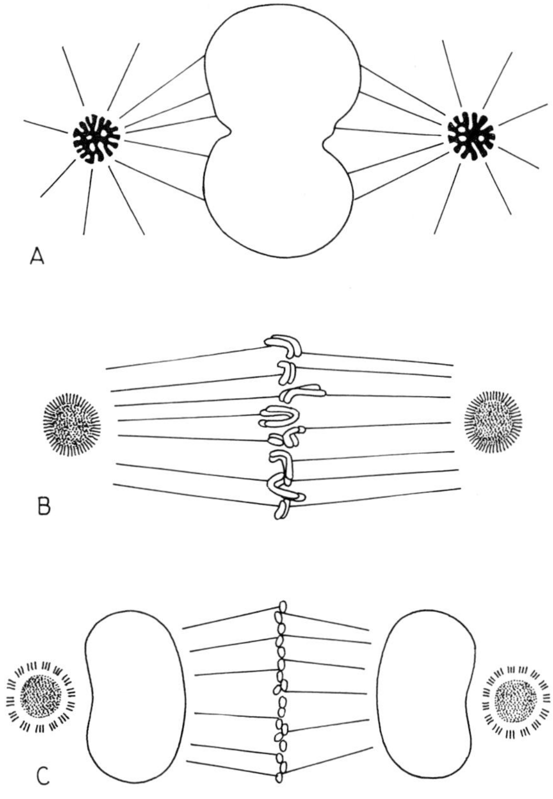

Fig. 4.7 A–C. Behavior of the blepharoplast during spermatogenesis in the fern, *Marsilea*. (A) Prophase of the 9th division: the blepharoplasts are at the spindle poles and act as MTOC. The MT come into close relation with the nuclear membrane. (B) At metaphase and anaphase of the same division, the blepharoplast remains at the poles, but the MT are no longer oriented toward it. The blepharoplast swells and begins its transformation into procentrioles, which form at its surface. (C) At telophase, the blepharoplast is close to the nuclei. The basal bodies continue to grow and become separated from the central material (cf. Fig. 4.6) (redrawn after Hepler [83])

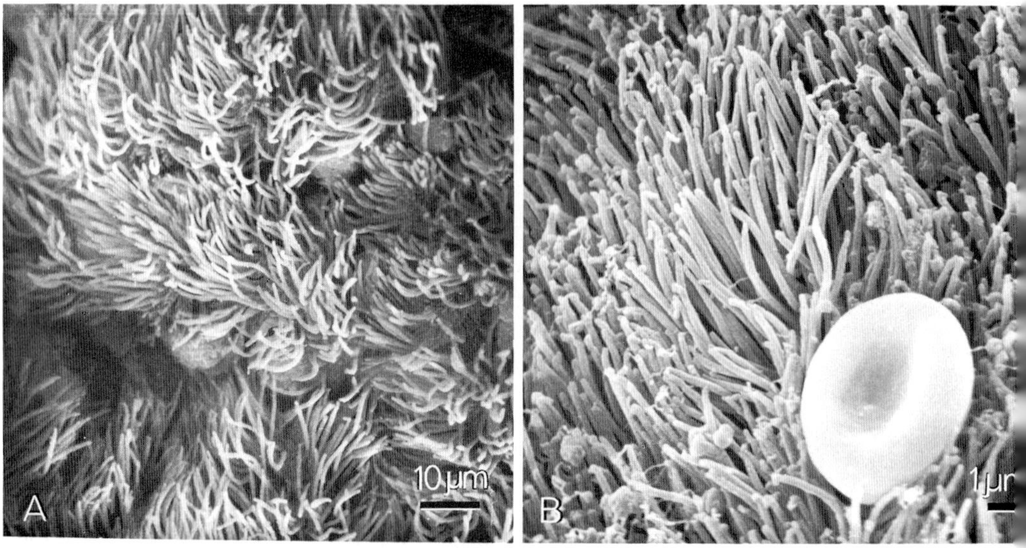

Fig. 4.8 A, B. Scanning electron microscope (SEM) aspect of the ciliated epithelium of the bronchus of a dog. (A) The low power view shows the ciliary bends which are related to the metachronal wave of ciliary beatings. (B) shows irregular bends at the tips of the cilia which are partly artefactual, and may correspond to the region where the stiffening action of the MT is not active. The round body is a red blood cell

4.3 Cilia and Flagella

The two terms are nearly synonyms: flagella will only be used here for the long and specialized cilia playing a role in cell motility, as seen in flagellates and flagellated gametes such as spermatozoa. The ultrastructure of cilia, and the presence in most of them of a particular grouping of 9+2 MT was known before the concept of MT was born. The reason is that the MT of cilia (like those of centrioles) are stable, and once the modern techniques of electron microscopy (staining, sectioning) had been introduced, many details about ciliary structure became apparent. The early papers by Fawcett and Porter [51], Grassé [72], Manton [105], and Sotelo and Trujillo-Cenoz [157], to mention only a few, clearly indicated that in species belonging to widely different phyla, cilia had the same basic structure of nine peripheral doublets surrounding a central pair of two MT. The classic papers of Afzelius [2] and of Gibbons and Grimstone [67] indicated that this assembly of MT was most complicated, that special relations existed between the basal bodies—with their triplets of MT—and the cilia, that special links were provided between doublets and the central pair of MT, and that the cilia, at their base, were attached to the cell membrane. Peculiar fibrillar structures associated with the basal bodies were described. In 1963, negative staining methods demonstrated that the tubules of spermatozoa flagella could be dissociated into protofilaments: this was one of the first indications of the detailed ultrastructure of MT [122]. Early descriptions have been reviewed by André and Thiery [11], Fauré-Fremiet [49], Fawcett [50], and Sleigh [154, 155].

Since this period, the demonstration that the ciliary tubules were only a particular association of organelles present in all eucaryotic cells has brought a new impetus to work on ciliary structure and function, work which is reviewed in the monograph edited by Sleigh [155]. In this chapter the structure and chemical composition of cilia will be studied, and their growth and regeneration described. Some data on pathological cilia, which may help to throw light on the mechanisms of their movements, will also be reported. Cilia motility will be discussed in Chapter 7, for it is related to the general problem of motility associated with MT. Many sensory cells depend on cilia for the perception of external stimuli: this problem is closely linked with the role of MT in cell shape and nerve cells (cf. Chaps. 6 and 9).

Cilia are to be found in most biological phyla, with the sole exception of the phanerogames (which also have no centrioles). In metazoa many—if not all—cells may have one or two cilia ("oligocilia"; [64]), while specialized ciliated cells such as those of the respiratory tract or the oviduct grow several hundred cilia. The particular case of the male flagellated gametes will be briefly discussed, considering atypical cilia and the relations of structure and motility. The problems of ciliary coordination, wave mechanisms, metachronal waves, etc., will not be considered here (cf. [155]): our main interest is to understand why MT are grouped in such a peculiar way in cilia, and the relations between this grouping and motility from a general point of view.

4.3.1 Ciliary Structure

Many excellent descriptions of the ultrastructure of cilia have been published, and we will attempt here to give a general survey of the main structural details, which are almost identical in cilia of most species. Although cilia grow from basal bodies, these will be described later, as they show the transitions from centrioles to cilia. The morphology of cilia will be analyzed before giving a biochemical survey of the different proteins which take part in their structure.

4.3.1.1 The Nine Peripheral Doublets. The basic structure of a cilium is that of a cylindrical organelle, with a length which may vary from a few nm to as much as 200 μm (in the olfactory cells of the frog; [136]). At rest, this structure is cylindrical and more or less straight, rising perpendicularly from the cell surface where it is attached to a basal body and other accessory structures.

When observed in a transverse section the ciliary membrane, which is a continuation of the cell membrane and has the typical triple-layered structure, lies at a small distance of nine groups of doublets. These are formed by one complete MT (tubule A) which has 13 subunits [55, 78] and is united to a second, incomplete MT with only 10 subunits (tubule B); three subunits form a wall which is common to the two A and B tubules. A detailed analysis of the ultrastructure of these two tubules has been made by optical diffraction methods and computer analysis ([7]; cf. Chap. 2).

Ciliary MT do not seem to be all identical as in the 9+2 configuration, the two central MT may be extracted without affecting the outer nine [66].

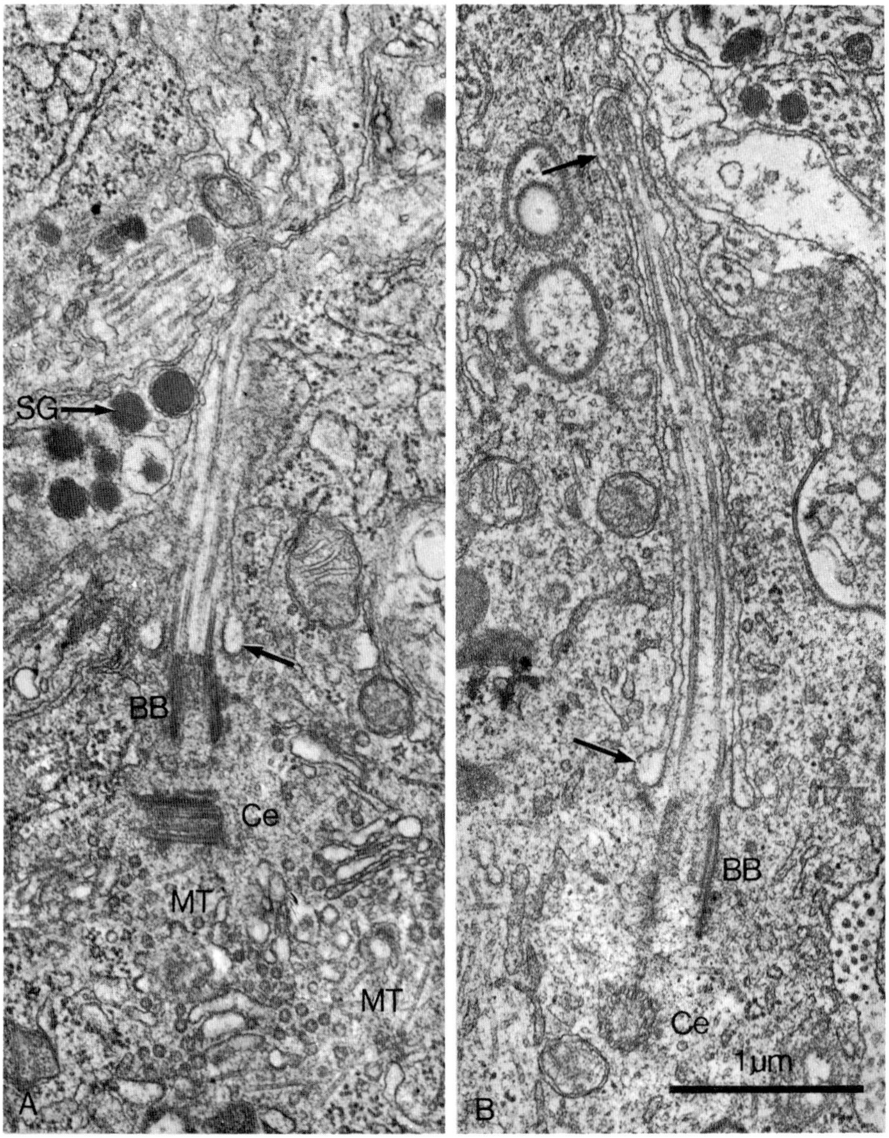

Fig. 4.9 A, B. Two aspects of cilia in cells (pituicytes) of the posterior lobe of the pituitary of the rat. (A) A single cilium is seen growing from a basal body (*BB*) lying close to a centriole (*Ce*) (diplosome), from which *MT* extend into the cytoplasm. Some of the peripheral and central doublets of the cilium are visible. The cilium grows into a ciliary vacuole, the limit of which is indicated by an *arrow*. *SG:* secretory granules in axons from the hypothalamic neurons. (B) Similar aspect, with a long cilium growing from a basal body (*BB*) lying close to a centriole (*Ce*). The limits of the long ciliary vacuole are apparent (*arrows*). The peripheral doublets are visible at the base of the cilium, and close to its extremity

Having demonstrated that subfibers A and B from outer doublets of sea-urchin sperm could be isolated by thermal fractionation, Stephens [163] concluded that they were made of two types of tubulin, A and B. However, later analysis demonstrated that both subfibers A and B contain equal amounts of tubulins α and β and thus appear chemically identical (cf. Chap. 2). The diffraction analysis by Amos et al. [7] indicates that the A and B tubules, while chemically identical, may behave

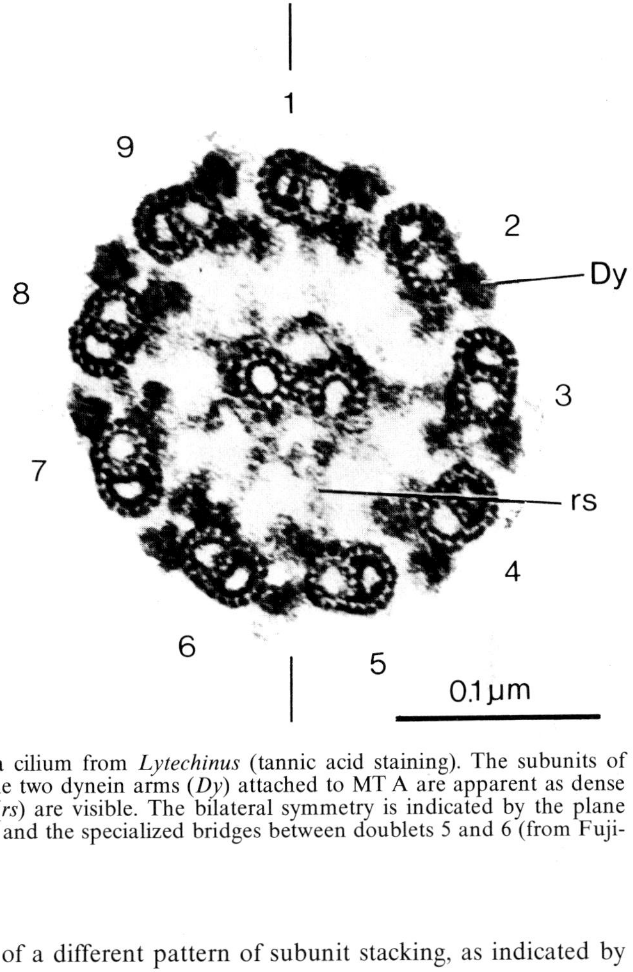

Fig. 4.10. Cross-section of a cilium from *Lytechinus* (tannic acid staining). The subunits of MT A and B are visible. The two dynein arms (*Dy*) attached to MT A are apparent as dense masses. The radial spokes (*rs*) are visible. The bilateral symmetry is indicated by the plane joining the two central MT, and the specialized bridges between doublets 5 and 6 (from Fujiwara and Tilney [55])

differentially as a result of a different pattern of subunit stacking, as indicated by Figure 4.11.

Another study on doublet MT has led to the conclusion that they are assembled from two different tubulins, named 1 and 2 [195]. From the electrophoretical data, these tubulins correspond to tubulins α and β: however, in the model proposed, the partition between MT A and B (cf. Chap. 4) would be assembled entirely from three subunits 1, the subunits 2 forming three rows in each MT. The proposed geometry is in agreement with the findings that the number of subunits in a MT is 13 (or an uneven number).

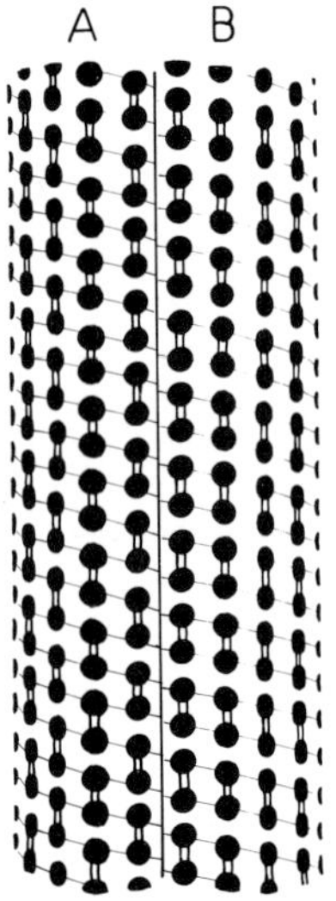

Fig. 4.11 A, B. Probable arrangement of dimers in a doublet MT in *Trichonympha*. The oblique lines are the 4.0 nm lattices; they are approximately continuous across the seam between MT A and B. In contrast, the 8.0 nm lattices are quite different in the two subfibers (redrawn from Amos and Klug [6])

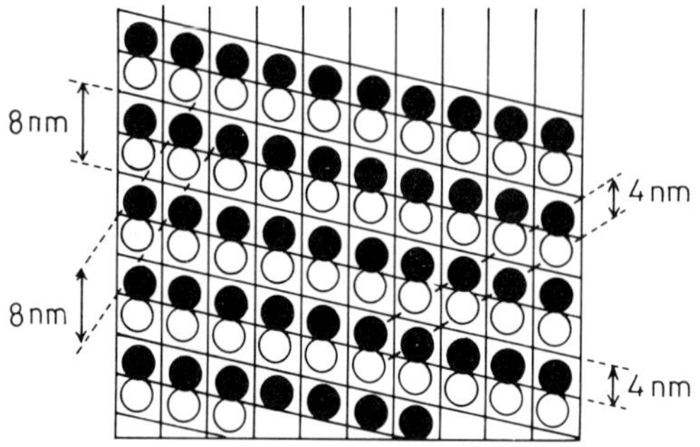

The same authors, having found by isoelectric focussing five different tubulin bonds, suggested that one of these may correspond to the partition MT, the remaining proteins forming two different types of dimers [195].

A confirmation of a particular composition of the partition MT in doublets has recently been published [98, 99]. The doublet tubules of sperm flagellae from two species (sea-urchin and scallop—*Pecten maximus*) were studied by three methods: electrophoresis of their proteins, extraction of some protein fractions, and electron microscopy after negative staining. The presence of many different components (28 in *Pecten*) confirms data from other authors; some of these components could be identified—the α and β tubulins, dynein. Extraction by sarkosyl, sulfocyanate or potassium iodide removed selectively some of these components, and their probable location could be controlled morphologically. It was shown first that the A tubule of the doublets contained a dense globule or fiber attached to its inside wall (Fig. 4.12): similar densities have been reported by several authors in the A tubules of doublets (cf. Chap. 2). Solubilization techniques showed that several proteins were associated with subfiber A; they may be located in this density. Further solubilization led to the isolation of ribbons of three protofilaments: there are reasons to believe that these form the limit between tubules A and B ([cf. 98, 99]; Fig. 4.12). Another interesting finding of this work is that in *Pecten*, solubilization by KSCN preferentially extracted β tubulin: this may indicate that the remaining three-ribbon association of protofilaments contains only α tubulin, or that a third type of tubulin may be present (cf. [22]). This clearly shows the complexity of tubulin doublets, which may explain some of the properties of these structures (resistance to poisons and physical agents, impossibility—so far—to assemble doublets in vitro).

This was discussed in Chapter 2. Tubules A and B are both left-handed, but the spatial relation of their α and β tubulins is different. The structure of the partition, made by three protofilaments, between tubules A and B, has been discussed recently [98, 99]. The differences between both tubules are further indicated by the fact

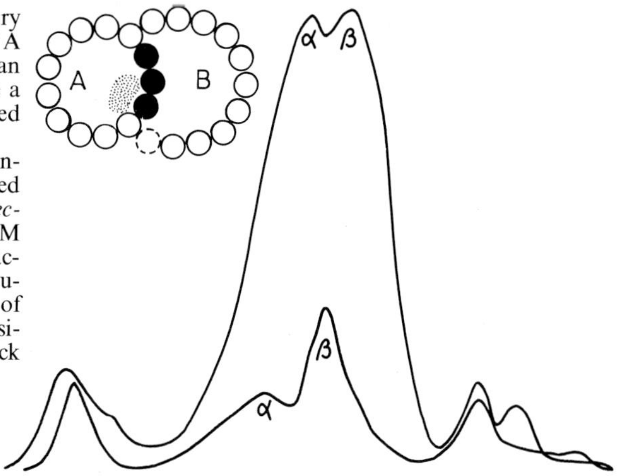

Fig. 4.12. Suggested structure of a ciliary doublet as seen in cross-section. MT A has 13 subunits, MT B 10 subunits and an 11th unit which may not necessarily be a protofilament. A dense globule is located close to the partition, in MT A. Preferential extraction of α tubulin: densitometer tracings of fast green-stained SDS gels of flagellar doublet MT of *Pecten* before and after extraction by 0.3 M KSCN. The material resistant to extraction contains about twice as much β tubulin as α, assuming equal dye-binding of both tubulins and that all the peak density represents tubulin (redrawn from Linck [98])

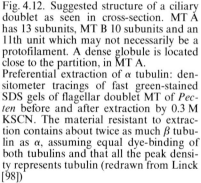

that when fragments of doublets are incubated in vitro with tubulin, the newly formed MT grow mainly from MT A [137]. Also, at the tips of cilia, subfiber B is shorter than subfiber A [146].

The two tubules A and B are in continuity with the corresponding centriolar MT; the plane linking their centers is oblique at an angle of about 10° to the tangent, pointing inward in the direction of MT A [184].

4.3.1.2 The Central Pair of MT. Most cilia have a 9+2 structure, and their axis is occupied by a pair of MT which imparts to the cilium a bilateral symmetry. These MT, which have the usual size and are formed of 13 subunits, start from above the basal body, and often end nearer the tip of the cilium than the doublet MT. The plane passing through the axis of these two MT is perpendicular to the plane of ciliary beating, and the relation of the central pair with movement is evident, from observations of 9+0 cilia, without central pair, which are often non-motile as in sensory organs [15], or display different patterns of motility, as in some insect spermatozoa [124].

The central pair is in close relation with a linking material which seems to envelop it, and also with the peripheral doublets, via the radial links oriented toward MT A.

4.3.1.3 Linkages and Appendices Between the Ciliary MT. These belong to at least three different categories.

a) *Dynein.* The important experiments of Gibbons [66, 67] on the chemical dissection of cilia (vide infra) demonstrated that a special protein, with the properties of an ATPase, was attached to the outer doublets. Its activation by ATP plays a role in ciliary movement. These "dynein arms" are fixed to MT A: an outer and an inner arm are clearly visible, and are connected to subfilaments 4–5 and 8–7 (cf. Fig. 4.10; [155, 186]). These projections have curved ends. They do not touch the neighboring doublet, although a link is visible between the inner arm and the next B tubule (interdoublet link; [185, 186]). Between doublets 5 and 6, there is often a more prominent double link, as if the dynein arms were longer, or fixed by some other

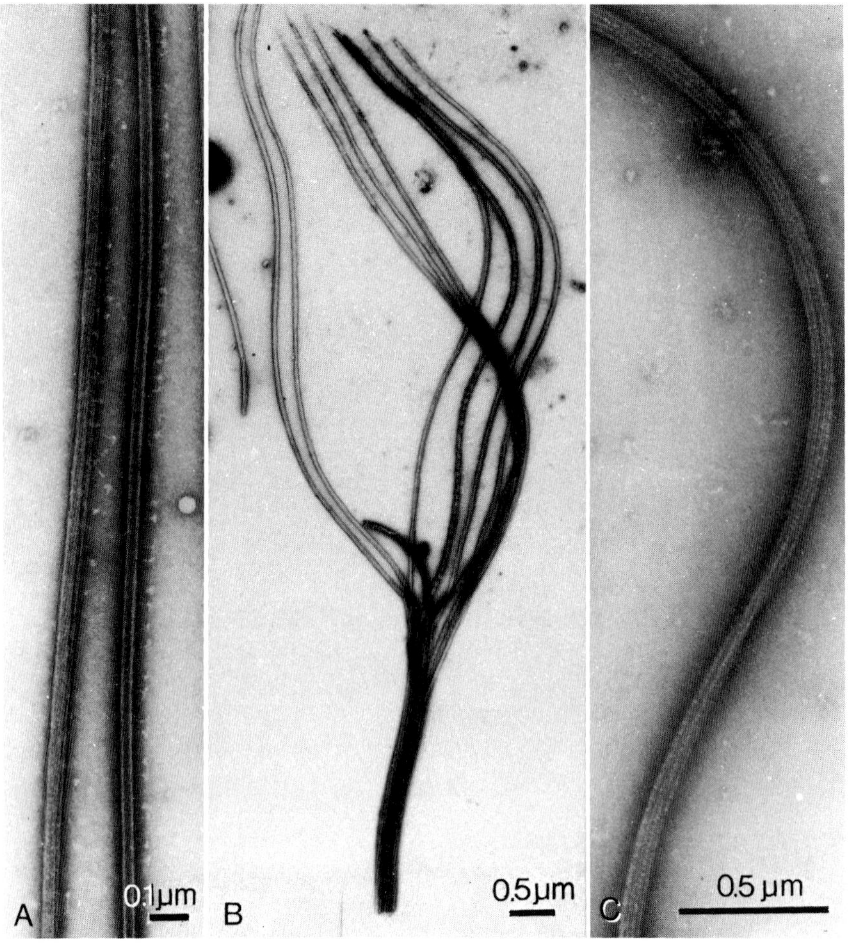

Fig. 4.13 A–C. Flagellar MT of *Chlamydomonas* (uranyl acetate negative stain). (A) Outer doublet MT showing radial links on MT A. (B) Whole axoneme with splayed distal end and squared-off proximal end. (C) Central pair MT with striated sheath (from Binder et al. [23])

Fig. 4.14. Diagrammatic representation of a cilium. *On the left,* the longitudinal section shows from tip to basis the following structures: (1) the cytoplasmic unitary membrane; (2) the ciliary doublets: the peripheral A tubules are longer than the incomplete B tubules; (3) the central pair of MT, connected by a sheath showing periodical densities; (4) the axial rods extending from the peripheral doublets to the central pair. They are grouped by three, and this periodicity may be related to that of the sheath surrounding the central pair of MT; (5) the granules indicated as local thickenings of the proximal part of the ciliary membrane; (6) below these, the ciliary necklace connected to the peripheral doublets; (7) below the central pair, the ciliary vesicle; (8) between the cilia and the basal body, the basal plate; (9) the basal body, with its nine triplets of MT, its alar sheets, its connections to the cell membrane and to the peripheral dense bodies, from which extend the cytoplasmic MT; (10) one ciliary rootlet, with its cross-striation. *On the right,* typical cross-sections at various levels show: (1) the aspect close to the tip, with some doublets and singlets, and the two central MT with their connections; (2) a typical cross-section, with the nine peripheral doublets, the two dynein arms attached to MT A, and the axial rods extending from the same MT toward the central pair. These end with a knob-like structure; (3) one possible aspect of the nexin links between the doublets; (4) the aspect at the level of the ciliary necklace, showing the connection between the membrane specializations and the nine doublets. These are surrounded by a dense material similar to that described around the triplets of centrioles; (5) the basal body triplets, linked by fibrils extending between the A MT and the peripheral dense bodies; (6) the cart-wheel and the connections between the triplets at the base of the basal body, showing the greater angulation (40°) of the triplet plane as compared to that of the ciliary doublets (10°)

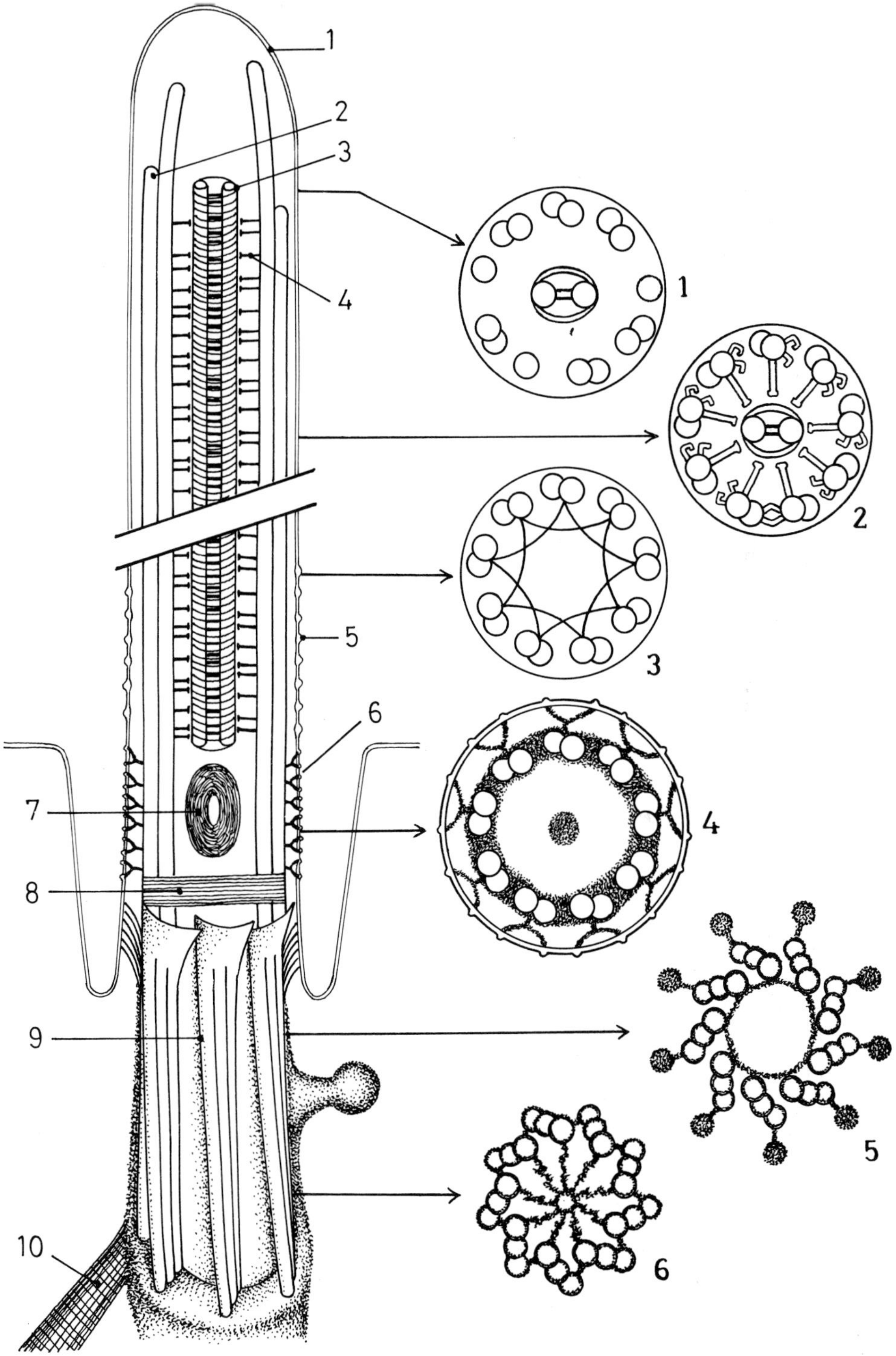
1
2
3
4
5
6
7
8
9
10
1
2
3
4
5
6

protein. This double link helps to determine the axial symmetry of the cilium, as it lies exactly opposite doublet 1, determining a plane which is perpendicular to that of the central doublet.

b) *Nexin.* The nine doublets are often seen to be linked by fine filaments which often form remarkable geometrical structures, such as three interlocked triangles (in *Chlamydomonas reinhardii,* [31]) or a nine-pointed star, as in the fern *Marsilea* [112] (Fig. 4.14). The fractionation of cilia shows that it is possible to obtain groups of nine doublets united by these filaments, which can be analyzed and demonstrated to be formed of a special protein, named *nexin* [161, 162]. It has a molecular weight of 150,000 – 165,000 daltons. It appears to unite the external triplets. It forms particularly complex patterns at the basal end of the cilium, in relation to central structures which are found there [94, 112]. It is not however evident that all such links, which do not seem to play a role in ciliary movements, contrary to dynein, are made of a single protein.

c) *Radial links (spokes).* These extend radially from subfiber A and end at a short distance from the central doublet and its sheath. They are easily apparent on negatively stained, isolated doublets, or in thin transverse sections of cilia [5, 32, 184]. They are more widely separated than the dynein arms; hence, they are not necessarily seen in transverse sections of cilia. They end in a knob-like structure. These links are often grouped by two or three, and their spacing varies from one species to another. The relation of this spacing with the structure of the sheath of the central doublet is important, for the radial links appear to provide a mechanism for guiding the movements of the MT during ciliary bending. The periodic alignment of the spokes becomes clearly visible in splayed cilia obtained, in *Tetrahymena,* by treating the cells with Triton X-100, which destroys the ciliary membrane, followed by critical point drying. The various ciliary MT flatten out on the support while retaining their relationships, and the spokes appear as densities aligned along the peripheral doublets [145]. This will be discussed in Chapter 7. The chemical nature of the radial links is unknown; it is possible that the terminal knob has a different composition from the links [86].

d) *The central sheath.* The two central tubules are linked together by a complex structure which has been called the central sheath, although it appears more to consist of ladder-like projections from the two MT [32]. According to Warner [186], these projections number four. They are much more closely spaced than the radial links, resembling more in this aspect the dynein arms of the peripheral doublets. The orientation of these links is a consequence of the bilateral symmetry of the cilium, two extending towards doublet 1, the two others towards doublet 5 and 6. There appears to be some dynamic linkage between the radial spokes and the sheath, which may change when the cilia bend (cf. Chap. 7).

4.3.1.4 Other Ciliary Structures. a) *Links between doublets and ciliary membrane.* Cross-sections of cilia often show, especially near the base of the cilium, delicate threads between the doublets and the membrane (cf. [31, 61, 176]). The topography of these structures has been revealed by the use of freeze fracturing and the study of replicas of the ciliary membranes. Two types of structures should be considered, and it is evident that large variations occur from species to species, and even within a single cell [150].

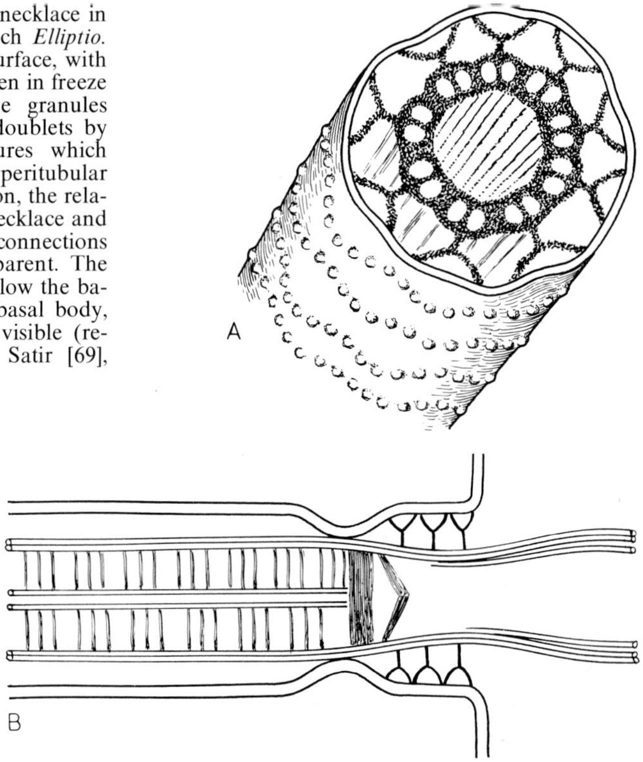

Fig. 4.15 A, B. The ciliary necklace in the gill of the lamellibranch *Elliptio*. (A) Aspect of the ciliary surface, with the necklace granules as seen in freeze fracture preparations. The granules are linked to the ciliary doublets by "champagne-glass" structures which are connected to dense peritubular material. (B) In cross-section, the relations between the ciliary necklace and the "champagne-glass" connections with the doublets are apparent. The necklace is located here below the basal plate, and above the basal body, the triplets of which are visible (redrawn from Gilula and Satir [69], slightly modified)

The *ciliary necklace* was discovered, in cilia of the lamellibranches *Elliptio* and *Mytilus,* by the technique of freeze etching, which reveals details of the ciliary membrane. Similar structures, differing in minor details, were found in cilia of species as different as the ameboflagellate *Tetramitus,* the sea-urchin embryon, the rat trachea and 9+0 cilia of cultured chick embryo fibroblasts [69]. The ciliary membrane shows a definite number of rows of particles located at the base of the cilium, above the basal body. These intramembranar particles (protein molecules?) are linked to the ciliary doublets by delicate connections which, in *Elliptio,* were described as having a "champagne-glass" appearance, the rim of the glass being connected with the membrane specializations (Fig. 4.15). Similar granules, located in plaque-like assemblies above the necklace, were described in *Tetrahymena* [198]. These "granule plaques" have been demonstrated to be located opposite the nine doublets, and form nine rows. In *Paramecium* there are three circles of five granules each, above the necklace, opposed to the doublets. A study of these structures after fixation in calcium-rich solutions has demonstrated that deposits of Ca^{2+} are electively located under the ciliary membrane, in the region of these granule plaques [131]. Although these do not appear to be present in all cilia, they may play an important role in ciliary motility and calcium exchanges.

As mentioned, ciliary necklaces have been found in many types of cells: in the brain of the rat, they form five or six strands perpendicular to the axis of the cilium, on faces A and B of the fractured ciliary membrane, appearing as a "standard feature" of cilia structure [171]. Similar granules, although much more numerous and more irregularly disposed, have been seen in the modified cilium of the photoreceptor cells of the rat retina [109]. All these observations indicate that the complexity of ciliary structures is even greater than was imagined a few years ago.

b) *Central body (axosome).* At the base of the cilium, at a level where there are no central doublets, a peculiar cylindrical structure may be observed [31], although in other cells, the central doublet may start from a globular "central body" or "axosome". This structure is attached to the outer doublets by fine fibers which may display the elaborate geometrical disposition mentioned above in the description of nexin, although the chemical nature of these links remains unknown. This central body is located close to the ciliary necklace [150]. The axosome may be linked with the inner doublets, as observed in *Didinium nasutum* [128]. It lies above, and close to, the basal plate. Its structure varies considerably from one species to another: cylindrical, rod-like, granular. It may, as indicated by its location, be related to the growth of the central pair of MT.

4.3.1.5 Basal Bodies. Basal bodies are still more complex than the cilia, for they are made of nine groups of MT, like those of centrioles, and several accessory structures which link them, on one side, to the cilium, on another to various anchoring filaments which extend into the cytoplasm. Moreover, their lumen shows several structures comparable to those already mentioned in centrioles. According to the plane of longitudinal sections, basal bodies appear as barrel-shaped organelles, or cylinders with slightly oblique walls. These aspects result from the twisted structure of the basal body, with all transitions from the typical pinwheel centriolar structure with oblique triplets, to that of the cilium, with doublets. Excellent three-dimensional reconstructions of the basal bodies of the *Rhesus* monkey oviduct have been published ([8]; Fig. 4.16).

a) *Triplet MT.* The structure of these is identical to that described in centrioles: one MT is complete, while the two others share common walls, probably made of three subunits. These triplets, when seen in a transverse section, are characteristically grouped like a pinwheel, the maximum angle with the tangent to circumference of the lumen being 40°, as observed at the base of the structure. Moving upward towards the cilium, this angle decreases progressively, to reach 10° where the triplets change abruptly to doublets, subtubule C stopping at the level of the basal plate or below. Moreover, these triplets are not always parallel with the axis of the basal body, and may have a pitch of 10 – 15° [8]. This geometry explains why it is often impossible, in a transverse section, to see with equal clarity the lumen of all nine triplets, and also why, in longitudinal sections, the basal body often shows a barrel shape. While the outside diameter of this structure remains more or less constant from one end to the other, the lumen enlarges (by the change of the "triplet angle") by about 56 nm between the proximal and the distal (ciliary) end [8]. This peculiar shape is probably maintained by the links already described in centrioles between tubules A and C, and by other internal structures which will now be described.

b) *Cart-wheel.* This structure, which is identical to that of centrioles, usually occupies the proximal end of the basal body, linking the triplets to a central granule through nine slightly curved fibrils. It is attached to fiber A and may play a role in

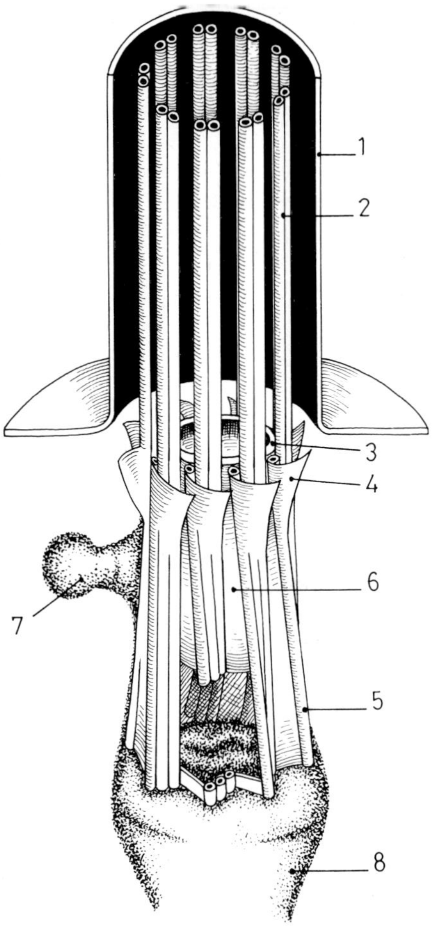

Fig. 4.16. Three-dimensional structure of the basal body of *Rhesus* monkey oviduct. The following structures are visible: (1) Cell membrane surrounding the cilium; (2) ciliary doublets (the central pair is not visible at this level); (3) apical ring structure (comparable to the basal plate, cf. Fig. 4.17); (4) alar sheets attached to the basal body triplets; (5) triplets with their 40° angulation and the links between MT A and C; (6) links between A and A MT of the triplets; (7) lateral basal "foot"; (8) ciliary rootlets (redrawn from Anderson [8])

the angulation of the triplets. In some basal bodies, the cart-wheel may extend all the length of the lumen [67, 181].

c) *Circular thread.* A thread of dense material has been observed in some basal bodies like in centrioles [128, 167]. As mentioned, this has been considered by some to be of nucleoprotein nature [76].

d) *Basal plate.* The distal extremity of the basal body, indicating the transition to the base of the cilium, is marked by the absence of tubules C (which may, at their distal extremity, sometimes show incomplete rings in the shape of a C; [197]) and by a complex structure, the basal plate. This is located in the same plane as the cell membrane (type I) or at some distance above it (type II) [128]. In transverse sections, it appears as a flat density in the lumen of the basal body. In fact, it is far more complex, and studies of isolated basal plates by negative staining [197] in *Tetrahymena pyriformis* show a delicate structure made of a web of fine filaments delimitating several holes disposed with a ninefold symmetry. This plate is in close relation to the basal ring, which is located outside the plate. It is from the plate—which represents the true beginning of the cilium—that the nine outer doublets start. It may persist when the cilium is shed.

e) *Transitional fibers (Alar sheets).* At the distal third of the basal body, extending outward, nine lamellae are observed. Three-dimensional reconstruction of these "alar sheets" [8] indicates that they are linked with the triplets, have the shape of trapezoidal sheets, enlarging toward the distal end of the basal body, and directed toward dense cytoplasmic masses which can be compared to the pericentriolar bodies. These, usually known as "transitional fibers", are of unknown chemical nature; they may help to link the basal body to surrounding structures and possibly to neighboring basal bodies or to the cytoplasmic membrane (cf. [31, 53, 61, 176]).

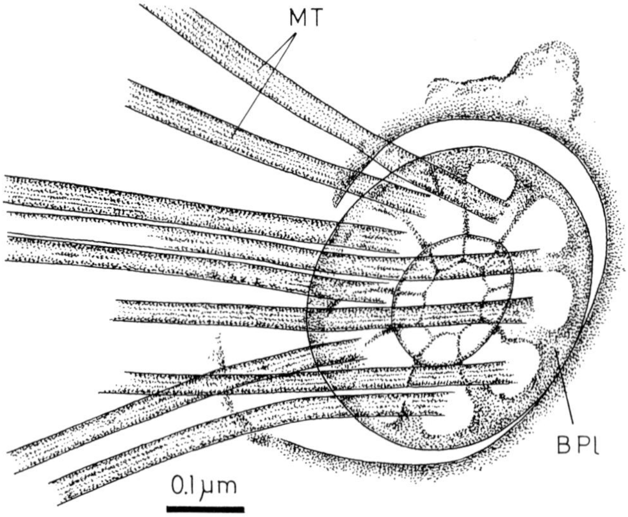

Fig. 4.17. Isolated basal body from oral apparatus of *Tetrahymena.* The nine doublet *MT* are attached to the complex basal plate (*BPl*), which shows a ninefold symmetry and a central cart-wheel structure (redrawn from Wolfe [197])

f) *Basal foot.* This is a structure which projects laterally from one side of the basal body. It is more or less club-shaped, and may show some delicate cross-striations [15, 65]. This structure is related to the plane of symmetry of the cilium, being located at the level of triplets 5 and 6 [53].

g) *Ciliary rootlets.* Extending from the side or the extremity of the basal body into the cytoplasm are structures which vary considerably from one cell to another and are made of cross-striated fibrils. These may, in some sensory cells [61], attain a considerable degree of complexity. They are found in most cells, and merge with the dense matrix which surrounds the triplet. The rootlets are particularly developed in protozoa: in ciliates they are specifically attached to a limited group of triplets. Their striations resemble that of collagen [128] [6]. The rootlets, most probably, have

[6] However, electrophoresis of isolated basal bodies from the gill epithelium of *Aequipecten irradians,* a mollusc, has indicated that rootlets were associated with a protein of molecular weight of 230,000 and 250,000 daltons, which has been named *ankyrin* [164]

mainly a mechanical role in attaching the basal bodies to other cytoplasmic structures or basal bodies, like in *Chlamydomonas* [137]. It should be noted that, although centrioles and basal bodies are closely related organelles, striated rootlets are not observed around centrioles, in contrast to the transitional fibers [119].

4.3.1.6 Relations of Cilia with Cytoplasmic MT. Basal bodies are related to centrioles; they are not in continuity with cytoplasmic MT, which are seen to start close to them, in relation with the rootlets and the transitional fibers. In ciliates, more elaborate groupings of MT are observed [71], displaying remarkable patterns geometrically related to the triplets of the basal bodies (*Chlamydomonas,* [137]). The basal bodies may be associated with long ribbons of closely linked MT, as observed in the ciliate *Condylostoma magnum* [128]. The complexity of these associations of basal bodies with radiating groups of MT may be very great, as shown in the choanoflagellate *Codosiga botrytis,* where radiating stacks of about 150 MT are found around the proximal part of the basal body [84].

4.3.2 The Biochemistry of Cilia

The morphological complexity of cilia indicates that various constituents are present in the tubules, the links and the membrane specializations. The possible existence of nucleic acids has also been the subject of many discussions.

The main components of cilia, as isolated in a more or less purified form, will be considered here: tubulin, dynein, nexin, and nucleic acids. Ankyrin was mentioned above [164]. This list is certainly far from covering the whole subject of cilia chemistry, but is sufficient to understand the relations of cilia with tubulins and "ordinary" MT, and mechanisms involved in ciliary movements (studied in Chap. 7).

4.3.2.1 Tubulins. Early contributions [33, 187] on the fractionation of isolated cilia from *Tetrahymena pyriformis* indicated that a particular protein—later to be identified as tubulin—was present, and also that a fraction with an ATPase activity could be isolated—this was to be named dynein [68]. The techniques of separating these compounds and confirming their location by electron microscopy were perfected by Gibbons [66] in experiments of "chemical dissection" of cilia. In treating pellets of isolated cilia of *Tetrahymena* either by digitonin or by dialysis, it was found that the central MT were solubilized first. In a second step, the arms extending from the outer doublets disappeared (it was in this fraction that dynein was found). This second step was demonstrated to be reversible, and adding the "dynein" fraction to the isolated doublets, the outer arms again became apparent, while some "fluffy" unidentified material was found coating the MT. Sperm tails of *Arbacia punctulata,* treated similarly, also showed that the outer doublets were more resistant than the central pair of MT.

This differential solubility of the central MT and the peripheral doublets was confirmed by several authors [47, 195] and led Stephens [160, 161] to the conception that two different types of tubulin (A and B) were present. In a study of the solubilization of ciliary tubules from *Chlamydomonas,* after sarcosyl treatment, it was found that the detergent destroyed the tubules in a definite sequence: (a) one of the two

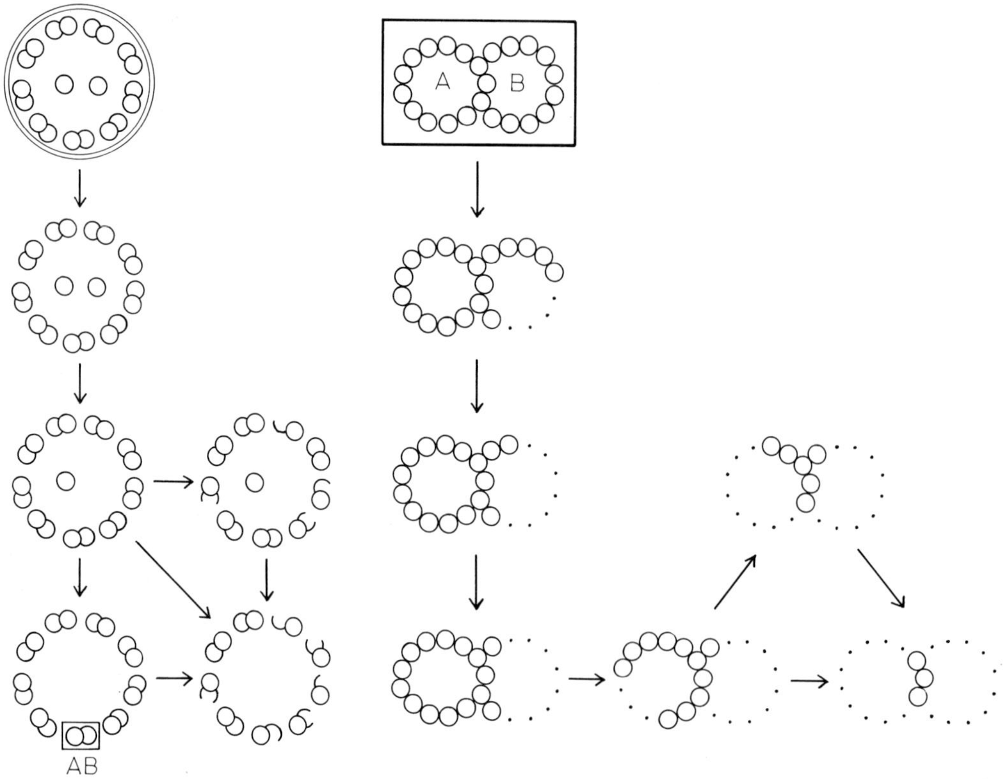

Fig. 4.18. Sequence of solubilization of MT from 9+2 flagella of *Chlamydomonas* after treatment with sarkosyl. After removal of the membranes, one of the central pair is removed, and then the B tubules of the doublet break down (*left*). *On the right,* the changes of the MT are indicated: the progressive destruction of the doublet MT by increasing action of sarkosyl finally leaves only the three partition protofilaments (cf. also Fig. 4.12) (redrawn from Witman et al. [194], simplified)

central MT, (b) the remaining central MT and the outer wall of tubule B, (c) the remaining tubule B, (d) the outer wall of tubule A. There remained then a more resistant fraction, which appeared to be made of the three subunits shared by tubules A and B [194]. Electrophoretic studies of the solubilized proteins indicated their tubulin nature, and two proteins, with MW of 56,000 and 53,000 daltons respectively were isolated [194]. These observations led to the conception, mentioned in Chapter 2, that the wall was made of tubulin 1, which was associated with tubulin 2 in tubules A and B. Isoelectric focusing demonstrated in this same work that more than two proteins may be present, as five bands were detected [195].

While the stability of MT and doublets is clearly different, it was shown by electrophoresis that the proteins extracted from the outer and the central MT of *Tetrahymena* cilia were identical and showed the typical migration velocity of tubulin [47]. A comparison of tubulins from *Chlamydomonas* flagella and mammalian brain suggested that two tubulins were present in the flagella and that only one of these was identical with that of brain [120]. Other results, described in Chapter 2, indicate however that ciliary tubulins are identical with those found in brain, although the stability of the assemblies may differ [162]. The different structural lattice found in MT A and B may be responsible for the greater resistance of the doublets (cf. [6]).

Tubulin is known for its capacity of binding colchicine (cf. Chap. 5), and this property is found in cilia, although lacking in the 9+2 axoneme of *Aequipecten irradians:* it has been concluded that a major protein component of the ciliary membrane, fixing colchicine, could be a modified form of tubulin [163]: this result should be compared with other data on the close relations between tubulins (or MT?) and cell membranes (cf. Chap. 3).

4.3.2.2 Dynein. This protein extracted from cilia and presenting an ATPase activity was named by Gibbons and Rowe [68]. Two fractions, with sedimentation coefficients of 30 and 14S, were extracted from cilia by the techniques mentioned above. The 30S protein is apparently a polymer of the 14S one [66]. Electron microscopic observations indicated that the arms of the external doublets had this ATPase activity, and reconstruction experiments demonstrated that dynein, added to doublets without lateral side-arms, became bound to them. It was observed (and this is important for a proper understanding of ciliary structure, cf. [55, 73]) that "the arms appeared to have returned with remarkable precision to the same position that they had in the intact cilia" [66]. The activation of dynein by ATP would generate the mechanical force conducting to ciliary beating, a problem which will be studied in Chapter 7. Dynein has been isolated from various types of cilia and flagellae, such as *Pecten irradians* [160] and mammalian brain [24, 63] (however with a smaller ATPase activity). The presence of dynein attached to MT is related to fundamental problems of MT physiology: heavy molecular weight proteins associated with tubulins, phosphorylation by proteins extracted with tubulins, origin of the forces necessary to explain movements linked with MT. The description of non-motile human spermatozoa without dynein arms is further evidence of the role of this ATPase ([3]; cf. Chap. 11). Detailed biochemical information about dynein is to be found in Stephens [162].

The role of dynein in cell motility is well demonstrated by studies on the various types of spermatozoa in cecidomyid flies. Some, with a 9+0 structure, are devoid of dynein arms and immotile. Others have a large number (up to 1,000) of doublets. These have only one dynein arm each: their motility is rudimentary, compared to flagella with two dynein arms [13]. Motility is thus independent of the ninefold symmetry but requires the presence of dynein arms.

4.3.2.3 Nexin. In his work on the dissection of cilia, Gibbons [66] noticed that in cilia from which the membrane and the central pair were dissolved, the nine doublets maintained their relations, although no bridges or spokes were visible. The integrity of the cylindrical groups was maintained by a "thin filament running around the inside of the cylinder from one A subfiber to the next" [66]. In further work along the same lines, this thread which was found to link the A tubules—even after the destruction of all other structures, including the B tubules—was named nexin, from the Latin *Nexus,* "meaning a tie binding together members of a group" [161]. Nexin can be fractionated with the outer ciliary MT and is found to have a MW of 160,000 daltons.

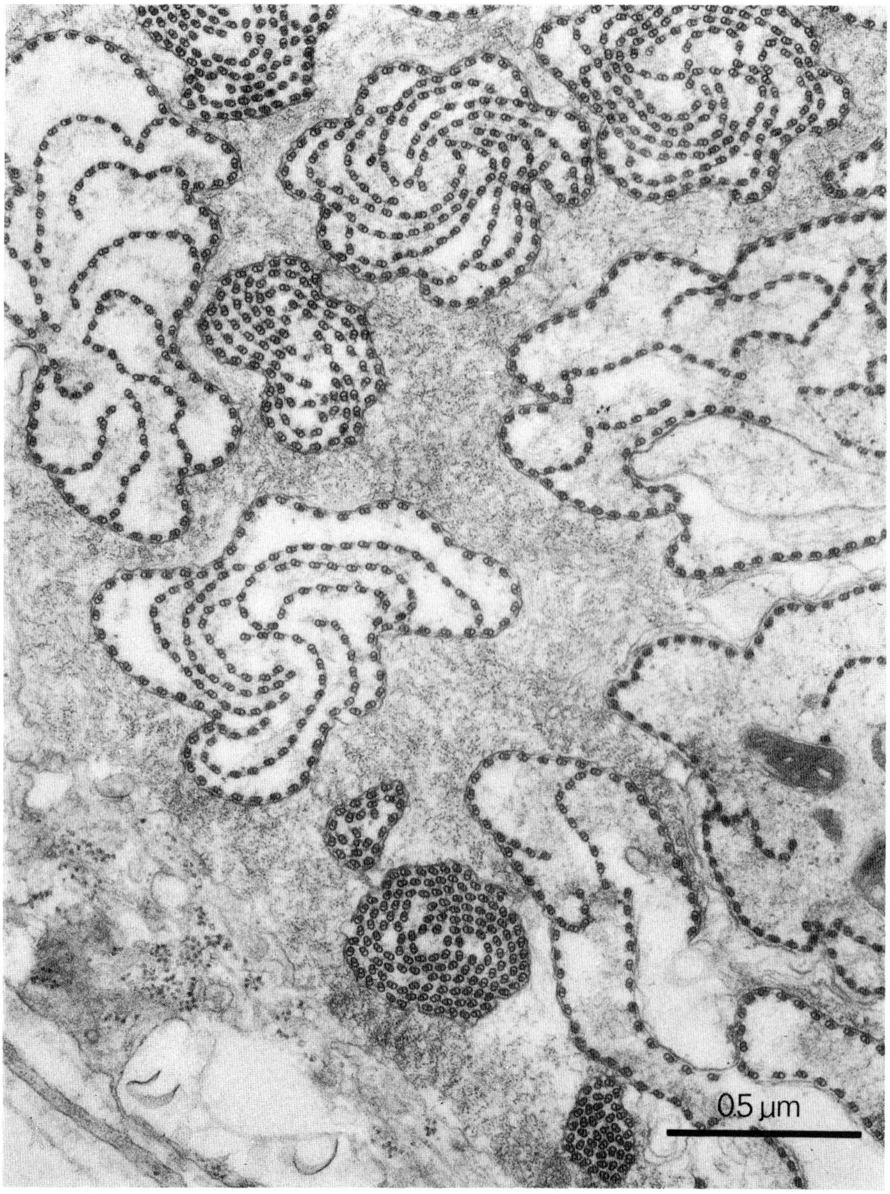

Fig. 4.19. Spermatozoon of *Monarthropalpus buxi*. The ciliary structures are made of a great number of doublets, connected by one visible arm (dynein). The A MT of each doublet appears more dense and may contain an additional protein (from Baccetti et al. [14])

4.3.2.4 Nucleic Acids. Much has been written on the "self-replicating" properties of basal bodies. As with centrioles, it can only be said that new basal bodies are often formed in the vicinity of old ones. The problem of nucleic acid in basal bodies has already been discussed in relation to the chemistry of centrioles. The careful study of regeneration of the oral band in *Stentor coeruleus,* where thousands of new basal bodies are assembled in 3 h, indicated that no ^{3}H-thymidine incorporation takes place, and that regeneration is not prevented by DNA poisons such as ethidium bromide, mitomycin C, or hydroxyurea [199]. In this model, on the contrary, RNA synthesis is detected during cilia regeneration; the hypothesis mentioned above that RNA may act as a structural unit in centrioles and basal bodies deserves further study [77].

4.3.3 Ciliary Birth, Growth, and Regeneration

The formation of cilia is preceded by the multiplication of centrioles which will become basal bodies. The changes from centrioles to basal bodies and the resulting growth of cilia, the problems of ciliary regeneration observed after deciliation, and some quantitative aspects of ciliary growth will be considered here. The regulatory influence which plays a role in all of these steps will be evoked.

4.3.3.1 Formation of Basal Bodies. The steps in the multiplication of cilia in various cells have already been described; these are very similar whatever the type of cell, and may take place in the cytoplasm (a) in relation to already formed centrioles ("kinetosomal mode"), (b) without any relation to other organelles, except a dense body called "deuterosome" ("deuterosomal mode"; [128, 156]), (c) de novo in cells without centrioles. The assembled centrioles, with their nine groups of triplets, migrate toward the cell membrane, where ciliogenesis will take place, with the formation of a "ciliary vesicle" deriving from the smooth endoplasmic reticulum, into which grows the cilium [108]. Once a centriole has become linked to the cell membrane, the cilium will grow from the external MT A and B (doublets) and two new MT will appear in its axis, together with all the other complex links described above. The growth of cilium, that is to say the assembly of its tubulin molecules, proceeds from the distal end, as shown in *Naegleria* [59] and *Chlamydomonas* [141]. Similarly, when flagella or cilia shorten, disassembly starts at the tip of the MT [31, 139].

Basal bodies do not necessarily arise from centrioles, and in *Chlamydomonas* they are assembled close to the cell membrane in the following sequence [30]: (a) the tubular hub of the cart-wheel, (b) nine spokes, (c) the A tubules, (d) the B and C tubules. This is similar to the sequence described for centrioles [9] and indicates the importance of the cart-wheel for imparting the ninefold symmetry.

The role of basal bodies for tubulin assembly has been studied in *Chlamydomonas.* It is possible, by treatment with EDTA and polyethylene glycol, to separate from the cell the whole basal body – axoneme complex with the two cilia. Negative staining of these preparations gives clear illustrations of the formation of the basal body from an "annulus" of nine granules from which the MT will grow [70]. Moreover, when such preparations are incubated in the presence of tubulin, under conditions favoring the assembly of MT, these are seen to grow rapidly, mainly from the distal end, except if the concentration of tubulin is high, when MT also grow proximally [139].

Basal bodies, in ciliates such as *Tetrahymena,* do not necessarily grow cilia, and a detailed quantitative study has indicated that, according to the moment in the cell cycle, from 17 to 46% of them remain "naked", without cilia [114]. This shows that although the cilium grows fast, the time elapsing between the formation of basal body and the growth of the cilium may be as long as the cell cycle [114].

4.3.3.2 Regeneration after Regression or Deciliation. The regeneration of cilia can take place in a few hours, and entails complex cytoplasmic changes, as many new proteins must be synthesized: several steps of this process may be arrested by an inhibitor of protein synthesis, such as cycloheximide. On the other hand, the regeneration of the tubulin part of the cilia may take place without protein synthesis, and is related to the presence in the cell of a pool of tubulin [12, 115]. It is also apparent that whenever possible, the cell retains the tubulin, as observed when, before cell division and the formation of new flagella, the old ones are "retracted" into the cell. Some of these problems have already been mentioned in Chapter 3.

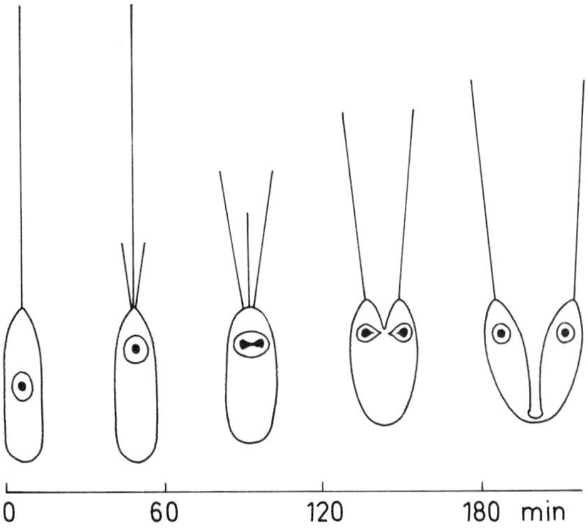

Fig. 4.20. Flagellar development in *Peranema trichophorum* at mitosis. The "old" flagella is gradually retracted and resorbed while two new flagella grow during cell division (redrawn from Tamm [169])

In a careful study of the flagellar cycle in the flagellate *Peranema trichophorum,* which has one long flagellum, this is seen to shorten at the onset of cell division, while two new flagella start growing, and will reach their normal size at the end of division. While the growth obeys a complex logarithmic pattern, the retraction of the "old flagellum" takes place in less then one hour. The two processes—shortening and lengthening—appear as fundamentally different. It is probable that the old flagellar material (tubulin) is re-used, although experimental removal of the flagella in a predivision cell was not found to slow down the growth of the new flagellae [169]. An important conclusion of this work is that differential controls may exist in the same cell, even in closely related regions: this has already been noted when describing the assembly of tubulins in the cilia formation in *Naegleria* [59].

In protozoa, different techniques have been devised which lead to a rapid deciliation: in *Stentor coeruleus,* for instance, various agents such as urea, sucrose or salt water lead to a shedding of thousands of cilia within less than one minute [107]. This is followed by a complete regeneration from the persisting basal bodies and also from newly formed ones, with a reconstitution of the complex ciliary patterns within 8 h. This regeneration is inhibited, reversibly, by colchicine [116], griseofulvin, VLB, podophylloxin, and other MT poisons [106], while metabolic inhibitors such as dinitrophenol, actinomycin D, cycloheximide, mitomycin, and hydroxyurea are without effect, indicating that no new synthesis of proteins is necessary.

In other species, cycloheximide may inhibit regeneration, as shown in *Chlamydomonas* and *Tetrahymena* [140, 141]: the assembly of the ciliary tubulins took place at the extremity of the cilium, as shown by radioautography (cf. Chap. 3). In *Tetrahymena*, regeneration takes place after deciliation by EDTA; it is prevented by colchicine, which does not affect protein synthesis, and also by cycloheximide, indicating that for cells in the logarithmic phase of growth there is no pool of tubulins and that these must be synthesized shortly before their assembly [134]. In contrast, in the non-growing condition, ciliary regeneration does not imply any protein synthesis, and proceeds from a tubulin pool [115]. However, in starved cells, cycloheximide does inhibit regeneration: it seems that the synthesis of new molecules of tubulin is initiated by the deciliation [75]. Regeneration, as observed by scanning electron microscopy, is rapid and completed in 90 min [175].

In *Chlamydomonas*, the action of cycloheximide on flagellar regeneration is complex, for although the cell is known to contain a sufficient pool of tubulin for the regeneration of its two flagellae, this was decreased when cycloheximide was present during the first 60 min after deflagellation. Synthesis of some morphogenetic protein different from tubulin may be the limiting step [48].

Slightly different results were obtained, always on *Tetrahymena*, in cells deprived of amino acids. The regeneration of the oral apparatus was studied; the destruction of the cilia took place without any evidence of autophagy, indicating that material from resorbed cilia may be used anew. The stages of ciliary regression were observed, and shown to follow in the opposite order the steps of assembly: loss of the external wall of tubule B, then of tubule A. This is similar to the results obtained by solubilization by detergents (cf. also the sterile mutant of *Drosophila* described in 3.7).

No synchronism in the destruction of the nine doublets was observed [191]. Regeneration of the oral apparatus could take place with no more than 6% of the tubulins being synthesized, and this regeneration, which was completed in 2 h, utilized tubulin from a cytoplasmic pool. However, synthesis of other proteins which may play a part in ciliary structures or in assembly, may be necessary [192]. In this species, it appears that the control of ciliary resorption plays a great role in the morphogenesis of the cell. It is also apparent from labeling experiments [115] that the tubulin subunits of the cilia are in a condition of dynamic exchange, subunits appearing to be replaced without the MT structures being altered. This fact is important for any study of the factors which confer on some MT, such as those of basal bodies and centrioles, a great stability, as compared to the cytoplasmic and mitotic MT [190].

4.3.3.3 Experimental Ciliogenesis. From the experiments mentioned above, cilia appear to be in a dynamic state—quite apart from their movements, which will be studied in Chapter 7. Their growth is rapid, as shown in *Naegleria*, where at 28 °C a cilium comprizing about 1,500,000 polypeptides may be assembled in 30 min, from a pool of tubulin different from that present in the ameba stage [59].

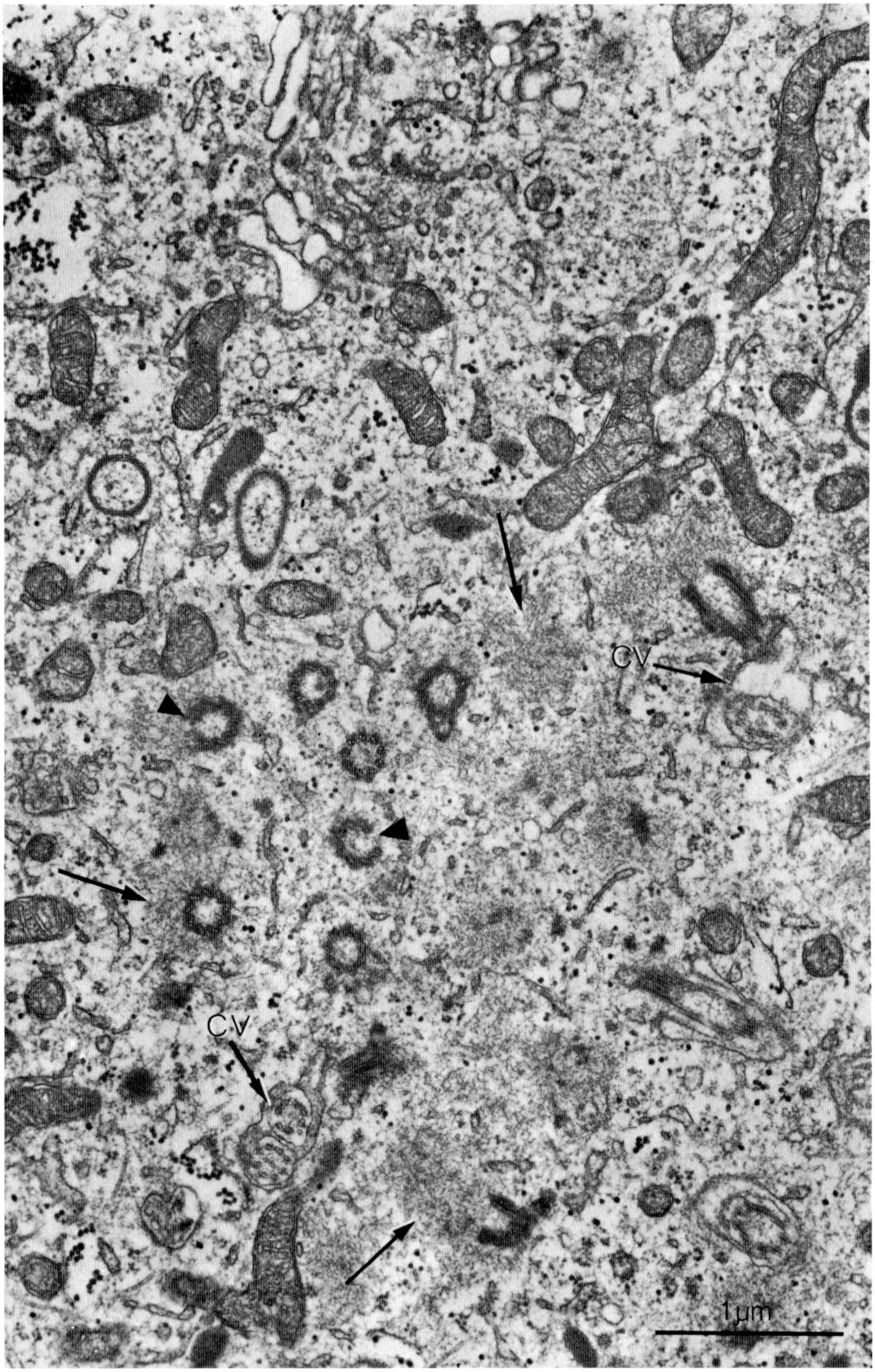
CV
CV
1 µm

Cilia play an important role in the genital tract, and in Chapter 2 references on the hormonal control of ciliogenesis have been mentioned. The balance between ciliated and non-ciliated cells is influenced by estrogens, which increase the number of ciliated cells: this was discussed above in relation to centriole and basal body multiplication and growth [9, 10, 25].

Several observations, apparently without any link between the techniques used, indicate that in experimental conditions cells which normally have no cilia or only one cilium ("oligocilia" [64]) may develop many new centrioles, basal bodies, and cilia. Ciliogenesis appears to be a general property of cells, although it is often restricted by controls which are as yet unknown. Chinese hamster fibroblasts, treated in vitro by the colchicine derivative, colcemid (0.06 μg/ml), developed after 2 h basal bodies with ciliary vesicles, and small cilia were observed. When these cells were returned to a normal medium, the growth of the cilia was resumed, and more than 25% of the centrioles (as opposed to 4% in normal cells) displayed ciliogenetic activity. The authors suggested that colcemid had induced "a specific... differentiation" [166]. Pargylin, a monoamine-oxidase inhibitor, leads to an active ciliogenesis in various cells, mainly glial, of the cat's brain [111]. Many cilia, of a normal 9+2 structure, are seen to arise from normal basal bodies in some regions, in particular the habenula and the mamillary bodies. These results have been confirmed [171].

Pituicytes, specialized glial cells of the posterior pituitary, which normally have only two centrioles and often a single cilium, may develop many new centrioles and cilia [87]. The mechanism and causes of this change—which can take place in rats chronically intoxicated by the tubulin poison VCR—remain obscure, although a possible link with pituitary activity cannot be ruled out at this moment [52].

4.3.4 Atypical and Pathological Cilia

It is not surprizing that the complex structure of the basal body and cilium may fail to be built correctly. However, considering the tremendous numbers of cilia which are built in so many species, the abnormalities are so rare that at one time it was supposed that a mutation affecting the structure of the cilia would always be lethal. This is not the case, and much effort is directed now toward the study of such mutants, for they may throw light on the problems of ciliary assembly and physiology. The cilia devoid of dynein arms recently described in man [2, 3] will be studied in Chapter 11.

Already in 1964, a mutant of *Chlamydomonas reinhardii* with immotile flagella devoid of the central pair of MT was described [133], and it was suggested that ciliary motility and bending needed these two central MT. The description of cilia with a 6+0 structure in the male gametes of the gregarin *Lecudina tuzetae* [151] is of par-

Fig. 4.21. Rat pituitary, posterior lobe. A pituicyte showing an active formation of basal bodies, leading to that of cilia. The animal had received repeated intraperitoneal injections of VCR (total dose 0.85mg/kg body weight in two weeks). Most basal bodies, showing typical triplets, are associated with dense zones of the cytoplasm (*arrows*). Some bodies appear incomplete, C-shaped (*arrow heads*). The cilia are growing into ciliary vesicles (*CV*) (unpublished, from [52])

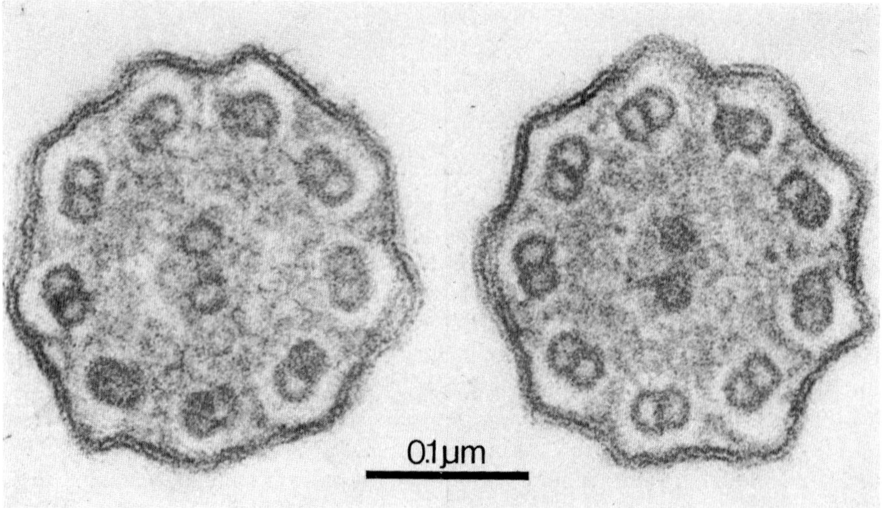

Fig. 4.22. Cross-sectioned cilia from the respiratory epithelium of patient: most of the cilium is occupied by a dense matrix, and the majority of doublets are devoid of dynein arms, and appear surrounded by a clear space. The cilia and the spermatozoa were immotile (original document, by courtesy of Afzelius)

ticular interest from this point of view. In these cells, the very long flagellum extends from one side of the cell to the other, which it crosses before protruding as a motile structure. The abnormal flagellae had six doublets, apparently with dynein arms, and linked by relatively long peripheral filaments, without any central pair of MT. A normal cart-wheel was present in the basal body, centered by a tubular structure; it showed more or less a sixfold symmetry. The flagellum was motile, although much slower than normally, with a cycle time of about 2 s. This indicates that links

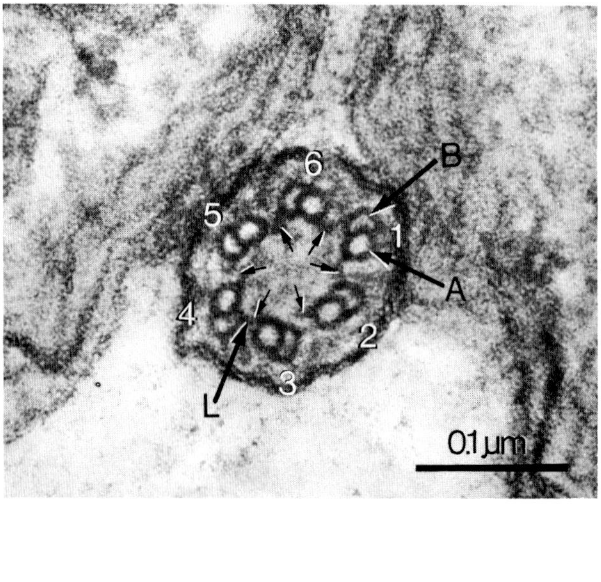

Fig. 4.23. Flagellar axoneme of 6+0 pattern in the male gamete of the gregarin *Lecudina tuzetae*. The six doublets have a normal structure. *Arrows* indicate the location of lateral arms (dynein?). A special link is visible (*L*) between doublets 3 and 4 (from Schrével and Besse [151])

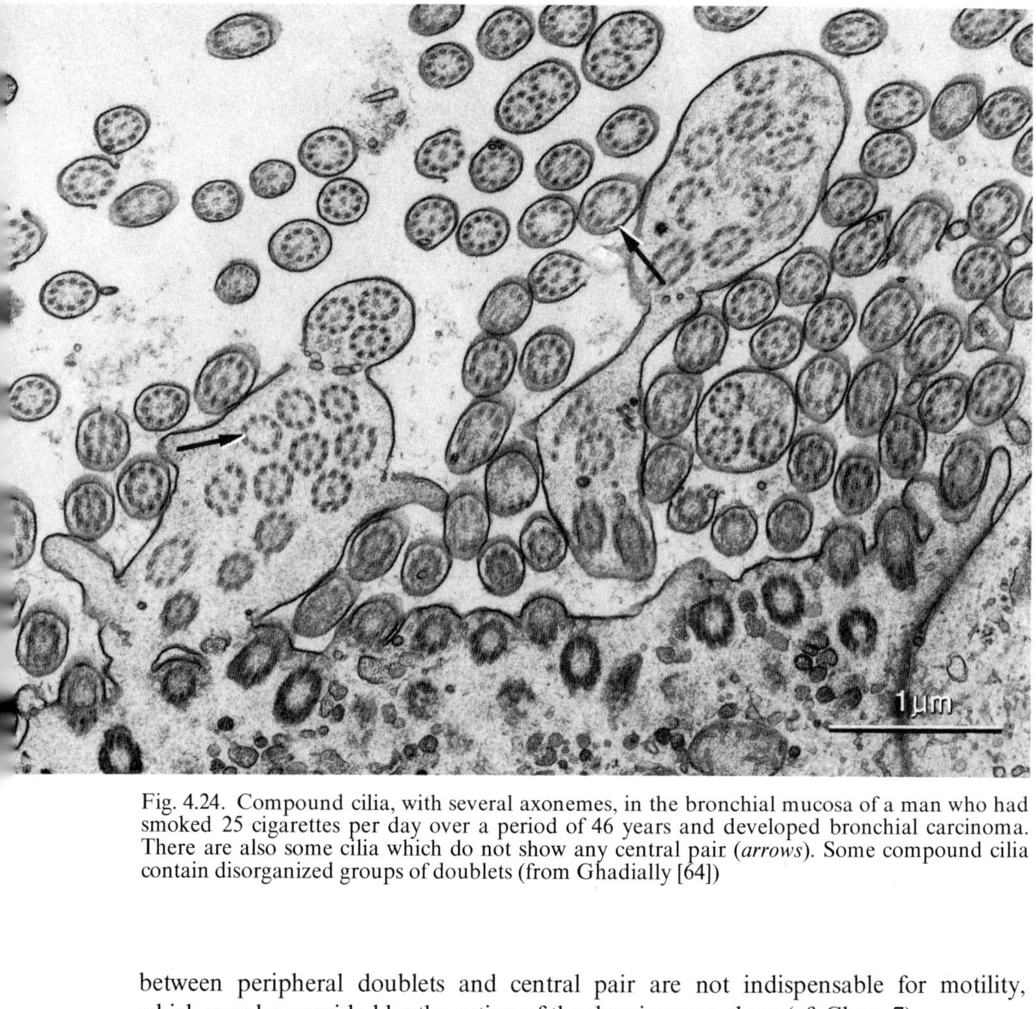

Fig. 4.24. Compound cilia, with several axonemes, in the bronchial mucosa of a man who had smoked 25 cigarettes per day over a period of 46 years and developed bronchial carcinoma. There are also some cilia which do not show any central pair (*arrows*). Some compound cilia contain disorganized groups of doublets (from Ghadially [64])

between peripheral doublets and central pair are not indispensable for motility, which may be provided by the action of the dynein arms alone (cf. Chap. 7).

Another mutant is that observed in *Chlamydomonas reinhardii,* where in the strain *bald-2* the basal body is unable to form triplet MT. These cells are nearly devoid of cilia, and the few basal bodies observed show a ring of nine singlet MT. A cart-wheel, although rather indistinct, is present, and the MT are linked to each other by fine fibers. Short and atypical cilia may arise from these structures, but they never show more than nine peripheral singlet MT. These cilia seem to be unstable structures. Other MT in the cell are quite normal, and the *bald-2* mutation apparently only affects the formation of doublet and triplet MT, suggesting strongly that other factors than tubulin are necessary. It should be remembered that when doublet or triplet structures are incubated in the presence of tubulin, single MT have been demonstrated to grow from MT A, but never doublets (cf. Chap. 2). Another observation on this strain is that rootlets are present near the basal body, but are never attached to it, indicating that this attachment needs the presence of tubules B and C.

No doubt similar findings may help to understand the relations between the different ciliary structures, and another mutant has been mentioned [196] which has no radial spokes and whose flagella are paralyzed. Many other peculiar flagellar structures have also been described in spermatozoa. For instance, the central pair of MT may be replaced, in the spermatozoa of plathelminths, by a central core with a complex structure, more or less helical, which seems quite different from the sheath surrounding the central pair of MT in the 9+2 configuration [174].

In pathological conditions, abnormal cilia may be found: two principal changes have been described (cf. [64]). In the nasal mucosa of piglets infected with *Bordetella bronchiseptica* swollen cilia, with less than the normal complement of MT and a large cytoplasm, are present; the cytoplasmic bulge may contain circular arrangements of MT. These changes lead to the destruction of the cilia [46]. Similar changes have been found in the bronchial mucosa of man, besides 9+0 cilia, and so-called compound cilia, that is to say groups of up to 20 axonemes with a normal complement of 9+2 MT each enveloped in a single membrane [4]. Cilia with a "swollen" cytoplasm have also been observed in the hamster respiratory tract under the influence of various carcinogenic compounds. All these changes thus appear to affect the ciliary cytoplasm more than the MT themselves. The striking ciliary lesions observed in the gill of a freshwater mussel under the influence of serum from patients affected with cystic fibrosis (mucoviscidosis), with reversible destruction of the ciliary MT, deserve further investigation [1].

4.4 Other Complex Associations of MT: Axopodia and Axostyles

In protists, many complex associations of MT are found: some of these have been described in relation to the patterned MT groupings in Chapter 3, other will be studied in relation to MT poisons (Chap. 5) and cell movements (Chap. 7). These structures are particularly visible in radiolaria and heliozoa, where they form long extensions radiating from the center of the cell, and playing an active role in the capture of nutrients. They are usually associated with some central body or axoplast (centroplast) from which they radiate. In other cells, they are attached to the nuclear membrane. From several points of view, these structures resemble centrioles and cilia although they differ from these by their much greater lability, either under the influence of MT poisons or during the normal cellular activity or the cell cycle. However, let us not forget that in ciliates the ciliary MT may be resorbed and reutilized after disassembly.

The terms "axostyles" and "axopodia" will be used here, indicating that these are long straight extensions of the cell wall, the second term suggesting a fact which is apparent in some species, that the axopodium plays a role in cell locomotion. Axostyles may also be put to use for the differentiation of a stalk, as in the pseudoheliozoan *Clathrulina elegans* [16]. Other complex MT structures such as the cytopharyngeal baskets of ciliates like *Nassula* [180] will not be considered here. The term "axoneme" should be restricted to the axial groupings of MT; it is also used to describe the MT assembly in cilia. "Axostyle" is preferable, to avoid any confusion with ciliary structures.

4.4.1 Axoplasts

All axopodia appear to arise either from the nuclear membrane or from an axoplast [85]. This is a dense structure, showing no differentiation into MT, located either at the central part of the cell, or close to the nucleus and somewhat excentrical [27]. From this structure, groups of MT linked together in complex patterns are seen to radiate, although all MT are not necessarily attached to the axoplast. In the heliozoan *Raphidiophrys* the axonemal MT have a most precise pattern (cf. Chap. 3): they radiate from a centroplast which has no definite fine structure, showing an inner sphere of dense material surrounded by an outer cottony material from which the MT, already neatly patterned, radiate in all directions [177]. In the radiolaria *Cystidosphaeria reticulata,* the axoplast is deeply embedded in the nucleus. Several groups of MT radiate from this amorphous mass, often crossing the nucleus, which is perforated by fine canaliculi [27, 28]. In the heliozoan *Gymnosphaera albida,* where the MT of the axopodia are linked in an hexagonal network (cf. Chap. 3), the axoplast is central and spherical. The axonemes arise from this structure, enveloped at their base by an electron-dense fibrillar material [90]. A more complex axoplast is described in the centroheliozoan *Heterophrys marina.* This "centroplast" is made of a central sphere, comprizing a disk of electron-dense material sandwiched between two half-spheres of granular substance. This is surrounded by a narrow clear zone, and around this are located the groups of axonemes with their radiating MT [17]. This complex structure thus appears as a true MTOC, and the groupings of the basal structures of the radiating MT resemble the multiple basal bodies grouped around the central spherical "blepharoplast" of the fern *Marsilea* [112].

4.4.2 Relation with the Nuclear Membrane

When axopodia are related to the nuclear membrane, very complex differentiations may be found, as described in the radiolaria *Sticholonche zanclea* Hertwig: the parallel MT are fixed to a dense body, maintained in position by fine fibrils and capable of sliding, like a joint, within a densification of the nuclear membrane. This species moves by the synchronous displacement of numerous axopodia, which have been compared to oars ([29, 85]; cf. Chap. 7). However, these rigid structures may rapidly dissociate into wavy fibrils, by a complete disassembly of their MT [85]. In the heliozoan *Actinosphaerium (Echinosphaerium) nucleofilum,* the axonemes are also attached to the nuclear envelope, although all their MT do not extend as far as the nucleus. The complex patterns described in this species (p. 106) appear to result from "self-assembly" regulated by a definite set of rules [142], governed however by the geometry of the central parallelogram (cf. Chap. 3). The pseudoheliozoan *Clathrulina elegans* has a particularly interesting cell cycle from the point of view of MT: it may exist as a motile cell with two flagellae, which becomes a heliozoan with a complex skeleton, growing a stalk made of many MT, which matures into a hollow structure. While the flagellated form has only a few MT outside its cilia, the fixed form has about 1000 MT which form a rigid rod extending from the outer nuclear membrane. In this species, the nuclear membrane acts as a nucleation site [16]. Similar relations of axopodia with the nucleus have been observed in the heliozoan *Actinophrys sol* [118].

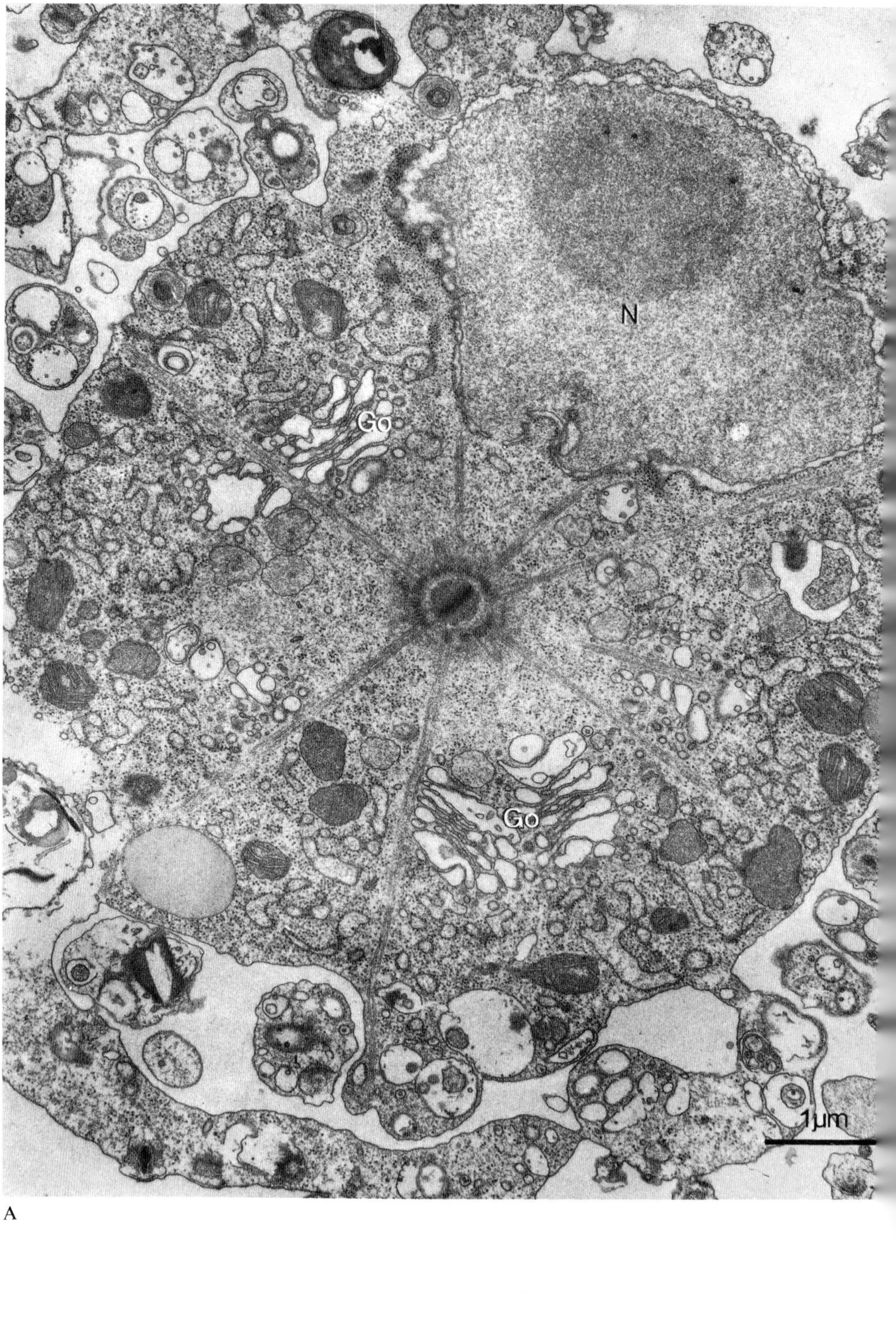

A

B

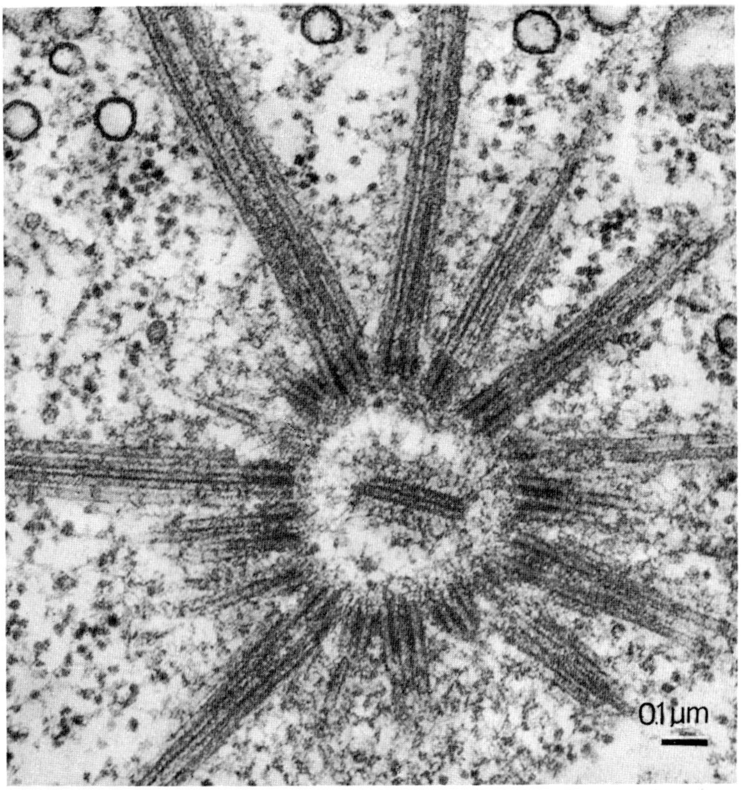

Fig. 4.25 A, B. (A) *Heterophrys marina* (centrohelidian). Cross-section passing through the centroplast. The axonemes radiate from the cortex of the centroplast. *N*, nucleus. *Go*, Golgi apparatus (dictyosomes). (B) *Heterophrys marina*. The axonemes radiate around the centroplast. They originate, in groups of six, in the dense cortex of the centroplast; other MT arise outside this region (cf. Chap. 3) (from Bardele [17])

4.4.3 Some Properties of Axopodia

Axopodia are not only interesting for the geometrical patterns of their MT; they also display a remarkable plasticity similar to that of some flagella and to the mitotic apparatus. They are excellent tools for studies of MT poisons and assembly and disassembly mechanisms. They show a group of properties which are found in most studies of MT organelles: they evidently play a mechanical role, in motility and in the capture of preys, while displaying a considerable plasticity, as evidenced by their rapid disassembly in various conditions.

a) *Growth, assembly, and disassembly.* The axopodia of *Actinosphaerium nucleofilum,* like the MT of the mitotic apparatus, are destroyed by exposure to a temperature of 4 °C. When the cells are replaced at 20 °C, the axonemes are reformed almost immediately [178]. In *Heterophrys marina,* a centrohelidian heliozoa, the axopodia may retract very rapidly (from 2 to 5 µm in about 20 msec). This results from a complete disassembly of their subunits; assembly takes longer, at a rate of about

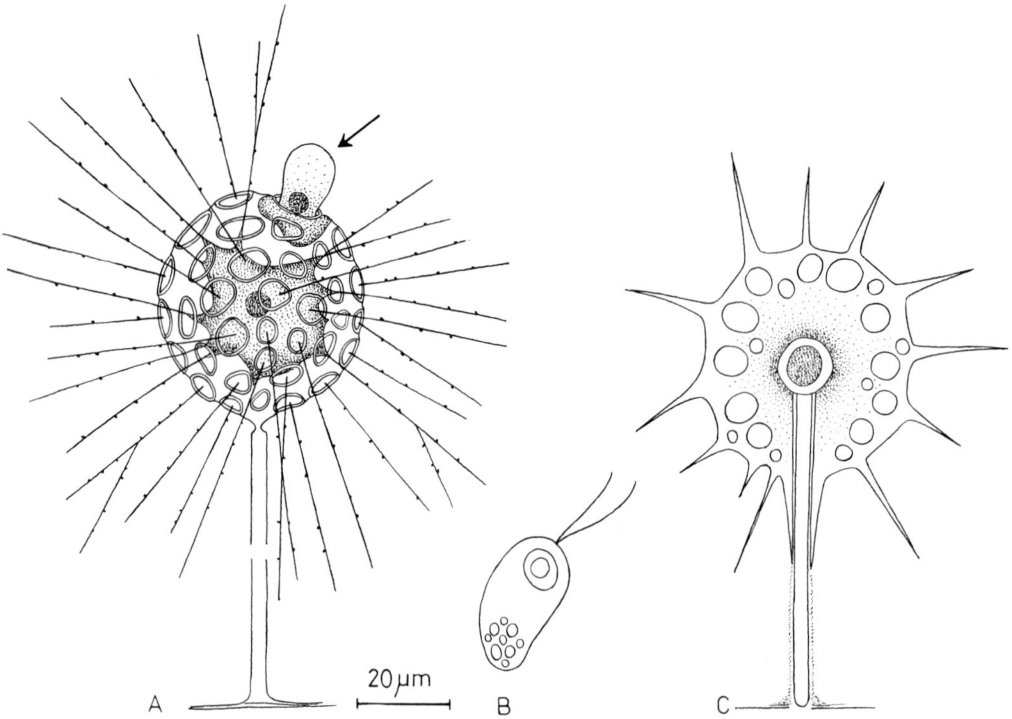

Fig. 4.26 A–C. Cell cycle of *Clathrulina elegans.* (A) Encapsulated heliozoan-like stage, which reproduces by the formation of ameboid cells (*arrow*) which develop two flagella. (B) Flagellated swarmer. (C) The ameboid or the flagellated form changes into a naked heliozoan that grows one large pseudopod which becomes the stalk, and later secretes its capsule. MT are involved in the structure of the axonemes in (A) and (C), the formation of cilia in (B), and the stalk in (C) (redrawn from Bardele [16], simplified)

15 µm/min [35]. In *Actinophrys sol,* the axopodia, when the cell captures small flagellates, shorten at the rate of 100 µm in less than 1 sec. They are disassembled from the tips toward the base. After destruction by cold, they grow at a rate of 100 µm/min. Colchicine shortens them at a speed of 50 µm/min: this is always the result of a disassembly, never of a true contraction [118]. The axopodia of *Actinophrys sol,* when their distal end is irradiated by UV light, retract at a rate of about 15 µm/min [117]. Their regrowth is more rapid than after amputation, indicating the probable reutilization of tubulins. These results should be compared to the spindle changes after UV irradiation (cf. Chap. 10).

In this respect, the MT of axopodia resemble much more those of the mitotic apparatus ("labile MT") than those of cilia.

b) *Cell movement.* The shortening of the axopodial MT exerts on the surrounding structure a pulling force which has been estimated to be as great as 10^{-5} dyne, that is to say an instantaneous force greater than that of muscle [35]. Like cilia, on the other hand, axopodia may bend: in *Saccinobaculus* a contractile wave can be seen to propagate from one end to the other of the axopodium: a sliding movement of some MT in relation to others appears partially to explain this contraction [102], which would resemble the "sliding filament" models of ciliary movement (Chap. 7) and mitosis (Chap. 10). In *Pyrsonympha vertens,* the axopodia may be animated by rhythmic beatings which have been compared to ciliary movements, although here each axopodium is made of about 2,000 MT [95]. The role of axopodia in the motility (crawling movements) of the radiolaria *Sticholonche zanclea* has already been mentioned [29, 85].

Other resemblances to ciliary movements are indicated by the activation of isolated axopodia of the flagellate *Cryptocercus punctulatus* by ATP, and by the demonstration that several proteins other than tubulin are present in these structures, in particular a high molecular weight component comparable to dynein, which may be located in the regularly spaced "bridges" between the MT [113].

These questions will be discussed again in relation to cell motility. According to McIntosh [102], the axostyle of *Saccinobaculus* "may represent one of the simplest biological engines composed of MT". Axonemes are probably far from being so simple, although they certainly represent one of the most promising structures for the study of relations between MT and movement.

References

1. Adshead, P. C., Martinez, J. R., Kilburn, K. H., Hess, R. A.: Ciliary inhibition and axonemal microtubule alterations in freshwater mussels. Ann. N. Y. Acad. Sci. **253,** 192 – 212 (1975)
2. Afzelius, B. B.: Electron microscopy of the sperm tail. Results obtained with a new fixative. J. Biophys. Biochem. Cytol. **5,** 269 – 278 (1959)
3. Afzelius, B. A., Eliasson, R., Johnsen, Ø., Lindholmer, C.: Lack of dynein arms in immotile human spermatozoa. J. Cell Biol. **66,** 225 – 232 (1975)
4. Ailsby, R. L., Ghadially, F. N.: Atypical cilia in human bronchial mucosa. J. Path. **109,** 75 – 78 (1973)
5. Allen, R. D.: A reinvestigation of cross-sections of cilia. J. Cell Biol. **37,** 825 – 831 (1968)
6. Amos, L. A., Klug, A.: Arrangement of subunits in flagellar microtubules. J. Cell Sci. **14,** 523 – 550 (1974)
7. Amos, L. A., Linck, R. W., Klug, A.: Molecular structure of flagellar microtubules. In: Cell motility (eds.: R. Goldman, T. Pollard, J. Rosenbaum), pp. 847 – 868. Cold Spring Harbor Lab. 1976
8. Anderson, R. G. W.: The three-dimensional structure of the basal body from the rhesus monkey oviduct. J. Cell Biol. **54,** 246 – 265 (1972)
9. Anderson, R. G. W., Brenner, R. M.: The formation of basal bodies (centrioles) in the rhesus monkey oviduct. J. Cell Biol. **50,** 10 – 34 (1971)
10. Anderson, R. G. W., Brenner, R. M.: Estrogen-stimulated oviduct of the Rhesus monkey in organ culture. In: Electron Microscopic Concepts of Secretion (ed.: M. Hess), pp. 85 – 97. New York: John Wiley and Sons 1975
11. André, J., Thiery, J.-P.: Mise en évidence d'une sous-structure fibrillaire dans les filaments axonématiques des flagellés. J. Micros. **2,** 71 – 80 (1973)
12. Auclair, W., Siegle, B. W.: Cilia regeneration in the sea urchin embryo: evidence for a pool of ciliary proteins. Science **154,** 913 – 915 (1966)
13. Baccetti, B., Dallai, R.: The spermatozoon of Arthropoda. XXVII. Uncommon axoneme patterns in different species of the Cecidomyid flies. J. Ultrastruct. Res. **55,** 50 – 69 (1976)
14. Baccetti, B., Dallai, R., Giusti, F., Bernini, F.: The spermatozoon of arthropoda. XXV. A new model of tail having up to 170 doublets: *Monarthropalpus Buxi.* Tissue and Cell **6,** 269 – 278 (1974)
15. Barber, V. C.: Cilia in sense organs. In: Cilia and Flagella (ed.: M. A. Sleigh), pp. 403 – 436. London, New York: Acad. Press 1974
16. Bardele, C. F.: Cell cycle, morphogenesis, and ultrastructure in the pseudoheliozoan *Clathrulina elegans.* Z. Zellforsch. **130,** 219 – 242 (1972)
17. Bardele, C. F.: The fine structure of the centrohelidian heliozoan *Heterophrys marina.* Cell Tissue Res. **161,** 85 – 102 (1975)

18. Bernhard, W., de Harven, E.: L'ultrastructure du centriole et d'autres éléments de l'appareil achromatique. Verhandl. 4. int. Kongr. Elektronmikroskopie, pp. 217 – 227. Berlin: Springer 1960
19. Berns, M. W., Rattner, J. B.: Irradiation of the centriolar region in mitotic potorous cells with a Laser microbeam. J. Cell Biol. **67,** 30 a (1975)
20. Bessis, M., Breton-Gorius, J., Thiery, J. P.: Centriole, corps de Golgi et aster des leucocytes. Etude au microscope électronique. Rev. Hémat. **13,** 363 – 386 (1958)
21. Bessis, M., Breton-Gorius, J.: Rapports entre noyau et centrioles dans les granulocytes étalés. Rôle des microtubules. Nouv. Rev. Fr. Hématol. **7,** 601 – 620 (1967)
22. Bibring, T., Baxandall, J., Denslow, S., Walker, B.: Heterogeneity of the alpha subunit of tubulin and the variability of tubulin within a single organism. J. Cell Biol. **69,** 301 – 312 (1976)
23. Binder, L. I., Dentler, W. L., Rosenbaum, J. L.: Assembly of chick brain onto flagellar microtubules from *Chlamydomonas* and sea urchin sperm. Proc. Natl. Acad. Sci. USA **72,** 1122 – 1126 (1975)
24. Borisy, G. G., Marcum, J. M., Olmsted, J. B., Murphy, D. B., Johnson, K. A.: Purification of tubulin and associated high molecular weight proteins from porcine brain and characterization of microtubule assembly in vitro. Ann. N. Y. Acad. Sci. **253,** 107 – 132 (1975)
25. Brenner, R. M.: Hormonal control of cilia renewal in the primate oviduct: ultrastructural studies. Progr. Gynecol. **6,** 77 (1970)
26. Buckland-Niks, J. A.: The fine structure of the spermatozoon of *Littorina* (Gastropoda Prosobranchia) with special reference to sperm motility. Z. Zellforsch. **144,** 11 – 29 (1973)
27. Cachon, J., Cachon, M.: Le système axopodial des Radiolaires Sphaerodiés. II. Les Périaxoplastidiés. III. Les Cryptoaxoplastidiés (Anaxoplastidiés). IV. Les fusules et le système rhéoplasmique. Arch. Protistenk. **114,** 291 – 307 (1972)
28. Cachon, J., Cachon, M.: Les systèmes axopodiaux. Année Biol. **13,** 523 – 560 (1974)
29. Cachon, J., Cachon, M., Tilney, L. G., Tilney, M. S.: Movement generated by interactions between the dense material at the ends of microtubules and non-actin-containing microfilaments in *Sticholonche zanclea.* J. Cell Biol. **72,** 314 – 338 (1977)
30. Cavalier-Smith, T.: Organelle development in *Chlamydomonas reinhardii.* Ph. D. Thesis, University of London 1967
31. Cavalier-Smith, T.: Basal body and flagellar development during the vegetative cell cycle and the sexual cycle of *Chlamydomonas reinhardii.* J. Cell Sci. **16,** 529 – 556 (1974)
32. Chasey, D.: Observations on the central pair of microtubules from the cilia of *Tetrahymena pyriformis.* J. Cell Sci. **5,** 453 – 458 (1969)
33. Child, F. M.: The characterization of the cilia of *Tetrahymena pyriformis.* Exp. Cell Res. **18,** 258 – 267 (1959)
34. Dalcq, A.: Le centrosome. Bull. Acad. Belg. Cl. Sci. 5e série **50,** 1408 – 1449 (1964)
35. Davidson, L.: Contractile axopodia of the centrohelidan heliozoan *Heterophrys marina.* J. Cell Biol. **59,** 71 a (1973)
36. Defoor, P. H., Stubblefield, E.: Effects of actinomycin D, amethopterin, and 5-fluoro-2′-deoxyuridine on procentriole formation in Chinese hamster fibroblasts in culture. Exp. Cell Res. **85,** 136 – 142 (1974)
37. de Harven, E.: The centriole and the mitotic spindle. In: Ultrastructure in Biological Systems (eds.: A. J. Dalton, F. Haguenau), Vol. 3, pp. 197 – 227. New York, London: Acad. Press 1968
38. de Harven, E., Bernhard, W.: Etude au microscope électronique de l'ultrastructure du centriole chez les Vertébrés. Z. Zellforsch. **45,** 378 – 498 (1956)
39. de Thé, W.: Cytoplasmic microtubules in different animal cells. J. Cell Biol. **23,** 265 – 275 (1974)
40. Dingle, A. D.: Control of flagellum number in *Naegleria.* Temperature-shock induction of multiflagellate cells. J. Cell Sci. **7,** 463 – 481 (1970)
41. Dingle, A. D., Fulton, C.: Development of the flagellar apparatus of *Naegleria.* J. Cell Biol. **31,** 43 – 54 (1966)
42. Dippell, R. V.: Effects of nuclease and protease digestion on the ultrastructure of *Paramecium* basal bodies. J. Cell Biol. **69,** 622 – 637 (1976)
43. Dirksen, E. R.: The presence of centrioles in artificially activated sea urchin eggs. J. Biophys. Biochem. Cytol. **11,** 244 – 247 (1961)
44. Dirksen, E. R.: Centriole morphogenesis in developing ciliated epithelium of the mouse oviduct. J. Cell Biol. **51,** 286 – 302 (1971)
45. Dirksen, E. R., Staprans, I.: Tubulin synthesis during ciliogenesis in the mouse oviduct. Developm. Biol. **46,** 1 – 13 (1975)

46. Duncan, J. R., Ramsey, F. K.: Fine structural changes in the porcine nasal ciliated epithelial cell produced by *Bordetella bronchiseptica* rhinitis. Am. J. Path. **47,** 601 – 612 (1965)
47. Everhart, L. P. Jr.: Heterogeneity of microtubule proteins from *Tetrahymena* cilia. J. Mol. Biol. **61,** 745 – 748 (1971)
48. Farrell, K. W.: Flagellar regeneration in *Chlamydomonas reinhardtii:* evidence that cycloheximide pulses induce a delay in morphogenesis. J. Cell Sci. **20,** 639 – 654 (1976)
49. Faure-Fremiet, E.: Cils vibratiles et flagelles. Biol. Rev. **36,** 464 – 536 (1961)
50. Fawcett, D. W.: Cilia and flagella. In: The Cell (eds.: J. Brachet, A. E. Mirsky), Vol. 2, pp. 217 – 298. New York, London: Acad. Press 1961
51. Fawcett, D. W., Porter, K. R.: A study of the fine structure of ciliated epithelia. J. Morphol. **94,** 221 – 282 (1954)
52. Flament-Durand, J., Hubert, J. P., Dustin, P.: Centriolo- and ciliogenesis in the rat's pituicytes under the influence of microtubule poisons in vitro. Exp. Cell Res. **99,** 435 – 437 (1976)
53. Flock, A., Duvall, A. J.: The ultrastructure of the kinocilium of the sensory cells in the inner ear and lateral line organs. J. Cell Biol. **25,** 1 – 8 (1965)
54. Friedländer, M., Wahrman, J.: Giant centrioles in neuropteran meiosis. J. Cell Sci. **1,** 129 – 144 (1966)
55. Fujiwara, K., Tilney, L. G.: Substructural analysis of the microtubule and its polymorphic forms. Ann. N. Y. Acad. Sci. **253,** 27 – 50 (1975)
56. Fulton, C.: Amebo-flagellates as research partners. The laboratory biology of *Naegleria* and *Tetramitus.* In: Methods in Cell Physiology (ed.: D. M. Prescott), Vol. 4, pp. 341 – 476. New York: Acad. Press 1970
57. Fulton, C.: Centrioles. In: Origin and Continuity of Cell Organelles (eds.: J. Reinert, H. Ursprung) (Results and Problems in Cell Differentiation, Vol. 2), pp. 170 – 221. Berlin, Heidelberg, New York: Springer 1971
58. Fulton, C., Dingle, A. D.: Basal bodies, but not centrioles, in *Naegleria.* J. Cell Biol. **51,** 826 – 836 (1971)
59. Fulton, C., Kowit, J. D.: Programmed synthesis of flagellar tubulin during cell differentiation in *Naegleria.* Ann. N. Y. Acad. Sci. **253,** 318 – 322 (1975)
60. Fulton, C., Simpson, P. A.: Selective synthesis and utilization of flagellar tubulin. The multi-tubulin hypothesis. In: Cell Motility (eds.: R. Goldman, T. Pollard, J. Rosenbaum), pp. 987 – 1006. Cold Spring Harbor Lab. 1976
61. Gaffal, K.-P., Bassemir, U.: Vergleichende Untersuchung modifizierter Cilienstrukturen in den Dendriten mechano- und chemosensitiver Rezeptorzellen der Baumwollwanze *Dysdercus* und der Libelle *Agrion.* Protoplasma **82,** 177 – 202 (1974)
62. Gall, J. G.: Centriole replication. A study of spermatogenesis in the snail *Viviparus.* J. Biophys. Biochem. Cytol. **10,** 163 – 193 (1961)
63. Gaskin, F., Kramer, S. B., Cantor, C. R., Adelstein, R., Shelanski, M. L.: A dynein-like protein associated with neurotubules. FEBS Lett. **40,** 281 – 286 (1974)
64. Ghadially, F. N.: Ultrastructural Pathology of the Cell. A text and atlas of physiological and pathological alterations in cell fine structure. London-Boston: Butterworths 1975
65. Gibbons, I. R.: The relationship between the fine structure and direction of beat in the gill cilia of a lamellibranch mollusk. J. Biophys. Biochem. Cytol. **11,** 179 – 205 (1971)
66. Gibbons, I. R.: Chemical dissection of cilia. Arch. Biol. (Liège) **76,** 317 – 352 (1965)
67. Gibbons, I. R., Grimstone, A. V.: On flagellar structure in certain flagellates. J. Biophys. Biochem. Cytol. **7,** 697 – 716 (1960)
68. Gibbons, I. R., Rowe, A. J.: Dynein: a protein with adenosine triphosphatase activity from cilia. Science **149,** 424 – 426 (1965)
69. Gilula, N. B., Satir, P.: The ciliary necklace. A ciliary membrane specialization. J. Cell Biol. **53,** 494 – 509 (1972)
70. Gould, R. R.: The basal bodies of *Chlamydomonas reinhardii.* Formation from probasal bodies, isolation, and partial characterization. J. Cell Biol. **65,** 65 – 74 (1975)
71. Grain, J., de Puytorac, P.: Particularités ultrastructurales des cinétosomes et de leurs annexes fibrillaires chez quelques Ciliés Astomes *Hoplitophryidae.* J. Microsc. **19,** 231 – 246 (1974)
72. Grassé, P. P.: L'ultrastructure de *Pyrsonympha vertens* (Zooflagellata, Pyrsonymphina): les flagelles et leur coaptation avec le corps, l'axostyle contractile, le paraxostyle, le cytoplasme. Arch. Biol. **67,** 595 – 609 (1956)
73. Grimstone, A. V., Klug, A.: Observations on the substructure of flagellar fibres. J. Cell Sci. **1,** 351 – 362 (1966)
74. Gross, P. R., Spindel, W.: Heavy water inhibition of cell division: an approach to mechanism. Ann. N. Y. Acad. Sci. **90,** 500 – 522 (1960)

75. Guttman, S. D., Gorovsky, M. A.: Cilia regeneration in starved *Tetrahymena.* J. Cell Biol. **67,** 149 a (1975)
76. Hartman, H.: The centriole and the cell. J. Theoret. Biol. **51,** 501 – 510 (1975)
77. Hartman, H., Puma, J. P., Gurney, T.: Evidence for the association of RNA with the ciliary basal bodies of *Tetrahymena.* J. Cell Sci. **16,** 241 – 260 (1974)
78. Hausmann, K., Hinsen, H.: Zur Feinstruktur der Cilientubuli von *Paramecium caudatum* nach Negativ-staining. J. Microsc. **15,** 107 – 110 (1972)
79. Heath, I. B.: Effect of antimicrotubule agents on growth and ultrastructure of Fungus *Saprolegnia ferax* and their ineffectiveness in disrupting hyphal microtubules. Protoplasma **85,** 147 – 176 (1975)
80. Heath, I. B.: Colchicine and colcemid binding components of Fungus *Saprolegnia ferax.* Protoplasma **85,** 177 – 192 (1975)
81. Heidemann, S. R., Kirschner, M. W.: Induction of aster formation in eggs of *Xenopus laevis* by isolated basal bodies. J. Cell Biol. **67,** 164 a (1975)
82. Henneguy, L. F.: Sur les rapports des cils vibratiles avec les centrosomes. Arch. Anat. Microsc. **1,** 481 – 496 (1897)
83. Hepler, P. K.: The blepharoplast of *Marsilea;* its de novo formation and spindle association. J. Cell Sci. **21,** 361 – 390 (1976)
84. Hibberd, D. J.: Observations on the ultrastructure of the choanoflagellate *Codosiga botrytis* (Ehr.) with special reference to the flagellar apparatus. J. Cell Sci. **17,** 191 – 220 (1975)
85. Hollande, A., Cachon, J., Cachon, M., Valentin, J.: Infrastructure des axopodes et organisation générale de *Sticholonche zanclea* Hertwig (Radiolaire *Sticholonchidae*). Protistologica **3,** 155 – 166 (1967)
86. Hopkins, J. M.: Subsidiary components of the flagella on *Chlamydomonas reinhardii.* J. Cell Sci. **7,** 823 – 839 (1970)
87. Hubert, J. P., Flament-Durand, J., Dustin, P.: Centrioles and cilia multiplication in the pituitary of the rat after furosemid and colchicine treatment. I. The posterior lobe. Cell Tissue Res. **149,** 349 – 361 (1974)
88. Jacobson, D. N., Johnke, R. N., Adelman, M. R.: Studies on motility in *Physarum polycephalum.* In: Cell Motility (eds.: R. Goldman, T. Pollard, J. Rosenbaum), pp. 749 – 770. Cold Spring Harbor Lab. 1976
89. Jadin, J. B., Willaert, E.: Trois cas de méningoencéphalite amibienne primitive à *Naegleria gruberi* observés à Anvers (Belgique). Protistologica **8,** 95 – 100 (1972)
90. Jones, C. W.: The pattern of microtubules in the axonemes of *Gymnosphaera albida* Sassaki: evidence for 13 protofilaments. J. Cell Sci. **18,** 133 – 156 (1975)
91. Kalnins, V. I., Porter, K. R.: Centriole replication during ciliogenesis in the Chick tracheal epithelium. Z. Zellforsch. **100,** 1 – 30 (1969)
92. Kato, K. H., Sugiyama, M.: On the de novo formation of the centriole in the activated sea urchin egg. Developm. Growth Differ. **13,** 359 – 366 (1971)
93. Kowit, J. D., Fulton, C.: Programmed synthesis of tubulin for the flagella that develop during cell differentiation in *Naegleria gruberi.* Proc. Natl. Acad. Sci. USA **71,** 2877 – 2881 (1974)
94. Lang, N. J.: Electron microscopy of the *Volvocacae* and *Astrephenomacae.* Am. J. Bot. **50,** 280 – 300 (1963)
95. Langford, G. M., Inoue, S., Sabran, I.: Analysis of axostyle motility in *Pyrsonympha vertens.* J. Cell Biol. **59,** 185 a (1973)
96. Lenhossek, M. von: Über Flimmerzellen. Verh. Anat. Ges. **12,** 106 – 128 (1898)
97. Lin, H., Chen, I-Li: Development of the ciliary complex and microtubules in the cells of rat subcommisural organ. Z. Zellforsch. **96,** 186 – 205 (1969)
98. Linck, R. W.: Flagellar doublet microtubules: fractionation of minor components and α-tubulin from specific regions of the A-tubulin. J. Cell Sci. **20,** 405 – 440 (1976)
99. Linck, R. W.: Fractionation of minor component proteins and tubulin from specific regions of flagellar-doublet microtubules. In: Cell motility (eds.: R. Goldman, T. Pollard, J. Rosenbaum), pp. 869 – 890. Cold Spring Harbor Lab. 1976
100. McGill, M, Brinkley, B. R.: Human chromosomes and centrioles as nucleating sites for the in vitro assembly of microtubules from bovine brain tubulin. J. Cell Biol. **67,** 189 – 199 (1975)
101. McGill, M., Highfield, D. P., Monahan, T. M., Brinkley, B. R.: Effects of nucleic acid specific dyes on centrioles of mammalian cells. J. Ultrastruct. Res. **57,** 43 – 53 (1976)
102. McIntosh, J. R.: The axostyle of *Saccinobaculus.* II. Motion of the microtubule bundle and a structural comparison of straight and bent axostyles. J. Cell Biol. **56,** 324 – 339 (1973)
103. McNitt, R.: Centriole ultrastructure and its possible role in microtubule formation in an aquatic fungus. Protoplasma **80,** 91 – 108 (1974)

104. Maller, J. L., Poccia, D. P.: Spindle formation and parthenogenesis by centriole fractions from sperm. J. Cell Biol. **67,** 258 a (1975)
105. Manton, I.: Observations on the microanatomy of the spermatozoid of the bracken fern (*Pteridium aquilinum*). J. Biophys. Biochem. Cytol. **6,** 413 – 418 (1959)
106. Margulis, L.: Microtubules and evolution. In: Microtubules and Microtubule Inhibitors (eds.: M. Borgers, M. De Brabander), pp. 3 – 18. Amsterdam, Oxford: North-Holland. New York: Am. Elsevier 1975
107. Margulis, L., Banerjee, S., Kelleher, J. K.: Assay for antitubulin drugs in live cells: oral regeneration in *Stentor coeruleus.* In: Microtubules and Microtubule Inhibitors (eds.: M. Borgers, M. De Brabander), pp. 453 – 470. Amsterdam, Oxford: North-Holland. New York: Am. Elsevier 1975
108. Martinez, P., Daems, W. T.: Les phases précoces de la formation des cils et le problème de l'origine du corpuscule basal. Z. Zellforsch. **87,** 46 – 68 (1968)
109. Matsusaka, T.: Membrane particles of the connecting cilium. J. Ultrastruct. Res. **48,** 305 – 312 (1974)
110. Mazia, D., Harris, P. J., Bibring, T.: The multiplicity of the mitotic centers and the time-course of their duplication and separation. J. Biochem. Biophys. Cytol. **7,** 1 – 20 (1960)
111. Milhaud. M., Pappas, G. D.: Cilia formation in the adult cat brain after pargyline treatment. J. Cell Biol. **37,** 599 – 609 (1968)
112. Mizukami, I., Gall, J.: Centriole replication. II. Sperm formation in the fern, *Marsilea* and the cycad, *Zamia.* J. Cell Biol. **29,** 97 – 111 (1966)
113. Mooseker, M. S., Tilney, L. G.: Isolation and reactivation of the axostyle. Evidence for a dynein-like ATPase in the axostyle. J. Cell Biol. **56,** 13 – 26 (1973)
114. Nanney, D. L.: Patterns of basal body addition in ciliary rows in *Tetrahymena.* J. Cell Biol. **65,** 503 – 512 (1975)
115. Nelsen, E. M.: Regulation of tubulin during ciliary regeneration in non-growing *Tetrahymena.* Exp. Cell Res. **94,** 152 – 158 (1975)
116. Neviackas, J. A., Margulis, L.: The effect of colchicine on regenerating membranellar cilia in *Stentor coeruleus.* J. Protozool. **16,** 165 – 171 (1969)
117. Ockleford, C. D.: Ultraviolet light microbeam irradiation of the microtubules in single heliozoan axopodia. Exp. Cell Res. **93,** 127 – 135 (1975)
118. Ockleford, C. D., Tucker, J. B.: Growth, breakdown, repair and rapid contraction of microtubular axopodia in the heliozoan *Actinophrys sol.* J. Ultrastruct. Res. **44,** 369 – 387 (1973)
119. O'Hara, P. T.: Spiral tilt of triplet fibers in human leukocyte centrioles. J. Ultrastruct. Res. **31,** 195 – 198 (1970)
120. Olmsted, J. B., Witman, G. B., Karlson, K., Rosenbaum, J. L.: Comparison of the microtubule proteins of neuroblastoma cells, brain and *Chlamydomonas* flagella. Proc. Natl. Acad. Sci. USA **68,** 2273 (1971)
121. Outka, D. E., Kluss, B. C.: The ameba-to-flagellate transformation in *Tetramitus rostratus.* II. Microtubular morphogenesis. J. Cell Biol. **35,** 323 – 346 (1967)
122. Pease, D. C.: The ultrastructure of flagellar fibrils. J. Cell Biol. **18,** 313 – 326 (1963)
123. Phillips, D. M.: Fine structure of *Sciara coprophila* sperm. J. Cell Biol. **30,** 499 – 518 (1966)
124. Phillips. D. M.: Structural variants in Invertebrate sperm flagella and their relationship to motility. In: Cilia and Flagella (ed.: M. A. Sleigh), pp. 379 – 402. New York, London: Acad. Press 1974
125. Phillips. S. G., Rattner, J. B.: Dependence of centriole formation on protein synthesis. J. Cell Biol. **70,** 9 – 19 (1976)
126. Pickett-Heaps, J. D.: Aspects of spindle evolution. Ann. N. Y. Acad. Sci. **253,** 352 – 361 (1975)
127. Pitelka, D. R.: Centriole replication. In: Handbook of Molecular Cytology (ed.: A. Lima de Faria), pp. 1200 – 1218. Amsterdam, London: North-Holland. New York: Wiley Intersci. Division 1969
128. Pitelka, D. R.: Basal bodies and root structures. In: Cilia and Flagella (ed.: M. A. Sleigh), pp. 437 – 470. New York, London: Acad. Press 1974
129. Pitelka, D. R., Child, F. M.: The locomotor apparatus of Ciliates and Flagellates: relations between structure and function. In: Biochemistry and Physiology of Protozoa (ed.: S. H. Hutner), Vol. 3, pp. 131 – 198. London: Acad. Press 1964
130. Pittam, M. D.: Studies of an amoebo-flagellate, *Naegleria gruberi.* Quart. J. Microsc. Sci. **104,** 513 – 529 (1963)
131. Plattner, H.: Ciliary granule plaques: membrane-intercalated particle aggregates associated with Ca^{2+} binding sites in *Paramecium.* J. Cell Sci. **18,** 257 – 270 (1975)

132. Rafalko, J. S.: Cytological observations on the amoebo-flagellate, *Naegleria gruberi* (Protozoa). J. Morphol. **81**, 1 – 44 (1947)
133. Randall, J., Warr, J. R., Hopkins, J. M., McVittie, A.: A single-gene mutation of *Chlamydomonas reinhardii* affecting motility: a genetic and electron microscopic study. Nature (London) **203**, 912 – 914 (1964)
134. Rannestad, J., Willimans, N. E.: The synthesis of microtubule and other proteins of the oral apparatus in *Tetrahymena pyriformis*. J. Cell Biol. **50**, 709 – 720 (1971)
135. Rattner, J. B., Phillips, S. G.: Independence of centriole formation and DNA synthesis. J. Cell Biol. **57**, 359 – 372 (1973)
136. Reese, T. S.: Olfactory cilia in the frog. J. Cell Biol. **25**, 209 – 230 (1965)
137. Ringo, D. L.: The arrangement of subunits in flagellar fibers. J. Ultrastruct. Res. **17**, 266 – 277 (1967)
138. Robbins, E. G., Jentzsch, G., Micall, A.: The centriole cycle in synchronized HeLa cells. J. Cell Biol. **36**, 329 – 338 (1968)
139. Rosenbaum, J. L., Binder, L. I., Granett, S., Dentler, W. L., Snell, W., Sloboda, R., Haimo, L.: Directionality and rate of assembly of chick brain tubulin onto pieces of neurotubules, flagellar axonemes, and basal bodies. Ann. N. Y. Acad. Sci. **253**, 147 – 177 (1975)
140. Rosenbaum, J., Carlson, A.: Cilia regeneration in *Tetrahymena* and inhibition by colchicine. J. Cell Biol. **40**, 415 – 425 (1969)
141. Rosenbaum, J. L., Moulder, J. E., Ringo, D. L.: Flagellar elongation and shortening in *Chlamydomonas*. The use of cycloheximide and colchicine to study the synthesis and assembly of flagellar proteins. J. Cell Biol. **41**, 600 – 619 (1969)
142. Roth, L. E., Pihlaja, D. J., Shigenaka, Y.: Microtubules in the heliozoan axopodium. I. The gradion hypothesis of allosterism in structural proteins. J. Ultrastruct. Res. **30**, 7 – 37 (1970)
143. Sachs, M. I., Anderson, E.: A cytological study of artificial parthenogenesis in the sea urchin *Arbacia punctulata*. J. Cell Biol. **47**, 140 – 158 (1970)
144. Sagan, L.: On the origin of mitosing cells. J. Theoret. Biol. **14**, 225 – 274 (1967)
145. Sale, W. S., Satir, P.: Splayed *Tetrahymena* cilia. A system for analyzing sliding and axonemal spoke arrangements. J. Cell Biol. **71**, 589 – 605 (1976)
146. Sandoz, D., Boisvieux-Ulrich, E.: Ciliogénèse dans les cellules à mucus de l'oviducte de caille. I. Etude ultrastructurale chez la caille en ponte. J. Cell Biol. **71**, 449 – 459 (1976)
147. Sandoz, D., Boisvieux-Ulrich, E., Laugier, C., Brard, E.: Ciliogénèse dans les cellules à mucus de l'oviducte de caille. II. Contrôle hormonal. J. Cell Biol. **71**, 460 – 471 (1976)
148. Satir, P.: The present status of the sliding microtubule model of ciliary motion. In: Cilia and Flagella (ed.: M. A. Sleigh), pp. 131 – 142. London, New York: Acad. Press 1974
149. Satir, P., Satir, B.: A model for ninefold symmetry in α-keratin and cilia. J. Theoret. Biol. **7**, 123 – 128 (1964)
150. Sattler, C. A., Staehelin, L. A.: Ciliary membrane differentiations in *Tetrahymena pyriformis*. *Tetrahymena* has four types of cilia. J. Cell Biol. **62**, 473 – 490 (1974)
151. Schrével, J., Besse, C.: Un type flagellaire fonctionel de base 6+0. J. Cell Biol. **66**, 492 – 507 (1975)
152. Schuster, F. L.: An electron microscope study of the amoebo-flagellate, *Naegleria gruberi* (Schardinger). I. The amoeboid and flagellate stages. J. Protozool. **10**, 297 – 313 (1963)
153. Schuster, F. L.: Ultrastructure of mitosis in the amoebo-flagellate *Naegleria gruberi*. Tissue and Cell **7**, 1 – 12 (1975)
154. Sleigh, M. A.: The Biology of Cilia and Flagella. Oxford: Pergamon Press 1962
155. Sleigh, M. A. (Ed.): Cilia and Flagella. London, New York: Acad. Press 1974
156. Sorokin, S. P.: Reconstruction of centriole formation and ciliogenesis in mammalian lungs. J. Cell Sci. **3**, 207 – 230 (1968)
157. Sotelo, J. R., Trujillo-Cenoz, O.: Electron microscope study of the kinetic apparatus in animal sperm cells. Z. Zellf. **48**, 565 – 601 (1958)
158. Staprans, I., Dirksen, E. R.: Microtubule protein during ciliogenesis in the mouse oviduct. J. Cell Biol. **62**, 164 – 174 (1974)
159. Steinman, R. M.: An electron microscopic study of ciliogenesis in developing epidermis and trachea in the embryo of *Xenopus laevis*. Am. J. Anat. **122**, 19 – 56 (1968)
160. Stephens, R. E.: Thermal fractionation of outer fiber doublet microtubules into A- and B-subfiber components: A- and B-tubulin. J. Mol. Biol. **47**, 353 – 363 (1970)
161. Stephens, R. E.: Isolation of nexin—the linkage protein responsible for maintenance of the ninefold configuration of flagellar axonemes. Biol. Bull. **139**, 438 (1970)
162. Stephens, R. E.: Enzymatic and structural proteins of the axoneme, In: Cilia and Flagella (ed.: M. A. Sleigh), pp. 39 – 78. London, New York: Acad. Press 1974

163. Stephens, R. E.: Structural chemistry of axoneme. Evidence for chemically and functionally unique tubulin dimers in outer fibers. In: Molecules and Cell Movement (eds.: S. Inoue, R. E. Stephens), pp. 181 – 206. New York: Raven Press 1975
164. Stephens, R. E.: The basal apparatus. Mass isolation from the molluscan ciliated gill epithelium and a preliminary characterization of striated rootlets. J. Cell Biol. **64**, 408 – 420 (1975)
165. Stockinger, L., Cireli, E.: Eine bisher unbekannte Art der Zentriolenvermehrung. Z. Zellforsch. **68**, 733 – 740 (1965)
166. Stubblefield, E., Brinkley, B. R.: Cilia formation in Chinese hamster fibroblasts in vitro as a response to colcemid treatment, J. Cell Biol. **30**, 645 – 652 (1966)
167. Stubblefield, E., Brinkley, B. R.: The architecture and function of the mammalian centriole. In: The Origin and Fate of Cell Organelles (ed.: K. B. Warren), pp. 175 – 218. New York: Acad. Press 1967
168. Szollosi, D.: The structure and function of centrioles and their satellites in the jellyfish *Phialidium gregarium*. J. Cell Biol. **21**, 465 – 479 (1964)
169. Tamm, S. L.: Flagellar development in the protozoan, *Peranema trichophorum*. J. Exp. Zool. **164**, 163 – 186 (1967)
170. Tamm, S. L.: Free kinetosomes in Australian flagellates. I. Types and spatial arrangement. J. Cell Biol. **54**, 39 – 55 (1972)
171. Tani, E., Ikeda, K., Nishiura, M., Higashi, N.: Specialized intercellular junctions and ciliary necklace in rat brain. Cell Tissue Res. **151**, 57 – 68 (1974)
172. Taylor, A. C.: The centrioles and microtubules. Am. Nat. **99**, 267 – 278 (1965)
173. Thiery, J. P.: Mise en évidence d'une sous-structure fibrillaire dans les filaments axonématiques des flagelles. J. Microsc. **2**, 71 – 80 (1963)
174. Thomas, M. B.: The structure of the 9 + 1 axonemal core as revealed by treatment with trypsin. J. Ultrastruct. Res. **52**, 409 – 422 (1975)
175. Thompson, G. A. Jr., Bauch, L. C., Walker, L. F.: Non-lethal deciliation of *Tetrahymena* by a local anesthetic and its utility as a tool for studying cilia regeneration. J. Cell Biol. **61**, 253 – 256 (1974)
176. Thornhill, R. A.: The ultrastructure of the olfactory epithelium of the lamprey *Lampetra fluviatilis*. J. Cell Sci. **2**, 591 – 602 (1967)
177. Tilney, L. B.: How microtubule patterns are generated. The relative importance of nucleation and bridging of microtubules in the formation of the axoneme of *Raphidiophrys*. J. Cell Biol. **51**, 837 – 854 (1971)
178. Tilney, L. G., Porter, K. R.: Studies on the microtubule in Heliozoa. II. The effect of low temperature on these structures in the formation and maintenance of the axopodia. J. Cell Biol. **34**, 327 – 343 (1967)
179. Todd, S. R.: Effects of high hydrostatic pressure on transformation in *Naegleria gruberi*. Symp. Soc. Exp. Biol. **26**, 485 – 486 (1972)
180. Tucker, J. B.: Initiation and differentiation of microtubule patterns in the ciliate *Nassula*. J. Cell Sci. **7**, 793 – 821 (1970)
181. Tucker, J. B.: Development and deployment of cilia, basal bodies, and other microtubular organelles in the cortex of the ciliate *Nassula*. J. Cell Sci. **9**, 539 – 568 (1971)
182. Turner, F. R.: An ultrastructural study of plant spermatogenesis. Spermatogenesis in *Nitella*. J. Cell Biol. **37**, 370 – 393 (1968)
183. van Assel, S., Brachet, J.: Formation de cytasters dans les œufs de Batraciens sous l'action de l'eau lourde. J. Embryol. Exp. Morphol. **15**, 143 – 151 (1966)
184. Warner, F. D.: The fine structure of the ciliary and flagellar axoneme. In: Cilia and Flagella (ed.: M. A. Sleigh). pp. 11 – 38. London, New York: Acad. Press 1974
185. Warner, F. D.: Ciliary intermicrotubule bridges. J. Cell Sci. **20**, 101 – 114 (1976)
186. Warner, F. D., Satir, P.: The structural basis of ciliary bend formation. Radial spoke positional changes accompanying microtubule sliding. J. Cell Biol. **63**, 35 – 63 (1974)
187. Watson, M. R., Alexander, J. B., Silverster, N. R.: The cilia of *Tetrahymena pyriformis*: fractionation of the isolated cilia. Exp. Cell Res. **33**, 112 – 129 (1964)
188. Weisenberg, R. C., Rosenfeld, A. C.: In vitro polymerization of microtubules into asters and spindles in homogenates of surf clam eggs. J. Cell Biol. **64**, 42 – 53 (1975)
189. Went, H. A.: The behaviour of centrioles and the structure and formation of the achromatic figure. Protoplasmatologia **6**, 1 – 109 (1966)
190. Williams, N. E.: Regulation of microtubules in *Tetrahymena*. Intern. Rev. Cytol. **41**, 59 – 86 (1975)
191. Williams, N. E., Frankel, J.: Regulation of microtubules in *Tetrahymena*. I. Electron microscopy of oral replacement. J. Cell Biol. **56**, 441 – 457 (1973)

192. Williams, N. E., Nelson, E. M.: Regulation of microtubules in *Tetrahymena.* II. Relation between turnover of microtubule proteins and microtubule dissociation and assembly during oral replacement. J. Cell Biol. **56,** 458 – 465 (1973)
193. Willmer, E. N.: Amoeba-flagellate transformation. Exp. Cell Res. **8,** (Suppl.) 32 – 46 (1961)
194. Witman, G. B., Carlson, K., Berliner, J., Rosenbaum, J. L.: *Chlamydomonas* flagella. I. Isolation and electrophoretic analysis of microtubules, matrix, membranes, and mastigonems. J. Cell Biol. **54,** 507 – 539 (1972)
195. Witman, G. B., Carlson, K., Rosenbaum, J. L.: *Chlamydomonas* flagella. II. The distribution of tubulins 1 and 2 in the outer doublet microtubules. J. Cell Biol. **54,** 540 – 555 (1972)
196. Witman, G. B., Fay, R., Plummer, J.: *Chlamydomonas* mutants: evidence for the roles of specific axonemal components in flagellar movement. In: Cell Motility (eds.: R. Goldman, T. Pollard, J. Rosenbaum), pp. 969 – 986. Cold Spring Harbor Lab. 1976
197. Wolfe, J.: Structural analysis of basal bodies of the isolated oral apparatus of *Tetrahymena.* J. Cell Sci. **6,** 679 – 700 (1970)
198. Wunderlich, F., Speth, V.: Membranes in *Tetrahymena.* I. The cortical pattern. J. Ultrastruct. Res. **41,** 258 – 269 (1972)
199. Younger, K. B., Banerjee, S., Kelleher, J. K., Winston, M., Margulis, L.: Evidence that synchronized production of new basal bodies is not associated with DNA synthesis in *Stentor coeruleus.* J. Cell Sci. **11,** 621 – 637 (1972)

Chapter 5 Microtubule Poisons

5.1 Introduction

A small number of chemicals, principally colchicine, colchicine derivatives and the *Vinca* (*Catharanthus*) alkaloids, are capable of binding specifically to tubulin, preventing its assembly into MT. It is now indispensable to give further details about the action of these drugs, and of other chemicals which may have, under various conditions, similar effects. In the next chapters, colchicine and other poisons will often be mentioned, in relation to cell movements, secretion, and mitosis; their applications in various fields of medicine will be studied in Chapter 11. Here, the specific mechanisms of tubulin binding of these poisons, and the ways in which they alter MT functions will be analyzed. However, great care must be taken not to forget that some pharmacological effects of these drugs are unrelated to MT poisoning. As MT play such a diverse role in many cell functions, it is not always easy to find out which effect is mediated via changes in the MT. One approach to this, which has been followed by several authors, is to study the effects on a definite cell function of a series of chemically unrelated MT poisons: if they are similar, and can also be imitated by physical means, such as cold, high hydrostatic pressures, or by deuterium oxide, it can with some safety be concluded that the effects are mediated through MT poisoning and not by non-specific actions.

Colchicine is by far the best known of all MT poisons. It is not our purpose to detail the many results which have been obtained with this substance and some closely related molecules since the publication of our 1955 monograph [90]. At that time, few effects other than mitosis inhibition were known, while the numerous papers published in the last ten years deal mainly with changes unrelated to mitosis. Again, only the fundamentals of colchicine action will be discussed here.

The *Vinca* alkaloids, and several other MT poisons which may be of interest, either as tools for understanding better the problems of MT assembly, or as specific poisons of specialized cells, will be considered next. This chapter will be concluded by a study of the poisoning of MT by physical agents and heavy water. In many ways, these imitate colchicine, and help to understand the reasons for the differential resistance of MT and the significance of their stability or lability.

5.2 Colchicine

A historical account of the discovery of the properties of the *Colchicum* alkaloids has been given in Chapter 1. Before analyzing the manifold effects discovered since

1955, it is of interest to recall some early results which only became understood after the discovery of the role of MT in cell functions other than mitosis.

5.2.1 Actions Unrelated to Mitosis, Before 1955

Several hints that colchicine could affect intermitotic cells were known many years ago. The following, mentioned in our monograph [90] may be listed:

a) The first experiments on plants demonstrated that root-tips growing in a colchicine solution, besides mitotic abnormalities including polyploid cells, showed peculiar swellings which were named "C tumors". It was found that after inhibiting mitoses of *Allium fistulosum* by X-irradiation, the typical colchicine tumors could still appear: a phytohormone effect was suggested [154].

b) Effects on plant cell wall formation had been described, the most spectacular being the alteration of the regular hexagonal pattern of the alga *Hydrodictyon* [106] (a change recently rediscovered, cf. [171]; cf. Chap. 6).

c) The formation of stomata in plants involves complex differentiations related to mitosis and to the action of MT on cell shape: alterations had been observed early in colchicine research [264].

c) Colchicine was also long known to be a nervous poison, as an intracerebral injection in the cat resulted in a rapid death, in contrast to the action of parenteral injections of similar doses [78]. Other nervous disturbances had been mentioned, and were to find some explanation after the discovery of tubulin in neurons.

e) Several effects on hormone secretion had been observed and found no explanation from the known properties of the alkaloid (cf. Chap. 8).

These facts, and many others, suggested that the pharmacology of colchicine was complex, and indicated that, particularly in plants, cell changes unrelated to mitosis or mitotic poisoning could be observed. It was however thought that the spindle proteins were the main receptors for colchicine, for in 1955 it could not be suspected that the same proteins influenced cell shape and motility. It may be worthwhile to recall one of the last sentences of the 1955 monograph:

> "Colchicine ... must be considered a singular substance. Not only does it possess remarkable side-effects, such as its action on gout, the colchicine-leukocytosis, its action on the nervous system and on neuromuscular contraction, its induction of specific malformations in embryos; it is also the most efficient and active of all mitotic poisons known". "It is probably more than mere chance that the unique structure of this tropolone derivative is associated with so many physiological activities."

Tubulin was to be one answer to these facts, although one should be careful today not to conclude too quickly that any action of colchicine is necessarily mediated through MT poisoning. The doses used must always be carefully considered, and effects on MT can only be demonstrated with certainty when a clear-cut association of the colchicine and the tubulin molecules is evident. It is for this reason that the introduction of ^{3}H-colchicine in 1965 [250] was of such importance.

5.2.2 The Chemistry of Colchicine and its Derivatives

Colchicine is one of the numerous substances extracted from the meadow saffron, *Colchicum automnale*, and from various other plants of the colchicum family and related groups [90]. *Gloriosa superba* contains as much as 0.3% of colchicine as compared to 0.5% for *C. automnale*, and colchicine from this source has recently been tested for its action on mitosis [186]. The chemistry of *Colchicum* extracts is most complex, and numerous chemicals have been isolated from this species. Colchicine was purified as early as in 1883 [283], although the complete elucidation of its chemical structure was only solved in 1950 after extensive work by Cook et al. (cf. [90]). A de-

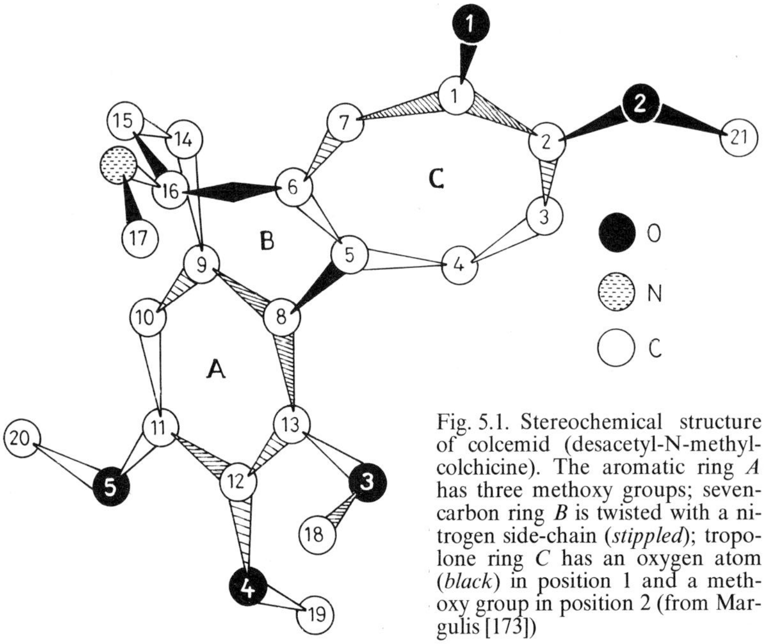

Fig. 5.1. Stereochemical structure of colcemid (desacetyl-N-methylcolchicine). The aromatic ring *A* has three methoxy groups; seven-carbon ring *B* is twisted with a nitrogen side-chain (*stippled*); tropolone ring *C* has an oxygen atom (*black*) in position 1 and a methoxy group in position 2 (from Margulis [173])

tailed stereochemical analysis of this complex molecule was published in 1975 [173, 174].

Colchicine is a tropolone derivative, with three rings: one aromatic ring (A) with three methoxy groups, one twisted seven-carbon ring (B) with a substituted amino group, and another seven-carbon, tropolone ring (C) with one oxygen and one methoxy group. The principal derivatives of colchicine, of interest as mitotic and MT poisons, are given in Table 5.1.

In a review of antimitotic agents it was concluded in 1963 [84] that the following structures were required for activity: at least one methoxy group on ring A; a seven-membered ring C (the lumicolchicines, obtained by the action of ultraviolet light on colchicine, are inactive, a fact which has been put to use in several studies of colchicine activity); a methoxy group on ring C—which may be replaced by a thiomethyl group. It had been thought that the amino group of ring B should be esterified, but it has since been found that desacetyl-aminocolchicine is ten times more active than the parent compound on a murine mast-cell tumor [224]. A group led by Lettré [153] has studied the actions of several halogenated derivatives, of which the two most potent mitotic poisons, ten times more active than colchicine, are N-fluoro-acetyl-desacetylcolchicine and N-fluoro-acetyl-14-methyl-thio-14-demethoxy-desacetylcolchicine. Moreover, a chloroacrylonitrile derivative of N-desacetylcolchicine has been found to be 100 times more active as a mitotic poison, and to combine with tubulin on the "colchicine site" [109].

Desacetyl-N-methyl-colchicine (colcemid) has been widely used as a MT poison and in chemotherapy: its action does not fundamentally differ from that of colchicine.

The three-dimensional structure of colchicine shows a planar A ring, and a twisted ring B [173]: one expects that the reason for the specificity of action on tubulin may be explained stereochemically. One important feature of the molecule is the three methoxy groups of ring A: podophyllotoxin, which is known to become attached to the same site of the tubulin molecule [42, 43], also has an aromatic ring with three methoxy groups. The increased activity of some amino-substituted compounds indicates that this group also plays a role in chemical activity. The position of oxy and methoxy groups on ring C also has its importance, as *iso*-colchicine, where their positions are inverted, has an antimitotic activity 100 times lower than that of colchicine [90], while colchiceine, without methoxy group on ring C, is inactive ([90]; Table 5.1).

In a study of simple derivatives of colchicine Fitzgerald [95], studying the inhibition of spindle formation in HeLa cells, and that of tubulin assembly in vitro, de-

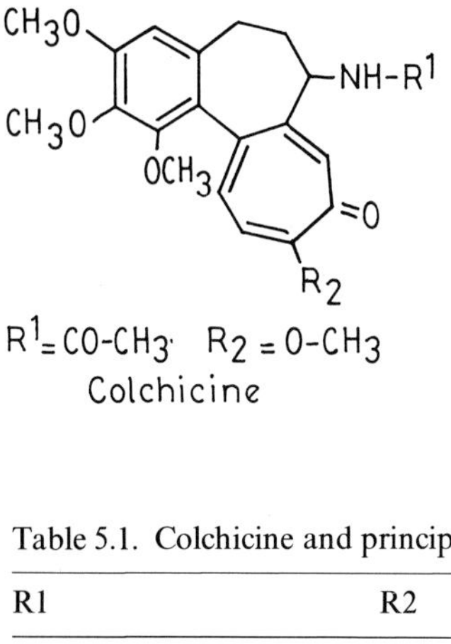

$R^1 = CO\text{-}CH_3$, $R_2 = O\text{-}CH_3$
Colchicine

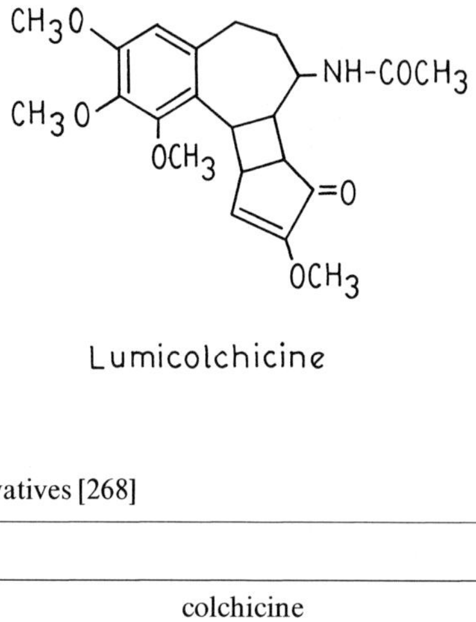

Lumicolchicine

Table 5.1. Colchicine and principal derivatives [268]

R1	R2	
$CO-CH_3$	$O-CH_3$	colchicine
CH_3	$O-CH_3$	desacetyl-N-methyl-colchicine (demecolchicine, colcemid)
H_3	$O-CH_3$	desacetyl-amino-colchicine
$CO-CH_3$	$-CH_2-CH_2=N<$	colchicamides
$CO-CH_3$	$S-CH_3$	thiocolchicine [153]
$CO-CH_2-F$	$S-CH_3$	fluorthiocolchicine [153]
$CO-CH_2-Cl$	$S-CH_3$	chlorthiocolchicine [153]
$CO-CH_2-F$	$O-CH_3$	N-fluoro-desacetylcolchicine [153]
$CO-CH_3$	OH	colchiceine
$CH_2-CHCl-CN$	$O-CH_3$	chlorocyanoethyl derivative of N-desacetyl-colchicine [109]
H	$O-CH_3$	trimethylcolchicinic acid
$CO-CH_2-Br$	$O-CH_3$	bromocolchicine [225]

monstrated the activity of two interesting compounds. Allocolchicine, which has a slightly modified ring C, is as active as colchicine (cf. Table 5.2); more interestingly, a compound with an open ring B [2-methoxy-5-(2′,3′,4′-trimethoxyphenyl)tropone] was about as active, indicating that the amino-substituted group of ring B was not necessary for action. A related substance, but with no methoxy groups on ring A, was less active in vivo and showed an inhibition of less than 30% of tubulin assem-

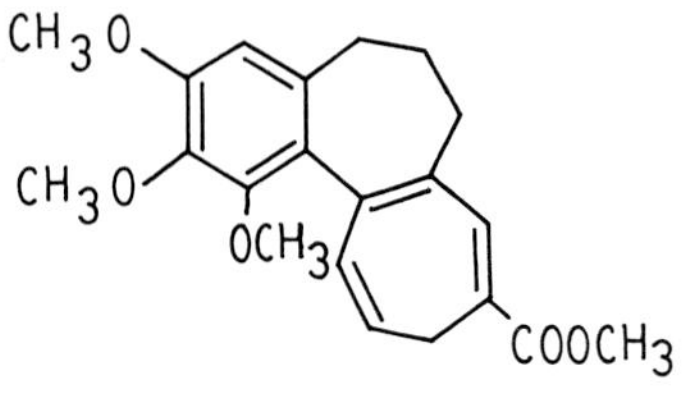

Allocolchicine

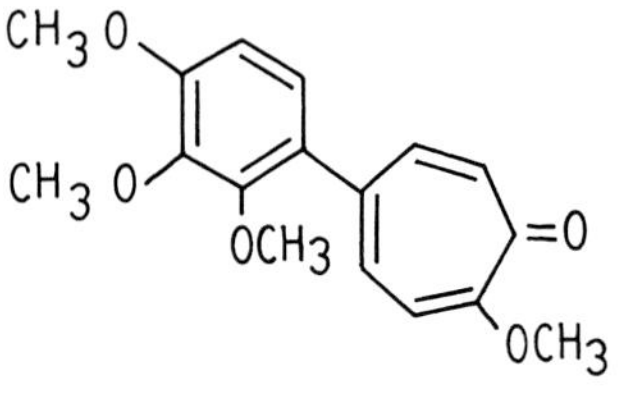

2 methoxy-5-(2′,3′,4′-trimethoxy-phenyl) tropone

Table 5.2. Inhibition of tubulin polymerization (assembly) and antimitotic activity of colchicine and related compounds (from Fitzgerald [94])

Compound	Half-maximal concentration inhibitory for assembly (M)	Mitotic inhibition (HeLa cells)	
		maximum inhibition concentration (M)	per cent of metaphases at maximum inhibition
Colchicine	2.3×10^{-7}	10^{-7}	20
Desacetamidocolchicine	1.9×10^{-7}	10^{-8}	21
Isocolchicine	$>10^{-4}$	= control	
Allocolchicine	1.9×10^{-7}	10^{-7}	21
2-methoxy-5(2′3′4′-trimethoxyphenyl)tropone	4.0×10^{-7}	10^{-7}	22

bly in vitro. Moreover, this compound (2-methoxy-5-phenyltropone) did not prevent the trimethoxy-derivative from inhibiting tubulin assembly, confirming the importance of the methoxy groups of ring A. The actions in vivo and in vitro were parallel except for desacetamido-colchicine, more active on mitosis than on assembly.

The isotopically labeled colchicine derivatives have already been mentioned: ^{14}C labeling of the methoxy group of ring C has helped to study colchicine metabolism in man [260] and colchicine binding [103], and the fundamental work of Taylor and his group was done with colchicine labeled with tritium on the same methoxy group of ring C [250].

Many attempts at total synthesis of colchicine have been made, and success appears to have been achieved, starting from a substituted 1,3-diphenylpropane, leading to desacetamido-isocolchicine, which has the ring structure of colchicine but without the amino-substituted group on ring B. This substance can then be transformed to (±)-colchicine [140].

5.2.3 The Linkage of Colchicine and Tubulin

This may be considered one of the important problems of MT research. It has long been known that colchicine acts at very low concentrations (less than 10^{-6} M) on some MT, in particular those of the mitotic spindle. The demonstration of the association between colchicine and a specific protein was the turning-point in the discovery of tubulin. However, many important questions remain without clear answers: what is the chemical nature of the bond between the alkaloid and the protein; can tubulin become linked with colchicine when organized in MT; may already formed MT be altered by colchicine and other similar poisons; why are some MT apparently resistant; why are the concentrations of the drug which are effective so different from one species to another?

Some of these questions may seem obvious, as it has been repeatedly written that MT are destroyed by colchicine. However, this apparent destruction may be explained by an inhibition of assembly, if MT were in a dynamic condition with a rapid turnover [269]. Nevertheless, when MT are destroyed by agents such as high hydrostatic pressure or cold, they disappear rapidly, in a few minutes, and this is difficult to explain other than by a disassembly of MT. Many experiments show MT, in cells of various types, vanishing under the influence of colchicine, even in low concentrations. This is difficult to explain if one admits that "colchicine does not bind to the intact microtubule" [236].

This opinion is mainly based on the demonstration that the outer doublets of sea-urchin sperm bind less than 0.001 mol of colchicine per mol of tubulin, while in the same conditions soluble tubulin would have bound 1 mol of colchicine per mol of tubulin [274]. This appears to be true for *doublets* (and probably for the centriolar triplets). However, doublets are known to be resistant to colchicine in most conditons, and results obtained from these structures, which are apparently stabilized (by HMW proteins?), do not necessarily apply to single MT.

At this point, the following problems will be considered: what is the nature of the colchicine – to – tubulin bond? What action has colchicine on MT? What are the fibrillary structures which have been described by several authors in cells intoxicated by the drug?

It should be understood that the answers to several of these problems apply also to MT poisons which act on different tubulin receptors, and that the problem of tubule lability can only be seen in a true perspective at the end of this monograph (cf. Chaps. 10 and 12).

5.2.3.1 The Tubulin – Colchicine Bond. Early observations demonstrated that colchicine was bound to tissues or structures which contained a special protein, named tubulin, and in which electron microscopy demonstrated the presence of MT [34, 93, 210, 274]. However, the techniques used did not indicate that MT—assembled tubulins—fixed colchicine. The discovery that the crystalline structures induced in cells by VLB, which are composed of tubulin (albeit assembled differently from MT), fixed colchicine [145] was an indication that in some conditions assembled tubulin could be stoichiometrically bound to colchicine. This led to the conclusion that each dimer of tubulin (MW = 110,000) has a single site for colchicine binding (which is also the receptor for another MT poison, podophyllotoxin), another for the *Vinca* alkaloids, and two others for the guanine nucleotides [42].

Another group of experiments conducted in vitro, after the discovery that tubulin could assemble in the absence of Ca^{2+} (cf. Chap. 2), indicated that colchicine prevented this assembly, when added to the assembling tubulin solution (Fig. 5.2; [35]). Ultraviolet light, inactivating colchicine to lumicolchicine, permitted the assembly to proceed (Fig. 5.3).

The colchicine binding affinity of tubulin has been studied, although this is made difficult by the fact that in vitro tubulin is not stable (except when combined in crystalline form with VLB): the affinity constants must take this instability into account. The colchicine – tubulin complex has a half-life of about 36 h [101]. The affinity constant has been found to be more than ten times greater than first indicated. The binding is relatively slow and during this time some of the tubulin molecules will denature. The rate constants of association and dissociation were found to be respectively $k = 0.37 \times 10^6$/M/h and $k_{-1} = 0.009$/h. The conclusion was reached that "the tubulin conformation required for assembly is incompatible with the conformation required for colchicine binding", "or that the colchicine binding site is buried within the microtubule" [236].

It had been shown that irradiation of the colchicine tubulin complex (changing the alkaloid to lumicolchicine) destroyed part of the binding site [6]. This has been confirmed and used as a technique for understanding the binding reaction: while

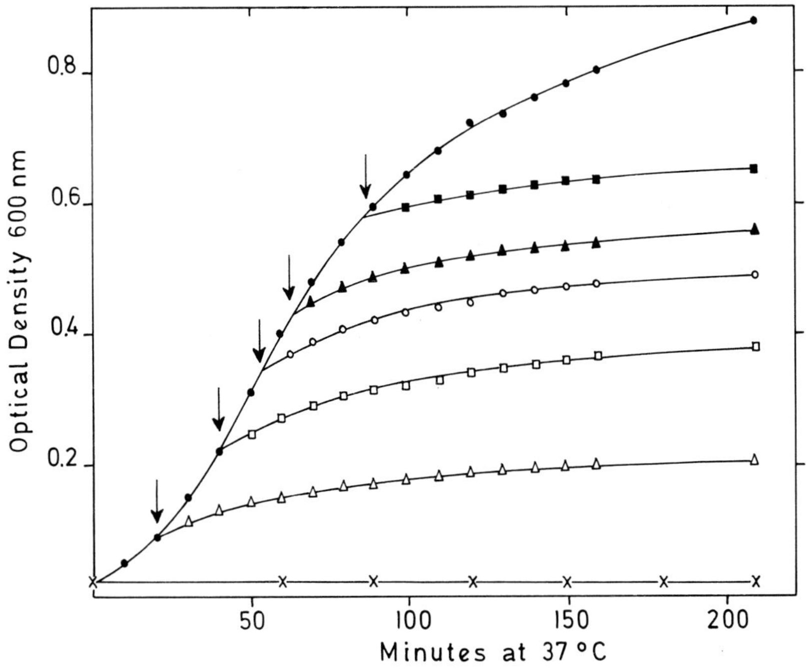

Fig. 5.2. Inhibition by colchicine of porcine brain tubulin assembly in vitro, monitored by turbidimetry. 100 μM colchicine was added at 0, 20, 40, 50, 60, and 90 min (*arrows*) after raising the temperature to 37 °C, inhibiting the assembly. The *line with closed circles* refers to assembly without colchicine (redrawn from Borisy et al. [35])

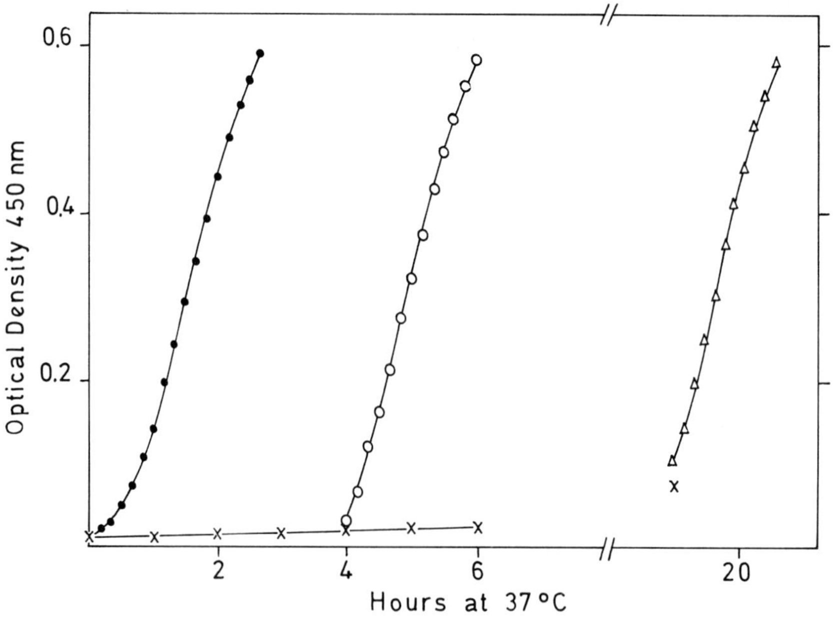

Fig. 5.3. Reversal of inhibition of MT assembly in vitro from porcine brain tubulin by transformation of colchicine to lumicolchicine, as measured by turbidimetry. Microtubule protein was prepared with 100 μM colchicine, which prevented assembly. A sample without colchicine was incubated at 37 °C at zero time (*black dots*). The others were irradiated by long-wave ultraviolet light at 4 (*open dots*) and 19 h (*triangles*), changing the colchicine to inactive lumicolchicine and permitting assembly (redrawn from Borisy et al. [35])

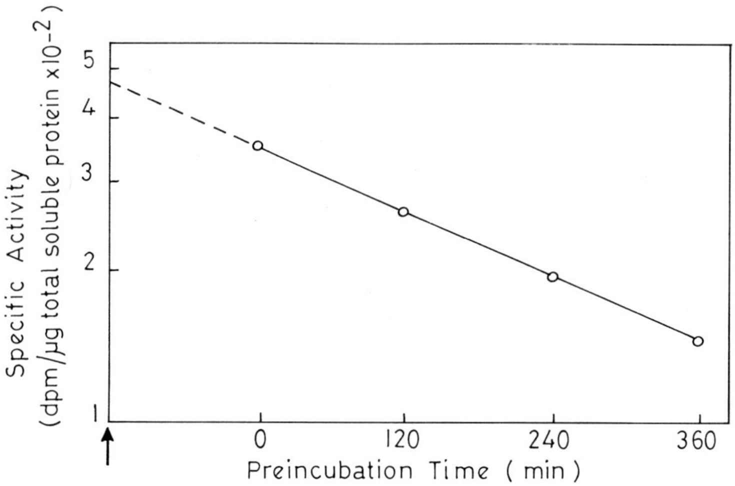

Fig. 5.4. First-order decay of colchicine-binding activity of tubulin. Chick embryo brain tubulin was incubated at 37 °C. At the times indicated, aliquots were incubated with 2×10^{-6} M acetyl-^{3}H-colchicine for 2 h, at 37 °C. Bound colchicine was estimated by the gel filtration method. Extrapolation to include the time of colchicine incubation yields the initial binding of tubulin. The half-time is calculated from the slope of the line (redrawn from Wilson [269])

the site is destroyed by 366 nm light, a (phospho)lipid complex of colchicine is liberated. This may be part of the tubulin molecule, combining with an intermediary product in the formation of lumicolchicine [44]. The problem of colchicine binding is further complicated by the conclusions of Kirschner et al. ([139]; Chap. 2) that two forms of tubulin, differing in their affinity for colchicine, may be present in cells.

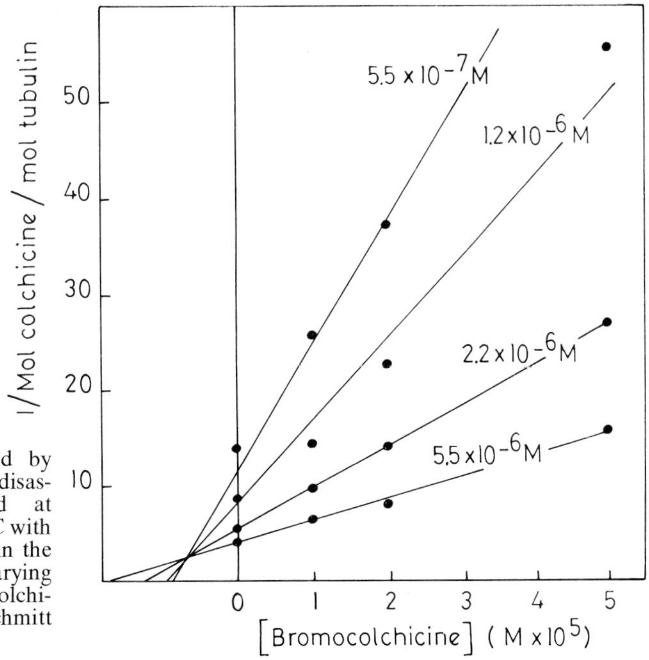

Fig. 5.5. Tubulin, purified by two cycles of assembly–disassembly, was incubated at 0.4 mg/ml for 2 h at 37 °C with 2×10^{-6} M ^{3}H-colchicine in the presence or absence of varying concentrations of bromocolchicine (redrawn from Schmitt and Atlas [225])

The progressive decay with time of the colchicine-binding property of tubulin is decreased by VLB in the case of the tubulin of *Strongylocentrotus* eggs, which however has an association constant for colchicine which is ten times lower than that of chick brain tubulin. This fact explains the lower fixation of colchicine on VLB crystals from sea-urchins [201].

An interesting finding is that bromocolchicine binds covalently to tubulin, but in different proportions for the two isomers: 30% for the α subunit, 70% for β tubulin. It inhibits competitively colchicine binding, which could take place on more than one site, the irreversible binding site being located on α tubulin [225].

Contrary to the findings with tubulin of various species (sea-urchin, porcine brain), no ^{3}H-colchicine binding could be found in post-ribosomal supernatants from yeasts, two species of protozoa, and plant cells [50]: these results would be evidence of a heterogeneity of tubulins in evolution, a conclusion which is acceptable (cf. Chap. 2), but is in contradiction to many experiments on the action of colchicine in protozoa and plants.

While colchicine has been so widely used for obtaining polyploid plant cells, through its action on mitosis, it is remarkable that plant MT appear to be more resistant to colchicine: the concentrations required to affect cell division in various species are as high as 10^{-5} and even 10^{-3} M. The chemical relations between plant tubulins and colchicine remain poorly understood (cf. review in [110]). This apparent resistance of plant MT is also observed with other poisons, and it is not evident that differences in permeability may explain the necessity of such high concentrations of drugs [111]. The protein binding colchicine has been extracted, and is similar to animal tubulin, in particular by the fact that colchicine binding is enhanced by VLB. Some of the substances present in plant extracts may have an inhibitory action on colchicine binding, and the problem is made more complex by the fact that in some conditions the extracts also bind lumicolchicine, which does not affect MT [112].

Colchicine, in aqueous solvants, becomes fluorescent when combined with brain tubulin [27]. Podophyllotoxin competitively decreases this fluorescence. A study of various colchicine derivatives shows that the tropolone ring and ring A are involved. A study of the thermodynamics of colchicine binding by this method yielded results comparable to those mentioned in Chapter 2 (enthalpy of binding: 10 kcal/mol, entropy: + 62 eu) [27]. This technique has also led to the conclusion that the binding of 1-anilino-8-naphtalene sulfonate to brain tubulin, which increases more than 100fold the fluorescence of the protein, takes place on a different site from colchicine [28]. The very large increase of colchicine fluorescence at 430 nm (300fold) when bound to tubulin shows a linear relation with concentration up to 2 mg/ml [7].

Curiously, colchicine binding is increased in vitro in the presence of rabbit antisera against chick brain tubulin; the decay rate of binding is however not affected [8].

5.2.3.2 Destruction of MT by Colchicine. While the binding of colchicine prevents, either in vivo or in vitro, the assembly of tubulin subunits into MT, many facts appear to indicate that colchicine is capable of destroying already formed MT. In the literature on MT, the expression "destruction of MT by colchicine" is to be found over and over again, many authors having concluded hastily that the absence of MT in colchicine-treated cells meant that they had been "destroyed" by the alkaloid.

A contradiction is apparent between several of the experimental data and the fact that tubulin subunits bind colchicine much more strongly than MT. This conclusion [271, 272] has often been based on the study of *doublets*, which belong to structures—cilia, flagella, centrioles—which are known to be resistant to colchicine [20, 21]. Other resistant MT, which are not in doublet form, are those of the marginal bundle of nucleated erythrocytes and thrombocytes [20, 21].

The singlet MT of nerve cells (neurotubules) are more resistant to colchicine than those of other cells (cf. Chap. 9). Their turn-over is probably quite slow, assembly taking place in the neuronal body, and disassembly at the extremity of axons and dendrites, the MT being transported in the extensions of the nerve cell by the slow component of axonal flow (cf. Chap. 9).

One interpretation which has often been presented is that only MT which are in a condition of dynamic renewal could be destroyed by colchicine—their disappearance being in fact the consequence of a block in their assembly.

While this may explain the great sensitivity of mitotic spindle MT, which are known to be in a condition of rapid change and renewal (cf. the "northern lights" phenomenon as described in Chap. 10), it is more difficult to understand the rapid disappearance (or disassembly) of MT in leukocytes, platelets, and axopodia. In all these cells, the MT are singlets, even if linked by various types of bonds: the most resistant MT are mainly found in cilia and centrioles, which are only affected by MT poisons during their regeneration (vide infra) and appear completely resistant once assembled, as mentioned in Chapter 4. The resistance of basal bodies is similar: all studies on deciliation show that the basal bodies are not affected and rapidly grow new cilia. Colchicine has been shown to affect ciliary regeneration, as observed in *Stentor* [187] and *Tetrahymena* [205]: this is explained by the fixation of colchicine on the tubulin pool necessary for cilia assembly.

It could be argued that the doses of colchicine necessary to destroy already formed MT are far greater than those which prevent their assembly: this is possible, as inhibition of assembly may result from inactivation of a small percentage of the tubulin molecules present. As suggested by Wilson [269], the assembly could be prevented "by the presence of a colchicine . . . molecule bound at the growing end of a microtubule". This would be possible if the growth of MT only took place at one of their ends by the addition one by one of subunits (monomers or dimers) to the MT, a fact which is far from being demonstrated. Moreover, the doses of colchicine which destroy MT are not necessarily much larger than those which are known to inhibit MT growth, and these differences may result from other factors (permeability of the cells, size of the tubulin pools, influence of temperature and ionic conditions).

In vitro, however, while not affecting assembled MT, colchicine (0.1×10^{-3} M) does destroy the 30 S component (rings) that is an intermediate in MT assembly ([194]; cf. Chap. 2).

The conclusion that the colchicine site is not accessible in MT [272, 275] is only apparently in contradiction to many facts from the literature, where already formed MT are rapidly seen to vanish under the influence of the alkaloid. These observations can perhaps all be explained by the dynamic condition and the relatively rapid renewal of MT. In contrast, in doublets and perhaps better in triplets, a complete resistance to colchicine is observed in all types of cells. This suggests either that the colchicine site is protected (by some stabilizing protein?) or that the doublet configuration (in which the tubulin dimers are differently oriented, as demonstrated by Amos and Klug ([5]; cf. Chaps. 2 and 4), has a stabilizing effect. Evidently, the thermodynamics of this stability deserve further notice, for this property is not limited to colchicine, and applies to all MT poisons and to the action of cold and high hydrostatic pressures.

5.2.3.3 Assembly of MT in the Presence of Colchicine. In dealing with ciliogenesis, it has already been mentioned that in experimental conditions colchicine did not appear to disturb the formation of new centrioles, basal bodies, and cilia. The following facts should be kept in mind and show the different behavior of these complex assemblies of MT as compared to most single MT:

a) After treatment with colcemid, Chinese hamster fibroblasts grow an increased number of cilia [247].

b) In homogenates of surf clam eggs, functional MTOC and centrioles were present even after colchicine treatment, suggesting the presence of tubulin "in a state . . . insensitive to colchicine" [267].

c) In a study of the regeneration of cilia in *Tetrahymena*, normal cilia grew in 6 h, although ^{3}H-colchicine was present in the cytoplasm, in excess in relation to the amount of tubulin [280]. It is suggested that these newly formed cilia are formed of a special kind of tubulin, which may have lost its capacity for binding colchicine, in agreement with the conclusions of Kirschner et al. [139] on the existence of two states of tubulin differing by their colchicine-binding properties (cf. Chap. 2).

d) A multiplication of centrioles has also been observed in fibroblasts treated with the MT poison VCR [143].

e) In rat pituicytes, the formation of basal bodies and cilia was not affected by a single dose of colchicine (400 μg) or by repeated injections of VCR [85, 86].

5.2.4 Fibrillary Cytoplasmic Changes

In many cells an increased number of cytoplasmic fibrils is visible when MT are destroyed by colchicine. This was first noticed in the anterior horn neurons of rabbits injected either with colchicine or VLB [277, 278, 279] and in tissue cultures of neurons, suggesting a transformation of MT into filaments.

However, it was also noticed that similar fibrils could appear after treating neurons by aluminium salts, which do not destroy MT [45]. These filaments, with a diameter of 10 nm, were compared with similar structures found in pathological conditions in man (cf. Chap. 11). Such fibrils were observed in cultured nerve cells [282]

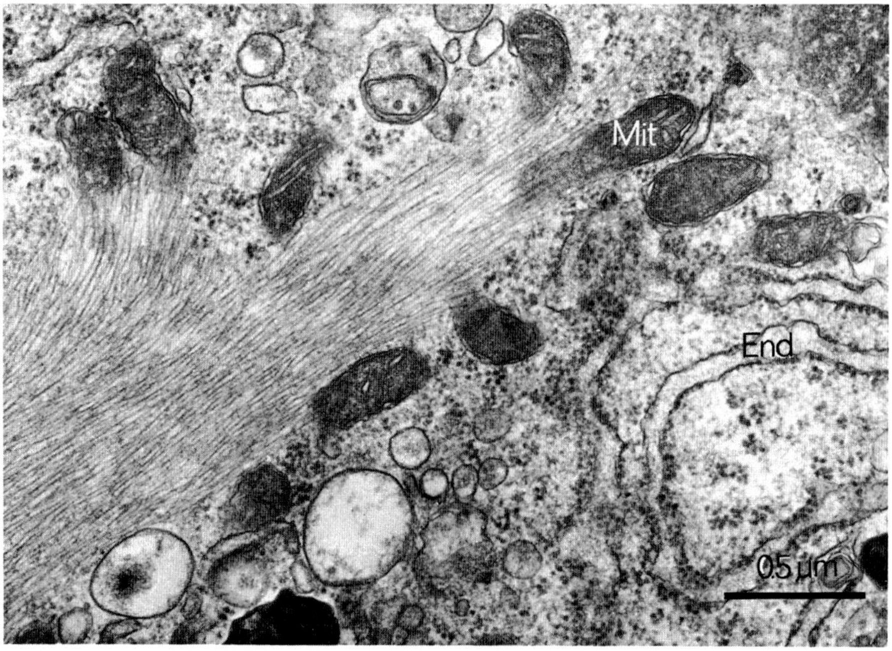

Fig. 5.6. MO cell (mouse embryonal cell) treated with colchicine 1 μg/ml for 24 h. Fibrillar changes of the cytoplasm. *Mit* mitochondria; *End* endoplasmic reticulum (from De Brabander [70])

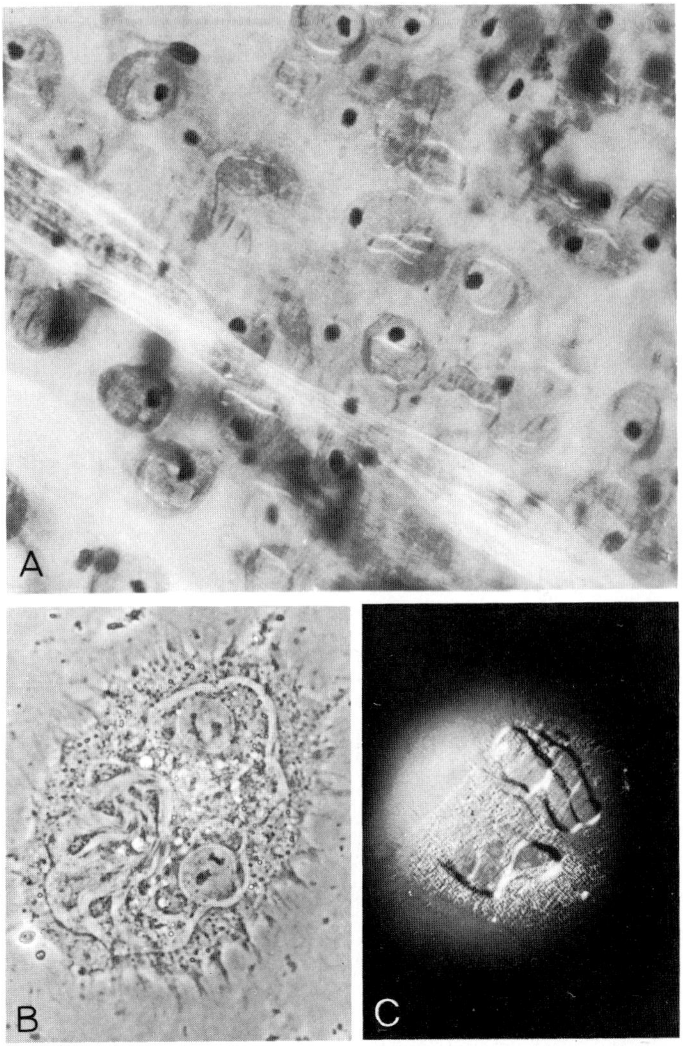

Fig. 5.7 A–C. (A) 5-day-old muscle culture from chick treated for 24 h with colcemid 10^{-6} M, viewed through polarizing microscope. The metaphase-arrested cells are recognizable by their dark pycnotic nuclei. Each of these cells shows a positively birefringent band, generally continuous. This band is made of 10 nm fibrils. (B) Two muscle cells have been treated first with cytochalasin B and then for 40 h with colcemid (10^{-6} M). The cells are binucleated and show twisted, continuous clear bands made of 10 nm fibrils (phase contrast). (C) The same cells under the polarizing microscope, showing the birefringence of the 10 nm filaments (from Croop and Holtzer [62])

and in fibroblasts after destruction of their MT by colchicine (5 to 40 μg/ml). These structures, which are birefringent, form a sheath around the nucleus [104].

Similar accumulations of filaments have been described in fibroblasts after treatment with colchicine and other poisons: however, these 10 nm filaments grow more and more numerous during 48 h, and when the cells are replaced in a normal medium, they decrease quite slowly, while MT reappear within 4 h. Cycloheximide completely prevents their formation: they may also appear when tubulin has been crystallized intracellularly under the action of *Vinca* alkaloids [71].

In post-mitotic myotubes, cultured in vitro, large bands of such birefringent fibrils appear under the influence of 10^{-6} M colcemid. They have a diameter of 10 nm, and assemble even in the presence of cycloheximide, making up as much as 20% of the cell volume. These fibrillar "cables" are still more apparent in cells treated with cytochalasin B. Their nature and possible relations to actomyosin are not clear, although they resemble the other fibrillary changes described in cells treated with MT poisons [62]. As mentioned above, in other cells cycloheximide inhibits similar changes. Holtzer et al. [125] have reviewed the various types of cells in which the 10 nm filaments may appear, sometimes in very large quantities. They are not stained by fluorescent antitubulins. Attempts to purify their constituent proteins have yielded two bands, one with a molecular weight of about 58,000 daltons. Although these filaments may represent up to 25% of the cell volume, their nature remains undefined.

In the axons of the crayfish nerve cord, warming destroys the MT; if the cooled preparation is incubated with colchicine, masses of aggregated filaments of 5 – 6 nm are formed. They are quite different from the VLB crystals [121]. It should be indi-

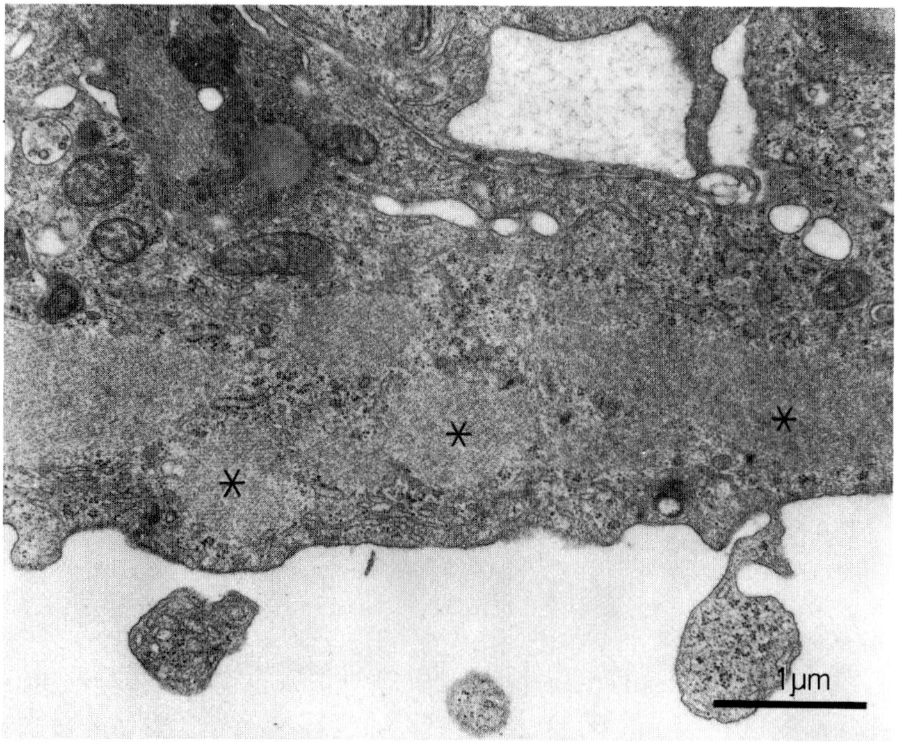

Fig. 5.8. Rat. Intraventricular injection of 100 μg colchicine over 1 h, after cannulation. Formation of crystalline inclusions (✱) in the tanycytes of the brain third ventricle. The crystals occupy the apical pole of the cell; no more MT are visible (from Schechter et al. [223])

cated that in this animal, neither low temperature (0 °C) nor colchicine destroys MT.

There are at least two other references to *paracrystalline* inclusions after colchicine treatment. In rat's tanycytes (glial cells located close to the ependyma) intraventricular injection of the alkaloid destroyed the MT and crystals comparable to those found after VLB were observed [223]. In quite different conditions, in leaflets of the moss *Sphagnum* treated for 10 h with a 1/1000 solution of colchicine, cytoplasmic paracrystals were found, although the MT did not disappear entirely, except in mitotic cells [227].

In studies on the tubulin polymorphs assembled in vitro (Chap. 2), colchicine was found to lead to the formation of filamentous polymers, about 17 nm in diameter (larger than the filaments described above), in the presence of 1 mM EGTA at pH from 5.8 to 6.4. When twisted tubulin ribbons formed in the presence of calcium ions were treated by colchicine, filamentous polymers were observed [175].

5.2.5 Modifications of MT Side-Arms

In spermatogenesis, in various species, the elongating nucleus of spermatids is surrounded by a "manchette" of MT which may play a role in the shaping of the nucleus (cf. Chap. 6). Colcemid destroys these MT in adult rats (6 h after an injection of 5 μg/g of body weight). When this destruction is complete, rounded dense bodies, much less numerous than the MT, are observed close to the nucleus: these may be formed of disassembled tubulin, and resemble the dense bodies described in centriole formation (Chap. 4). Other changes are the presence of C microtubules (cf. Chap. 2) and modifications of the inter-MT links, which become considerably longer than in control animals.

These abnormal bridges, which reach a length of 16 nm, twice the average size, were thought to be formed of MT subunits [162]. This is in agreement with several indications that some of the lateral expansions of MT are modified by cold or MT poisons: in HeLa cells, interMT bridges disappear after one hour at 4 °C [30], and in the testis of *Gerris remigis* Say the curving lateral expansions between MT are destroyed by colchicine. These are more or less comparable to C MT [255].

5.2.6 Resistance to Colchicine

The doses of colchicine which are active in various species differ largely, from 10^{-6} M to 10^{-2} M. This is related to the varying fragility of MT — those of the mitotic spindle being very readily destroyed, those of some axopodia needing much larger concentrations. It depends also on the cell permeability to the alkaloid, as demonstrated many years ago in the protist *Amoeba sphaeronucleus*, where mitotic arrest is only obtained by injecting the drug into the cytoplasm [56]. Permeability factors also play a role in mammalian cells, as evidenced by the study of a colchicine-resistant line of Chinese hamster cells [155].

In 1952, it had been shown that large doses of colchicine (10 – 20 mg/kg of weight) did not affect the mitoses of the golden hamster (*Mesocricetus auratus*)

[195]. Later, it was demonstrated that resistance to podophyllotoxin also existed (which binds to the same site of the tubulin molecule) [75]. However, larger doses of both drugs were effective: for colchicine, 50 mg/kg was necessary to arrest all mitoses in the bone marrow within 6 h. On the other hand, the spindle MT could be destroyed by doses of sodium cacodylate similar to those active in the rat [75].

The resistance of the hamster has been correlated with the metabolism of colchicine: tubulin from hamsters polymerizes in vitro like that of other rodents, and is equally sensitive to colchicine [95]. In the mouse colchiceine is formed, but not in the hamster [229]. On the other hand, the liver of the hamster converts about four times more colchicine to other metabolites, but this cannot explain the animal's resistance, as the difference of lethal doses between the hamster and the mouse or the rat is 100fold [229]. The resistance is not explained by an increased biliary excretion [126, 127].

Resistant clones of cells have been selected from tissue cultures of Chinese hamster ovary cells, resistance being measured by the number of surviving cells after colchicine treatment. Several mutants lines displaying resistance were studied, and demonstrated to have a reduced uptake of ^{3}H-colchicine. However, the colchicine-binding properties of cell extracts were normal. Hence, resistance was correlated with a relative impermeability of the cell toward the alkaloid [155]. These mutant lines were resistant to VLB and to colcemid, and also to the quite unrelated drug actinomycin D. Resistance was decreased by the action of the detergent Tween 80 [217]. The role of permeability has been confirmed in resistant lines from mouse and Syrian hamsters: these appear with a high frequency and are not true mutants. They revert to a sensitive condition when cultured in the absence of colchicine. Resistance was also present for VLB, actinomycin and puromycin, and can only be explained in terms of a decreased cell permeability. The natural resistance of the hamster cannot result from the same mechanism: it has been suggested either that an extra binding pool of tubulin is present [181] or that a surface site binds colchicine and protects the intracellular tubulin. This is in agreement with several recent observations of colchicine-binding proteins (tubulins?) in cell membranes. Other problems of genetic resistance to MT poisons will be considered in Chapter 11.

It is evident that the varying sensitivity toward colchicine and the drug resistance of some cells deserve further study. The natural resistance of the hamster, with its cross-resistance with podophyllotoxin, is of particular interest for a study of the binding site of tubulin for these two poisons.

5.2.7 Colchicine Pharmacology

It is fundamental for research on whole animals to understand properly which actions of colchicine are related to MT poisoning, and which are not. This is far from easy, and all recent work on colchicine, demonstrating its action on so many different cell processes, and often conducted with widely differing dosages, and on diverse animal species, makes it difficult to assess the effects truly related to MT, particularly if morphological data do not clearly show marked changes of these organelles. For instance, the central nervous toxicity of colchicine may result from alterations of neurotubules, although considerable neuronal changes may take place under the influence of colchicine without apparent MT involvement (cf. Chap. 9).

Colchicine toxicity was well known in the 19th century and, as explained in Chapter 1, led to the discovery of its action on cell division and later on MT. Colchicine poisoning in man will be discussed in Chapter 11.

Colchicine pharmacology was analyzed in our 1955 monograph [90] and a good review of recent work has been published by Creasey [61]. It may be worthwhile to summarize rapidly the principal pharmacological effects of colchicine as known in 1955, apart from its therapeutic action in gout. The effects which are clearly connected with mitotic MT poisoning, such as diarrhea and blood changes, will not be mentioned here.

a) *Nervous system*: the intracerebral injection of colchicine in the cat leads to a rapid death, with vasomotor changes and respiratory paralysis.

b) *Striated muscle*: a sustained increase in contractile force has been observed with large doses of colchicine (1.6×10^{-2} M), with a marked increase of glycolysis.

c) *Liver*: steatosis and cell damage have been described in the mouse, mainly after repeated injections. Intranuclear inclusions (resulting probably from cytoplasmic invagination) appear after daily injections in mice [149].

d) *Blood*: apart from the leukocytosis and thrombocytosis (cf. Chap. 11) and bone marrow aplasia after large doses, colchicine may affect the blood-clotting time.

The metabolism of colchicine has been repeatedly studied and the first use of ^{14}C-colchicine was to follow its location in the body. Colchicine is excreted unchanged in urine [9], and remains for some time in many organs of patients dying of an overdose. The persistence of the alkaloid was known from observations on hibernating bats, which do not appear to be affected until awakened and warmed [90]. Similarly, in the frog, a considerable difference in toxicity is found in relation to the temperature: in normal conditions a large dose of colchicine (50 mg) is excreted unchanged in the urine without any noticeable disturbance; in contrast, if the animal

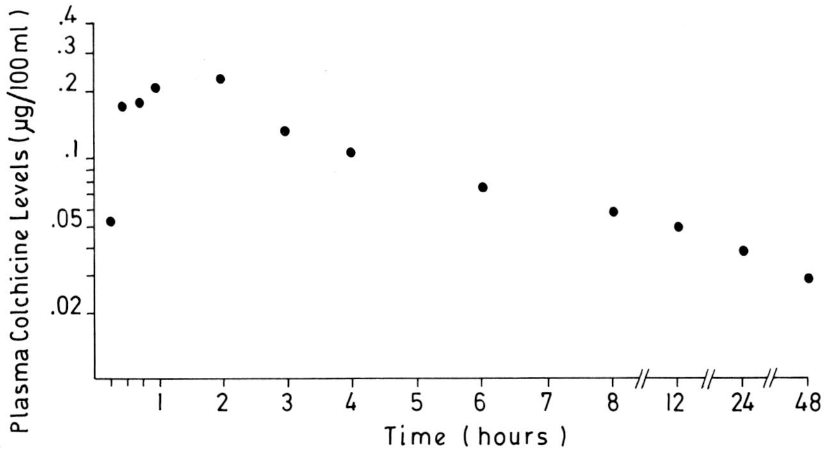

Fig. 5.9. Colchicine levels in plasma of ten human subjects, after oral administration of 1.0 mg. The curve represents the averaging ot two populations: four patients had peak colchicine levels at 30 min after ingestion, four at 2 h, and two others at 45 min and 1 h respectively (redrawn from Wallace and Ertel [260], slightly modified) (By permission of Grune and Stratton, Inc.)

is warmed to 32 °C, death results in a few days with evidence of nervous paralysis. In *Rana esculenta* the lethal dose at 15 – 20 °C is as high as 1.2 – 2.0 g/kg of weight, while at 30 – 32 °C it is similar to that found in the mouse, i.e., 2 – 4 mg/kg [cf. 90].

Some more recent results on the fate of colchicine in the body, its excretion and metabolism may be mentioned. In human volunteers, the metabolism of 1 mg methoxy ^{14}C labeled colchicine administered orally was studied. The plasma concentration reached a peak between one and two hours, and fell slowly within the next eight hours (Fig. 5.9). The highest mean concentration for all subjects was 0.323 ± 0.73 μg/100 ml. After an intravenous injection, colchicine is cleared from the plasma with a half-time of less than 20 min, indicating its rapid penetration into cells [260]. In rats, the biliary excretion of ^{3}H-colchicine has been studied, and compared to that of hamsters, dogs, and rabbits. The rat excreted the largest quantity and 68% were found in the feces in 48 h after intravenous administration. Colchicine is concentrated in the liver and excreted in the bile against a bile-plasma gradient, indicating that this is an active process. Twenty minutes after injections of ^{3}H-colchicine the liver and the kidney contained the highest radioactivity, and the lowest concentration was found in the brain [126, 127]. In man, therapeutic doses of ^{3}H-colchicine also showed a short plasmatic half-life (19.3 min), while the concentration in the cytoplasm of leukocytes was 3 to 17 times higher than that in the plasma [92].

A new method of study of colchicine metabolism, which holds great promise, is the application of radioimmunological techniques. It has been possible to prepare antibodies in rabbits injected with a bovine serum albumin conjugate of N-desacetylthiocolchicine. This appears to have a good specificity, as only N-desacetylthiocolchicine, N-desacetyl-N-methylcolchicine, and colchicine showed a slight cross-reaction (less than 8% of that of colchicine; [36]).

The conclusion that colchicine is not metabolized in the body is in contradiction to several observations, indicating that in the liver microsome fraction it undergoes oxidative demethylation, however in varying amounts according to the animal species. Ring A is demethylated, and only in the rat and the mouse, which are most sensitive to colchicine, is ring C demethylated, leading to the formation of colchiceine [229].

There are some hints of an action of colchicine on nucleic acids. A decrease in cytoplasmic RNA and an increase in nucleolar ribonucleoprotein has been found in Ehrlich ascites tumor cells [158], indicating a decreased synthesis of RNA, also found with VLB and to a lesser extent with VCR [60, 61]. An interaction of colchicine with DNA has been suggested as a explanation of the optical rotation of mixtures of colchicine and DNA [128]. A depression of DNA, RNA and protein synthesis has been related in *Tetrahymena* after a prolonged action of colchicine [148]. Another effect related to nucleoproteins (although not specific, as lumicolchicine is equally active) is the inhibition of nucleoside transport across membranes of mammalian cells (at concentrations of $10^{-4} - 10^{-5}$ M [25, 182]).

5.2.8 Colchicine Antagonists

Although in several experimental conditions the action of colchicine on MT (principally on spindle MT) appears to be somewhat antagonized by various agents, these effects do not show any evident specificity, and may in some conditions result from a stabilizing action on the MT.

The antimitotic action of γ-hexachlorocyclohexane or gammexane on plant mitoses led to the discovery that meso-inositol, whose chemical structure is closely related, acted as an antagonist. More surprizing was the finding that meso-inositol (and neither D-inositol nor D-sorbitol) could antagonize the actions of colchicine on root-tips and prevent the formation of the typical bulbous growths. A similar antagonism was demonstrated in rat fibroblast cultures, where the recovery from the action of colchicine on mitoses was more rapid in the presence of *meso*-inositol. Other sugars and related molecules did not show this effect [184].

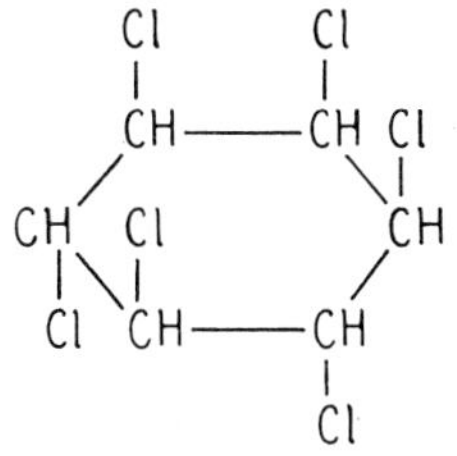

γ-HEXACHLORO-
CYCLOHEXANE
(Gammexane)

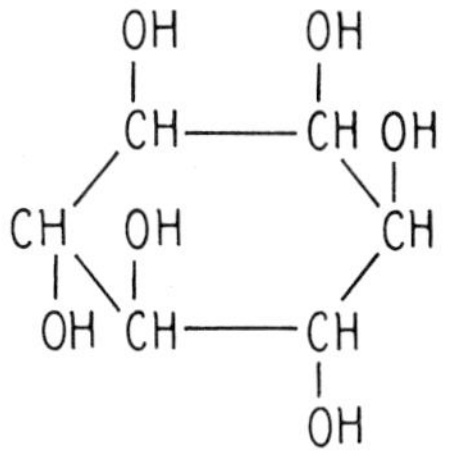

MESO-INOSITOL

The discovery of the role of ATP in muscle contraction led to speculations about its possible role in spindle dynamics, and a decrease of the antimitotic action of colchicine in the presence of large doses of ATP was reported ([152]; cf. Chap. 10). The recent knowledge about the role of guanosine phosphoric esters in tubulin assembly, and the possible phosphorylation of tubulin (cf. Chap. 2) suggest that ATP may help to stabilize the MT, and only indirectly antagonize colchicine.

A quite different mode of action is that of the hormone melatonin, which has been shown to act as a MT poison (vide infra), inhibiting the oral regeneration of *Stentor coeruleus* [13]. An antagonism of melatonin and colchicine on the spindle birefringence had been mentioned [163]. Melatonin, which binds to MT, was shown later, at concentrations which are without other effect on *Stentor*, to protect this ciliate against death by colcemid. Some effects of melatonin can conversely be antagonized by colchicine. Melatonin would be bound to MT at a different site from colchicine and increase MT resistance to the alkaloid [13, 96].

Another hormone may affect the survival of cells arrested in metaphase by colchicine or colcemid which, in mammals, are unable to survive more than a few hours. This antagonism has been shown in the mouse where the mitotic increase, resulting from the destruction of spindle MT, lasts several hours longer in adrenalectomized animals. The effect of adrenalectomy could be prevented by cortisone, which apparently decreases by indirect metabolic changes the resistance of arrested mitoses [83] (Fig. 5.10).

In plant cells, a new antagonist of colchicine has been recently mentioned, *pemoline* or 5-phenyl-2-imino-4-oxazoleidine. In *Vicia fava* seeds treated with 0.005% solutions of colchicine, no increase of metaphases could be observed for four hours in the presence of 3×10^{-4} g/ml of pemoline [214]. This also protects the cells against the effects of caffeine, which has rather different actions from colchicine, while inhibiting mitosis. The action of pemoline has been compared to that of giberellic acid, which also shows effects antagonistic to colchicine in plants [97]. The growth effects of gibberelin, which promotes cell expansion, are reversed by colchicine, through its effect on MT [222, 237]. Colchicine acted also as an inhibitor of the

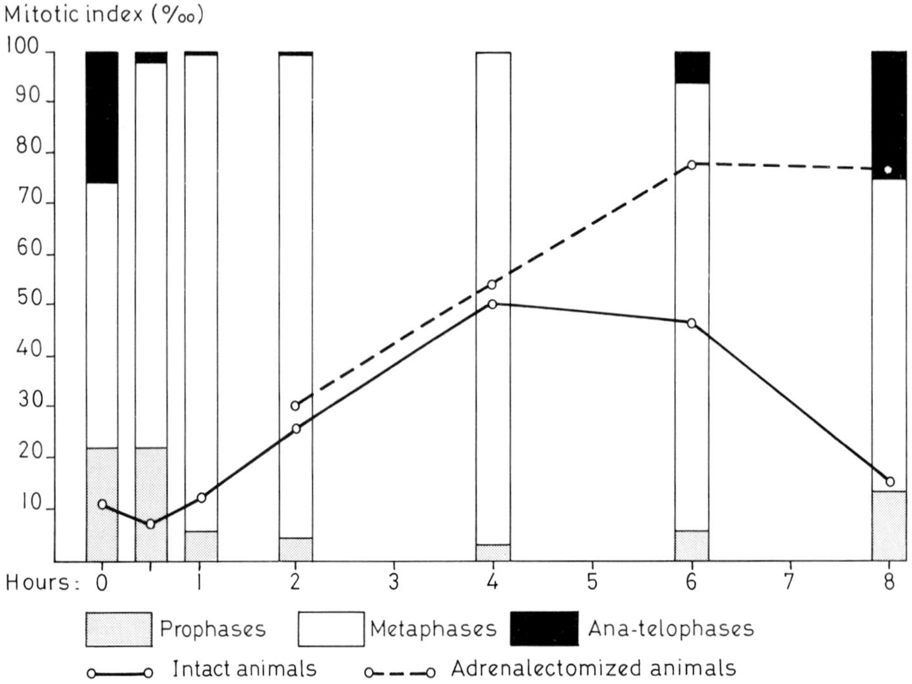

Fig. 5.10. Variations of the mitotic index and the percentage of mitotic stages in the bone marrow of normal (O —— O) and adrenalectomized (O - - - - O) mice after a single injection of 1 mg/kg of body weight of colcemid. The mitotic arrest lasts longer after adrenalectomy; the fall of the mitotic index after 6 h in intact animals is largely the consequence of cell necrosis, which is less pronounced after adrenalectomy. The columns refer to the intact animals (slightly modified, from Dustin [83])

action of indolacetic acid on cell elongation [237], demonstrating complex interrelations between MT and cell shape (cf. Chap. 6).

It can be concluded that the antagonistic actions which have been related are poorly specific and probably quite indirect. No true antagonist of colchicine has yet been found.

5.3 The *Catharanthus (Vinca)* Alkaloids

Originating from Madagascar, *Catharanthus roseus* (L.) G. Don (*Vinca rosea* L.) is a plant cultivated in warm climates, which has brought to phytochemists a wealth of new interesting chemicals. The recently published monograph by Taylor and Farnsworth [251] gathers much information on the plant, the chemical structure of its components, their action, and their use in cancer chemotherapy. Studied first because extracts of *Catharanthus* were supposed, in popular medicine, to be effective in diabetes (which proved false), the discovery of their marrow-depressing actions led to the observation of their mitotic poisoning effects which are linked with a destruction of spindle MT similar to that described for colchicine.

Contrary to colchicine, which has found its widest use as an agent for the production of polyploid plants, apart from its role in MT research, the two main indol alkaloids from *Catharanthus*, VLB and leurocristine or VCR, have important applications in the treatment of neoplastic diseases (cf. Chap. 11).

These two complex molecules—which only differ slightly—have many similar properties, while their chemotherapeutic effects are different. In this chapter, the fundamental problems of their fixation on tubulins and their action on MT will be studied, and in particular the crystalloid tubulin inclusions which result from their action. Some information on pharmacological actions unrelated to MT poisoning and on antagonists of their action will also be summarized.

The complex structure of these bifunctional indol derivatives (cf. [249]) results from the linking of a catharanthine molecule to a vindoline molecule. It is interesting to mention that although the general structure differs widely from that of colchicine, there is one analogy in the presence of two or three methoxy groups. The stereochemical structure of the molecule has not been published: while it does not combine to the same site as colchicine, it would be interesting to know if its shape has some resemblance to that of the tropolone alkaloid.

The *Catharanthus* alkaloids are active at very small concentrations, a fact which may explain their success in cancer chemotherapy (cf. Chap. 11). The assembly of porcine brain tubulin is prevented by 4.3×10^{-7} M VLB, a concentration consistent with the levels observed in vivo [198]. The mitoses of cells from a human cervix carcinoma cultures in vitro are arrested at concentrations of 10^{-9} g/l [65].

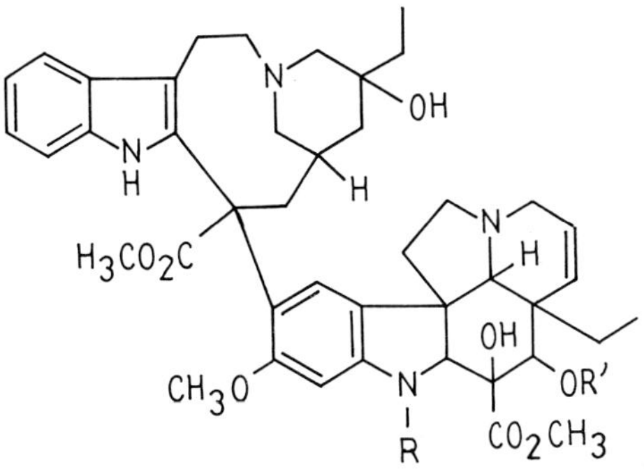
Vinblastine

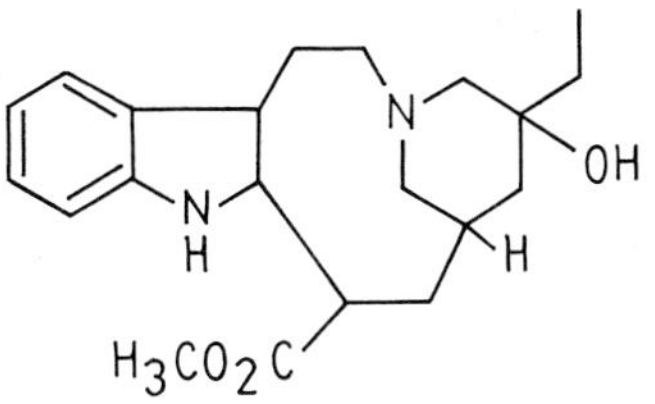
Catharantine

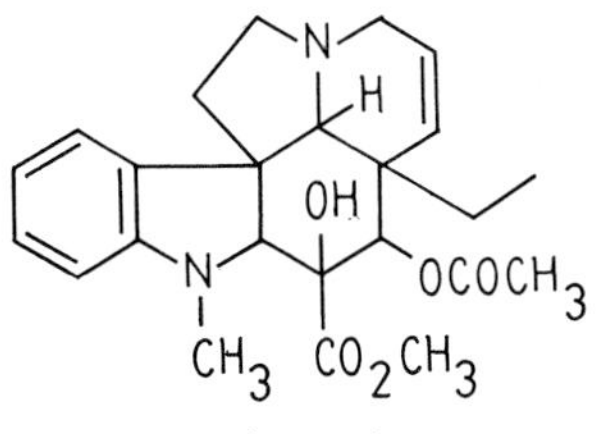
Vindoline

VBL: R = $-CH_3$ R′ = $-CO-CH_3$
VCR: R = $-COH$ R′ = $-CO-CH_3$

The study of molecules similar to VLB and VCR brings some information about the relation between their structure and activity. Desacetylvinblastine (which has only two methyl groups on the vindoline moiety of the molecule) is as active—as measured by the dose required to arrest mitoses in hamster cells in vitro—as VLB and VCR, while leurosidine, which differs by a slight change of the catharanthine part of the molecule, is 100 times less active. Catharanthine is slightly active (1,000 times less than VLB), while vindoline is completely inactive [271]. The concentrations required for a 50% mitotic arrest in HeLa cells are:

desacetyl VLB:	7.5×10^{-8} M
leurosidine:	7.5×10^{-6} M
catharanthine:	5×10^{-5} M.

The degree of stabilization of colchicine-binding activity (vide infra), which is an important property of VLB and VCR, follows the same order of activity [271].

5.3.1 Action of VLB and VCR on MT

Although VLB and VCR have different effects in cancer chemotherapy (cf. Chap. 11), their relations with MT are similar, and on the next pages these two drugs will be studied together. Many of the problems which have been met with colchicine will be encountered again, and the *Vinca* alkaloids bring further evidence of linkage with the tubulin molecule on the one hand, and apparent destruction of MT on the other. In vitro, freshly assembled MT are however resistant to VLB, while rapidly destroyed by the addition of calcium ions [270]. One main difference from colchicine is that VLB and VCR may aggregate the tubulin molecules in the cytoplasm or in vitro, leading to the formation of paracrystalline structures which may be isolated and have been used for obtaining pure tubulin (cf. Chap. 2).

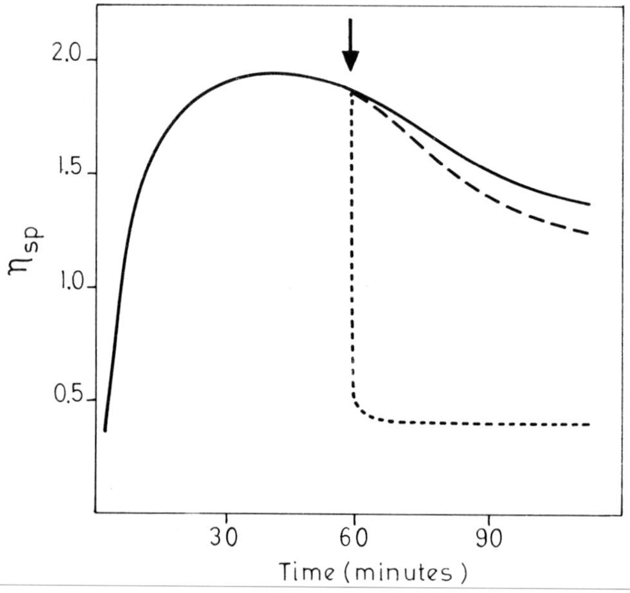

Fig. 5.11. Inability of VLB to depolymerize in vitro freshly assembled embryonic chick brain MT. Viscosity measurements. *Upper curve:* control. — — — —: 5×10^{-5} M VLB added at 60 min. - - - -: 5 mM excess calcium chloride added at 60 min: rapid decrease of viscosity indicating destruction of MT (redrawn from Wilson et al. [270])

The early morphological studies indicated a rapid destruction of spindle MT by VLB and VCR, leading to arrested mitoses similar to those found after colchicine [40, 53, 63, 64, 99, 102, 130].

Mitotic arrest could last longer than after colchicine—up to 20 h [40]. As in colchicine-arrested mitoses, centrioles are not affected: their reduplication proceeds normally [102] but they are often located at the center of the mitotic figure [131, 284]. In cells recovering from the action of VLB (Earle L cells) "large numbers of MT and centrioles" were observed [142].

All later experiments indicated that not only the spindle MT, but MT in various cells, independent of mitosis, became invisible under the action of VLB or VCR: in neurons, after intrathecal injection of VCR [228]; in neurons and oligodendroglial cells, after intracerebral implantation of VLB pellets in the rat [123]; in the rat kidney glomeruli, after intravenous VLB [256]; in the rat retina ganglion cells, after intravitreous injection [47, 48], etc. Other similar observations will be mentioned in later chapters, as the *Vinca* alkaloids have been used like colchicine as tools for the destruction of MT.

Like colchicine, this action may be followed, in many cells, by the accumulation of fibrils about 10 nm wide which are not believed to originate from MT [123, 131, 143, 150, 278] (HeLa cells, L fibroblasts, lymphoma cells, neurons).

All these changes are often—but not always—related to the formation of paracrystalline structures which are a feature of these drugs.

5.3.2 Binding of VLB and VCR to Tubulin and Crystal Formation

Studying the action of relatively low concentrations of VLB and VCR (from 1×10^{-5} M to 4×10^{-4} M) on L-strain fibroblasts and human leukocytes, Bensch and Malawista [23] were the first to describe crystalline bodies as long as 8 μm in some cells. The crystals were already visible in less than one hour, and appeared first in the pericentriolar region. Preincubation with colchicine (4×10^{-4} M for 24 h) did not prevent their formation: the importance of this observation will be underlined below. Properly oriented sections showed that the crystalline arrays were formed of hexagonally packed tubular structures, with a diameter of about 27 – 28 nm. Some isolated MT with a diameter of about 28 nm (*macro*tubules) were also seen by these authors, who concluded that the crystals were formed of tubulin. This was soon confirmed by experiments in which solutions of tubulin were incubated with *Vinca* alkaloids. Purified tubulin from brain treated with 10^{-4} M VLB formed complex precipitates with ladder-like and irregularly coiled profiles, much less regular than the crystals seen intracellularly [23]. The tubulin nature of the crystals was demonstrated by chromatography [170] and electrophoresis and confirmed by immunological methods [185]. Similar crystals were found in starfish oocytes treated with VLB (not with VCR); they were birefringent and uniaxial [166]. Crystalline inclusions were described in the following months in various cells (frog leucocytes, neurons, tissue cultures). It is now known that in any cell containing a pool of tubulin, this paracrystalline assembly of tubulin can be obtained either with VLB or VCR.

These crystals were important from several points of view: they led to a method of purification of tubulin (cf. Chap. 2) and provided material for crystallographic studies. The structure of the crystals in relation to that of tubulins and MT has turned out however to be difficult to solve (cf. Chap. 2).

The crystals induced by VLB in the cytoplasm of unfertilized eggs of *Strongylocentrotus purpuratus*, are remarkably stable (in fact much more stable than any other tubulin preparations), even in the presence of divalent cations [41]. Their formation is enhanced by heavy water—which in other conditions stabilizes MT (vide infra). They made up about 2% of all proteins of L cells, and showed some ATPase activity [41]. Electron microscopical observations have shown that in various cells, the crystals may be closely associated with other cytoplasmic structures, in particular ribosomes. For instance, in cell cultures treated at low temperatures with VLB, ribosomes "crystallize" helically around tubular structures [168]. These complex assemblies were inhibited by puromycin, cycloheximide, and actinomycin D, indicating the necessity of protein synthesis [144].

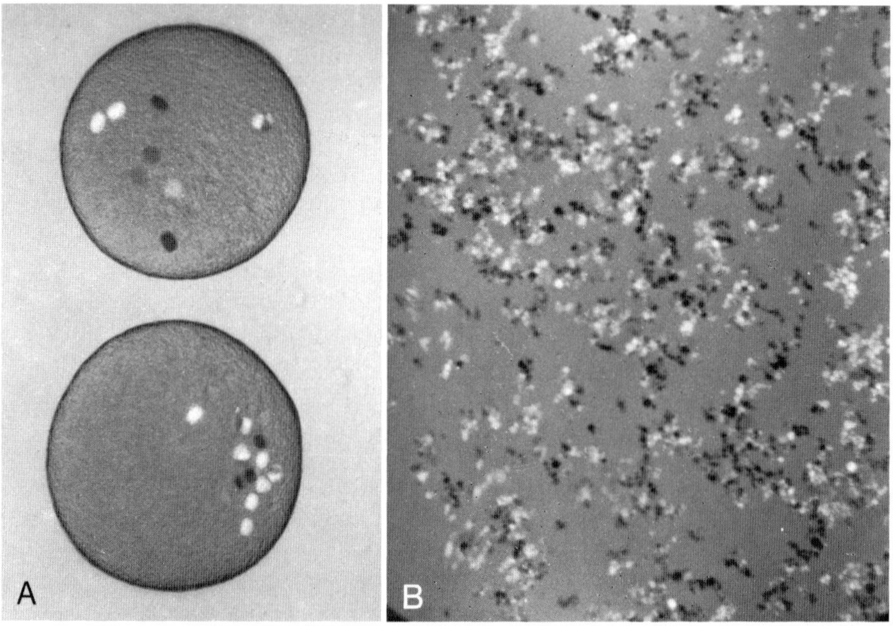

Fig. 5.12 A, B. VLB crystals in polarized light. (A) VLB crystals in living sea-urchin eggs. (B) Isolated VLB crystals (original document, by courtesy of H. Sato)

Two problems need further research: what are the links associating tubulins in these crystalline structures? What is the exact geometry of these assemblies? An important answer to the first question was the finding that the tightly bound GTP (on the N site) of tubulin was released when VLB induced crystals from tubulin solutions [26]: the nucleotide on the E site was not released. These results suggested that the nucleotides played a major role in MT assembly, although it remained impossible to know whether the longitudinal or the lateral bonds between tubulin molecules (or dimers) were affected. The VLB crystals contain equal amounts of α and β tubulins, evidence that MT are heterodimers (cf. Chap. 2). A precise study of the amounts of drugs bound by the crystals has also provided important data about the tubulin molecule: as they bind one mol of VLB per dimer and one mol of colchicine (on a different site, as the crystals bind colchicine as readily, or more so, than intact tubulin) and two mol of guanosine nucleotide, it was clear that at least four sites, of three different types, were present on the tubulin dimer [42]. It was also demonstrat-

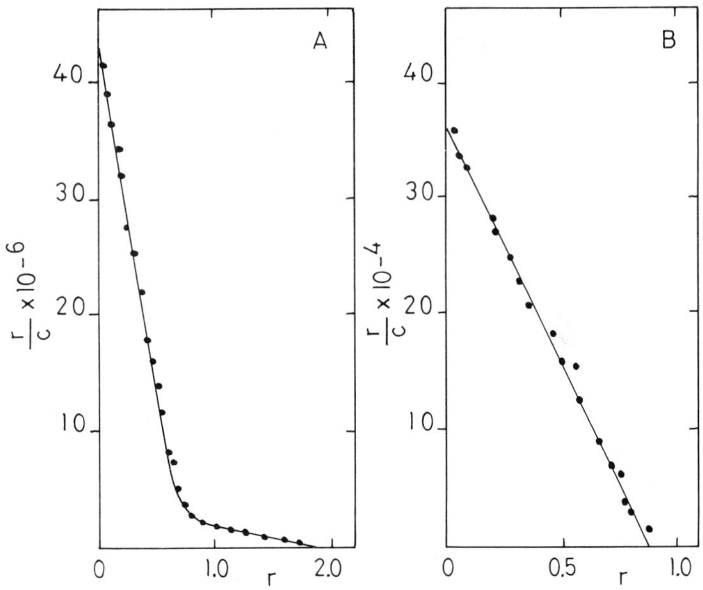

Fig. 5.13 A, B. ^{3}H-VLB binding to rat tubulin. Incubation at 37 °C for 30 min (Scatchard plots). (A) Freshly prepared solution (0.2 μM) in 0.25 M sucrose, pH 6.8. (B) Aged solution devoid of colchicine-binding properties. The non-linearity of plot (A) indicates the presence of two or more binding sites. When tubulin preparations were stored at 4 °C until they lost any affinity for colchicine (cf. Fig. 5.4) the low affinity site for VLB persisted with a stoichiometry of 0.85 mol of VLB per mol of tubulin (B) (redrawn from Bhattacharyya and Wolff [29])

ed by this technique that podophyllotoxin competes with colchicine, indicating that it is probably fixed on the same binding site. There is thus evidence that the crystals are made of tubulins α and β: the remaining problem is to understand how they are assembled, and why the indole alkaloids alone have this molecular action.

Further details about the binding of VLB to tubulin have been published by Bhattacharyya and Wolff [29]: rat brain tubulin has two binding sites, a high affinity one, with a constant of 6.2×10^6/M, and a low affinity one (8×10^4/M). The first corresponds to the half-maximal concentration of VLB needed to prevent assembly, the second to that needed to aggregate tubulin. Both sites approch a 1 : 1 stoichiometry. The high affinity site, like that for colchicine, decays rapidly. It is protected against decay by colchicine, while VLB also protects the colchicine site. At these concentrations, the effects of VLB are specific, while at concentrations of 10^{-3} M and higher, other proteins are also precipitated.

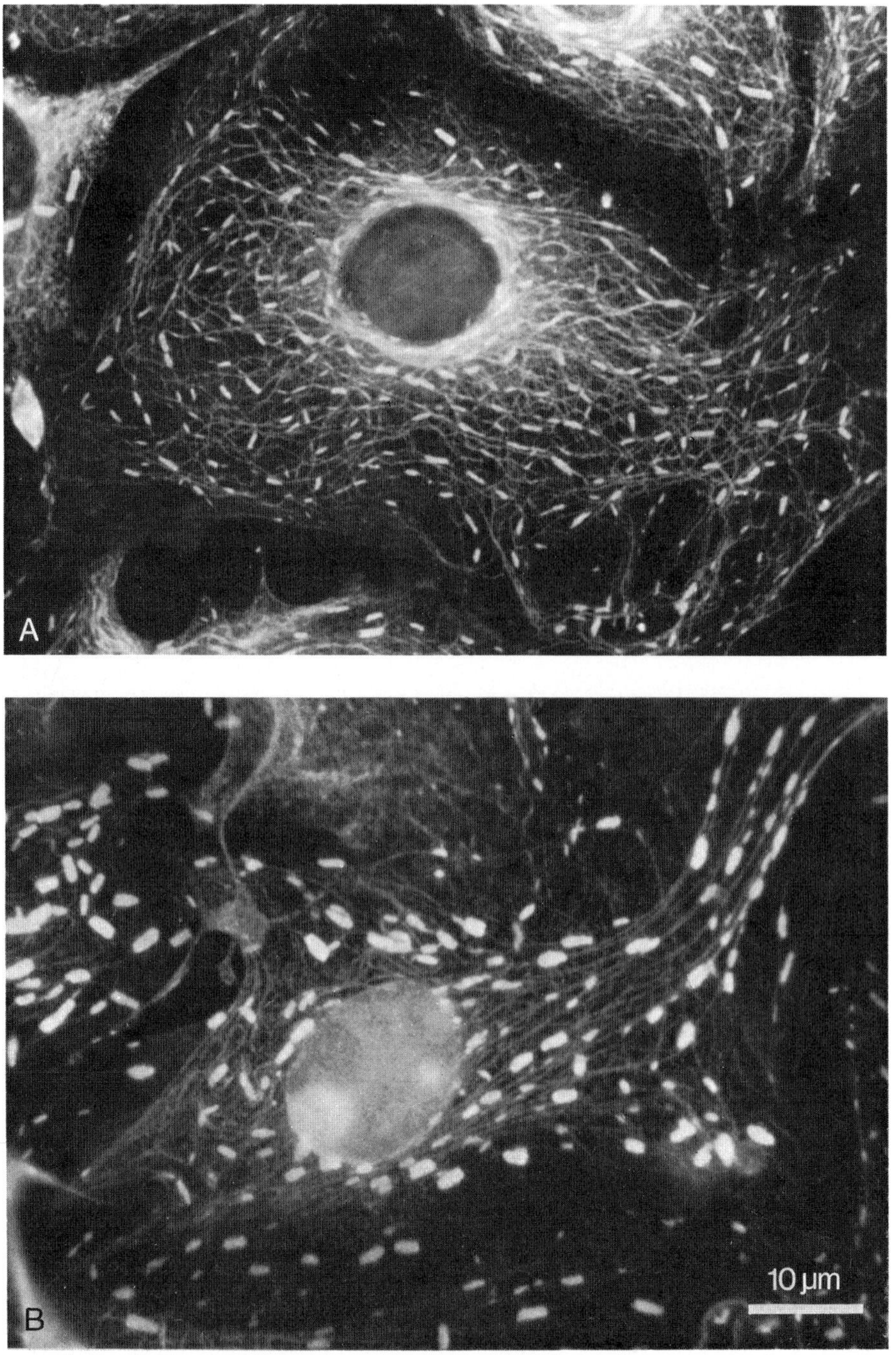

Fig. 5.14 A, B. Immunofluorescent staining of MT and VLB crystals in rat kangaroo Pt-K2 cells after simultaneous incubation with 10^{-5} M vinblastine sulfate (A) and 10^{-5} M colcemid (B). Time: 9 – 15 min, 37 °C. The crystals are formed along the cytoplasmic MT and then enlarge (from Sato et al. [219])

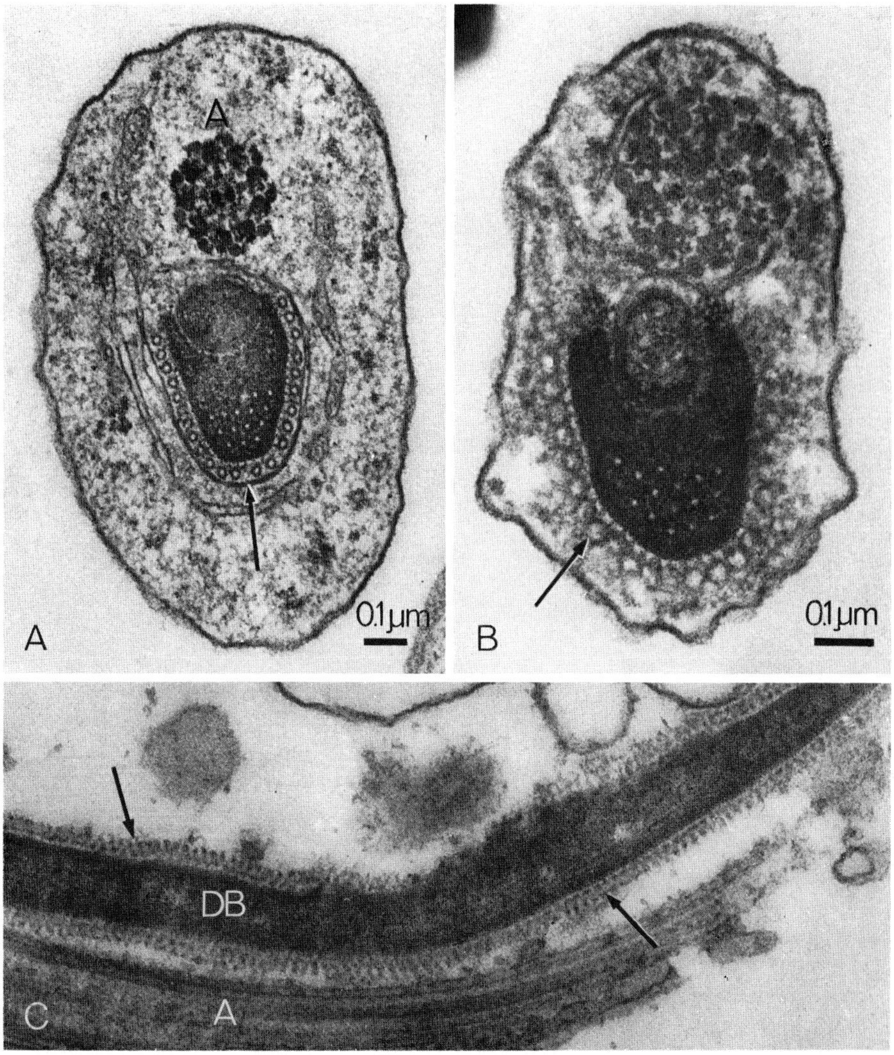

Fig. 5.15 A–C. (A) Transverse section of non-treated *Nephrotoma* spermatid showing the flagellar axoneme with 9+2 symmetry (*A*), the perimitochondrial dense body and the sheath of MT around this body (*arrow*). (B) Spermatid treated for 1 h at room temperature in Ringer solution with 1 mM VLB sulfate. The MT located around the dense body now appear as circular profiles (*arrow*). The flagellar axoneme is not modified. (C) Longitudinal section of a *Nephrotoma* spermatid after 1 h treatment with 1 mM VLB sulfate. The dense body (*DB*) is surrounded by helical structures resulting from the modification of the MT (*arrows*). The flagellar MT are not affected (*A*) (from Behnke and Forer [22])

The *Vinca* crystals are readily stained by immunofluorescence [265] and immunohistochemical techniques [76], a fact which indicates that the antigenic sites of tubulins are accessible.

The study of the crystal structure is closely linked with that of the "macrotubules" mentioned above. Although it is not at all evident that all large tubules which have

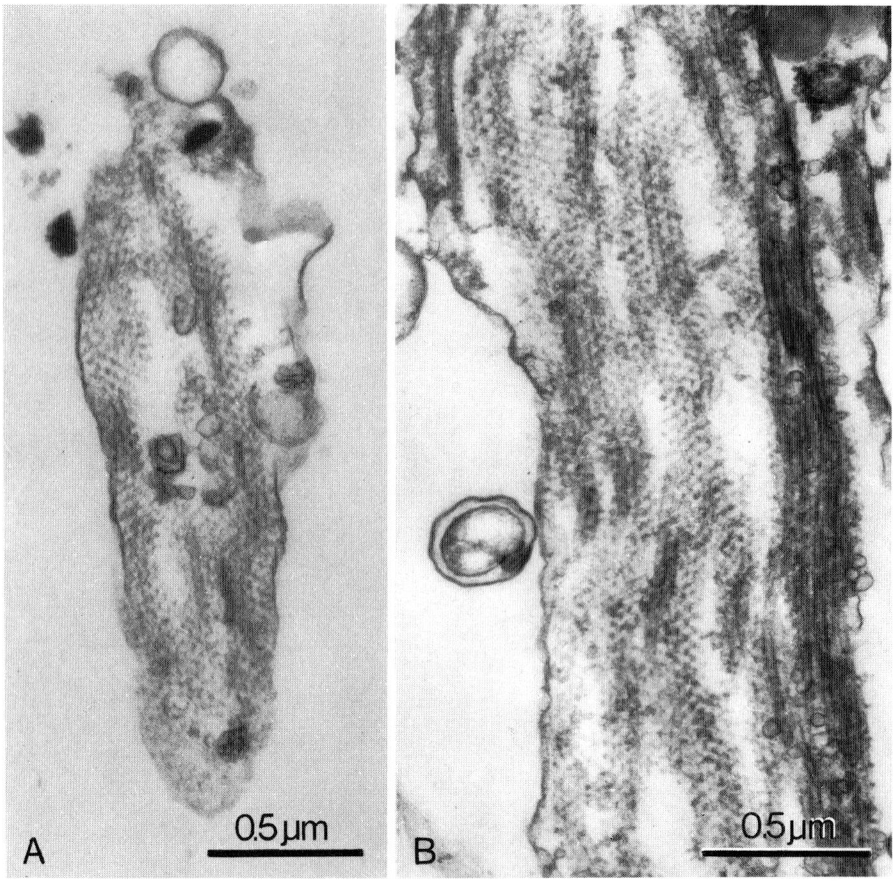

Fig. 5.16 A, B. Longitudinal sections through the axopodial tip (A) and base (B) of *Echinosphaerium nucleofilum*, after treatment for 10 min with 30 mM $MgCl_2$. Helical structures made of MT subunits (cf. similar structures observed after VLB and VCR, Figs. 5.15 and 5.18) (from Shigenaka et al. [240])

been called "macrotubules" are identical, a careful study of their formation and ultrastructure with the help of negative staining methods has shown, in spermatids of the crane fly (*Nephrotoma suturalis*), that the normal MT surrounding the nucleus ("manchette", cf. Chap. 6) change, under the action of VLB, into helical structures [21]. This observation confirms that made in vitro by similar negative staining techniques, of a loose spiral arrangement of filaments in the VLB crystals [169]. It is interesting to mention that in the axonemes of heliozoa, similar helical structures may be induced by Mg^{2+} [240].

The molecular structure of the *Vinca* paracrystals has been carefully studied in two recent publications. Crystals resulting from the action of VLB on the unfertilized eggs of the sea-urchin *Echinus esculentus* have been observed after osmium acid fixation (as they are damaged by the conventional glutaraldehyde fixatives) with an electron microscope provided with a goniometer stage. Electron diffraction from the crystals and optical diffraction patterns obtained from micrographs were compared. Three different patterns were observed: (a) a network similar to that mentioned by Bensch and Malawista [23] and resembling a hexagonal packing of tubules about 24 – 28 nm wide, which is considered to result from the intersection of three linear arrays, forming more or less hexagonal tubes separated by small triangles; (b) parallel rows with a beaded appearance, separated by oblique "rungs" regularly spaced;

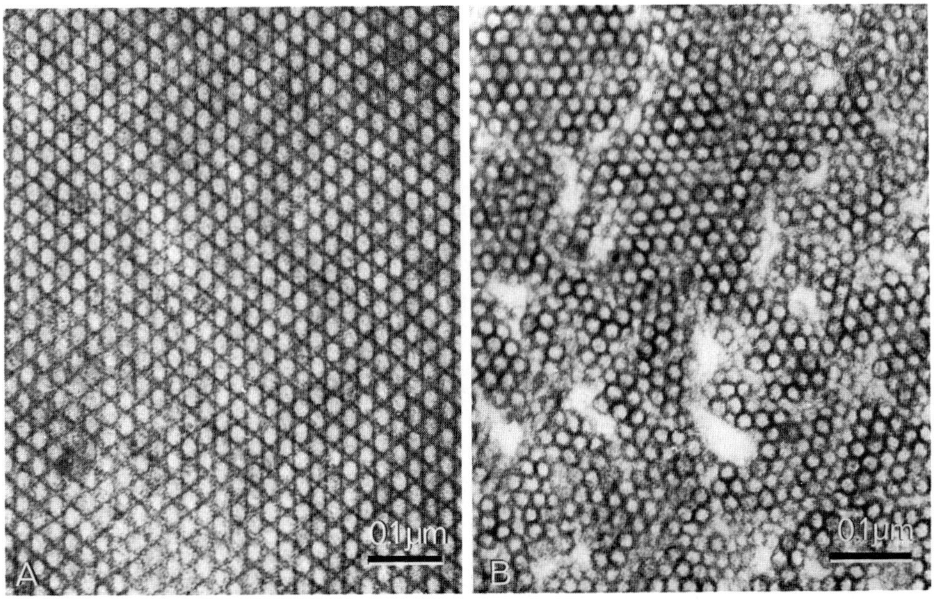

Fig. 5.17 A, B. Eggs of the sea-urchin *Echinus esculentus* incubated with 0.1 mM VLB at 12 °C for 18 h. The paracrystals which formed were isolated and sectioned after fixation. Two different crystalline lattices are apparent, whether the eggs are incubated at low density (less than 100 eggs/ml, LD) or high density (over 250 eggs/ml, HD). (A) LD crystal. The star pattern appears to be generated by the intersection of three sets of linear arrays. (B) HD crystal. The aspect here is that of a close, nearly hexagonal packing of tubules, with a diameter of 23 nm, corresponding to that of the normal MT in sea-urchins (from Starling [244])

(c) parallel rows of continuous dense lines with closely spaced rungs. The hexagonal pattern could not arise from a packing of enlarged MT, and the paracrystals would be formed directly from the tubulin pool and not from MT. The enhancement of VLB crystal formation by other mitotic inhibitors (colchicine, colcemid, podophyllin, griseofulvin, iso-propyl-N-phenylcarbamate) has been confirmed [243, 244, 245].

Fujiwara and Tilney [100] have compared the structures of VLB crystals from unfertilized eggs of sea-urchins and macrotubules from *Echinosphaerium* after low temperature treatment. Tannic acid was found to stabilize the structure of the crystals, which then resist glutaraldehyde fixation better. Confirming earlier work (cf. Chap. 2) macroT were found, like the crystals, to form in the presence of colchicine or colcemid; cross-sections failed to reveal any definite substructure similar to that seen in MT, and this is an argument for a model in which the protofilaments are at

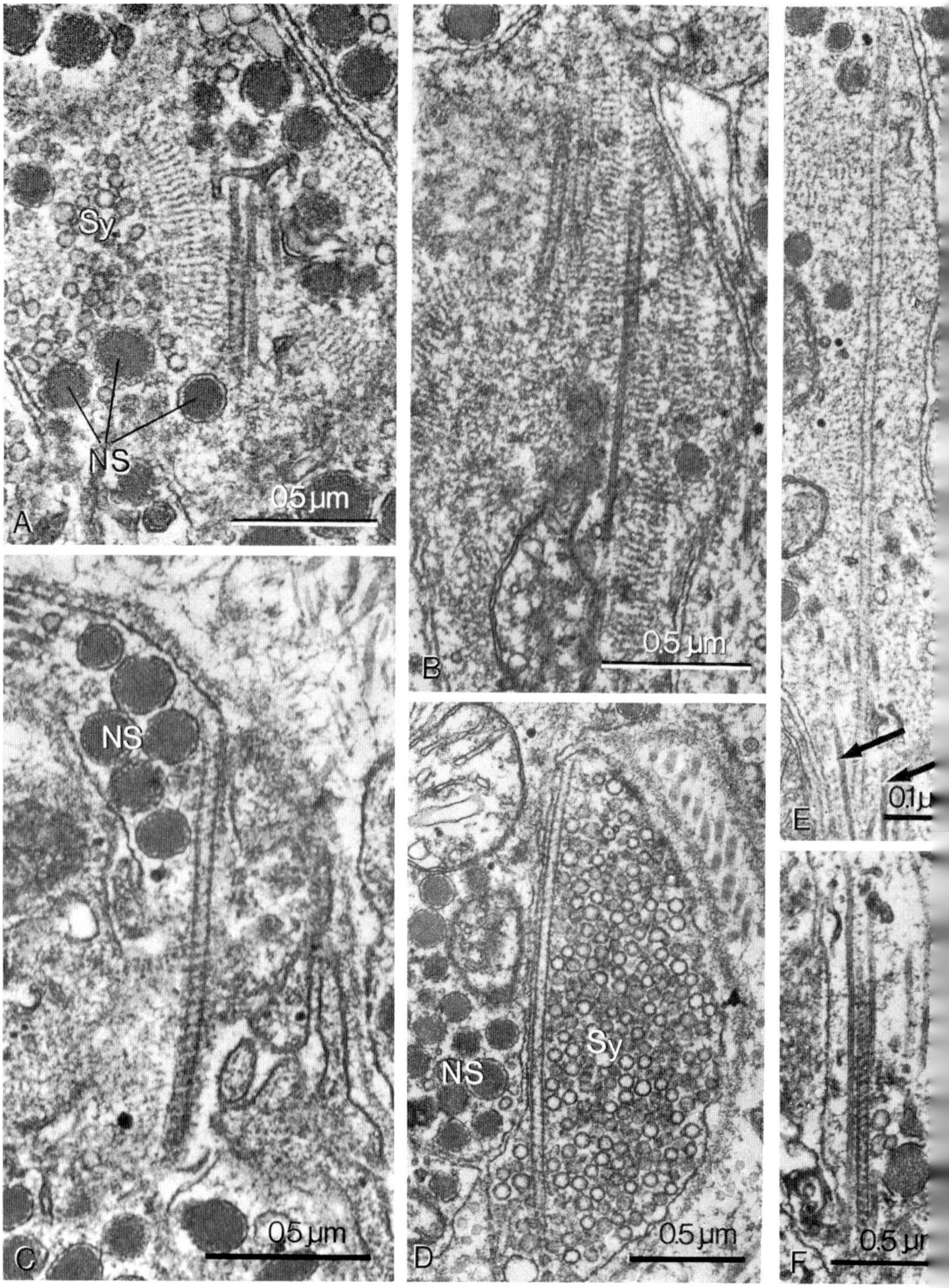

Sy
NS
0.5 µm
A
0.5 µm
B
NS
0.5 µm
C
NS
Sy
0.5 µm
D
E
0.5 µ
F

45° to the tubule axis. The crystals have the shape of hexagonal plates. They are birefringent and their optical axis is perpendicular to the hexagonal faces. Sections cut parallel to this face show hexagonally packed tubules of 32 nm diameter, with a wall thickness of 5 nm. A detailed analysis of the electron micrographs and of various planes of sections leads to a model in which the tubulin subunits are arranged in two parallel helices, 180° out of phase one with another. The helices are made of strands of heterodimers (cf. Fig. 2.27). The colchicine site is free, contrary to MT [100]. The model which is presented explains many properties of VLB crystals. One point of interest is that each tubular structure is formed of a double helix. The study of macroT induced by VCR indicates that these may present a spiral structure, although some aspects resemble the "decorated" MT described in Chapter 2 [85]. The relations between this model and that of MT, as discussed in Chapter 2, remain difficult to explain, unless one admits that no MT to crystal transition takes place, the crystals being formed from the tubulin pool. However, the observations on *Nephrotoma* [22] show clearly a change of MT to loosely coiled structures.

Such action of *Vinca* alkaloids on preformed MT has been confirmed in an important paper on VCR, VLB and desacetyl-VLB amide, and tubulin polymerized in vitro. All three components, at low doses ($1-10\times10^{-6}$ M), disassembled MT within a few minutes, and led to the formation of spiral structures made of longitudinally assembled protofilaments. The spirals were resistant to cold and to Ca^{2+}; they were disassembled by NaCl 0.25 M. These results, which confirm other observations, clearly indicate that VLB and VCR induce structural changes in already assembled MT [117]. Hence spiral structures, which are probably closely related to the paracrystalline inclusions, can arise from MT and not only from unassembled tubulin.

In autonomic nerves (where MT are supposed to be in a "stable" condition), VLB (10^{-4} M) may destroy in 15 min all MT, with the formation of paracrystalline structures. This is not prevented by 2.5×10^{-4} M colchicine, which destroys about 80% of the MT [254].

The *Vinca* crystals are further evidence of the polymorphism of tubulin assemblies (cf. Chap. 2). They are not always identical to the structures described above, and much more complex patterns are found in the optic nerve terminals in the superior colliculus of the rabbit, after intraocular injection of VLB. These are formed of regularly spaced roughly cylindrical structures, which display in cross-section a

Fig. 5.18 A–F. Formation of macroT, "decorated" MT, and paracrystalline inclusions in the posterior pituitary of a young rat, after 2 h incubation at 37 °C in a 10^{-5} M solution of VCR in Locke's solution. *NS:* neurosecretory granules; *Sy:* synaptoid vesicles. (A) One normal and one large MT in close relation with a VCR inclusion with a ladder-like structure. The "macroT" shows a helical "decoration", and may be a double tubule (cf. Chap. 2). (B) Within one large VCR crystal, a MT appears to continue into a large "decorated" tubule. Another large tubule is visible above. (C) "Decorated" macroT located close to a cell membrane. The helical external structure is clearly visible. (D) A "decorated" macroT located close to the cell membrane in a nerve extremity containing a large number of synaptoid vesicles. (E) A single long macroT crossing a large VCR crystalline inclusion. Two normal MT are visible (*arrows*). (F) Close relations between two apparently normal MT and one macroT displaying a helical "decoration". The helix appears to be oriented in two directions, according to which side of the structure is in focus. It is important to note that the external limit of the macroT appears as a smooth line, contrary to what would be expected if this structure were made of a normal MT around which a helical group of protofilaments was wound (unpublished documents, after Dustin et al. [85])

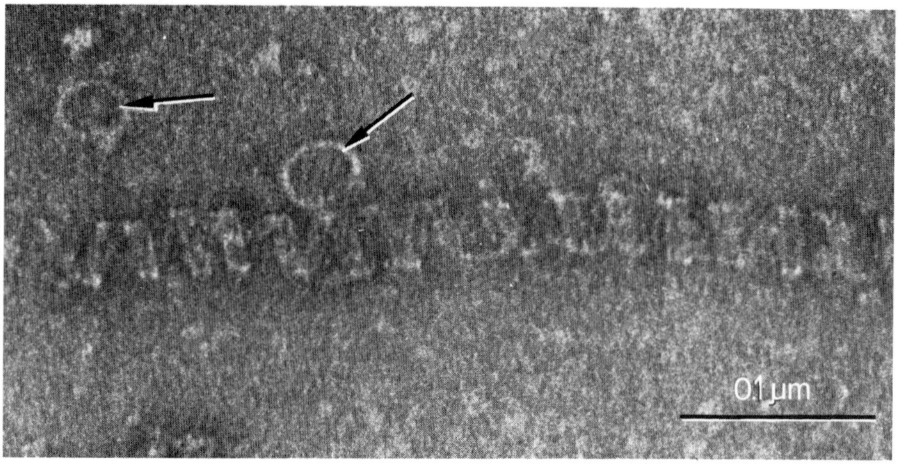

Fig. 5.19. Presynaptic matrix material from guinea-pig treated in vitro for 1 h in 0.5 mM VLB sulfate at pH 6.5, 37 °C. Formation of a double helix (negative stain). *Arrows:* tubulin rings (from Kadota et al. [132])

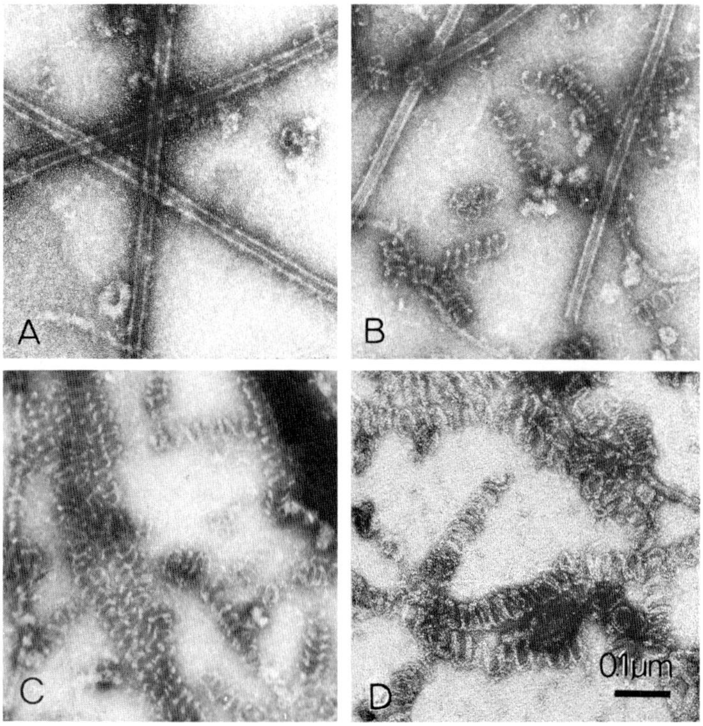

Fig. 5.20 A–D. Action of VLB on MT formed in vitro. Tubulin (1.0 mg/ml; 8×10^{-6} M) was polymerized in 0.5 ml of reassembly buffer containing 0.5×10^{-3} M of GTP at 37 °C, and then treated with VLB for 5 min. (A) 1×10^{-6} VLB. (B) 10×10^{-6} VLB. (C) 10×10^{-6} VLB, then 3×10^{-3} M $CaCl_2$ for 5 min. (D) 10×10^{-6} VLB, then cooled to 0 °C for 5 min. The samples were stained with 2% uranyl acetate (from Himes et al. [117])

pentagonal symmetry, with a central tubule 25 nm wide surrounded by five half-tubules 15 nm in diameter. As VLB can precipitate proteins other than tubulins, this may explain these peculiar structures, which are only found in the axon terminals, close to the synaptic vesicles, which appear unchanged [46, 47, 48]. In these experiments, as in many others, the nerve MT (neurotubules) were destroyed by the large quantities of VLB injected (1 – 500 μg).

The binding of VLB to cells has been studied in blood platelets, which concentrate, after injection of large doses, a high proportion of VLB in the circulating blood. VLB is known to destroy the circular MT of these cells. Using ^{3}H-VLB, the uptake was found to be rapid at 37 °C; the drug was not metabolized or degraded by the platelets. In 90 min, the cellular concentration reached a plateau and 70% of the isotope was bound to the cells. The uptake is slower at low temperatures. About 0.3 μg of VLB is bound by 10^9 platelets; the binding curve is hyperbolic. The fixation is reversible. In agreement with other authors, colchicine was found to increase the binding of ^{3}H-VLB [230]. Other effects of *Catharanthus* alkaloids on blood platelets will be considered in Chapter 11.

5.3.3 Assembly of Tubulin in the Presence of VLB or VCR

Results comparable to those related for colchicine indicate that tubulin assembly may in some conditions take place in the presence of *Vinca* alkaloids. In mice injected with 0.1 μg/g of weight VLB, the cytoplasmic MT of the spermatogonia decreased in number, while the centrioles—numbering four—became located at the center of the mitotic figure. A dense, "fuzzy" material was found around these organelles. One day after the injection, new MT were seen to grow out of one end of the centrioles, and in some places new MT could be seen in the center of the centriolar structure. The dense juxtacentriolar structures were compared to the deuterosomes (cf. Chap. 4; [284]). In another experiment, repeated injections of VCR (10 μg per day, up to a total dose of 0.85 mg/kg of body weight in about two weeks) did not affect the formation of centrioles and cilia in the pituicytes of rats [85].

5.3.4 Other Actions of *Catharanthus* Alkaloids

Like colchicine, VLB and VCR have many cellular actions, some of which may be explained by their action on MT, while others are apparently quite independent of MT. It is often difficult to know whether some effects are or are not linked to MT poisoning. Such instances are neurotoxicity, which is one of the complications of VCR therapy, and which may result from changes of neuronal MT. One instance (discussed in Chap. 11) is the syndrome of inappropriate antidiuretic hormone secretion (cf. [193] and Chap. 11). Another is the remarkable platelet-increasing action of VLB. Lastly, the antimitotic actions of VLB and VCR do explain some of their therapeutic properties in cancer chemotherapy, although a comparison with colchicine indicates that other properties must be invoked to understand the results obtained [74].

The various biochemical effects which have been attributed to the *Catharanthus* alkaloids have been reviewed [60], and only some of these will be mentioned here.

Independent of the antimitotic (spindle) action of VCR and VLB, RNA and DNA synthesis is depressed in various cells [59]. A significant depression of DNA polymerase has been found in erythroblasts [213]. In *Tetrahymena* the incorporation of ^{3}H-uridine and ^{3}H-thymidine was decreased. Protein synthesis may also be inhibited [60]. The action on nucleic acid synthesis may be important in relation to the chemotherapeutic actions of these drugs.

5.3.5 Antagonists and Resistance

There are no animal species naturally resistant to the alkaloids, as the hamster is to colchicine. In selected lines of Ehrlich ascites tumor, resistance to four drugs—daunomycin, adriamycin, VCR, and VLB—has been selected [66]; these cells have different cytological features from the other sublines [114]. It had been shown already in 1968 [59] that a resistant line of the same tumor and a much lower uptake of ^{3}H-VLB. Resistance has been studied in a haploid line of frog cells in tissue culture. While resistance of podophyllin is associated with resistance to colchicine (which binds on the same site), strains resistant to VLB, tolerating doses more than 30 times larger than the parental strain, showed no cross-resistance to podophyllin [98]. In these cells, resistance does not appear to be related to a decreased permeability to the drugs.

Early work indicated that glutamic acid and tryptophan protected cells against the mitotic-arresting effect of VLB [63]. Glucose, D-ribose, and cortisone were also found to decrease the antineoplastic action of VLB [77]; as mentioned above, in mammals cortisone increases the number of degenerating cells after mitotic arrest [82]. Glutamic acid acts as a competitive inhibitor of the uptake of VLB by leucocytes and this may explain its antagonistic action [61].

5.3.6 Metabolism

The binding of VLB to blood platelets has already been mentioned. A study of the metabolism of ^{3}H-VLB in the dog [61] has shown a rapid initial clearance phase, followed by a slow elimination from the blood. Platelets and leukocytes concentrate VLB and may act as reservoirs for the drug. The initial phase may be related to the fixation of VLB in tissues containing large pools of tubulin. VLB also binds to plasma proteins, in particular globulins, which may fix as much as 75% (cf. [79]). The major path of excretion is through the bile, with some possible enterohepatic circulation and urinary excretion [61].

5.4 Podophyllin and Related Compounds

One chemical binds to the same site of the tubulin molecule as colchicine: podophyllotoxin [42]. Cross-resistance of the Syrian hamster to these two drugs had already suggested a similarity of action [75].

The resin of *Podophyllum peltatum* L. (May apple, mandrake) and various preparations from this plant have a strong emetic and cathartic activity, which was already known in Indian popular medicine. The resin was introduced commercially in 1850 and widely used as a cathartic, replacing calomel, and also for a variety of other purposes. Following a report on the treatment of *condylomata acuminata* (benign skin tumors which, by the way, have also been favorably influenced by colchicine) [90, 138, 248], a large amount of research work on this drug and its chemical constituents was started, and already in 1954, in a review article [134], nearly 400 references to the literature were mentioned. At that time, it had been demonstrated that podophyllin modified cell division like colchicine, by a destruction of the spindle fibers. It was rapidly shown that the biological activity of this drug was complex, and inhibition of cytochrome oxidase and cell respiration was described [263]. Progressively, podophyllotoxin and several related compounds became used in cancer chemotherapy; their complex action on mitosis and cell respiration has brought favorable results (cf. [176]).

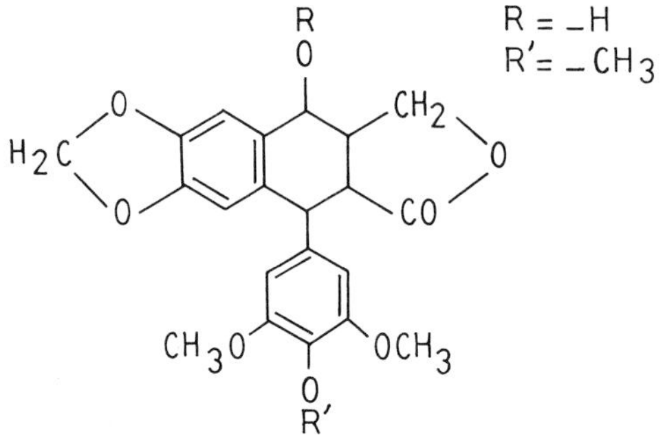

Podophyllotoxin

The chemical structure of podophyllotoxin and peltatins, which have similar cytological effects, is different from that of colchicine; one notices however the presence of three methoxy groups like in the colchicine molecule, which may explain the fixation on the same tubulin site.

Picropodophyllin is a stereoisomer of podophyllotoxin and is biologically inactive [134] (this is comparable to the inactivity of iso-colchicine). Podophyllic acid, which has an opened lactone ring, is also without biological action. A related substance, 4′-demethyl-epipodophyllotoxin-B-D-ethylidene-glucoside or VM 26 (with a substituted glucoside chain attached at R, and a hydrogen at R′), is used in cancer chemotherapy and affects cells during the mitotic cycle by another mechanism, leading to premitotic blocking [176].

Podophyllotoxin competes with colchicine, as shown on tubulin crystals from sea-urchin eggs, outer doublet MT, and chick embryo brain tubulin [201, 272]. With soluble forms of tubulin, the affinity of podophyllotoxin appears to be twice as great as with colchicine; the reverse is true for VLB crystals [270]. While the binding site is the same, the mechanisms are different: podophyllotoxin binds more rapidly, its binding is not temperature-dependent like that of colchicine, and is more rapidly reversible [58]. The binding of podophyllotoxin to rat brain tubulin has an association rate ten times higher than that of colchicine, according to recent results

(3.8×10^{6}/M/h). About 0.8 mol of podophyllotoxin is bound per mol of tubulin dimer, and this reaction is endothermic (43 cal/deg/mol). Tropolone, which inhibits colchicine binding, does not affect the podophyllotoxin – tubulin reaction. Colchicine and podophyllotoxin both appear to have two sites of attachment, sharing the one located on the trimethoxyphenyl moiety [58].

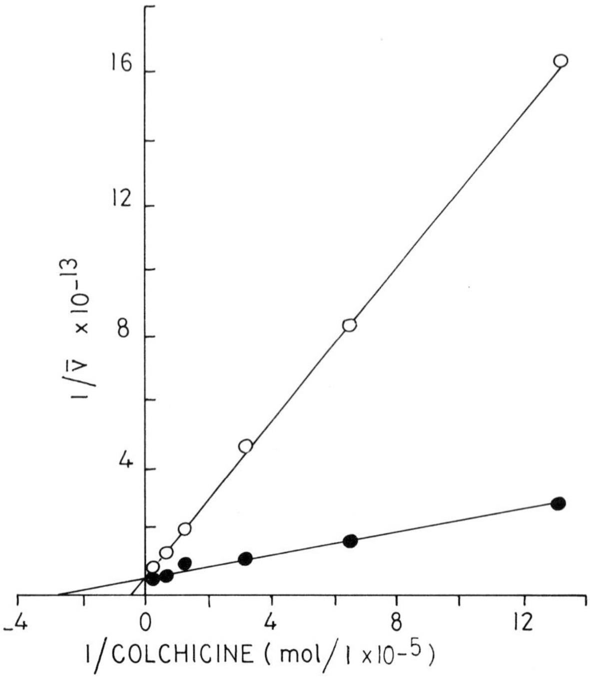

Fig. 5.21. Competitive inhibition of colchicine binding by podophyllotoxin, at 37 °C. Sea-urchin egg supernate (7.0 mg protein per ml) was incubated for 8 h with different concentrations of labeled colchicine in the presence (*open circles*) or absence (*closed circles*) of 4.0×10^{-6} M podophyllotoxin. The data were corrected for loss of colchicine binding activity during incubation (half-time at 37 °C = 12.8 h). V̄: mol of bound colchicine per µg total soluble protein (redrawn from Pfeffer et al. [201])

The podophyllin derivatives are thus most interesting for a study of the attachment site of colchicine and the relation between structure and activity. Their biological action, which is used in cancer chemotherapy, results only partly from their MT poisoning action, which has been observed on a wide variety of cells [134].

Some of the podophyllin derivatives which are used in cancer chemotherapy, such as 4′-demethylepipodophyllotoxin-ethylidene-β-D-glucoside, do not inhibit MT assembly in vitro, and act on the cell metabolism by interfering with nucleoside transport—a property which has also been mentioned for colchicine [156]. It also induces single-strand breaks in DNA [157].

5.5 Griseofulvin

This antibiotic, isolated from *Penicillium griseofulvum* Dierckx in 1939 [199], was found to arrest mitoses in metaphase when large doses (100 – 200 mg/kg of weight) were injected intravenously to rats. Various tissues were affected, and similar effects found in transplanted tumors and in plant cells [24]. The spindle inhibition in cells of *Pisum* root-tips was less effective than that of colchicine [200].

Griseofulvin has a structure which differs from that of other MT poisons, although one again finds several methoxy groups.

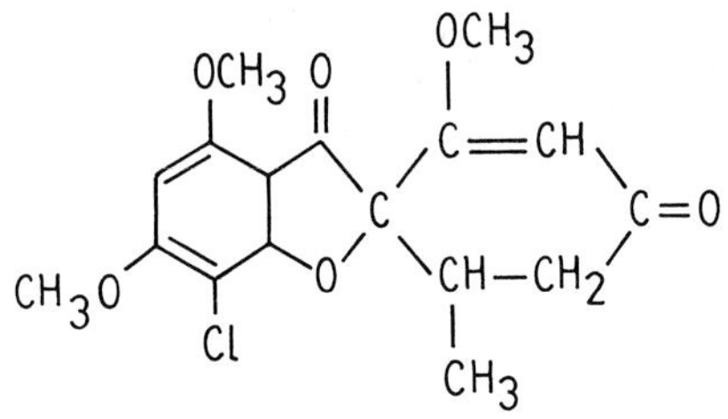

Griseofulvin

Like VLB, griseofulvin may reversibly destroy the spindle MT [165]. On cultures of hamster fibroblasts, doses of 4×10^{-5} M have identical effects to those of colchicine at 1.6×10^{-7} M, but they are more reversible. However, all MT are not destroyed and they appear disoriented; some groups of 6 – 7 nm microfibrils are also observed [2]. Griseofulvin (which is widely used in medicine for its antifungal action; [23]) has other effects related to MT: it destroys the MT of polymorphonuclear leukocytes [188] and, together with many other MT poisons (VLB, podophyllotoxin, β-peltatin, colchicine, melatonin), inhibits oral regeneration in *Stentor* [172], capping in lymphocytes [281], and the movement of melanin granules in frog melanocytes [164]. It is however ineffective in other conditions where MT poisons like colchicine and VLB are active. Griseofulvin was thought not to prevent MT polymerization in vitro [273]; more recent results indicate that at a concentration of $20 - 200 \times 10^{-6}$ M it does prevent brain tubulin assembly, but this action is not apparent in a glycerol-containing solution [211]. Blocked mitoses of HeLa cells show that many MT remain intact, normally fixed to the kinetochores and to the poles [107].

Immunofluorescent studies of mouse 3T3 cells show however that MT are destroyed by 5×10^{-5} M griseofulvin. Typical arrested metaphases are observed in HeLa cells: these show a diffuse staining with fluorescent antitubulins [266].

The mode of action of griseofulvin has been compared to that of isopropyl-N-phenylcarbamate (vide infra; cf. [272]). Studies on the inhibitory action of griseofulvin, at concentrations of $20 - 200 \times 10^{-6}$ M, on the in vitro assembly of sheep brain tubulin, have indicated that this drug does not bind to the tubulin dimer. Its action on assembly can be explained by a specific binding to the MAPs which are necessary for tubulin assembly (cf. Chap. 2). This appears to be the first demonstration of an inhibition of MT formation by an action on the MAPs [212].

In the myxomycete *Physarum polycephalum* mitosis is inhibited and the nuclei grow very large, probably by polyploidy. Bundles of MT were found in the nuclei of cells treated with 20 μg/ml of griseofulvin. These semi-crystalline arrays were so large that they distended the nuclear membrane (in this species, the spindle is intranuclear and VLB-precipitable protein has been identified in the nucleus). Some nuclei also contained a fibrillar material which may be of MT origin [108].

Griseofulvin has been shown to have antiinflammatory action in man, a fact which may be secondary to its action on polymorphonuclear MT [188]. It has been tested in acute cases of gout [242, 261].

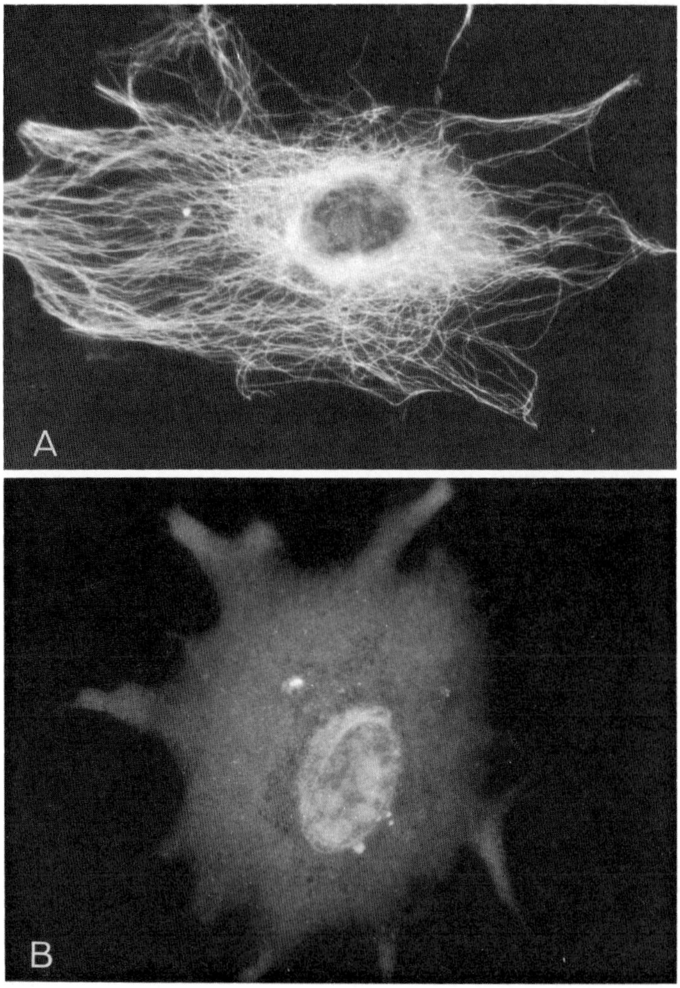

Fig. 5.22 A, B. Griseofulvin interacts with MT. (A) Mouse 3T3 cells, immunofluorescence staining of MT. (B) Loss of MT after 18 h griseofulvin 2×10^{-4} M (same technique) (from Weber et al. [266]) (Copyright Academic Press, Inc. London)

Griseofulvin is not the only substance of plant origin with antimitotic effects probably related to MT: several others mentioned in 1955 [90] would deserve more research, such as patulin, chelidonine [151], cryptopleurine, sanguinarine, protoanemonin.

5.6 Halothane and Other Anesthetic Drugs

Narcotic drugs have long been known to influence mitosis by affecting the spindle: in 1926, chloral hydrate was found to arrest at metaphase root-tip mitoses of *Vicia fava* [258]. After an intensive series of studies on the mitoses of *Allium* root-tips, the Swedish author Östergren [196, 197] presented a general theory based on the relations between the liposolubility of various organic compounds and their action on the spindle. The theory did not apply to colchicine, which had a far more potent action than could be predicted (cf. [90]). However, several new lines of research do indicate that several anesthetic drugs may influence MT.

Chloral hydrate—which has other actions on MT such as acting as a deciliating agent in *Paramecium* [136]—was studied on the endosperm mitoses of *Haemanthus katherinae* [10] and on the eggs of *Pleurodeles* [234]. In these last cells, spindle and astral fibers disappeared after 4 h in 0.1 M solutions. Electron microscopy showed that disoriented fragments of MT were visible and fibrillar structures arising perhaps from MT disintegration.

Halothane (1,1,1-trifluoro-2,2-chlorobromethane) was found to destroy neurotubules, and this observation led to a proposed model for general anesthesia in which MT played a central role [4]. The fact that the axopodia of *Actinophrys sol* were also destroyed by this drug, and by other anesthetics (methoxyflurane, chloroform, cyclopropane, nitrous oxide), and this in proportion to their anesthetic properties, appeared to confirm this idea [3]. However, the action of concentrated solutions may be the result of permeability changes, and studies on the isolated rabbit vagus nerve indicate that the number of MT is actually increased by 3 – 10 mM solutions of halothane [119]. In the nervous system of the crayfish, a reversible alteration of MT could be observed (at 40 °C); moreover, macroT of 42 nm in diameter and C-MT were observed after prolonged treatment. No macroT were formed if colchicine was present, and this drug destroyed the macroT formed by halothane. In the crayfish similar macroT could be formed by digitonin, hyaluronidase and low temperatures [120]. MacroT appear in one minute in 5 mM halothane. Negative staining indicates that they are formed by a helical arrangement of protofilaments [118].

The mechanism of formation of macroT under the influence of halothane has been studied in the crayfish cord. Rapidly and reversibly, MT are disassembled into protofilaments and ribbons. These become twisted, and the assembly of two such ribbons leads to the formation of macroT in which the protofilaments are at an angle of 30 – 40° to the axis of the tubule. The ribbons, stained by tannic acid, were found to be formed of only 12 protofilaments (cf. Chap. 2). This action is shared by other anesthetic drugs, such as methoxyflurane, isoflurane, enflurane, trichlorethylene [118]. This change of MT is comparable to that described above for VLB, and confirms that drugs may rapidly modify already assembled MT.

As halothane fails to modify axonal transport in the rabbit vagus nerve [135], and does not depolymerize MT in the mouse optic nerve [220], it is difficult to explain its anesthetic effects as a consequence of MT poisoning [221]. Halothane has many other cytotoxic effects, on lysosomes and smooth endoplasmic reticulum [55].

Nitrous oxide has been studied on mitosis in HeLa cells: these were exposed to N_2O under a pressure of 80 lb per square inch. Reversible inactivation of the spindles resulted in typical "colchicine metaphases". However, the MT were not destroyed and the centrioles remained normally located [38].

The local anesthetic *lidocaine* (xylocaine) is known to inhibit rapid axonal transport, which is related to MT (cf. Chap. 9). A decrease of the number of MT in the rabbit vagus nerve was observed only with doses above 0.4%, acting for about one hour. This was reversible, but the MT changes did not coincide with the modifications of axonal transport [51]. On the other hand, lidocaine does prevent the repolymerization of rat brain tubulin fractions, and this effect is reversible. Counts indicate that the length of the MT was not modified, while their number was decreased. The local anesthetics procaine and etidocaine have similar effects. Lidocaine would combine reversibly with non-polymerized tubulin, and prevent the assembly of MT. The anesthetic potency is not proportional to these effects [113].

All these data, compared to the facts assembled by Östergren [197], indicate that MT may be affected by relatively large doses of substances with anesthetic properties. They certainly do not prove that anesthesia is a consequence of MT alterations in the nerve cells; it is more probably, like the effects on mitosis, the consequence of non-specific cytoplasmic changes, perhaps related to those induced by physical agents such as cold rather than by highly specific poisons such as colchicine.

5.7 Isopropyl-N-Phenylcarbamate (IPC)

This substance, like other carbamates such as urethane (ethylcarbamate), has complex actions on dividing cells and DNA synthesis. It is widely used as a herbicide, and may modify MT, as well as inhibit cilia regeneration in *Stentor* [13]. In the endosperm cells of *Haemanthus katherinae* mitosis is severely disturbed and micronuclei are formed: this does not result from a destruction of MT but from their disorientation [116]. In vitro, IPC does not prevent the assembly of brain tubulin, and studies with isopropyl-N-3-chloro-phenylcarbamate (CIPC) marked by ^{14}C show that this molecule does not bind to tubulin [18]. The pluricentric mitoses, in a green alga, are shown to result from a separation of the polar MTOC in several functional subunits. In the protist *Ochromonas,* IPC and CIPC do not affect intact MT, but prevent their assembly after mechanical deflagellation. The reassembly of rhizoplast MT after pressure disassembly is also prevented, although some macroT may be formed. All these effects are reversible; they differ from those of colchicine. It is suggested that IPC binds to tubulin subunits and changes their conformation, permitting only reassembly in the form of macroT [39]. These results are however contradicted by studies on the assembly in vitro of chick brain tubulin: this is not prevented by IPC at concentrations from $10^{-4} - 10^{-3}$ M. The structure of the MT appeared quite normal and no macroT could be observed. It is suggested that the action of IPC may be on MTOC or on interMT bridges [57].

5.8 Melatonin

The finding that a naturally occurring hormone of vertebrates, melatonin, a tryptophane derivative secreted by the pineal gland, could affect MT, is interesting as an indication that the control of MT assembly may be mediated by relatively simple molecules present in tissues.

It was first observed that this hormone was antagonistic to colchicine on melanocytes [163]. This antagonism was confirmed on quite another model, the cilia regeneration of *Stentor:* colchicine protected the cells against the inhibition of cilia regeneration caused by melatonin, and melatonin protected the organism against the lethal effects of colchicine [12]. On the other hand, the two drugs inhibited synergis-

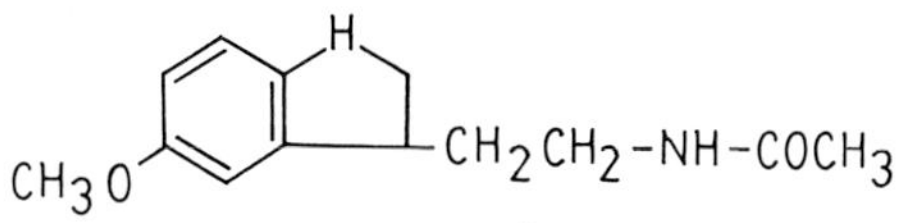

Melatonin

tically the regeneration of the oral band, suggesting that both bound to MT protein. In HeLa and KB cells, melatonin (10^{-3} M) alone has no action on mitosis; however, preincubation for one hour in melatonin (10^{-5} M) antagonizes the action of colchicine (10^{-7} M) on mitosis [96]. An antagonism between melatonin and IPC on mitosis has also been mentioned [129]. The hypothesis that colchicine (or colcemid) may bind to the same site as melatonin has been proposed; contrary to colchicine, which would not remain bound to tubulin during assembly, melatonin may be incorporated in growing tubules "with no deleterious effects" [12]. The inhibition of colchicine binding to tubulin has been confirmed [133]. In plant cells (*Allium cepa* root-tips) melatonin, at a concentration of 8×10^{-4} M, arrests mitoses in metaphase. Serotonin and N-acetylserotonin are inactive, although closely related chemically. A methoxy group on an aromatic ring about 7 Å distant from an electronegative atom, which would ensure the linking with the protein (tubulin), is proposed as a base of MT poisoning [14]. Further evidence of a relation between colchicine and melatonin is the fact that ^{3}H-colchicine binding to brain tubulin can be completely inhibited by high concentrations (5 mM) of melatonin [276]. These results are contradicted by Poffenbarger and Fuller [203], who did not find any effect of melatonin on the in vitro assembly of bovine brain MT, and no inhibition of ^{3}H-colchicine binding or increase in the mitotic index in Chinese hamster ovary cells: for these authors, melatonin cannot be considered a MT poison or a mitotic inhibitor.

The biological implications of these findings in relation to the hormonal action of melatonin are interesting, although the concentrations of the hormone used in many of these experiments are far above those met under physiological conditions.

5.9 Sulfhydryl Binding Drugs and Metals

Some of the first observations of arrested mitoses by inactivation of the spindle were made with *arsenical* derivatives (arsenious oxide, sodium cacodylate) and opened the way for the discovery of the morphologically similar action of colchicine [80, 202]. The accumulation of arrested metaphases in mice injected with cacodylate is, at the light microscopical level, identical with that following colchicine (cf. Chap. 1). In agreement with the early work of Rapkine [206] on the role of -SH groups in spindle dynamics, this action was explained as a result of the binding of arsenic to these groups [81]. It was further noticed that substances with -SH functions, such as dimercaptopropanol ("British anti-lewisite") and dithiodiethylcarbamate had similar effects on mitosis, suggesting that a correct balance between reduced and oxidized -SH groups was necessary for spindle function [82]. The importance of -SH in the spindle was further substantiated by the experiments of Mazia [177], who stabilized the mitotic apparatus by dithiodiglycol before isolating it. The theoretical basis of this method was to protect S-S bonds during spindle isolation. This was confirmed by the fact that the isolated mitotic apparatus was dissolved by *p*-mercurybenzoate, which combines to -SH groups. This work found a chemical confirmation in the dem-

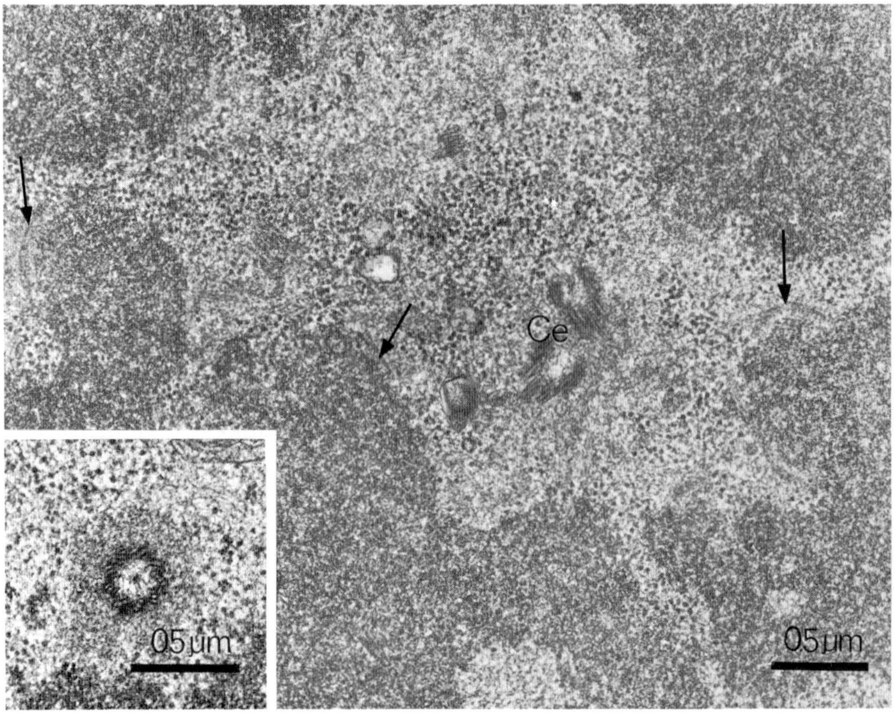

Fig. 5.23. Mouse intestinal mitosis 3 h after a single injection of 2 mg/g of sodium cacodylate. Centrally located centrioles (*Ce*). Absence of MT. Irregularly located kinetochores (*arrows*). *Inset:* Guinea-pig intestinal mitosis arrested 3 h after 2 mg/g of sodium cacodylate. A centriole surrounded by a granular dense substance (disassembled tubulin?)

onstration that tubulin effectively contains several sulfhydryl residues [115, 147, 209] (cf. Chap. 2).

A more recent poison of MT assembly is diamide $[(CH_3)_2NCON=NCON(CH_3)_2]$, which combines with two -SH groups, and completely inhibits cell division (cf. Chap. 10). Diamide at the concentration of 3×10^{-4} M prevents the in vitro assembly of brain tubulin; this effect is reversed by dithiothreitol, which reduces the SS links. Diamide may act either directly on the tubulin molecule or on glutathione, which would be involved in the assembly of the spindle MT [179].

Apart from a small number of papers on the action of heavy metals on mitoses in plants and animals, no modern ultrastructural work had been done on the action of arsenic, and it was not known whether it destroys or alters MT.

Recent personal observations have indicated that in mice and guinea-pigs, sodium cacodylate (2 mg/g of body weight) destroys within less than one hour all MT from intestinal mitoses, leading to a typical "metaphase arrest". It is interesting to mention that Luftig et al. [159] have pointed out that the use of cacodylate in buffer solutions for electron microscopy may lead to an extensive disassembly of MT. They have recommended a buffer more favorable for maintaining the integrity of MT, and comparable to that used in studies of MT assembly in vitro (PIPES buffer). Figure 2.1 is an illustration of the excellent results obtained with this procedure.

Some other effects of heavy metals may be mentioned here.

Organomercurials, such as methylmercury and phenylmercury, have been studied on various types of cells, and their action has been recently reviewed [252]. In HeLa cells, chronic inhibition of cell growth by doses of 0.03 – 0.3 mg/l led to colchicine-type mitotic figures. On plant cells (*Allium cepa*) similar effects are found; it seems that phenylmercury is more readily bound to kinetochore MT and methylmercury to pole-to-pole MT [252]. In *Tetrahymena pyriformis*, cilia regeneration is inhibited by methylmercuric chloride at concentrations much lower than colchicine (10^{-7} – 10^{-8} M). The outer fibers of cilia of *Tetrahymena* contain about 7.5 sulfhydryl residues per molecule of tubulin [209]. The binding of ^{3}H-methylmercuric chloride to tubulin proteins has been confirmed in the eggs of *Lytechinus pictus* [252]. Mercury compounds harm many other cytoplasmic structures, such as the mitochondria, and inhibit many -SH-containing enzymes. In *Tetrahymena* MT are not necessarily destroyed by toxic doses of mercuric chloride. On a quite different material—the dorsal root ganglion of the frog—methylmercury is found to destroy MT. In vitro, it depolymerizes MT at a concentration of 50 – 100 nM/ml [1]; the presence of seven -SH groups per molecule and their role in tubulin assembly was confirmed [147].

Some other metals may modify MT. A solution of copper sulphate (10^{-5} M) destroys in ten minutes the axopodia of the heliozoan *Actinosphaerium nucleofilum*. When they regenerate, instead of the regular spiral-like arrangement described in Chapter 3, multiple spirals are formed. Nickelous ions (10^{-5} M) cause similar retraction of axopodia. Another early effect of copper is to detach the axonemes from the nuclear membrane (cf. Chap. 4). C-MT are also observed. The mechanism of action of the divalent copper and nickel ions is not explained. The rapidity of the change, qualified as "cataclysmic", may result from the binding of a few copper ions to some tubulin subunits, with an action at distance on the neighboring, unpoisoned molecules of tubulin. These results are compared to the known harmful consequences for the nervous system of increased copper concentrations as found in Wilson's disease [216].

The action of heavy metal ions on the axonemes of *Echinosphaerium nucleofilum* have been studied quantitatively; according to their concentration, they may have a stabilizing action (at very low concentrations) or a destructive action on MT. The strength of this effect is in the order $Hg^{2+} > Cd^{2+} > Zn^{2+}$. Degradation of MT, with formation of filamentous structures, occurs with 2×20^{-6} M solutions of Hg^{2+}, lower concentrations having in contrast a stabilizing action [238].

Yet another metal which may bind to tubulin is *zinc*: this stabilizes the neurotubules of rat peripheral nerves, which become more resistant to osmic acid fixation.

However, C-shaped tubules are also found [141]. The formation of sheets of tubulin molecules in vitro in the presence of Zn ions has been mentioned in Chapter 2.

It is evident that the metals may affect MT by various pathways, either by combining with tubulin, or by poisoning cell enzymes necessary for MT metabolism. The fixation on -SH groups of tubulin, glutathione or other -SH proteins is one of these modes of action, particularly evident for arsenic and probably mercury compounds.

5.10 Other MT Poisons

The list of substances capable of modifying MT in different cells is very long: already twenty years ago, the number of drugs which had been found to inhibit the mitotic spindle was considerable [84, 90]. However, few drugs may destroy or inhibit MT at low concentrations without harming other cell functions. Some recently described MT poisons have been grouped in Table 5.3, which is far from being complete and has left aside many substances active only at high concentrations. Attention has mainly been given to animal cells, and to drugs with promising new approaches to the study of tubulin assembly and disassembly.

Some substances destroy MT by non-specific protein denaturation: digitonin and urea come into this category. They do demonstrate that MT may be rapidly destroyed without damage to other cell structures, indicating their particular fragility.

Chlorpromazine is interesting, for it has been suggested that some of its actions on the central nervous system may be mediated by changes in neurotubules. This deserves further studies, for nervous system MT are particularly stable (cf. Chap. 9).

The dynamics of chlorpromazine binding to brain tubulin have been further studied. Two sites are present, one molecule of the drug binding strongly, and 8 – 9 weakly. The higher affinity site involves hydrophobic bonds, like the binding of chlorpromazine to serum albumin. The mitotic arrest is related to this reaction, while the inhibition of fast axonal transport requires larger doses (1×10^{-3} M) [122].

Table 5.3. Various MT poisons: some recent results

Drug	Tissue or cell	EM	Results	Reference
Acenaphtene	*Haemanthus katherinae* endosperm	+	modification of mitotic MT	Bajer [10]
Barbiturates	MT in vitro		ultrastructure and polymerization	Edström et al. [89]
Chlorpromazine	frog sciatic nerve, in vitro	+	inhibition of axonal transport; decrease of MT	Edström et al. [87]
	hypothalamo-hypophysial system of the rat	+	partial arrest of axonal flow; no modifications of MT	Edström et al. [88]
	brain tubulin (guinea-pig)	–	selective precipitation of tubulin	McGuire et al. [161]
	brain tubulin (mice)		inhibition of assembly and colchicine binding; disassembly of MT	Cann and Hinman [52]

Table 5.3. continued (1)

Drug	Tissue or cell	EM	Results	Reference
Diamide		+	destruction of MT mitotic arrest	Mellon and Rebhun [179, 180]
Digitonin	*Allium* root-tips	+	formation of macro T,	Olah and Hanzely [192]
			mitotic abnormalities	Olah and Bozzola [191]
5-fluoro-pyrimidine-2-one	human liver cells in tissue culture	–	metaphase arrest of mitosis	Oftebro et al. [190]
Formaldehyde, formamide	eggs of *Pleurodeles waltii*	–	multipolar spindles	Sentein [232]
Formaldehyde, glutaraldehyde	eggs of *Triturus helveticus*		disappearance of spindle and astral fibers	Sentein [233]
Isoprenaline	rat parotid	+	arrested mitoses with central centrioles	Radley [204]
Maytansine (from *Maytenus serrata*)	sea-urchin eggs CHO cells in culture	+	destroys MT (5×10^{-8} M) prevents in vitro assembly by destroying rings ($100 \times$ more potent than VCR in eggs)	Schnaitman et al. [226] Kupchan et al. [146] Remillard et al. [208]
Mebendazole methyl-(5-benzoyl-1H-benzimidazol-2-yl) carbamate	*Ascaris suum, Syngamus trachea, Taenia taeniaeformis* (cysticerci)	+	destruction of cytoplasmic MT	Borgers and De Nollin [31]
Oncodazole methyl [5-(2-thienyl-carbonyl)-1H-benzimidazol-2-yl] carbamate	fibroblast cultures neoplastic cells in tissue culture	+	destruction of MT formation of bundles of 10 nm filaments	De Brabander et al. [72]
Quinoline	eggs of *Triturus helveticus*	–	destruction of spindle and astral fibers	Sentein [231]
Rotenone	Chinese hamster cells		metaphase arrest	Meisner and Sorensen [178]
	Chinese hamster ovary cells in vitro	+	metaphase arrest, inhibition of 3H-colchicine binding to brain tubulin	Brinkley et al. [37] Barham and Brinkley [16]
Trifluralin (trifluoro 2,6-dinitro-N,N-dipropyl *p* toluidine)			oral morphogenesis in *Stentor coeruleus*	Bartels and Hilton [18] Banerjee et al. [11]
Urea	heliozoan axopodia	+	opening and destruction of MT	Shigenaka et al. [239]
	Allogromia	+	rapid disassembly; formation of pseudocrystalline inclusions	McGee-Russel [160]
	spermatozoa of *Rhynchosciara* sp.	+	progressive destruction of MT	Shay [235]
Withaferin (lactone from *Withania somnifera* Dun)	HeLa cells in tissue culture	–	metaphase arrest of mitosis	Shohat [241]

An important group of MT poisons are benzimidazole derivatives, which were first introduced as fungicides. The three most interesting are methyl benzimidazol-2-yl carbamate (MBC or carbendazin), mebendazole and oncodazole

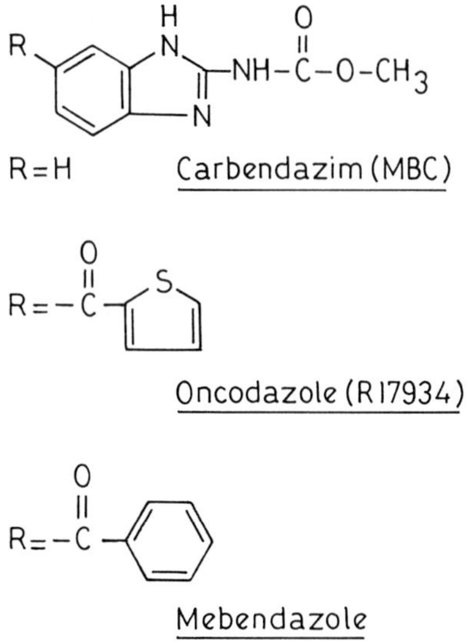

[69]. Abnormal chromatin configurations had been observed in *Aspergillus nidulans* treated with MBC, and an interference with spindle activity was suspected [67]. Further investigations confirmed that this antimitotic action resembled that of colchicine; however, MBC did not affect in vitro assembly of porcine brain tubulin into MT, although it was shown to combine, in fungi, with a 110,000 daltons protein [68]. MBC was also shown to destroy selectively the MT of parasitic worms, without affecting those of the host [31, 32, 33]. The closely related compound, *oncodazole* or R 17934 [methyl 5-(2-thienylcarbonyl)benzimidazol-2-yl carbamate] arrests mammalian mitoses in vitro within 10 min at a concentration of 1 μg/ml, with disappearance of spindle MT. This effect is reversible, and synergic with colchicine, VLB, and VCR. The in vitro assembly of tubulin is also prevented. When the drug is applied longer, many bundles of microfilaments appear in the cytoplasma, like those

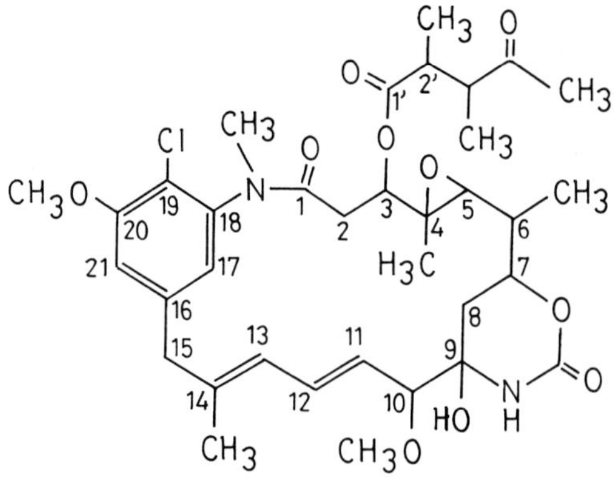

MAYTANSINE

described with colchicine [70, 72, 73]. The binding to tubulin is in a mol : mol ratio, and may inhibit colchicine binding [124].

Maytansine, found in alcoholic extract from *Maytenus ovatus* Loes, is a macrolide with a structure similar to the rifamycins [146], but devoid of any action on nucleic acid synthesis. It irreversibly inhibits mitosis of sea-urchin eggs at the low dose of 6×10^{-8} M. It also inhibits in vitro formation of MT from tubulin solutions. It does not appear to affect the replication of centrioles, as multicentric mitoses were observed. Maytansine added to polymerized tubulin suspensions in vitro rapidly decreased their turbidity. It may act by combining to sulfhydryl groups of tubulin [208]. It causes a rapid breakdown of MT, within minutes in vitro, and destroys within 15 min VLB crystalline inclusions in the eggs of *Lytechnicus*: the macrolide would become fixed on a site of the tubulin molecule which is not buried by assembly into MT [207], which may be the VLB site [167]. This inhibitor is interesting as it shows in vivo an activity 100 times greater than VCR, while in vitro it is somewhat less powerful than colchicine.

Rotenone, a heterocyclic substance with two methoxy groups attached to an aromatic ring, is a substance of plant origin which rapidly destroys the MT of mitotic cells. The ultrastructural changes are identical with those of colchicine. The drug is also an inhibitor of cell respiration (cf. Chap. 10; [17]).

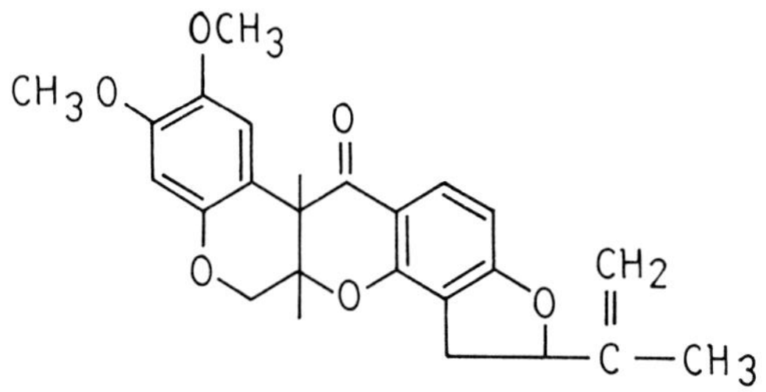

ROTENONE

Another new MT poison is *steganacin*, a lactone extracted from *Steganotaenia araliacea* Höchst. Low concentrations (7×10^{-6} M) completely prevent tubulin assembly in vitro, and inhibit cell division. This molecule, which has a trimethoxybenzene ring, competes with colchicine and probably binds to the same site [262].

It is thus evident that MT may be destroyed (or their assembly prevented) by many drugs, although structures like centrioles and cilia are resistant. A limited number of substances are known to combine specifically with tubulin, like colchicine, the *Catharanthus* alkaloids, benzimidazole derivatives, maytansine. A better understanding of the molecular action of these poisons on the tubulin molecule, of the possible existence of tubulins with different sensitivities, and of the relation between drug fixation and tubulin assembly, will be the aim of future work.

5.11 Comparison with the Action of Physical Agents and Heavy Water

The destruction of MT by cold and *high hydrostatic pressure* has been mentioned in previous chapters, and has brought important information on the thermodynamics of MT assembly (Chap. 2) [218]. The rapid destruction of various types of MT by hydrostatic pressure is an indication of the instability of MT. This is not only found in dividing cells, where the spindle MT might be considered as being in a dynamic condition, continuously assembled and disassembled: high hydrostatic pressures may destroy even the most stable of MT. The central ciliary MT of *Tetrahymena* disappear (from the top to the basis) in less than 10 minutes under a pressure of 7,500 or 10,000 psi [137]; the doublet MT of *Tetrahymena* are also destroyed and the basal bodies resorbed, although the oral MT, except in periods of growth, are resistant [183]. In contrast, neurotubules from nerves of *Rana pipiens* are resistant to pressures as high as 10,000 psi [7] and MT assembled from purified tubulin of beef brain are not modified by similar pressures [189]. This is in contradiction to more recent results on rat brain MT, where the pressure effects were found to be temperature-dependent: under 25 °C a complete dissociation of MT in vitro was obtained at pressures of 500 atm. [91]. Another instance of "resistant" MT is that of the spindle midbody at telophase (cf. Chap. 10): this is not affected, contrary to other cytoplasmic MT, by pressures of 10,000 psi. Centrioles are also resistant (in HeLa cells) [105]. The thermodynamic considerations derived from pressure-induced depolymerization have been discussed in Chapter 2 (cf. 2.5.3.1). It may be concluded that various degrees of MT stability are observed, and that these are closely parallel to the stability towards MT poisons.

The action of *cold* is similar. MT, of the mitotic spindle, are rapidly disassembled at low temperatures [15, 215] (cf. Chap. 2). Blood platelet MT disappear at 0 °C; when they reassemble after rewarming, C-MT may be observed [19]. MacroT (which may have some relations with C tubules) are found when the axopods of *Actinosphaerium*, which lose all their MT at 4 °C, are rewarmed [253]. Here again, cold mimicks closely the effects of specific poisons such as colchicine and VLB or VCR.

An *increased temperature* may also affect MT: heat-shock treatment was, before the discovery of colchicine as a polyploidizing agent, one of the frequently used techniques for doubling the number of chromosomes (cf. [90]). The MT of the ventral nerve cord of the crayfish (*Procambarus clarkii*), which are resistant to colchicine, are rapidly destroyed at 40 °C; this effect is reversible [120]. In the sea-urchin sperm tails, the B MT of the outer doublets are also selectively destroyed at 40 °C [246].

The effects on MT of *ultraviolet* beams have brought results which will be discussed in Chapter 10.

Heavy water has opposite effects on MT: it increases their stability, perhaps their number, and appears to favorize their assembly. D_2O has often been used in comparison with other MT poisons to ascertain that MT were involved in various cell functions (cf. Chap. 8). The thermodynamics of this effect have been studied [54]. The artificial production by D_2O of newly formed centrioles in activated eggs has been mentioned in Chapter 4 [257]. In mitotic cells of wheat meristems, the number of spindle fibers appears increased by D_2O [49].

In conclusion, the effects of MT poisons can be imitated in many cells by high hydrostatic pressures and low temperatures; in a few cells, on the contrary, by heat. Heavy water has an opposite stabilizing action on MT while modifying and usually depressing the cellular activities associated with MT activity.

---

[7] cf. note on pressure units, Chap. 1

References

1. Abe, T., Haga, T., Kurokawe, M.: Blockage of axoplasmic transport and depolymerisation of reassembled microtubules by methyl mercury. Brain Res. **86,** 504 – 508 (1975)
2. Adair, G. M.: Antimitotic action of griseofulvin. J. Cell Biol. **63,** 2 a (1974)
3. Allison, A. C., Hulands, G. H., Nunn, J. F., Kitching, J. A., MacDonald, A. C.: The effects of inhalational anaesthetics on the microtubular system in *Actinosphaerium nucleofilum*. J. Cell Sci. **7,** 483 – 499 (1970)
4. Allison, A. C., Nunn, J. F.: Effects of general anaesthetics on microtubules: a possible mechanism of anaesthesia. Lancet **II,** 1326 – 1329 (1968)
5. Amos, L. A., Klug, A.: Arrangement of subunits in flagellar microtubules. J. Cell Sci. **14,** 523 – 550 (1974)
6. Amrhein, N., Filner, P.: Sensitization of colchicine binding protein to ultraviolet light by bound colchicine. FEBS Lett. **33,** 139 – 142 (1973)
7. Arai, T., Okuyama, T.: Fluorometric assay of tubulin – colchicine complex. Ann. Biochem. **69,** 443 – 450 (1975)
8. Aubin, J. E., Subrahmanyan, L., Kalnins, V. I., Ling, V.: Antisera against electrophoretically purified tubulin stimulate colchicine-binding activity. Proc. Natl. Acad. Sci. **73,** 1246 – 1249 (1976)
9. Back, A., Walaszek, E. J.: Fixation of radioactive colchicine by Ehrlich ascites carcinoma cells. Nature **172,** 202 – 203 (1953)
10. Bajer, A.: Chromosome movement and fine structure of the mitotic spindle. Symp. 22 Soc. Exp. Biol. Aspects of Cell Motility, pp. 285 – 310. Cambridge Univ. Press 1968
11. Banerjee, S., Kelleher, J. K., Margulis, L.: The herbicide trifluralin is active against microtubule-based oral morphogenesis in *Stentor coeruleus*. Cytobios **12,** 171 – 178 (1975)
12. Banerjee, S., Kerr, V., Winston, M., Kelleher, J. K., Margulis, L.: Melatonin: inhibition of microtubule-based oral morphogenesis in *Stentor coeruleus*. J. Protozool. **19,** 108 – 112 (1972)
13. Banerjee, S., Margulis, L.: Reversible inhibition of cilia regeneration in *Stentor coeruleus* by isopropyl N-phenyl-carbamate. Nature **224,** 180 (1969)
14. Banerjee, S., Margulis, L.: Mitotic arrest by melatonin. Exp. Cell. Res. **78,** 314 – 318 (1973)
15. Barber, H. N., Callan, H. G.: The effects of cold and colchicine on mitosis in the newt. Proc. Roy. Soc. London B **131,** 258 – 271 (1943)
16. Barham, S. S., Brinkley, B. R.: Action of rotenone and related respiratory inhibitors on mammalian cell division. I. Cell kinetics and biochemical aspects. Cytobios **15,** 85 – 96 (1976)
17. Barham, S. S., Brinkley, B. R.: Action of rotenone and related respiratory inhibitors on mammalian cell division. II. Ultrastructural studies. Cytobios **15,** 97 – 110 (1976)
18. Bartels, P. G., Hilton, J. L.: Comparison of trifluralin, oryzalin, pronamide, propham, and colchicine treatment on microtubules. Pestic. Biochem. Physiol. **3,** 462 – 472 (1973)
19. Behnke, O.: Incomplete microtubules observed in mammalian blood platelets during microtubule polymerization. J. Cell Biol. **34,** 697 – 701 (1967)
20. Behnke, O.: Microtubules in disk-shaped blood cells. Intern. Rev. Exp. Path. **9,** 1 – 92 (1970)
21. Behnke, O., Forer, A.: Evidence for four classes of microtubules in individual cells. J. Cell Sci. **2,** 169 – 192 (1967)
22. Behnke, O., Forer, A.: Vinblastine as a cause of direct transformation of some microtubules into helical structures. Exp. Cell Res. **73,** 506 – 508 (1972)
23. Bensch, K. G., Malawista, S. E.: Microtubular crystals in mammalian cells. J. Cell Biol. **40,** 95 – 106 (1969)
24. Bent, K. J., Moore, R. H.: The mode of action of griseofulvin. In: Biochemical Studies of Antimicrobial Drugs (eds.: B. A. Newton, P. E. Reynolds), pp. 82 – 110. London: Cambridge Univ. Press 1966
25. Berlin, R. D.: Temperature dependence of nucleoside transport in rabbit alveolar macrophages and polymorphonuclear leukocytes. J. Biol. Chem. **248,** 4724 – 4730 (1973)
26. Berry, R. W., Shelanski, M. L.: Interactions of tubulin with vinblastine and guanosine triphosphate. J. Mol. Biol. **71,** 71 – 80 (1972)
27. Bhattacharyya, B., Wolff, J.: Promotion of fluorescence upon binding of colchicine to tubulin. Proc. Natl. Acad. Sci. USA **71,** 2627 – 2631 (1974)
28. Bhattacharyya, B., Wolff, J.: The interaction of 1-anilino-8-naphthalene sulfonate with tubulin a site independent of the colchicine-binding site. Arch. Biochem. Biophys. **167,** 264 – 269 (1975)

29. Bhattacharyya, B., Wolff, J.: Tubulin aggregation and disaggregation – mediation by 2 distinct vinblastine-binding sites. Proc. Natl. Acad. Sci. USA **73,** 2375 – 2378 (1976)
30. Bhisey, A. N., Freed, J. J.: Cross-bridges on the microtubules of cooled interphase HeLa cells. J. Cell Biol. **50,** 557 – 561 (1971)
31. Borgers, M., De Nollin, S.: Ultrastructural changes in *Ascaris suum* intestine after mebendazole treatment in vivo. J. Parasitol. **61,** 110 – 122 (1975)
32. Borgers, M., De Nollin, S., De Brabander, M., Thienpont, D.: Influence of the anthelmintic mebendazole on microtubules and intracellular organelle movement in nematode intestinal cells. Am. J. Vet. Res. **36,** 1153 – 1166 (1975)
33. Borgers, M., De Nollin, S., Verheyen, A., De Brabander, M., Thienpont, D.: Effects of new anthelmintics on the microtubular system of parasites. In: Microtubules and Microtubule Inhibitors (eds.: M. Borgers, M. De Brabander), pp. 497 – 508. Amsterdam-Oxford: North-Holland. New York: Elsevier 1975
34. Borisy, G. G., Taylor, E. W.: The mechanism of action of colchicine. Binding of colchicine-^{3}H to cellular protein. J. Cell Biol. **34,** 524 – 534 (1967)
35. Borisy, G. G., Olmsted, J. B., Klugman, R. A.: In vitro aggregation of cytoplasmic microtubule subunits. Proc. Natl. Acad. Sci. USA **69,** 2890 – 2894 (1972)
36. Boudene, C., Duprey, F., Bohuon, C.: Radioimmunoassay of colchicine. Biochem. J. **151,** 413 – 426 (1975)
37. Brinkley, B. R., Barham, S. S., Barranco, S. C., Fuller, G. M.: Rotenone inhibition of spindle microtubule assembly in mammalian cells. Exp. Cell Res. **85,** 41 – 46 (1974)
38. Brinkley, B. R., Rao, P. N.: Nitrous oxide: effects on the mitotic apparatus and chromosome movement in HeLa cells. J. Cell Biol. **58,** 96 – 106 (1973)
39. Brown, D. L., Bouck, G. B.: Microtubule biogenesis and cell shape in *Ochromonas.* III. Effects of the herbicidal mitotic inhibitor isopropyl *N*-phenylcarbamate on shape and flagellum regeneration. J. Cell Biol. **61,** 514 – 536 (1974)
40. Bruchowsky, N., Owen, A. A., Becker, A. J., Till, J. E.: Effects of vinblastine on the proliferative capacity of L cells and their progress through the division cycle. Cancer Res. **25,** 1232 – 1237 (1965)
41. Bryan, J.: Vinblastine and microtubules. I. Induction and isolation of crystals from sea urchin oocytes. Exp. Cell Res. **66,** 129 – 136 (1971)
42. Bryan, J.: Definition of three classes of binding sites in isolated microtubule crystals. Biochemistry **11,** 2611 – 2615 (1972)
43. Bryan, J.: Biochemical properties of microtubules. Fed. Proc. **33,** 152 – 157 (1974)
44. Bryan, J.: Preliminary studies on affinity labeling of the tubulin – colchicine binding site. Ann. N. Y. Acad. Sci. **253,** 247 – 259 (1975)
45. Bunge, R. P., Bunge, M. B.: A comparison of neuronal changes following colchicine treatment with observations on other conditions involving the accumulation of neurofilaments. J. Neuropath. Exp. Neurol. **28,** 169 (1969)
46. Bunt, A. H.: Protein synthesis in ganglion cells of the rabbit retina after intravitreous injection of vinblastine. Invest. Ophthal. **12,** 467 – 469 (1973)
47. Bunt, A. H.: Effects of vinblastine on microtubule structure and axonal transport in ganglion cells of the rabbit retina. Invest. Ophthal. **12,** 579 – 590 (1973)
48. Bunt, A. H.: Paracrystalline inclusions in optic nerve terminals following intraocular injection of vinblastine. Brain Res. **53,** 29 – 40 (1973)
49. Burgess, J., Northcote, D. H.: Action of colchicine and heavy water on the polymerization of microtubules in wheat root meristem. J. Cell Sci. **5,** 433 – 451 (1969)
50. Burns, R. G.: ^{3}H-colchicine binding. Failure to detect any binding to soluble proteins from various lower organisms. Exp. Cell Res. **81,** 285 – 292 (1973)
51. Byers, M. R., Fink, B. R., Kennedy, R. D., Middaugh, M. E., Hendrickson, A. E.: Effects of lidocaine on axonal morphology, microtubules, and rapid transport in rabbit vagus nerve in vitro. J. Neurobiol. **4,** 125 – 144 (1973)
52. Cann, J. R., Hinman, N. D.: Interaction of chlorpromazine with brain microtubule subunit protein. Mol. Pharmacol. **11,** 256 – 267 (1975)
53. Cardinali, G., Cardinali, G., Enein, M. A.: Studies on the antimitotic activity of leurocristine (vincristine). Blood **21,** 102 – 110 (1963)
54. Carolan, R. M., Sato, H., Inoue, S.: Further observations on the thermodynamics of the living mitotic spindle. Biol. Bull. **131,** 385 (1966)
55. Chang, L. W., Dudley, A. W. Jr., Lee, Y. K., Katz, J.: Ultrastructural changes in the kidney following chronic exposure to low levels of halothane. Am. J. Path. **78,** 225 – 242 (1975)
56. Comandon, J., Defonbrune, P.: Action de la colchicine sur *Amoeba sphaeronucleus.* C. R. Soc. Biol. Paris **136,** 410 – 411; 423; 460 – 461; 746 – 747; 747 – 748 (1942)

57. Coss, R. A., Bloodgood, R. A., Brower, D. L., Pickett-Heaps, J. D., McIntosh, J. R.: Studies on the mechanism of action of isopropyl N-phenylcarbamate. Exp. Cell Res. **92,** 394–398 (1975)
58. Cortese, F., Bhattacharyya, B., Wolff, J.: Podophyllotoxin as a probe for the colchicine-binding site of tubulin. J. Biol. Chem. **252,** 1134–1140 (1977)
59. Creasey, W. A.: Modifications in biochemical pathways produced by the *Vinca* alkaloids. Cancer Chemother. Rep. **52,** 501–507 (1968)
60. Creasey, W. A.: Biochemistry of dimeric *Catharanthus* alkaloids. In: The *Catharanthus* Species: Botany, Chemistry, Pharmacology, and Clinical Use (eds.: W. I. Taylor, N. R. Farnsworth) New York: Marcel Dekker 1975
61. Creasey, W. A.: *Vinca* alkaloids and colchicine. Handb. Exp. Pharmakol. **38/2,** 670–694 (1975)
62. Croop, J., Holtzer, H.: Response of myogenic and fibrogenic cells to cytochalasin B and to colcemid. I. Light microscope observations. J. Cell Biol. **65,** 271–285 (1975)
63. Cutts, J. H.: The effect of Vincaleukoblastine on dividing cells in vivo. Cancer Res. **21,** 168–172 (1961)
64. Cutts, J. H., Beer, C. T., Noble, R. L.: Biological properties of Vincaleukoblastine, an alkaloid in *Vinca rosea* Linn. with reference to its antitumor action. Cancer Res. **20,** 1023–1031 (1960)
65. Dahl, W. N., Ottebro, R., Pettersen, E. O., Brustad, T.: Inhibitory and cytotoxic effects of oncovin (vincristine sulfate) on cells of human line NHIK 3025. Cancer Res. **36,** 3101–3105 (1976)
66. Danø, K.: Development of resistance to adriamycin (NSC-123127) in Ehrlich ascites cells in vivo. Cancer Chemother. Rep. **56,** 321–326 (1972)
67. Davidse, L. C.: Antimitotic activity of methyl benzimidazol-2-yl carbamate (MBC) in *Aspergillus nidulans*. Pestic. Biochem. Physiol. **3,** 317–325 (1973)
68. Davidse, L. C.: Antimitotic activity of methyl benzimidazol-2-yl carbamate in fungi and its binding to cellular protein. In: Microtubules and Microtubule Inhibitors (eds.: M. Borgers, M. De Brabander) pp. 483–495. Amsterdam: North-Holland. New York: Am. Elsevier 1975
69. Davidse, L. C., Flach, W.: Differential binding of methyl benzimidazol-2-yl carbamate to fungal tubulin as a mechanism of resistance to this antimitotic agent in mutant strains of *Aspergillus nidulans*. J. Cell Biol. **72,** 174–193 (1977)
70. De Brabander, M.: Onderzoek naar de rol van microtubuli in gekweekte cellen met behulp van een nieuwe synthetische inhibitor van tubulin-polymerisatie. Thesis, Brussels 1977
71. De Brabander, M., Aerts, F., Van de Veire, R., Borgers, M.: Evidence against interconversion of microtubules and filaments. Nature **253,** 119–210 (1975)
72. De Brabander, M., Van de Veire, R., Aerts, F., Geuens, G., Brogers, M., Desplenter, L., De Cree, J.: Oncodazole (R 17934): a new anti-cancer drug interfering with microtubules. Effects on neoplastic cells cultured in vitro and in vivo. In: Microtubules and Microtubule Inhibitors (eds.: M. Borgers, M. De Brabander), pp. 509–521. Amsterdam-Oxford: North-Holland. New York: Am. Elsevier 1975
73. De Brabander, M. J., Van de Veire, R. M. L., Aerts, F., Borgers, M., Janssen, P. A. J.: The effects of methyl 5-(2-thienyl-carbonyl)-1H-benzimidazol-2-yl carbamate (R 17934; NSC 238 159), a new synthetic antitumoral drug interfering with microtubules, on mammalian cells cultured in vitro. Cancer Res. **36,** 1011–1018 (1976)
74. Deconti, R. C., Creasey, W. A.: Clinical aspects of the dimeric *Catharanthus* alkaloids. In: The *Catharanthus* Alkaloids. Botany, Chemistry, Pharmacology, and Clinical Use (eds.: W. I. Taylor, N. R. Farnsworth), pp. 237–278. New York: Marcel Dekker 1975
75. de Harven, E.: A propos de la résistance du hamster doré (*Mesocricetus auratus*) à certains poisons stathmocinétiques. Bull. Cl. Sci. Acad. Roy. Belg. 5e série, **41,** 1056–1060 (1955)
76. De Mey, J., Hoebeke, J., De Brabander, M., Geuens, G., Joniau, M.: Immunoperoxidase visualisation of microtubules and microtubular proteins. Nature **264,** 273–275 (1976)
77. Di Marco, A., Soldati, M., Gaetani, M.: Ricerche sull'azione della "Vincaleukoblastine sulphate" in associazione ad altre sostanze. Tumori **47,** 278–288 (1961)
78. Dixon, W.: A Manual of Pharmacology. London: Arnold 1906
79. Donigian, D. W., Owellen, R. J.: Interaction of vinblastine, vincristine and colchicine with serum proteins. Biochem. Pharmacol. **22,** 2113–2120 (1973)
80. Dustin, A. P.: L'action des arsenicaux et de la colchicine sur la mitose. La stathmocinèse. C. R. Ass. Anat. **33,** 204–212 (1938)
81. Dustin, P.: Some new aspects of mitotic poisoning. Nature **159,** 794 (1947)

82. Dustin, P. Jr.: Mitotic poisoning at metaphase and -SH proteins. Exp. Cell Res. Suppl. **1,** 153 – 155 (1949)
83. Dustin, P.: The quantitative estimation of mitotic growth in the bone marrow of the rat by the stathmokinetic (colchicinic) method. In: The Kinetics of Cellular Proliferation (ed.: F. Stohlman), pp. 50 – 56. New York: Grune and Stratton 1959
84. Dustin, P. Jr.: New aspects of the pharmacology of antimitotic agents. Pharmacol. Rev. **15,** 449 – 480 (1963)
85. Dustin, P., Hubert, J. P., Flament-Durand, J.: Centriole and cilia formation in rat pituicytes after treatment with colchicine and vincristine. Studies in vivo and in vitro. In: Microtubules and Microtubule Inhibitors (eds.: M. Borgers, M. De Brabander), pp. 289 – 296. Amsterdam-Oxford: North-Holland. New York: Am. Elsevier 1975
86. Dustin, P., Hubert, J. P., Flament-Durand, J.: Action of colchicine on axonal flow and pituicytes in the hypothalamopituitary system of the rat. Ann. N. Y. Acad. Sci. **253,** 670 – 684 (1975)
87. Edström, A., Hansson, H. A., Norström, A.: Inhibition of axonal transport in vitro in frog sciatic nerves by chlorpromazine and lidocaine. A biochemical and ultrastructural study. Z. Zellforsch. **143,** 53 – 70 (1973)
88. Edström, A., Hansson, H.-A., Norstrom, A.: The effect of chlorpromazine and tetracaine on the rapid axonal transport of neurosecretory material in the hypothalamo-neurohypophysial system of the rat. A biochemical and ultrastructural study. Z. Zellforsch. **143,** 71 – 92 (1973)
89. Edström, A., Hansson, H.-A., Larsson, H., Wallin, M.: Effects of barbiturates on ultrastructure and polymerization of microtubules in vitro. Cell Tissue Res. **162,** 35 – 48 (1975)
90. Eigsti, O. J., Dustin, P. Jr.: Colchicine in Agriculture, Medicine, Biology, and Chemistry. Ames, Iowa: Iowa State Coll. Press 1955
91. Engelborghs, Y., Heremans, K. A. H., De Maeyer, L. C. M., Hoebeke, J.: Effect of temperature and pressure on polymerisation equilibrium of neuronal microtubules. Nature **259,** 686 – 688 (1976)
92. Ertel, N., Omokoku, B., Wallace, S. L.: Colchicine concentrations in leukocytes. Arthritis Rheum. **12,** 293 (1969)
93. Feit, H., Barondes, S. H.: Colchicine-binding activity in particulate fractions of mouse brain. J. Neurochem. **17,** 1355 – 1364 (1970)
94. Fitzgerald, T. J.: Molecular features of colchicine associated with antimitotic activity and inhibition of tubulin polymerization. Biochem. Pharmacol. **25,** 1383 – 1388 (1976)
95. Fitzgerald, T. J., Mayfield, D. G.: Effect of colchicine on polymerization of tubulin from rats, mice, hamsters and guinea-pigs. Experientia **32,** 83 – 84 (1976)
96. Fitzgerald, T. J., Veal, A.: Melatonin antagonizes colchicine-induced mitotic arrest. Experientia **32,** 372 – 373 (1976)
97. Fragata, M.: The mitotic apparatus, a possible site of action of gibberellic acid. Naturwissenschaften **57,** 139 (1970)
98. Freed, J. J., Ohlsson-Wilhem, B. M.: Cultured haploid cells resistant to antitubulins. In: Microtubules and Microtubule Inhibitors (eds.: M. Borgers, M. De Brabander), pp. 367 – 378. Amsterdam-Oxford: North-Holland. New York: Am. Elsevier 1975
99. Frei, E. III, Whang, J., Scoggins, R. B., Scott, E. J. van, Rall, D. P., Ben, M.: The stathmokinetic effect of vincristine. Cancer Res. **24,** 1918 – 1925 (1964)
100. Fujiwara, K., Tilney, L. G.: Substructural analysis of the microtubule and its polymorphic forms. Ann. N. Y. Acad. Sci. **253,** 27 – 50 (1975)
101. Garland, D., Teller, D. C.: A reexamination of the reaction between colchicine and tubulin. Ann. N. Y. Acad. Sci. **253,** 232 – 238 (1975)
102. George, P., Journey, L. J., Goldstein, M. N.: Effect of vincristine on the fine structure of HeLa cells during mitosis. J. Nat. Cancer Inst. **35,** 355 – 375 (1965)
103. Gillespie, E.: Colchicine binding in tissue slices. Decrease by calcium and biphasic effect of adenosine-3′,5′-monophosphate. J. Cell Biol. **50,** 544 – 549 (1971)
104. Goldman, R. D.: The role of three cytoplasmic fibers in BHK-21 cell motility. I. Microtubules and the effects of colchicine. J. Cell Biol. **51,** 752 – 762 (1971)
105. Goode, D., Salmon, E. D., Maugel, T. K., Bonar, D. B.: Microtubule disassembly and recovery from hydrostatic pressure treatment of mitotic HeLa cells. J. Cell Biol. **67,** 138 a (1975)
106. Gorter, C.: De invloed van colchicine op den groei van den celwand van wortelharen. Proc. Kon. Nederl. Akad. Wetensch. **48,** 3 – 12 (1945)
107. Grisham, L. M., Wilson, L., Bensch, K. G.: Antimitotic action of griseofulvin does not involve disruption of microtubules. Nature **244,** 294 – 296 (1973)

108. Gull, K., Trinci, A. P. J.: Ultrastructural effects of griseofulvin on the myxomycete *Physarum polycephalum*. Inhibition of mitosis and the production of microtubule crystals. Protoplasma **81,** 37 – 48 (1974)
109. Hammond, S., Bryan, J.: Microtubule affinity labels. J. Cell Biol. **55,** 103 a (1972)
110. Hart, J. W., Sabnis, D. D.: Colchicine and plant microtubules: a critical evaluation. Curr. Adv. Plant. Sci. **26,** 1095 – 1104 (1976)
111. Hart, J. W., Sabnis, D. D.: Colchicine binding activity in extracts of higher plants. J. Exp. Bot. **27,** 1353 – 1360 (1976)
112. Hart, J. W., Sabnis, D. D.: Binding of colchicine and lumicolchicine to components in plants extracts. Phytochemistry **15,** 1897 – 1902 (1976)
113. Haschke, R. H., Byers, M., Fink, B. R.: Effects of lidocaine on rabbit brain microtubular protein. J. Neurochem. **22,** 837 – 844 (1974)
114. Hasholt, L., Danø, K.: Cytogenetic investigations on an Ehrlich ascites tumor, and four sublines resistant to daunomycin, adriamycin, vincristine and vinblastine. Hereditas **77,** 303 – 310 (1974)
115. Hauser, M.: Differentielles Kontrastverhalten verschiedener Mikrotubulisysteme nach Mercury Orange-Behandlung. Cytobiologie **6,** 367 – 381 (1972)
116. Hepler, P. K., Jackson, W. T.: Isopropyl-*N*-phenylcarbamate affects spindle microtubules orientation in dividing endosperm cells of *Haemanthus katherinae* Baker. J. Cell Sci. **5,** 727 – 743 (1969)
117. Himes, R. H., Kersey, R. N., Heller-Bettinger, I., Samson, F. E.: Action of the *Vinca* alkaloids vincristine, vinblastine, and desacetyl vinblastine amide on microtubules in vitro. Cancer Res. **36,** 3798 – 3802 (1976)
118. Hinkley, R. E. Jr.: Microtubule – macrotubule transformations induced by volatile anesthetics. Mechanism of macrotubule assembly. J. Ultrastruct. Res. **57,** 237 – 250 (1976)
119. Hinkley, R. E. Jr., Green, L. S.: Effects of halothane and colchicine on microtubules and electrical activity of rabbit vagus nerves. J. Neurobiol. **2,** 97 – 106 (1971)
120. Hinkley, R. E., Samson, F. E.: Anesthetic-induced transformation of axonal microtubules. J. Cell Biol. **53,** 258 – 263 (1972)
121. Hinkley, R. E., Samson, F. E. Jr.: The effects of an elevated temperature on colchicine, and vinblastine on axonal microtubules of the crayfish (*Procambarus clarkii*). J. Exp. Zool. **188,** 321 – 336 (1974)
122. Hinman, N. D., Cann, J. R.: Reversible binding of chlorpromazine to brain tubulin. Mol. Pharmacol. **12,** 769 – 777 (1976)
123. Hirano, A.: Neurofibrillary changes in conditions related to Alzheimer's disease. In: Alzheimer's Disease and Related Conditions (eds.: G. E. W. Wolstenholme, M. O'Connor), pp. 185 – 201. London: Churchill 1970
124. Hoebeke, J., van Nijen, G., De Brabander, M.: Interaction of oncodazole (R 17934), a new antitumoral drug, with rat brain tubulin. Biochem. Biophys. Res. Commun. **69,** 319 – 324 (1976)
125. Holtzer, H., Fellini, S., Rubinstein, N., Chi, J., Strahs, K.: Cells, myosins and 100-A filaments. In: Cell Motility (eds.: R. Goldman, T. Pollard, J. Rosenbaum), pp. 823 – 840. Cold Spring Harbor Lab. 1976
126. Hunter, A. L., Klaassen, C. D.: Biliary excretion of colchicine. J. Pharmacol. Exp. Therap. **192,** 605 – 617 (1975)
127. Hunter, A. L., Klaassen, C. D.: Biliary excretion of colchicine in newborn rats. Drug Metab. Dispos. **3,** 530 – 535 (1975)
128. Ilan, J., Quastel, J. H.: Effect of colchicine on nucleic acid metabolism during metamorphosis of *Tenebrio molitor* L. and in some mammalian tissues. Biochem. J. **100,** 448 – 457 (1966)
129. Jackson, W. T.: Regulation of mitosis. II. Interaction of isopropyl-n-phenyl carbamate and melatonin. J. Cell Sci. **5,** 745 – 755 (1969)
130. Johnson, I. S., Armstrong, J. G., Gorman, M., Burnett, J. P. Jr.: The *Vinca* alkaloids: a new class of oncolytic agents. Cancer Res. **23,** 1390 – 1427 (1963)
131. Journey, L. J., Burdman, J., George, P.: Ultrastructural studies on tissue culture cells treated with vincristine. Cancer Chemother. Rep. **52,** 509 – 518 (1968)
132. Kadota, T., Kadota, A., Gray, E. G.: Coated-vesicle shells, particle/chain material, and tubulin in brain synaptosomes. An electron microscope and biochemical study. J. Cell Biol. **69,** 608 – 621 (1976)
133. Kelleher, J. K., Johnson, E., Winston, M., Banerjee, S., Margulis, L.: Melatonin: induces mitotic arrest and inhibits colchicine-binding to microtubule protein. J. Cell Biol. **59,** 164 a (1973)
134. Kelly, M. G., Hartwell, J. L.: The biological effects and the chemical composition of podophyllin. A review. J. Nat. Cancer Inst. **14,** 967 – 1010 (1954)

135. Kennedy, R. D., Fink, B. R., Byers, M. R.: The effect of halothane on rapid axonal transport in the rabbit vagus. Anesthesiology **36,** 433 – 443 (1972)
136. Kennedy, J. R. Jr., Brittingham, E.: Fine structure changes during chloral hydrate deciliation of *Paramecium caudatum*. J. Ultrastruct. Res. **22,** 530 – 545 (1968)
137. Kennedy, J. R., Zimmerman, A. M.: The effects of high hydrostatic pressure of the microtubules of *Tetrahymena pyriformis*. J. Cell Biol. **47,** 568 – 576 (1970)
138. King, L. S., Sullivan, M.: Effects of podophyllin and of colchicine on normal skin, on condyloma acuminatum and on verruca vulgaris. Pathologic observations. Arch. Path. **43,** 373 – 386 (1947)
139. Kirschner, M. W., Williams, R. C., Weingarten, M., Gerhart, J. C.: Microtubules from mammalian brain: some properties of their depolymerization products and a proposed mechanism of assembly and disassembly. Proc. Natl. Acad. Sci. USA **71,** 1159 – 1187 (1974)
140. Kotani, E., Miyazaki, F., Tobinaga, S.: A new synthesis of the alkaloid (±)-colchicine. J. chem. Sci. **8,** 300 (1974)
141. Krammer, E. B., Zenker, W.: Effekt von Zinkionen auf Struktur und Verteilung der Neurotubuli. Acta Neuropath. **31,** 59 – 69 (1975)
142. Krishan, A.: Time-lapse and ultrastructure studies on the reversal of mitotic arrest induced by vinblastine sulfate in Earle's L cells. J. Nat. Cancer Inst. **41,** 581 – 596 (1968)
143. Krishan, A., Hsu, D.: Observations on the association of helical polyribosomes and filaments with vincristine-induced crystals in Earle's L-cell fibroblasts. J. Cell Biol. **43,** 553 – 563 (1969)
144. Krishan, A., Hsu, D.: Vinblastine-induced ribosomal complexes. Effect of some metabolic inhibitors on their formation and structure. J. Cell Biol. **49,** 927 – 931 (1971)
145. Krishan, A., Hsu, D.: Binding of colchicine-^{3}H to vinblastine- and vincristine-induced crystals in mammalian tissue culture cells. J. Cell Biol. **48,** 407 – 409 (1971)
146. Kupchan, S. M., Komoda, Y., Court, W. A., Thomas, G. J., Smith, R. M., Karim, A., Gilmore, C. J., Haltiwanger, R. C., Bryan, R. F.: Maytansine, a novel antileukemic ansa macrolide from *Maytenus ovatus*. J. Am. Chem. Soc. **94,** 1354 – 1356 (1972)
147. Kuriyama, R., Sakai, H.: Viscometric demonstration of tubulin polymerization. J. Biochem. **75,** 463 – 472 (1974)
148. Kuzmich, M. J., Zimmerman, A. M.: The action of mercaptoethanol on cell division in synchronized *Tetrahymena*. J. Protozool. **19,** 129 – 132 (1972)
149. Lambers, K.: Über Organveränderungen bei chronischer Colchicine-Vergiftung. Virchows Arch. **321,** 88 – 100 (1951 – 1952)
150. Lapis, K.: Vincaleukoblastine-induced changes in the ultrastructure of ascites tumour cells. Acta Morph. Acad. Sci. Hung. **16,** 65 – 83 (1968)
151. Lettré, H.: Über Mitosegifte. Ergebn. Physiol. **46,** 379 – 452 (1950)
152. Lettré, H., Albrecht, M.: Über die Abhängigkeit der Colchicine-Wirkung von der Adenosintriphosphorsäure. Naturwissenschaften **38,** 547 (1951)
153. Lettré, H., Dönges, K.-H., Barthold, K., Fitzgerald, T. J.: Synthese neuer Colchicin-Derivate mit hoher antimitotischer Wirksamkeit. Liebigs Ann. **758,** 185 – 189 (1972)
154. Levan, A.: The macroscopic colchicine effect: a hormonic action? Hereditas **26,** 262 – 276 (1942)
155. Ling, V., Thompson, L. H.: Reduced permeability in CHO cells as a mechanism of resistance to colchicine. J. Cell. Physiol. **83,** 103 – 116 (1974)
156. Loike, J. D., Horwitz, S. B.: Effects of podophyllotoxin and VP-16-213 on microtubule assembly in vitro and nucleoside transport in HeLa cells. Biochemistry **15,** 5435 – 5442 (1976)
157. Loike, J. D., Horwitz, S. B.: Effect of VP-16-213 on the intracellular degradation of DNA in HeLa cells. Biochemistry **15,** 5443 – 5448 (1976)
158. Love, R.: Studies on the cytochemistry of nucleoproteins. I. Effect of colchicine on the ribonucleoproteins of the cell. Exp. Cell Res. **33,** 216 – 231 (1964)
159. Luftig, R. B., McMillan, P. N., Weatherbee, J. A., Weihing, R. R.: Increased visualization of microtubules by an improved fixation procedure. J. Histochem. Cytochem. **25,** 175 – 187 (1977)
160. McGee-Russel, S. M.: Rapid disassembly and reassembly of labile microtubules in *Allogromia* by urea treatment, with formation of pseudocrystalline arrays. J. Cell Biol. **55,** 170 a (1972)
161. McGuire, J., Quinn, P., Knutton, S.: The selective precipitation of tubulin by chlorpromazine. J. Cell Biol. **63,** 217 a (1974)
162. MacKinnon, E. A., Abraham, P. J., Svatek, A.: Long link induction between the microtubules of the manchette in intermediate stages of spermiogenesis. Z. Zellforsch. **136,** 447 – 460 (1973)

163. Malawista, S. E.: On the action of colchicine. The melanocyte model. J. Exp. Med. **122,** 361 – 384 (1965)
164. Malawista, S. E.: Microtubules and the movement of melanin granules in frog dermal melanocytes. Ann. N. Y. Acad. Sci. **253,** 702 – 710 (1975)
165. Malawista, S. E., Sato, H., Bensch, K. G.: Vinblastine and griseofulvin reversibly disrupt the living mitotic spindle. Science **160,** 770 – 771 (1968)
166. Malawista, S. E., Sato, H.: Vinblastine produces uniaxial, birefringent crystals in starfish oocytes. J. Cell Biol. **42,** 596 – 599 (1969)
167. Mandelbaum-Shavit, F., Wolpert-Defilippes, M. K., Johns, D. G.: Binding of maytansine to rat brain tubulin. Biochem. Biophys. Res. Commun. **72,** 47 – 54 (1976)
168. Maraldi, N. M., Simonelli, L., Pettazzoni, P., Barbieri, M.: Ribosome crystallization. II. Ribosome and protein crystallization in hypothermic cell cultures treated with vinblastine sulfate. J. Submicrosc. Cytol. **2,** 51 – 68 (1970)
169. Marantz, R., Shelanski, M. L.: Structure of microtubular crystals induced by vinblastine in vitro. J. Cell Biol. **44,** 234 – 238 (1970)
170. Marantz, R., Ventilla, M., Shelanski, M. L.: Vinblastine-induced precipitation of microtubule protein. Science **165,** 498 – 499 (1969)
171. Marchant, H. J., Pickett-Heaps, J. D.: The effect of colchicine on colony formation in the algae *Hydrodictyon, Pediastrum* and *Sorastrum*. Planta **116,** 291 – 300 (1974)
172. Margulis, L.: Microtubules and evolution. In: Microtubules and Microtubule Inhibitors. (eds.: M. Borgers, M. De Brabander) pp. 3 – 18 Amsterdam-Oxford: North-Holland. New York: Am. Elsevier 1975
173. Margulis, T. N.: Structure of the mitotic spindle inhibitor colcemid. N-desacetyl-N-methyl-colchicine. J. Am. Chem. Soc. **96,** 899 – 901 (1974)
174. Margulis, T. N.: X-ray analysis of microtubule inhibitors. In: Microtubules and Microtubule Inhibitors (eds.: M. Borgers, M. De Brabander), pp. 67 – 78. Amsterdam-Oxford: North-Holland. New York: Am. Elsevier 1975
175. Matsumura, F., Hayashi, M.: Polymorphism of tubulin assembly. In vitro formation of sheet, twisted ribbon and microtubule. Biochem. Biophys. Acta **453,** 162 – 175 (2976)
176. Mathé, G., Schwarzenberg, L., Pouillart, P., Oldham, R., Weiner, R., Jasmin, C., Rosenfeld, C., Hayat, M., Misset, J. L., Musset, M., Schneider, M., Amiel, J. L., de Vassal, F.: Two epipodophyllotoxin derivatives, VM 26 and VP 16213, in the treatment of leukemias, hematosarcomas and lymphomas. Cancer **34,** 985 – 992 (1974)
177. Mazia, D.: Mitosis and the physiology of cell division. In: The Cell. Biochemistry, Physiology, Morphology (eds.: J. Brachet, A. E. Mirsky), Vol. 3, pp. 77 – 412. New York-London: Acad. Press 1961
178. Meisner, H. M., Sorensen, L.: Metaphase arrest of Chinese Hamster cells with rotenone. Exp. Cell Res. **42,** 291 – 295 (1966)
179. Mellon, M. G., Rebhun, L. I.: Sulfhydryls and in vitro polymerization of tubulin. J. Cell Biol. **70,** 226 – 238 (1976)
180. Mellon, M., Rebhun, L. I.: Studies on the accessible sulfhydryls of polymerizable tubulin. In: Cell Motility (eds.: R. Goldman, T. Pollard, J. Rosenbaum), pp. 1149 – 1164. Cold Spring Harbor Lab. 1976
181. Minor, P. D., Roscoe, D. H.: Colchicine resistance in mammalian cell lines. J. Cell Sci. **17,** 381 – 396 (1975)
182. Mizel, S. B., Wilson, L.: Nucleoside transport in mammalian cells. Inhibition by colchicine. Biochemistry **11,** 2573 – 2577 (1972)
183. Moore, K. C.: Pressure-induced regression of oral apparatus microtubules in synchronized *Tetrahymena*. J. Ultrastruct. Res. **41,** 499 – 518 (1972)
184. Murray, M. R., de Lam, H., Chargaff, E.: Specific inhibition by meso-inositol of the colchicine effect on rat fibroblasts. Exp. Cell Res. **11,** 165 – 177 (1951)
185. Nagayama, A., Dales, S.: Rapid purification and the immunological specificity of mammalian microtubular paracrystals possessing ATPase activity. Proc. Natl. Acad. Sci. USA **66,** 464 – 471 (1970)
186. Narain, P., Raina, S. N.: Cytological assay of C-mitotic potency of colchicine obtained from *Gloriosa superba* L. Cytologia **40,** 751 – 757 (1975)
187. Neviackas, J. A., Margulis, L.: The effect of colchicine on regenerating membranellar cilia in *Stentor coeruleus*. J. Protozool. **16,** 165 – 171 (1969)
188. Norberg, B.: Cytoplasmic microtubules and radialsegmented nuclei (Rieder cells). Effects of osmolality, ionic strength, pH, penetrating non-electrolytes, griseofulvin, and a podophyllin derivative. Scand. J. Haemat. **7,** 445 – 454 (1970)
189. O'Connor, T. M., Houston, L. L., Samson, F.: Stability of neuronal microtubules to high pressure in vivo and in vitro. Proc. Natl. Acad. Sci. USA **71,** 4198 – 4202 (1974)

190. Oftebro, R., Grimmer, Ø., Øyen, T. B., Laland, S. G.: 5-fluoropyrimidin-2-one, a new metaphase arresting agent. Biochem. Pharmacol. **21,** 2451 – 2456 (1972)
191. Olah, L. V., Bozzola, J.: Effect of digitonin on cell division. IV. Interrelation between nucleus and phragmoplast. Cytologia **37,** 365 – 376 (1972)
192. Olah, L. V., Hanzely, L.: Effect of digitonin on cellular division. V. The distribution of microtubules. Cytologia **38,** 55 – 72 (1973)
193. Oldham, R. K., Pomeroy, T. C.: Vincristine-induced syndrome of inappropriate secretion of antidiuretic hormone. South. Med. J. **65,** 1010 – 1012 (1972)
194. Olmsted, J. B., Marcum, J. M., Johnson, K. A., Allen, C., Borisy, G. G.: Microtubule assembly: some possible regulatory mechanisms. J. Supramol. Struct. **2,** 429 – 450 (1974)
195. Orsini, M. W., Pansky, B.: The natural resistance of the Golden Hamster to colchicine. Science **115,** 88 – 89 (1952)
196. Östergren, G.: Colchicine mitosis, chromosome contraction, narcosis and protein chain folding. Hereditas **30,** 429 – 467 (1944)
197. Östergren, G.: Narcotized mitosis and the precipitation hypothesis of narcosis. Coll. Int. Centre Nat. Rech. Sci. **26,** 77 – 88 (1951)
198. Owellen, R. J., Hartke, C. A., Dickerson, R. M., Hains, F. O.: Inhibition of tubulin-microtubule polymerization by drugs of the *Vinca* alkaloid class. Cancer Res. **36,** 1499 – 1503 (1976)
199. Oxford, A. E., Raistrick, H., Simonart, P.: Studies in the biochemistry of microorganisms. LX. Griseofulvin $C_{18}H_{17}O_6Cl$, a metabolic product of *Penicillium Griseofulvin* Dierckx. Biochem. J. **33,** 240 – 248 (1939)
200. Paget, G. E., Walpole, A. L.: The experimental toxicology of griseofulvin. Arch. Derm. **81,** 750 – 757 (1960)
201. Pfeffer, T. A., Asnes, C. F., Wilson, L.: Properties of tubulin in unfertilized sea urchin eggs. Quantitation and characterization by the colchicine-binding reaction. J. Cell Biol., **69,** 599 – 607 (1976)
202. Piton, R.: Recherches sur les actions caryoclasiques et caryocinétiques des composés arsenicaux. Arch. Intern. Méd. Exp. **5,** 355 – 411 (1929)
203. Poffenbarger, M., Fuller, G. M.: Is melatonin a microtubule inhibitor? Exp. Cell Res. **103,** 135 – 142 (1976)
204. Radley, J. M.: Ultrastructure of mitotic arrest induced by isoprenaline in rat parotid acinar cells. J. Cell Sci. **16,** 309 – 332 (1974)
205. Rannestad, J., Willimans, N. E.: The synthesis of microtubule and other proteins of the oral apparatus in *Tetrahymena pyriformis*. J. Cell Biol. **50,** 709 – 720 (1971)
206. Rapkine, L.: Sur les processus chimiques au cours de la division cellulaire. Ann. Physiol. Physicochim. Biol. **7,** 382 – 418 (1931)
207. Rebhun, L. I., Nath, J., Remillard, S. P.: Sulfhydryls and regulation of cell division. In: Cell Motility (eds.: R. Goldman, T. Pollard, J. Rosenbaum), pp. 1343 – 1366. Cold Spring Harbor Lab. 1976
208. Remillard, S., Rebhun, L. I., Howie, G. A., Kupchan, S. M.: Antimitotic activity of the potent tumor inhibitor Maytansine. Science **189,** 1002 – 1005 (1975)
209. Renaud, F. L., Rowe, A. J., Gibbons, K. R.: Some properties of the proteins forming the outer fibers of cilia. J. Cell Biol. **36,** 79 – 90 (1968)
210. Robbins, E., Shelanski, M.: Synthesis of a colchicine-binding protein during the HeLa cell life cycle. J. Cell Biol. **43,** 371 – 373 (1969)
211. Roobol, A., Gull, K., Pogson, C. I.: Inhibition by griseofulvin of microtubule assembly in vitro. FEBS Lett. **67,** 248 – 251 (1976)
212. Roobol, A., Gull, K., Pogson, C. I.: Evidence that griseofulvin binds to a microtubule associated protein. FEBS Lett. **75,** 149 – 153 (1977)
213. Roodman, G. D., Hutton, J. J., Bollum, F. J.: DNA polymerase activities during erythropoiesis. Effects of erythropoietin, vinblastine, colcemid, and daunomycin. Exp. Cell Res. **91,** 269 – 278 (1975)
214. Röper, W.: Antagonistic effects of pemoline to colchicine and caffeine. Experientia **31,** 1200 – 1201 (1975)
215. Roth, L. E.: Electron microscopy of mitosis in amebae. III. Cold and urea treatments: a basis for tests of direct effects of mitotic inhibitors on microtubule formation. J. Cell Biol. **34,** 47 – 59 (1967)
216. Roth, L. E., Shigenaka, Y.: Microtubules in the Heliozoan Axopodium. II. Rapid degradation by cupric and nickelous ions. J. Ultrastruct. Res. **31,** 356 – 374 (1970)
217. Rozenblat, V. A., Serpinskaia, A. S., Stavrovskaia, A. A.: The increase of sensitivity of cells, resistant to colchicine, by means of non-ionogen detergent twin 80. Proc. Acad. Sci. USSR **215,** 208 – 210 (1974)

218. Salmon, E. D.: Pressure-induced depolymerization of spindle microtubules. II. Thermodynamics of in vivo spindle assembly. J. Cell Biol. **66,** 114 – 127 (1975)
219. Sato, H., Ohnuki, Y., Fujiwara, K.: Tubulin immunofluorescence in mitotic spindle and unpolymerized spindle subunits. In: Mitosis. Facts and Questions (eds.: M. Little, N. Paweletz, C. Petzelt, H. Ponstingl, D. Schroeter, H.-P. Zimmermann) p. 239. Berlin-Heidelberg-New York: Springer 1977
220. Saubermann, A. J., Gallagher, M. L.: Mechanisms of general anesthesia: failure of pentobarbital and halothane to depolymerize microtubules in mouse optic nerve. Anesthesiology **38,** 25 – 28 (1973)
221. Saubermann, A. J., Gallagher, M. L.: Allison and Nunn revisited. Anesthesiology **39,** 357 (1973)
222. Sawhney, V. K., Srivastava, L. M.: Gibberellic acid induced elongation of lettuce hypocotyls and its inhibition by colchicine. Can. J. Bot. **52,** 259 – 264 (1974)
223. Schechter, J., Yancey, B., Weiner, R.: Response of tanycytes of rat median eminence to intraventricular administration of colchicine and vinblastine. Anat. Rec. **184,** 233 – 249 (1976)
224. Schindler, R.: Desacetylamino-colchicine: a derivative of colchicine with increased cytotoxic activity in mammalian cell cultures. Nature **196,** 73 – 74 (1962)
225. Schmitt, H., Atlas, D.: Specific affinity labelling of tubulin with bromocolchicine. J. Mol. Biol. **102,** 743 – 758 (1976)
226. Schnaitman, T., Rebhun, L. I., Kupchan, S. M.: Antimitotic acitivity of the antitumor agent, maytansine. J. Cell Biol. **67,** 388 a (1975)
227. Schnepf, E., Deichgräber, G.: The effects of colchicine, ethionine, and deuterium oxide on microtubules in young *Sphagnum* leaflets. A quantitative study. Cytobiologie **13,** 341 – 353 (1976)
228. Schochet, S. S. Jr., Lampert, P. W., Earle, K. M.: Neuronal changes induced by intrathecal vincristine sulfate. J. Neuropath. Exp. Neurol. **27,** 645 – 658 (1968)
229. Schönharting, M., Mende, G., Siebert, G.: Metabolic transformation of colchicine. II. The metabolism of colchicine by mammalian liver microsomes. Hoppe-Seyler's Z. Phys. Chem. **355,** 1391 – 1399 (1974)
230. Secret, C. J., Hadfield, J. R., Beer, C. T.: Studies on the binding of ^{3}H-vinblastine by rat blood platelets in vitro. Effects of colchicine and vincristine. Biochem. Pharmacol. **21,** 1609 – 1624 (1972)
231. Sentein, P.: Action de la quinoline sur les mitoses de segmentation des œufs d'Urodèles: le blocage de la centrosphère. Chromosoma **32,** 97 – 134 (1970)
232. Sentein, P.: Formamide, formaldéhyde et acide formique. Trois substances à action antimitotique différente. Bull. Ass. Anat. **56,** 712 – 720 (1971)
233. Sentein, P.: Action of glutaraldehyde and formaldehyde on segmentation mitoses. Inhibition of spindle and astral fibers, centrospheres blocked. Exp. Cell Res. **95,** 233 – 246 (1975)
234. Sentein, P., Ates, Y.: Action de l'hydrate de chloral sur les mitoses de segmentation de l'œuf de Pleurodèle. Etude cytologique et ultrastructurale. Chromosoma **45,** 215 – 244 (1974)
235. Shay, J. W.: Electron microscope studies of spermatozoa of *Rhynchosciara* sp. I. Disruption of microtubules by various treatments. J. Cell Biol. **54,** 598 – 608 (1972)
236. Sherline, P., Leung, J. T., Kipnis, D. M.: Binding of colchicine to purified microtubule protein. J. Biol. Chem. **250,** 5481 – 5486 (1975)
237. Shibaoka, H.: Involvement of wall microtubules in gibberellin promotion and kinetin inhibition of stem elongation. Plant Cell Physiol. **15,** 255 – 264 (1974)
238. Shigenaka, Y.: Microtubules in Protozoan cells. II. Heavy metal ion effects on degradation and stabilization of the Heliozoan microtubules. Annot. Zool. Jpn. **49,** 164 – 176 (1976)
239. Shigenaka, Y., Roth, L. E., Pihlaja, D. J.: Microtubules in the heliozoan axopodium. III. Degradation and reformation after dilute urea treatment. J. Cell Sci. **8,** 127 – 152 (1971)
240. Shigenaka, Y., Tadokoro, Y., Kaneda, M.: Microtubules in Protozoan cells. I. Effects of light metal ions on the Heliozoan microtubules and their kinetic analysis. Annot. Zool. Jpn. **48,** 227 – 241 (1975)
241. Shohat, B.: Effect of withaferin A on cells in tissue culture. Z. Krebsforsch. **80,** 97 – 102 (1973)
242. Slonim, R. R., Howell, D. S., Brown, H. E. Jr.: Influence of griseofulvin upon acute gouty arthritis. Arthritis Rheum. **5,** 397 – 404 (1962)
243. Starling, D.: Two ultrastructurally distinct tubulin paracrystals induced in sea-urchin eggs by vinblastine sulphate. J. Cell Sci. **20,** 79 – 90 (1976)
244. Starling, D.: The effects of mitotic inhibitors on the structure of vinblastine-induced tubulin paracrystals from sea-urchin eggs. J. Cell Sci. **20,** 91 – 100 (1976)

245. Starling, D., Burns, R. G.: Ultrastructure of tubulin paracrystals from sea urchin eggs, with determination of spacings by electron and optical diffraction. J. Ultrastruct. Res. **51**, 261 – 268 (1975)
246. Stephens, R. E.: Thermal fractionation of outer fiber doublet microtubules into A- and B-subfiber components: A- and B-tubulin. J. Mol. Biol. **47**, 353 – 363 (1970)
247. Stubblefield, E., Brinckley, B. R.: Cilia formation in Chinese hamster fibroblasts in vitro as a response to colcemid treatment. J. Cell Biol. **30**, 645 – 652 (1966)
248. Sullivan, M., King, L. S.: Effects of resin of podophyllum on normal skin, condylomata acuminata and verrucae vulgaris. Arch. Dermatol. Syphilol. **56**, 30 – 45 (1947)
249. Svoboda, G. H., Blake, D. A.: The phytochemistry and pharmacology of *Catharanthus roseus* (L.) G. Don. In: The *Catharanthus* Alkaloids. Botany, Chemistry, Pharmacology, and Clinical Use (eds.: W. I. Taylor, N. R. Farnsworth) New York: Marcel Dekker 1975
250. Taylor, E. W.: The mechanism of colchicine inhibition of mitosis. I. Kinetics of inhibition and the binding of H^3-colchicine. J. Cell Biol. **25**, 145 – 160 (1965)
251. Taylor, W. I, Fransworth, N. R. (eds.): The *Catharanthus* Alkaloids. Botany, Chemistry and Clinical Use. New York: Marcel Dekker 1975
252. Thrasher, J. D.: The effect of mercuric compounds on dividing cells. In: Drugs and the Cell Cycle (eds.: A. M. Zimmerman, G. M. Padilla, I. L. Cameron), pp. 25 – 48. New York-London: Acad. Press 1973
253. Tilney, L. G., Porter, K. R.: Studies on the microtubules in Heliozoa. II. The effect of low temperature on these structures in the formation and maintenance of the axopodia. J. Cell Biol. **34**, 327 – 343 (1967)
254. Tomlinson, D. R., Bennett, T.: An ultrastructural examination of the action of vinblastine on microtubules, neurofilaments and muscle filaments in vitro. Cell Tissue Res. **166**, 413 – 420 (1976)
255. Turner, F. R.: Modified microtubules in the testis of the water strider, *Gerris remigis* (Say). J. Cell Biol. **53**, 263 – 270 (1972)
256. Tyson, G. E., Bulger, R. E.: Effect of vinblastine sulfate on the fine structure of cells of the rat renal corpuscule. Am. J. Anat. **135**, 319 – 344 (1972)
257. Van Assel, S., Brachet, J.: Formation de cytasters dans les œufs de Batraciens sous l'action de l'eau lourde. J. Embryol. Exp. Morph. **15**, 143 – 151 (1966)
258. Van Regemoorter, D.: Les troubles cinétiques dans les racines chloralosées et leur portée pour l'interprétation des phénomènes normaux. Cellule **37**, 43 – 73 (1926)
259. Wallace, S. L.: Colchicum: the panacea. Bull. N. Y. Acad. Med. **49**, 130 – 135 (1973)
260. Wallace, S. L., Ertel, N. H.: Preliminary report: plasma levels of colchicine after oral administration of a single dose. Metabolism **22**, 749 – 754 (1973)
261. Wallace, S. L., Nissen, A. W.: Griseofulvin in acute gout. New Engl. J. Med. **266**, 1099 – 1101 (1962)
262. Wang, R. W.-J., Rebhun, L. I., Kupchan, S. M.: Antimitotic and antitubulin activity of the tumor inhibitor steganacin. J. Cell Biol. **70**, 335 a (1976)
263. Waravdekar, V. S., Paradis, A. D., Leiter, J.: Enzyme changes induced in normal and malignant tissues with chemical agents. III. Effects of acetylpodophyllotoxin-ω-pyridinium chloride on cytochrome oxidase, cytochrome c, succinoxidase, succinic dehydrogenase and respiration of sarcoma 37. J. Natl. Cancer Inst. **14**, 585 – 592 (1953)
264. Weber, F.: Spaltöffnungsapparat-Anomalien colchicinierter *Tradescentia*-Blätter. Protoplasma **37**, 556 – 565 (1943)
265. Weber, K.: Visualization of tubulin-containing structures by immunofluorescence microscopy: cytoplasmic microtubules, mitotic figures and vinblastine-induced paracrystals. In: Cell Motility (eds.: R. Goldman, T. Pollard, J. Rosenbaum), pp. 403 – 418. Cold Spring Harbor Lab. 1976
266. Weber, K., Wehland, J., Herzog, W.: Griseofulvin interacts with microtubules both in vivo and in vitro. J. Mol. Biol. **102**, 817 – 830 (1976)
267. Weisenberg, R. C., Rosenfeld, A. C.: In vitro polymerization of microtubules into asters and spindles in homogenates of surf clam eggs. J. Cell Biol. **64**, 42 – 53 (1975)
268. Werner, D., Dönges, K. H.: Chemie und Wirkungsmechanismus des Colchicins und seiner Derivate. Planta Med. **22**, 306 – 315 (1972)
269. Wilson, L.: Microtubules as drug receptors: pharmacological properties of microtubule protein. Ann. N.Y. Acad. Sci. **253**, 213 – 231 (1975)
270. Wilson, L., Anderson, K., Chin, D.: Nonstoichiometric poisoning of microtubule polymerization: a model for the mechanism of action of the *Vinca* alkaloids, podophyllotoxin and colchicine. In: Cell Motility (eds.: R. Goldman, T. Pollard, J. Rosenbaum), pp. 1051 – 1064. Cold Spring Harbor Lab 1976
271. Wilson, L., Anderson, K., Creswell, K.: On the mechanism of action of vinblastine. J. Cell Biol. **63**, 373 a (1974)

272. Wilson, L., Bamburg, J. R., Mizel, S. B., Grisham, L. M., Creswell, K. M.: Interaction of drugs with microtubule proteins. Fed. Proc. **33**, 158 – 166 (1974)
273. Wilson, L., Bryan, J.: Biochemical and pharmacological properties of microtubules. Adv. Cell Mol. Biol. **3**, 21 – 72 (1975)
274. Wilson, L., Friedkin, M.: The biochemical events of mitosis. II. The in vivo and in vitro binding of colchicine in Grasshopper embryos and its possible relation to inhibition of mitosis. Biochemistry **6**, 3126 – 3135 (1967)
275. Wilson, L., Meza, I.: The mechanism of action of colchicine. Colchicine binding properties of sea urchin sperm tail outer doublet tubulin. J. Cell Biol. **58**, 709 – 720 (1973)
276. Winston, M., Johnson, E., Kelleher, J. K., Banerjee, S., Margulis, L.: Melatonin: cellular effects on live stentors correlated with the inhibition of colchicine-binding to microtubule protein. Cytobios **9**, 237 – 243 (1974)
277. Wisniewski, H., Shelanski, M. L., Terry, R. D.: Experimental colchicine encephalopathy. I. Induction of neurofibrillary degeneration. Lab. Invest. **17**, 577 – 587 (1967)
278. Wisniewski, H., Shelanski, M. L., Terry, R. D.: Effects of mitotic spindle inhibitors on neurotubules and neurofilaments in anterior horn cells. J. Cell Biol. **38**, 224 – 229 (1968)
279. Wisniewski, H., Terry, R. D.: Neurofibrillar pathology J. Neuropath. Exp. Neurol. **19**, 163 – 176 (1970)
280. Wunderlich, F., Heumann, H.-G.: In vivo reassembly of microtubules in the presence of intracellular colchicine. Cytobiologie **10**, 140 – 151 (1974)
281. Yahara, I., Edelman, G. M.: Electron microscopic analysis of the modulation of lymphocyte receptor mobility Exp. Cell Res. **91**, 125 – 142 (1975)
282. Yamada, K. M., Spooner, B. S., Wessells, N. K.: Ultrastructure and function of growth cones and axons of cultured nerve cells. J. Cell Biol. **49**, 614 – 635 (1971)
283. Zeisel, S.: Über das Colchicin. Monatsh. Chem. **7**, 557 – 596 (1886)
284. Zuckerberg, C., Solari, A. J.: Centriolar changes induced by vinblastine sulphate in the seminiferous epithelium of the mouse. Exp. Cell Res. **76**, 470 – 475 (1973)

Chapter 6 Cell Shape

6.1 Introduction

It should be clear now that MT have two contradictory properties: they are rigid structures, and act as supporting skeletons of many cellular differentiated organelles, such as axopodia, cilia, cytopharyngeal baskets; on the other hand, in many conditions, they may disassemble into their molecules of tubulin within a short time, and display a considerable lability which is apparently surprizing for a skeletal structure. MT are also associated with many intracellular movements, in mitosis, neuroplasmic flow, saltatory movements, secretion, phagocytosis. One of the principal problems about MT function is this relation between "skeletal" and "muscular" activities: are these both directly related to tubulins, or are the MT only supports for other proteins playing a role in movement, as in cilia and flagella, where dynein is the true agent of motion?

Many facts indicate that MT, notwithstanding their lability, are used by diverse cells as supporting structures, and their relation with cell shape must be discussed before any study of MT-associated motility, which may be a more indirect property of MT. It is of course impossible to cover entirely all this field and this chapter will be limited to cells in which shape and MT are intimately connected, the destruction of the MT leading to a change of the cell shape. In these instances, MT are not apparently related to other cell functions; however, as will become evident later, it is often difficult to dissociate the structural and the dynamic aspects of MT functions. For instance, MT give to the axopodia of heliozoa their rigidity, but it is also along the axoneme that particles are seen to migrate. In nerves, MT are associated with cytoplasmic movements, and will be studied in Chapter 9, although there are good reasons to think that MT have also a supporting role.

6.2 MT in Disk-Shaped Blood Cells

The relations between cell shape and MT are particularly apparent in nucleated red blood cells and platelets. This has been the subject of an excellent review [14]. Only the principal facts will be summarized here. The fibrillar ultrastructural nature of the marginal bundle of erythrocytes of vertebrates other than mammals had been suspected since the early observations of Meves [79] (cf. Chap. 1). In a short note, Fawcett [37] observed that the marginal band of toadfish erythrocytes was formed of fibrils with a tubular aspect, about 30 nm in diameter. He suggested that they maintained the shape of the cells. After the discovery of MT, it was demonstrated that in

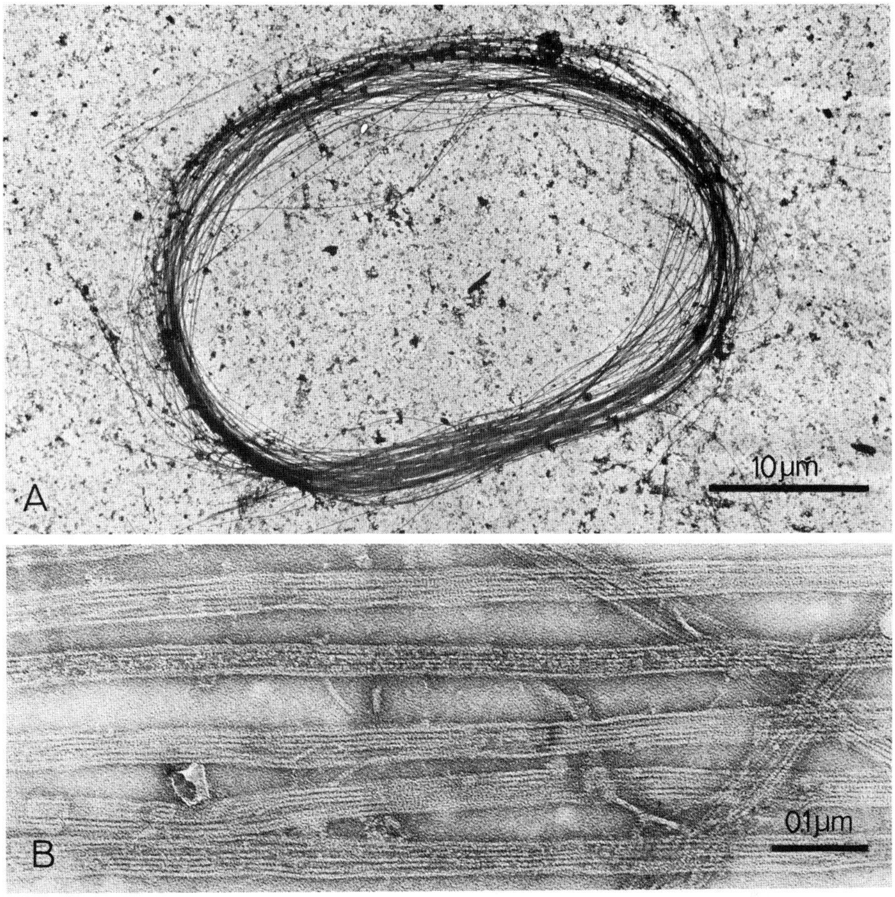

Fig. 6.1 A, B. Marginal band of the erythrocytes of *Triturus cristatus*. (A) Isolated marginal band from erythrocytes (citrated blood hemolyzed by sodium deoxycholate). The numerous MT form a ribbon-shaped band. (B) Isolated MT from the marginal band negatively stained. The protofilaments, showing globular subunits, are apparent (from Bertolini and Monaco [17])

all groups of cold-blooded vertebrates the bundle was made of tubular elements which were compared to the recently discovered MT [77]. The relation of this structure with the shape of the erythrocytes was stressed [40] and the observation that in the non-nucleated but oval red blood cells of the camelidae similar peripheric MT were present [8] was further evidence of their relation to shape, as the red blood cells of other mammals have no such structure [8]. The number of MT in the marginal bundle of nucleated red blood cells varies considerably, from 5 to more than 40 [14] even within the same species [106]. In the duck, about 30 – 40 MT, lying out in one sheet, are found in transverse sections at each pole of the erythrocytes. These MT are sometimes linked by "side-arms" [8].

[8] Observations of red blood cells of camelidae (camel and llama) have not confirmed the findings reported in the literature [8]: MT are present, like in other mammals, during the differentiation of red blood cells, but do not appear to be a constant feature of adult cells [46]

In amphibia, the number of MT appears species-specific (118 in *Triturus cristatus*, 281 in *Necturus maculosa*, 395 in *Amphiuma tridactylum*) (these figures referring of course to the numbers of cross-sections of MT seen in electron microscopy: the actual number may be smaller if some MT complete several revolutions around the cell). In these cells, the MT would maintain the cell shape [108]. These facts suggest one question: how is the number and the location of these MT regulated after each mitosis of primitive erythroblasts?

The marginal band of erythrocytes of the newt can be isolated after citrate hemolysis: it is formed by about 130 MT, the protofilaments of which can be readily observed after negative staining. The band, when free from the cell, becomes circular and flexible: it cannot alone explain the shape of the cells [17].

The numbers of MT in the marginal bands appear to be related to the size of the erythrocytes, as demonstrated in a comparative study of 23 species. The number varies from less than 10 to more than 400, and the largest cells—like those of *Salamandra* and *Lepidosiren*—have the greatest number of MT. A nearly linear relation was demonstrated between cell diameter and number of MT expressed on a logarithmic scale (Fig. 6.2; [46]). The MT would be required to maintain the ellipsoidal and biconvex shape of the cells.

In mammals, a study of erythropoiesis in the rabbit showed that in erythroblasts 16 to 18 sections of MT are found at the periphery. In adult cells, no more MT are found, except in some reticulocytes where short segments of MT may be seen. It is suggested that the MT are eliminated from the cells at the same time as the nucleus

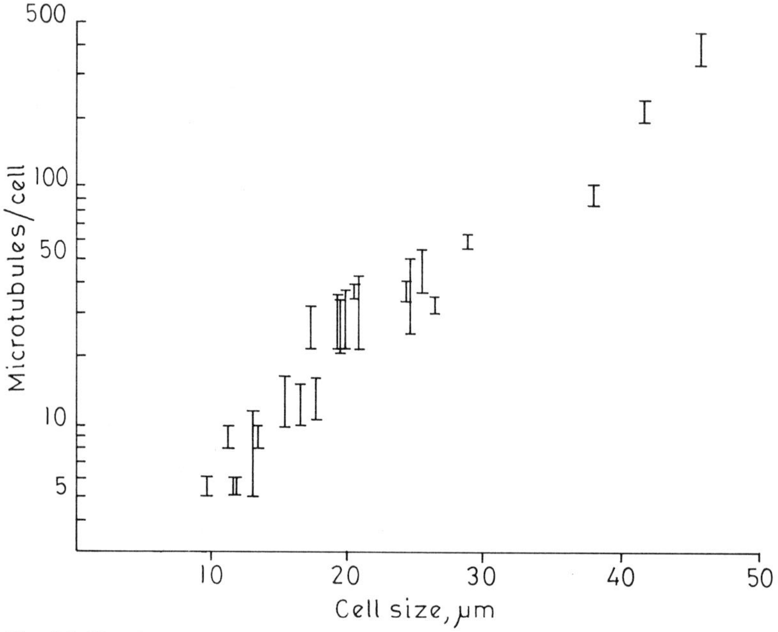

Fig. 6.2. Numbers of MT in nucleated red blood cells of various species of vertebrates: relation to cell size. There is a linear correlation between the average erythrocyte length and the number of MT (on a logarithmic scale). The results include those of the authors and data from the literature (redrawn from Goniakowska-Witalińska and Witaliński [46])

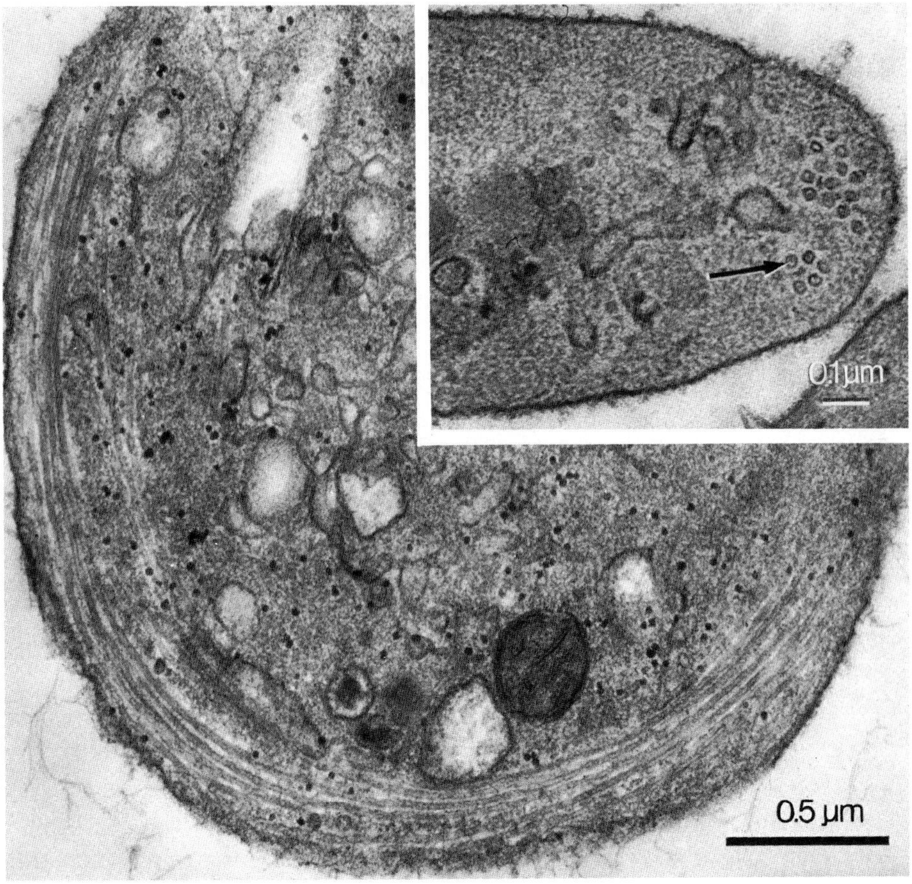

Fig. 6.3. Aspects of the marginal band of blood platelets. The MT maintain the disk shape of the platelet, and are grouped close to the membrane. *Inset:* A section in a plane perpendicular to this figure shows about 20 MT forming the marginal band. Some of these (*arrow*) show a central density (cf. Chap. 2; from Behnke [15])

[48]. The persistence of telophasic MT as Cabot's rings has been mentioned in Chapter 1 [18].

It is interesting, from the point of view of evolution, that the primitive erythroblasts of mammals, in the fetal liver, have a marginal band of about 50 MT. In erythroblasts of the adult, MT are present, but without this peripheral location, which appears like an ontogenic recapitulation [120].

The relation of the marginal bundle of chick erythrocytes to cell shape has been carefully analyzed by studying the numbers of MT at various stages of maturation, and the effect of MT destruction by colchicine, VCR, or cold [9]. The number of MT decreases during maturation, nearly parallel to the decrease in the number of ribosomes. The young red blood cells have a flattened spheroidal shape, and later become disk-shaped, with a sharp rim. When MT are disassembled by cold (about 1 h at 0 °C), the cells assume a spheroidal shape. When warmed and treated with colchicine or VCR, they remain spheroidal and do not become flat, contrary to controls. However, the erythrocytes which had their mature, flat, elliptic shape before disassembly of the MT retained their shape, indicating that the presence of MT was necessary for the acquisition of the normal shape, but not for maintaining it [9]. It was also observed that the MT of young red blood cells were more sensitive to cold than those of adult cells, perhaps because the first were in a more dynamic condition. Two important conclusions can be drawn from this work: (a) MT are necessary for the proper shaping of the cells; (b) once the adult shape is obtained, MT may be destroyed without affecting it. This indicates that MT act more like a scaffolding than a skeleton: similar facts will be described in quite different cells.

The blood *platelets* of mammals are also disk-shaped cells with a peripheral ring of MT, and their shape depends on the integrity of these MT. This relation with cell shape was mentioned in the first descriptions of this structure [19, 55, 101]. A thorough study and a review of this subject has been published by Behnke [12, 13, 14]. The mammalian platelets are formed from the cytoplasm of the megacaryocytes, which divides into regular fragments. While the megacaryocytes (polyploid cells) have many MT without any particular order [13] and a important pool of tubulin as evidenced by crystal formation after VCR treatment [16], the MT have a definite location in the platelets. Sections perpendicular to the plane of the disk-shaped cells show a ring of from 5 to 20 sections of MT. These appear to form some kind of structural unit, as shown by their preservation as a bundle after destruction of the cytoplasm. On the other hand, bridges appear to maintain this bundle by uniting the MT and keeping the MT bent. Like in red blood cells, it is not certain whether there are as many MT as cross-sections, for a few MT may be wound around the cell in several circles (Fig. 6.4). These MT have a typical substructure, with about 12 protofilaments [12, 124, 125].

What happens to blood platelets when these MT are destroyed? This may be obtained quite simply by keeping platelets at 0 °C: all MT disappear, and the cells become irregular or spherical. If they are warmed up to 37 °C, two possibilities are

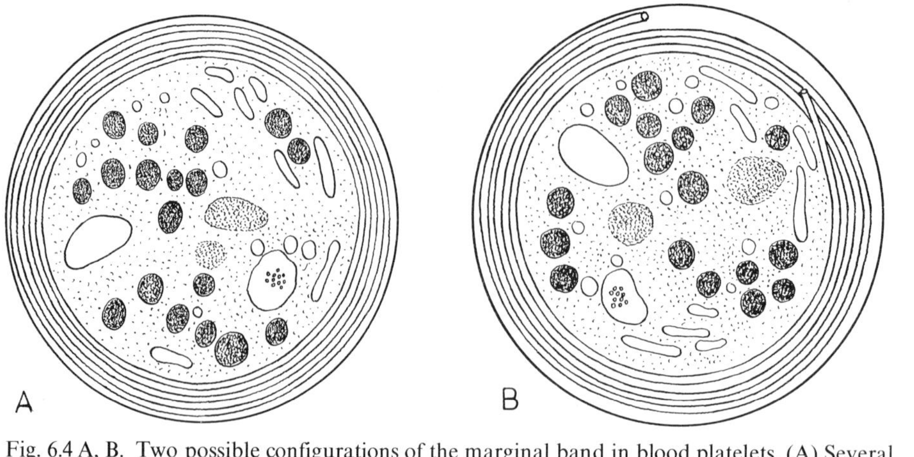

Fig. 6.4 A, B. Two possible configurations of the marginal band in blood platelets. (A) Several parallel circular MT. (B) A smaller number (or only one) of spirally circled MT (redrawn from Behnke [14])

open: either the MT reform as in the normal platelet, which resumes its disk shape, or the MT reassemble parallel to one another in a central rod-like structure, and the platelet becomes spindle-shaped [14]. Here again is evidence of the lability of the MT and of their relation to the shape of the cell. However, as it is mentioned that the MT are under tension, and become straight when broken, one must imagine some external force curving them into the marginal bundle, perhaps through the ac-

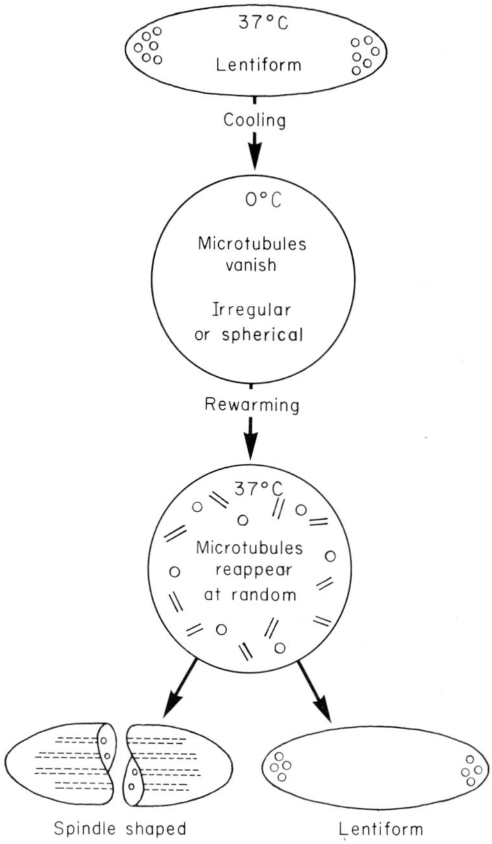

Fig. 6.5. Diagrammatic presentation of the effects of cooling and rewarming on blood platelet MT and platelet shape (from Behnke [14])

tivity of the interMT bridges. The shape is clearly related to the MT but other dynamic factors must bend the MT. The fact that sodium fluoride does not prevent the reassembly of the MT but prevents the formation of disk-shaped platelets may be an indication of this; however, sodium cyanide or monoiodoacetate do not prevent the formation of the marginal bundle [14].

Platelets treated in vitro by colchicine or colcemid lose their MT: however, large doses of colchicine are needed: with 100 μg/ml, only 50% of the MT are destroyed in 30 min, and to obtain a destruction of 90% a concentration of 1 mg/ml is needed [41, 126]. Similar results are obtained with VLB and VCR with the formation of paracrystals in the platelet cytoplasm, indicating that a pool of tubulin is present [124 – 127]. Apart from the destruction of MT, the general structure of the cells and their organelles is only slightly affected. The binding of VLB to platelets is enhanced by colchicine [105]; this binding may play a role in the transport of the *Vinca* alcaloids in the blood.

Platelets contain other fibrillar structures, related to the actomyosin-like protein thrombosthenin, which plays an important role in clot retraction. An interaction between thrombosthenin and MT in giving the platelets their disk shape cannot be ruled out, in the same way as dynein plays an evident role in maintaining the proper shape of cilia. As Behnke concludes [14]: "It may be argued that the discoid shape of platelets . . . and the arrangement of MT in a marginal bundle is a secondary phenomenon that is determined by the shape of the cell."

During blood clotting, the platelets undergo rapid changes related to the contractile activity of their MF (thrombosthenin): in the contracted condition, the MT are still clearly visible but occupy a more central or random location [14]. In rats and rabbits, the aggregation of platelets by ADP or contact with collagen is not modified by previous destruction of their MT by colchicine. The bleeding time is however slightly increased [62]. A study of the contraction of platelets devoid of MT has shown some slight changes, although they may form a normal clot [126]. The contraction of platelets is not affected by the destruction of their MT by colchicine, although the centripetal movement of their organelles is decreased [127]. However, if platelets are incubated with relatively large doses of colchicine ($0.01 - 0.1 \times 10^{-3}$ M), a retardation of clot retraction is apparent. This is related to the delay in the formation of pseudopods containing MT and MF, these last structures being involved in the contraction of platelets and clot retraction [30]. In later stages of blood clotting, the platelet organelles undergo a progressive destruction which involves the MT.

6.3 Nuclear and Cytoplasmic Shaping in Spermatogenesis

The role of MT in the shaping of spermatozoa is evident, and the multiple variants of the male sex gametes are composed, structurally, of a few organelles—acrosome, nucleus, mitochondria, flagella—the last being the organ of motility (cf. Chap. 7). The shaping of this highly specialized cell, throughout the animal and vegetal kingdom, results from a gradual condensation and deformation of the nucleus, which is often related to the location of MT. It is impossible in the limits of this study to embark on a comparative study of spermatogenesis, which has undergone in evolution some remarkable changes, some male gametes being only formed of a nucleus with a sheath of cytoplasm full of MT, others having multiple flagellae, some even having lost all ciliary structure. Two recent reviews cover several of these fields [32, 38]. The main problem which will be discussed here is the relation between the MT, in particular the MT "manchette", and the shaping of the nucleus during spermatogenesis. As in disk-shaped cells, it will again be apparent that MT act both as a scaffolding and as an active shaping element. Some remarkable relations between chromatin condensation and MT will also be demonstrated.

The structure known as manchette (= cuff) is a more or less cylindrical grouping of MT which surrounds the nucleus during some types of spermatogenesis. Whatever the future shape of the nucleus, the manchette is found in species as different

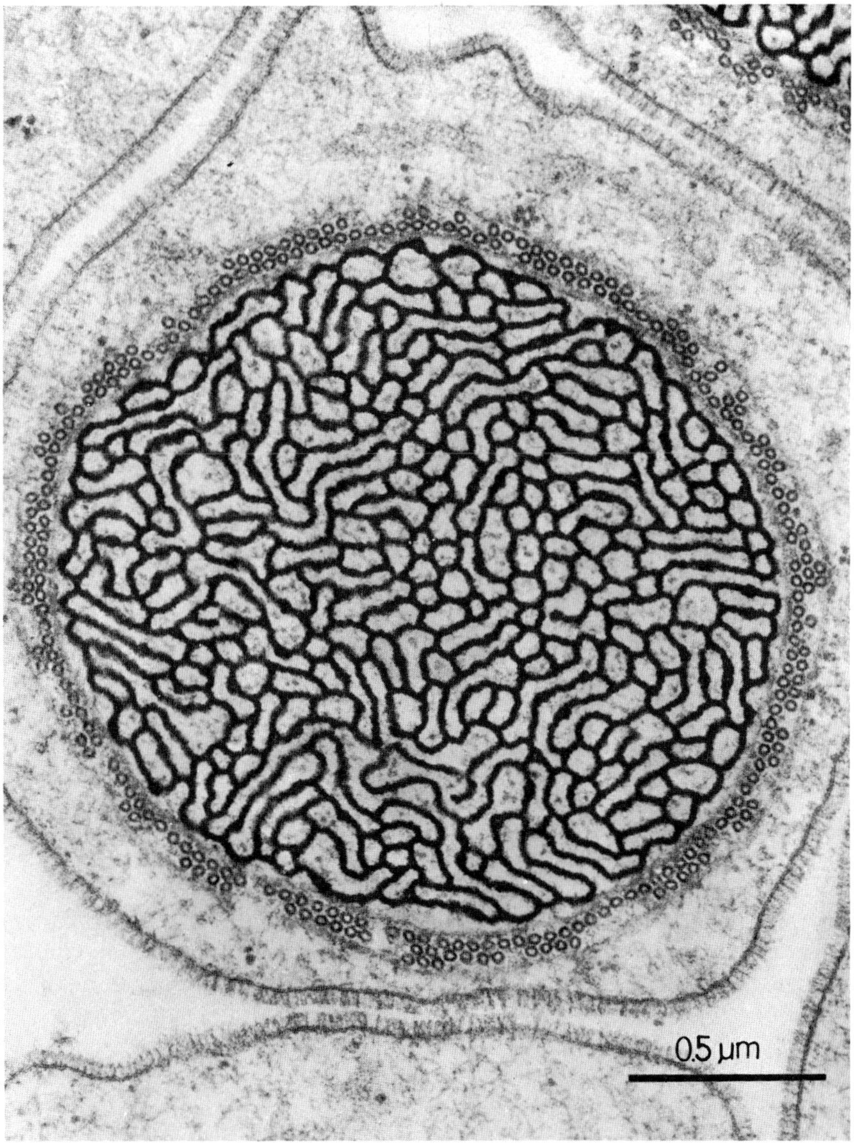

Fig. 6.6. Spermatid of the grasshopper *Acrida lata*. The nucleus has a paracrystalline structure. The single-layered nuclear envelope is surrounded by two or three layers of MT (from Yasuzumi et al. [131])

as mammals, birds, insects, worms, arthropods, and gymnosperm plants. These MT are closely packed, in one or several rows, often attached side by side by bridges. Their direction varies from one species to another, and varies also in relation to the stage of nuclear differentiation: it may be longitudinal, slightly helical, or tightly helical. Considered in the earliest works [43] as a skeletal structure, the manchette is helical in the grasshoppers *Melanoplus differentialis differentialis* (Thomas) [68] and

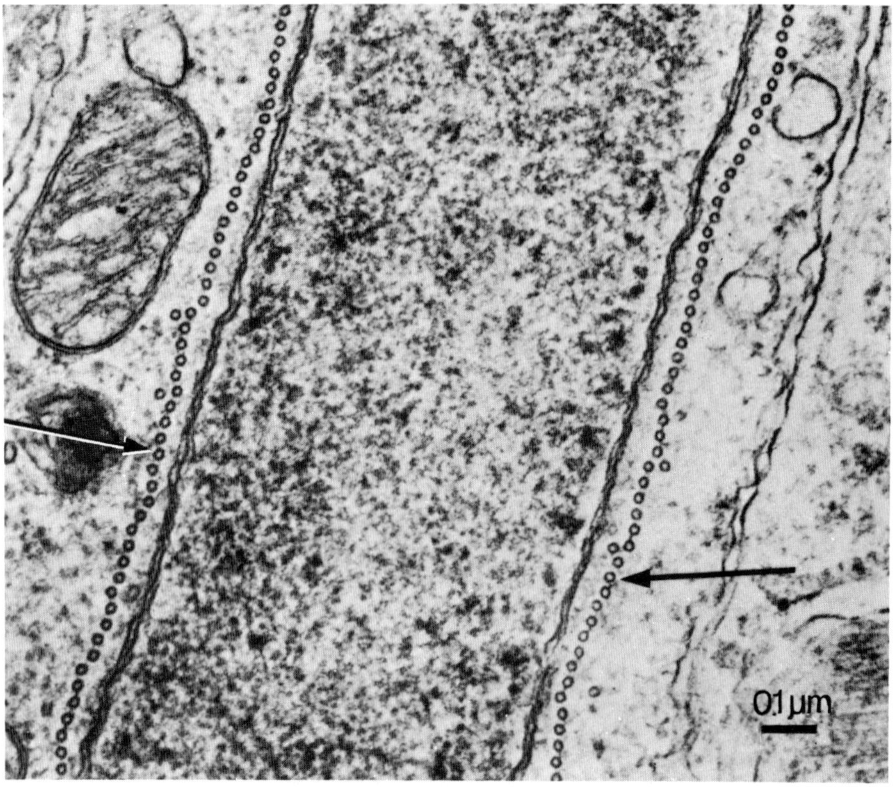

Fig. 6.7. Spermatid of the rooster. The MT (*arrows*) form a sheath ("manchette") along the elongating nucleus. From serial sections, it can be demonstrated that this structure is made of a few very long, spirally wrapped MT (from McIntosh and Porter [75])

Acrida lata [131], in the domestic fowl [74], in the coccid *Steatococcus tuberculatus* Morrison [80], in the worm *Tubifex tubifex* [42], in the huridinea [34]; straight, by contrast, in the charophyte *Nitella missourensis* Allen [118], to mention only a few instances.

One of the most important studies on the role of this helical grouping of MT is that of McIntosh and Porter [75]. In the adult rooster, spectacular modifications of the manchette are observed in various stages of spermatozoan differentiation. In spermatids, the MT circle the nucleus in a plane perpendicular to the axis of the cell. The radius of these circles was seen to decrease with progressive lengthening and condensation of the nucleus. The single layer of MT was studied in serial sections, and demonstrated to consist in fact of two interlocked helices made of two very long MT each, circling around the nucleus, and united at intervals of about 60 nm along their length by connections. This structure persisted until the nucleus had attained its normal length, and then, within a short time, these MT disassembled completely, to be replaced by others—probably made from the reassembly of the same subunits. These, parallel to the nuclear axis, may participate in the final bending of the nucleus. The authors suggested that the connections or bridges between the helical MT were active in sliding the two helices past one another, tightening the MT spiral and hence compressing and elongating the nucleus, the spacing between the MT remaining constant at about 30 nm. As will be shown below, there is however evidence from other species that the nuclear shaping may take place without any action of manchette MT.

Other observations suggest a role of the manchette MT in the shaping of spermatozoa. In the grasshopper, the manchette is helical and straightens when the nucleus elongates; it disappears once the elongation is ended [68]. In rodents, the number of MT which form several perinuclear layers is typical of each species, although it varies during spermatozoan maturation. In the Chinese hamster, for instance, the number of MT profiles varies from 589 to 1099 and later decreases to 425 as the nucleus matures [94, 95]. The relation between manchette and nuclear shape is also indicated by the nuclear deformation observed in the mutant t-gene mouse. In the t^{w2} allele, the MT become disorganized in late spermatids and this results in abnormally shaped nuclei [35]. Similar facts have been described in the mutant *ms* (3) 10R in *Drosophila*, in which thc nuclci do not elongate and the manchette MT are missing while normal MT are found in the flagellae. This mutation may be imitated by colcemid or VLB treatment: the manchette MT are disorganized, and the nucleus, in which chromatin condenses normally, fails to elongate [128, 129].

In some exceptional types of spermatozoa, the MT evidently play several roles, as seen in the coccid *Steatococcus tuberculatus* Morrison: here the spermatozoa are formed of a condensed nucleus and a double "helical meshwork" of about 80 MT: these not only shape the cell, but are also the cause of its motility, as no other organelles are present (no centrioles, no acrosome, no mitochondria) [80].

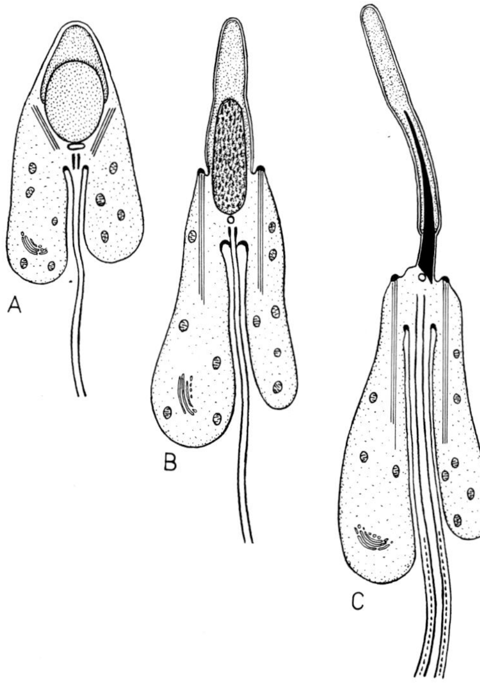

Fig. 6.8 A–C. Spermiogenesis in the guinea-pig. Three stages of the position of the nuclear ring and the manchette. (A) The MT are obliquely disposed at the rear end of the nucleus. (B) The nucleus is elongating. The MT extend backward, and are attached to a circular thickening of the cytoplasmic membrane (nuclear ring). (C) The nucleus has completely condensed. At this stage, the MT manchette is located at a distance from the nucleus, and clearly cannot play any role in the shaping of the nucleus (redrawn from Fawcett et al. [39])

While the conclusions of McIntosh and Porter [75] may seem to apply to various types of spermatzoa in many different groups, they are contradicted by several facts. In the spermatogenesis of the guinea-pig, for instance, groups of MT are close to the rear end of the nucleus. This elongates considerably, while the MT, which are attached to a circular thickening of the cytoplasm, become located in the cytoplasm which surrounds the flagellum, far from the condensing nucleus: there is no evident

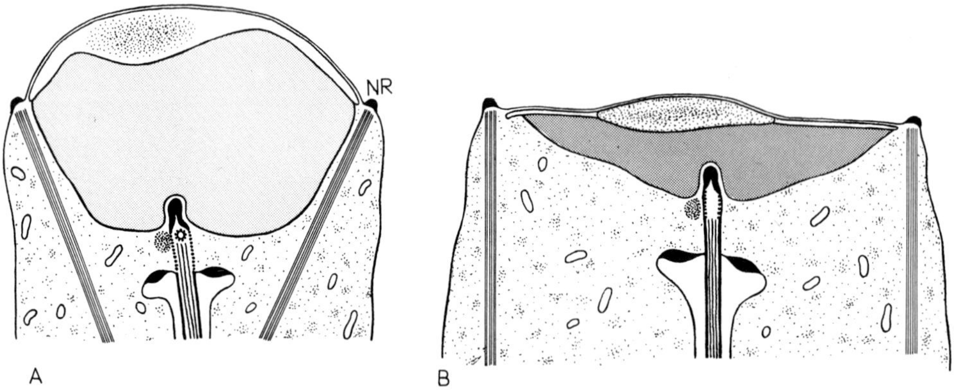

Fig. 6.9 A, B. Spermiogenesis in the opossum *Caluromys.* Nuclear ring and manchette. The nucleus is disk-like, and the MT are attached to a specialized region of the cell membrane (nuclear ring, *NR*). While they may be active in shaping the cytoplasm, their position in (B) indicates that the shape of the nucleus has no relation with the MT (cf. Fig. 6.8). A shows an early, and B a late stage of condensation of the spermatozoan nucleus (redrawn from Fawcett et al. [39])

relation between nuclear shape and MT ([39]; Fig. 6.6). Similarly, in the spermatogenesis of the Opossum, the nucleus becomes flattened in a direction perpendicular to the MT, which extend at the edge of the nucleus, from the nuclear ring toward the extremity of the cell. The nucleus assumes when mature a complex spiral shape which is not related to the location of the MT [39, 93]. Further, in the chaetognaths *Sagitta setosa* and *Spadella cephaloptera* MT are distributed at random in early

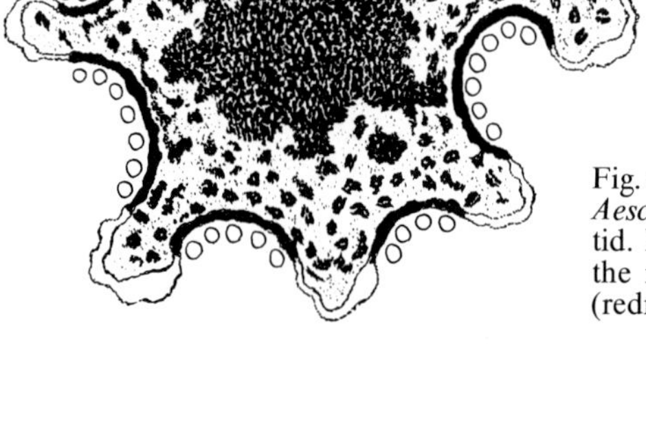

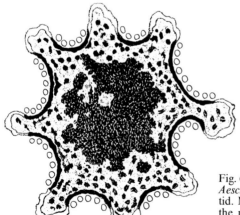

Fig. 6.10. Spermiogenesis in the dragon-fly *Aeschna grandis.* Transverse section of a spermatid. MT are present in each nuclear furrow, and the nuclear envelope is denser in these regions (redrawn from Kessel [67], slightly modified)

spermatids and are no longer present when in later stages the nucleus condenses; it is clear that the chromatin condensation, like in the marsupials [39], is not determined by cytoplasmic factors but by intrinsic properties of the nucleus [119].

In the spermatogenesis of the scorpions *Hadrurus hirsutus* and *Vejous spinigerus* the nucleus elongates and condenses as in other types of spermatogenesis. However, the complete absence of any MT around the nucleus during this shaping, while they

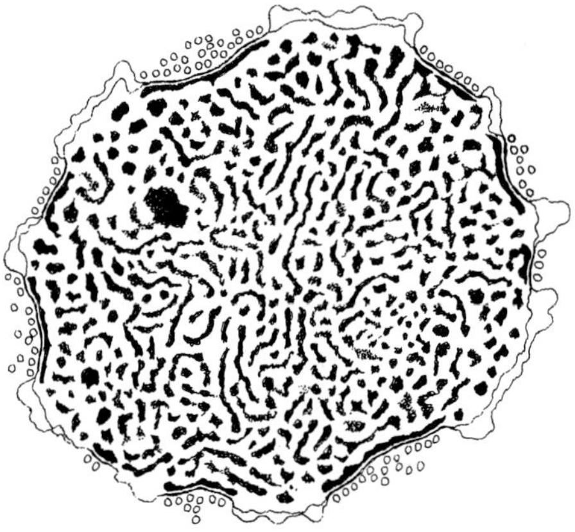

Fig. 6.11. Spermatogenesis of *Bacillus rossius.* Cross-section of the spermatid nucleus. MT are preferentially located close to the zones of nuclear envelope where the chromatin is layered (redrawn from Baccetti et al. [5])

are present in the cytoplasm of spermatogonia, precludes a role in nuclear shaping, which appears to result from intrinsic properties of the chromatin [89].

Several interesting observations indicate that dynamic, maybe metabolic interrelations may exist between the MT and the chromatin condensation. In the dragonfly, *Aeschna grandis,* MT play a role in nuclear shaping; they are located in deep infoldings of the nuclear membrane ([67]; Fig. 6.10). Similar findings in *Bacillus rossius* ([5, 6]; Fig. 6.11) and in *Lumbricus terrestris* ([73]; Fig. 6.12) indicate that the MT may participate in the elimination of the nucleoplasm (Fig. 6.9). An incomplete ring of MT surrounds the nucleus, and in its gaps the nuclear membrane and the nucleoplasm, without chromatin, are seen to herniate [1]. The manchette may also intervene in the elimination of unwanted cytoplasm in the last stages of spermato-

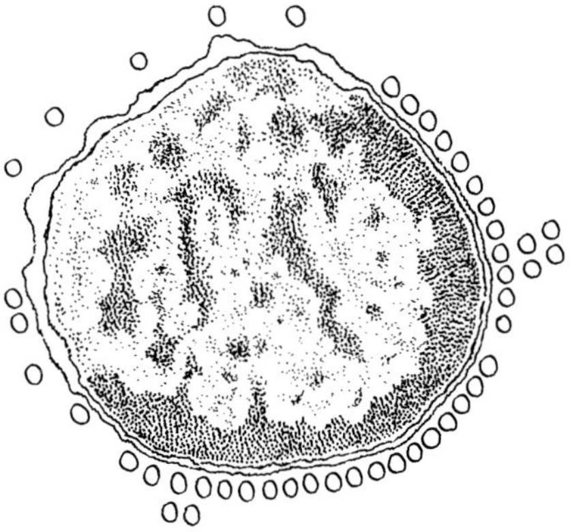

Fig. 6.12. Spermiogenesis of *Lumbricus terrestris.* Oblique section near the nuclear base. On the *right,* the chromatin is condensed and the MT are close together. On the *left,* there are few MT and the chromatin is loose (redrawn from Lanzavecchia and Lora Lamia Donin [73], slightly modified)

zoon maturation: the MT provide "a specialized mechanism that maintains cell form and organisation" ... "while all cell constituents outside the (MT) scaffolding are discarded" [1]. This role of MT as a scaffolding which helps the cell to locate its various constituents is similar to that described above for the nucleated erythrocytes, and many other examples may be mentioned.

The relations between MT and chromatin, as seen in *Lumbricus terrestris*, suggest that the MT may influence the intranuclear activities, and indirectly change the nuclear shape, without acting by a mechanical compression of the nucleus [73]. A similar "inductive" action of perinuclear MT on chromatin condensation has been described in *Tubifex tubifex* and another annelids, where a spiral manchette is present, and a relation between the MT and the nuclear membrane is evident: the two unit membranes of the nuclear envelope adhere to one another only close to the MT [42]. In *Drosophila*, there is also a "distinctive association of MT with endoplasmic reticulum and nuclear envelope", and the chromatin is seen to adhere to the nuclear membrane at regular spaces which have the same periodicity as the perinuclear MT [109]. These associations may be related to the shaping of the nucleus [113].

A detailed study of the spermatogenesis in hirudinea, where a helical sheath of MT envelops the maturing nucleus, indicates that there is a correspondence between the direction of the MT helices and the helical condensation of the chromatin. When the nucleus matures, it assumes a doubly helical shape, which originates at the level of the acrosome, also enveloped by a helical group of MT. Later, when the nucleus is finally shaped, the manchette disappears. This study, although apparently pointing to definite relations between MT and spiral shaping of the nucleus, also indicates an effect of the MT on chromatin condensation which "should be considered of inductive and not of mechanical nature". The only inductive system with a spiral structure is evidently that of the MT; however, the authors do not carry this reasoning one step further, and inquire into the inductive actions which may explain the helical disposition of the MT themselves [34].

The destruction of the manchette by MT poisons has been mentioned. In mouse spermatozoa, it resists for one hour at 0 – 4 °C and is destroyed by colchicine and VLB (10^4 M), while the 9+2 MT of the flagellum remains unaffected. In contrast, pepsin was found to destroy the flagellar MT in 60 min while not modifying the MT of the manchette [130].

In conclusion, this review of some studies on the shaping of spermatozoan nucleus shows that the relations with MT vary considerably from one species to another, although a "manchette" of MT is found in widely different species. The conception of MT acting as a mechanical instrument of constriction cannot be generalized; anyhow, it would not be the MT themselves which provide the shaping force, but the interMT bridges. Interesting relations have been noted between MT and chromatin condensation. Shape and MT are evidently related, but this relation is essentially dynamic. MT sometimes appear as a scaffolding which is discarded once its function is ended, and also as a structure endowed with properties of motility and even contractility. These must be related to other proteins than those of the MT. All these observations show that MT are far from being supporting structures comparable to a cytoskeleton; their instability and their rapid disassembly indicate that their function is more dynamic than static.

6.4 Morphogenesis and Cell Growth

The form of the embryo is closely related to that of its cells and tissues, and many facts indicate that close relations exist between MT and morphogenetic shaping. However, these processes are essentially dynamic, and involve not only orientation and asymmetrical growth of cells, but also cell movements and displacements of cellular membranes. It is thus often difficult to know whether the MT and their location are the cause or the consequence of shaping, and whether MT act as skeleton, scaffolding, framework—to quote some terms mentioned in the literature—or as active organelles involved in deformation of the cells. This may result either from a pushing of the cell membranes as a consequence of MT lengthening, or from more complex actions involving the MT side-arms and cytoplasmic movements controlled by MT-associated proteins. It is also often difficult to know the respective roles of MT and MF, which are ubiquitous structures of a contractile nature (actin, myosin, cf. Chap. 7). MT poisons are of some help in analyzing these actions; however, the changes in shape which result from their action on mitotic growth further complicate this analysis. The shape of the elongated eggs of insects is related to MT located close to the cell membrane, with which they are connected by small desmosomes. This stiff circumferential meshwork, by resisting tensions stresses, could explain the elongation of the egg [117].

In the formation of the neural folds in amphibia, a marked cell elongation takes place, and MT become oriented parallel to the long axis of the cell; hence the idea that they may play a role in this asymmetrical growth [122]. However, in the neural tube of *Hyla regilla* and *Xenopus laevis*, a dense meshwork of fine MF is located at the apical pole of the neural cells, while the MT are randomly oriented throughout the cytoplasm, sometimes in bundles. The contraction of the MF appears to be more important than the MT in the folding of the neural tube [7]. A study of *Xenopus* embryos treated with VLB (250 μg/ml) confirms that the destruction of the MT prevents the cell elongation which results from the orientation of the MT in the long axis of the cells and also from the contractile action of MF [64, 65]. In the same species, this action of MT has been interpreted in a more complex manner. MT do not appear to behave as skeletal structures, and the elongation of the cell may be explained by several hypotheses: (a) either the MT, by lenthening, push apart the two cell extremities; this would imply that the MT run all the length of the cell, which is not certain; (b) the MT could also slide in relation to one another; however, the distance between them seems too great for this; (c) a third possibility, in agreement with other findings on the relations between MT and intracellular transport, is that the MT play a role in neural plate elongation by a "displacement or transport of cytoplasm towards the extending base of the cell" [25].

In the neural tube of the chick, similar observations have been made. The destruction of MT by cold or colchicine decreases the cell asymmetry, indicating that MT are implicated in the maintenance of cell shape [54]. It should be recalled that the idea that MT are related to the development of asymmetrical cell shapes was proposed in 1966 by Porter [93]. Studies with VLB indicate that this drug prevents neurulation in the chick, although the interpretation of these results is made more complex by the fact that the MF were also disrupted [66]. Other recently described actions of VLB on the morphogenesis of the limbs in the chick are related to the mitotic poisoning by this drug [70]; they should be compared to older works on the production by colchicine of various types of abnormal embryonic development and monstrosities in the chick (cf. [36]).

In the gastrulation of *Arbacia punctulata* the MT of the primary mesenchyme, which arise from the pericentriolar bodies in close relation to the single cilia of these cells, elongate and appear to form "a framework which operates to shape the cells".

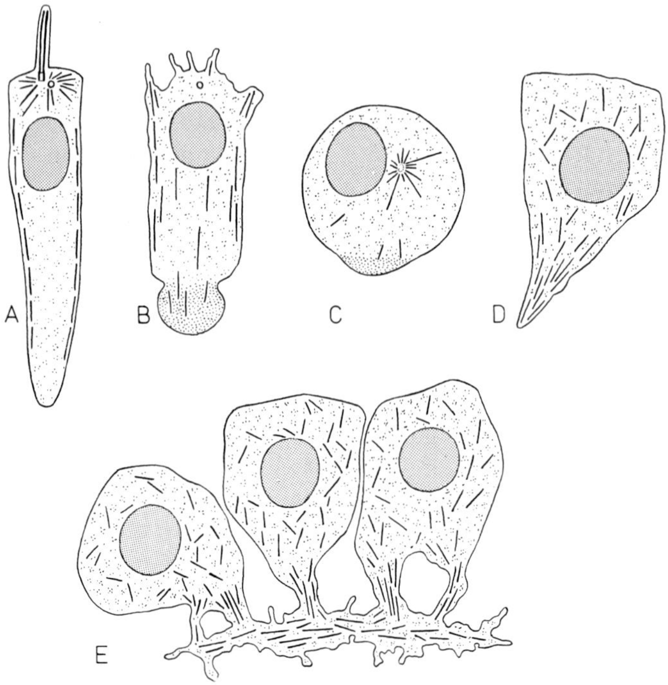

Fig. 6.13 A–E. Distribution of MT (represented as *black lines*) in the differentiation of the primary mesenchyme of *Arbacia punctulata.* The MT are oriented parallel to the direction of cell asymmetry at each stage. In (A) and (C), they are also seen to radiate from the centriolar zone (redrawn from Gibbins et al. [44], slightly modified)

However, an important problem remains, that of the control of MT formation: it is suggested that sites of MT formation are sequentially activated leading to an "orderly progression of shape" [44]. The invagination of the archenteron in this species does not take place when the MT are destroyed either by colchicine or by high hydrostatic pressures. The contractile processes would be related to MF—a conclusion which has since been widely confirmed. The MF are also affected by high hydrostatic pressures [111].

In the lens, during the synthesis of the specific proteins (crystallins), considerable cell elongation takes place. The relation with parallely oriented MT was mentioned in 1964 [26]. The role of hormones, and in particular insulin, in this growth was indicated in Chapter 3. The close relation between MT and cell elongation is demonstrated by the fact that colchicine and MT poisons prevent the normal cellular differentiation. This has been observed in the squid, where colchicine ($10^{-3}-10^{-4}$ M) reversibly destroys in 24 h all MT [2, 3]. In the chick, while cycloheximide slows down cell growth gradually, colchicine destroys the MT without affecting protein synthesis. This inhibits completely the stimulatory action of insulin [91]. These studies, which have been conducted in vitro, indicate that VLB has a similar effect. The inhibition of growth could not be explained by a decrease of the population pressure of the tightly packed cells, and it is concluded that these experiments "suggest the importance of MT, although their role remains elusive" [90]. It should be noted that in this model again the parallelism between the MT and the cell shape does not explain in a satisfactory way what is the motive force for elongation, and how the MT are related to this force.

The lens MT can be demonstrated by immunofluorescent staining of bovine lens cells cultured in vitro with antibodies from rabbits injected with brain or platelet tubulins. The pattern of staining, which is quite different from that obtained with anti-actin antibodies, disappears after colchicine (2 mg/ml) or after overnight incubation at 4 °C [31].

6.5 The Shaping of Nerve Cells

Neurons with their extensive ramifications—dendrites and axones—contain a large amount of tubulin and, as described in Chapter 2, the brain of vertebrates has become the favorite source of this protein. Numerous MT—often called "neurotubules"—are present in all the cytoplasmic extensions of neurons, in which they play various roles: their implication in the transport of metabolites and organelles will be discussed in Chapter 9. Nearly all sensory cells—in the retina, the olfactory epithelium, the ear, the lateral line, and various types of mechanochemical receptors—rely on MT which arise from more or less modified ciliary structures. These will be considered after the study of neuroplasmic flow, for a relation between transport functions and sensory activity is not impossible (Chap. 9).

Some relations between the formation of MT and the shaping of neurons under the action of nerve growth factor (NGF) have been mentioned in Chapter 3. This is antagonized, in vitro, by colcemid, and the formation of MT appears necessary for the elongation of nerve processes [54, 98]. However, nerve cells also contain a large number of fibrillar structures—neurofilaments (NF)— which also participate closely in the formation of dendrites and axones. The function of NF remains quite mysterious, while that of MT has clear relations with transport.

It is interesting to note that MT are associated with the structure of myelin, which is, in the central nervous system, an extension of the membrane of oligodendrocytes, and in peripheral nerves, of Schwann cells. These very large cytoplasmic extensions wrap round nerves in a tightly apposed series of cytoplasmic membranes which constitute the myelin sheath [60]. The myelin layers are separated at the Ranvier nodes and the so-called Schmidt-Lanterman incisures, which are in fact helical zones of Schwann cell cytoplasm. Within each turn of these helices a single MT is observed [53]. This has been found in various species, and appears to be a constant

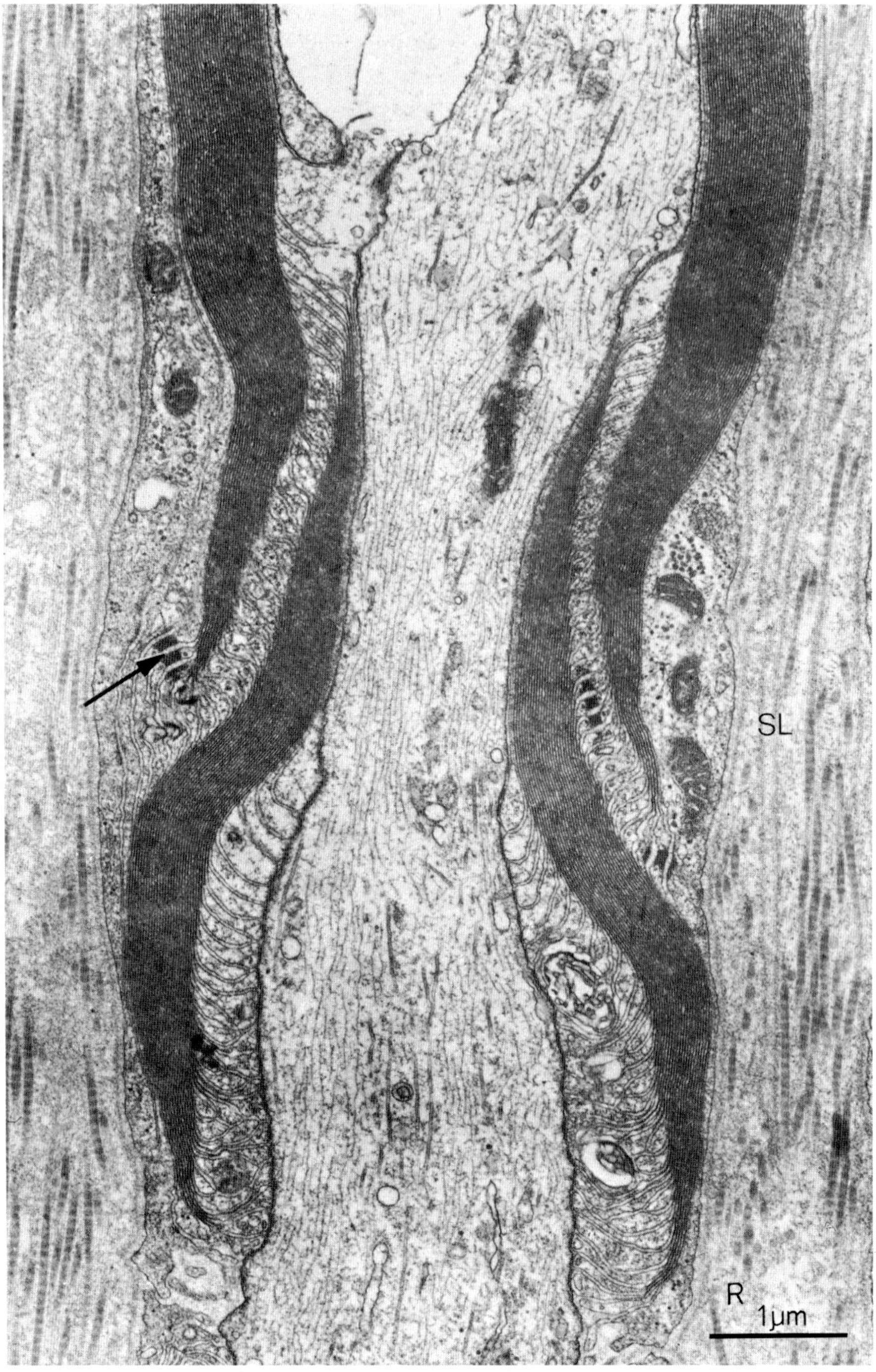

Fig. 6.14. Man. Normal sural nerve. Ranvier node (*R*) and Schmidt-Lanterman cleft (*SL*). The MT are hardly visible at this magnification. Densities between the layers of myelin in the cleft are apparent (*arrow*)

structure. The problem mentioned about platelets and red blood cells is apparent here: if this MT has a cytoskeletal function, how can its helical bending be explained? In fact, the Schmidt-Lanterman incisures and the Ranvier nodes are regions of cytoplasmic transport from the cell body of the Schwann cell to the sheaths of myelin, and these MT may have a more dynamic role than that of a skeleton.

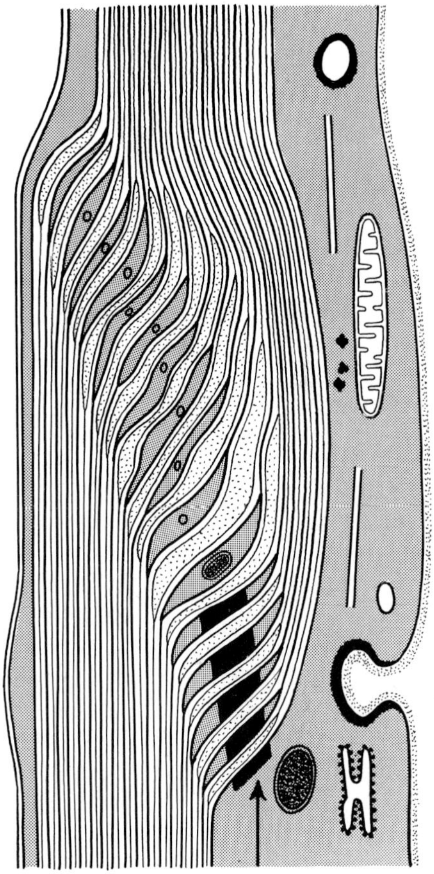

Fig. 6.15. Diagrammatic structure of a Schmidt-Lanterman cleft. The myelin sheaths, made by the circular wrapping of the Schwann cell membrane, separate, leaving a helical band of Schwann cell cytoplasm which connects the external and the internal zones of the myelin sheath. The band is occupied by a MT which may have a helical shape. Some of the myelin membranes are held together by a cytoplasmic condensation (*arrow*) (redrawn from Hall and Williams [53])

In the shrimp, *Penaeus japonicus*, a giant nerve fiber has a most remarkable structure: Schwann cells surround the axon, and the myelin sheath proper is separated from these by a nearly empty space. The Schwann cell cytoplasm contains a "tremendous" number of tightly packed MT, which have a supporting function [52].

In sensory cells, MT are often found in very elongated cell processes, and two theories have been put forward: for some [4] they play a role in the transport of the sensory stimuli toward the cell body; for others, they provide a skeleton for the long and narrow cell processes, which may be formed by a single or a small number of MT surrounded by a thin sheet of cytoplasm [4, 96]. These problems will be discussed in Chapter 9 in relation to neuroplasmic transport.

6.6 Other Relations of MT and Cell Shape in Metazoa

Behnke [14] has given a table of several papers describing the involvement of MT in the determination or maintenance of cell shape. While in some instances MT appear as rigid skeletal structures, in other cells their relation with shape is mainly dynamic, as demonstrated by the reversible changes in shape when the MT are disassembled by high hydrostatic pressures or MT poisons. A few instances of these relations of MT and shape in metazoan cells will be mentioned here. The use of immunohistochemical techniques for visualizing MT has clearly indicated the relations between the cell shape and the cytoplasmic MT. When mouse 3T3 cells become attached to a glass surface and spread, the growth of the MT from the perinuclear (perhaps the centriolar zone) parallels the flattening of the cell. The MT extend from the cytocenter to the cell membrane, and maintain the cell shape [87]. Similar observations, showing the loss of polarity of fibroblasts in vitro when MT are disassembled by various MT poisons (cf. Chap. 5), confirm the idea proposed by Porter in 1966 [93] that cell asymmetry and polarity are related to the cytoplasmic MT skeleton (cf. [33]).

6.6.1 Connective Tissue Cells

When treated with 5 – 40 µg/ml of colchicine, the shape of BHK-21 fibroblasts changes: they resemble more epithelial cells, and numerous microfibrils appear in the juxtanuclear region. Cell locomotion is disturbed, although the movements of the cell surface, which are related to MF, are not affected [45, 121].

Cultures of muscle cells also undergo morphological changes when treated with low concentrations (10^{-8}) of colchicine which do not affect cell division, the myotubes taking on the aspect of "myosacs" [21]. In tail muscle regeneration in *Rana pipiens* tadpoles, MT are oriented longitudinally. Lateral interactions between them may be active in maintaining the bipolar form of the cells. In the youngest cells, the MT radiate from pericentriolar satellites. Sliding interactions between the MT are suggested [123]. This shows again the dynamic participation of MT in cell elongation. MT have also been described as "cytoskeleton" in the smooth muscle cells of human aorta [99].

The shape of the nucleus in mammalian polymorphonuclear leukocytes results from the interaction of MT radiating from the centrosphere and the pericentriolar bodies with the nuclear membrane [20]. When the MT are destroyed by colchicine, the nucleus reverts to a less lobulated shape. In these cells, MT are involved in locomotion, diapedesis and phagocytosis (cf. Chap. 8). Their participation in the movements of cell surface receptors ("capping") is mentioned in Chapters 3 and 11. When polymorphonuclear leukocytes are treated by sodium oxalate, their nuclei assume a special type of "radial segmentation", known to hematologists as Rieder's cells [85, 86]. This effect is inhibited by colcemid, VCR [63, 84], and cold (4 °C) [83]. This oxalate-induced segmentation has been used as a test for antibubulin substances [86]. It can also be observed in lymphocytes and monocytes. This shape may be observed spontaneously in leukemic cells, where the centrioles occupy a central position: it results from the radial disposition of the MT and some contractile process associated with them [110]. It would be interesting to know whether the changes in calcium content of the leukocyte cytoplasm induced by sodium oxalate may not influence the assembly of MT or the contraction of proteins of the actomyosin type.

6.6.2 Epithelial and Glandular Cells

MT may act either as a network, keeping in place the cell organelles, or as a scaffolding, which will determine the cell shape. The first action has been observed in cells with a complex stratification of organelles (mitochondria, Golgi apparatus, lysosomes): in the thyroid of the cream hamster, this stratification disappears when MT are destroyed by VCR in vitro (6×10^{-4} M): within 4 h, the Golgi saccules move about the cell, the endoplasmic reticulum spaces lose their elongated shape, and the lysosomes enlarge. The centrioles are not affected [69].

In flattening epithelial cells of the skin of the new-born rat, numerous long MT, with a clear "exclusion zone", are visible in the stratum granulosum: they are arranged in two orthogonal layers. Although these MT are not closely spaced, it has been suggested that they could, by a contractile peristaltic action, influence the flattening of the differentiating cell [23]. This idea is similar to that mentioned above for erythrocytes or platelets.

In liver cells cultivated in vitro, the destruction of the MT leads to a disorganization of the cell organelles with a dispersion of the Golgi apparatus [33]. Several other examples of changes of the Golgi structures after MT poisoning will be mentioned in Chapter 8 (cf. also Chap. 3).

6.7 Cell Shape in Protozoa

In protozoa, MT have been put to use in many different ways. They are not only necessary for mitosis and ciliary beating, but also for cell shaping and the formation of differentiated organs, which act as "cell anchoring devices, teeth, legs, fishing rods, stirrups or cell elongation machines" [77].

The shape of the protozoal cells or of some of these appendages depends largely on the integrity of the MT. Although they may be used in a mechanical manner, for instance as "legs", these rigid structures by the rapid disassembly of MT, may be retracted in a few minutes, only to be reassembled later more slowly. Several instances of this have been given in the preceding chapters.

Clearly, while the MT assemblies in unicellulars have through evolution become comparable to the skeletal structures of pluricellulars, they remain in a dynamic condition—except for centrioles and basal bodies—and the relation between shape and MT is very different from that of a fixed and rigid skeleton. A problem which has already been considered, and which arises again here, is that of the spatial determinism of the complex MT assemblies which determine the shape of the cell. One answer to this is given by the studies on the flagellate *Ochromonas* [22, 24]. The overall form of this protist depends on two groups of MT. When these are destroyed, either by high hydrostatic pressures or by colchicine, the cells become rounded,

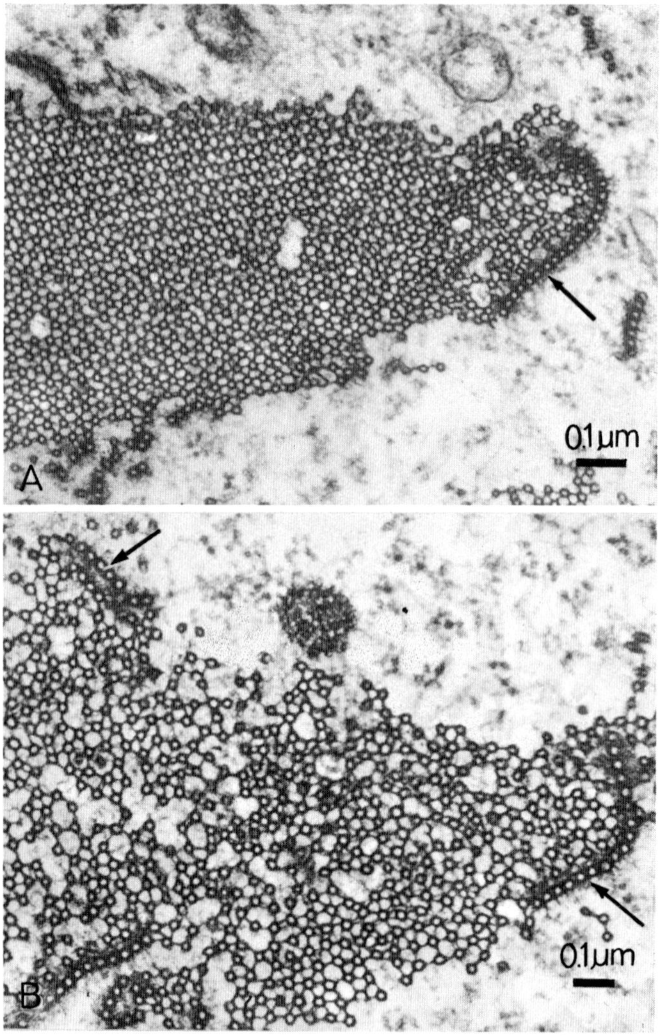

Fig. 6.16 A, B. Rods made of hexagonally linked MT in the cytopharynx of the ciliate *Nassula*. (A) Cross-section of a rod in an organism treated with colchicine during the second hour of rod development. The rod is attached to a lamella (*arrow*) made of MT joined side by side. There are some irregularities in the hexagonal lattice. (B) The same rod cut 8 µm lower, showing several gaps in the lattice. The rod lamellae are indicated by the *arrows* (from Tucker et al. [116]) (Copyright Academic Press, Inc. New York)

while the cilia are not affected. The destruction of MT is gradual and always proceeds from the posterior to the anterior part of the cell. The reassembly of the MT is related to two specific nucleating sites. While the shape of the cell requires the integrity of the MT, the location of the MTOC is the true determinant of cellular asymmetry.

The role of the axonemes of heliozoa and radiolaria has been described in Chapter 3. It is evident that the rigidity of the long cytoplasmic extensions of heliozoa such as *Actinosphaerium* is related to the presence of a structured group of MT, the axoneme, which acts as a support for the cytoplasm. It is probably also involved (cf. Chap. 7) in the movements of food particles which are captured by the axopodia and carried toward the cell body [28].

In some radiolaria, like *Sticholonche zanclea*, the numerous rigid axopodia move synchronously and are used by the cell as oars in its swimming movements. These structures, with their hexagonal lattice of MT, are rigid, but various chemicals and physical stimuli may lead to their almost complete retraction, only their most basal part remaining, the MT disappearing or being replaced by lamellar or fibrillar structures [27, 61]. This is reversible, the basal structure acting probably as a MTOC.

More complex structures result from the assembly of numerous MT which appear to have a skeletal function, like the cytopharyngeal baskets of ciliates such as *Nassula* [114, 115]. The MT are in a hexagonal pattern, which is apparently defined by the properties of an initiating site. The tubules attain a greater length when their number has been reduced by a heat-shock, suggesting that they are formed from a tubulin pool [115].

In fixed species, MT are used in the formation of rigid stalks. In the pseudoheliozoan *Clathrulina elegans*, a hollow stalk made mainly of MT is formed when the cell becomes fixed (cf. Chap. 4, Fig. 4.26). This property of MT is also found in the oocytes of a freshwater muscle of the species *Anodonta*, where the cells grow a stalk containing parallel MT [11].

These few examples indicate that cellular shape and MT assembly are closely related, and that in several cases a “skeletal” role of MT is apparent. However, the possibility for MT to reassemble in the same patterns after experimental disassembly indicates that the shape is determined by the poorly understood orienting factors which guide the assembly of MT.

6.8 Cell Shape in Plants

Before the discovery of MT, observations on changes of cell walls in plants treated with colchicine, and some extensive modifications of cell shape, as observed in the alga *Hydrodictyon* [47], had been mentioned [36]. These were the first indications that colchicine could have effects unrelated to mitosis. A disorientation of cellulose fibers was found in the alga *Nitella* after colchicine [49, 50], and shortly afterwards, Ledbetter and Porter [74] described plant MT for the first time and found that they were often parallel to cellulose fibers.

This relation between the orientation and the location of MT and the structure of the plant cell wall will be examined here. Other problems of interest in plant cells, such as the role of MT in the orientation and location of the phragmoplast, will be studied in Chapter 10. An extensive review of these problems was published a few years ago [58]. From this the following relations between cell wall formation and MT may be summarized:

1. MT are oriented parallel to cellulose fibers and may control their orientation in primary and secondary wall formation, in various types of plants.

2. The destruction of MT does not prevent the formation of the cellulose fibers, but may lead to an incorrect orientation.

3. Strong correlations between MT and cellulose fibers are found in sieve elements, stoma guard cells, secondary walls of xylem cells.

4. MT are clustered in regions where the cell walls will thicken, and appear to anticipate and control this thickening. Colchicine leads to an irregular formation of the secondary wall of xylem [57].

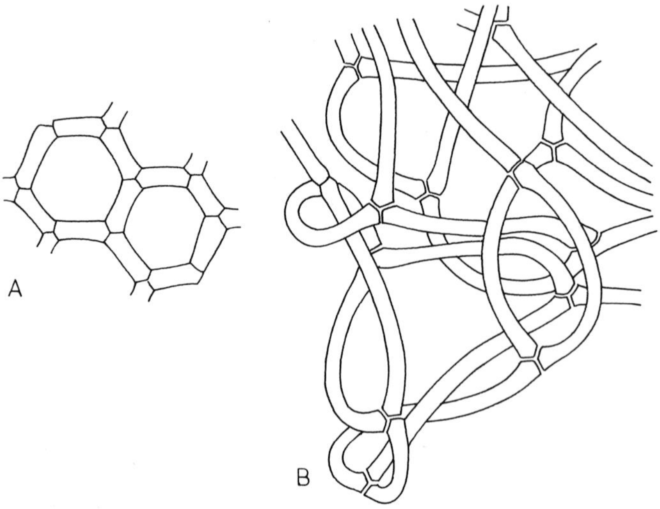

Fig. 6.17 A, B. *Hydrodictyon.* (A) The normal network of this alga. (B) After colchicine treatment, the cells are abnormally elongated and the network is deformed, although the intercellular links are not modified (after C. Gorter, in Eigsti and Dustin [36])

5. The relation between MT and cell wall may be indirect, MT possibly generating a flow, toward the membrane, of cell organelles (Golgi vesicles) or metabolites indispensable for cellulose synthesis and assembly.

6. The mechanism by which MT become aligned "is much less clear" [58]. A few specific observations in this field may be mentioned. In wheat cells, colchicine destroys the mitotic MT and the MT aligned parallel to the cell wall, which becomes thickened [92]. In a review on plant MT [82] these were considered highly labile structures existing in a dynamic equilibrium with a pool of subunits; they "play a directive role in morphogenesis, either by controlling the movement and positioning of other cellular structures or by serving as cytoskeletal agents". This quotation, underlining the ambiguity of MT function, remains correct today. In the pollen in *Haemanthus katherinae* the MT are parallel to the long axis of the cell during its elongation. When they are destroyed either by colchicine or isopropyl phenylcarbamate (cf. Chap. 5), the cells become spherical [102]. A quantitative study of MT in the growing leaves of *Sphagnum* shows that after mitosis they become aligned along the cell wall. They are more numerous close to thickened regions of this wall. While they are considered "cytoskeletal elements", their action is indirect and involves a lifting of the cytoplasm from the wall and the local accumulation of Golgi-derived material [104].

A similar mechanism has been suggested in the guard cells of *Allium*: observations after colchicine poisoning indicate that they help to locatize materials necessary for the cell wall deposition and the formation of cellulose fibrils [88].

A similar explanation has been presented in theoretical form: MT could play the role of a track along which cellulose synthetase is carried, laying down cellulose fibers parallel to the MT [56]. The orientation of the MT is certainly complex, and studies on the azuki bean (*Azukia angularis*) have shown that in cells treated with indolacetic acid (IAA) the MT are randomly oriented, while after IAA and kinetin the MT are parallel to the cell axis, but transverse to this if IAA is associated with gibberellin. This substance thus controls the direction of cell growth by acting on the MT [107].

The function of MT in the algae *Hydrodictyon reticulatum* and two other species has again been demonstrated (cf. [47]): the network is disorganized by colchicine, indicating that MT may have a cytoskeletal function [76]. Similar observations have been made on the coleoptile of *Avena*, where MT showing 13 subunits may be found in the cellulose wall. This is the consequence of the extracellular location of the peripheral MT, which pass from the cytoplasm into vacuoles and further into the wall. They play a part in the orientation of cellulose fibers, similar to that observed during the orientation of the phragmoplast at the end of mitosis (cf. Chap. 10; [59]).

Studies on the wall of the green alga *Oocystis solitaria* have shown definite relations between the pattern of cellulose microfibrils and MT. The fibrils form several layers, each oriented at 90° to the other. The fibrils are laid down parallel to MT of the cortical cytoplasm, which are located close to the cell membrane and surrounded by a clear "exclusion zone" (cf. Chap. 2). It is suggested that by successive steps of disassembly and reassembly, the MT change their direction before a new cellulose wall is formed, assuring the parallelism between MT and microfibrils. When MT are destroyed by colchicine (10^{-2} M) the typical pattern does not develop, and the cells remain spherical [97, 100].

These results, like those described in protists and metazoa, always indicate a dynamic relation between MT and shape, and if several authors mention the "cytoskeleton" function of MT, these always remain a labile assembly of subunits, determined by morphogenetic factors which remain largely unknown. The interesting and striking relation between MT and cellulose walls—which are at a relatively long distance from the peripheral band of MT—recalls the relations described above for the chromatin location in spermatogenesis.

6.9 MT with Structural and Mechanical Functions

In some cases, MT are not only a support for cell structures, but are put to use by the cell to carry mechanical loads, acting more or less as tendons; this provides some interesting insight into the mechanical strength of MT.

The relations between MT and cell wall have been mentioned in Chapter 3: in the examples which will be considered here, the linkages are particularly evident.

In insects, several instances of cells with attachments of MT to the cell wall are known, such as the MT-associated cell ridges described in the sternal gland of the termite *Zootermopsis nevadensis* (Hagen) [103] and the epithelial cells of the ejaculatory duct of orthoptera, where the cytoplasm contains about 70% of its volume of parallel MT which are attached to the apical pole of the cell. Several of these MT, as seen in *Locusta migratoria*, show a central density. The attachment of the MT takes place at specialized structures considered hemidesmosomes [10]. In these cells, the function of the MT remains unknown, and the hypothesis that they may play a part in water transport has been proposed. Similar MT with central densities, filling literally all the cytoplasm of connective cells in the rectum of the blowfly *Calliphora erythrocephala* (Meig.), may have comparable functions [51].

The role of MT in the epithelial cells located between muscle cells and the cuticle of insects is that of the transmission of mechanical stresses. In *Calpodes ethlius* (lepidoptera) and *Rhodnius prolixus* (hemiptera), the muscle cells which terminate at the Z line, are attached to epithelial cells which contain large numbers of parallel MT

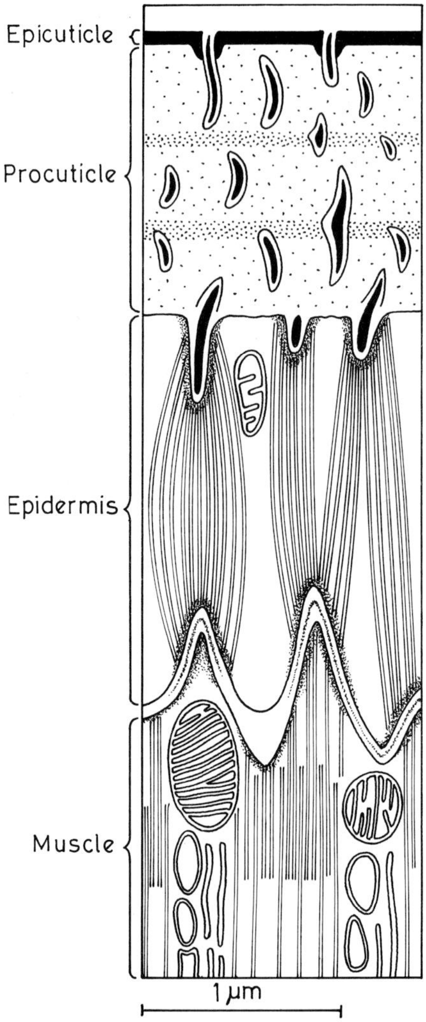

Fig. 6.18. General view of muscle attachment to cuticle in *Apterygota.* The procuticle is traversed by muscle attachment fibers (*black*), which originate in conical hemidesmosomes of the epithelial cells, to which bundles of MT are attached. At the base of these cells, the MT are attached to desmosomes which link the epidermal cells to the muscle (redrawn from Caveney [29], slightly modified)

which, at the other end of the cell, close to the cuticle, are attached to hemidesmosomes. These epidermal cells, or "tendinous cells", have a mechanical function, and the orientation of the MT is related to the pulling force exerted on them and through them [72]. Similar structures have been described in the hypopharyngeal cavity of *Blabera craniifer* (Burm) (dictyoptera): the epidermis cells, with a great number of MT attached to hemidesmosomes, have in this peculiar structure a skeletal function, preventing the collapse of the hypopharynx by food particles. The relations between the cells and the cuticle are comparable to those found in tendon cells [81]. In several species of apterygota, similar types of muscle attachment to the cuticle have been described. The MT link up with conical hemidesmosomes, and are related to fibers which are fixed to the cuticle. The bundles of MT provide a link between differentiated parts of the muscle cell membrane, and the complex structures of the procuticle (muscle attachment fibers). Here also, MT with a denser inner zone have been noticed. The MT, at the moment of moulting, are not destroyed, but are less linear; they apparently grow in length before the formation of a new cuticle [29].

Such structures are found in other invertebrates like the tendon cells of balanidae (*Balanus balanoides* and *B. improvisus*), where huge amounts of MT are present in the epidermal cells, and are attached at each end to specialized zones of the cell membrane. During moulting, the aligned MT are disorganized. It is of interest to note that cells not attached to muscle, and observed in the integument, may also contain a very large number of MT, the function of which remains unknown [71].

Much remains to be learnt about these "tendinous" MT: their mode of formation in the developing cell, their relation with other organelles (centrioles, as in the gland of *Zootermopsis* [103]), and their mechanical resistance. Evidently, MT not only play a role in the shape of these specialized epidermal cells, but have been put to use as stable, force-transmitting organelles. This is apparently so different from the unstable MT as seen in the mitotic spindle or axonemes that one may wonder whether all these structures are identical. The presence of a central density has been described in various other cells (cf. Chap. 2); the close relations between MT and cell membrane have already beeen mentioned above, and may explain some problems of cell secretion, as described in Chapter 8. Further research on the chemical composition of these structural MT (and in general, on the MT of invertebrates) is required.

References

1. Anderson, W. A., Weissman, A., Ellis, R. A.: Cytodifferentiation during spermiogenesis in *Lumbricus terrestris*. J. Cell Biol. **32,** 11 – 26 (1967)
2. Arnold, J. M.: Squid lens development in compounds that affect microtubules. Biol. Bull. **131,** 383 (1966)
3. Arnold, J. M.: On the occurence of microtubules in the developing lens of the squid *Loligo pealii*. J. Ultrastruct. Res. **14,** 534 – 539 (1966)
4. Atema, J: Stimulus transmission along microtubules in sensory cells: a hypothesis. In: Microtubules and Microtubule Inhibitors (eds.: M. Borgers, M. De Brabander), pp. 247 – 257. Amsterdam-Oxford: North Holland 1975
5. Baccetti, B., Burrini, A. G., Dallai, R., Pallini, V., Periti, P., Piantelli, F., Rosati, E., Selmi, G.: Structure and function in the spermatozoon of *Bacillus rossius*. The spermatozoon of Arthropoda, XIX. J. Ultrastruct. Res. Suppl. **12,** 5 – 73 (1973)

6. Baccetti, B., Dallai, R., Giusti, F., Bernini, F.: The spermatozoon of arthropoda. XXV. A new model of tail having up to 170 doublets: *Monarthropalpus* Buxi. Tissue Cell **6,** 269 – 278 (1974)
7. Baker, P. C., Schroeder, T. E.: Cytoplasmic filaments and morphogenetic movements in the amphibian neural tube. Dev. Biol. **15,** 432 – 450 (1967)
8. Barclay, N. E.: Marginal bands in duck and camel erythrocytes. Anat. Rec. **154,** 313 (1966)
9. Barrett, L. A., Dawson, R. B.: Avian erythrocyte development: microtubules and the formation of the disk shape. Dev. Biol. **36,** 72 – 81 (1974)
10. Bassot, J. M., Martoja, R.: Données histologiques et ultrastructurales sur les microtubules cytoplasmiques du canal éjaculateur des Insectes Orthoptères. Z. Zellforsch. **74,** 145 – 181 (1966)
11. Beams, H. W., Sekhon, S. S.: Electron microscope studies on the oocyte of the fresh-water mussel (*Anodonta*), with special reference to the stalk and mechanism of yolk deposition. J. Morphol. **119,** 477 – 502 (1966)
12. Behnke, O.: Incomplete microtubules observed in mammalian blood platelets during microtubule polymerization. J. Cell Biol. **34,** 697 – 701 (1967)
13. Behnke, O.: An electron microscope study of the rat megakaryocyte. II. Some aspects of platelet release and microtubules. J. Ultrastruct. Res. **26,** 111 – 129 (1969)
14. Behnke, O.: Microtubules in disk-shaped blood cells. Intern. Rev. Exp. Path. **9,** 1 – 92 (1970)
15. Behnke, O.: Microtubules et microfilaments. Triangle **13,** 7 – 16 (1974)
16. Behnke, O., Pedersen, T. N.: Ultrastructural aspects of megacaryocyte maturation and platelet release. In: Platelets. Production, Function, Transfusion and Storage (eds.: M. G. Baldini, S. Ebbe), New York-San Francisco-London: Grune and Stratton 1974
17. Bertolini, B., Monaco, G.: The microtubule marginal band of the newt erythrocyte. J. Ultrastruct. Res. **54,** 59 – 67 (1976)
18. Bessis, M.: Living blood cells and their ultrastructure. Berlin-Heidelberg-New York: Springer 1973
19. Bessis, M., Breton-Gorius, J.: Les microtubules et les fibrilles dans les plaquettes étalées. Nouv. Rev. Fr. Hématol. **5,** 657 (1965)
20. Bessis, M., Breton-Gorius, J.: Rapports entre noyau et centrioles dans les granulocytes étalés. Rôle des microtubules. Nouv. Rev. Fr. Hématol. **7,** 601 – 620 (1967)
21. Bischoff, R., Holtzer, H.: The effect of mitotic inhibitors on myogenesis in vitro. J. Cell Biol. **36,** 111 – 128 (1968)
22. Bouck, G. B., Brown, D. L.: Microtubule biogenesis and cell shape in *Ochromonas.* I. The distribution of cytoplasmic and mitotic microtubules. J. Cell Biol. **56,** 340 – 359 (1973)
23. Branson, R. J.: Orthogonal arrays of microtubules in flattening cells of the epidermis. Anat. Rec. **160,** 109 – 121 (1968)
24. Brown, D. L., Bouck, G. B.: Microtubule biogenesis and cell shape in *Ochromonas.* III. Effects of the herbicidal mitotic inhibitor isopropyl-N-phenylcarbamate on shape and flagellum regeneration. J. Cell. Biol. **61,** 514 – 536 (1974)
25. Burnside, B.: Microtubules and microfilaments in newt neurulation. Dev. Biol. **26,** 416 – 441 (1971)
26. Byers, B., Porter, K. R.: Oriented microtubules in elongating cells of the developing lens rudiment after induction. Proc. Natl. Acad. Sci. USA. **52,** 1091 – 1099 (1964)
27. Cachon, J., Cachon, M.: Les systèmes axopodiaux. Année Biol. **13,** 523 – 560 (1974)
28. Cachon, J., Cachon, M.: Rôle des microtubules dans les courants cytoplasmiques des axopodes. C. R. Acad. Sci. Paris **280,** 2341 – 2343 (1975)
29. Caveney, S.: Muscle attachment related to cuticle architecture in *Apterygota.* J. Cell Sci. **4,** 541 – 559 (1969)
30. Chao, F. C., Shepro, D., Tullis, J. L., Belamarich, F. A., Curby, W. A.: Similarities between platelet contraction and cellular motility during mitosis: role of platelet microtubules in clot retraction. J. Cell Sci. **20,** 569 – 588 (1976)
31. Crawford, N., Trenchev, P., Castle, A. G., Holborow, E. J.: Platelet tubulin and brain tubulin antibodies: immunofluorescence of cell microtubules. Cytobios **14,** 121 – 130 (1975)
32. Dalcq, A.: Cytomorphologie normale du testicule et spermatogénèse chez les Mammifères. Mém. Acad. Roy. Med. Belg. **46,** 1 – 834 (1973)
33. De Brabander, M.: Onderzoek naar de rol van microtubuli in gekweekte cellen met behulp van een nieuwe synthetische inhibitor van tubulin-polymerisatie. Thesis, Brussels 1977
34. Donin, C. L. L., Lanzavecchia, G.: Morphogenetic effects of microtubules. III. Spermiogenesis in Annelida Hirudinea. J. Submicrosc. Cytol. **6,** 245 – 259 (1974)

35. Dooher, G. B., Bennett, D.: Abnormal microtubular systems in mouse spermatids associated with a mutant gene at the T-locus. J. Embryol. Exp. Morph. **32,** 749 – 762 (1974)
36. Eigsti, O. J., Dustin, P. Jr.: Colchicine—in Agriculture, Medicine, Biology, and Chemistry. Ames, Iowa: Iowa State College Press 1955
37. Fawcett, D. W.: Electron microscopic observations on the marginal band of nucleated erythrocytes. Anat. Rec. **133,** 379 (1959)
38. Fawcett, D. W.: The mammalian spermatozoon. Dev. Biol. **44,** 394 – 436 (1975)
39. Fawcett, D. W., Anderson, W. A., Phillips, D. M.: Morphogenetic factors influencing the shape of the sperm head. Dev. Biol. **26,** 220 – 251 (1971)
40. Fawcett, D. W., Witebsky, F.: Observations on the ultrastructure of nucleated erythrocytes and thrombocytes, with particular reference to the structural basis of their discoidal shape. Z. Zellforsch. **62,** 785 – 806 (1964)
41. Feo, C., Breton-Gorius, J.: Action comparée d'un dérivé de la colchicine et de la podophylline sur la blocage des mitoses et la disparition des microtubules. Nouv. Rev. Fr. Hématol. **8,** 827 – 840 (1968)
42. Ferraguti, M., Lanzavecchia, G.: Morphogenetic effects of microtubules. I. Spermiogenesis in Annelida Tubificidae. J. Submicrosc. Cytol. **3,** 121 – 138 (1971)
43. Flechon, J. E., Courot, M.: Structure microtubulaire de la manchette des spermatides chez le taureau et le bélier. J. Micros. **5,** 50 a (1966)
44. Gibbins, J. R., Tilney, L. G., Porter, K. R.: Microtubules in the formation and development of the primary mesenchyme in *Arbacia punctulata.* I. The distribution of microtubules. J. Cell Biol. **41,** 201 – 226 (1969)
45. Goldman, R. D.: The role of three cytoplasmic fibers in BHK-21 cell motility. I. Microtubules and the effects of colchicine. J. Cell Biol. **51,** 752 – 762 (1971)
46. Goniakowska-Witalińska, L., Witaliński, W.: Evidence for a correlation between the number of marginal band microtubules and the size of vertebrate erythrocytes. J. Cell Sci. **22,** 397 – 402 (1976)
47. Gorter, C.: De invloed van colchicine op den groei van den celwand van wortelharen. Proc. Kon. Nederl. Akad. Wetensch. **48,** 3 – 12 (1945)
48. Grasso, J. A.: Cytoplasmic microtubules in Mammalian erythropoietic cells. Anat. Rec. **156,** 397 – 413 (1966)
49. Green, P. B.: Mechanism for plant cellular morphogenesis. Science **138,** 1404 – 1405 (1962)
50. Green, P. B.: On mechanisms of elongation. In: Cytodifferentiation and Macromolecular Synthesis (ed.: M. Locke) pp. 203 – 234. New York: Acad. Press. 1963
51. Gupta, B. L., Berridge, M. J.: Fine structural organization of the rectum in the Blowfly, *Calliphora erythrocephala* (Meig.) with special reference to connective tissue, tracheae and neurosecretory innervation in the rectal papillae. J. Morphol. **120,** 23 – 81 (1966)
52. Hama, K.: The fine structure of the Schwann cell sheath of the nerve fiber in the shrimp (*Penaeus japonicus*). J. Cell Biol. **31,** 624 – 632 (1966)
53. Hall, S. M., Williams, P. L.: Studies on the "incisures" of Schmidt and Lanterman. J. Cell Sci. **6,** 767 – 791 (1970)
54. Handel, M. A., Roth, L. E.: Cell shape and morphology of the neural tube: implications for microtubule function. Dev. Biol. **25,** 78 – 95 (1971)
55. Haydon, G. B., Taylor, A.: Microtubules in hamster platelets. J. Cell Biol. **26,** 673 – 676 (1965)
56. Heath, I. B.: A unified hypothesis for the role of membrane bound enzyme complexes and microtubules in plant cell wall synthesis. J. Theoret. Biol. **48,** 445 – 450 (1974)
57. Hepler, P. K., Fosket, D. E.: The role of microtubules in vessel member differentiation in *Coleus.* Protoplasma **72,** 213 – 236 (1971)
58. Hepler, P. K., Palewitz, B. A.: Microtubules and microfilaments. Ann. Rev. Plant Physiol. **25,** 309 – 362 (1974)
59. Heyn, A. N. J.: Intra- and extracytoplasmic microtubules in coleoptiles of *Avena.* J. Ultrastruct. Res. **40,** 433 – 457 (1972)
60. Hirano, A., Dembitzer, H. M.: A structural analysis of the myelin sheath in the central nervous system. J. Cell Biol. **34,** 555 – 567 (1967)
61. Hollande, A., Cachon, J., Cachon, M., Valentin, J.: Infrastructure des axopodes et organisation générale de *Sticholonche zancela* Hertwig (Radiolaire Sticholonchidae). Protistologica **3,** 155 – 166 (1967)
62. Hovig, T.: L'ultrastructure des plaquettes sanguines en relation avec leur fonction hémostatique. Nouv. Rev. Hématol. **9,** 496 – 505 (1969)
63. Ito, S.: Study on the in vitro Rieder cell. Scand, J. Haematol. **12,** 355 – 365 (1974)
64. Karfunkel, P.: The activity of microtubules and microfilaments during neurulation in *Xenopus laevis.* J. Cell Biol. **47,** 102 a (1970)

65. Karfunkel, P.: The role of microtubules and microfilaments in neurulation in *Xenopus*. Dev. Biol. **25,** 30 – 56 (1971)
66. Karfunkel, P.: The activity of microtubules and microfilaments in neurulation in the chick. J. Exp. Zool. **181,** 289 – 302 (1972)
67. Kessel, R. G.: The association between microtubules and nuclei during spermiogenesis in the dragonfly. J. Ultrastruct. Res. **16,** 293 – 304 (1966)
68. Kessel, R. G.: An electron microscope study of spermiogenesis in the grasshopper with particular reference to the development of microtubular systems druing differentiation. J. Ultrastruct. Res. **18,** 677 – 694 (1967)
69. Ketelbant-Balasse, P., Neve, P.: New ultrastructural features of the Cream Hamster thyroid with special reference to the second kind of follicle. Cell Tissue Res. **166,** 49 – 63 (1976)
70. Kieny, M.: Effets de la vinblastine sur la morphogénèse du pied de l'embryon de poulet. Aspects histologiques. J. Embryol. Exp. Morphol. **34,** 609 – 632 (1975)
71. Koulish, S.: Microtubules and muscle attachment in the integument of balanidae. J. Morphol. **140,** 1 – 14 (1973)
72. Lai-Fook, J.: The structure of developing muscle insertions in Insects. J. Morphol. **123,** 503 – 528 (1967)
73. Lanzavecchia, G., Lora Lamia Donin, A.: Morphogenetic effects of microtubules. II. Spermiogenesis in *Lumbricus terrestris*. J. submicrosc. Cytol. **4,** 247 – 260 (1972)
74. Ledbetter, M. C., Porter, K. R.: A "microtubule" in plant fine structure. J. Cell Biol. **19,** 239 – 250 (1963)
75. McIntosh, J. R., Porter, K. R.: Microtubules in the spermatids of the domestic fowl. J. Cell Biol. **35,** 153 – 173 (1967)
76. Marchant, H. J., Pickett-Heaps, J. D.: The effect of colchicine on colony formation in the algae *Hydrodictyon, Pediastrum* and *Sorastrum*. Planta **116,** 291 – 300 (1974)
77. Margulis, L.: Microtubules and evolution. In: Microtubules and Microtubule Inhibitors. (Eds.: M. Borgers, M. De Brabander) pp. 3 – 18. Amsterdam-Oxford: North-Holland 1975
78. Maser, M. D., Philpott, C. W.: Marginal bands in nucleated erythrocytes. Anat. Rec. **150,** 365 – 381 (1964)
79. Meves, F.: Gesammelte Studien an den roten Blutkörperchen der Amphibien. Arch. mikrosc. Anat. **77,** 465–540 (1911)
80. Moses, M. J., Wilson, M. H.: Spermiogenesis in an iceryne coccid, *Steatococcus tuberculatus* Morrison. An electron microscope study. Chromosoma **30,** 373 – 429 (1970)
81. Moulins, M.: Etude ultrastructurale d'une formation de soutien épidermo-conjonctive inédite chez les insectes. Z. Zellforsch. **91,** 112 – 134 (1968)
82. Newcombe, E. H.: Plant microtubules. Ann. Rev. Plant Physiol. **20,** 253 – 288 (1969)
83. Norberg, B.: The influence of temperature on the formation of radial-segmented nuclei in human lymphocytes and monocytes. Scand. J. Haematol. **5,** 255 – 263 (1968)
84. Norberg, B.: Cytoplasmic microtubules and radial-segmented nuclei (Rieder cells). Effects of vinblastine, sulfhydryl reagents, heparin and caffeine. Scand. J. Haematol. **6,** 312 – 318 (1969)
85. Norberg, B.: Segmentation radiale du noyau des cellules polynucléées du sang et de la moelle osseuse (Formation de cellules de Rieder). Nouv. Rev. Fr. Hématol. **9,** 191 – 198 (1969)
86. Norberg, B., Uddman, R.: The oxalate-induced radial segmentation of the nuclei of lymphocytes and monocytes from peripheral blood. A possible screening test for metaphase-blocking agents. Blut **26,** 261 – 267 (1973)
87. Osborn, M., Weber, K.: Tubulin-specific antibody and the expression of microtubules in 3T3 cells after attachment to a substratum. Further evidence for the polar growth of cytoplasmic microtubules in vivo. Exp. Cell Res. **103,** 331 – 340 (1976)
88. Palewitz, B. A., Hepler, P. K.: Cellulose microfibril orientation and cell shaping in developing guard cells of *Allium*: the role of microtubules and ion accumulation. Planta. **132,** 71 – 94 (1976)
89. Phillips, D. M.: Nuclear shaping in absence of microtubules in scorpion spermatids. J. Cell Biol. **62,** 911 – 916 (1974)
90. Piatigorsky, J.: Lens cell elongation in vitro and microtubules. Ann. N. Y. Acad. Sci. **253,** 333 – 347 (1975)
91. Piatigorsky, J. Rothschild, S. S., Wilberg, M.: Stimulation by insulin of cell elongation and microtubule assembly in embryonic chick-lens epithelia. Proc. Natl. Acad. Sci. USA. **70,** 1195 – 1198 (1973)

92. Pickett-Heaps, J. D.: The effects of colchicine on the ultrastructure of dividing plant cells, xylem wall differentiation and distribution of cytoplasmic microtubules. Dev. Biol. **15,** 206 – 236 (1967)
93. Porter, K. R.: Cytoplasmic microtubules and their functions. In: Ciba Foundation Symposium in Principles of Biomolecular Organization. pp. 308 – 345. London: Churchill 1966
94. Rattner, J. B.: Nuclear shaping in marsupial spermatids. J. Ultrastruct. Res. **40,** 498 – 512 (1972)
95. Rattner, J. B., Brinckley, B. R.: Ultrastructure of mammalian spermatogenesis. III. The organization and morphogenesis of the manchette during rodent spermiogenesis. J. Ultrastruct. Res. **41,** 209 – 218 (1972)
96. Reese, T. S.: Olfactory cilia in the frog. J. Cell Biol. **25,** 209 – 230 (1965)
97. Robinson, D. G., Grimm, I., Sachs, H.: Colchicine and microfibril orientation. Protoplasma **89,** 375 – 380 (1976)
98. Roisen, F. J., Murphy, R. A.: Neurite development in vitro. II. The role of microfilaments and microtubules in dibutyryl adenosine 3′5′-cyclic monophosphate and nerve growth factor stimulated maturation. J. Neurobiol. **4,** 397 – 412 (1973)
99. Sachs, E. S., Daems, W. T.: Microtubules in human aortic intimal cells. Z. Zellforsch **73,** 553 – 558 (1965)
100. Sachs, H., Grimm, I., Robinson, D. G.: Structure, synthesis and orientation of microfibrils. I. Architecture and development of the wall of *Oocystis solitaria.* II. The effect of colchicine on the wall of *Oocystis solitaria.* Cytobiologie **14,** 49 – 60; 61 – 74 (1976)
101. Sandborn, E. B., Le Buis, J. J., Bois, P.: Cytoplasmic microtubules in blood platelets. Blood **27,** 247 – 252 (1966)
102. Sanger, J. M., Jackson, W. T.: Fine structure study of pollen development in *Haemanthus katherinae* Baker II. Microtubules and elongation of the generative cells. J. Cell Sci. **8,** 303 – 316 (1971)
103. Satir, P., Stuart, A. M.: A new apical microtubule-associated organelle in the sternal gland of *Zootermopsis nevadensis* (Hagen), Isoptera. J. Cell Biol. **24,** 277 – 283 (1965)
104. Schnepf, E.: Microtubulus-Anordnung und -Umordnung, Wandbildung und Zellmorphogenese in jungen *Sphagnum*-Blättchen. Protoplasma **78,** 129 – 144 (1973)
105. Secret, C. J., Hadfiled, J. R., Beer, C. T.: Studies on the binding of ^{3}H-vinblastine by rat blood platelets in vitro. Effects of colchicine and vincristine. Biochem. Pharmacol. **21,** 1609 – 1624 (1972)
106. Sekhon, S. S., Beams, H. W.: Fine structure of the developing Trout erythrocytes and thrombocytes with special reference to the marginal band and the cytoplasmic organelles. Am. J. Anat. **125,** 353 – 374 (1969)
107. Shibaoka, H.: Involvement of wall microtubules in gibberellin promotion and kinetin inhibition of stem elongation. Plant Cell Physiol. **15,** 255 – 264 (1974)
108. Small, J. V.: The significance of the marginal band microtubules. J. Ultrastruct. Res. **38,** 207 – 208 (1972)
109. Stanley, H. P., Bowman, J. T., Romrell, L. J., Reed, S. C., Wilkinson, R. F.: Fine structure of normal spermatid differentiation in *Drosophila melanogaster.* J. Ultrastruct. Res. **41,** 433 – 466 (1972)
110. Stenstam, M., Mecklenburg, C. von, Norberg, B.: The ultrastructure of spontaneous radial segmentation of the nuclei in bone marrow cells from 3 patients with acute myeloid leukemia. Scand. J. Haematol. **15,** 63 – 71 (1975)
111. Tilney, L. G., Cardell, R. R. Jr.: Factors controlling the reassembly of the microvillus border of the small intestine of the salamander. J. Cell Biol. **47,** 408 – 422 (1970)
112. Tilney, L. G., Gibbins, J. R.: Microtubules in the formation and development of the primary mesenchyme in *Arbacia punctulata.* II. An experimental analysis of their role in development and maintenance of cell shape. J. Cell Biol. **41,** 227 – 250 (1969)
113. Tokuyasu, K. T.: Dynamics of spermiogenesis in *Drosophila melanogaster.* IV. Nuclear transformation. J. Ultrastruct. Res. **48,** 284 – 303 (1974)
114. Tucker, J. B.: Fine structure and function of the cytopharyngeal basket in the Ciliate *Nassula.* J. Cell Sci. **3,** 493 – 514 (1968)
115. Tucker, J. B.: Morphogenesis of a large microtubular organelle and its association with basal bodies in the Ciliate *Nassula.* J. Cell Sci. **6,** 385 – 429 (1970)
116. Tucker, J. B., Dunn, M., Pattisson, J. B.: Control of microtubule pattern during the development of a large organelle in the ciliate *Nassula.* Dev. Biol. **47,** 439 – 453 (1975)
117. Tucker, J. B., Meats, M.: Microtubules and control of insect egg shape. J. Cell Biol. **71,** 207 – 217 (1976)
118. Turner, F. R.: An ultrastructural study of plant spermatogenesis. Spermatogenesis in *Nitella.* J. Cell Biol. **37,** 370 – 393 (1968)

119. Van Deurs, B.: Chromatin condensation and nuclear elongation in the absence of microtubules in Chaetognath spermatids. J. Submicros. Cytol. **7,** 133 – 138 (1975)
120. Van Deurs, B., Behnke, O.: The microtubule marginal band of mammalian red blood cells. Z. Anat. **143,** 43 – 48 (1973)
121. Vasiliev, J. M., Gelfand, I. M: Effects of colcemid on morphogenetic processes, and locomotion of fibroblasts. In: Cell Motility. (Eds.: R. Goldman, T. Pollard, J. Rosenbaum) pp. 279 – 304. Cold Spring Harbor Lab. 1976
122. Waddington, C. H., Perry, M. M.: A note on the mechanisms of cell deformation in the neural folds of the amphibia. Exp. Cell Res. **41,** 691 – 693 (1966)
123. Warren, R. H.: Microtubular organization in elongating myogenic cells. J. Cell Biol. **63,** 550 – 566 (1974)
124. White, J. G.: Effects of colchicine and *Vinca* alkaloids on human platelets. I. Influence on platelet microtubules and contractile function. Am. J. Path. **53,** 281 – 292 (1968)
125. White, J. G.: Effects of colchicine and *Vinca* alkaloids on human platelets. II. Changes in the dense tubular system and formation of an unusual inclusion in incubated cells. Am. J. Path. **53,** 447 – 463 (1968)
126. White, J. G.: Effects of colchicine and *Vinca* alkaloids on human platelets. III. Influence on primary internal contraction and secondary aggregation. Am. J. Path. **54,** 467 – 478 (1969)
127. White, J. G.: Physico-chemical dissection of platelet structural physiology. In: Platelets. Structure, Function, Transfusion and Storage. (Eds.: M. G. Baldini, S. Ebbe) pp. 235 – 252. New York-San Francisco-London: Grune and Stratton 1974
128. Wilkinson, R. F., Stanley, H. P., Bowman, J. T.: Genetic control of spermiogenesis in *Drosophila melanogaster*. The effects of abnormal cytoplasmic microtubule populations in mutant *ms*(3)10R and its colcemid-induced phenocopy. J. Ultrastruct. Res. **48,** 242 – 258 (1974)
129. Wilkinson, R. F., Stanley, H. P., Bowman, J. T.: Role of microtubules in nuclear shaping of *Drosophila* spermatids. J. Cell Biol. **63,** 372 a (1974)
130. Wolosewick, J. J., Brayn, J. H. D.: Characterization of manchette microtubules in the mouse. J. Cell Biol. **59,** 367 a (1973)
131. Yasuzumi, G., Mataono, Y., Asai, T., Nagasaka, M., Yasuzumi, F.: Spermatogenesis in animals as revealed by electron microscopy. XXII. Development of nuclei and cytoplasmic microtubules in the Grasshopper spermatids. Z. Zellforsch. **115,** 543 – 552 (1971)

Chapter 7 Cell Movement

7.1. Introduction

Intimate relations exist between cellular locomotion, intracellular displacements of various organelles, and MT. The role of MT in the movements of cilia is evident, and their association with the displacements of chromosomes at mitosis has been at the origin of some of the most important contributions on MT. However, when this complex subject is approached more closely, the relations of MT and motion appear more complex and often indirect. The study of the mechanics of ciliary bending provides one of the best examples of the role of other proteins, establishing connections between MT, in the dynamic changes of these structures.

Recent work on other cytoplasmic fibrillar structures, grouped under the general term of microfilaments (MF), has brought important knowledge about the factors of cell motility. MF—structures measuring between 5 and 10 nm in diameter—are by no means a single entity, and new techniques—in particular immunofluorescence staining and the treatment of glycerinated cells with heavy meromyosin (HMM)—have brought evidence that proteins of the acto-myosin group are present in many if not all cells, and are the true agents of contractility in the most general sense, of cell locomotion, and perhaps even of the complex intracellular motions of secretory granules and chromosomes. The main problem which needs a better understanding is that of the relations between MF and MT: it is linked with the nature of the proteins associated with tubulins, in particular the heavy molecular weight fraction. In the preceding chapter, MT were seen to act as supports in cell shapes where they were bent and maintained in this condition by other components; in cell motion, MT act similarly as supports, closely associated with the true "contractile proteins", actin and myosin. The possibility that MT assembly and disassembly may by itself lead to intracellular movements has been suggested in mitosis, as indicated by changes of location of chromosomes when MT are disassembled by various drugs (cf. Chap. 10).

In this chapter, after some explanations about the nature and the role of MF, ciliary motion will be considered first. While it concerns particularly complex structures, the results obtained provide good insight into the rules which may underlie the relations of MT with motility. The role of MT in several intracellular movements will be studied in other chapters: secretion entails the transport of granules toward the cell membrane and their excretion: some steps of this transport cannot take place if the integrity of MT is affected (cf. Chap. 8). Chapter 9 will be devoted to the movements of cytoplasm and various organelles in the dendrites and axones of neurons: this neuroplasmic flow is an interesting model for the study of the role of MT.

The movements of chromosomes at mitosis are of course closely related to MT: they will be analyzed in Chapter 10.

The subject treated here has been discussed at several important meetings, and the advances made in this field are clearly apparent from a comparison of the monographs edited by Allen and Kamiya [3], Inoué and Stephens [60], and Goldman *et al.* [48]. Ciliary motion is thoroughly discussed in the book edited by Sleigh [112], while a Ciba Foundation book on locomotion of cells brings information on various aspects of MT activity [26]. Other contributions on this extensive subject can be found in the review by Mohri [82], and the conferences edited by Soifer [113], and Borgers and De Brabander [15].

7.2 Microtubules and Microfilaments

Before considering the role of MT in cell movements, other fibrillar constituents of cytoplasm must be described. Some confusion existed in this field before a proper definition of MT, which were often mentioned as "fibrillar" structures. It was also believed for some time that the subunits of MT could reassemble into fibrils. This conclusion, which appeared to agree with the substructure of MT protofilaments (cf. Chap. 2), was based on observations on colchicine-treated cells and on neuron pathology. The formation, after colchicine, of large bundles of filaments measuring about 10 nm in diameter, has been described in Chapter 5.

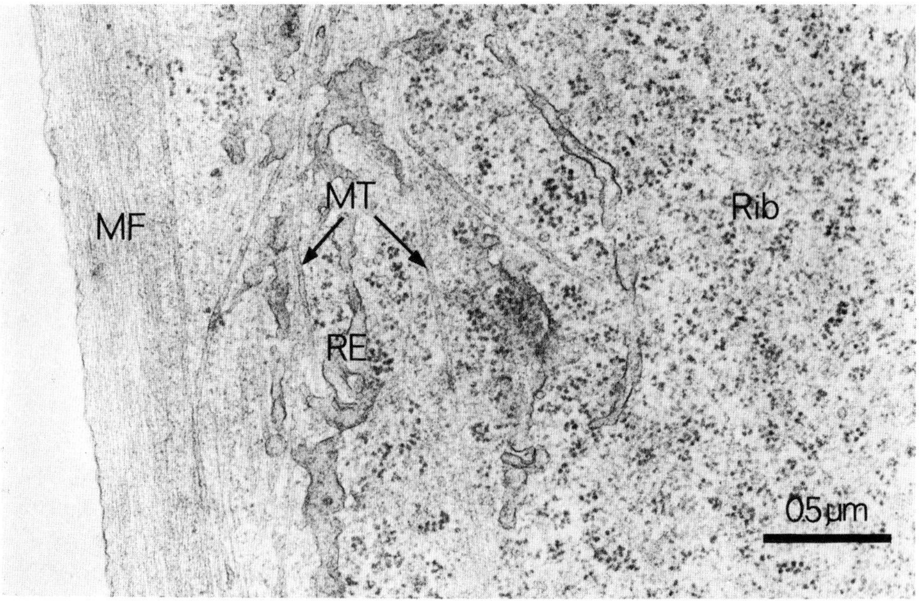

Fig. 7.1. Mouse embryonic cell: microfilaments (*MF*) located immediately under the cell membrane (cell web) and MT in the cytoplasm which contains numerous ribosomes (*Rib*) and some endoplasmic reticulum vesicles (*RE*) (from De Brabander [30])

These fibrils do not originate from the tubulin pool [31]. Similar bundles of fibrils can be induced by aluminium [126], which is not a MT poison.

In normal cells, fibrils measuring between 5 and 7 nm in diameter are often present, and are considered to be functionally associated with MT. The expression "microtubular – microfilament system" is often used, in particular in papers on the mechanisms of secretion (cf. Chap. 8). Considerable advances have been made in this field, demonstrating the actin nature of many of these fibrils. The techniques which have been used, at the cytological level, are the specific fixation of heavy meromyosin (HMM), and immunofluorescence. Both techniques yield comparable results, although they may require a preliminary extraction of some of the cytoplasmic proteins and rupturing of the membranes by glycerol or acetone, to increase the permeability of the cell. HMM attaches specifically to actin fibers, displaying their polarity, as the HMM molecules bind by one of their extremities, giving typical arrow-head patterns showing an identical orientation of the HMM molecules all along the MF [61, 89]. The results of this difficult but excellent technique for electron microscopy have been completely confirmed by immunofluorescence methods using anti-actin antibodies (cf. [68, 124]).

Actin appears today to be an almost universal component of cells, and most—but not all—MF are made of this protein. Other proteins which play a role in muscle contraction, such as myosin, α-actinin [68], and troponin, are also found in non-muscle cells. However, the ratio of actin and myosin, which is small in muscle, is quite different in other cells, as shown by Table 7.1 [89].

This difference may result in a much smaller force, but a much greater shortening than in muscle [89]. Other differences between muscle and non-muscle actin, and the associated proteins, certainly exist. However, it is apparent—and the fact is essential for a proper understanding of all cell movements which are associated with MT—that a similar mechanism of molecular sliding of actin and myosin fibers is present in many cells, and is responsible for various movements. The possibility of contraction by actin alone has also been considered [117]. The actin filaments, as shown by immunofluorescence and by electron microscopy, are often attached to cell membranes or form, close to the cell limits, a "web" of fibrils located immediately under the cell membrane. In microvilli, the MF which are readily observed in electron microscopy are specifically "decorated" by HMM, and show a polarity which explains the retraction of these differentiations [117].

Table 7.1. Contractile protein contents in various cells (from Pollard [89])

Protein system	Percent total protein	Concentration		Molar ratio
		mg/g	μmol/g	actin : myosin
Actin				
Acanthameba	14	10.5	250	193
Human platelets	10	10	240	109
Rabbit skeletal muscle	19	37	900	6.2
Myosin				
Acanthameba	0.3	0.2	1.3	
Human platelets	1	1	2.2	
Rabbit skeletal muscle	35	66	144	

Actin has been identified in many different cells, and the reader is referred to the reviews by Bairati [4], Goldman *et al.* [48], and Tilney [117]. Some locations of actin may be mentioned here: neural plate [5], cleavage furrow or contractile ring [110], blood platelets (where it was described under the name of "thrombostenin") [9], neuroblastoma cells [25], red blood cells [118], and brain [63, 69]. The possible role of actin in mitosis will be discussed in Chapter 10. It is thus evident that many

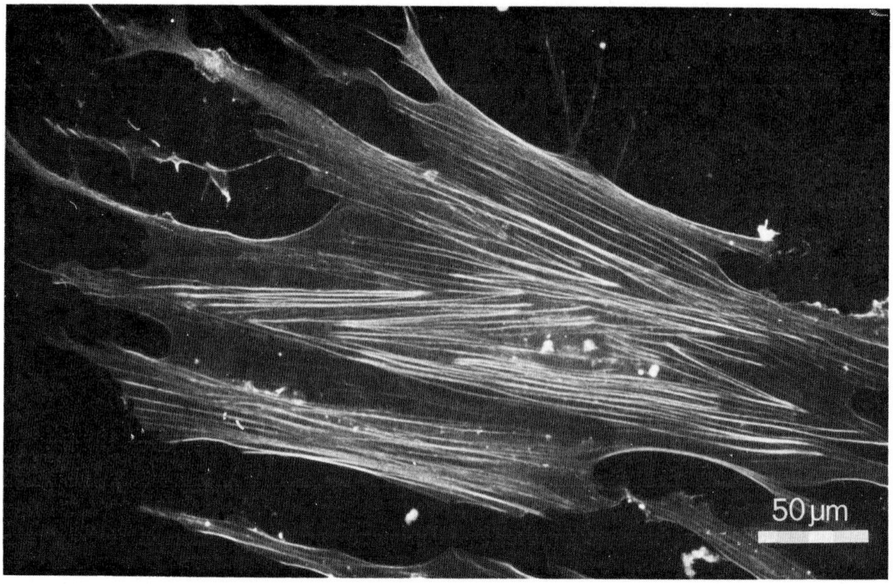

Fig. 7.2. Immunofluorescence staining of actin in human fibroblasts. Contrary to the MT (cf. Fig. 2.3) the bundles ("cables") of actin filaments are usually straight and often extend from one side of the cell to the opposite cell membrane (from Lazarides and Lindberg [68])

forms of cell motility are related to contractile proteins of the actin-myosin type, and one problem which will have to be discussed is that of their relations with MT.

Another aspect of MF physiology has been the discovery that *cytochalasin* B, a substance extracted from several types of moulds, may more or less specifically modify these fibrillary proteins [23]. Its action is complex, and it is certainly not as specific a poison as, for instance, colchicine for MT. It may also affect cell membranes. Cytochalasin [9] has been often used to modify, reversibly, the cell MF, and is an interesting tool for the study of cell motility.

Immunofluorescence studies with anti-actin antibodies show that in mouse fibroblasts in vitro the pattern of actin bundles is rapidly disorganized by cytochalasin B, while MT are not affected [124].

In cilia and flagella the presence of *dynein*, a protein with ATPase activities similar to myosin [42], has been described in Chapter 4. This protein apparently acts

[9] While cytochalasin B has no effect on tubulin assembly, the closely related cytochalasin A inhibits tubulin and actin assembly in vitro and decreases the colchicine binding of tubulin, through an apparently irreversible action on -SH groups of proteins [53]

alone, the MT—or some MT-associated proteins—playing the role of actin in a more or less similar sliding filament mechanism. Cilia contain other more or less fibrillar structures: the spokes which unit the central pair and the peripheral doublets, and the *nexin* filaments which hold together the doublets. These proteins do not appear to have any "contractile" function.

Other structures which may be readily mistaken for contractile MF are the fibrillar proteins playing a part in cell structure: the best known are the tonofilaments of epithelial cells, the fibrils attached to desmosomes, and the glial filaments of the brain. The fibrillar rootlets of basal bodies should also be mentioned. Last, pathological fibrils of a protein nature, often showing a "twisted" structure, are those of amyloid (not a definite chemical entity, but a family of fibrillar proteins with a β-pleated structure, cf. [45]), and the peculiar filaments found in neurons of Alzheimer's disease and in some so-called "slow virus" infections, such as kuru (cf. Chap. 11; [126, 127]).

Thus, many types of fibrillary proteins found in cell have a diameter of about 5 to 10 nm: they belong to different categories, the principal ones being actin (probably associated with a small number of myosin molecules), structural fibrils (tonofilaments), and pathological changes, such as those described after colchicine and those found in several diseases. Although cytochalasin may be of some help, any identification of these proteins, apart from by biochemical means, needs the use of immunofluorescent staining methods or of meromyosin binding.

7.3 An Introductory Note on Mechanisms of Movement Associated with MT

Before proceeding to an analysis of models of relations between MT and movements, it should be made clear that two radically different possibilities may be encountered. The first, which has been mentioned in Chapter 4 in relation to the assembly and disassembly of MT—which can take place quite rapidly—, has often been considered in studies of mitosis: if MT are assembled from a tubulin pool close to centrioles or MTOC, their growth may play a direct role in the separation of these structures. The elongation of the anaphase spindle in many types of mitosis (cf. Chap. 10) is apparently the result of new tubulin subunits being added to polar MT. This is similar to the growth of cilia, flagella, and axonemes (cf. Chap. 4).

Movements resulting from disassembly are evident, such as the retraction of cilia or, still more spectacular, that of axonemes. However, these consist in a disappearance of organized structures, and it is not evident that the active displacement of cell structures could result from disassembly alone.

Other forms of movement associated with MT are quite different. While MT have no contractility, they are closely realted to rapid intracytoplasmic movements and rapid displacement of cell structures, leading to cell locomotion, phagocytosis, intracellular flow. Ciliary motion, which will now be considered, results from the interaction between MT and dynein acting as an ATPase, the movement depending on the sliding of one MT (or a doublet MT) in relation to another. The "sliding filament" mechanism, which has already been mentioned in relation to MT and cell shape (Chap. 6), may be generalized, following the early model of mitosis proposed by McIntosh et al. [74]. In this case, movement is generated by the interaction of MT with other proteins which establish, between closely associated MT, shearing forces leading to the sliding, one past another, of different MT, or to the sliding of cytoplasmic particles along MT. As will become evident, this last mechanism appears to be the most important in the various types of movement associated with MT.

7.4 Ciliary Movement

The complex structure of cilia and flagella governs their rapid undulating movements; structural and biochemical studies have led to a good understanding of the main principles of this movement, which is closely related to the facts described in Chapter 4: the nine doublets, the two central MT, the double dynein arms between the doublets, the radial spokes, the nexin links, the sheath around the central pair of MT. The basal body need not be considered here, as it has been shown that cilia isolated without the basal body are still capable of beating regularly. The complex structure of the basal body plays a role of course in ciliogenesis, and assures also a mechanical link between cilia and cytoplasm.

Two authors have particularly contributed to explain the mechanism of ciliary beating: Gibbons and Satir, the first by a careful chemical analysis of the structural components of cilia [41 – 44], the second by demonstrating the validity of the "sliding filament" model [100 – 103]. A clear relation between movement and structure was demonstrated by studies on cilia of *Anodonta cataracta*, where the bilateral symmetry is marked by the position of the two central MT and by the special link between doublets 5 and 6. The effective beating stroke takes place in a plane perpendicular to that passing by the two central MT [41]. The work of Gibbons [43, 44] leading to the isolation of *dynein* has already been described. This protein is an ATPase with a great specificity, as other nucleotides cannot act as substrates, except in cells capable of phosphorylating them to ATP or transphosphorylating other triphosphate nucleotides to ATP, as observed in *Tetrahymena* [99]. The action of ATP and dynein was studied on isolated flagella from which the cell membrane had been removed by Triton X-100. In the absence of ATP, the flagella enter a condition of rigor, comparable to that of muscle in similar conditions; this is reversible: slowly if low concentrations of ATP are added, completely and leading to a renewal of ciliary beating when a normal concentration of ATP is present. The biochemical relation between the quantity of ATP used, the number of molecules of dynein, and the energy needed for ciliary movement have been figured out and indicate that the mechanism does not differ fundamentally from that of other contractile structures [43].

When isolated cilia, without their cytoplasmic membrane, are treated with weak solutions of trypsin, and ATP is added, the sliding movement of some doublets in relation to others becomes spectacular: probably as a result of the destruction of the nexin bridges, some of the doublets slide rapidly in relation to the others, the whole structure increasing rapidly in length. These changes were observed in bull and human sperm flagella, extracted with Triton X-100, digested briefly with trypsin at pH 9.0, and reactivated by ATP in the presence of Mg^{2+} ([70, 116]; Fig. 7.3).

Further indirect evidence of the role of dynein is that in sea-urchin sperm, the removal of the outer arms by a solution of KCl did not change the aspect of beating but slowed down its rhythm. About 60% of the dynein was lost by this technique [44]. The fact that trypsin treatment, which destroyed the radial spokes, did not affect motility, indicates that these structures do not play an active part in movement; their role is probably to maintain a certain degree of rigidity and elasticity of the whole cilium or flagellum. It may be more complex, as *Chlamydomonas* mutants without radial spokes show a slower bending movement [128], and mutants without central tubules are immobile. However, a flagella of 6+0 structure, observed in the male gamete of the gregarin *Lecudina tuzetae,* is capable of beating, but its movements are particularly slow [109]. On the other hand, pathological human spermatozoa lacking dynein arms are immobile ([1]; cf. Chap. 11).

Many other authors have confirmed these facts (cf. reviews in Inoué and Stephens [60], and Sleigh [114]). Isolated axonemes of *Chlamydomonas* become active in the presence of ATP; after trypsination, the sliding induced by ATP leads to a separation of some groups of doublets ([2]; Fig. 7.3). When, in the same species, the two flagellae are isolated with their basal bodies, activation by ATP (or ADP, which is phosphorylated by this species to ATP) induces coordinated movements of the flagellae, indicating the intrinsic regulation of this activity. In this experiment, the ciliary membranes were intact; they may play some coordinating role [59].

The sliding filament concept implies that the length of the MT does not vary and that, by an interaction between doublets mediated by dynein, the MT of one side slide in relation to the opposite ones. As a consequence, sections through the extremities of cilia should show less MT on the convex than on the concave side of the bend, even if this bend is located at some distance from the tip of the cilium. This

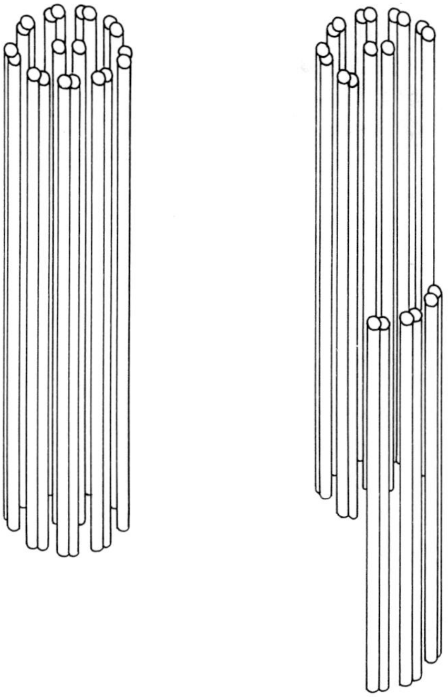

Fig. 7.3. Action of ATP on demembranated ciliary axonemes. The rapid and considerable increase in length after addition of ATP is considered to result from a sliding movement of one group of doublets in relation to the others (redrawn from Satir [103])

was demonstrated in the ciliated cells of the mussels *Elliptio* and *Anodonta* by Satir [100, 101, 102]. The observation was made easier by the fact that the doublets become singlets close to their distal extremity: hence, on the concave side of the bend, doublets could be found, while on the opposite side, in transversal sections, only singlets were visible. The motions of the doublets relative to the central pair of MT could also be analyzed in detail by measuring the spacings between the spokes and the central pair [122]. In *Elliptio* these spokes, when observed in longitudinal preparations of cilia, are grouped by three with a 86 nm repeat. Their relation during ciliary bending, to the lateral projections, extending from the central MT and spaced at about 14 nm, indicates that the spokes are capable of bending to a certain angle, and are not firmly attached to the central tubules: they undergo, during bending, a cycle of intermittent attachment. These studies show moreover the remarkably precise geometrical dispostion of the various ciliary constituents, suggesting strongly that definite relations exist between the molecular structure of the MT with their 4 nm periodicity, the location of the radial spokes, and the lateral projections attached to the central MT. It is also possible, on biochemical grounds, that not all the dynein is present in the dynein arms, and that some similar molecule may be located, perhaps at the spoke head-sheath interface. Warner and Satir [123] conclude by presenting the hypothesis "that the radial spokes are a main part of the transduction mechanism that converts interdoublet sliding into local bending".

Other complexities of the bending problem are indicated by Gibbons in a recent contribution [43]: flagellae, in a condition of rigor, may readily undergo, under the effect of mechanical forces, a rotation of 180°; also, ultrastructural analysis of the aspects of ciliary MT observed with a high voltage electron microscope (1 MeV) shows that some twist is present along the cilia in the regions of bending. The dynamics of the cross-bridge sliding filament model have been studied with the aid of a computer, and the results agree well with the known data [16]; the study of these models suggests however "a wide range of other possible mechanisms for control of the active progress and regulation of the flagellar wave". This last aspect has not been discussed here, for it would be beyond the scope of this book; the dynamics of cilia movements and metachronal waves involve complex problems of hydrodynamics and perhaps intracellular regulatory processes (cf. [112]).

Other types of flagellar movements are known and indicate that movement can be created by MT with few other structures: in the armored scale insect *Parlatoria oleae* (Homoptera, Coccoidea) sperm bundles, animated by undulating movements, are made of nuclei surrounded by many MT with no 9+2 structure, and no other organelles: apparently, the MT move in relation to one another to create the movement, perhaps through the action of interMT bridges (dynein?). Similar spermatozoa are known in 10 species of coccid insects; links do exist between neighboring MT, as mentioned in Chapter 4 [95, 96].

7.5 Intracellular Displacements and Motion

The movements associated with cell activity are diverse, and may lead to displacements and locomotion of the whole cell, of some particular regions of the cell surface, as observed in phagocytosis and secretion (Chap. 8), or displacements of various particles or organelles inside the cytoplasm. This last category will be considered first, for its study throws some light on the complex interactions between MT and other fibrillar structures; it is the best introduction to the study of granule movements as observed in exo- and endocytosis, and of continuous flow as found in nerves (Chap. 9), and even of chromosome movements during mitosis. In all these dynamic changes close relations are observed between the moving objects (granules, organelles, foreign particles) and MT, while the mechanisms of motion almost certainly involve, in most cases, other proteins of a contractile nature (actomyosin-like). The central problem which requires a proper answer is that of the spatial and dynamic associations between MT and force-generating structures.

7.5.1 Saltatory Motion

Apart from slow deformations and displacements of cell structures, two different types of motion should be clearly separated: cell streaming and saltatory motions. Cell streaming, as observed in cyclosis of plant cells, is independent of MT and will not be studied here. In the Alga *Nitella*, for instance, MT are located in regions of the cell where no streaming takes place and they could not be the driving force [86]. Saltatory motions, in contrast, often appear to require MT or MF: they are defined by sudden, rapid, oriented displacements of any small intracytoplasmic particles, movements which by their amplitude and rapidity are clearly different from brownian motion and cannot be explained by simple thermodynamic agitation. Rebhun [92, 93, 94], who has shown the significance of these movements in various cell activities such as mitosis, pigment granule movements, axonal flow, has given the following definition (for reviews, see [64, 94]):

a) The movements are discontinuous, with sudden changes in velocity: from rest, particles may move distances of up to 30 μm in a few seconds, at a speed of up to 5 μm/sec, and then stop suddenly.

b) Two closely located particles, separated by about 1 μm, may show different movements.

c) All particles in a given cell are not affected, and while small granules may show saltatory movements, cell organelles may remain motionless.

d) The velocity of movement appears to be uniform for all the distance traversed, an observation which implies that a continuous force is applied.

As these movements may be observed in cells with ameboid motion, it was imagined earlier that pseudopod formation could result from an assembly – disassembly mechanism probably involving MT, while saltatory movements would imply contractions of MT [32]: this last conclusion is now known to be incorrect. However, the relation with MT was demonstrated by the effects of colchicine, VLB, and podophyllotoxin, which rapidly stop this type of movement. In cultured fibroblasts, studied by cinematography, all saltations stop in the presence of 10^{-4} M colchicine [36]; puromycin is without any effect. The orientation of saltatory motions of diverse bodies (lysosomes, carbon particles, lipids) is parallel to that of MT. However, the origin of the motile force remains unknown [37]. The possible role of MF, and of proteins of the actin – myosin type, has been suggested, that of MT being "one of support of the mechanochemical transducer and probably of the establishment of cellular form and internal pathway chanelling" [94].

The generality of this type of discontinuous, rapid, intracellular movement will appear further in this and the next chapter. It is the basis of many intracellular movements, and the movements of mitosis (Chap. 10) may be considered a special case of saltatory movement of particularly large structures, the chromosomes.

7.5.2 The Movement of Pigment Granules

Melanin granules, which are large particles, undergo rapid movements related to adaptive changes of skin color. Other pigmented granules (rhodopsin), as observed for instance in fish scales, move similarly, sometimes in a rhythmic fashion [90]. In mammals, melanin granules are injected by melanoblasts into the cytoplasm of epithelial cells of the skin: this transfer, which is stimulated by ultraviolet light in the so-called "immediate tanning reaction", may take place in a few minutes and also involves MT and MF [62]. We will limit this study to the problem of melanophores and the reversible expansion and contraction of their pigment granules, as observed in cold-blooded vertebrates. The analysis of this model is illuminating, for we are faced with the rapid displacement of large numbers of relatively big and heavy granules, under the influence of chemical stimuli which are well known, displacement which can be determined at will, even in isolated fragments of skin or scales of fish. The role of MT is recognized by all authors, but two different mechanisms have been proposed. Either movement is generated by reversible cycles of assembly and disassembly of MT, or MT are only a support for the motive action of other proteins, the pigment granules "sliding" along the MT.

From the start, this problem was studied with the help of colchicine, and it was during work on the mode of action of this drug in gout that Malawista [77], before

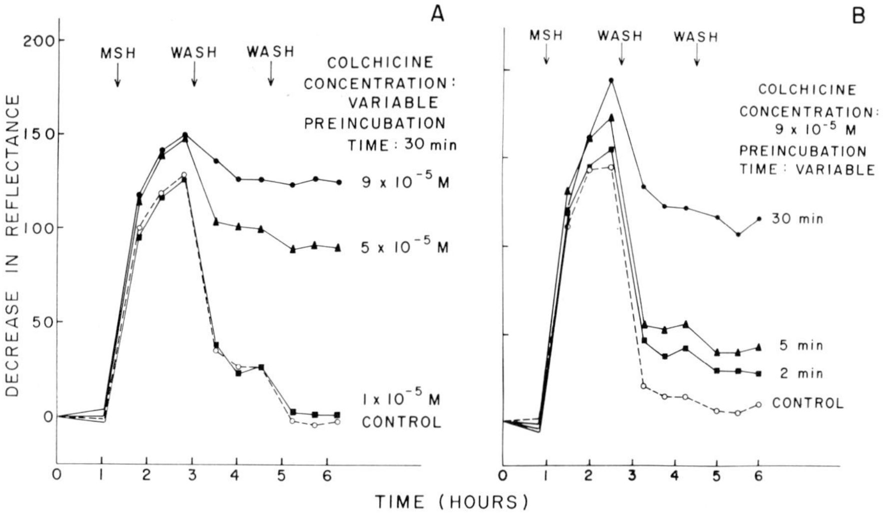

Fig. 7.4 A, B. Action of colchicine on the movement of pigment granules of *Rana pipiens* melanophores after action of melanophore stimulating hormone (*MSH*). Colchicine prevents the return to normal after blackening. This effect is proportional to the dose (A) and to the duration of action (B) (redrawn from Malawista [77])

any mention of MT, used melanophores of the frog's skin as a model. The expansion of the pigment in these cells—the pigment granules moving outward in the stellate expansions of the cytoplasm—can be provoked by pituitary melanin-stimulating hormone (MSH) in vitro; this action is reversible, and after washing, the pigment granules reassemble in the center of the cell, the skin becoming whiter. Colchicine, at doses of 5×10^{-5} M, counteracts this centripetal movement of pigment granules, and this is related to concentration and duration of action. It is potentiated by high hydrostatic pressures [79]. Similar effects were found with other darkening agents, such as caffeine or ATP. Colchicine alone, at a concentration of 9×10^{-5} M, produces a gradual, irreversible darkening after several hours of action. This is counteracted by the lightening agent, melatonin, the pineal gland hormone (cf. Chap. 5). These results were first interpreted in terms of modifications of cell viscosity, as no ultrastructural observation had been made. It was shown later [78] by the same author that the MT poisons, colcemid, VLB, and VCR had similar effects, although VLB was found to be 100 times more active in producing an irreversible darkening of the skin (9×10^{-7} M). Griseofulvin was found to be active but only when

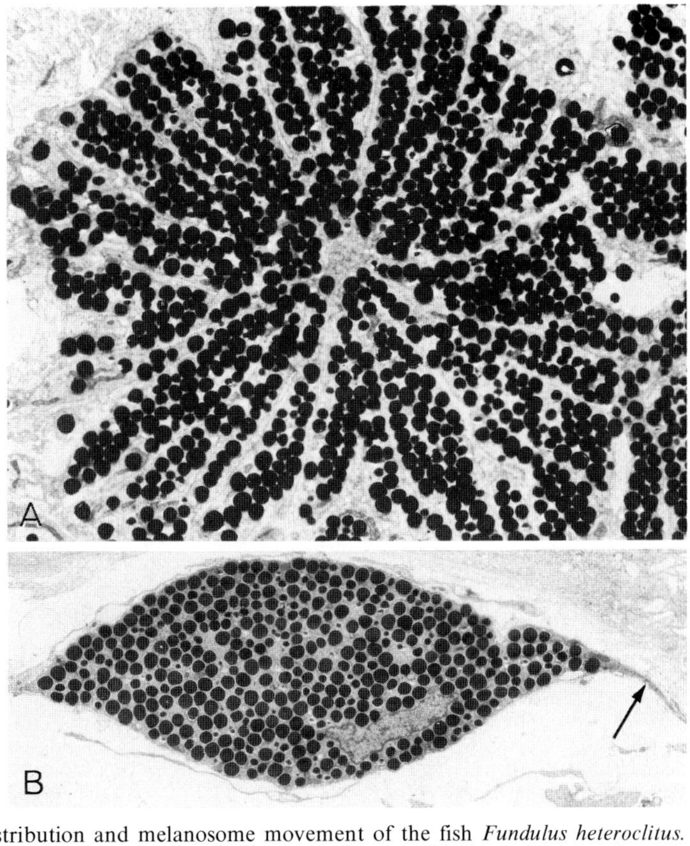

Fig. 7.5 A, B. MT distribution and melanosome movement of the fish *Fundulus heteroclitus*. (A) Horizontal section of a melanophore with dispersed pigment. (B) Vertical section through the central part of a melanophore almost completely aggregated. The melanosomes are gathered in the center near the nucleus while the cell processes appear collapsed (*arrow*) (from Schliwa [104])

present in the medium, contrary to colchicine, VLB, or VCR, the action of which persisted even after they had been removed from the incubation medium. This suggested that MT are involved in aggregation, and not in dispersion, as their destruction or poisoning led to the dispersion of the pigment granules.

Table 7.2. Action of MT poisons on melanocytes

Reference	Animal species	Poison	Dose	Action
Malawista, 1965 [77]	*Rana*	colchicine	9×10^{-5}	Darkening. Prevents contraction after MSH and other darkening agents
Malawista, 1971 [78]	*Rana*	colchicine VLB, VCR griseofulvin	$2.5-5 \times 10^{-5}$ M $2.5-5 \times 10^{-5}$ M $2-5 \times 10^{-5}$ M	Darkening. VLB is the most active
Wikswo and Noyales, 1972 [125]	*Fundulus heteroclitus*	colchicine	5×10^{-3} M	Isolated scales: destruction of MT, increase of MF. Decreased contraction by epinephrine
Lambert and Crowe, 1973 [65]	*Uca pugilator*	colchicine	3 mg/ml	Inhibition of aggregation. No effect on dispersion
Robison and Charlton, 1973 [97]	*Palaemonetes vulgaris*	colchicine VLB	1×10^{-3} M	Destruction of MT, without action on pigment expansion
Murphy and Tilnea, 1974 [85]	Teleost erythrophores	colchicine VLB cold, hydrostatic pressure	1×10^{-3} M	Total inhibition of granule movement, without destruction of MT
Schliwa and Bereiter-Hahn, 1974 [107]	*Pterophyllum scalare*	VLB colchicine	0.2×10^{-3} M 5×10^{-3} M	Inhibition of pigment movement
Castrucci, 1975 [24]	*Tilapia melanopleura*	colchicine VLB	1×10^{-5} M 1×10^{-5} M	Inhibition of pigment movement (melano- and erythrophores)
Murphy, 1975 [84]	*Gymnocorymbus*	VLB	1×10^{-5} M	VLB arrests granule movement and destroys MT
		colchicine	2×10^{-3} M	Colchicine retards movement without destroying MT
Moellmann and McGuire, 1975 [81]	*Rana*	VLB	1×10^{-5} M	Inhibition of lightening after cytochalasin B
Fingerman *et al.* 1975 [34]	*Palaemonetes vulgaris*	colchicine	25×10^{-3} M	Inhibition of pigment aggregation (erythrophores)
Schliwa and Bereiter-Hahn, 1975 [108]	*Pterophyllum scalare*	VLB	0.3×10^{-3} M	No change in the pressure required for dispersion
Lambert and Fingerman, 1976 [66]	*Uca pugilator*	colchicine lumicolchicine	0.25×10^{-3} M	Inhibition of aggregation. Lumicolchicine is as active as colchicine

A rather large number of publications have dealt with similar experiments, and their main results are summarized in Table 7.2. Besides the various MT poisons, cytochalasin has been shown to inhibit the movement of pigment granules. The role of cAMP, acting as the second messenger of MSH on the cells, has been demonstrated, and will not be discussed here. It leads to a dispersion of pigment granules, i.e., a darkening [12, 72, 76, 87, 107], mimicking the effect of MSH. It has also been claimed that cAMP may modify the numbers of MT in the melanocytes, decreasing the MT while the number of MF increases [80, 107]. This is similar to the action of ultraviolet light [62]. Ionic changes also affect the aggregation of pigment, Na^+ being a darkening agent and K^+ a lightening one [24].

Melanophores from different species of vertebrates and invertebrates contain MT and MF. These are linked with the movement of the large pigment granules. The MF may be comparable to the actomyosin fibrils visible in various cells. Some authors describe an increase of similar fibrils after colchicine treatment, but these may be of a quite different nature, and similar to those mentioned in Chapter 5. Tables 7.2 and 7.3 summarize the main findings on the action of MT poisons and cytochalasin B on various types of melanophores. Of course, one should be careful before comparing pigment cells in insects or crustacea, melanophores or erythrophores of teleost scales, and melanocytes of frog's skin. These cells are of a different embryonic origin and it would be surprising if the same mechanisms were at play in all cases. While hormonal control exists both in invertebrates and vertebrates, the nature of the hormones is quite different; however, in both cases, cAMP may be the messenger of hormone action.

In the retina of teleosts, adaptation to light is provided by large movements of pigment granules, which slide in relation to the photoreceptors, as studied in the grey snapper, *Lutjanus griseus*. This migration of pigment granules along cytoplasmic processes, which contain MT and actin filaments, may cover a distance of up to 75 μm. The MT are very difficult to fix, and their lability suggests that their disassembly may play a role in the pigment movement, although other mechanisms involving actin filaments cannot be excluded [18, 19].

From Table 7.2 it appears that MT poisons prevent the movement of the pigment granules in both directions, although when acting alone they favorize the pigment expansion. Thus, MT appear to be mainly important for pigment aggregation.

Table 7.3. Action of cytochalasin B on melanocytes

Reference	Animal species	Action
McGuire et al., 1972 [72]	*Rana pipiens*	Aggregation of pigment after darkening by MSH (antagonist of melanin-dispersing MF)
Lambert and Crowe, 1973 [65]	*Uca pugilator*	Inhibition of pigment translocation (aggregation and dispersion)
Magun, 1973 [76]	*Rana*	Inhibition of pigment, dispersion (after MSH, theophyllin, dbcAMP). No inhibition of aggregation
Robison and Charlton, 1973 [97]	*Palaemonetes vulgaris*	Inhibition of pigment dispersion without destruction of MF (10 μg/ml)
Moellman and McGuire, 1975 [81]	*Rana*	Aggregation of pigment (antagonized by VLB)

There are fewer results with cytochalasin B, which has a complex action affecting MF (Table 7.3): in mammals, it clearly has an aggregation effect and may act as an antagonist of the darkening effect of VLB [81]. In all these experiments, it is far from evident that the actions of MT poisons and cytochalasin B result from a destruction or disassembly of MT or MF respectively, and the possible relations between these structures and pigment granule movement must now be discussed. The recently demonstrated fact that lumicolchicine (which does not modify MT) is as active as colchicine in the crab, *Uca pugilator*, should be kept in mind [66].

Two different theories have been proposed; they are contradictory, but they result from observations on different types of pigmented cells.

7.5.2.1 The Sliding Filament Theory. As mentioned above, saltatory motion, of which pigment granule displacements are an example, is characterized by oriented movements at a uniform speed, as if the particles were guided by a constant force applied during all the time of motion. It could thus be imagined that the pigment granules slide along the MT, possibly moved by interactions with contractile proteins attached to these structures. Two difficulties are at once apparent: the movement takes place in two directions, and links between pigment and MT are not observed [11, 51]. One may imagine that two populations of MT are present, radial ones centered on the centrioles, and necessary for aggregation, and peripheral ones, perhaps linked to the cell membrane, and used in pigment dispersal [11, 90]. Although the absence of links between MT and pigment granules has been mentioned by several authors [106, 125], these structures are difficult to fix properly and anyhow, the granules must be attached to some kind of support, for it is evident that they have no autonomous properties of motility.

The "sliding filament" theory has been defended by Murphy and Tilney [85], and their conclusions are based on the following observations: (1) movements and alignment of granules disappear under the action of several MT poisons (physical and chemical) while the MT are relatively resistant; (2) MT are needed both for centripetal and centrifugal motions; (3) the MT are fixed and the granules slide along them: the number of MT does not change when the cell contracts or expands its pigment; (4) there are not two different populations of MT; (5) pigment granules are connected to the MT by labile links; (6) no relationship is apparent between MF and pigment movement. These observations are based on the study of melano- and erythrophores of three species of Fish. The authors conclude that "the granules slide along the MT through active interaction", and suggest that some dynein-like protein may be involved.

In a more recent contribution, the stability of the MT of *Fundulus* is stressed, as they resist 2 mM of colchicine and pressures of 10,000 psi [84]. The impairment of pigment movement by antitubulin drugs, even without any destruction of MT, is compared with similar facts observed in nerve cells (cf. Chap. 9).

7.5.2.2 Movement by Assembly – Disassembly of MT. All modern research has shown that MT may rapidly be disassembled into tubulin molecules (or dimers) and reassembled: many instances have been described in relation to tubulin biochemistry (Chap. 2) and the growth and retraction of cilia or axopods (Chap. 4). Porter, in studies on the melanophores of *Fundulus heteroclitus* and the erythrophores of *Ho-*

locentrus ascensionis, concluded that pigment movement may be closely related to disassembly of MT—leading to central accumulation of pigment, thus to whitening—and assembly, with growth of MT toward the periphery at the same time as the pigment granules move out and the cell darkens. It is important to note that the two movements have different velocities: aggregation is more rapid (about 30 μm/min) while dispersion takes longer and is less uniform (about 100 μm/min).

Such differences are similar to those described in retraction and regrowth of axopodia in protozoa. During dispersion, the movements are more irregular, more "saltatory" than during the rapid aggregation. When the erythrophores were studied in the contracted condition, the peripheral cytoplasm appeared nearly empty, no more MT being visible outside the pigmented center. No bridges between granules and MT are apparent [90].

Further studies of these cells have shown however that the MT were not destroyed when pigment granules moved into the contracted condition, but were pushed close to the cell membrane in the regions devoid of pigment granules; the "empty" appearance of the cytoplasm had suggested, in sections oriented in the plane of the cell, that no more MT were present ([91] and Porter, personal communication).

In a series of studies on fish melanophores (*Pterodophyllum scalare*), Schliwa and Bereiter-Hahn [105 – 108] have reached similar conclusions, adding however that some MT-independent contractile system is probably present in these cells. Pigment aggregation is accompanied by a decrease in the numbers of MT and an increase of MF. Some C-MT may be observed [106]. MT turnover and melanin displacement occur simultaneously, disassembly being the fastest motion (darkening in 40 sec by Na^+, whitening in 15 sec by K^+). Sulfhydryl groups may play a role (cf. Chaps. 2 and 5), as evidenced by the dispersive effect of parachloromercurybenzoic acid [105]. Pigment dispersion is prevented by VLB, acting as a MT poison; however, this may be counteracted by cAMP [107], which has no action on aggregation. These results would indicate that cAMP facilitates the assembly of MT, which would be the moving force for pigment dispersion. In a more recent work [108], the use of mechanical compression on pigment granule movements suggest the presence of a contractile component in the cell, independent of MT. Its chemical nature remains to be characterized. These results "rule out the sliding mechanism" of pigment movement [108]; they can be explained if long bridges (about 30 nm) link the granules with the MT.

While it is possible that in some types of melanophores the aggregation of pigment takes place at the same time as a disassembly of MT, and vice versa, it remains difficult to understand how these two changes are mechanically linked: if it is relatively easy to imagine that the outgrowth of MT may push the granules toward the periphery of the cell, how could disassembly provide the centripetal force? (hence the idea of some other "contractile" protein and the possible role of MF). A similar question will appear in the study of mitosis: whether disassembly of MT can, alone, move chromosomes.

7.5.3 Other Intracellular Movements Linked With MT

Many different types of cytoplasmic organelles or ingested particles move in the cell close to MT, suggesting again relations between MT and motility. The following instances indicate that MT are either guides, or pathways, or propulsive structures, in various cells; in any case, they are associated with molecules involved in motility.

7.5.3.1 Ribosomes. In the insect *Notonecta glauca glauca*, the oocytes receive a flow of ribosomes from a "trophic" zone, through long cellular expansions or nutritive tubes. These are about 15 μm wide and contain about 30,000 MT which are about 2 μm long. These structures are birefringent, this property disappearing after colchicine or cooling to less than 2 °C [75]. The tubes contain only, apart from MT, innumerable ribosomes which become incorporated into the maturing eggs. The MT have a clear exclusion zone ([115]; cf. Fig. 2.9); it may contain minute filaments, bridging the space between MT and ribosomes. In these cells, MT "have one definite function: they facilitate the flow of ribosomes", "keeping channels open for their transport". It is however not proven that the MT exert a motive force on these organelles.

7.5.3.2 Nuclei. Close connections between MT and the nuclear membrane have been mentioned above (Chaps. 3 and 6). In hamster cells infected with parainfluenza virus SV5 multinucleated syncytia demonstrate a migration of nuclei, at a speed of about 1 – 2 μm/min. They form linear arrays between which birefringent cytoplasm is observed, containing MT and MF of 8 nm diameter. When such cells are treated with colchicine, the MT are no longer visible, the MF appear to be more numerous, and the nuclei are located without order. Linear groups of nuclei can be isolated from these cells and are associated with MT, which seem to delineate channels along which the nuclei move [57]. Another instance of the guiding role of MT in nuclear migration has been described in heterokaryons between macrophages and melanocytes, where it is disrupted by colchicine [50]. Here again, as in melanocytes, it is difficult to know whether the MT act as guides or if they provide some motive force.

7.5.3.3 Viruses. The relations between viruses and MT have been described in Chapter 3. The role of MT in virus transport in the cells is of great importance, in particular in the nervous system, where several viruses (*herpes simplex*, for instance) are known to migrate in relation to the direct neuroplasmic flow (cf. Chap. 9). Reoviruses associate closely with both neurotubules and spindle tubules, evidence of the similarity (or identity) of these two types of MT [49]. Adenoviruses bind to MT, mainly at their edge [71]; they also bind to VLB paracrystals in HeLa cells [29]. It is possible that MT participate in the movement of adenoviruses from the cell surface toward the nucleus. The fact that the viruses are intimately attached to the MT is difficult to reconcile with active transport, unless the MT were in a condition of movement by assembly and disassembly.

7.5.3.4 Food Particles. MT play a prominent role in the feeding of protozoa, and this very elaborate mode of endocytosis is related to the simpler forms of endocytosis found in metazoa, which will be described in the next chapter.

A remarkable type of transport of food particles is provided by the *Suctoria*, a group of sessile tentacle-bearing ciliates which often feed on other free-swimming ciliates engulfed through the activity and movements of complex tentacles [7]. *Toko-*

Fig. 7.6. General aspect of a suctorian, showing the location of the tentacles with a terminal knob (redrawn from Grell [52])

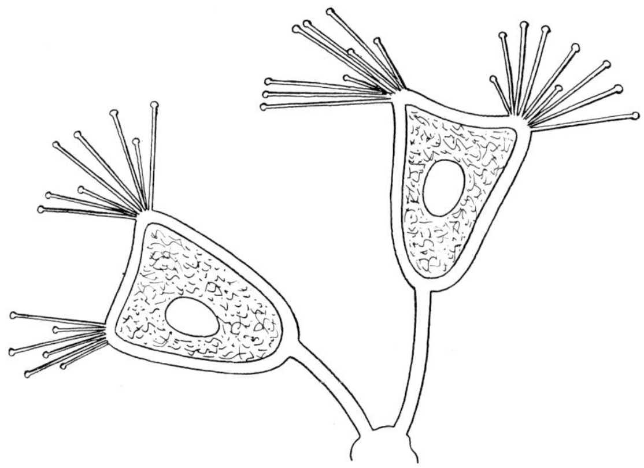

phyra infusionum has been the most thoroughly studied. The long tentacles of this species stick to their prey, which is engulfed by the formation, in the axis of the tentacle, of a channel leading to the body of the host. This channel, as in other species of suctoria, is limited by a complex structure, comprizing at least two more or less circular groups of MT. Similar structures are found in other species, such as *Dendrocometes paradoxus* [6], *Acineta tuberosa* Ehrenberg [8], *Choanophrya infundibulifera* Hartog [54], *Rhyncheta cyclopum* Zenker [55], and *Discophrya* sp. [28].

The following description, with minor changes, applies to all these species. The tentacle, in the resting condition, is a long cylindrical structure often with a swollen extremity. It is supported by a double microtubular structure, forming a circular, exterior row of MT which at a special region called the sleeve, are attached by lateral links, and a more complex inner structure made of several obliquely disposed groups of MT, also linked together, and attached by tenuous links to the external group. The number of these inner groups is variable; from about 30 in *Choanophrya* [54] to seven in *Tokophyra* [119, 120]. These inner groups have a ribbon-like appearance, and are more or less oblique in relation to the axis of the tentacle (Fig. 7.7).

When in the feeding position, the membrane is closely connected to the membrane of the prey. This is carried through a channel, extending the whole length of the tentacle, which becomes shorter and wider and is capable of ingesting large particles such as whole *Tetrahymena*. In this position, the inner group of MT extends toward the extremity, where it curves out rather sharply (about 140° in *Tokophyra*), while the MT, pushed laterally by the wide canal, are in close relation to the cytoplasmic membrane lining this pathway. The inner MT, in most species, show lateral arms which are mainly pointed toward the lumen, although some connections do exist between the two concentric groups of MT. These complex structures and their relations during resting and feeding are illustrated by Figure 7.7. It should be pointed out that, like so many MT structures, all these tubules are more or less oblique, and this is most apparent in the feeding position, when the inner MT curve out around the opening of the tentacle. The inner MT are also helical, as clearly shown in model reconstitutions of *Tokophyra* [119].

So, here we have a most dynamic cellular structure, capable of transporting large particles and undergoing, when it has captured some food, important changes in the position of the two main groups of MT. These are not attached, at their base, to

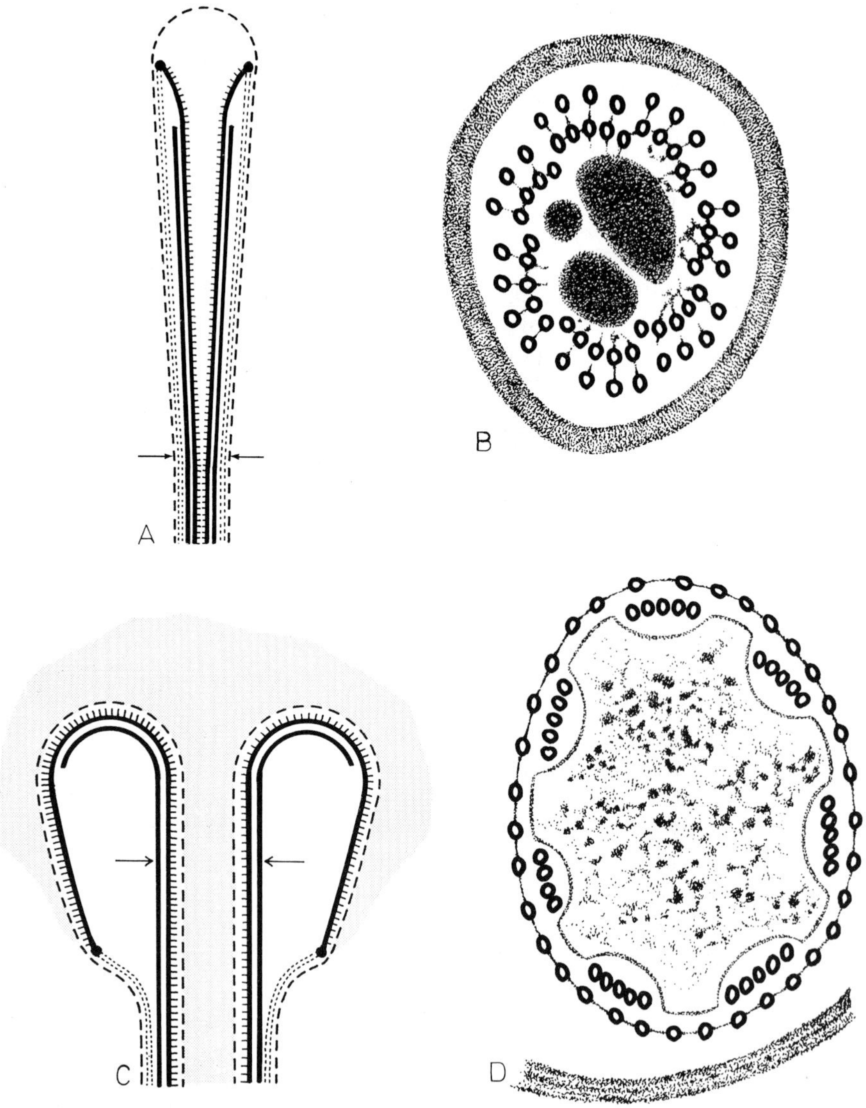

Fig. 7.7 A–D. Movements and position on the MT of the tentacles of *Tokophyra* at rest and during feeding. (A) Longitudinal section. The tentacle shows two rows of MT (represented here as *heavy lines*). The innermost ones show regularly spaced projections turned toward the center. Between the MT and the cytoplasmic membrane, groups of fibrils (*dotted lines*) are attached at the extremity to a circular band to which the internal MT are also attached. (B) Cross-section at the level indicated by the *arrows* in (A). The two rows of MT and their connections are represented here at a higher magnification. The external MT are linked to the central ones by bridges, while the central MT show "free" extensions turned toward the axis of the tentacle. (C) The tentacle is ingesting the cytoplasm of a relatively large protozoan (*grey*). The MT are curved outward and down by the traction of the filaments on the circular band. A large cavity is formed in the tentacle axis, along which the prey is ingested. (D) Cross-section of (C) at the level of the *arrows*. The center is occupied by the prey's cytoplasm. The internal MT form regularly spaced bundles (their lateral expansions may play some role in the sliding movement of the prey toward the cell body). The external MT are widely separated, but maintained in place by bridges (redrawn after Tucker [119])

any fixed structure; they do show, at least for the exterior group, a close connection with a circular rim of dense material located at the distal end. This ring is connected with a circular sheath of "epiplasmic" material, ill defined, which is present all along the tentacle and which may have a contractile function [119]. How can the relative motions of MT and cytoplasmic structures be explained? Certainly not by any contraction of the MT themselves, as was thought at one time [98]. As two groups of MT are present, with bridges between them, it is possible to imagine that these structures are in a dynamic condition, like the dynein bridges in cilia, and that the two groups of MT slide in relation to one another. This mechanism was proposed for *Dendrocometes paradoxus*, the transport of the prey resulting from "repeated sliding back and forth of the inner MT ribbons", as a result of cyclic bridging and release of the MT arms and the membrane of the tentacle [6]. A similar opinion was expressed in a study of *Choanophrya infundibulifera:* however, the connections between the inner MT and the cell membrane limiting the channel in which the prey is transported did not appear to be present in all suctoria, and a possible action of "unresolved cytoplasmic microfilaments" was suggested [54].

In more recent studies, it was first shown that in *Tokophyra* the links between the MT could stretch considerably during feeding, reaching a length eight times greater than in the resting condition. The arms of the inner MT were imagined to play a role in moving the food particles, without any movement of the MT in relation to one another [119]. A more refined study of the changes in the MT during feeding has led to the conclusion that there is no evidence that the two groups of MT slide past one another: the spectacular changes observed during feeding can be explained, as suggested above, by a contraction of structures located in the tentacle pellicle, and which attach to the rim at the extremity of the inner group of MT. This evaginates during feeding as observed in other species, and these changes can be compared to the opening of an umbrella, where the MT represent the rigid, skeletal elements, and the cytoplasmic pellicle the motive element. On the other hand, the sidearms of the inner MT may help the centripetal cytoplasmic flow which has been observed close to the alimentary channel [120].

The studies on this remarkable group of protozoa lead thus to the conclusion that MT are mainly skeletal structures, capable of being displaced (and even deformed) by outside forces located in the cytoplasm. Links between the MT ensure the rigidity of the whole structure (and the rigidity of the long tentacles) and may help to move cytoplasm and perhaps the food particles. Movement is thus the result of contractile proteins associated with the MT and not of an intrinsic property of these organelles.

Other studies on movements in protozoa—which it is not possible to study in detail here—show relations between contractile elements and MT structures in which sliding filament motions are observed. The study of the contraction of the large protist *Stentor coeruleus* demonstrates links between contractile fibers (myonemes) and complex associations of MT (the *km* fibers) which are attached to the basal bodies of the very numerous cilia of this organism. During contraction, the myonemes shorten, and the groups of MT slide in relation to one another, although they are linked by definite "bridges". These cannot be considered stable, and may be "mechanicochemical units", although the contractile elements are the myonemes [58]. These MT are stable components, resistant to colchicine and to cytochalasin.

7.5.4 Axoneme-Associated Movements

The contractile motions of intracellular axostyles made of many MT, as observed for instance in *Pyrsonympha* [35], have many points of resemblance with ciliary movement. Several different types of motion associated with the long axopodia of heliozoa have been described; some are related to MT, others apparently not:

a) Retraction: as mentioned in Chapter 4, this may be very rapid (1000 μm/sec, [88]) and results from the sudden disassembly of all the MT of the axopodium. It may be followed by elongation, which is slower and linked to the formation of new MT. These movements are used in the capture of prey by the axopodia.

b) Flexion: axopodia, in *Actinophrys sol,* may be slightly bent and elastically resist an externally applied force. A more severe bending leads to a local change, and this bend is seen to move gradually toward the tip of the axopod [88].

c) Beating: in several species, the axopodia are put to use for crawling movements of the cell, as described above (Chap. 4; [21, 56]). Rhythmic beatings, comparable to those of cilia and of the intracellular axonemes of *Pyrsonympha* [35, 67], may involve the sliding of MT in relation to one another [73]; dynein or a similar type of ATPase appears to take part in this activity [13, 14, 83]. Various particles, food or specialized structures such as kinetocysts, may move in both directions along the axopodia of heliozoa: it is this type of movement which will be discussed now.

This translation of particles is of the "saltatory" type, and fulfils the criteria mentioned above [35]. However, the particles traveling along the axopodia of *Echinosphaerium nucleofilum* move in both directions, and they are often located too far from the axoneme, without any attachment to the MT, for any acceptable "sliding" theory. Their course is also straight, while the spiral structure of the axoneme (cf. Chap. 4) would suggest a more complex pathway. Although these movements are not prevented by cytochalasin B, the fibrils present in the cytoplasm may provide the motive force [35]. A microcinematographic study of these movements has also indicated that they are independent of the MT. They are visible in articifial "axopodia" made by extension of the cytoplasm, and are not affected by colchicine [33].

However, in a recent study of the regenerating MT of the axopodia of a radiolaria, *Thalassicola nucleata* Huxley, the newly formed MT displayed remarkable sidearms with a nodular extremity, which are rather similar to the radial links described in cilia and flagella: these arms may, by an alternating movement, provide the force necessary for the movements of the neighboring cytoplasm and particles [20]. Once again, in this hypothesis, the movement bears some relation to the MT, but the motive force is provided by another structure attached to the MT. These appear, in the axopodia of heliozoa and radiolaria, much more as skeletal elements than as the motors of cytoplasmic displacement. However, they are peculiar supporting structures, capable of retraction in a few seconds.

Saccinobaculus, an intestinal parasite of wood-eating roaches, is particularly well suited for studies of the contractile properties of axostyles [13]. These may be isolated, and their contraction studied in vitro. The axoneme is made of many MT, which are linked in two directions, sometimes forming a hexagonal lattice. This structure, when contracted, assumes a helical shape; this can be induced by ATP (from 2×10^{-6} to 1×10^{-2} M), while GTP is weakly active and other nucleosides inactive. Divalent cations are required, Mg^{2+} ($1 - 5 \times 10^{-3}$ M) giving the best results, and competing with Ca^{2+}, which alone fails to induce any motility. The ratio of Ca^{2+}/Mg^{2+} must be larger than 1 : 3. Undulatory movements, lasting several minutes, may be obtained in these conditions. These results, confirming earlier studies on the same species [83], show the necessity of free sulfhydryl groups for motility, as indicated by the inhibitory effects of parachloro-mercurybenzoic acid and mersalyl, which are reversed by cysteine. Sodium fluorate also inhibits all motility. Electron microscopical studies of the isolated axonemes show that bridging is more evident in the contracted than in the extended condition; these bridges are completely extracted by 0.5 M KCl, which reduces the ATPase activity associated with the axoneme. Anti-MT drugs such as colchicine, VLB, and podophyllotoxin, have no effect on the structure or the motility of the axonemes. It is concluded that contractility is associated with an ATPase which requires ATP in a millimolar concentration, and Mg^{2+} ions. A model of the mechanochemical action of the bridges between MT has been presented, taking into account these biochemical requirements and the formation of temporary links between MT, producing movement by sliding motions of the MT. While the nature of the ATPase is not known, the resemblance between these movements and those of cilia is striking, although the exact mechanism of action of the cross-bridges remains poorly understood [13]. This model is of interest in relation to other MT-associated movements.

The possibility that other contractile proteins different from actomyosin may play a role is indicated by several studies on protozoa. In the large species, *Sticholonche zanclea,* where the axopodia extend outward from specialized zones of the nuclear membrane (cf. Chap. 4), and act as oars to propel the organism, the axonemes are made of a dense hexagonal pattern of closely linked MT which are attached, at the base of the axopodia, to a dense material which turns in a depression of the nuclear membrane. The movements of this structure, comparable to a joint, result from contractions of thin filaments (about 3 nm) attached to the membrane and the axoneme. These do not fix heavy meromyosin, and apparently belong to another group of contractile proteins. In this cell, the MT are essentially supportive structures, giving rigidity to the long axopodia [21].

7.6 Cell Motility and Locomotion

Movement of whole cells, like ameboid motion, is the result of complex interactions between cytoplasm, contractile proteins, and MT. Actin, myosin, and tropomyosin have been demonstrated in various cells, either biochemically or by immunofluorescence (cf. [124]). In an early paper on the movements of rat embryo cells in tissue culture, the 7.5 nm fibrils, often attached to the cell membrane, were already considered the source of energy for intra- and extracellular movements and cell shape [17].

Several studies with colchicine indicated that the destruction or disassembly of MT could change the aspect of cell motion without suppressing ameboid activity. In HeLa cells treated by 10^{-5} M colchicine, saltatory motion is arrested although membrane motion, and the formation and extension of microvilli, are not modified [36].

Fibroblasts and macrophages, under the effect of MT poisons, do not become immotile, but lose the polarized aspect of their motion: their cytoplasm maintains its contractility, but locomotion in a definite direction is not seen any more. Ruffling becomes evenly distributed on all the cell surface, instead of being mainly active at the leading edge [10, 30, 39, 121].

The migration of monocytes in vitro, in Boyden chambers, is enhanced by low concentrations of colchicine or VLB: this has been supposed to result from a facilitated deformation of the cells secondary to MT destruction [27]. More complex mechanisms may however play a role in the control of chemotaxis of granulocytes. In studies with the Boyden chamber, human polymorphonuclears have been shown to require Ca^{2+} and Mg^{2+} for optimal chemotaxis, and doses of colchicine and cytochalasin which decreased the cell movements also decreased the release of Ca^{2+} from the cells. In these experiments, the chemotactic factor was the fraction C5a of complement, which was found to increase significantly the number of MT in the centriolar region. This would be favored by the release of Ca^{2+}. The mechanical work of chemotaxis would be provided by MF (actin ?), while the MT would play the role of vectors of motion [40].

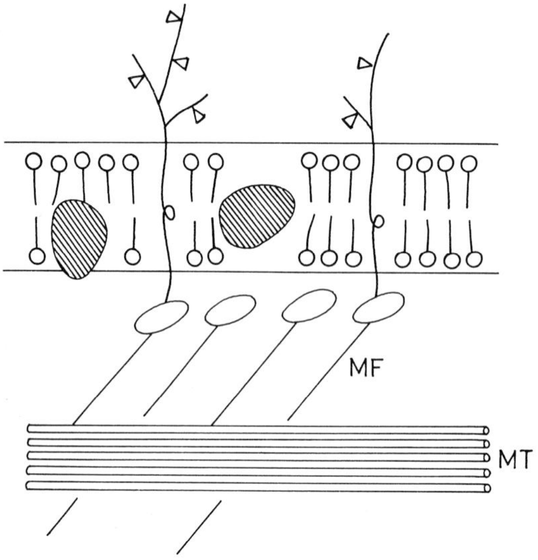

Fig. 7.8. Model of modulation of surface receptors by MT and MF: in this model MT form, at a short distance from the cell membrane, a support on which MF (actomyosin?) become loosely attached at one end, the other being fixed to the extremity of long protein molecules which cross the cell membrane [showing its double lipidic layers and containing large protein molecules (*hatched*)]. These molecules may act as receptors for lectins, antibodies, etc. Their lateral movement would require the integrity of both MF and MT ("microfilament–microtubular system") (redrawn from Yahara and Edelman [129])

In quite different cells, gliocytes, colchicine was observed, in tissue culture, to destroy the MT without affecting cell motility, while cytochalasin destroyed reversibly the 5 nm MF [114] and arrested all ameboid motion. Similar results have been described with various types of cells (cf. [46, 47]).

The relations between MT and MF have been the subject of several recent reviews and monographs [38, 48, 60, 64]. It becomes more and more evident that motion results from an interaction between actin and myosin (although the possibility of actin – actin interactions should not be excluded), and that MT poisons modify cell motility indirectly, by destroying the MT which maintain the general structure and polarity of the cell. Most cells appear now to be contractile, and the MF seem to be the main agents of movement: this conclusion is not in contradiction to what has been said above, and more and more, we are led to a conclusion which was already that of Buckley and Porter in 1967, and which was summarized, in concluding a work on macrophage motility, by the sentence: "The appearances fit the view that microfibrils are contractile and microtubules skeletal in function" [22].

The facts which will be studied in Chapter 8 in relation to endo- and exocytosis are in agreement with this conclusion; it will remain to be seen whether the movements of chromosomes at mitosis can be explained without assuming that the spindle MT provide the energy for mitosis.

References

1. Afzelius, B. A., Eliasson, R., Johnsen, Ø., Lindholmer, C.: Lack of dynein arms in immotile human spermatozoa. J. Cell Biol. **66,** 225 – 232 (1975)
2. Allen, C., Borisy, G. G.: Flagellar motility in *Chlamydomonas*: reactivation and sliding in vitro. J. Cell Biol. **63,** 5 a (1974)
3. Allen, R. D., Kamiya, N. (eds.): Primitive Motile Systems in Cell Biology. New York-London: Acad. Press 1964
4. Bairati, A.: Comparative ultrastructure of ectoderm-derived filaments. Boll. Zool. **39,** 283 – 308 (1972)
5. Baker, P. C., Schroeder, T. E.: Cytoplasmic filaments and morphogenetic movements in the amphibian neural tube. Dev. Biol. **15,** 432 – 450 (1967)
6. Bardele, C. F.: Cell cycle, morphogenesis, and ultrastructure in the pseudoheliozoan *Clathrulina elegans*. Z. Zellforsch. **130,** 219 – 242 (1972)
7. Bardele, C. F.: Transport of materials in the suctorian tentacle. Symp. Soc. Exp. Biol. **27,** 191 – 208 (1974)
8. Bardele, C. F., Grell, K. G.: Elektronmikroskopische Beobachtungen zur Nahrungsaufnahme bei dem Suktor *Acineta tuberosa* Ehrenberg. Z. Zellforsch. **80,** 108 – 123 (1967)
9. Behnke, O., Kristensen, B. I., Weilsen, L. E.: Electron microscopical observations on the actinoid and myosinoid filaments in blood platelets. J. Ultrastruct. Res. **37,** 351 – 369 (1971)
10. Bhisey, A. N., Freed, J. J.: Ameboid movement induced in cultured macrophages by colchicine or vinblastine. Exp. Cell Res. **64,** 419 – 429 (1971)
11. Bickle, D., Tilney, L. G., Porter, K. R.: Microtubules and pigment migration in the melanophores of *Fundulus heteroclitus* L. Protoplasma **61,** 322 – 345 (1966)
12. Bitensky, M. W., Keirns, C. N., Keirns, J. J.: The amphibian melanocyte: microtubules, cyclic AMP, and organelle translocation. Ann. N. Y. Acad. Sci. **253,** 685 – 691 (1975)
13. Bloodgood, R. A.: Biochemical basis of axostyle motility. Cytobios **14,** 101 – 120 (1975)
14. Bloodgood, R. A., Miller, K. R.: Freeze-fracture of microtubules and bridges in motile axostyles. J. Cell Biol. **62,** 660 – 671 (1974)
15. Borgers, M., De Brabander, M. (eds.): Microtubules and Microtubule Inhibitors. Amsterdam-Oxford: North-Holland; New York: Am. Elsevier 1975
16. Brokaw, C. J.: Cross-bridge behaviour in a sliding filament model for flagella. In: Molecules and Cell Movement (eds.: S. Inoué, E. E. Stephens), pp. 165 – 180. New York: Raven Press; Amsterdam: North-Holland 1975
17. Buckley, I. K., Porter, K. R.: Cytoplasmic fibrils in living cultured cells. A light and electron microscope study. Protoplasma **64,** 349 – 380 (1967)
18. Burnside, M. B.: Possible roles of microtubules and actin filaments in retinal pigmented epithelium. Exp. Eye Res. **23,** 257 – 275 (1976)
19. Burnside, B.: Microtubules and actin filaments in teleost visual cone elongation and contraction. J. Supramol. Struct. **5,** 257 – 275 (1976)
20. Cachon, J., Cachon, M.: Rôle des microtubules dans les courants cytoplasmiques des axopodes. C. R. Acad. Sci. (Paris) D **280,** 2341 – 2344 (1975)
21. Cachon, J., Cachon, M., Tilney, L. G., Tilney, M. S.: Movement generated by interactions between the dense material at the ends of microtubules and non-actin-containing microfilaments in *Sticholonche zanclea*. J. Cell Biol. **72,** 314 – 338 (1977)
22. Carr, I.: The fine structure of microfibrils and microtubules in macrophages and other lymphoreticular cells in relation to cytoplasmic movement. J. Anat. **112,** 383 – 390 (1972)

23. Carter, S. B.: Effects of cytochalasins on mammalian cells. Nature (London) **213,** 261 – 264 (1967)
24. Castrucci, A. M. de L.: Chromatophores of the teleost *Tilapia melanopleura*. II. The effect of chemical mediators, microtubule-disrupting drugs and ouabain. Comp. Biochem. Physiol. **50,** 457 – 462 (1975)
25. Chang, C. M., Goldman, R. D.: The localization of actin-like fibers in cultured neuroblastoma cells as revealed by heavy meromyosin binding. J. Cell Biol. **57,** 867 – 874 (1973)
26. Ciba Foundation Symposium 14. Locomotion of Tissue Cells. Amsterdam-London-New York: Elsevier, Excerpta Medica, North-Holland Associated Scientific Publishers 1973
27. Crispe, I. N.: The effect of vinblastine, colchicine and hexylene glycol on migration of human monocytes. Exp. Cell Res. **100,** 443 – 446 (1976)
28. Curry, A., Butler, R. D.: The ultrastructure, function and morphogenesis of the tentacle in *Discophrya* sp. (Suctorida) cileatea. J. Ultrastruct. Res. **56,** 164 – 176 (1976)
29. Dales, S., Chardonnet, Y.: Early events in the interaction of adenoviruses with HeLa cells. IV. Association with microtubules and the nuclear pore complex during vectorial movement of the inoculum. Virology **56,** 465 – 483 (1973)
30. De Brabander, M.: Onderzoek naar de rol van microtubuli in gekweekte cellen met behulp van een nieuwe synthetische inhibitor van tubulin-polymerisatie. Thesis, Brussels 1977
31. De Brabander, M., Aerts, F., van de Veire, R., Borgers, M.: Evidence against interconversion of microtubules and filaments. Nature **253,** 119 – 120 (1975)
32. Dupraw, E.J.: The organization of honey bee embryonic cells. I. Microtubules and amoeboid activity. Dev. Biol. **12,** 53 – 71 (1965)
33. Edds, K.: Particle movements in artificial axopodia of *Echinosphaerium nucleofilum*. J. Cell Biol. **59,** 88 a (1973)
34. Fingerman, M., Fingerman, S. W., Lambert, D. T.: Colchicine, cytochalasin B, and pigment movements in ovarian and integumentary erythrophores of the prawn, *Palaemonetes vulgaris*. Biol. Bull. **149,** 165 – 177 (1975)
35. Fitzharris, T. P., Bloodgood, R. A., McIntosh, J. R.: The effect of fixation on the wave propagation of the protozoan axostyle. Tissue and Cell **4,** 219 – 225 (1972)
36. Freed, J. J., Bhisey, A. N., Lebowitz, M. M.: The relation of microtubules and microfilaments to the motility of cultured cells. J. Cell Biol. **39,** 46 a (1968)
37. Freed, J. J., Lebowitz, M. M.: The association of a class of saltatory movements with microtubules in cultured cells. J. Cell Biol. **45,** 334 – 354 (1970)
38. Gail, M.: Time lapse studies on the motility of fibroblasts in tissue culture. In: Locomotion of Tissue Cells. Ciba Foundation Symposium 14, pp. 287 – 310. Amsterdam-London-New York: Elsevier, Excerpta Medica, North-Holland 1973
39. Gail, M. H., Boone, C. W.: Effect of colcemid on fibroblast motility. Exp. Cell Res. **65,** 221 – 227 (1971)
40. Gallin, J. I., Rosenthal, A. S.: Regulatory role of divalent cations in human granulocyte chemotaxis. Evidence for an association between calcium exchanges and microtubule assembly. J. Cell Biol. **62,** 594 – 609 (1974)
41. Gibbons, I. R.: The relationship between the fine structure and direction of beat in the gill cilia of a lamellibranch mollusk. J. Biophys. Biochem. Cytol. **11,** 179 – 205 (1961)
42. Gibbons, I. R.: The organization of cilia and flagella. In: Molecular Organization of Biological Function. (ed.: J. M. Allen), pp. 211 – 237. New York: Evanston; London: Harper and Row 1967
43. Gibbons, I. R.: Molecular basis of flagellar motility in sea urchin spermatozoa. In: Molecules and Cell Movement. (eds.: S. Inoué, R. E. Stephens) pp. 207 – 232, New York: Raven Press; Amsterdam: North-Holland 1975
44. Gibbons, B. H., Gibbons, I. R.: The effect of partial extraction of dynein arms on the movement of reactivated sea-urchin spermatozoa. J. Cell Sci. **13,** 337 – 358 (1973)
45. Glenner, G. G., Page, D. L.: Amyloid, amyloidosis, and amyloidogenesis. Intern. Rev. Exp. Path. **15,** 2 – 93 (1976)
46. Goldman, R. D., Berg, G., Bushnell, A., Cheng Ming Chang, Dickerman, L., Hopkins, N., Miller, M. L., Pollack, R., Wang, E.: Fibrillar systems in cell motility. In: Locomotion of Tissue Cells. Ciba Foundation Symposium 14, pp. 83 – 108. Amsterdam-London-New York: Elsevier, Excerpta Medica, North-Holland 1973
47. Goldman, R. D., Knipe, D. M.: Functions of cytoplasmic fibers in non-muscle cell motility. Cold Spring Harbor Symp. Quant. Biol. **27,** 523 – 534 (1972)
48. Goldman, R., Pollard, T., Rosenbaum, J. (eds.): Cell Motility. Cold Spring Harbor Lab. 1976

49. Gonatas, N. K., Margolis, G., Kilham, L.: Reovirus type III encephalitis: observations of virus-cell interactions in neural tissues. II. Electron microscopic studies. Lab. Invest. **24,** 101 – 109 (1971)
50. Gordon, S., Cohn, Z.: Macrophage-melanocyte heterokaryons. I. Preparation and properties. J. Exp. Med. **131,** 981 – 1003 (1970)
51. Green, L.: Mechanism of movements of granules in melanocytes of *Fundulus heteroclitus.* Proc. Natl. Acad. Sci. USA. **59,** 1179 – 1186 (1968)
52. Grell, K. G.: Protozoology. Berlin-Heidelberg-New York: Springer 1973
53. Himes, R. H., Houston, L. L.: The action of cytochalasin A on the in vitro polymerization of brain tubulin and muscle G-actin. J. Supramol. Struct. **5,** 81 – 90 (1976)
54. Hitchen, E. T., Butler, R. D.: Ultrastructural studies of the commensal suctorian, *Choanophrya infuldibulifera* Hartog. I. Tentacle structure, movement and feeding. Z. Zellforsch. **144,** 37 – 57 (1973)
55. Hitchen, E. T., Butler, R. D.: The ultrastructure and function of the tentacle in *Rhyncheta cyclopum* Zenker (Ciliatea, Suctoria) J. Ultrastruct. Res. **46,** 279 – 295 (1974)
56. Hollande, A., Cachon, J., Cachon, M., Valentin, J.: Infrastructure des axopodes et organisation générale de *Sticholonche zanclea* Hertwig (Radiolaire Sticholonchidae). Protistologica **3,** 155 – 166 (1967)
57. Holmes, K. V., Choppin, P. W.: On the role of microtubules in movement and alignment of nuclei in virus-induced syncytia. J. Cell Biol. **39,** 526 – 543 (1968)
58. Huang, B., Pitelka, D. R.: The contractile process in the ciliate *Stentor coeruleus.* I. The role of microtubules and filaments, J. Cell Biol. **57,** 704 – 728 (1973)
59. Hyams, J., Borisy, G. G.: The isolation and reactivation of the flagellar apparatus of *Chlamydomonas reinhardi.* J. Cell Biol. **63,** 150 a (1974)
60. Inoué, S., Stephens, R. E. (eds.): Molecules and Cell Movement. New York: Raven Press; Amsterdam: North-Holland 1975
61. Ishikawa, H., Bischoff, R., Holtzer, H.: Formation of arrowhead complexes with heavy meromyosin in a variety of cell types. J. Cell Biol. **43,** 312 – 328 (1969)
62. Jimbow, K., Pathak, M. A., Fitzpatrick, T. B.: Effect of ultraviolet on the distribution pattern of microfilaments and microtubules and on the nucleus in human melanocytes. Yale J. Biol. Med. **46,** 411 – 426 (1973)
63. Johnson, L. S., Sinex, F. M.: On the relationship of brain filaments to microtubules. J. Neurochem. **22,** 321 – 326 (1974)
64. Komnick, H., Stockem, W., Wohlfarth-Botterman, K. E.: Cell motility: mechanisms in protoplasmic streaming and ameboid movement. Intern. Rev. Cytol. **34,** 169 – 252 (1973)
65. Lambert, D. T., Crowe, J. H.: Colchicine and cytochalasin B; effect on pigment granule translocation in melanophores of *Uca pugilator* (Crustacea; Decapoda). Comp. Biochem. Physiol. **46,** 11 – 16 (1973)
66. Lambert, D. T., Fingerman, M.: Evidence of a non-microtubular colchicine effect in pigment granule aggregation in melanophores of the fiddler crab, *Uca pugilator.* Comp. Biochem. Physiol. C. **53,** 25 – 28 (1976)
67. Langford, G. M., Inoué, S., Sabran, I.: Analysis of axostyle motility in *Pyrsonympha vertens.* J. Cell Biol. **59,** 185 a (1973)
68. Lazarides, E., Lindberg, U.: Actin is the naturally occurring inhibitor of deoxyribonuclease I. Proc. Natl. Acad. Sci. U.S.A. **71,** 4742 – 4746 (1974)
69. Lebeux, Y. J., Willemot, J.: An ultrastructural study of the microfilaments in rat brains by means of heavy meromyosin labelling. I. The perikaryon, the dendrites and the axon. II. The synapses. Cell Tiss. Res. **160,** 1 – 36; 37 – 68 (1975)
70. Lindemann, C. B., Gibbons, I. R.: Adenosine triphosphate-induced motility and sliding of filaments in mammalian sperm extracted with triton X-100. J. Cell Biol. **65,** 147 – 162 (1975)
71. Luftig, R. B., Weihing, R. R.: Adenovirus binds to rat brain microtubules in vitro. J. Virol. **16,** 696 – 706 (1975)
72. McGuire, J., Moellmann, G., McKeon, F.: Cytochalasin B and pigment granule translocation. Cytochalasin B reverses and prevents pigment granule dispersion caused by dibutyryl cyclic AMP and theophylline in *Rana pipiens* melanocytes. J. Cell Biol. **52,** 754 – 758 (1972)
73. McIntosh, J. R., Ogata, E. S., Landis, S. C.: The axostyle of *Saccinobaculus.* I. Structure of the organism and its microtubule bundle. J. Cell Biol. **56,** 304 – 323 (1973)
74. McIntosh, J. R., Hepler, R. K., Wie, D. G. van: Model for mitosis. Nature **224,** 659 – 663 (1969)
75. Macgregor, H. C., Stebbings, H.: A massive system of microtubules associated with cytoplasmic movement in telotrophic ovarioles. J. Cell Sci. **6,** 431 – 449 (1970)

76. Magun, B.: Two actions of cyclic MAP on melanosome movement in frog skin. Dissection by cytochalasin B. J. Cell Biol. **57,** 845 – 858 (1973)
77. Malawista, S. E.: On the action of colchicine. The melanocyte model. J. Exp. Med. **122,** 361 – 384 (1965)
78. Malawista, S. E.: The melanocyte model. Colchicine-like effects of other antimitotic agents. J. Cell Biol. **49,** 848 – 855 (1971)
79. Malawista, S. E., Asterita, H., Marsland, D.: Potentiation of the colchicine effect on frog melanocytes by high hydrostatic pressure. J. Cell. Physiol. **68,** 13 – 17 (1966)
80. Moellmann, G., McGuire, J., Lerner, A. B.: Intracellular dynamics and the fine structure of melanocytes. With special reference to the effects of MSH and cyclic AMP on microtubules and 10-nm filaments. Yale J. Biol. Med. **46,** 337 – 360 (1973)
81. Moellmann, G., McGuire, J.: Correlation of cytoplasmic microtubules and 10-nm filaments with the movement of pigment granules in cutaneous melanocytes of *Rana pipiens.* Ann. N. Y. Acad. Sci. **253,** 711 – 722 (1975)
82. Mohri, H.: The function of tubulin in motile systems. Bioch. Biophys. Acta **456,** 85 – 127 (1976)
83. Mooseker, M. S., Tilney, L. G.: Isolation and reactivation of the axostyle. Evidence for a dynein-like ATPase in the axostyle. J. Cell Biol. **56,** 13 – 26 (1973)
84. Murphy, D. B.: The mechanism of microtubule-dependent movement of pigment granules in teleost chromatophores. Ann. N. Y. Acad. Sci. **253,** 692 – 701 (1975)
85. Murphy, D. B., Tilney, L. G.: The role of microtubules in the movement of pigment granules in teleost melanophores. J. Cell Biol. **61,** 757 – 779 (1974)
86. Nagai, R., Rebhun, L. I.: Cytoplasmic microfilaments in streaming *Nitella* cells. J. Ultrastruct. Res. **14,** 571 – 589 (1966)
87. Novales, R. R., Fujii, R.: A melanin-dispersing effect of cyclic adenosine monophosphate on *Fundulus* melanophores. J. Cell. Physiol. **75,** 133 – 135 (1970)
88. Ockleford, C. D., Tucker, J. B.: Growth, breakdown, repair, and rapid contraction of microtubular axopodia in the heliozoan *Actinophrys sol.* J. Ultrastruct. Res. **44,** 369 – 387 (1973)
89. Pollard, T. D.: Functional implications of the biochemical and structural properties of cytoplasmic contractile proteins. In: Molecules and Cell Movement (eds.: S. Inoué, R. E. Stephens) pp. 259 – 286. New York: Raven Press 1975
90. Porter, K. R.: Microtubules in intracellular locomotion. In: Locomotion of Tissue Cells. Ciba Foundation Symposium 14, pp. 149 – 170. Amsterdam-London-New York: Elsevier, Excerpta Medica, North-Holland 1973
91. Porter, K. R.: Motility in cells. In: Cell Motility (eds.: R. Goldman, T. Pollard, J. Rosenbaum), pp. 1 – 23. Cold Spring Harbor Lab. 1976
92. Rebhun, L. I.: Saltatory particle movements and their relation to the mitotic apparatus. In: The Cell in Mitosis (ed.: L. Levine), pp. 67 – 106. New York: Acad. Press 1963
93. Rebhun, L. I.: Saltatory particle movements in cells. In: Primitive Motile Systems in Cell Biology (eds.: R. D. Allen, N. Kamiya), pp. 503 – 525. New York-London: Acad. Press 1964
94. Rebhun, L. I.: Polarized intracellular particle transport: saltatory movements and cytoplasmic streaming. Intern. Rev. Cytol. **32,** 93 – 139 (1972)
95. Robison, W. G.: Microtubules in relation to the motility of a sperm syncytium in an armored scale insect. J. Cell Biol. **29,** 251 – 265 (1966)
96. Robison, W. G.: Microtubular patterns in spermatozoa of coccid insects in relation to bending. J. Cell Biol. **52,** 66 – 83 (1972)
97. Robison, W. G., Charlton, J. S.: Microtubules, microfilaments, and pigment movement in the chromatophores of *Palaemonetes vulgaris* (Cristacea) J. Exp. Zool. **186,** 279 – 304 (1973)
98. Rudzinska, M. A.: The fine structure and function of the tentacle in *Tokophyra infusionum.* J. Cell Biol. **25,** 459 – 477 (1965)
99. Saavedra, S., Renaud, F.: Studies on reactivated cilia. I. The utilization of various nucleoside triphosphates during ciliary movement. Exp. Cell Res. **90,** 439 – 442 (1975)
100. Satir, P.: Structure and function of cilia and flagella. Protoplasmatologia **3,** 1 – 52 (1965)
101. Satir, P.: Studies on cilia. II. Examination of the distal region of the ciliary shaft and the role of filaments in motility. J. Cell Biol. **26,** 805 – 834 (1965)
102. Satir, P.: Studies on cilia. III. Further studies on the cilium tip and a "sliding filament" model of ciliary motility. J. Cell Biol. **39,** 77 – 94 (1968)
103. Satir, P.: The present status of the sliding microtubule model of ciliary motion. In: Cilia and Flagella (ed.: M. A. Sleigh) London and New York: Acad. Press 1974
104. Schliwa, M.: Microtubule distribution and melanosome movements in fish melanophores. In: Microtubules and Microtubule Inhibitors. (eds.: M. Borgers, M. De Brabander) pp. 215 – 228. Amsterdam-Oxford: North-Holland 1975

105. Schliwa, M., Bereiter-Hahn, J.: Pigment movements in fish melanophores: morphological and physiological studies. II. Cell shape and microtubules. Z. Zellforsch. **147,** 107 – 125 (1973)
106. Schliwa, M., Bereiter-Hahn, J.: Pigment movements in fish melanophores: morphological and physiological studies. III. The effects of colchicine and vinblastine. Z. Zellforsch. **147,** 127 – 148 (1973)
107. Schliwa, M., Bereiter-Hahn, J.: Pigment movements in fish melanophores: morphological and physiological studies. IV. The effect of cyclic adenosine monophosphate on normal and vinblastine treated melanophores. Cell Tissue Res. **151,** 423 – 432 (1974)
108. Schliwa, M., Bereiter-Hahn, J.: Pigment movements in fish melanophores: morphological and physiological studies. V. Evidence for a microtubule-independent contractile system. Cell Tissue Res. **158,** 61 – 74 (1975)
109. Schrével, J., Besse, C.: Un type flagellaire fonctionnel de base 6+0 J. Cell Biol. **66,** 492 – 507 (1975)
110. Schroeder, T. E.: The contractile ring. I. Fine structure of dividing mammalian (HeLa) cells and the effects of cytochalasin B. Z. Zellforsch. **109,** 431 – 449 (1970)
111. Schroeder, T. E.: Dynamics of the contractile ring. In: Molecules and Cell Movement (eds.: S. Inoué, R. E. Stephens) pp. 305 – 334. New York: Raven Press; Amsterdam: North-Holland 1975
112. Sleigh, M. A. (ed.): Cilia and Flagella. London-New York: Acad. Press 1974
113. Soifer, D., (ed.): The Biology of Cytoplasmic Microtubules. Ann. N. Y. Acad. Sci. **253,** (1975)
114. Spooner, B. S., Yamada, K. M., Wessells, N. K.: Microfilaments and cell locomotion. J. Cell Biol. **49,** 595 – 613 (1971)
115. Stebbings, H., Bennett, C. E.: The sleeve element of microtubules. In: Microtubules and Microtubule Inhibitors (eds.: M. Borgers, M. De Brabander), pp. 35 – 45. Amsterdam-Oxford: North-Holland; New York: Elsevier 1975
116. Summers, K.: ATP-induced sliding of microtubules in bull sperm flagella. J. Cell Biol. **60,** 321 – 324 (1974)
117. Tilney, L. G.: Nonfilamentous aggregates of actin and their association with membranes. In: Cell Motility (eds.: R. Goldman, T. Pollard, J. Rosenbaum), pp. 513 – 528. Cold Spring Harbor Lab. 1976
118. Tilney, L. G., Detmers, P.: Actin in erythrocyte ghosts and its association with spectrin. Evidence for a non-filamentous form of these two molecules in situ. J. Cell Biol. **66,** 508 – 520 (1975)
119. Tucker, J. B.: Microtubule arms and cytoplasmic streaming and microtubule bending and stretching of intertubule links in the feeding tentacle of the suctorian ciliate *Tokophyra.* J. Cell Biol **62,** 424 – 437 (1974)
120. Tucker, J. B., Mackie, J. B.: Configurational changes in helical microtubule frameworks in feeding tentacles of the suctorian ciliate *Tokophyra.* Tissue and Cell **7,** 601 – 612 (1975)
121. Vasiliev, J. M., Gelfand, I. M., Domnina, L. V., Ivanova, O. Y., Komm, S. G., Olshevskaja, L. V.: Effect of colcemid on the locomotory behaviour of fibroblasts. J. Embryol. Exp. Morph. **24,** 625 – 640 (1970)
122. Warner, F. D.: Cross-bridge mechanisms in ciliary motility: the sliding – bending conversion. In: Cell Motility (eds.: R. Goldman, T. Pollard, J. Rosenbaum) pp. 891 – 914. Cold Spring Harbor Lab. 1976
123. Warner, F. D., Satir, P.: The structural basis of ciliary bend formation. Radial spoke positional changes accompanying microtubule sliding. J. Cell Biol. **63,** 35 – 63 (1974)
124. Weber, K.: Visualization of tubulin-containing structures by immunofluorescence microscopy: cytoplasmic microtubules, mitotic figures and vinblastine-induced paracrystals. In: Cell Motility (eds.: R. Goldman, T. Pollard, J. Rosenbaum), pp. 403 – 418. Cold Spring Harbor Lab. 1976
125. Wikswo, M. A., Novales, R. R.: Effect of colchicine on microtubules in the melanophores of *Fundulus heteroclitus.* J. Ultrastruct. Res. **41,** 189 – 201 (1972)
126. Wisniewski, H., Terry, R. D.: An experimental approach to the morphogenesis of neurofibrillary degeneration and the argyrophilic plaque. In: Alzheimer's disease and related conditions (eds.: G. E. Wolstenholme, M. O'Connor), pp. 223 – 240. Ciba Foundation Symposium. London: Churchill 1970
127. Wisniewski, H., Terry, R. D.: Neurofibrillary pathology. J. Neuropath. Exp. Neurol. **19,** 163 – 176 (1970)
128. Witman, G. B., Fay, R., Plummer, J.: *Chlamydomonas* mutants: evidence for the roles of specific axonemal components in flagellar movement. In: Cell Motility (eds.: R. Goldman, T. Pollard, J. Rosenbaum), pp. 969 – 986. Cold Spring Harbor Lab. 1976
129. Yahara, I., Edelman, G. M.: Electron microscopic analysis of the modulation of lymphocyte receptor mobility. Exp. Cell Res. **91,** 125 – 142 (1975)

Chapter 8 Secretion, Exo- and Endocytosis

8.1 Introduction

The first mention of the word "microtubules" was made by Slautterback [103] in a description of secretory cells. However, it was only seven years later that Lacy *et al.* [53] suggested that MT play a role in the secretory activity of the B-cells of the Langerhans islets. They showed that several agents known to modify MT (colchicine, VLB, VCR, and deuterium oxide) inhibited the release of insulin by these cells when stimulated by glucose. Similar findings concerning the thyroid gland were made in 1970 by two groups working independently [79, 122]. As a matter of fact, interest in colchicine and MT was revived in our laboratory by the research in these two fields carried on in Brussels [69, 78]. Since then, a large number of contributions have shown relations between MT and various modes of cell secretion, and have demonstrated that the integrity of MT is necessary for the cell activities related to the release of secretory granules.

This work covers many types of secretion, and it is far from certain that the role of MT is the same in all. In processes which may involve cell motility, membrane changes, cAMP, hormones, and ionic changes, the possible non-specific actions of MT poisons must not be forgotten, especially when large doses are used, or when MT do not appear to be destroyed or disassembled.

In this chapter, the secretion of the Langerhans islets will be discussed first, for it provides a well-studied example of liberation of large granules which move in the cell in relation to MT. The thyroid gland provides quite another type of activity, where phagocytosis of thyroglobulin represents the first step in the liberation of the thyroid hormones. This indicates clearly that phagocytosis (endocytosis) may be closely related to secretion (exocytosis): for this reason, the second part of this chapter will be devoted to polymorphonuclear leukocytes. The liberation of the enzymes of their specific granules—which are lysosomes—is similar to secretion, and cannot be separated from the opposite activity, phagocytosis. Both imply some intervention of contractile proteins (actin ?) and of MT.

Several other aspects of intra- or transcellular transport will be grouped at the end of this chapter, as they are influenced by MT poisons; they may be only indirectly related to secretion. Neurosecretion will be discussed in Chapter 9.

8.2 Endocrine Secretion

8.2.1 The Langerhans Islets

The endocrine tissues of the pancreas have a complex activity, as they secrete two hormones with opposite actions, insulin and glucagon. The B-cells, secreting insulin, have been most thoroughly studied, either in whole animals [23], in tissue culture [41, 82], or in isolated islets maintained in a functional condition by perifusion. The results obtained by this last technique are the most interesting, for the influence of stimulants or inhibitors on the steps of insulin release may be followed and excellent electron microscopic observations are possible.

The liberation of insulin comprises several steps, which are similar to those observed in other merocrine cells, but are particularly well suited for research because of the possibilities of measuring small quantities of insulin, and the good knowledge of the various factors, such as glucose and Ca^{2+}, which stimulate the secretion. The secretory granules are formed in the Golgi region of the B-cells and transported toward the cell membrane in relation to numerous cytoplasmic MT. Before being excreted by emiocytosis—the granules opening at the cell surface by fusion of their membranes with those of the cell—they must cross a submembranous zone which contains a dense web of microfilaments. Their exact nature is still being discussed, but actin has been demonstrated to be present in the islets [28].

When the islets are stimulated by glucose, insulin is liberated in two steps, a rapid, early phase followed by a more prolonged secretion. From the first experiments with MT poisons [55], it was suggested that the early phase consisted in a rapid release of granules already associated with MT or MF, while the later phase implied the transport of newly formed granules within the cell and their association with MT. This was indicated by the fact that VCR had an inhibitory action on glucose-stimulated secretion only after 90 min, while deuterium oxide inhibited insulin release reversibly and immediately [69]. Colchicine (10^{-3} M) was found not to inhibit the first phase, either after glucose or tolbutamide stimulation [55].

While electron microscopy demonstrated topographical relations between MT and insulin secretory granules, the further study of various MT inhibitors clearly confirmed that complex mechanisms were involved (some granules are so close to MT that two authors [34], suggested that they may empty their content *into* the MT, the distal extremities of which, being attached to the cell membrane (cf. Chap. 3), would provide an outlet for the secretion. This hypothesis has received no confirmation).

In an important series of papers, a group comprizing inter alia Malaisse and Orci have enlarged on the first studies of Lacy (cf. reviews in [52, 54, 68, 100]). Their main findings may be summarized in the following way:

a) The secretion of insulin after glucose stimulation is increased by cytochalasin, which modifies the MF of the subapical cell web [115] without affecting the basal release of insulin. The action of cytochalasin, which is reversible, could be to facilitate emiocytosis by modifying, at the secretory pole of the cell, actin-like MF which would provide the motive force for the transport of the granules and their extrusion [117].

b) Colchicine, at smaller doses than those mentioned above (2×10^{-5} M), actually increases the early and late glucose-stimulated secretion of insulin. This is also true after stimulation by the sulfonylurea derivative, gliclazide, and in this condition even doses of 10^{-3} M are still stimulatory [114]. These results can be interpreted if two modalities of transport and extrusion of granules exist, one being independent of any MT activity [105, 116]. This is confirmed by the fact that glucose and gliclazide stimulate insulin release by islets previously treated by D_2O and cytochalasin [116].

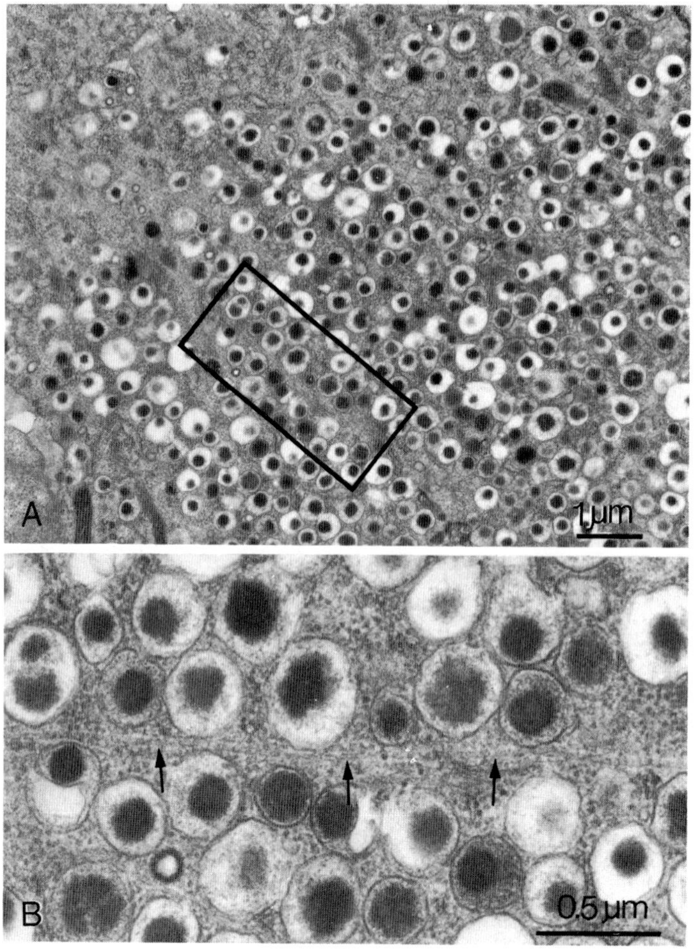

Fig. 8.1 A, B. Monolayer culture of neonatal rat pancreas. (A) Low power magnification of a B-cell. (B) MT, indicated by *arrows,* between two rows of secretory granules (from Malaisse et al. [68])

Glucose also increases the amount of assembled tubulin in the islets, as demonstrated by a new technique of dosage of free and polymerized tubulin [88].

c) Similar findings have been made with VCR: if it acts during less than 30 min, it increases secretion, and becomes inhibitory only after 2 h: this suggests a role of MT in the slow release of secretory granules [18]. This result is in agreement with the observations made on whole animals, where VLB decreases the basal secretion of insulin while having no action on the secretion stimulated by sulfonylurea: in this last condition, MT would play only a minor role [23].

d) The islets of spiny mice (*Acomys cahirinus*) contain less tubulin than in normal animals, as indicated by the smaller volume of VCR-induced crystalline inclusions [70]. An impairment of the MT could afford one explanation of the diabetes-

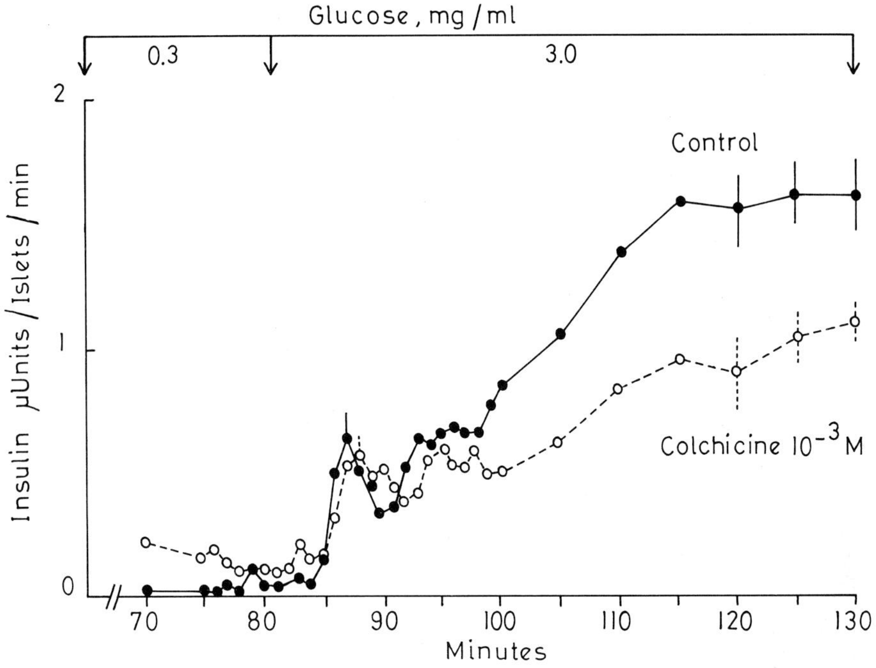

Fig. 8.2. Effect of colchicine on the first and second phase of glucose-induced insulin secretion in perifused Langerhans islets of the rat. Colchicine (10^{-3} M) does not inhibit the first phase, but produces a significant inhibition of the second phase (redrawn from Lacy et al. [55])

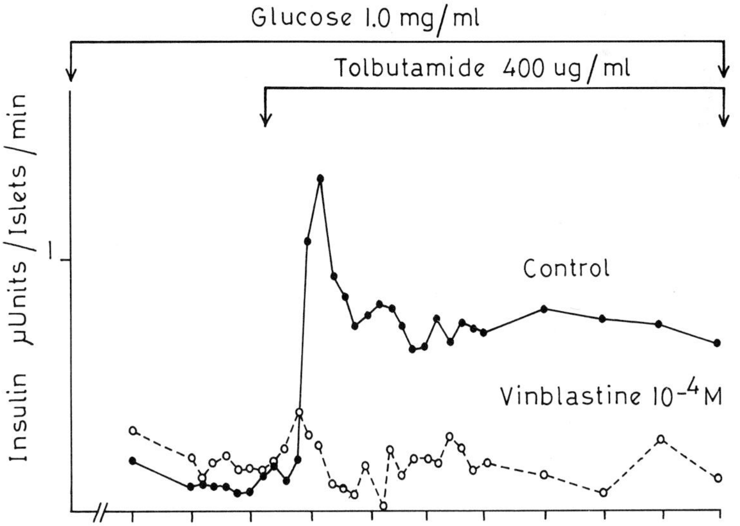

Fig. 8.3. Effect of VLB on tolbutamide-induced insulin secretion in perifused rat Langerhans islets. The first phase of secretion is completely suppressed, as is the elevated basal secretion following perifusion with tolbutamide (redrawn from Lacy et al. [55])

like syndrome found in these animals. Their islets do react normally to glucose after cytochalasin [90] and an initial decreased sensitivity to glucose may explain the functional defect.

These results, as discussed recently [68], lead to the conclusion that the secretion of insulin (stimulated physiologically by glucose) involves two steps: a rapid phase related to the liberation of the granules located close to the cell membrane, in the web of MF, which would imply the action of actin-like molecules without any intervention of MT; and a slow phase, necessitating the transport of other granules from the Golgi zone to the cell surface. This would take place along the MT acting as channels [45] along which the granules would slide, possibly through the action of MT-associated contractile proteins. This conclusion is similar to that described for the transport of melanin granules (Chap. 7) and gives to the MT the role of a scaffolding linked with the granule displacement. Some of the actions of MT poisons may result from the disruption of this scaffolding, while others (specially with large doses) may result from effects either on MF or on cell membranes.

The relations of tubulins to glucose-stimulated secretion may be more complex, as dosage of tubulin in the Langerhans islets of rats by a radioimmunological method has indicated that its synthesis was stimulated by glucose and by cAMP, and decreased by fasting. In fact, in the islets, tubulin appears as a most actively synthetized protein, quantitatively comparable to proinsulin synthesis, as demonstrated by ^{3}H-leucine incorporation. cAMP would at the same time stimulate tubulin synthesis and MT assembly in these cells [88].

It is remarkable that the secretion of glucagon by the A-cell is apparently also linked with MT but obeys different laws. It has not been so extensively studied: colchicine and VLB disrupt the MT of A-cells and increase the secretion of glucagon in guinea-pig islets incubated in vitro [20]. Colchicine (10^{-3} – 10^{-5} M) destroys in vitro the MT of islets A- and B-cells of guinea-pigs and disperses the Golgi apparatus of the cells, a fact which may indirectly affect the secretory activity [76]. Cytochalasin is without effect on glucagon secretion in duct-ligated pancreases of rats, while it potentiates arginine-induced secretion, also stimulated by colchicine and VLB (10^{-4} M) and inhibited by deuterium oxide. These results suggest, like in B-cells, the implication of MT and MF in secretion, but do not explain the stimulatory action of MT poisons [57].

The perifused Langerhans islet is certainly a most interesting tool for the study of cell secretion, and from all facts collected in the last few years, it is clear that MT play an important role in continued secretion, with other proteins (probably actin) providing the motive force.

8.2.2 The Thyroid

The secretory activity of the thyroid is more complex than that of other endocrine glands, as it takes place in two different steps: a specific protein, thyroglobulin, is synthetized in the endoplasmic reticulum of the thyroid cells, iodinated at the cell surface, and stored as colloid in the extracellular vesicles of the gland. The hormones (triiodotyronine and thyroxine) are liberated from the thyroglobulin and secreted in the bloodstream by a complex mechanism in which thyroglobulin is captured

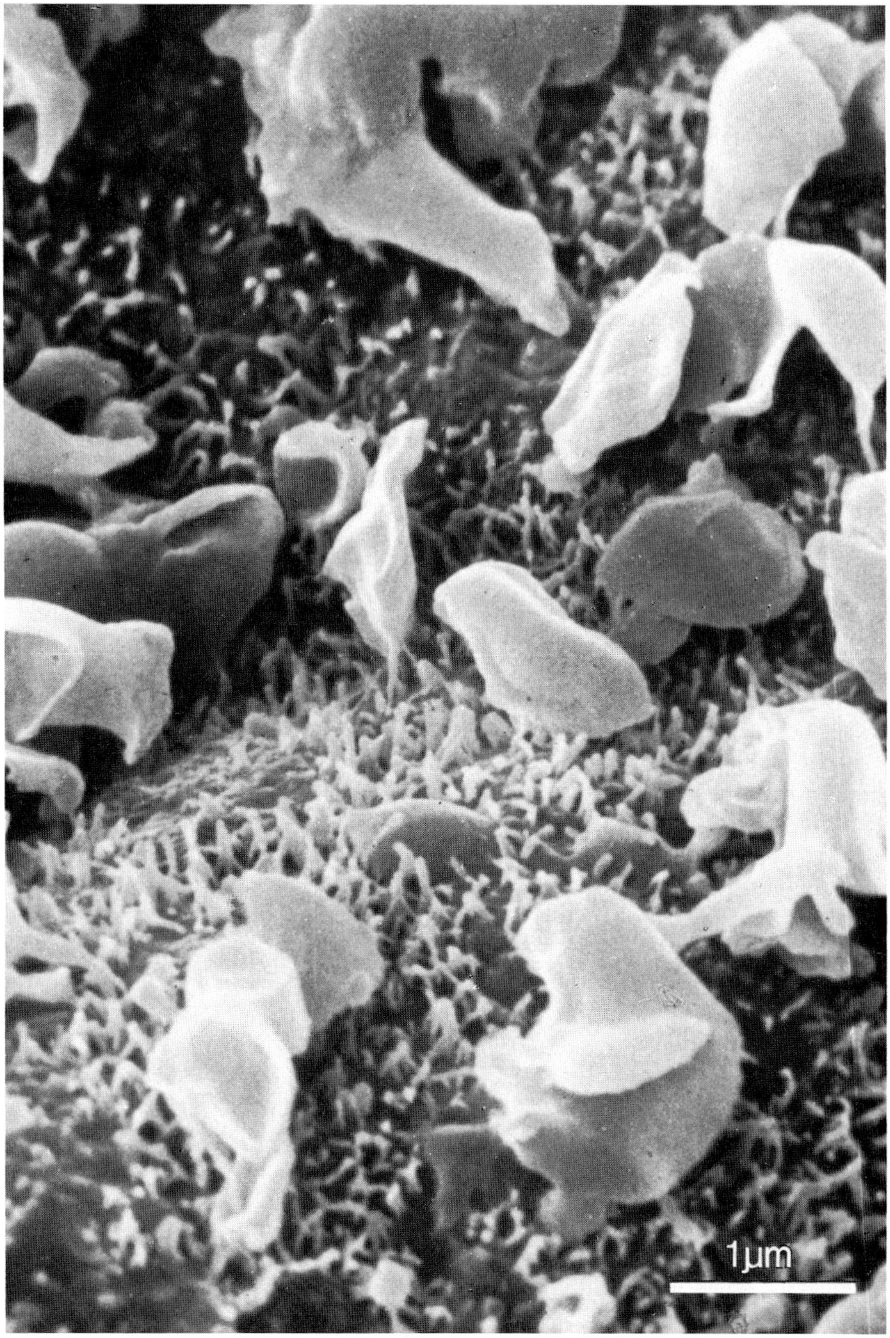

Fig. 8.4. Scanning electron microscopy of the apical poles of dog thyroid cells studied in vitro 15 min after a stimulation by 1.5 mU/ml of bovine TSH. Large lamellipodia arise from the apex of the cells, between the microvilli, and capture by phagocytosis droplets of colloid (not visible here, as the preparation has been washed before examination). This rapid phase of colloid capture is inhibited by MT poisons and by cytochalasin (Ketelbant-Balasse et al. [48])

by the cells, transported to lysosomes, and hydrolyzed, the released hormones diffusing towards the cell membrane and the blood. Thyroid cells (which have been specially studied in mammals from the functional point of view) contain numerous MT which are located in all the cell body, without any particular relation to centrioles. Microfilaments measuring between 5 and 10 nm form an apical web close to the cell membrane, similar to that mentioned above in pancreatic islet cells. This is closely related to the fibrils present in the numerous microvilli which cover the apex of the cell (cf. review, [27]).

When the cells are stimulated by the specific pituitary hormone (TSH) or by cAMP, which acts as a second messenger, the capture of droplets of thyroglobulin at the apical pole is increased: this activity, which resembles phagocytosis, may involve MT or MF or both. Two groups of workers published in 1970 their first findings on this subject [79, 122]. Both works were inspired by the findings on insulin secretion, and indicated that MT were involved. In slices of dog's thyroid, incubated in vitro, and where secretion was measured by the release of butanol-extractable ^{131}I (comprizing ^{131}iodide plus ^{131}labeled thyroid hormones), it was found that colchicine (1×10^{-2} M), VLB (25×10^{-6} M), and VCR (0.2×10^{-3} M) inhibited hormonal secretion and prevented the formation of the apical droplets which result from phagocytosis of colloid by thyroid cells [79]. Moreover, deuterium oxide was found to have a similar action, while puromycin did not prevent droplet formation. It was also observed that the MT poisons decreased the number of visible MT, with paracrystalline structure formation after VCR. It was concluded that the action of these agents known to modify MT indicated that MT and possibly also MF were needed for colloid phagocytosis and hormone secretion. The second group of workers [122, 124] reached similar conclusions, after studying the action of colchicine and VLB on the thyroid of mice incubated in vitro: the MT poisons blocked the stimulatory action of TSH on secretion at doses between 10^{-7} M and 10^{-6} M for colchicine, 10^{-7} M and 10^{-6} M for VLB. However, while the TSH stimulation secretion was inhibited, the basal rate of secretion, measured by ^{131}I release, was not affected. The significance of this result will appear later. These authors demonstrated also the presence of a 6S colchicine binding protein (tubulin) in their thyroids. Secretion stimulated by dbcAMP was also inhibited, and it was concluded that endocytosis of colloid required the integrity of MT [122]. However, as cytochalasin B (0.5 – 3.0 μg/ml) blocked colloid phagocytosis, it appeared that both MT and MF were involved [123].

These results were confirmed by both groups [78, 124], and other MT poisons were tested in a thorough investigation of this problem [78, 126]. The relatively slow action of colchicine was explained by the slow uptake of the drug by the thyroid, related probably to its binding to MT protein. A high correlation was found between MT poisoning and action of other poisons: podophyllotoxin was about as effective as colchicine, while griseofulvin was inactive. Colchiceine was much less active than colchicine or colcemid, and colchicoside was inactive [126].

The phagocytosis of colloid droplets after TSH stimulation lends itself very well to three-dimensional study with the scanning electron microscope. In dog thyroid slices (prepared by thyroxin injections so as to be in a basal condition) TSH added in vitro or injected in vivo has a spectacular effect. Within a few minutes, large lamellipodia or ruffles grow from the apical pole of the thyroid cells. They curve and appear to capture large drops of colloid before retracting into the cell [47, 48]. The formation of these lamellipodia, which resemble the surface changes of other phagocytic cells, is inhibited by MT poisons and cytochalasin B. However, only large doses of colchicine [78] prevent their formation and may have other non-specific effects ([36]; cf. Chap. 5); the effective doses of VLB are smaller and may be more specific.

Other activities of the thyroid cell—formation of hydrogen peroxide, oxidation of glucose—are depressed by 10^{-5} M colchicine [14]. The action of VLB may also prove to be more complex than the simple formation of crystalloid inclusions. It does not influence the thyroid secretion of mice except after TSH stimulation, when VLB is given two hours before TSH. Moreover, when large doses of TSH (more than 1 mU) were given, the crystalloid structures were seen to disappear, MT were again found, and ruffles were visible at the apical pole. Similar changes of VLB crystals were observed after dbcAMP stimulation, while VLB was found to have no inhibitory action on pseudopod formation stimulated by dbcAMP [22]. These results suggest that cAMP plays an active role in maintaining the integrity of MT (cf. Chap. 2). It should be noted that in other models, for instance the Langerhans islets, it has also been mentioned that stimulation could increase the numbers of MT [88], and in the thyroid of the mouse, an increase of the number of MT has been observed after 70 min incubation in vitro with 3 mU of TSH [126].

More recently, VLB was demonstrated to arrest within one hour the response of mouse thyroids to TSH, and to suppress the formation of colloid droplets in the cells. It was concluded that MT were necessary for cAMP to initiate the formation of these droplets [121].

One difficulty in the study of MT poisons and thyroid secretion is that several other metabolic pathways may be modified, in particular if rather large doses are used. Deuterium oxide may influence MF. Cytochalasin B, also, may have side-effects not related to MF [36].

In recent studies on the effects of chronic stimulation of thyroids by TSH (in the dog and the guinea-pig) it has been found that secretion apparently takes place without any formation of lamellipodia, suggesting that another pathway, perhaps micropinocytosis, not affected by MT or MF poisons, plays a role in secretion. However, when such thyroids are treated in vitro by a large single dose of TSH, ruffle formation with phagocytosis of thyroglobulin is visible (Ketelbant-Balasse, personal communication).

In summary, the liberation of the thyroid hormones follows the capture by the cell of thyroglobulin; this can take place in a basal state or in chronic stimulation by TSH, by a mechanism which remains to be defined and is insensitive to MT poisons and to cytochalasin B. On the contrary, phagocytosis of colloid droplets represents a mechanism of rapid capture of relatively large amounts of thyroglobulin and takes place in acute secretory responses. Studies on human thyroids lead to similar conclusions [47].

In this organ, it appears more and more evident that both MT and MF are necessary for the rapid secretion associated with phagocytosis. MT have the role of a structural scaffolding, maintaining the shape and the location in the thyroid cells of various organelles, as observed in the cream hamster [80]. The MF located in the cytoplasm and close to the pseudopods probably provide the motive force in the rapid formation of the apical ruffles. They may also act in the transport of the colloid vesicles toward the lysosomes, where the hormones are separated from the protein support. Like in pancreatic cells, a close cooperation between MT and contractile proteins (actin?) is apparent.

8.2.3 Other Endocrine Glands

These have not been studied as thoroughly as pancreas and thyroid, and many reports are either contradictory, negative, or positive with doses of MT poisons for which non-specific effects cannot be excluded.

8.2.3.1 Pituitary (Anterior Lobe). The various types of cells of this organ — the posterior lobe of which will be studied in Chapter 9 under the heading of neurosecretion — contain MT, as demonstrated by the preparation of this protein from bovine anterior pituitary [101], formation of typical VCR crystals in prolactin and growth-hormone (STH) secreting cells of the rat pituitary [51], and by serial section studies of the relations of MT with secretory granules and cell organelles in the prolactin cells in the rat [118].

Some effects of MT poisons on pituitary secretion are listed in Table 8.1.

From these results, it is apparent that the synthesis of hormones is not impaired, as an increase of secretory granules has been observed [85]. Colchicine binds to pituitary slices and, if it is allowed to act for several hours, may destroy the MT [102]. The basal release of various hormones does not appear to be affected in most experiments, and when the stimulated secretion is depressed, it is only slightly so. All hormones do not seem to be equally affected. The conclusion of MacLeod et al. [66] that "MT are not significantly involved in prolactin and growth hormone secretion" is perhaps excessive. However, it is apparent that the basal rate of hormone secretion is usually not modified by MT poisons, while a depression of stimulated secretion has been repeatedly observed. Considering the complexity of the pituitary, with its varied cells and multiple hormones, it appears possible that, like in the pancreas and the thyroid, MT are only indispensable when a peak secretion of hormones is required.

8.2.3.2 Parathyroid. Bovine parathyroid glands may be stimulated to secrete parathormone (PTH) in vitro by vitamin A (35 μg/ml): this effect is inhibited for three hours by VLB (8 μg/ml), and stimulated by cytochalasin B, indicating that MT or "microtubular-like" proteins facilitate PTH secretion [12, 13]. However, VLB does not depress low Ca^{2+} stimulated secretion in the same conditions [12]. The cytochalasin increased release of PTH is inhibited by VLB, a fact which may involve changes in the endoplasmic reticulum and PTH synthesis more than MT modifications (although typical paracrystals were observed, [13]). A quantitative stereological study of MT in the rat after administration of colchicine (2.5 mg/kg of body weight) showed a disappearance of MT after three hours, an increased density of secretory granules and a blockage of hormone release. In contrast, the administration of phosphorus, which increases the secretion, increases the amount of MT in the parathyroid cells. These results indicate some form of participation of MT in parathormone secretion [93].

Table 8.1. Action of MT poisons of secretion in the anterior pituitary

Reference	Animal species	MT poison	Dose	Results
Ewart and Taylor, 1971 [24]	rat	colchicine	0.1 mM	No inhibition of STH secretion after dbcAMP stimulation
Kraicer and Milligan, 1971 [50]	rat, in vitro	colchicine	–	Inhibition of K^+ stimulated ACTH release
Lockhart-Ewart and Taylor, 1971 [64]	rat	colchicine	–	Growth hormone: no inhibition
Pelletier and Bornstein, 1972 [85]	rat, tissue culture	colchicine	10^{-6} M, 24 h	Accumulation of secretory granules; disappearance of MT (somatotrophic cells only)
Temple et al., 1972 [110]	rat, in vitro	colchicine	5×10^{-5} M	No inhibition of secretion after 5 μg/ml TRH
Howell, 1973 [43]	rat, in vitro	colchicine VLB		No action on cAMP stimulated secretion of ACTH, TSH, and STH
Labrie et al., 1973 [51]	rat, in vitro	VCR	10^{-5} M	Inhibition of STH and prolactin release
Gautvik and Tshjian, 1973 [29]	culture of rat pituitary tumor	colchicine	5×10^{-6} M	Decrease of prolactin and STH secretion after stimulation by TRH or hydrocortisone
MacLeod et al., 1973 [66]	rat, in vitro	colchicine VCR	10^{-5} M	Decreased synthesis and release of prolactin No action on STH secretion (inhibition by VCR 10^{-4} M)
Sundberg et al., 1973 [109]	rat, in vitro	colchicine	10^{-3} M – 10^{-6} M	Enhancement of stimulated release of LH, FSH, TSH and GH. Increase of basal rate of GH secretion
Sheterline et al., 1975 [12]	heifer, in vitro	colchicine	10^{-5} M	40% inhibition of GH release after 3-isobutyl-1-methyl-xanthine stimulation

8.2.3.3 Adrenal Cortex. There is little information on the possible role of MT in adrenal steroid hormones secretion (the activity of the adrenal medulla will be studied in Chap. 9, as it is closely related to neurosecretion). VLB has been mentioned as stimulating the secretion of steroid hormones in the mouse [111]. In the rat, the formation of numerous crystalline inclusions after the injection of a considerable dose (50 mg/kg) of VLB, while MT disappear, indicates that a relatively large amount of tubulin is present in the cells [49].

8.3 Exocrine Secretion

In comparison with the large number of papers on endocrine glands, few studies on the possible role of MT in exocrine secretion have been published. The probable reason is the greater difficulty in measuring slight changes in secretory activity, rather than any fundamental difference between the mechanisms involved.

8.3.1 Pancreas

The effects of colchicine have been studied in the release of amylase by fragments of guinea-pig pancreas in vitro. As one hour preincubation with 10^{-4} M colchicine did not change the basal secretion, and did not affect the release after stimulation by bethanechol or pancreozymin, it was concluded that MT did not play a role in the stimulus – secretion coupling in these cells [4]. In a series of similar experiments, the action of colchicine (10^{-3} M), VCR (6×10^{-5} M), and deuterium oxide was studied on amylase secretion by fragments of rat pancreas perifused in vitro, and stimulated by pancreozymin. Colchicine and heavy water inhibited the secretion, while VCR was slightly active. The doses used are high and may have effects other than on the MT, and no definite conclusions could be reached [95]. An inhibition of pilocarpin-stimulated amylase secretion in mice 5 h after VLB was found to be proportional to the dose, however at high concentrations (from 0.4 to 400 mg/kg). The integrity of MT may be necessary for pancreatic secretion, but other actions (on amino acid metabolism, or the anticholinergic effect of VLB) should also be considered [77]. In the rat, similar results have been obtained with colchicine and VLB in vitro. Colchicine (10^{-3} – 10^{-5} M) decreases by 50% the discharge of secretory granules, while VLB (10^{-4} M) leads to an inhibition of 90% which lasts for over 3 h. The arrest of the intracellular transport results in an accumulation of crystalline proteins in the endoplasmic reticulum. At larger doses, colchicine (10^{-2} M) and VLB (10^{-4} M) also depress protein synthesis. The exact role of MT and probably of MF remains poorly understood [99]. The intracellular transport appears more affected than the discharge of secretory granules, and it is concluded that it is "difficult to decide . . . if unspecific inhibition of energy production . . . or MT function" may explain these facts [46].

8.3.2 Salivary Glands

The carbamylcholine-induced secretion of submaxillary gland mucin in the rat is inhibited by colchicine (10^{-5} M) [97], and a study of α amylase release from rat parotid slice in vitro has shown that the results depended on the quantity of dbcAMP present. With small doses of this nucleotide, which are without effect on amylase secretion, colchicine (10^{-4} M) and cytochalasin (52×10^{-6} M) increase the secretion. At higher dbcAMP concentrations (10^{-4} M) colchicine was without effect. It is suggested that dbcAMP may have a stabilizing action on MT, and at low concentrations promote the reassembly of MT, stimulating the secretion. The ultrastructure of these glands was not described [8]. In both these papers, the influence of cAMP, tubulin phosphorylation, and Ca^{2+} uptake have been considered, and the relation between MT and secretion appears rather indirect.

8.3.3 Mammary Gland

Fragments of mammary glands of lactating rabbits incubated in vitro have been treated by 10^{-6} M colchicine, and the transfer of labeled proteins (^{3}H-leucine) measured: the rate of migration of the granules between the endoplasmic reticulum and the Golgi vesicles is slowed, and after more than 2 h of incubation there is a marked increase of radioactive proteins in these structures. Colchicine inhibits almost completely the extrusion of the labeled proteins. However, it is not certain that MT are present in these cells [81]. Other results were obtained in the goat by the technique of intramammary infusion of colchicine. An injection of 1 mg of colchicine decreases by 67% the milk secretion at 36 h. This is normal again at 48 h. There is a dose relationship between the yield of milk and the dose of colchicine: 5 mg depresses the secretion to 40% of normal. Similar effects were obtained with VCR, although the cellular basis of these results remains unknown [84].

From these results with exocrine glands, it is evident that the action of MT poisons is difficult to understand at this time and deserves further analysis by in vitro studies combined with ultrastructural observations, similar to those which have yielded good results in the study of pancreatic islets.

8.4 Leukocytes: Phagocytosis and Exocytosis

The polymorphonuclear leukocytes of vertebrates are specialized cells, containing specific granules closely related to lysosomes. Their function is to protect the tissues by their phagocytic activity. They are very motile, with properties of chemotaxis which enable them to accumulate in injured tissues. From several points of view, they may be compared to primitive unicellular organisms, such as amebae, the specific granules being similar to secretion droplets, capable of opening into phagocytic, digestive vacuoles or "phagosomes". In some conditions, they may release their enzymes into the surrounding medium.

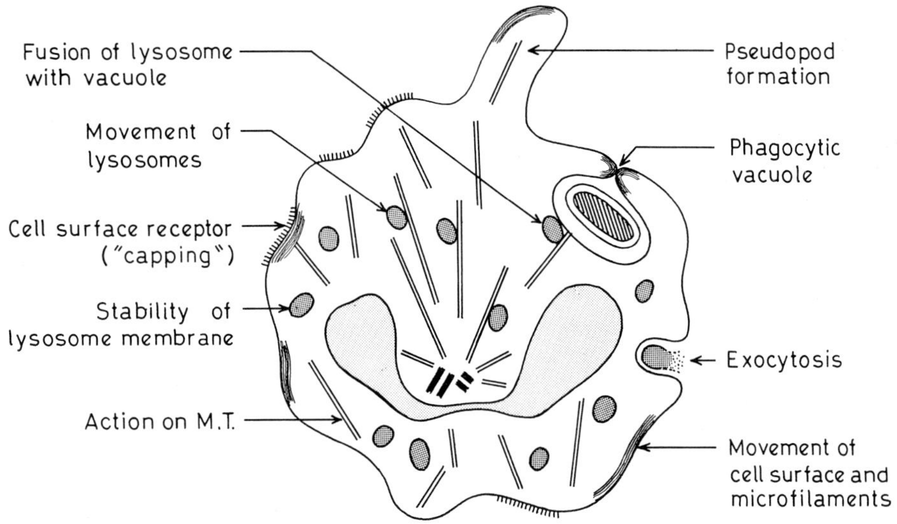

Fig. 8.5. Possible sites of action of MT poisons on the phagocytic activity of polymorphonuclear leukocytes

This double aspect of movement, closely linked with phagocytosis, and the secretion of lysosomal enzymes into vacuoles, are reminiscent of the activities described above in the thyroid cells, where phagocytosis precedes, like in leukocytes, the intracellular, lysosomal digestion of colloid, leading to the release of the specific hormone. Leukocytes also release secretory products, but their activity is short-lived, the cells undergoing lysis after the dissolution of their granules.

It is not surprizing to find many papers indicating that MT are related to phagocytosis and the movement and displacements within the cells of the specific granules. Since colchicine has so long been known as a cure for gout, a disease in whose acute crises the release of enzymes from polymorphonuclears plays a great role, it is natural that many explanations have been proposed for the therapeutic action of the alkaloid.

This problem will be treated in Chapter 11. MT poisons may also interfere with the liberation of histamine and other substances from the granules of mastcells and basophilic leukocytes, one of the first steps in the inflammatory reaction. They also affect the activity of macrophages originating from the blood monocytes. It is easier to deal separately with these three types of cell: polymorphonuclears (mainly neutrophil), macrophages, and mast cells.

8.4.1 Polymorphonuclears

Colchicine, in low doses (about 10^{-6} M), has been found to interfere with chemotaxis, formation of phagosomes, and release of enzymes into these vacuoles (cf. [71]). This could be related to the destruction of polymorphonuclear MT by colchicine [72, 74]. At concentrations of 5×10^{-5} M, colchicine and VLB were shown to affect the phagocytosis of bacteria by human leukocytes in the same way: the capture of the microorganisms was not prevented, although the formation of phagosomes and the degranulation of the cells was inhibited [72]. MT would be indispensable for the fusion of the specific granules with the phagocytic vacuoles, either by action on their membrane, or better, by preventing the intracellular movements of the granules.

Further information about the movements of granules was provided by work with cytochalasin B, which facilitates (by modifying the cell surface or the submembranar web of MF) the liberation of the granule contents in the medium, changing the leukocytes into secretory cells. The release of β-glucuronidase (a lysosomal enzyme) is decreased by colchicine and VLB; the MT would play a role in transporting the granules toward the cell surface [42]. These results were confirmed, and an important additional fact was noticed: an increase in cytoplasmic cAMP also prevented the release of the granules [120]. This observation has led to important conclusions which will be mentioned below.

This action of cytochalasin B on phagocytosis and exocytosis has shown the role of Ca^{2+} in the contraction of the cellular web involved in phagocytosis, and perhaps also in the polymerization of MT [128]. This has been compared with the variations of Ca^{2+} during mitosis [1]. In human polymorphonuclears treated with cytochalasin B, colchicine (10^{-6} M) inhibited the release of granules, while cGMP had the opposite effect. cAMP, theophyllin, and prostaglandin E_1 were also inhibitory. This was studied after the action on the leukocytes of a constituent of complement (C5a) which

Table 8.2. Mean MT number in complement-stimulated human neutrophils [a] (from Weissmann et al., [119])

Treatment	N	MT	p vs control
Fresh serum + 0.25 M epsilon-aminocaproic acid (EACA)	27	11.9 ± 0.8	<0.001
Zymosan treated serum [b] + EACA (*control*)	16	30.3 ± 2.6	–
cAMP (10^{-3} M) + theophyllin (10^{-3} M), 30 min + ZTS 1 min	15	22.3 ± 2.0	< 0.01
cGMP (10^{-5} M), 4 min, ZTS 1 min	19	47.1 ± 1.7	< 0.01

[a] All cells were pretreated with 5 μg/ml of cytochalasin B for 10 min
[b] Homologous serum + 0.25 M EACA incubated for 20 min at 37° C with 1 mg/ml zymosan (ZTS)

stimulates the liberation of leukocyte enzymes (after cytochalasin A) [32]. These opposing actions of cAMP and cGMP have led to the formulation [42, 119] of the so-called "ying-yang" hypothesis of enzyme release, further substantiated by the fact that phorbol myristate, a powerful stimulator of cGMP, increases granule release and the number of leukocyte MT [33]. Tables 8.2 and 8.3 demonstrate the relations between the numbers of leukocyte MT and these factors. Carbamylcholine, which increases the amount of cGMP, and acts as a cholinergic agent, may be antagonized by the anticholinergic agent, atropin. In these experiments, D_2O, contrary to colchicine, increases enzyme release [119]. These results have been discussed in relation to the possible action of cAMP and cGMP on MT assembly, and the protein-kinase activity which has been described as associated with MT (cf. Chap. 2).

In studies of secretory cells, it has been mentioned that D_2O, while stabilizing the MT, inhibits their action, as does colchicine. However, in leukocytes, like in mast cells, D_2O appears to stimulate granule release [32]. In leukocytes, D_2O alone does not induce any enzyme release, but increases it after complement stimulation. This effect is antagonized partially by colchicine [130].

Several other papers confirm that colchicine, even in low concentrations, inhibits intracellular movements in polymorphonuclear leukocytes, and the transport of lysosomes toward phagocytic vacuoles or toward the cell surface (after cytochalasin B pretreatment): they indicate close similarities between leukocytes and secretory cells and a similar action of MT on transport of secretory granules.

Table 8.3. Modification of complement-mediated enzyme release from cytochalasin-treated neutrophils (from Weissmann et al., [119])

Compounds added before exposure to C5a	Percent enzyme activity released in supernatant	
	β-glucuronidase	LDH
Control	26.4	4.8
cAMP (10^{-3} M) + theophyllin (5×10^{-4} M)	18.2	5.1
colchicine (5×10^{-6} M)	14.8	4.2
carbamylcholine (5×10^{-6} M)	30.1	4.9
cGMP (5×10^{-6} M)	33.4	4.6
phorbolmyristate acetate (1.5×10^{-8} M)	38.2	5.0

Larger doses (6×10^{-3} M) of colchicine are necessary to depress particle uptake by human polymorphs [58]; in a review on phagocytosis, MT have been considered "levers to amplify the movements by actin", this molecule being the real source of motility [108].

These findings may be related to the action of colchicine in gout [73], the inhibition of the fusion of the lysosomal granules with phagocytic vacuoles explaining, possibly, the anti-inflammatory action of colchicine and VLB.

Another action of colchicine on polymorphonuclear leukocytes in man (and also on mononuclear phagocytes) is to increase the liberation of the endogeneous pyrogen, the protein which, by acting on the central nervous system, induces fever. The mechanism of its liberation remains unknown, and it was expected that colchicine may have an inhibitory action: on the other hand, leukocytes incubated in vitro with concentrations of colcemid and VLB higher than 10^{-6} M, liberated more pyrogen, as measured by its action on rabbits. Lumicolchicine was inactive. The secretion of pyrogen, which takes place in several hours, is prevented by cycloheximide. The mechanism of action remains unexplained; the doses used are about ten times larger than those used in therapeutics, where colchicine has never been mentioned as occasioning fever [6].

In Chapter 3, the relations of MT to capping in polymorphonuclear leukocytes were mentioned, concanavalin A having been shown to increase the number of MT in the human leukocytes, and to lead to selective discharge of specific granules.

8.4.2 Macrophages

The mononuclear phagocytes are blood-born cells which are closely related to polymorphonuclears and assume various macrophagic activities in the tissues. Their MT are also destroyed by colchicine (10^{-5} M) and the saltatory movements of ingested bacteria are arrested; phagocytosis however, like in polymorphs, is not prevented [5]. In contrast, the role of MF in phagocytosis has been demonstrated [92]. In mouse peritoneal macrophages colchicine (10^{-6} M) prevents the release of acid phosphatase, while lumicolchicine is without any effect; colchicine does not affect the pinocytosis of ^{3}H-sucrose. It is not evident that MT are necessary for the fusion of lysosomes and phagocytic vacuoles [86, 87]. While colchicine destroys the MT at the concentration of 10^{-6} M in two hours, it does not interfere with the transfer of acid phosphatase to phagosomes after uptake of polyvinyltoluene [86, 87]: MT "are unlikely to play a critical role in ... fusion of lysosomes with endosomes". In these cells, colchicine may also depress protein synthesis and amino acid uptake without modifying the intracellular digestion of bacteria [86, 87].

8.4.3 Mastocytes

These cells, which are closely related to the basophil leukocytes of blood, are fixed elements which may release, in the first steps of inflammation, histamine, serotonin, and heparin. Early observations indicated that the shape of peritoneal mast cells was modified by large doses (1 mg/kg) of colchicine in the rat, while other mitotic poisons (podophyllotoxin, sodium cacodylate, chelidonin) were without action [83]. This was one of the first indications of an action of colchicine on cell shape in intermitotic vertebrate cells (cf. Chap. 5). Histamine release induced by polymixin B and other drugs was depressed by colchicine (5×10^{-4} M) in rat peritoneal mastocytes [30, 113]. VLB and griseofulvin had similar effects. On the other hand (like in polymorphonuclears) D_2O alone (at concentrations from 40 to 90%) caused histamine release, and potentiated—at 35% concentration—the histamine release by compounds such as 48/80, polymixin B, or reserpine [113]. In histamine release by basophil leukocytes in allergy, colchicine also had a depressing effect [62]. In human basophils, histamine release after the action of IgE was enhanced by heavy water, which antagonized colchicine [31]. It may be worthwhile mentioning in this context, recalling the facts observed in leukocytes, that Ca^{2+} elicits the liberation of the granules of peritoneal mast cells [44].

However, in a recent study of histamine release induced by polymyxin in rat mastocytes, colchicine (10^{-5} M) was found to destroy the MT completely, leading to changes in cell shape, without depressing more than 20% of the histamine secretion [56].

8.5 Other Cell Activities Related to Secretion or Transport

Many investigators, applying colchicine or other MT poisons to various cells, have observed inhibitory effects on secretion or transport of molecules into the medium. These results are often of a preliminary nature: they indicate the complexities of the pharmacology of antiMT agents, without necessarily demonstrating the participation of MT in these forms of transport.

8.5.1 Liver Cells

In a recent review, colchicine was mentioned as decreasing the secretion by the liver cells of the following proteins: albumin, prothrombin, fibrinogen, and very low density lipoproteins (VLDL) [25, 26], while biliary secretion and formation of biliary cholesterol and acids were not impeded. In the perfused mouse liver, colchicine (85×10^{-6} M) and VCR (2.5×10^{-6} M) decreased the secretion of triglycerides. In the cells treated by VCR, besides amorphous inclusions made probably of precipitated tubulin, numerous vacuoles containing VLDL particles are visible [60, 61]. Similar vacuoles are seen after perfusion with colchicine (85×10^{-6} M). Colchicine, VLB, and VCR were found to inhibit strongly the release of newly synthetized proteins in the perfused mouse liver, while protein synthesis was not impaired [59]. This would be the consequence of an arrest of the transport of intracellularly synthetized proteins toward the cell surface, for which MT would be indispensable. Even small polypeptides appear to be transported by the same mechanism. Lipoproteins are affected similarly, and in the rat, colchicine decreases by 80% the secretion of VLDL and by 20% that of HDL: this is not the result of a decreased synthesis and would confirm that MT are necessary for transport and excretion [106]. Biliary lipid secretion is not altered. A possible action of colchicine on the cell membranes is suggested, as the administered doses are rather high (0.5 mg/100 g) and may have non-specific effects [107]. For instance, the transport and incorporation of uridine, hypoxanthine, and choline in Novikoff hepatoma cells is inhibited by colcemid in vitro, at concentrations of 10×10^{-6} M, but this is considered a non-specific effect comparable to the transport inhibition of nucleosides [89] by colchicine (cf. Chap. 5).

In rats, the secretion of coagulation factors V and VII is decreased, as well as the formation of triglycerides; these components appear to accumulate in the cytoplasm of the liver cells in Golgi-derived vesicles. Lumicolchicine is without effect [35]. However, the large doses injected (0.5 mg of colchicine/100 g body weight) suggest that these effects may not be specific or related to MT.

In a study of the inhibition of protein secretion by rat liver slices, colchicine was found to act at the step prior to the fusion of the Golgi vesicles with the plasma membrane. Lumicolchicine has no similar effect. However, the authors are uncertain about the role of MT in these changes, as membrane effects cannot be ruled out [94].

Colchicine (25×10^{-6} M/100 g) causes an inhibition of secretion of glycoproteins by the rat with an accumulation, in the Golgi region, of proteins, conjugated to sialic acid and galactose residues [3].

8.5.2 Fibroblasts and Collagen Secretion

The general effects of MT poisons on secretion are confirmed by several results on collagen-secreting cells. In chick embryo frontal bone, and mouse fibroblasts, colchicine and VLB (10^{-6} M) rapidly inhibit the incorporation of ^{14}C-proline into collagen: this substance is transported to the cell membrane, to be secreted by vesicles whose movement is impaired by MT poisons [19]. Similarly, in chick embryo tendons intracellular ^{14}C-collagen increases while secretion is decreased by about 70% in 2 h by 10^{-6} M doses of colchicine [16]. In embryonic chick cranial bone, procollagen synthesis and secretion are inhibited by colchicine, while the Golgi apparatus disperses into many vesicles. These sometimes contain filamentous inclusions: the action of the MT poisons would disturb the movements of the Golgi vesicles, the secretory changes being secondary to this disorganization of the cell ([21]; cf. Chap. 6). Similar findings have been described in isolated chondrocytes from fetal guinea-pigs. Doses of 10^{-5} M of colchicine or VLB lead to a disappearance of MT, a dissociation of the Golgi complex, and a clustering of lysosomes. These changes are slowly reversible. The relation between MT and secretion cannot be defined, although the changes of the Golgi system may result in a dysfunction leading to a decrease in collagen formation and deposition of intercellular matrix [75].

The inhibition of collagen formation by colchicine has also been observed in pathological conditions, such as liver cirrhosis in rats after carbon tetrachloride intoxication [96]. This has led to therapeutic attempts in man (cf. Chap. 11).

8.5.3 Adipocytes and Lipid Metabolism

Various steps of lipid metabolism may be affected by colchicine: its action on the liberation of lipoproteins from the liver cells has been mentioned above. In the adipocytes from the epididymis of rats, physiological concentrations of insulin increase the number of MT. These MT disappear after 5×10^{-5} M colchicine. MT would be required to "direct" in the cell the metabolism of insulin [104]. The inhibition of insulin-stimulated metabolism of glucose to fatty acids has been confirmed [65]. On the other hand VCR (10×10^{-6} M) has no effect. Colchicine also impairs the liberation of free fatty acids (FFA) from epididymis adipocytes in the presence of epinephrine or theophyllin. Glucose metabolism was not modified; the sensitive step would be the extrusion of FFA [98].

An inhibition of glucose oxidation in fat cells treated with high concentrations of colchicine (10^{-3} or 10^{-2} M) has been described: this inhibition is immediate and results from an action on glucose transport across the cell membrane. It is as apparent at 4 °C as at 25 °C, a fact which leads to the conclusion that this is an effect unrelated to MT [11].

Fetal-guinea pig chondrocytes secrete less proteoglycans, and these have a smaller molecular size than usual after colchicine and VLB (10^{-5} M). Heavy water, in contrast, increases secretion while inhibiting synthesis. These effects may be the consequence of the dispersion of the Golgi complex resulting from the absence of MT [63].

Colchicine also affects the liberation from endothelial cells of the "clearing factor" or lipoprotein lipase in the blood. This is released after intravenous injection of heparin, and colchicine, like VLB, decreases this activity by half in rats. An action on cell membranes has been suggested; the doses of colchicine (0.5 mg/100 g body weight) and VLB (1.0 mg/100 g) are quite high [9]. Similar changes were found in the rat heart, where the synthesis of the enzyme was not affected [10]. MT poisons would mainly affect the transport of lipoprotein lipase from the myocardial cells to the endothelia, where it is present in a functional state [7]. The liberation of the clearing factor from fat cells is also inhibited by colchicine [15].

A decrease of chylomicron release from the intestinal epithelium of the rat after fat ingestion may also be a non-specific effect of colchicine (0.5 mg/kg; [2]).

8.5.4 Other Effects of MT Poisons on Transport

Some effects of colchicine which do not appear to be linked with MT have been mentioned in Chapter 5, in particular the inhibition of nucleotide transport across cell membranes. The following references are further indications of the manifold action of MT poisons, not necessarily related to alterations of MT:

a) Bone resorption in organ culture by parathyroid hormone or active metabolites of vitamin D, such as 1,25-(OH) D3, is blocked by low concentrations of colchicine (10^{-7} – 10^{-5} M): this is related to modifications of the brush border of osteoclasts, which are comparable to those induced by calcitonin [91]. Colchicine, possibly through this mechanism, can produce hypocalcemia in rats [39]. These results may be related to the fact that synthesis and release of collagenase by fibroblasts is increased by small doses of colchicine [37].

b) The secretion of IgM antibody by mouse spleen cells in response to an injection of sheep red blood cells is depressed by colchicine (2×10^{-5} M) and VCR (6×10^{-3} M). Loss and disaggregation of MT in the immunocytes was observed. The precise role of MT was however not clear; cytochalasin had no effect [112].

c) The liberation of lung surfactant by hamster lung slices incubated in vitro, measured by the release in the medium of ^{14}C-phosphatidylcholine, was decreased by 60% by colchicine and VLB, either added to the medium or injected 4 h before killing the animals. The fact that a syndrome of respiratory distress has been mentioned in one case of voluntary overdosage with colchicine [40] would confirm the hypothesis that MT may be involved in the exocytosis of the surfactant-laden granules of the type II pneumocytes [17].

d) The formation of the antiviral protein interferon, by fibroblasts of the human foreskin, is inhibited by both colchicine and VLB, at the concentration of 10^{-4} M. Colchicine would act before secretion, on one step of interferon synthesis, while VLB would more specifically inhibit secretion [38].

e) Osmotic water flow across the toad urinary bladder in response to vasopressin is inhibited by 10^{-5} M colchicine, only on the serosal side. Colchicine did not affect water transport in response to other stimuli (urea, mannitol, amphotericin B). The site of action of colchicine remains unknown [127].

It can be concluded that effects of MT poisons on various types of transport reveal the complexities of colchicine and VLB pharmacology, and cannot be considered to result from MT damage without careful verification. Most facts gathered in this chapter indicate that links exist between MT and the transport of secretory granules or the steps required for secretion, but the definite relations of MT with secretion often remain poorly understood. In the next chapter, the problem of the transport of neurosecretory granules will demonstrate links with MT, although also of an indirect nature. Far too often, actions of MT poisons have been ascribed to MT without any careful consideration of the mechanisms by which these could regulate the secretory processes.

References

1. Allison, A. C.: The role of microfilaments and microtubules in cell movement, endocytosis and exocytosis. In: Locomotion of Tissue Cells. Ciba Foundation Symposium Amsterdam-London-New York: Elsevier, Excerpta Medica, North Holland 1973
2. Arreaza-Plaza, C. A., Bosch, V., Otayek, M. A.: Lipid transport across the intestinal epithelial cell. Effect of colchicine. Biochim. Biophys. Acta **431,** 297 – 302 (1976)
3. Banerjee, D., Manning, C. P., Redman, C. M.: The in vivo effect of colchicine on the addition of galactose and sialic acid to rat hepatic serum glycoproteins. J. Biol. Chem. **251,** 3887 – 3892 (1976)
4. Benz, L., Eckstein, B., Matthews, E. K., Williams, J. A.: Control of pancreatic amylase release in vitro: effects of ions, cyclic AMP, and colchicine. Brit. J. Pharmac. **46,** 66 – 77 (1972)
5. Bhisey, A. N., Freed, J. J.: Ameboid movement induced in cultured macrophages by colchicine or vinblastine. Exp. Cell Res. **64,** 419 – 429 (1971)
6. Bodel, P.: Colchicine stimulation of pyrogen production by human blood leukocytes. J. Exp. Med. **143,** 1015 – 1026 (1976)
7. Borensztajn, J., Rone, M. S., Sandros, T.: Effects of colchicine and cycloheximide on the functional and non-functional lipoprotein lipase fractions of rat heart. Biochim. Biophys. Acta **398,** 394 – 400 (1975)
8. Butcher, F. R., Goldman, R. H.: Effect of cytochalasin B and colchicine on α-amylase release from rat parotid tissue slices. Dependence of the effect on $N^6,O^{2'}$-dibutyryl adenosine 3′,5′-cyclic monophosphate concentration. J. Cell Biol. **60,** 519 – 522 (1974)
9. Chajek, T., Stein, O., Stein, Y.: The effect of microtubule inhibitors on degradation of serum lipoproteins. In: Microtubules and Microtubule Inhibitors (eds.: M. Borgers, M. De Brabander), pp. 207 – 213. Amsterdam-Oxford: North-Holland 1975

10. Chajek, T., Stein, O., Stein, Y.: Interference with the transport of heparin-releasable lipoprotein lipase in the perfused rat heart by colchicine and vinblastine. Biochim. Biophys. Acta **388,** 260 – 267 (1975)
11. Cheng, K., Katsoyannis, P. G.: The inhibition of sugar transport and oxidation in fat cell ghosts by colchicine. Biochem. Biophys. Res. Commun. **64,** 1069 – 1075 (1975)
12. Chertow, B. S., Williams, G. A., Baker, G. R., Surbaugh, R. D., Hargis, G. K.: The role of subcellular organelles in hormone secretion: the interactions of calcium, vitamin A, vinblastine and cytochalasin B in PTH secretion. Exp. Cell Res. **93,** 388 – 394 (1975)
13. Chertow, B. S., Buschmann, R. J., Henderson, W. J.: Subcellular mechanisms of parathyroid hormone secretion: ultrastructural changes in response to calcium, vitamin A, vinblastine, and cytochalasin B. Lab. Invest. **32,** 190 – 200 (1975)
14. Chiraseveenuprapund, P., Rosenberg, I. N.: Effects of colchicine on the formation of thyroid hormone. Endocrinology **94,** 1086 – 1093 (1974)
15. Cryer, A., McDonald, A., Williams, E. R., Robinson, D. S.: Colchicine inhibition of the heparin-stimulated release of clearing-factor lipase from isolated fat-cells. Biochem. J. **152,** 717 – 720 (1975)
16. Dehm, P., Prockop, D. J.: Time lag in the secretion of collagen by matrix-free tendon cells and inhibition of the secretory process by colchicine and vinblastine. Biochem. Biophys. Acta **264,** 375 – 383 (1972)
17. Delahunty, T. J., Johnston, J. M.: The effect of colchicine and vinblastine on the release of pulmonary surface active material. J. Lipid Res. **17,** 112 – 116 (1976)
18. Devis, G., Obberghen, E. van, Somers, G., Malisse, F., Orci, L., Malaisse, W. J.: Dynamics of insulin release and microtubular – microfilamentous system. II. Effects of vincristine. Diabetologia **10,** 53 – 60 (1974)
19. Dieglemann, R. F., Peterkofsky, B.: Inhibition of collagen secretion from bone and cultured fibroblasts by microtubular disruptive drugs. Proc. Natl. Acad. Sci. USA. **69,** 892 – 896 (1972)
20. Edwards, J. C., Howell, S. L.: Effects of vinblastine and colchicine on the secretion of glucagon from isolated guinea-pig islets of Langerhans. FEBS Lett. **30,** 89 – 92 (1973)
21. Ehrlich, H. P., Ross, R., Bornstein, P.: Effects of antimicrotubular agents on the secretion of collagen. A biochemical and morphological study. J. Cell Biol. **62,** 390 – 405 (1974)
22. Ekholm, R., Ericson, L. E., Josefsson, J. O., Melander, A.: In vivo action of vinblastine on thyroid ultrastructure and hormone secretion. Endocrinology **94,** 641 – 649 (1974)
23. Ericson, L. E., Lindquist, I.: Effect of vinblastine in vivo on ultrastructure and insulin releasing capacity of the B-cell following sulphonylurea and isopropyl-noradrenaline. Diabetologia **11,** 467 – 474 (1975)
24. Ewart, R. B. L., Taylor, K. W.: The regulation of growth hormone secretion from the isolated rat anterior pituitary in vitro. The role of adenosine 3′:5′-cyclic monophosphate. Biochem. J. **124,** 815 – 826 (1971)
25. Feldman, G., Maurice, M.: Microtubules, microfilaments et sécrétion cellulaire. Biol. Gastroenterol. **8,** 269 – 274 (1975)
26. Feldman, G., Maurice, M., Sapin, C., Benhamou, J. P.: Inhibition by colchicine of fibrinogen translocation in hepatocytes. J. Cell Biol. **67,** 237 – 242 (1975)
27. Fujita, H.: Fine structure of the thyroid gland. Intern. Rev. Cytol. **40,** 197 – 280 (1975)
28. Gabbiani, G., Malaisse-Lagae, F., Blondel, B., Orci, L.: Actin in pancreatic islet cells. Endocrinology **95,** 1630 – 1635 (1974)
29. Gautvik, K. M., Tshjian, A. H. Jr.: Effects of cations and colchicine on the release of prolactin and growth hormone by functional pituitary tumor cells in culture. Endocrinology **93,** 793 – 799 (1973)
30. Gillespie, E., Levine, R. J., Malawista, S. E.: Histamine release from rat peritoneal mast cells: inhibition by colchicine and potentiation by deuterium oxide. J. Pharm. Exp. Ther. **164,** 158 – 165 (1968)
31. Gillespie, E., Lichtenstein, L. M.: Histamine release from human leukocytes: studies with deuterium oxide, colchicine, and cytochalasin B. J. Clin. Invest. **51,** 2941 – 2947 (1972)
32. Goldstein, I., Hoffstein, S., Gallin, J., Weissmann, G.: Mechanisms of lysosomal enzyme release from human leukocytes: microtubule assembly and membrane fusion induced by a component of complement. Proc. Natl. Acad. Sci. USA. **70,** 2916 – 2920 (1973)
33. Goldstein, R. C., Schwabe, A. D.: Prophylactic colchicine therapy in familial Mediterranean fever. Ann. Intern. Med. **81,** 792 – 794 (1974)
34. Gomez-Acebo, J., Garcia Hermida, O.: Morphological relations between β-secretory granules and the microtubular – microfilament system during sustained insulin release in vitro. J. Anat. **114,** 421 – 438 (1973)
35. Gratzl, M., Schwab, D.: The effect of microtubular inhibitors on secretion from liver into blood plasma and bile. Cytobiologie **13,** 199 – 210 (1976)

36. Grenier, G., van Sande, J., Willems, C., Nève, P., Dumont, J. E.: Effects of microtubule inhibitors and cytochalasin B on thyroid metabolism in vitro. Biochemie **57,** 337 – 342 (1975)
37. Harris, E. D., Jr., Krane, S. M.: Effects of colchicine on collagenase in cultures of rheumatoid synovium. Arthritis Rheum. **14,** 669 – 684 (1971)
38. Havell, E. A., Vilcek, J.: Inhibition of interferon secretion by vinblastine. J. Cell Biol. **64,** 716 – 718 (1975)
39. Heath, D. A., Palmer, J. S., Aurbach, G. D.: The hypocalcemic action of colchicine. Endocrinology **90,** 1589 – 1593 (1972)
40. Hill, R. N., Spragg, R. G., Wedel, M. K., Moser, K. M.: Acute respiratory distress syndrome associated with colchicine intoxication. Ann. Intern. Med. **83,** 523 – 524 (1975)
41. Hilwig, I.: The influence of colchicine on endocrine pancreatic cells in vitro. Z. Zellforsch. **132,** 263 – 272 (1972)
42. Hoffstein, S., Zurier, R. B., Weissmann, G.: Mechanisms of lysosomal enzyme release from human leukocytes. III. Quantitative morphologic evidence for an effect of cyclic nucleotides and colchicine on degranulation. Clin. Immunol. Immunopathol. **3,** 201 – 217 (1974)
43. Howell, S. L.: Secretion of growth hormone in the rat. Postgrad. Med. J. **49,** (Suppl. 1), 127 – 131 (1973)
44. Kagayama, M., Douglas, W. W.: Electron microscope evidence of calcium-induced exocytosis in mast cells treated with 48/80 or the ionophores A-23187 and X-537A. J. Cell Biol. **62,** 519 – 525 (1974)
45. Kern, H. F.: Fine structural distribution of microtubules in pancreatic B cells of the rat. Cell Tissue Res. **164,** 261 – 270 (1975)
46. Kern, H. F., Seybold, J., Bieger, W.: Discussion paper on inhibition of secretory processes in the rat exocrine pancreatic cell by microtubule inhibitors. In: Stimulus – Secretion Coupling in the Gastro-intestinal Tract (eds.: R. Case, H. Goebell), pp. 85 – 88. Cambridge: MIT Press 1976
47. Ketelbant-Balasse, P., Nève, P.: Morphological modifications of apical surfaces of thyroid cells in different functional conditions. In: Scanning Electon Microscopy (eds.: O. Johari, I. Corvin), pp. 761 – 768. Chicago: IIT Res. Institute 1974
48. Ketelbant-Balasse, R., Rodesch, P., Nève, Pasteels, J. M.: Scanning electron microscope observations of the apical surface of dog thyroid cells. Exp. Cell Res. **79,** 111 – 119 (1973)
49. Kovacs, K., Horvath, E., Szabo, S., Dzau, V. J., Chang, Y. C., Feldmann, D., Reynolds, E. S.: Effect of vinblastine on the fine structure of the rat adrenal cortex. Horm. Metab. Res. **7,** 365 – 366 (1975)
50. Kraicer, J., Milligan, J. V.: Effect of colchicine on in vitro ACTH release induced by high K^+ and hypothalamus-stalk-median eminence extract. Endocrinology **89,** 408 – 412 (1971)
51. Labrie, F., Gauthier, M., Pelletier, G., Borgeat, P., Lemay, A., Gouge, J. J.: Role of microtubules in basal and stimulated release of growth hormone and prolactin in rat adenohypophysis in vitro. Endocrinology **93,** 903 – 914 (1973)
52. Lacy, P. E.: Endocrine secretory mechanisms. Am. J. Path. **79,** 170 – 188 (1975)
53. Lacy, P. E., Howell, D., A., Young, C., Fink, J.: New hypothesis of insulin secretion. Nature **219,** 1177 – 1179 (1968)
54. Lacy, P. E., Malaisse, W., J.: Microtubules and beta cell secretion. Recent Prog. Horm. Res. **29,** 199 – 228 (1973)
55. Lacy, P. E., Walker, M. M., Fink, C. J.: Perifusion of isolated rat islets in vitro: Participation of the microtubular system in the biphasic release of insulin. Diabetes **21,** 987 – 997 (1972)
56. Lagunoff, D., Chi, E. Y.: Effect of colchicine on rat mast cells. J. Cell Biol. **71,** 182 – 195 (1976)
57. Leclercq-Meyer, V., Marchand, J., Malaisse, W. J.: Possible role of a microtubular – microfilamentous system in glucagon secretion. Diabetologia **10,** 215 – 224 (1974)
58. Lehrer, R. I.: Effects of colchicine and chloramphenicol on the oxidative metabolism and phagocytic activity of human neutrophils. J. Infect. Dis. **127,** 40 – 48 (1973)
59. Le Marchand, Y., Patzelt, C., Assimacopoulos-Jeannet, F., Loten, E. G., Jeanrenaud, B.: Evidence for a role of the microtubular system in the secretion of newly synthesized albumin and other proteins by the liver. J. Clin. Invest. **53,** 1512 – 1517 (1974)
60. Le Marchand, Y., Singh, A., Assimacopoulos-Jeannet, F., Orci, L., Rouiller, C., Jeanrenaud, B.: A role for the microtubular system in the release of very low density lipoproteins by perfused mouse livers. J. Biol. Chem. **248,** 6862 – 6870 (1973)
61. Le Marchand, Y., Singh, A., Patzelt, C., Orci, L., Jeanrenaud, B.: In vivo and in vitro evidences for a role of microtubules in the secretory processes of liver. In: Microtubules and Microtubule Inhibitors (eds.: M. Borgers, M. De Brabander), pp. 153 – 164. Amsterdam-Oxford: North-Holland 1975

62. Levy, D. A., Carlton, J. A.: Influence of temperature on the inhibition by colchicine of allergic histamine release. Proc. Soc. Exp. Biol. Med. **130,** 1333 – 1336 (1969)
63. Lohmander, S., Moskalewski, S., Madsen, K., Thyberg, J., Friberg, U.: Influence of colchicine on the synthesis and secretion of proteoglycans and collagen by fetal guinea-pig chondrocytes. Exp. Cell Res. **99,** 333 – 345 (1976)
64. Lockhart-Ewart, R. B. L., Taylor, K. W.: The regulation of growth hormone secretion from the isolated rat anterior pituitary in vitro. Biochem. J. **124,** 815 – 826 (1971)
65. Loten, E. G., Jeanrenaud, B.: Effects of cytochalasin B, colchicine and vincristine on the metabolism of isolated fat cells. Biochem. J. **140,** 185 – 192 (1974)
66. MacLeod, R. M., Lehmeyer, J. E., Bruni, C.: Effect of anti-mitotic drugs on the in vitro secretory activity of mammotrophs and somatrophs and on their microtubules. Proc. Soc. Exp. Biol. Med. **144,** 259 – 267 (1973)
67. Malaisse, W. J., Leclercq-Meyer, V., van Obberghen, E., Somers, G., Devis, G., Ravazzola, M., Malaisse-Lagae, F., Orci, L.: The role of the microtubular – microfilamentous system in insulin and glucagon release by the endocrine pancreas. In: Microtubules and Microtubule Inhibitors (eds.: M. Borgers, M. De Brabander), pp. 143 – 152. Amsterdam-Oxford: North-Holland 1975
68. Malaisse, W. J., Malaisse-Lagae, F., Obberghen, E. van, Somers, G., Devis, G., Ravazzola, M., Orci, L.: Role of microtubules in the phasic pattern of insulin release. Ann. N. Y. Acad. Sci. **253,** 630 – 652 (1975)
69. Malaisse-Lagae, F., Greider, M. H., Malaisse, W. J., Lacy, P. E.: The stimulus – secretion coupling of glucose-induced insulin release. IV. The effect of vincristine and deuterium oxide on the microtubular system of the pancreatic beta cell. J. Cell Biol. **49,** 530 – 535 (1971)
70. Malaisse-Lagae, F., Ravazzola, M., Amherdt, M., Gutzeit, A., Stauffacher, W., Malaisse, W. J., Orci, L.: An apparent abnormality of the B-cell microtubular system in spiny mice (*Acomys cahirinus*) Diabetologia **11,** 71 – 76 (1975)
71. Malawista, S. E.: Colchicine: a common mechanism for its anti-inflammatory and anti-mitotic effects. Arthritis Rheum. **11,** 191 – 197 (1968)
72. Malawista, S. E.: Vinblastine: colchicine-like effects on human blood leukocytes during phagocytosis. Blood **37,** 519 – 529 (1971)
73. Malawista, S. E.: Microtubules and the mobilization of lysosomes in phagocytizing human leukocytes. Ann. N. Y. Acad. Sci. **253,** 738 – 749 (1975)
74. Malawista, S. E., Bensch, K. G.: Human polymorphonuclear leukocytes: demonstration of microtubules and effect of colchicine. Science **156,** 521 – 522 (1967)
75. Moskalewski, S., Thyberg, J., Lohmander, S., Friberg, U.: Influence of colchicine and vinblastine on the Golgi complex and matrix deposition in chondrocyte aggregates. An ultrastructural study. Exp. Cell Res. **95,** 440 – 454 (1975)
76. Moskalewski, S., Thyberg, J., Friberg, U.: In vitro influence of colchicine on the Golgi complex in A- and B-cells of guinea-pig pancreatic islets. J. Ultrastruct. Res. **54,** 304 – 317 (1976)
77. Nevalainen, T. J.: Cytotoxicity of vinblastine and vincristine to pancreatic acinar cells. Virchows Arch. B. Cell Path. **18,** 119 – 128 (1975)
78. Nève, P., Ketelbant-Balasse, P., Willems, C., Dumont, J. E.: Effect of inhibitors of the microtubules and microfilaments on dog thyroid slices in vitro. Exp. Cell Res. **74,** 227 – 244 (1972)
79. Nève, P., Willems, C., Dumont, J. E.: Involvement of the microtubule – microfilament system in thyroid secretion. Exp. Cell Res. **63,** 457 – 460 (1970)
80. Nève, P., Wollman, S. H.: Ultrastructure of the thyroid gland of the cream hamster. Anat. Rec. **171,** 81 – 98 (1971)
81. Ollivier-Bousquet, M., Denamur, R.: Inhibition par la colchicine de la sécrétion des protéines du lait. C. R. Acad. Sci. Paris D **276,** 2183 – 2186 (1973)
82. Orci, L., Blondel, B., Malaisse-Lagae, F., Ravazzola, M., Wollheim, C., Malaisse, W. J., Renold, A. E.: Cell motility and insulin release in monolayer culture of endocrine pancreas. Diabetologia **10,** 382 (1974)
83. Padawer, J.: Effect of colchicine and related substances on the morphology of peritoneal mast-cells. J. Natl. Cancer Inst. **25,** 731 – 747 (1960)
84. Patton, S.: Reversible suppression of lactation by colchicine. FEBS Lett. **48,** 86 – 88 (1974)
85. Pelletier, G., Bornstein, M. B.: Effect of colchicine on rat anterior pituitary gland in tissue culture. Exp. Cell Res. **70,** 221 – 223 (1972)
86. Pesanti, E. L., Axline, S. G.: Colchicine effects on lysosomal enzyme induction and intracellular degradation in the cultivated macrophage. J. Exp. Med. **141,** 1030 – 1046 (1975)
87. Pesanti, E. L., Axline, S. G.: Phagolysosome formation in normal and colchicine-treated macrophages. J. Exp. Med. **142,** 903 – 913 (1975)

88. Pipeleers, D. G., Pipeleers-Marichal, M. A., Kipnis, D. M.: Microtubule assembly and the intracellular transport of secretory granules in pancreatic islets. Science **191**, 88 – 89 (1976)
89. Plagemann, P. G. W., Erbe, J.: Inhibition of transport systems in cultured rat hepatoma cells by colcemid and ethanol. Cell **2**, 71 – 74 (1974)
90. Rabinovitch, A., Gutzeit, A., Kikuchi, M., Cerasi, E., Renold, A. E.: Defective early phase insulin release in perifused pancreatic islets of spiny mice (*Acomys cahirinus*) Diabetologia **11**, 457 – 466 (1975)
91. Raisz, L. G., Holtrop, M. E., Simmons, H. A.: Inhibition of bone resorption by colchicine in organ culture. Endocrinology **92**, 556 – 562 (1973)
92. Reaven, E. P., Axline, S. G.: Subplasmalemmal microfilaments and microtubules in resting and phagocytizing cultivated macrophages. J. Cell Biol. **59**, 12 – 27 (1973)
93. Reaven, E. P., Reaven, G. M.: Quantitative ultrastructural study of microtubule content and secretory granule accumulation in parathyroid glands of phosphate-treated and colchicine-treated rats. J. Clin. Invest. **56**, 49 – 55 (1975)
94. Redman, C. M., Banderjee, D., Howell, K., Palade, G. E.: Colchicine inhibition of plasma protein release from rat hepatocytes. J. Cell Biol. **66**, 42 – 59 (1975)
95. Robberecht, P.: L'adaptation des hydrolases et la sécrétion exocrine du pancréas du rat. Thesis, Brussels 1974
96. Rojkind, M., Kershenobich, D.: Effect of colchicine on collagen, albumin and transferrin synthesis by cirrhotic rat liver slices. Biochim. Biophys. Acta **378**, 415 – 423 (1975)
97. Rossignol, B., Herman, G., Keryer, G.: Inhibition by colchicine of carbamylcholine induced glycoprotein secretion by the submaxillary gland. A possible mechanism of cholinergic induced protein secretion. FEBS Lett. **21**, 189 – 194 (1972)
98. Schimmel, R. J.: Inhibition of free fatty acid mobilization by colchicine. J. Lipid Res. **15**, 206 – 210 (1974)
99. Seybold, J., Bieger, W., Kern, H. F.: Studies on intracellular transport of secretory proteins in the rat exocrine pancreas. II. Inhibition by antimicrotubular agents. Virchows Arch. A. Path. Anat. **368**, 309 – 327 (1975)
100. Sharp, G. W. G., Wollheim, C., Muller, W. A., Gutzeit, A., Truehart, P. A., Blondel, B., Orci, L., Renold, A. E.: Studies on the mechanism of insulin release. Fed. Proc. **34**, 1537 – 1548 (1975)
101. Sheterline, P., Schofiled, J. G.: Endogenous phosphorylation and dephosphorylation of microtubule-associated proteins isolated from bovine anterior pituitary. FEBS Lett. **56**, 297 – 302 (1975)
102. Sheterline, P., Schofiled, J. G., Mira, F.: Colchicine binding to bovine anterior pituitary slices and inhibition of growth hormone release. Biochem. J. **148**, 453 – 460 (1975)
103. Slautterback, D. B.: A fine tubular component of secretory cells. Am. Soc. Cell Biol. Abstr. 199 (1961)
104. Soifer, D., Braun, T., Hechter, O.: Insulin and microtubules in rat adipocytes. Science **172**, 269 – 270 (1971)
105. Somers, G., Obberghen, E. van, Devis, G., Ravazzola, M., Malaisse-Lagae, F., Malaisse, W. J.: Dynamics of insulin release and microtubular – microfilamentous system. III. Effect of colchicine upon glucose-induced insulin secretion. Europ. J. Clin. Invest. **4**, 299 – 306 (1974)
106. Stein, O., Sanger, L., Stein, Y.: Colchicine-induced inhibition of lipoprotein and protein secretion into the serum and lack of interference with secretion of biliary phospholipids and cholesterol by rat liver in vivo. J. Cell Biol. **62**, 90 – 103 (1974)
107. Stein, Y., Stein, O.: Lipoprotein synthesis, intracellular transport and secretion in liver. In: Atherosclerosis III (eds.: G. Schettler, A. Weizel), pp. 652 – 657, Berlin-Heidelberg-New York: Springer 1974
108. Stossel, T. P.: Phagocytosis: recognition and ingestion. Semin. in Hematol. **12**, 83 – 116 (1975)
109. Sundberg, D. K., Krulich, L., Fawcett, C. P., Illner, P., Maccann, S. M.: The effect of colchicine on the release of rat anterior pituitary hormones in vitro. Proc. Soc. Exp. Biol. Med. **142**, 1097 – 1100 (1973)
110. Temple, R., Williams, J. A., Wilber, J. F., Wolff, J.: Colchicine and hormone secretion. Bioch. Biophys. Res. Commun. **46**, 1454 – 1461 (1972)
111. Temple, R., Wolff, J.: Stimulation of steroid secretion by antimicrotubular agents. J. Biol. Chem. **248**, 2691 – 2698 (1973)
112. Teplitz, R. L., Mazie, J. C., Gerson, I., Barr, K. J.: The effects of microtubular binding agents on secretion of IgM antibody. Exp. Cell Res. **90**, 392 – 400 (1975)
113. Tolone, G., Bonasera, L., Parrinello, N.: Histamine release from mast cells: role of microtubules. Experientia **30**, 426 (1974)

114. Obberghen, E. van, Devis, G., Somers, G., Ravazzola, M., Malaisse-Lagae, F., Malaisse, W. J.: Dynamics of insulin release and microtubular – microfilamentous system. IV. Effect of colchicine upon sulphonylurea-induced insulin secretion. Europ. J. Clin. Invest. **4,** 307 – 312 (1974)
115. Obberghen, E. van, Somers, G., Devis, G., Vaughan, G. D., Malaisse-Lagae, F., Orci, L., Malaisse, W. J.: Dynamics of insulin release and microtubular – microfilamentous system. I. Effect of cytochalasin B. J. Clin. Invest. **52,** 1041 – 1051 (1973)
116. Obberghen, E. van, Somers, G., Devis, G., Ravazzola, M., Malaisse-Lagae, F., Orci, L., Malaisse, W. J.: Dynamics of insulin release and microtubular – microfilamentous system. VI. Effect of D_2O. Endocrinology **95,** 1518 – 1528 (1974)
117. Obberghen, E. van, Somers, G., Devis, G., Ravazzola, M., Malaisse-Lagae, F., Orci, L., Malaisse, W. J.: Dynamics of insulin release and microtubular – microfilamentous system. VII. Do microfilaments provide motive force for translocation and extrusion of beta-granules? Diabetes **24,** 892 – 901 (1975)
118. Warchol, J. B., Herbert, D. C., Williams, M. G., Rennels, E. G.: Distribution of microtubules in prolactin cells of lactating rats. Cell Tissue Res. **159,** 205 – 213 (1975)
119. Weissmann, G., Goldstein, I., Hoffstein, S., Tsung, P. K.: Reciprocal effects of cAMP and cGMP on microtubule-dependent release of lysosomal enzymes. Ann. N. Y. Acad. Sci. **253,** 750 – 762 (1975)
120. Weissmann, G., Zurier, R. B., Hoffstein, S.: Leukocytic proteases and the immunologic release of lysosomal enzymes. Am. J. Path. **68,** 539 – 559 (1972)
121. Williams, J. A.: In vitro studies on the nature of vinblastine inhibition of thyroid secretion. Endocrinology **98,** 1351 – 1358 (1976)
122. Williams, J. A., Wolff, J.: Possible role of microtubules in thyroid secretion. Proc. Natl. Acad. Sci. USA **67,** 1901 – 1908 (1970)
123. Williams, J. A., Wolff, J.: Cytochalasin B inhibits thyroid secretion. Biochem. Biophys. Res. Commun. **44,** 422 – 425 (1971)
124. Williams, J. A., Wolff, J.: Colchicine-binding protein and the secretion of thyroid hormone. J. Cell Biol. **54,** 157 – 165 (1972)
125. Wolff, J., Bhattacharyya, B.: Microtubules and thyroid hormone mobilization. Ann. N. Y. Acad. Sci. **253,** 763 – 770 (1975)
126. Wolff, J., Williams, J. A.: The role of microtubules and microfilaments in thyroid secretion. Rev. Prog. Horm. Res. **29,** 229 – 286 (1973)
127. Yuasa, S., Urakabe, S., Kimura, G., Shirai, D., Takamitsu, Y., Orita, Y., Abe, H.: Effect of colchicine on the osmotic water flow across the toad urinary bladder. Biochim. Biophys. Acta **408,** 277 – 282 (1975)
128. Zurier, R. B., Hoffstein, S., Weissmann, G.: Cytochalasin B: effect on lysosomal enzyme release from human leukocytes. Proc. Natl. Acad. Sci. USA **70,** 844 – 848 (1973)
129. Zurier, R. B., Hoffstein, S., Weissmann, G.: Mechanisms of lysosomal enzyme release from human leukocytes. I. Effect of cyclic nucleotide and colchicine. J. Cell Biol. **58,** 27 – 41 (1973)
130. Zurier, R. B., Weissmann, G., Hoffstein, S.: Mechanisms of lysosomal enzyme release from human leukocytes. II. Effects of cAMP and cGMP, autonomic agonists, and agents which affect microtubule function. J. Clin. Invest. **53,** 297 – 309 (1974)

Chapter 9 Neurotubules: Neuroplasmic Transport, Neurosecretion, Sensory Cells

9.1 Introduction

The MT of neurons, or "neurotubules", were observed in 1956 by Palay [121] and are a constituent of all nerve cells, in invertebrates and vertebrates. As described in Chapter 2, brain has become one of the favorite sources of tubulin for biochemical studies, and this can be readily understood in view of the large number of MT in dendrites and axons. The MT of glial cells, in the central nervous system and in the nerves, may be associated with some complex structures, such as the Ranvier nodes and the Schmidt-Lanterman incisures (cf. Chap. 6).

Most of the problems associated with the function of MT appear in a study of nerve cells, in particular their role in maintaining the shape of long cytoplasmic extensions, and their relations with the transport, along the axons and the dendrites, of various organelles and granules. The considerable size of the neurons, the long distances between the "active" part of the cell, the perinuclear cytoplasm where the main syntheses take place, and the extremities of the axons, explain the importance of transport. This is often described as "axonal" transport or flow, but as it is also present in dendrites (where its study is more difficult), the expression "neuroplasmic flow" is to be preferred. This is by no means a one-way movement of all constituents of the cytoplasm. The cellulifugal flow comprizes at least two different components, a fast one, covering several hundred mm per day, and a slow one, around 5 mm/day [10]. A reverse flow, towards the cell, is also observed.

In some neurons, transport involves secretory granules, and problems similar to those discussed in Chapter 8 will be met here under the heading of "neurosecretion". Quite another aspect of the role of MT in nerve cells is found in sensory cells. Many of these—in the eye, the ear, the olfactory organ—are modified ciliary cells, and the cilia may play some role in sensory transduction. Other cells, acting as mechanochemical transducers, have complex arrays of MT which are moved under the action of exterior stimuli, and convey some information to the cell body.

In this chapter, many problems which have been discussed in previous chapters will be met again, and the facts already gathered should help the reader to understand better the role of MT in shape, transport, and secretion in nerve cells, although several questions remain without a definite answer. A description of neurotubules and their changes under the action of MT poisons and physical effects, such as high hydrostatic pressures and cold, will be given first; the biochemistry of nerve MT has been studied in Chapter 2, as brain has often been used as a source of tu-

[10] Some authors prefer to express the speeds in μm/sec; all speeds will be converted here to mm/day, which appears more convenient (86.4 mm/day = 1 μm/sec)

bulin. The important problem of tubulin-associated proteins needs particular attention in relation to nerve cells, for it is closely related to the possible mechanisms of transport.

9.2 Morphology and General Properties

The MT of nerve cells are morphologically similar to those of other cells, although it would be hazardous to say that they are identical. For instance, they belong to the so-called "resistant" MT, that is to say that they are only slowly destroyed by cold, high pressures or MT poisons: whether this results from a particular composition or structure, or from a low turn-over and great metabolic stability is far from clear. Brain extracts, the starting point for most biochemical studies, contain high molecular weight proteins associated with MT (MAPs) as the lateral expansions of neurotubules (34; cf. Chap. 2).

MT of nerve cells often have central densities (cf. Chap. 2). Their principal peculiarity is their great length, which according to some authors approximates that of the axons [161]. However, while the total number of MT on cross-sections before and after a nerve bifurcation is approximately equal, some results are not in agreement with the constancy of this number, and it remains very possible that some MT may be shorter than the nerve, as shown in a thorough study of the crayfish nerve cord [104], which is a favorable material, as these cells do not contain any other fibrillar structures. This same work also indicates that there is an upper limit to the density of MT in nerves—about 450 per μm^2. No bifurcated MT have ever been observed.

The molecular structure of nerve MT is similar to that described in Chapter 2, although it is far from certain that the number of tubulin subunits is always the same: in the crayfish, the tannic acid staining technique clearly shows 13 subunits in MT of glial cells, while those of the closely associated neurons have only 12 subunits [19]. However, in other species, 13 subunits have been counted [65].

In neurons, and in particular in axons, MT are located side by side with other fibrils, the neurofilaments (NF). These fibrils are about 9.5 nm in diameter, somewhat larger than the fibrils (gliofilaments) seen in glial cells, which measure 7 nm. They appear to be made of protofilaments and to have a hollow center [170]. They often display prominent lateral expansions ("side-arms") similar to those of MT. The protein of NF, *filarin,* has been isolated: it is acidic, with a molecular weight of about 80,000 daltons. It differs completely from tubulin and does not seem to be related to actin. Its filaments are perhaps made of two protofibrils helically wound [73]. An interesting fact, studied in the frog [130] and in rodents [51], is the relation between the numbers of NF and MT and the thickness of the myelin sheaths. While unmyelinated fibers show mainly MT, myelinated fibers of peripheral nerves often contain more NF than MT, and this is proportional to the thickness of the myelin layer. The function of NF remains obscure and their relation to other MF requires further study. If MT were considered to have mainly a skeletal function, it could be argued that in myelinated fibers, the mechanical strength of the myelin would dispense with the structural role of MT. In the crayfish cord [141] the myelin layer is thin, there are no MF, and the number of MT is particularly high.

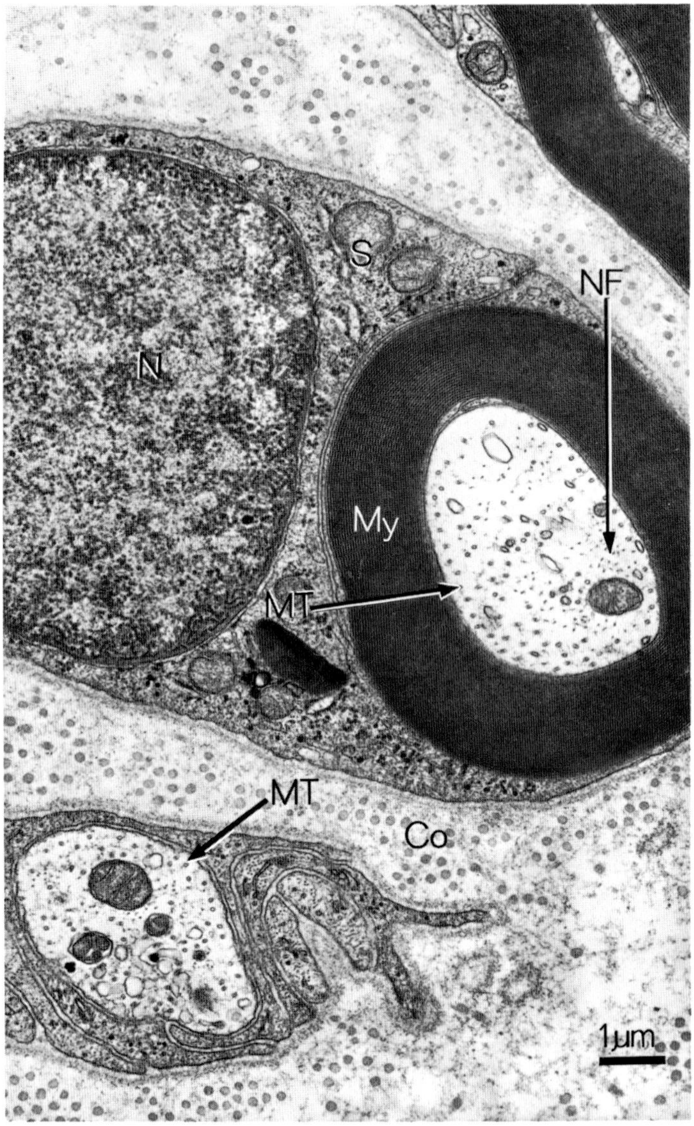

Fig. 9.1. Human sural nerve. Myelinated and non-myelinated fibers. *S* Schwann cell cytoplasm; *Co* collagen fibers; *My* myelin sheath; *N* nucleus; MT microtubules; NF neurofilaments

The association of MT and NF can be quite regular, as described in the nerve terminals of *Tinca tinca* L.: two types of MT are visible, some have a diameter of 28 nm, are made of ten subunits with a helicoidal disposition, and are surrounded by ten NF, while others have a diameter of 25.5 nm and nine subunits are surrounded by nine NF: this correspondence between the substructure of the MT and the connections with the NF indicates a definite function of this linkage. Longitudinal sections demonstrate a periodicity of the side-arms linking MT and NF. These

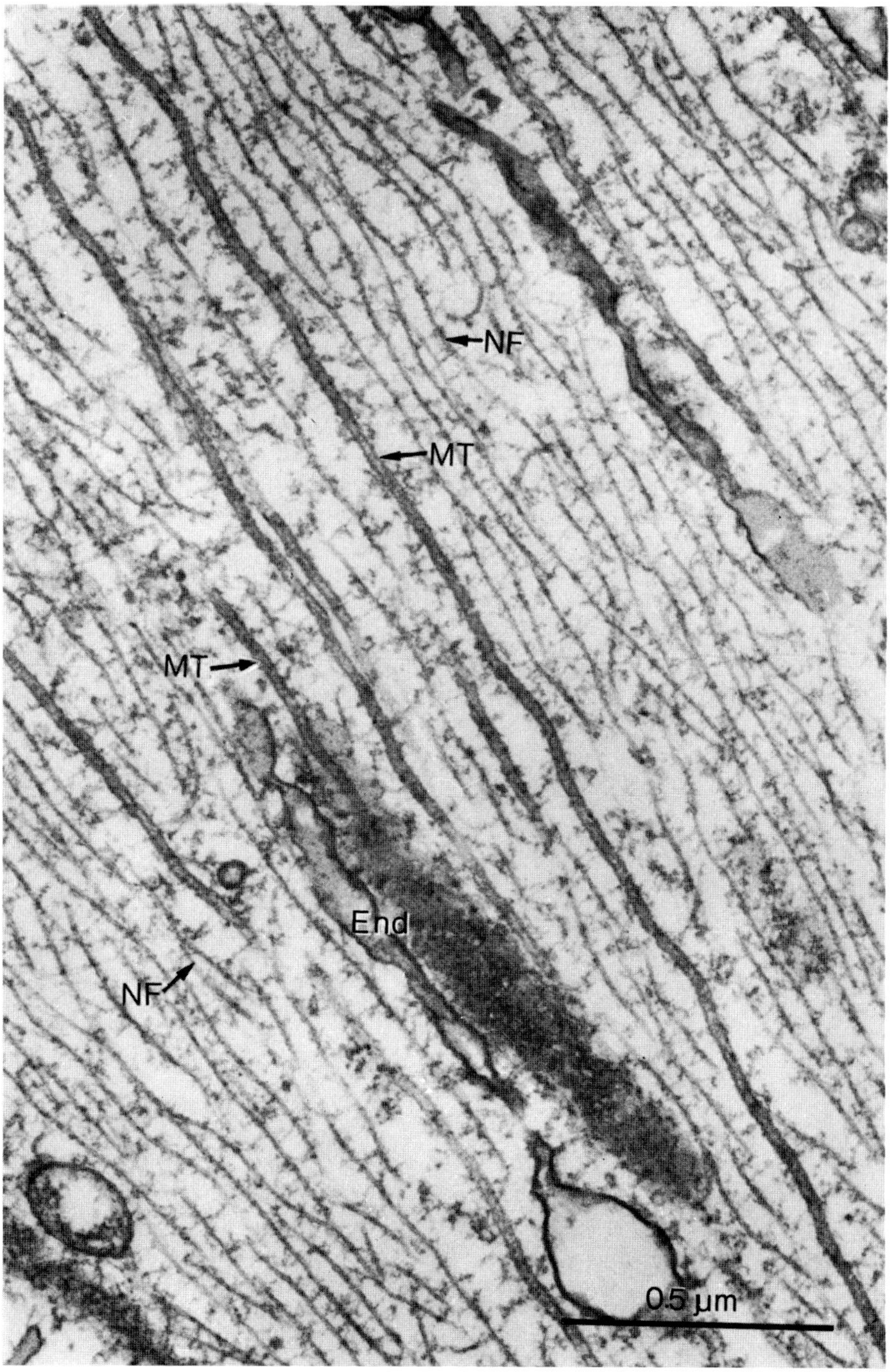

Fig. 9.2. Human sural nerve. Neurofilaments (*NF*) and microtubules (*MT*). Wispy filaments are attached to NF and MT alike. Elongated cisterns of the endoplasmic reticulum are also apparent (*End*)

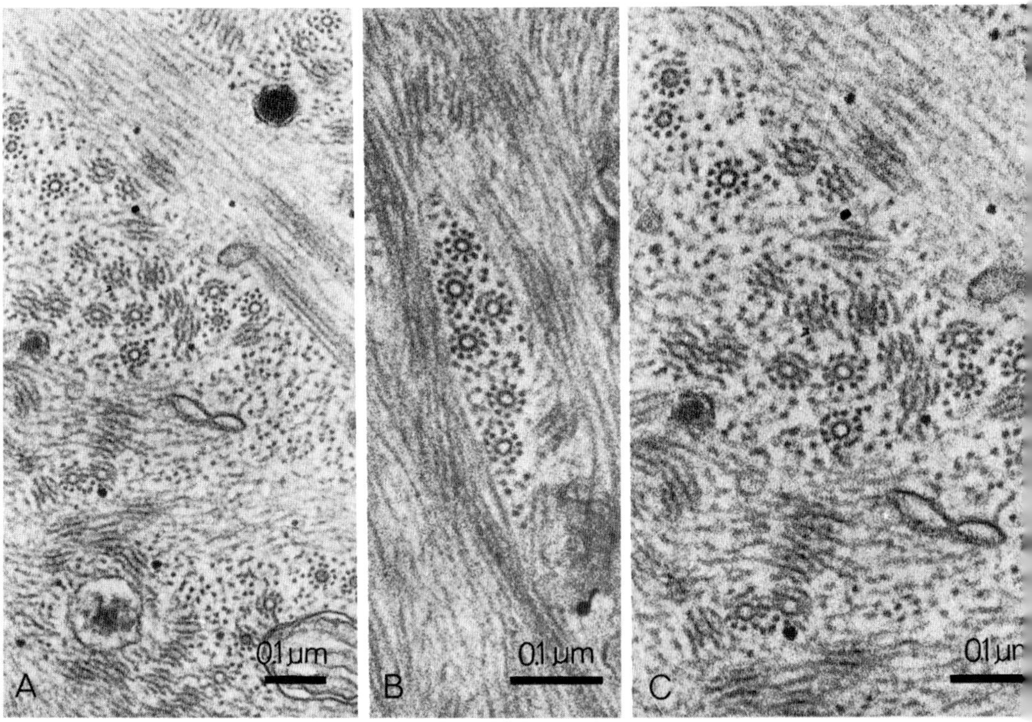

Fig. 9.3 A–C. Nervus terminalis of the tench (*Tinca tinca* L.). The NF, which far outnumber the MT, show definite relations with these in several places (from Bertolini et al. [11])

structures [11] are strikingly comparable to the periodic linkages which have been described (Chap. 4) between the central pair of MT and the radial links in cilia.

On the other hand, in the neurons of Clarke's nucleus of the kitten, MT appear regularly distributed in the dendrites, while NF are grouped in small "packets" sometimes associated with mitochondria or with endoplasmic reticulum. The filaments have a structural function, while the tubules are responsible for transport [150].

The MT of nerve cells, through their lateral expansions [18, 64, 92, 141, 171], may be interconnected and linked to other cytoplasmic organelles, an important observation in relation to the problems of neuroplasmic transport. Mitochondria—which are known to be carried along axons after originating in the neuronal cytoplasm—are often seen to be surrounded by MT in the rat [70]. A thorough study of the relations of MT and mitochondria in the large axons of *Petromyzon fluviatilis* ammocete larvae has clearly demonstrated that this is not a chance association; moreover, "bridges" are seen to link these two types of organelles [152]. In the same nerves, associations are also found between synaptic vesicles and MT, as already mentioned in Chapter 3: rosettes of these vesicles may be found around MT or aligned along MT [75, 151, 152]. The problem of the origin of these vesicles and similar structures and their possible transport along axons cannot be discussed here; it is probable that the vesicles are formed, not in the cell body, but close to the axon ending by a process of internalization of the cell membrane after the opening of the vesicles into the synaptic clefts.

Other cell organelles transported along the axons and dendrites become associated with MT, such as vesicles of the smooth endoplasmic reticulum. In the frog's brain, these are perpendicular to the MT which tunnel through their membranes [94].

Neurotubules also play a role in shaping nerve cells, as demonstrated by studies on tissue cultures and the changes of form induced by MT poisons; dbcAMP apparently increases the numbers of MT, and antagonizes the effects of colcemid on the shape of chick neurons [23, 138, 139]. This action of nucleotides is probably also linked with the action of nerve growth factor (NGF) (cf. Chap. 6). A dynein-like protein has been found associated with MT [55].

9.3 Experimental Changes of Neurotubules

The MT of nerve cells belong to the resistant category of MT. Some of the morphological changes which they undergo under the influence of drugs and physical agents will be summarized here, before studying the effects of these agents on neuroplasmic flow.

9.3.1 MT Poisons

One spectacular change observed in neurons treated with colchicine is the formation of large tangles of fibrillary material (cf. Chap. 5). These filaments, of 10 to 14 nm in diameter, with a hollow center, appear in rabbits after an intravenous injection of colchicine. It was first suggested that they were the result of a transformation of MT [165]. However, they could be induced by substances which are not known to interfere with MT, such as aluminium phosphate (Holt's adjuvant) [69, 84, 155]. In tissue cultures of neurons, 14 h after colchicine, the number of MT decreases and they progressively disappear, while these filaments accumulate [14]. Similar changes are found in the neurons of the anterior horn after VLB injection [166, 167] and in rabbits which received intracisternal or subarachnoid injections of relatively large doses of colchicine (0.1 mg), VLB (0.2 mg), or podophyllotoxin (0.4 mg). These changes were slowly reversible. Surprizingly, no VLB crystals were observed [167]. In the spinal ganglia of mice injected with 5 mg/kg body weight of VCR, a similar decrease in the numbers of MT with an increase of filaments was found, the axonal MT appearing more resistant than those of the neuronal body. These doses resulted in a paralysis of the hind limbs [77]. Similar findings were reported in tissue cultures of neurons exposed to VLB, VCR, or podophyllotoxin [127].

The formation of typical *Vinca* crystalline arrays of tubulin was reported for the first time after intracerebral implantation of VLB in the white matter [69]: the necessity of large doses is confirmed by the fact that most experiments on nerve cells have been done either in vitro or with local applications of MT poisons. Recently, typical VCR crystals have been found in axonal profiles after intrathecal injection of 20 μg of VCR [38, 47].

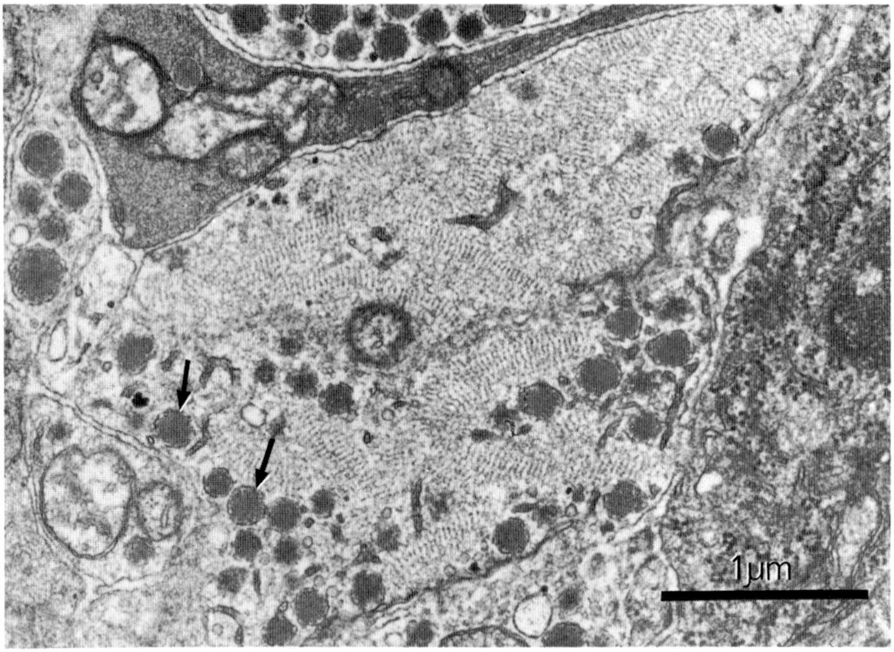

Fig. 9.4. Rat, aged 13 days. The posterior lobe of the pituitary was incubated for 2 h in Locke's solution containing VCR. An axonal ending, containing several dense neurosecretory granules (*arrows*), is almost entirely occupied by a VCR inclusion, showing a ladder-like structure. These observations indicate that a large amount of tubulin (either in the form of MT or not) is present in these axons

A quantitative study, demonstrating that colchicine acting in vitro on the vagus nerve of the rabbit may decrease the numbers of MT in the non-myelinated fibers, shows that this has only weak effects on the electrical activity of the isolated nerve: the number of MT decreases from 40.3 ± 2.9 per μm^2 to 16.1 ± 2.42. On the other hand halothane (10 mM), which blocks the electric activity, increases the number of MT, which reaches 58.0 ± 5.84 [66]. This result is in contradiction to the hypothesis of a "microtubular" explanation of the anesthetic properties of halothane (cf. Chap. 5; [2]). Other observations with colchicine are a glial activation after one to three daily injections of 20 to 40 µg in the rat, with some increase of cytoplasmic filaments. In the same paper "an increased number of pituicytes" is mentioned (containing) "centrioles from which tubuli radiated" (cf. Chap. 4; [61]). In tissue cultures of dorsal root ganglia of the chick, treatment with 0.05 µg/ml of colchicine inhibits neuronal growth and reduces the numbers of MT, while large numbers of 7 – 10 nm thick filaments become apparent. The reduction of 42% in the density of MT is attributed to disassembly of these structures [32]. The peculiar inclusions formed after intraocular injection of 10 to 500 µg of VLB in the rabbit have been described in Chapter 5 [15].

All these observations indicate that neurotubules may be destroyed by relatively large doses of MT poisons. The formation of large paracrystals after VLB or VCR suggests that a pool of non-assembled tubulin subunits is present in the nerve cells and axons. The formation of fibrillar structures is not specific for MT poisons, and evidence has been given that these are not formed by tubulin molecules.

9.3.2 Action of Physical Agents

Low temperatures affect the MT of neurons like those of other cells: in the nerve fibers of the toad, *Bufo arenarum,* MT disappear almost entirely after exposure, either in vivo or in vitro, to a temperature of 2 °C. This is reversible, as if the subunits of MT remained in place, perhaps by the action of the clear "exclusion zone" (cf. Chap. 2; [137]). On the other hand, NF are not affected by cold.

High hydrostatic pressures were first found to be without action on nerve MT, either in vivo or in vitro [118]. More recent data indicate that the results depend considerably on the temperature. Above 25 °C rat brain tubulin, polymerized in vitro, was only slightly affected, while a complete dissociation was obtained readily at a pressure of 500 atm at 15 °C [41]. These observations are comparable to those made on axonemes of heliozoa (cf. Chap. 5). This temperature effect has been discussed in relation to the thermodynamics of tubulin polymerization [42]. No studies on the action of pressure at lower temperatures on intact cells, as opposed to artificially polymerized tubulin, appear to have been published.

9.4 Neurotubules and Neuroplasmic ("Axonal") Flow and Transport

Much information on the transport of cell constituents along axons and dendrites has been gathered, and the purpose of this section will be to study the relations between this type of cell motility and the MT (neurotubules). Fundamentally, this is not different from the problems of melanin granule translocation and of intracellular movements of secretory granules, except that the scale is quite different. Moreover, transport in dendrites and axons is more complex, as in the same cell various rates of movement can be observed, and transport may also be bidirectional. Since the first observation of cytoplasmic movements in axons by Weiss in 1943 [162], several review articles have surveyed the extensive literature on this subject [33, 89, 95, 96, 97, 117, 129, 133, 158, 159]. The contribution by Price and Griffin [133] gives a detailed analysis of the recent work on this subject.

While, before the study of MT, axoplasmic flow appeared to be a property of neurons, as cells with very long cytoplasmic extensions in which various metabolites had to be carried from the cell body, modern work demonstrates that this is only one of the manifestations, albeit on a large scale, of the saltatory movements of cytoplasm (Chap. 7). The neurons provide particularly good experimental models for a better understanding of the complex relations between MT and movement.

9.4.1 Methods of Study; Fast, Slow, and Reverse Flow

Movements in nerves may be observed either in vivo or demonstrated indirectly. Direct observation, for instance in tissue cultures, shows that various small particles

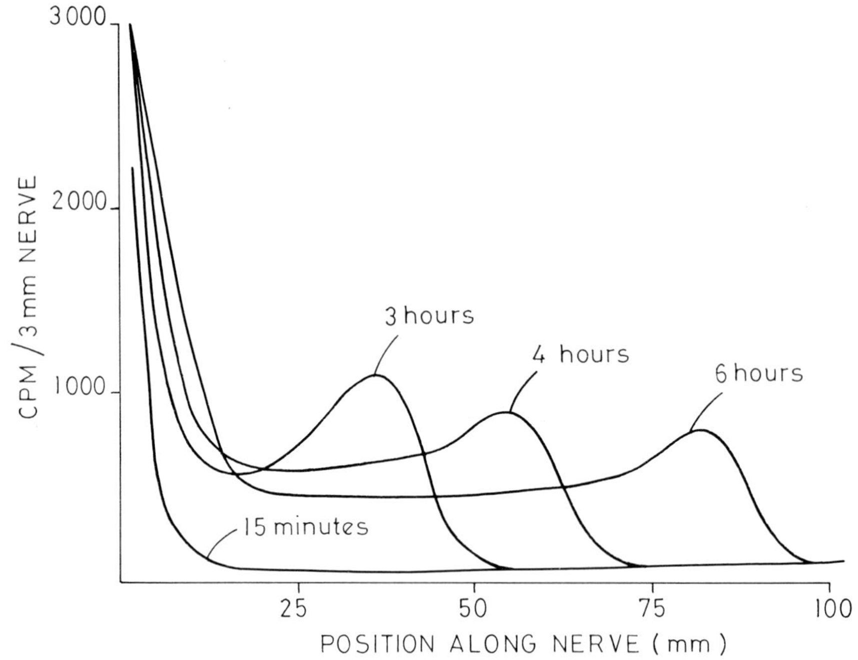

Fig. 9.5. Curves of fast anterograde neuroplasmic transport in the rat sciatic nerve after microinjection of ^{3}H-leucine into the L5 ventral horn. Three hours after injection, the labeled protein has moved about 50 mm, and at 6 h, nearly 100 mm. The mean rate of fast transport in this system is 385 ± 34 mm/day (redrawn from Griffin et al. [58], simplified)

Table 9.1. Some numerical data on neuroplasmic transport

Reference	Substances or structures transported	Fast transport	Slow transport
Dahlström and Haggendal, 1966 [29]	amine storage granules	250 mm/day	
Schmitt, 1968 [144]	neurosecretion (hypothalamus)	2,000 mm/day	
Ochs and Ranish, 1969 [115]	^{3}H-leucine, sciatic nerve of cat	401 ± 35 mm/day	
	^{32}P-orthophosphate		4.5 mm/day
Elam et al., 1970 [40]	sulfated mucopolysaccharides	+	
Lasek, 1970 [89]	axon proteins, MT, MF		0.4 – 3 mm/day
	catecholamine granules, mitochondria	100 – 500 mm/day	
Barondes, 1971 [7]	soluble proteins (MT)		+
	fucosylglycoproteins	+	
Smith, 1971 [49]	^{14}C-labeled proteins		1 – 2 mm/day
Chang, 1972 [23]	particles	620 mm/day	
Kirkpatrick, 1972 [83]	particles of less than 0.5 μm	86 – 174 mm/day	
Ambron et al., 1974 [3]	^{3}H-glycoproteins	50 mm/day	
Forman et al., 1975 [49]	organelles	93 mm/day	

appear to move along axons or dendrites. These movements are irregular, saltatory, and are not always in the same direction [10]. In axons of the bullfrog, the speed depends on the temperature; at 28 °C, particles move at about 93 mm/day (1 μm/sec), and 90% are seen to move towards the cell body [49]. The first observations on nerves were done by sectioning and ligating the axons: the proximal end showed a progressive swelling with an accumulation of particles and organelles (damming effect; [158, 173]). Later, injection of labeled proteins or inorganic molecules into the cytoplasm made it possible to follow the movements. This has been done with ^{3}H-leucine, ^{14}C-leucine, ^{32}P-orthophosphate, marked fucose, ^{3}H-proteins, and ^{45}calcium: although the techniques used are different, the results obtained indicate a satisfactory agreement in the speeds of transport.

While most authors agree that two modes of transport are present side by side, a slow component with a speed of a few mm/day, and a fast one varying, according to the different cells studied and the conditions of observation, between less than 100 and more than 1000 mm/day (Table 9.1), some authors have suggested that more than two different speeds may exist in the same cell: in the optic nerve of the rabbit, four speeds have been described: light particles move at two different speeds, 150 and 40 mm/day, mitochondria and lysosomes from 6 to 12 mm/day, and soluble proteins, comprizing free tubulin (colchicine binding protein), 2 mm/day [90, 164]. To facilitate our study, this may be simplified, and only two modes of transport considered—the fast, bidirectional and probably related to MT, and the slow, only centrifugal [96], which is that of the cytoplasm (and perhaps of the MT themselves).

While the physiology of neurons, with all cell machinery gathered close to the nucleus, suggests that centrifugal flow is the most important, for it alone can provide the very long cell extensions with various metabolites and some organelles, reverse flow has been repeatedly observed (cf. [58]). It is important in pathology, as it may carry toxins [132] or viruses toward the central nervous system. Some instances may be mentioned here: herpes virus along the axons of the sciatic nerve of the mouse [86], proteins in the regenerating hypoglossal nerve of the rabbit [87], nerve growth factor (cf. [21, 62, 154]), acetylcholinesterase in transected nerves (however at half the speed of the centrifugal movement; [97]) and horse radish peroxidase in peripheral nerves [119], the medulla of the cat [88], and the chick optic nerve [91]. The importance of retrograde transport, apart from the danger of carrying viruses to the central nervous system, is demonstrated by the changes which take place in the neuron following a section of its axon: it is probable that chemical information is carried, by the rapid mode, towards the cell body (cf. [96]). There are few experimental data on retrograde transport in relation to MT, but it is blocked by VLB, at larger doses, however, than orthograde transport, in the optic nerve and the colliculus of the rat [16, 17].

9.4.2 Nature of Transported Substances and Structures

Continuous displacements take place in the neuroplasm and soluble proteins, various small particles, cell organelles, and secretion products migrate along the axons and dendrites. The movements of several enzymes have also been followed (cf. Table 9.2)

Table 9.2. Fast and slow axonal transport: structures and organelles transported

Fast transport: >40 mm/day	Slow transport: ±2 mm/day
Acetylcholine	Axoplasm
Acetylcholinesterase	Choline acetyltransferase
Ca^{2+}	Cytoskeletal proteins
Dopa decarboxylase	Endoplasmic reticulum
Dopamine β-hydroxylase	Lysosomes
Fucosylglycoproteins	(Mitochondria)
Glycoproteins	Monoamine oxidase?
Mitochondria	Neurofibrils
Nerve growth factor (retrograde)	Neurotubules
Neurosecretory granules	Soluble proteins
(noradrenalin, posterior pituitary hormone-carrying granules)	Tubulin
Sulfated mucopolysaccharides	
Tubulin?	
Tyrosine hydroxylase	

9.4.2.1 Cytoplasm and Soluble Proteins. It is obvious that, as most ribosomes are located close to the nucleus, many proteins assembled in the cell body and/or in the Golgi apparatus must continuously migrate toward the extremities of the cell. However, it should be mentioned that the final fate of these proteins is not always clearly understood. As pointed out recently [96], if a protein is carried by the "slow" path, and has a relatively short life and a high turnover, it is not always easy to understand how it will eventually reach the most distal regions of the cell. There must certainly be some protein-synthetic activity at some distance from the cell body. The same is true for glycogen, and the problems raised by complex structures such as synaptic bodies, which are only seen at the extremities of axons or dendrites, are far from having found a satisfactory solution.

Among the "soluble" proteins which are carried—by the slow route—tubulin is of most interest for us. Colchicine injected in the eye of *Carassius aureus* was observed to move slowly in the optic nerve, indicating the slow transport of either tubulin or MT [56]. In the mouse, about 28% of the soluble proteins present in the nerve endings are tubulin, and ^{3}H-leucine labeling combined with VLB precipitation demonstrated that it was carried from the cell body; it is however possible that a fraction of this tubulin may be carried by the rapid route [43, 71]. The neurotubules themselves move at a slow rate (about 1 μm/min or 1.5 mm/day) and are imagined to be assembled continuously in the cell body and disassembled at the nerve endings [161]. This hypothesis appears satisfactory, although other possibilities may exist, for instance a status of equilibrium between tubulin molecules and MT all along the axons: this suggestion will be discussed in relation to one theory of axonal flow. Some proteins may be carried more rapidly, in particular fucosylglucoproteins [7, 13, 40].

9.4.2.2 Cell Organelles. Endoplasmic reticulum vesicles, mitochondria, and lysosomes move along the nerves. It has been figured out that about 1000 mitochondria are formed each day in the cell body, which explains the fact that a constriction of the axon rapidly leads to the local accumulation of several hundred mitochondria

[161]. The mitochondria of nerves have a rather rapid turnover and are destroyed in a few days (cf. [76]). The movements of mitochondria in nerves are difficult to study because of the lack of reliable markers [50]. While their transport along the axon has been described as slow [50, 82], studies of mitochondrial enzymes (hexokinase, glutamic dehydrogenase) have shown a rate of about 20 – 31 mm/day [122]. Transport may be bidirectional, for in sectioned isolated nerves mitochondria pile up at each end [96, 173]. In the frog sciatic nerve, incubation in colchicine (2×10^{-3} M in Ringer) and VLB (1×10^{-4} M) arrested the movements of mitochondria; this action was clearly related to the destruction of many MT, and impairment of mitochondrial transport was related to the numbers of MT [50]. This decrease in the numbers of MT was slow and only apparent after several hours of incubation. Neurofilaments and endoplasmic reticulum were not modified. An important observation made in these experiments was that a long incubation with cyanide and dinitrophenol—which were used by Ochs and Hollingsworth [114] to act as metabolic poisons of neuroplasmic flow—also decreased considerably the numbers of MT, albeit after ten hours of incubation.

The migration of various types of neurosecretory granules will be described further: suffice it to say that these large organelles appear mainly to be carried rapidly.

Vesicles of smooth endoplasmic reticulum are found in axons and dendrites: they may be carried and could possibly play a role in neuroplasmic flow [37]. As for the synaptic vesicles, these may be assembled at the nerve endings from soluble synaptosomal proteins carried along the axons, and there is some indication, from injections of colchicine in the brain of mice, that this transport may be slowed by the alkaloid and may perhaps depend on MT function [25]. However, the structure of the smooth endoplasmic reticulum of axons is still under discussion, and some authors think that a continuous system of vacuoles extends throughout the cell [37].

Close associations have been described, near the end of the nerves, between the synaptic vesicles and MT in the lamprey (*Petromyzon fluviatilis*), the vesicles being aligned along the MT, as mentioned in Chapter 3 ([75]; cf. also [12]).

9.4.2.3 Viruses. The retrograde transport of herpes virus plays an important role in the infection with *herpes simplex*, as the virus is carried from the nerves to the sensory ganglia, where it remains in an inactive condition between episodes of skin vesiculation [85, 86]. Other instances of associations between MT and viruses have been mentioned in Chapter 3. Poliomyelitis virus may also be carried from a sectioned nerve, by retrograde transport, to the spinal cord. No data on the possible role of MT have been published.

9.4.3 Action of MT Poisons on Neuroplasmic Flow

Many drugs interfering with MT or destroying them may arrest neuroplasmic movements, and the study of these results provides some information about the relation between the MT and movement. We will deal separately with colchicine, as it is the most specific and does not introduce other changes such as the crystalline formations produced by the *Catharanthus* alkaloids.

9.4.3.1 Colchicine. The published results, while indicating some action of colchicine, differ as to the type of transport inhibited (slow or fast) and the alterations of MT. This may result from the variety of techniques, animal species, and dosages used. Some of the most interesting results have been obtained by injecting colchicine in the eye of the rabbit and measuring the axonal flow in the optic nerve [78, 79, 80, 124]. An almost complete inhibition of rapidly migrating proteins was observed about 5 h after intraocular injection of 50 – 100 μg, which does not affect protein synthesis. The intracisternal injection of similar doses in the rabbit inhibited the rapid transport (300 – 400 mm/day) of proteins in the vagus and hypoglossal nerves [147]: a result as effective as that of a ligation of the nerve. Further work on the optic nerve indicated, however, that low doses of colchicine (1 – 10 μg), while completely arresting the fast transport, did not affect the slow displacements. Moreover, the MT appeared quite normal [79, 80]. Similar observations on the rat showed that only large doses of colchicine (about 1 mg) inhibited ^{3}H-proline fast axoplasmic transport unless the alkaloid was given 24 h before the amino acid, when 1 μg was effective. Lumicolchicine had no effect, while podophyllotoxin and VLB gave positive results. No study of the MT was reported [123]. Similar results have been obtained in the rabbit, by following the transfer of labeled ^{3}H-proline: while colchicine inhibits the transport and lumicolchicine does not, it is not certain that MT are affected [131].

In the chick, results which are sometimes in contradiction to those obtained in mammals have been reported: two authors have concluded that colchicine mainly affected the slow transport [74, 99]. This was observed by measuring the flow of ^{3}H-leucine in the sciatic nerve after injection of the drug in the spinal cord. The rapid flow was decreased when colchicine was injected up to three days before the amino acid. The action on the slow flow may be explained by an inhibition of the growth and movement of the MT themselves [99].

In *Xenopus laevis* sciatic nerves studied in vitro by dark-field microscopy, the saltatory motions of intraaxonal particles were inhibited within one hour by colchicine (and by VLB) at concentrations of or above 1×10^{-4} M. A progressive loss of MT was found by electron microscopy in nerves treated by 5×10^{-3} M colchicine, about 70% of the MT having disappeared within four hours. MT would be involved in the transport of large intraneural particles [60].

In the crayfish, it is also the slow component which is affected by cold or colchicine in the nerve cord [45], as measured by the movements of ^{3}H-proteins. The morphology of the MT is unchanged.

In another invertebrate, *Asterias glacialis,* two types of transport were also demonstrated, with speeds of 240 – 480 mm/day and 20 – 40 mm/day. They moved in two directions, centripetal and centrifugal. Injections of rather large doses of colchicine (0.5 mg) blocked the transport of ^{14}C-leucine marked proteins. Although the action of the drug was damaging and irreversible, MT and NF being destroyed, some slow flow was maintained [54].

9.4.3.2 VLB, VCR, Podophyllotoxin. The inhibition of fast axoplasmic transport by VLB in the rat has been mentioned above [79, 80]. In the crayfish (*Procambarus clarkii*) intraganglionic injection of VLB decreases the numbers of MT, although the block induced was detected as far as 10 mm from the site of injection, in regions "where the number and the morphology of MT appear unaltered" [44]. Paracrystalline inclusions were observed. These neurons have no MF. The inhibition of slow transport, whether by VLB, cold, or colchicine, takes place in cells containing apparently normal MT, and may last several days [44].

In the rat optic tract, VLB (8 μg intraocular) blocks transport [123], in agreement with other studies [98].

A study of axoplasmic transport in vitro in the rat sciatic nerve has shown a definite relation between MT-poisoning properties and inhibition: colchicine, podophyllotoxin and VLB are inhibitory, while lumicolchicine and picropodophyllin are more than 100 times less active. These last two compounds do not bind to tubulin. The exact role of MT remains however poorly explained, and the possibility that MT poisons may act on the axonal membranes (smooth endoplasmic reticulum or axolemma) should be further studied [125, 126].

9.4.3.3 Other Drugs Influencing Transport and MT. The discovery that local anesthetics, such as halothane, could modify MT, has led to many speculations on the mechanisms of anesthesia (cf. [2]). As recalled in Chapter 5, these are modern versions of the "narcotic" theory of Östergren [120] proposed to explain the action of drugs on the mitotic spindle. In 1968, the observation that saturated solutions of the local anesthetic, halothane, could destroy neurotubules, led to a general hypothesis on the mechanism of anesthesia [2] which was reformulated three years later [1]. However, halothane, even at 10 mM concentration, does not destroy the MT of the isolated rat vagus nerve: they are even slightly more numerous than in controls. The anesthetic action more probably results from membrane permeability changes [66]. In the crayfish, halothane does modify MT and leads to the formation of macroT (cf. Chap. 2; [67, 68]).

Another local anesthetic, lidocaine (xylocaine), inhibits rapid axonal transport in rabbit vagus nerves in vitro [46]. However, a morphological study, while demonstrating that large doses of lidocaine (0.6% for 75 min) could destroy up to 90% of the MT and lead to severe swelling of axons, and that intermediate doses (0.4% for 90 min, 0.6% for 45 min) did decrease the numbers of MT, indicated clearly that concentrations which effectively blocked impulse conduction and rapid axonal transport completely (0.4%, 10 – 20 min) did not modify the MT. It was even possible that the lowest concentration (0.4%) increased within 45 min the numbers of MT. Transport inhibition and MT changes thus appear to occur independently and by different mechanisms [20 a].

These results (cf. [81]) and a study of the optic nerve of mice treated by halothane or pentobarbital, which did not show any significant change in the number of MT during general anesthesia [140], have led to a rejection of the theory proposed in 1968 by Allison and Nunn [142, 143].

Other substances affecting the central nervous system have been suggested as acting on the neurotubules. The frog sciatic nerve shows a 50% inhibition of axonal transport of proteins under the action of chlorpromazine; the block is nearly total at the low concentration of 0.5 mM. These effects are reversible, although the number of MT is seen to decrease, while that of NF increases [39]. Similar changes have been obtained on this material with the local anesthetics lidocaine and tetracaine.

Other chemical inhibitors of axoplasmic transport are mescaline, which is interesting as a representative of a group of triphenyl methyoxy alkylamines—the role of the trimethoxyphenyl group in the pharmacology of colchicine and the *Catharanthus* alkaloids has been mentioned in Chapter 5 [123, 125, 126]—, and melatonin, the pineal gland hormone, a tryptophane derivative also studied in Chapter 5 [22].

A very powerful inhibitor of fast axoplasmic transport is batrachotoxin [117]: concentrations of 0.2×10^{-6} M are active on the cat sciatic nerve. The exact mechanism of action is not yet known, but may be related to the Ca^{2+}level in the nerve fibers, as the blocking effect is antagonized by placing the nerve preparation in a medium containing $35 - 125 \times 10^{-3}$ M Ca^{2+}.

The selective destructive action on MT of cyanide and dinitrophenol acting for many hours on frog's nerves has been mentioned above [50].

9.4.3.4 Action of Cold and Heavy Water. Close relations have been demonstrated between axoplasmic flow and nerve cell metabolism: these will be considered later. The rapid (410 mm/day) flow in the sciatic nerve of the cat was studied at various temperatures. Below 11 °C a complete block occurred, and no transport could be observed for up to 48 h. Two hypotheses were proposed: first, that at low temperatures the availability of ATP (which is necessary for transport, cf. [112]) is decreased; second, that MT could be disassembled [116]. However, in the crayfish, while exposure to 3 °C blocks transport, no evidence of disassembly of MT was noticed [45]. While there are indications that D_2O may enhance neurite growth in tissue culture [103], maybe by increasing the formation of MT, it has been shown to inhibit reversibly fast axonal flow in a sciatic nerve preparation of the frog [4].

9.5 Axonal Flow and Neurosecretion

Although not so long ago the idea of a neuron as a secretory cell appeared revolutionary, it is more and more evident that neurosecretion is a frequent activity, and plays a fundamental role not only in some specialized neuroendocrine organs, but in very general activities such as the liberation of catecholamines and the formation of the chemical mediators of synaptic transmission. As neurosecretory granules are readily observed, and as some of them may be labeled specifically with isotopes, their flow is an excellent object for the study of the mechanisms of transport. It is evident that this problem links the cytophysiology of neurons directly with that of secretory cells, and the observations studied in Chapter 8 may often apply to neurosecretory cells. Any theory of axoplasmic transport should apply to neurosecretory granules, and the relations between these and MT are of special interest for the general biologist.

Two principal neurosecretory systems have been studied: the transport and liberation of sympathicomimetic amines by the autonomous nervous system and the adrenal medulla, and the transport of neurosecretory material from the hypothalamic nuclei to the posterior pituitary. In both systems, the hormones are carried in linkage with large proteinaceous membrane-limited granules. Although much work has been done on these systems, the results in relation to the role of neurotubules are far from conclusive. Some positive results, with local applications of large doses of MT poisons, are open to criticism and may lack specificity. On the other hand, several observations indicate an arrest of neurosecretory granules without apparent changes of MT.

9.5.1 Adrenergic Fibers and Neurons

Catecholamines are transported, in nerves of the sympathetic system of vertebrates, associated with specific proteins. These are transported as granules by the fast component (250 mm/day) [29] of neuroplasmic flow. One of the first observations was that this flow was arrested by the local application, on sympathetic fibers of the rat, of a 20% solution of colchicine: the increased fluorescence indicated the accumulation of catecholamines above this zone [26, 27, 28, 30, 31]. Similar results were obtained by injecting colchicine (0.1 mM) and VLB (0.05 M) close to the axons (cf. [28]). The action on MT was not evident: after doses of 0.3 M colchicine, which are far above those which have a specific action, an increase of fibrils was found in the axons while the MT "seemed to be reduced" [72]. In the cat, it was also observed that in the hypogastric nerve, injections of colchicine (1–10 μg/ml) affected the transport of noradrenaline granules without destroying the MT [100]. In the chicken, intravenous injections of large doses of VLB (10 mg/kg body weight) arrested the transport of catecholamines. No evident change of MT was reported [8]. It is of interest to recall that such doses may have an effect on dopamine-β-hydroxylase, which is the final enzyme in epinephrine (adrenaline) synthesis [136].

In a recent report, colchicine and VLB are considered to interfere with the transport of amine storage granules and also of acetylcholine; the effective doses are however considerable (10^{-2} M colchicine, 10^{-3} M VLB) and the drugs are injected close to the nerve studied: such results do not bring any clear-cut evidence for the role of MT in the transport of neurotransmitters.

Another aspect of the possible role of MT is in the release, at the nerve terminals, of the neurotransmitters; this problem is more closely related to that of secretion, as it entails the opening of granules at the cell membrane (cf. Chap. 8). The release of catecholamines from bovine adrenal medullary cells after nicotine stimulation is inhibited by colchicine and VLB and potentiated by D_2O, indicating a probable role of MT [128]. In perfused rabbit adrenals, colchicine (0.5×10^{-3} M) inhibited the catecholamine release after acetylcholine and not after potassium stimulation. It is concluded that exocytosis does not seem to be mediated by MT, and the following sentence from these authors [35] may be quoted in full: "Colchicine appears to be yet another pharmacological agent whose usefulness as an analytical tool is compromized by its complexity of action." No morphological studies of these cells were described. The possibility of effects unrelated to MT is confirmed by studies of acetylcholine-induced adrenal secretion: the inhibition by colchicine and VLB are considered to result from an anticholinergic action of these drugs, unrelated to MT [134, 157]. Similar conclusions have been reached in studies on the liberation of norepinephrine from sympathetic endings in the heart. The inhibitory doses of colchicine (10^{-3} M) are higher than those which block axoplasmic transport, and it is unlikely that interference with MT plays a role [153]. The mechanism of catecholamine exocytosis is complex and cAMP and Ca^{2+} are important. Its blockage by colchicine (10^{-3} M) may reflect the atropin-like action of the alkaloid [169].

In conclusion, no definite evidence, either pharmacological or ultrastructural, indicates that MT play a role either in the transport or the release of neurotransmitters, and many other explanations of the experimental results are possible.

9.5.2 Hypothalamo–Hypophyseal Neurosecretion

The hormones of the posterior pituitary, in vertebrates, originate from neurons which are located in the hypothalamic region of the brain. The hormones or prohormones [53] are carried along the axons of the pituitary stalk in membrane-limited granules, which are released at the nerve endings in the posterior pituitary. Because of the high sulfur content of the antidiuretic hormone (ADH), labeling with ^{35}S is convenient, and makes it possible to follow the movements of the granules carried by the fast mode of flow. It is possible here to neglect some of the complexities of the system: the presence of at least two types of hypothalamic neurosecretory cells, their possible relations with the several hormones secreted [53], the role of the nervous "release factors" in controlling the rate of secretion, and the problems of exocytosis at the nerve endings, connected with the formation of "synaptoid" vesicles [105].

The axons of the pituitary stalk and posterior lobe contain large numbers of secretory granules, often packed densely in the so-called "Hering bodies", which are swellings of the axoplasm. MT and MF are present in these axons (which are unmyelinated). Their role in transport has been suspected for several years [9]. Subarachnoid injections of colchicine, in the rat, arrest the transport of ^{35}S-cysteine-labeled granules and destroy the MT of the supraoptic nucleus [107]. The transport of labeled secretion takes place at a speed of 50–70 mm/day [109]. In the rat, the complete arrest of ^{35}S-labeled secretory granule transport was observed after intracisternal injection of 0.2 mg colchicine. The labeling of the hypothalamic nuclei was not influenced, but 16 h after injecting ^{35}S-cysteine (intracisternally), no label was found in the posterior pituitary. However, considerable accumulations of neurosecretion granules were observed in the cell bodies and axons of the paraventricular nuclei, although the MT of these cells did not show the slightest change [48]. In further work on this subject, in rats which had been submitted to a forced secretion of ADH either by drinking water with 1% NaCl, or after injection of the powerful diuretic furosemid, a similar arrest of neurosecretion was found; besides normal granules, peculiar elongated structures were described, the exact signification of which remains mysterious [38]. Identical results have been obtained by intrathecal injection of 20 μg of VCR, with an impairment of neurosecretion movements, elongated structures, and also some accumulations of mitochondria and smooth endoplasmic reticulum. Neurotubules remain visible in the neurosecretory neurons [47], although no quantitative study of their numbers has been conducted.

These findings are in agreement with studies on the action of colchicine on the release of neurosecretion from the rat's posterior pituitary: colchicine not only blocked the rapid transport (indicated by ^{35}S-cysteine) but also decreased the amount of liberated hormone in response to potassium. Although granules accumulate in the axons, MT remain visible [108]. More recently, the transport arrest by colchicine with no destruction of MT has been confirmed [106].

Quite another approach to the possible role of MT in the hypothalamo–pituitary system has been the observation of rats with congenital diabetes insipidus, as compared to normal animals or rats receiving hypertonic saline for four days. Numerations of the MT in the pituitary stalk show a definite increase of the number of MT

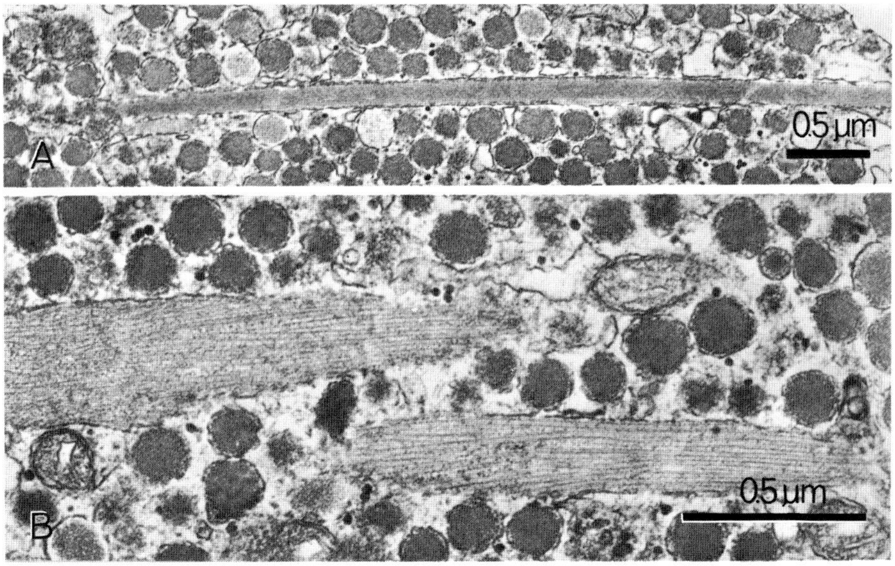

Fig. 9.6 A, B. Rat. Paraventricular nucleus 48 h after intrathecal injection of colchicine. (A) abnormal elongated neurosecretory granule. Radioautography with ^{35}S indicates that at this time the neuroplasmic transport of granules is arrested. (B) Abnormal fibrillary inclusions (original document, by courtesy of J. Flament-Durand)

after saline injections, i.e., after stimulation of ADH secretion, and a still more marked increase in the diabetes insipidus animals [57, 148]. The diameter of the axons increases in parallel fashion.

It may be concluded from these experiments that relations do exist between MT, as evidenced by the action of colchicine or VCR, and neurosecretion, both at the level of intraaxonal transport and perhaps of secretion itself (exocytosis). However, one fact remains evident: transport may be completely arrested, the granules accumulating in swollen axons, with no evident destruction of MT. It can be objected to this that no quantitative estimate of MT has been carried out in the experiments related above [47], but these facts are in agreement with the conclusions of many other authors that neurotubules belong to the "resistant" type of MT. Any theory of axoplasmic transport—and this may apply to other forms of secretory or melanin granule translocations—should explain how colchicine, in moderate concentrations, may arrest the transport function without altering MT morphology. As colchicine may be fixed on the MT or tubulin, one could imagine that this blocks some function which is indispensable for the intracellular transport of particles and secretory granules.

9.6 Theories of Neuroplasmic Flow

A good understanding of neuroplasmic transport would provide information on all other problems of intracellular displacements more or less linked with MT. Some of the theoretical models which have been proposed may be of general significance, while other authors have rejected any role of neurotubules in transport. The axons (and the dendrites, which have been much less studied) are evidently complex structures, and apart from the ground cytoplasm, which most probably moves slowly, the rapid form of transport, which will be discussed here, may be associated with at least three structures: MT, NF, and endoplasmic reticulum. The role of this last structure will be discussed later; what are the arguments for preferring the MT to the NF as organs associated with movement? Two appear particularly striking. Nerves do exist without any structure resembling NF: the crayfish provides the best instance, as mentioned above, of nerves with MT, rapid flow, and no NF. Another argument has still more importance, as the comparison can be made in different nerves of the same animal: axonal flow moves at the same fast rate (about 410 ± 50 mm/day; [112]) in motor, sensory, myelinated, non-myelinated, and neurosecretory axons. While the number of neurofilaments differs considerably in these nerves, the unmyelinated ones have few or no MF and many MT. Although in some nerves, geometrical relations have been described between MT and NF, this is not general, and cannot be used as a basis of a theory implying some interaction between MT and NF, as in some secretory cells (cf. Chap. 8). On the other hand, it should be made quite clear that this does not exclude the possibility of other filamentous proteins associated closely with the MT, as mentioned already in Chapter 2: the fact that purified nerve MT show a stoichiometrical relation with other proteins, and that the ultrastructure of MT assembled in vitro without the heavy molecular weight MAPs shows tubules with a smooth contour, without the wispy side-arms found when tubulin and the other proteins are present (cf. Chap. 2), is an indication that "filamentous" proteins are associated with MT in nerve cells. However, these are unrelated to NF, which differ from other types of cytoplasmic MF, for instance those which have been identified as actin (cf. Chap. 7).

9.6.1 Metabolic Requirements

In a series of papers, Ochs [110 – 117] has demonstrated the role of ATP in axoplasmic flow, as studied in the dorsal root ganglia and spinal nerves of cats after injection of ^{3}H-leucine. All transport is arrested within a short time (15 min) when the preparation is deprived of oxygen (in an atmosphere of N_2), or intoxicated by cyanide, azide, and dinitrophenol. In contrast, deprivation of glucose does not change the fast flow for several hours. The role of ATP is indicated by the fact that in the first 15 min of oxidation arrest, its amount falls by 50%. Glycolysis also plays a role in transport, but the inhibition by iodacetic acid does not arrest the flow until 1.5 to 2 h. Fluoroacetate, which blocks the citric acid cycle, also acts after 1.5 h. All these metabolic blocks are reversible, and flow may be resumed even after more than 1 h of anoxia. Hyperbaric oxygen is also inhibitory, after more than 1 h, at a moment

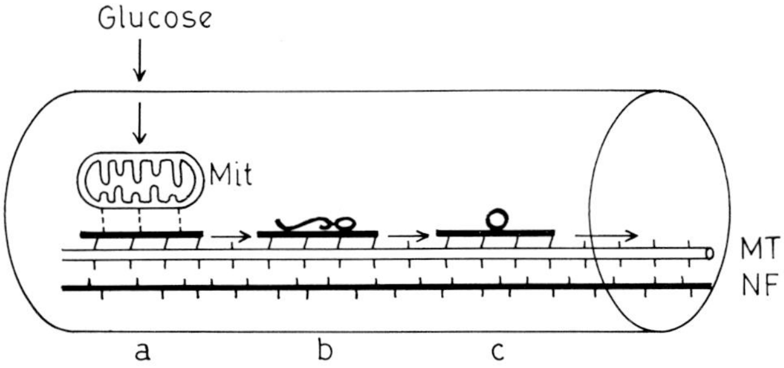

Fig. 9.7. Suggested model for axoplasmic transport. A carrier molecule (*thick line*) is transported along the *MT*, through the action of their side-arms. This carrier molecule transports various organelles (secretion granules, mitochondria, *Mit*). The transport requires ATP. The role of the neurofilaments (*NF*) remains unexplained (redrawn from Ochs [113], slightly modified)

when the amount of ATP is decreased. Substances which, on the other hand, modify nerve conduction, have no effect on axoplasmic flow. Oxalate is inhibitory; however, elimination of Ca^{2+} and Mg^{2+} by EDTA or EGTA has no consistent effect [112]. This may result from technical problems, as it has been shown that when nerve trunks were incubated in a calcium-free medium, transport was not affected, while an inhibition of 40 – 60% was found if the nerve cells were deprived of Ca^{2+} [36]. No ultrastructural changes were observed in these cells. It has been suggested that calcium ions may link proteins to the transport system, a fact which would explain why $^{45}Ca^{2+}$ is transported (in the frog) at the fast rate of 75 mm/day, like ^{3}H-proteins [59]. In the sciatic nerve of the chicken, studied in vitro in Krebs-Ringer medium from which ions were eliminated, it was shown that without K^+, Na^+, and Mg^{2+} the movement of small (0.5 μm) particles, as observed in phase-contrast, was not affected, while deprivation of Ca^{2+} by EGTA arrested all movement, without affecting the electric activity of the nerve [83].

In relation to the possible role of ATP, histochemical observations have shown that ATPase could be detected along NF and the bridges connecting them to MT, but not at the level of the MT themselves [168].

All these observations indicate that while rapid transport may persist in axons which are separated from the cell body, ATP is indispensable to provide the energy needed for movement. The problem which must now be solved is where ATP acts and what is the motor for the movements.

9.6.2 The Sliding Filament Theory

As MT in nerve cells are relatively stable structures, and as drugs like colchicine may modify axoplasmic movements without destroying MT, it has been suggested that particles or secretory granules have transitory connections with the MT through tubulin molecules: the particles could be imagined as sliding along the MT in a manner comparable to that mentioned in Chapter 8 for secretory granules [144]. As expressed by Schmitt [144], axonal flow is a "vital process in search of a mechanism," while the fibrous structures of neurons are "organelles in search of a function." NF, which have lateral expansions, may play a role in slow transport, and MT in fast flow, for instance in non-myelinated nerves, where NF are rare. The "sliding vesicle" mode of transport could, from a phylogenetic point of view, be older than the sliding filament mechanism of muscle contraction [144]. The MT, in this hypothesis, act as "conveyor-belts" [115]. This conception has been presented in a more sophisticated way by Ochs [111]: "transport filaments" (the nature of which is not defined) slide along the immobile MT through the action of "side-arms" attached to these (and which could be the heavy molecular weight proteins mentioned above). The particles, organelles, and molecules carried by the fast flow become attached to these transport molecules, ATP providing the energy for their sliding along the MT. This model explains how quite different substances may be carried at the same rate, as they become attached to the same carrier molecule. The cross-bridges would contain a $Mg^{2+}Ca^{2+}$ ATPase. NF would play no part. Although not stated, the carrier protein may be similar to tubulin—a fact which would explain how MT poisons arrest axoplasmic flow without disrupting the MT themselves—, or a protein of the actin-myosin group [55, 90]. In this second hypothesis, the action of colchicine (at low doses) remains mysterious.

9.6.3 Role of Assembly-Disassembly

In several instances of cellular movement, it has been suggested that cycles of assembly-disassembly of tubulin molecules (or dimers) could provide the motive force. In Chapter 10, the significance of this mechanism in mitotic movements will be discussed. Paulson and McClure [124, 125] have proposed a theory which may explain how the inhibitory actions of MT poisons on transport may take place without destruction of the MT. Their hypothesis is based on the idea that MT are in equilibrium with tubulin molecules. This is probably true in some cells, although in axons it may be that the association of subunits into MT takes place in the neuronal body, and the disassembly at the extremity of the axon, if, as mentioned above, MT move gradually along the nerve at the slow speed. There is no information on the possible reutilization of disassembled tubulin in nerve endings. However, the theory which is proposed here is based on the possibility of exchanges of subunits all along the MT, at single subunit "gaps" which are imagined as able to move all along the MT (cf. the gradion hypothesis of Roth et al. [140] for the axonemes of heliozoa, as explained in Chap. 3). The transport of particles would be somewhat linked to the displacement along the MT of a dislocation of its paracrystalline structure; the action of colchicine, VLB, and other drugs would not be on the tubules but on non-assembled subunits.

This theory has the advantage of explaining a block of transport without any apparent change in the shape of the MT, although it does not provide a definite link between the MT local change and the transported particles. However, a small change in the MT configuration could be imagined as interfering with the attachment of "side-arms" or other proteins which would provide the motive force for transport. It should also be noted that this theory, better than the sliding filament one, provides an explanation for retrograde flow, which is seldom considered in these studies.

9.6.4 Transport Without MT

A few authors deny MT any role in axonal flow, and this can be understood in view of the several experimental facts which show arrested transport with intact MT (and NF). A study of the rapid transport of ^{3}H-proline labeled proteins in the vagus nerve of rabbits has shown that the rapid transport (360 mm/day) can partly continue after exposure of the nerves to 7.5×10^{-3} M colchicine. The transport was observed by electron microscopic autoradiography. The relatively large dose of colchicine substantially reduces the number of MT (by more than 50%) without affecting the NF. The autoradiographic silver grains are mainly located close to the axoplasm, and profiles of smooth endoplasmic reticulum are seen to accumulate at the distal end of the nerve [20]. These results, which are in contradiction to several others, indicate a persistence of flow in the almost complete absence of MT, while it was mentioned above that flow could be arrested (by the same drug) without any alteration of MT.

It was concluded from these results that the marginal axoplasm, perhaps the axolemma, and the smooth endoplasmic reticulum may be involved in transport. A similar conclusion has been reached in a study of the transport of ^{3}H-amino acids and ^{3}H-fucose in nerves of the chicken (280 mm/day). About 80% of the proteins are carried by the slow transport, which comprizes the MT and NF themselves and the mitochondria. A study by high voltage electron microscopy (HVEM) of the nerves has shown that there is a continuity between the saccules of the smooth endoplasmic reticulum which are scattered all along the nerve: they provide continuous channels and these are the "preferential pathway for fast transport" [37].

The role of the Golgi apparatus should not be neglected, as its participation in colchicine-modified secretion has been mentioned in Chapter 8: the interrelations between Ca^{2+}, contractile proteins such as actin, and the endoplasmic reticulum and Golgi apparatus may be more important than MT in explaining the effects of colchicine on neuroplasmic flow [36, 63, 117].

It is evident from the summary of these publications that rapid transport and its relation to MT is far from being understood, and that the inhibitory effects of MT poisons remain mysterious, although axonal transport is apparently easy to study, as it persists in isolated fragments of cell and has definite metabolic requirements.

9.7 MT and Sensory Cells

Many sensory cells result from differentiations of ciliary cells, and cilia play an important role in the structure of the retina, the inner ear, the olfactory organs, and the mechanoreceptors of vertebrates and invertebrates. In some cells, cilia are located at the base of specialized structures such as the retina rods and cones, and appear to have mainly a structural role. In the inner ear of vertebrates, cilia are visible in the Corti organ and the sensory cells of the labyrinth. However, they are accompanied by other extensions of the cell membrane, the stereocilia, which do not contain MT and are structures more or less similar to microvilli. In some species, the Corti organ does not contain any true cilia, and the perception of sound is mediated through the stereocilia, which have a complex three-dimensional pattern. In the olfactory cells, the chemical influx affects cells which have very long, specialized ciliary structures; these cytoplasmic extensions may contain only a single MT, as described in the olfactory epithelium of the frog [135]. In contrast, in mechanoreceptors found in several invertebrates, numerous parallel MT form rigid structures which could transmit movements from the medium toward the cell body.

One could imagine that the passive bending of a cilium or of a bundle of MT could generate chemical changes in the cell, signals which would be transmitted to the nervous system. These changes could be similar but of reversed sign to those taking place in active bending and movements of cilia. This idea of a conformational change of MT structures leading to chemical reactions transmitted to the nervous system has been proposed by Atema [5, 6], and the MT changes compared to a lattice defect which would be carried along the MT, an idea similar to that described above for neuroplasmic transport ([125, 126]; review in [163]).

Some structures which strongly suggest a role of MT in mechanical transduction may be mentioned here. Mechanoreceptor cells of the bee show a modified ciliary extension which is embedded in the cuticle: this is formed of a cylinder of densely packed MT [156]. In the cockroach, sensitive "hairs" contain from 350 to 1000 MT: they lose their receptor properties after treatment with colchicine or VLB [101, 102]. A modified 9+0 cilium is at the base of the intracuticular extension of

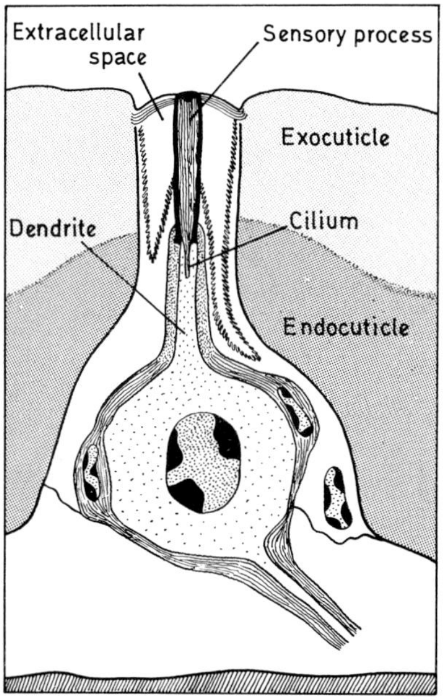

Fig. 9.8. Cockroach *campaniform sensilla* (diagrammatic representation): the sensory neuron, surrounded by supporting cells, has a single cilium from which extends, toward a specialized region of the exocuticle, a bundle of parallel MT which are embedded in a dense material and closely associated with the surface of the insect (redrawn from Moran et al. [101])

the sensory neuron. The MT have a dense central core and originate in a dense matrix, the nine ciliary doublets being located at the exterior of these sensory MT [102]. A similar relation of nine ciliary doublets with a central group of many MT is found in the olfactory sensory cells of several insects. The central MT may have a hexagonal pattern; they are linked by bridges to the membrane. At the bottom of this rigid structure may be located a "tubular body" which is compressed by the lateral displacements of the modified cilium, and would be the true receptor [52].

A theoretical discussion of the possible role of MT in these and other sensory cells has been presented by Atema [5, 6]. While the morphological relations between modified cilia and bundles of MT and various types of sensory perception are evident, the idea that the MT are an essential link in the transmission of stimuli, and not only mechanical supports for cells with specialized appendices, remains open to further study. As shown in Chapter 6, movement does not seem to arise from properties of the MT alone, but from modified relations between MT through the action of other proteins. If MT are capable of sensory transduction, and of originating signals when deformed, it is most probable that those other proteins will be involved. One should be careful not to jump too rapidly to conclusions: one instance of this, in the field of nerve cells, is the suggestion by Cronly-Dillon et al. [24] that MT are involved in memory fixation in goldfish: the animals received intracranial injections of 12.5 or 25 μg of colchicine in 50 μl saline, and this led to complete retrograde amnesia when studied 36 – 48 h later, while short-term memory did not appear to be affected. The role proposed for MT, axonal flow, and other factors is imaginative, although far from the hard facts known at this date about neuronal MT.

9.8 The Intranuclear Rods of Sympathetic Neurons

The assembly of MT in nuclei is by no means exceptional, and many instances will be described in relation to mitosis (Chap. 10). In sympathetic neurons of mammals, rod-like inclusions have been mentioned in the literature. In the cat, their number is related to the cellular activity. In a series of contributions [93, 145, 146] these structures have been demonstrated to consist of about 12 layers of microfilaments and of tubules. The filaments have a diameter of about 8 nm; and the tubules, which are slightly smaller than cytoplasmic MT, measure 17 – 18 nm in diameter. They are found in one layer at the periphery of the inclusion. Each inclusion, measuring from 0.4 to 0.8 μm in diameter comprizes about 800 filaments and 60 tubules. A thorough description of these inclusions, considered normal constituents of the sympathetic neurons, has been published ([145]; Fig. 9.9).

The physiological significance of these structures is indicated by the fact that electrical stimulation of the neurons leads to an increase of the number of nuclei showing these rodlets (up to ten times, in less than 30 min; [146]). Further studies have shown that this may take place in the presence of cycloheximide, and results apparently from the assembly of preexisting subunits. One suggestion that this could be comparable to the assembly of MT is brought by studies demonstrating that the injection of dbcAMP or theophyllin in the ganglia leads to a marked increase in the number of inclusions.

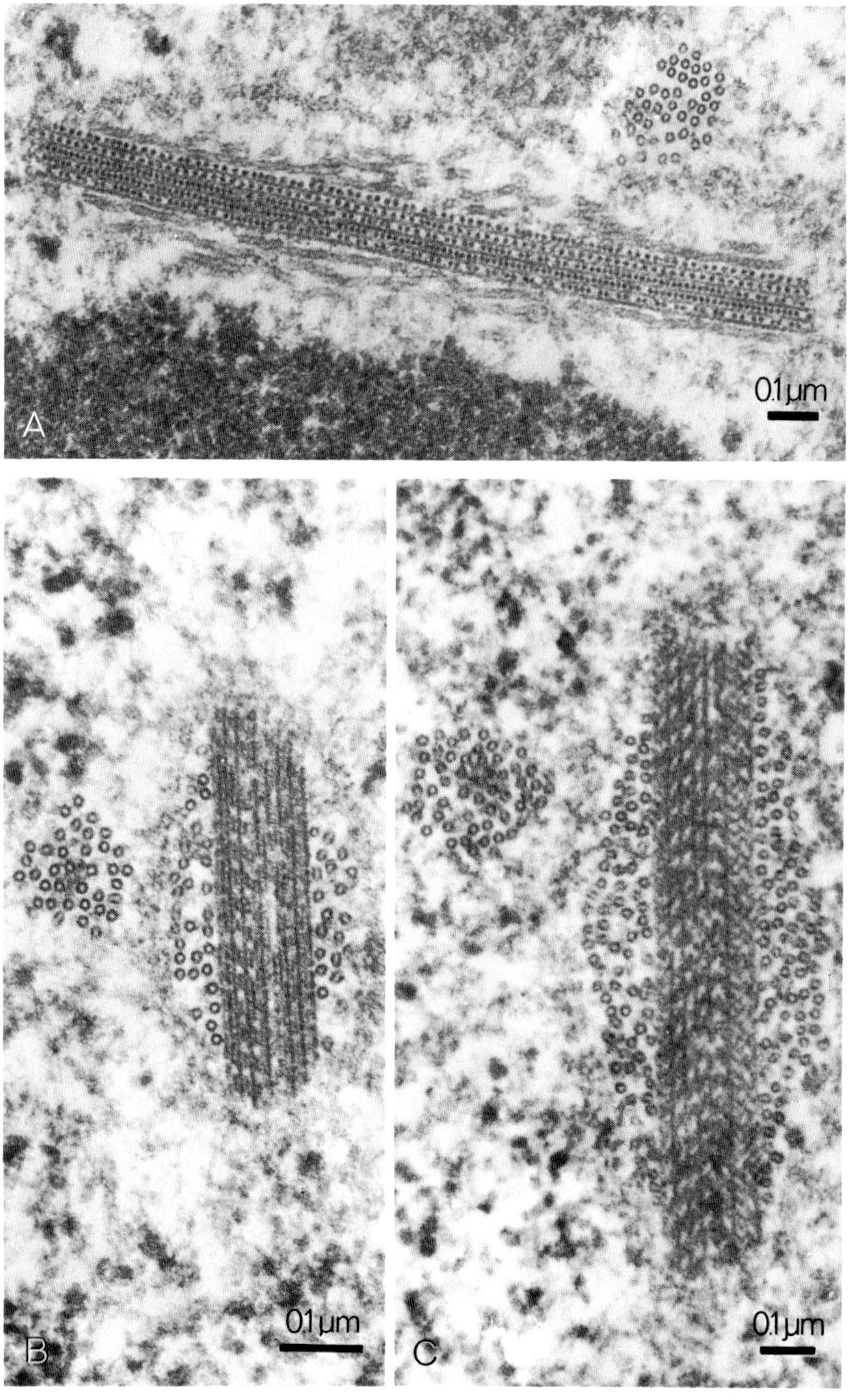
0.1 μm
A
0.1 μm
B
C
0.1 μm

These results should be completed, now that antiMT immunofluorescent antibodies are available, by the demonstration that the tubules surrounding the rodlets are truly identical with cytoplasmic MT. Further studies on other species of mammals should help to understand better the function of these inclusions.

References

1. Allison, A. C.: Function and Structure of Cell Components in Relation to Action of Anaesthetics. (eds.: T. C. Gray, J. F. Nunn) General Anaesthesia, Vol. 1. London: Butterworths 1971
2. Allison, A. C., Nunn, J. F.: Effects of general anaesthetics on microtubules: a possible mechanism of anaesthesia? Lancet **ii,** 1326 – 1329 (1968)
3. Ambron, R. T., Goldman, J. E., Schwartz, J. H.: Axonal transport of newly synthetized glycoproteins in a single identified neuron of *Aplysia californica.* J. Cell Biol. **61,** 665 – 675 (1974)
4. Anderson, K. E., Edström, A., Hanson, M.: Heavy water reversibly inhibits fast axonal transport of proteins in frog sciatic nerves. Brain Res. **43,** 299 – 302 (1972)
5. Atema, J.: Microtubule theory of sensory transduction. J. Theor. Biol. **38,** 181 – 190 (1973)
6. Atema, J.: Stimulus transmission along microtubules in sensory cells: a hypothesis. In: Microtubules and Microtubule Inhibitors (eds.: M. Borgers, M. De Brabander), pp. 247 – 257. Amsterdam-Oxford: North-Holland 1975
7. Barondes, S. H.: Slow and rapid transport of protein to nerve endings in mouse brain. Acta Neuropath. Suppl. **5,** 97 – 103 (1971)
8. Bennett, T., Cobb, J. L. S., Malmfors, T.: Fluorescence histochemical and ultrastructural observations on the effects of intravenous injections of vinblastine on noradrenergic nerves. Z. Zellforsch. **141,** 517 – 528 (1973)
9. Bergland, R. M., Tocack, R. M.: Microtubules and neurofilaments in axons of human pituitary stalk. Exp. Cell Res. **54,** 132 – 134 (1971)
10. Berlinrood, M., McGee-Russel, S. M., Allen, R. D.: Patterns of particle movements in nerve fibers “in vitro”. J. Cell Sci. **11,** 875 – 886 (1972)
11. Bertolini, B., Monaco, G., Rossi, A.: Ultrastructure of a regular arrangement of microtubules and neurofilaments. J. Ultrastruct. Res. **33,** 173 – 186 (1970)
12. Bird, M. M.: Microtubule-synaptic vesicle associations in cultured rat spinal cord neurons. Cell Tissue Res. **168,** 101 – 116 (1976)
13. Bondy, S. C., Madsen, C. J.: Development of rapid axonal flow in the chick embryo. J. Neurobiol. **2,** 279 – 286 (1971)
14. Bunge, R., Bunge, M.: Electron microscopic observations on colchicine-induced changes in neuronal cytoplasm. Anat. Rec. **160,** 323 (1968)
15. Bunt, A. H.: Effects of vinblastine on microtubule structure and axonal transport in ganglion cells of the rabbit retina. Investig. Ophthalmol. **12,** 579 – 590 (1973)
16. Bunt, A. H., Lund, R. D.: Vinblastine-induced blockage of orthograde and retrograde axonal transport of protein in retinal ganglion cells. Exp. Neurol. **45,** 288 – 297 (1974)
17. Bunt, A. H., Lund, R. D., Lund, J. S.: Retrograde axonal transport of horse-radish peroxidase by ganglion cells of the albino rat retina. Brain Res. **73,** 215 – 228 (1974)
18. Burton, P. R., Fernandez, H. L.: Delineation by lanthanum staining of filamentous elements associated with the surfaces of axonal microtubules. J. Cell Sci. **12,** 567 – 584 (1973)

Fig. 9.9 A – C. Microtubular and paracrystalline inclusions in the nuclei of sympathetic neurons from the cat. (A) A typical nuclear inclusion: it is made of two groups of perpendicular fibrils, and is surrounded by longitudinally cut MT. Another group of intranuclear MT is visible at right (from R. Seïte, unpublished). (B) A group of parallel lamellae surrounded by two groups of MT. Another group of MT is located to the left of the inclusions (from R. Seïte et al. [146]). (C) Another aspect of the nuclear inclusions, showing the close relations between the MT and the parallel sheets of rodlets (from R. Seïte, unpublished)

19. Burton, P. R., Hinkley, R. E.: Further electron microscopic characterization of axoplasmic microtubules of the ventral nerve cord of the crayfish. J. Submicrosc. Cytol. **6,** 311 – 326 (1974)
20. Byers, M. R.: Structural correlates of rapid axonal transport: evidence that microtubules may not be directly involved. Brain Res. **75,** 97 – 114 (1974)
20 a. Byers, M. R., Hendrickson, A. E., Fink, B. R., Kennedy, R. D., Middaugh, M. E.: Effects of lidocaine on axonal morphology, microtubules and rapid transport in rabbit vagus nerve in vitro. J. Neurobiol. **4,** 125 – 143 (1973)
21. Calissano, P., Levi, A., Alema, S., Chen, J., Levi-Montalcini, R.: Studies on the interaction of the nerve growth factor with tubulin and actin. In: Molecular Basis of Motility (eds.: L. M. G. Heilmeyer, J. C. Ruegg, T. Wieland), pp. 186 – 202. Berlin-Heidelberg-New York: Springer 1976
22. Cardinali, D. P., Freire, F.: Melatonin effects on brain. Interaction with microtubule protein, inhibition of fast axoplasmic flow and induction of crystalloid and tubular formations in the hypothalamus. Mol. Cell. Endocrin. **2,** 317 – 330 (1975)
23. Chang, C.: Effect of colchicine and cytochalasin B on axonal particle movement and outgrowth in vitro. J. Cell Biol. **55,** 37 a (1972)
24. Cronly-Dillon, J., Carden, D., Birks, C.: The possible involvement of brain microtubules in memory fixation. J. Exp. Biol. **61,** 443 – 454 (1974)
25. Crothers, S. D., McCluer, R. H.: Effect of colchicine on the delayed appearance of labelled protein into synaptosomal soluble proteins. J. Neurochem. **24,** 209 – 214 (1975)
26. Dahlström, A.: Effect of colchicine on transport of amine storage granules in sympathetic nerves of rat. Europ. J. Pharmacol. **5,** 111 – 113 (1968)
27. Dahlström, A.: The effects of drugs on axonal transport. In: New Aspects of Storage and Release Mechanisms of Catecholamines. Bayer Symposium 2, 20 – 36. Berlin-Heidelberg-New York: Springer 1970
28. Dahlström, A.: Axoplasmic transport (with particular respect to adrenergic neurons). Phil. Trans. Roy. Soc. London B **261,** 325 – 358 (1971)
29. Dahlström, A., Häggendal, J.: Studies on the transport and life-span of amine storage granules in a peripheral adrenergic neuron system. Acta Physiol. Scand. **67,** 278 – 288 (1966)
30. Dahlström, A., Häggendal, J.: Axonal transport of amine storage granules in sympathetic adrenergic neurons. In: Biochemistry of Simple Neuronal Models (eds.: E. Costa, E. Giocobini) Adv. in Bioch. Psychopharmacol. Vol. 2, pp. 65 – 93. New York: Raven Press; Amsterdam: North-Holland 1970
31. Dahlström, A., Heiwall, P. O., Häggendal, J., Saunders, N. R.: Effect of antimitotic drugs on the intraaxonal transport of neurotransmitters in rat adrenergic and cholinergic nerves. Ann. N. Y. Acad. Sci. **253,** 507 – 516 (1975)
32. Daniels, M. P.: Fine structural changes in neurons in nerve fibers associated with colchicine inhibition of nerve fiber formation in vitro. J. Cell Biol. **58,** 463 – 469 (1973)
33. Davison, P. F.: Axoplasmic transport: physical and chemical aspects. In: Neurosciences, Second Study Program (ed.: F. O. Schmitt), pp. 851 – 857. New York: Rockefeller University Press 1970
34. Dentler, W. L., Granett, S., Rosenbaum, J. L.: Ultrastructural localization of the high molecular weight proteins associated with in vitro-assembled microtubules. J. Cell Biol. **65,** 237 – 241 (1975)
35. Douglas, W. W., Sorimachi, M.: Colchicine inhibits adrenal medullary secretion evoked by acetylcholine without affecting that evoked by potassium. Brit. J. Pharmacol. **45,** 129 – 132 (1972)
36. Dravid, A. R., Hammerschlag, R.: Axoplasmic transport of proteins in vitro in primary afferent neurons of frog spinal cord: Effect of Ca^{++}-free incubation conditions. J. Neurochem. **24,** 711 – 718 (1975)
37. Droz, B., Rambourg, A., Koenig, H.: The smooth endoplasmic reticulum: structure and role in the renewal of axonal membrane and synaptic vesicles by fast axonal transport. Brain Res. **93,** 1 – 13 (1975)
38. Dustin, P., Hubert, J. P., Flament-Durand, J.: Action of colchicine on axonal flow and pituicytes in the hypothalamopituitary system of the rat. Ann. N. Y. Acad. Sci. **253,** 670 – 684 (1975)
39. Edström, A., Hansson, H. A., Norström, A.: Inhibition of axonal transport in vitro in frog sciatic nerves by chlorpromazine and lidocaine. A biochemical and ultrastructural study. Z. Zellforsch. **143,** 53 – 70 (1973)
40. Elam, J. S., Goldberg, J. M., Radin, N. S., Agranoff, B. W.: Rapid axonal transport of sulfated mucopolysaccharide proteins. Science **170,** 458 – 460 (1970)

41. Engelborghs, Y., Heremans, K. A. H., Hoebeke, J.: The effect of pressure on neuronal microtubules. In: Microtubules and Microtubule Inhibitors (eds.: M. Borgers, M. De Brabander), pp. 59 – 66. Amsterdam-Oxford: North-Holland 1975
42. Engelborghs, Y., Heremans, K. A. H., De Maeyer, L. C. M., Hoebeke, J.: Effect of temperature and pressure on polymerisation equilibrium of neuronal microtubules. Nature **259,** 686 – 688 (1976)
43. Feit, H., Dutton, G. R., Barondes, S. H., Shelanski, M. L.: Microtubule protein. Identification in and transport to nerve endings. J. Cell Biol. **51,** 138 – 147 (1971)
44. Fernandez, H. L., Burton, P. R., Samson, F. E.: Axoplasmic transport in the crayfish nerve cord. The role of fibrillar constituents of neurons. J. Cell Biol. **51,** 176 – 192 (1971)
45. Fernandez, H. L., Huneeus, F. C., Davison, P. F.: Studies on the mechanism of axoplasmic transport in the crayfish cord. J. Neurobiol. **1,** 395 – 409 (1970)
46. Fink, B. R., Kennedy, R. D.: Rapid axonal transport: effect of halothane anesthaesia. Anesthesiology **36,** 13 – 20 (1972)
47. Flament-Durand, J., Couck, A. M., Dustin, P.: Studies on the transport of secretory granules in the magnocellular hypothalamic neurons of the rat. II. Action of vincristine on axonal flow and neurotubules in the paraventricular and supraoptic nuclei. Cell Tissue Res. **164,** 1 – 9 (1975)
48. Flament-Durand, J., Dustin, P.: Studies on the transport of secretory granules in the magnocellular hypothalamic neurons. I. Action of colchicine on axonal flow and neurotubules in the paraventricular nuclei. Z. Zellforsch. **130,** 440 – 454 (1972)
49. Forman, D. S., Padjen, A. L., Siggins, G. R.: Movements of optically detectable intraaxonal organelle in vitro. J. Cell Biol. **67,** 119 a (1975)
50. Friede, R. L., Ho, K. C.: The relation of axonal transport of mitochondria with microtubules and other axoplasmic organelles. J. Physiol. **265,** 507 – 519 (1977)
51. Friede, R., Samorajski, T.: Axon caliber related to neurofilaments and microtubules in sciatic nerves of rats and mice. Anat. Rec. **167,** 379 – 387 (1970)
52. Gaffal, K. P., Bassemir, U.: Vergleichende Untersuchung modifizierter Cilienstrukturen in den Dendriten mechano- und chemosensitiver Rezeptorzellen der Baumwollwanze *Dysdercus* und der Libelle *Agrion.* Protoplasma **82,** 177 – 202 (1974)
53. Gainer, H., Sarne, Y., Brownstein, M. J.: Biosynthesis and axonal transport of rat neurohypophysial proteins and peptides. J. Cell Biol. **73,** 366 – 381 (1977)
54. Gamache, F. W. Jr., Gamache, J. F.: Changes in axonal transport in neurones of *Asterias vulgaris* and *Asterias forbesei* produced by colchicine and dimethyl sulfoxide. Cell Tissue Res. **152,** 423 – 436 (1974)
55. Gaskin, F., Kramer, S. B., Cantor, C. R., Adelstein, E., Shelanski, M. L.: A dynein-like protein associated with neurotubules. FEBS Lett. **40,** 281 – 286 (1974)
56. Grafstein, B., McEwen, B. S., Shelanski, M. L.: Axonal transport of neurotubule protein. Nature (London) **227,** 289 – 290 (1970)
57. Grainger, F., Sloper, J. C.: Correlation between microtubular number and transport activity of the hypothalamo-neurohypophyseal secretory neurons. Cell Tissue Res. **153,** 101 – 114 (1974)
58. Griffin, J. W., Price, D. L., Drachman, D. B., Engel, W. K.: Axonal transport to and from the motor nerve ending. Ann. N. Y. Acad. Sci. **274,** 31 – 45 (1976)
59. Hammerschlag, R., Dravid, A. R., Ciu, A. Y.: Mechanism of axonal transport: a proposed role for calcium ions. Science **188,** 273 – 274 (1975)
60. Hammond, G. R., Smith, R. S.: Inhibition of the rapid movements of optically detectable axonal particles by colchicine and vinblastine. Brain Res. **128,** 227 – 242 (1977)
61. Hansson, H. A., Sjöstrand, J.: Ultrastructural effects of colchicine on the hypoglossal and dorsal vagal neurons of the rabbit. Brain Res. **35,** 379 – 396 (1971)
62. Hendry, I. A., Stökel, K., Thoenen, H., Iversen, L. L.: The retrograde axonal transport of nerve growth factor. Brain Res. **68,** 103 – 121 (1974)
63. Hindelang-Gertner, C., Stoeckel, M. E., Porte, A., Stutinsky, F.: Colchicine effects on neurosecretory neurons and other hypothalamic and hypophyseal cells, with special reference to change in the cytoplasmic membranes. Cell Tissue Res. **170,** 17 – 42 (1976)
64. Hinckley, R. E. Jr.: Axonal microtubules and associated microfilaments stained by Alcian blue. J. Cell Sci. **13,** 753 – 762 (1973)
65. Hinckley, R. E., Burton, P. R.: Tannic acid staining of axonal microtubules. J. Cell Biol. **63,** 139 a (1974)
66. Hinckley, R. E. Jr., Green, L. S.: Effects of halothane and colchicine on microtubules and electrical activity of rabbit vagus nerves. J. Neurobiol. **2,** 97 – 106 (1971)
67. Hinckley, R. E., Samson, F. E.: Anesthetic-induced transformation of axonal microtubules. J. Cell Biol. **53,** 258 – 263 (1972)

68. Hinckley, R. E., Telser, A.: Halothane inhibition of neuroblastoma division and neurite formation. J. Cell Biol. **55,** 114 a (1972)
69. Hirano, A.: Neurofibrillary changes in conditions related to Alzheimer's disease. In: Alzheimer's Diseases and Related Conditions (eds.: G. E. W. Wolstenholme, M. O'Connor), pp. 185 – 201. Ciba Foundation Symposium. London: Churchill 1970
70. Hirano, A., Zimmerman, H. M.: Some new pathological findings in the central myelinated axon. J. Neuropath. Exp. Neurol. **30,** 325 – 336 (1971)
71. Hoffman, P. N., Lasek, R. J.: The slow component of axonal transport. Identification of major structural polypeptides of the axon and their generality among mammalian neurons. J. Cell Biol. **66,** 351 – 366 (1975)
72. Hökfelt, T., Dahlström, A.: Effects of two mitosis inhibitors (colchicine and vinblastine) on the distribution and axonal transport of noradrenaline storage particles, studied by fluorescence and electron microscopy. Z. Zellforsch. **119,** 460 – 482 (1971)
73. Huneeus, F. C., Davison, P. F.: Fibrillar proteins from squid axons. I. Neurofilament protein. J. Mol. Biol. **52,** 415 – 428 (1970)
74. James, K. A. C., Austin, L.: The binding in vitro of colchicine to axoplasmic proteins from chicken sciatic nerve. Biochem. J. **117,** 773 – 777 (1970)
75. Järlfors, U., Smith, D. S.: Association between synaptic vesicles and neurotubules. Nature (London) **224,** 710 – 711 (1969)
76. Jeffrey, P. L., James, K. A. C., Kidman, A. D., Richards, A. M., Austin, L.: The flow of mitochondria in chicken sciatic nerve. J. Neurobiol. **3,** 199 – 208 (1972)
77. Journey, L. J., Burdman, J., Whaley, A.: Electron microscopic study of spinal ganglia from vincristine-treated mice. J. Natl. Cancer Inst. **43,** 603 – 620 (1969)
78. Karlsson, J. O., Sjöstrand, J.: The effects of colchicine on the axonal transport in the optic nerve and tract of the rabbit. Brain Res. **13,** 617 – 619 (1969)
79. Karlsson, J. O., Sjöstrand, J.: Characterization of the fast and slow components of axonal transport in retinal ganglion cells. J. Neurobiol. **2,** 135 – 144 (1971)
80. Karlsson, J. O., Sjöstrand, J.: Transport of microtubular protein in axons of retinal ganglion cells. J. Neurochem. **18,** 975 – 982 (1971)
81. Kennedy, R. D., Kink, B. R., Byers, M. R.: The effect of halothane on rapid axonal transport in the rabbit vagus. Anesthesiology **36,** 433 – 443 (1972)
82. Khan, M. A., Ochs, S.: Slow axoplasmic transport of mitochondria (MAO) and lactic dehydrogenase in mammalian nerve fibers. Brain Res. **96,** 267 – 277 (1975)
83. Kirkpatrick, J. B., Palmer, S. M.: Ionic requirements for rapid particulate axoplasmic flow. Am. J. Pathol. **66,** 4 a (1972)
84. Klatzo, I., Wisniewski, H., Streicher, E.: Experimental production of neurofibrillary degeneration. I. Light microscopic observations. J. Neuropathol. Exp. Neurol. **24,** 187 – 199 (1965)
85. Kristensson, K.: Neural spread of herpes simplex virus. 7th Intern. Cong. Neuropathol. **2,** 329 – 332. Amsterdam: Excerpta Med. 1975
86. Kristensson, K., Lycke, E., Sjöstrand, J.: Transport of herpes simplex virus in peripheral nerves. Acta Physiol. Scand. (Suppl.) **357,** 14 – 14 (1970)
87. Kristensson, K., Lycke, E., Sjöstrand, J.: Retrograde transport of protein tracer in the rabbit hypoglossal nerve during regeneration. Brain Res. **45,** 175 – 181 (1972)
88. Kuypers, H. G. J. M., Masiky, V. A.: Retrograde axonal transport of horseradish peroxidase from spinal cord to brain stem cell groups in the cat. Neurosci. Lett. **1,** 9 – 14 (1975)
89. Lasek, R.: Protein transport in neurons. Intern. Rev. Neurobiol. **13,** 289 – 324 (1970)
90. Lasek, R. J., Hoffman, P. N.: The neuronal cytoskeleton, axonal transport and axonal growth. In: Cell Motility (eds.: R. Goldmann, T. Pollard, J. Rosenbaum), pp. 1021 – 1050. Cold Spring Harbor Lab. 1976
91. Lavail, J. H., Lavail, M. M.: The retrograde intraaxonal transport of horseradish peroxidase in the chick visual system: a light and electron microscopic study. J. Comp. Neurol. **157,** 303 – 358 (1974)
92. Le Beux, Y. L.: An ultrastructural study of the synaptic densities, nematosomes, neurotubules, neurofilaments and of a further three-dimensional filamentous network as disclosed by the E-PTA staining procedure. Z. Zellforsch. **143,** 239 – 272 (1973)
93. Leonetti, J., Seïte, R.: Influence du dibutyryl AMP cyclique et de la théophylline sur la fréquence des microtubules et des microfilaments nucléaires des neurones sympathiques. C. R. Acad. Sci. (Paris) D **281,** 423 – 426 (1975)
94. Lieberman, A. R.: Microtubule-associated smooth endoplasmic reticulum in the frog's brain. Z. Zellforsch. **116,** 564 – 577 (1971)
95. Lubińska, L.: Axoplasmic streaming in regenerating and in normal nerve fibers. Progr. Brain Res. **13,** 1 – 71 (1964)

96. Lubińska, L.: Axoplasmic flow. In: Intern. Rev. Morphol. Vol. 17, pp. 241 – 296 (eds.: C. C. Pfeiffer, J. R. Smythies). New York: Acad. Press 1975
97. Lubińska, L., Niermierko, S.: Velocity and intensity of bidirectional migration of acetylcholinesterase in transected nerves. Brain Res. **27,** 328 – 342 (1971)
98. McClure, W. O.: Effects of drugs on axoplasmic transport. Adv. Pharmacol. Chemother. **10,** 185 – 220 (1972)
99. McGregor, A. M., Komiya, Y., Kidman, A. D., Austin, L.: The blockage of axoplasmic flow of proteins by colchicine and cytochalasins A and B. J. Neurochem. **21,** 1059 – 1066 (1973)
100. Mayor, D., Tomlinson, D. R., Banks, P., Mraz, P.: Microtubules and the intra-axonal transport of noradrenaline storage (dense cored) vesicles. J. Anat. **111,** 344 – 345 (1972)
101. Moran, D. T., Chapman, K. M., Ellis, R. A.: The fine structure of cockroach campaniform sensilla. J. Cell Biol. **48,** 155 – 173 (1971)
102. Moran, D. T., Varela, F. G.: Microtubules and sensory transduction. Proc. Natl. Acad. Sci. USA **68,** 757 – 760 (1971)
103. Murray, M. R., Benitez, H. H.: Action of heavy water (D_2O) on growth and development of isolated nervous tissues. In: Ciba Foundation Symposium Growth of the Nervous System (eds.: E. W. Wolstenholme, M. O'Connor), pp. 148 – 178. London: Churchill 1968
104. Nadelhaft, I.: Microtubule densities and total numbers in selected axons of the crayfish abdominal nerve cord. J. Neurocytol. **3,** 73 – 86 (1974)
105. Nagasava, J., Douglas, W. W., Schulz, R. A.: Ultrastructural evidence of secretion by exocytosis and of "synaptic vesicle" formation in posterior pituitary glands. Nature (London) **227,** 407 – 409 (1970)
106. Norström, A.: Axonal transport in the hypothalamoneurohypophyseal system (HNS) of the rat. 7th Intern. Cong. Neuropath. **2,** 315 – 324. Amsterdam: Excerpta Med. 1975
107. Norström, A., Hansson, H. A., Sjöstrand, J.: Effects of colchicine on axonal transport and ultrastructure of the hypothalamoneurohypophyseal system of the rat. Z. Zellforsch. **113,** 271 – 293 (1971)
108. Norström, A., Hansson, H. A.: Effects of colchicine on release of neurosecretory material from the posterior pituitary gland of the rat. Z. Zellforsch. **142,** 443 – 464 (1973)
109. Norström, A., Sjöstrand, J.: Axonal transport of proteins in the hypothalamo-neurohypophyseal system of the rat. J. Neurochem. **18,** 29 – 40 (1971)
110. Ochs, S.: The dependence of fast transport in mammalian nerve fibers on metabolism. Acta Neuropathol. Suppl. **5,** 86 – 96 (1971)
111. Ochs, S.: Characteristics and a model for fast axoplasmic transport in nerve. J. Neurobiol. **2,** 331 – 346 (1971)
112. Ochs, S.: Energy metabolism and supply of $\sim$P to the fast axoplasmic transport mechanism in nerve. Fed. Proc. **33,** 1049 – 1058 (1974)
113. Ochs, S.: Axoplasmic transport. A basis for neural pathology. In: Peripheral Neuropathy (eds.: P. J. Dyck, P. K. Thomas, E. H. Lambert), pp. 213 – 230. Philadelphia-London-Toronto: Saunders 1975
114. Ochs, S., Hollingsworth, D.: Dependence of fast axoplasmic transport in nerve on oxidative metabolism. J. Neurochem. **18,** 107 – 114 (1971)
115. Ochs, S., Ranish, N.: Characteristics of the fast transport system in mammalian nerve fibers. J. Neurobiol. **2,** 247 – 261 (1969)
116. Ochs, S., Smith, C.: Low temperature slowing and cold-block of fast axoplasmic transport in mammalian nerves in vitro. J. Neurobiol. **6,** 85 – 102 (1975)
117. Ochs, S., Worth, R.: Batrachotoxin block of fast axoplasmic transport in mammalian nerve fibers. Science **187,** 1087 – 1089 (1975)
118. O'Connor, T. M., Houston, L. L., Samson, F.: Stability of neuronal microtubules to high pressure in vivo and in vitro. Proc. Natl. Acad. Sci. USA **71,** 4198 – 4202 (1974)
119. Olsson, Y., Kristensson, K.: Retrograde axonal transport in peripheral nerves and signals for chromatolysis. 7th Intern. Cong. Neuropathol. Budapest: Akadémiai Kiadó 1974
120. Östergren, G.: Narcotized mitosis and the precipitation hypothesis of narcosis. Coll Int. C. N. R. S. **26,** 77 – 88 (1951)
121. Palay, S. L.: Synapses in the central nervous system. J. Biophys. Biochem. Cytol. **2,** 193 – 201 (1956)
122. Partlow, L. M., Ross, C. D., Motwani, R., McDougal, D. B. Jr.: Transport of axonal enzymes in surviving segments of frog sciatic nerve. J. Gen. Physiol. **60,** 388 – 405 (1972)
123. Paulson, J. C., McClure, W. O.: Inhibition of axoplasmic transport by mescaline and other trimethoxyphenylalkylamines. Mol. Pharmacol. **9,** 41 – 50 (1973)
124. Paulson, J. C., McClure, W. O.: Microtubules and axoplasmic transport. Brain Res. **73,** 333 – 337 (1974)

125. Paulson, J. C., McClure, W. O.: Microtubules and axoplasmic transport. Inhibition of transport by podophyllotoxin: an interaction with microtubule protein. J. Cell Biol. **67,** 461 – 468 (1975)
126. Paulson, J. C., McClure, W. O.: Inhibition of axoplasmic transport by colchicine, podophyllotoxin, and vinblastine: an effect on microtubules. Ann. N. Y. Acad. Sci. **253,** 517 – 527 (1975)
127. Peterson, E. R.: Neurofibrillar alterations in cord-ganglion cultures exposed to spindle inhibitors. J. Neuropathol. Exp. Neurol. **28,** 168 (1969)
128. Poisner, A. M., Bernstein, J.: A possible role of microtubules in catecholamine release from the adrenal medulla. Effect of colchicine, *Vinca* alkaloids and deuterium oxide. J. Pharmacol. Exp. Ther. **177,** 102 – 108 (1971)
129. Pomerat, C., Hendelman, W. J., Raiborn, C. W., Massey, J. F.: Dynamic activities of nervous tissue in vitro. In: The Neuron (ed.: H. Hyden), pp. 119 – 178. New York: Acad. Press 1967
130. Potter, H. D.: The distribution of neurofibrils coextensive with microtubules and neurofilaments in dendrites and axons of the tectum, cerebellum and pallidum of the frog. J. Comp. Neurol. **143,** 385 – 410 (1971)
131. Price, M. T.: The effects of colchicine and lumicolchicine on the rapid phase of axonal transport in the rabbit visual system. Brain Res. **77,** 497 – 501 (1974)
132. Price, D. L., Griffin, J., Young, A., Peck, K., Stocks, A.: Tetanus toxin: direct evidence for retrograde intraaxonal transport. Science **188,** 945 – 947 (1975)
133. Price, D. L., Griffin, J. W.: Structural substrate of protein synthesis and transport in spinal motor neurons. In: Amyotrophic Lateral Sclerosis (eds.: J. M. Andrews, R. T. Johnson, M. A. B. Brazier), pp. 1 – 32. New York-San Francisco-London: Acad. Press 1976
134. Pumplin, D. W., McClure, W. O.: Effects of cytochalasin B and vinblastine on the release of acetylcholine from a sympathetic ganglion. Eur. J. Pharmacol. **28,** 316 – 325 (1974)
135. Reese, T. S.: Olfactory cilia in the frog. J. Cell Biol. **25,** 209 – 230 (1965)
136. Reid, J. L., Kopin, I. J.: The effects of ganglionic blockage, reserpine and vinblastine on plasma catecholamines and dopamine-β-hydroxylase in the rat. J. Pharmacol. Exp. Ther. **193,** 748 – 756 (1968)
137. Rodriguez Echandia, E. L., Piezzi, R. S.: Microtubules in the nerve fibers of the toad *Bufo arenarum* Hensel. Effect of low temperature on the sciatic nerve. J. Cell Biol. **39,** 491 – 497 (1968)
138. Roisen, F. J., Braden, W. G., Friedman, J.: Neurite development in vitro. III. The effects of several derivatives of cyclic AMP, colchicine and colcemid. Ann. N. Y. Acad. Sci. **253,** 545 – 561 (1975)
139. Roisen, F. J., Murphy, R. A.: Neurite development in vitro. II. The role of microfilaments and microtubules in dibutyryl adenosine 3′,5′-cyclic monophosphate and nerve growth factor stimulated maturation. J. Neurobiol. **4,** 397 – 412 (1973)
140. Roth, L. E., Pihlaja, D. J., Shigenaka, Y.: Microtubules in the heliozoan axopodium. I. The gradion hypothesis of allosterism in structural proteins. J. Ultrastruct. Res. **30,** 7 – 37 (1970)
141. Samson, F. E. Jr., Hinckley, R. E. Jr.: Neuronal microtubular systems. Anesthesiology **36,** 417 – 421 (1972)
142. Saubermann, A. J., Gallagher, M. L.: Allison and Nunn revisited. Anesthesiology **39,** 357 (1973)
143. Saubermann, A. J., Gallagher, M. L.: Mechanisms of general anesthesia: failure of pentobarbital and halothane to depolymerize microtubules in mouse optic nerve. Anesthesiology **38,** 25 – 29 (1973)
144. Schmitt, F. O.: Fibrous proteins—neuronal organelles. Proc. Natl. Acad. Sci. USA **60,** 1092 – 1101 (1968)
145. Seïte, R., Escaig, J., Couineau, S.: Microfilaments et microtubules nucléaires et organisation ultrastructurale des bâtonnets intranucléaires des neurones sympathiques. J. Ultrastruct. Res. **37,** 449 – 478 (1971)
146. Seïte, R., Mei, N., Vuillet-Luciani, J.: Effect of electrical stimulation on nuclear microfilaments and microtubules of sympathetic neurons submitted to cycloheximide. Brain Res. **50,** 419 – 423 (1973)
147. Sjöstrand, J., Frizell, M., Hasselgren, P. O.: Effects of colchicine on axonal transport in peripheral nerves. J. Neurochem. **17,** 1563 – 1570 (1970)
148. Sloper, J. C., Grainger, F.: Quantitation of microtubules in secretory neurons. In: Microtubules and Microtubule Inhibitors (eds.: M. Borgers, M. De Brabander), pp. 281 – 287. Amsterdam-Oxford: North Holland 1975
149. Smith, B. H.: Neuroplasmic transport in the nervous system of the cockroach *Periplaneta americana.* J. Neurobiol. **2,** 107 – 118 (1971)

150. Smith, D. E.: The location of neurofilaments and microtubules during the postnatal development of Clarke's nucleus in the kitten. Brain Res. **55,** 41 – 54 (1973)
151. Smith, D. S., Järlfors, U., Beraneck, R.: The organization of synaptic axoplasm in the lamprey (*Petromyzon marinus*) central nervous system. J. Cell Biol. **46,** 199 – 219 (1970)
152. Smith, D. S., Järlfors, U., Cameron, B. F.: Morphological evidence for the participation of microtubules in axonal transport. Ann. N. Y. Acad. Sci. **253,** 472 – 506 (1975)
153. Sorimachi, M., Oesch, F., Thoenen, H.: Effects of colchicine and cytochalasin B on the release of ^{3}H-norepinephrine from guinea-pig atria evoked by high potassium, nicotine and tyramine. Arch. Pharmacol. **276,** 1 – 12 (1973)
154. Stöckel, K., Paravicini, U., Thoenen, H.: Specificity of the retrograde axonal transport of nerve growth factor. Brain Res. **76,** 413 – 421 (1974)
155. Terry, R. D. Peña, C.: Experimental production of neurofibrillary degeneration. II. Electron mictroscopy, phosphatase histochemistry and electron probe analysis. J. Neuropath. Exp. Neurol. **24,** 200 – 210 (1965)
156. Thurm, U.: An insect mechanoreceptor. I. Fine structure and adequate stimulus. Cold Spring Harbor Symp. Quant. Biol. **30,** 75 – 82 (1965)
157. Trifaro, J. M., Collier, B., Lastoweka, A., Stern, D.: Inhibition by colchicine and by vinblastine of acetylcholine-induced catecholamine release from the adrenal gland: an anticholinergic action, not an effect on microtubules. Molec. Pharmacol. **8,** 264 – 267 (1972)
158. Weiss, P. A.: Neuronal dynamics. Neurosci. Res. Program Bull. **5,** 371 – 400 (1967). Reprinted in: Dynamics of Development. Experiments and Interferences (ed.: P. A. Weiss). New York-London: Acad. Press 1968
159. Weiss, P. A.: Neuronal dynamics and neuroplasmic ("axonal") flow. In: Cellular Dynamics of the Neuron (ed.: S. H. Barondes), pp. 3 – 34. New York-London: Acad. Press 1969
160. Weiss, P. A., Hiscoe, H. B.: Experiments on the mechanism of nerve growth. J. Exp. Zool. **107,** 315 – 395 (1948)
161. Weiss, P. A., Mayr, R.: Neuronal organelles in neuroplasmic ("axonal") flow. I. Mitochondria. II. Neurotubules. Acta Neuropath. Suppl. **5,** 187 – 197; 198 – 206 (1971)
162. Weiss, P., Taylor, A. C.: Impairment of growth and myelinization in regenerating nerve fibers subject to constriction. Proc. Soc. Exp. Biol. Med. **55,** 77 – 80 (1944)
163. Weiderhold, M. L.: Mechanosensory transfunction in "sensory" and "motile" cilia. Ann. Rev. Biophys. Bioengin. **5,** 39 – 62 (1976)
164. Willard, M., Cowan, W. M., Vazeles, P. R.: The polypeptide composition of intraaxonally transported proteins: evidence for four transport velocities. Proc. Natl. Acad. Sci. USA **71,** 2183 – 2187 (1974)
165. Wisniewski, H., Shelanski, M. L., Terry, R. D.: Experimental colchicine encephalopathy. I. Induction of neurofibrillary degeneration. Lab. Invest. **17,** 577 – 587 (1967)
166. Wisniewski, H., Terry, R. D., Shelanski, M. L.: Neurofibrillary degeneration of nerve cells after subarachnoid injection of mitotic spindle inhibitors. J. Neuropathol. Exp. Neurol. **28,** 168 – 169 (1969)
167. Wisniewski, H., Shelanski, M. L., Terry, R. D.: Effects of mitotic spindle inhibitors on neurotubules and neurofilaments in anterior horn cells. J. Cell Biol. **38,** 224 – 229 (1968)
168. Wolff, J. R., Wolff, A.: Is an ATP-ase the drive of the axonal flow? 7th Intern. Cong. Neuropathol. **2,** 305 – 308. Amsterdam: Excerpta Medica 1975
169. Wooten, G. F., Kopin, I. J., Axelrod, J.: Effects of colchicine and vinblastine on axonal transport and transmitter release in sympathetic nerves. Ann. N. Y. Acad. Sci. **253,** 528 – 534 (1975)
170. Wuerker, R. B.: Neurofilaments and glial filaments. Tissue Cell **2,** 1 – 9 (1970)
171. Yamada, K. M., Wessells, N. K.: Axon elongation. Effect of nerve growth factor on microtubule protein. Exptl. Cell Res. **66,** 346 – 352 (1971)
172. Zelená, J.: Bidirectional movements of mitochondria along axons of an isolated nerve segment. Z. Zellforsch. **92,** 186 – 196 (1968)
173. Zelená, J., Lubińska, L., Gutman, E.: Accumulation of organelles at the ends of interrupted axons. Z. Zellforsch. **91,** 200 – 219 (1968)

Chapter 10 The Role of MT in Mitosis

10.1 Introduction

The study of MT has been linked from the start with that of mitosis, even before the identification of spindle "fibers" as MT and the role of MT in accessory structures such as the centrioles became evident. The earliest work on colchicine cytotoxicity [175] illustrated the mitotic changes (Chap. 1), and modern colchicine research was the outcome of work on mitotic poisons [52, 137]. The importance of colchicine poisoning of mitosis for many years outshadowed the more fundamental problems, which could only find an answer after the purification of tubulin and the demonstration of the specific binding of colchicine to this protein (cf. Chaps. 2 and 5). Spindle poisons have for nearly fourty years played a considerable role in the study of mitotic growth, cytogenetics, and the production of polyploid or amphidiploid plants. These aspects of "applied,, research on mitosis with spindle poisons will not be considered here, and this chapter will be limited to the fundamental problems of the role of MT in mitotic movements.

Since the school of Mazia [151] showed that most of the proteins of the mitotic apparatus (MA), comprizing the spindle and related structures, were present before mitosis, the real importance of the MT observed by electron microscopy in the spindle and asters has been under discussion. These studies have given rise to a very large number of publications. No explanation of the role of MT is possible without an overall understanding of the chromosome movements and other cell changes. However, this is not a review of mitosis, and our attention will be fixed on the MT and the structures directly related to them, without considering the multiple other aspects of mitosis, such as its control, its metabolic requirements, the differential sensitivities of cells and organisms towards the spindle poisons, the mechanisms of chromosome formation and of DNA reduplication, the nucleolar cycle, etc.

On the other hand, any explanation of the role of MT in cell reduplication in eukaryotes should take into account not only the classical mitosis of histology textbooks (the mitosis of vertebrates in particular), but should be applicable to the many strange or "atypical" ones found in various species of animals, plants, and protozoa: some of these present illuminating aspects which lead to a clearer understanding of the role of MT. These structures being present in all eukaryotic mitoses—MT may have appeared in evolution when the nucleus became separated from the cytoplasm by a membrane—, it is natural to believe that some fundamental invariants must be present in all types of divisions. It may seem presuming to embark on such a generalization, but the task is worthwhile, for the discoveries of the last decades present to the student of mitosis a small number of facts which must be taken into account in any theoretical consideration of the role of MT. While some authors have, by painstaking research, analyzed in depth the mitosis of one or two species, the purpose of this chapter will be a tentative integration of results obtained on a large variety of cells: it will reveal some constant properties of MT and help to explain what their functions might be and how they may affect chromosome movements.

In preceding chapters, several properties of the spindle have been mentioned: its birefringence, and its changes under the effects of poisons and physical agents such as cold or high hydrostatic pressures. The problems of MT assembly and the thermodynamics of MT formation have been discussed in Chapter 2, while the structure of centrioles and other "microtubule organizing centers" or MTOC has been mentioned in Chapter 4. The various spindle poisons affecting MT have already been listed in Chapter 5, and Chapters 7 and 9 have provided some insight into the relations between MT and intracellular movements. The reader should thus be prepared to follow the dynamics of chromosome movement, although these are far more complex than any other type of transport, while considerably slower and requiring only minute amounts of energy (chromosomes move at a rate of about 1 μm/min, i.e., about 1.5 mm/day, a rate comparable to the slow neuroplasmic flow).

This field has been treated by many authors and the books by Schrader [209], Levine [135], Luykx [139], Bajer and Molè-Bajer [9], and the conferences edited by Soifer [216], Inoué and Stephens [113], and Goldman et al. [83] gather a large amount of information on the subject of mitotic MT. Other important reviews are those of Mazia [151], Went [226], de Harven [46], Nicklas [160], Forer [65], Hartmann and Zimmerman [92], Kubai [127], to mention only those directly related to the problems of MT and mitosis.

10.2 Some Aspects of the Evolution of Mitosis

In order to understand properly the multiple roles of MT in complex mitoses such as those of vertebrates, where several types of MT are involved apparently independently in various functions—centriolar structures, continuous fibers, kinetochore fibers, telophasic body—, simpler types of cell division should be described first. This leads to an attempt to classify the various types of mitosis on an evolutionary scale. The reader is referred for further information to a recent symposium [218] and to the papers by Pickett-Heaps [180, 182] and Kubai [127].

It may be useful, from the start, to attempt a definition of mitosis, from the point of view of chromosome movements, in order to grasp what is essential and what is not. Centrioles for instance, or centriole equivalents such as polar plaques or other MTOC, are absent from the mitoses of all higher plants and appear of secondary importance; the role of centrioles is mainly to provide the sister cells with basal bodies for cilia formation. Polar plaques, as found in many "primitive" mitoses, are true MTOC from which the mitotic MT are seen to arise. They may be related in some ways to the primitive kinetochores, as they appear as specializations of the nuclear membrane. Their presence does not appear indispensable for the formation of a spindle.

One fundamental property of mitosis is the *bipolarity* of its structure, which appears more or less closely related to the polarity of the spindle. Of course, multipolar mitoses do exist, but they have the same basic organization. Unipolar mitoses may exist in some special conditions (cf. the well-known case of *Sciara,* cf. [209]); they have special functions such as the elimination of some chromosomes, and need not be considered here. The twofold symmetry of the spindle would be useless if the chromosomes, after division, did not possess a similar symmetry, so that the two groups of sister chromosomes may move to the opposite poles. This last symmetry results from the duplication of chromatids, and in higher organisms from the twofold symmetry of the kinetochores, which become aligned with the spindle.

A second feature which is indispensable is longitudinal growth: in the simplest mitoses, elongation of the nucleus, which is related to the growth of polar MT, is the main change observed during division. It takes place in relation to the separation of the chromosomes, which may result from their attachment to the nuclear membrane, or from the relations between their kinetochores and MT. It will become clear that MT have a double function: nuclear elongation, and kinetochore separation, and this duality appears to be present in mitoses of all eukaryotes.

The mechanisms of mitosis normally lead to the formation of two daughter cells, and cytoplasmic movements are necessary for the last stages of telophase. At this step, where the role of contractile cytoplasmic proteins such as actin is apparent, the remnants of the mitotic MT may persist in the shape of a telophasic bundle, which is found in various cells, but has only a secondary role in mitosis; this structure is interesting because its MT show properties which are distinct from those of the other spindle MT.

The problem of imagining how this complex series of events appeared in evolution is formidable, but the study of some "primitive" mitoses gives some indications of possible steps. In prokaryotes the genetic information (genophore), after reduplication of the DNA molecule, is carried to the two daughter cells through attachment of the DNA to the cell membrane [181, 197]. The transition to eukaryotes separated the genome from the cell membrane by enclosing it in a nuclear envelope: it may be imagined that this also elongated with cell growth, and separated the chromosomes which would be linked to it. At some moment in evolution, this separation was helped by the formation of MT, linking the chromosomes to specialized zones of the nuclear membrane, and helping the elongating cell to divide in two, without error, its genetic information. Mitoses of such a type are known, and suggest that in the most primitive forms MT were intranuclear. At a later stage, the chromosome attachments to the nuclear envelope became more autonomous, and behaved as MTOC. Later still, they lost all relations with the envelope and became the kinetochores. These steps were linked with the formation of an extranuclear spindle with special fibers attached to the kinetochores (K-MT); at this moment, two types of spindle MT, continuous or polar (playing a role in elongation) and kinetochoric (indispensable for chromosome separation) were present, as in the mitosis of higher plants and animals.

10.3 Some Types of Mitosis

The evolutionary sequence sketched above may be illustrated by some examples which help to understand the role of MT. These have been chosen from the recent literature, placing special emphasis on the relations between the spindle structures and the mitotic movements. The reader should be aware that the variety of mitoses is considerable, and that unicellulars alone provide a wealth of interesting variants. The constancy of some features helps to clarify the role of MT. Their dual function—elongation and chromosome separation—will become more evident.

10.3.1 Intranuclear MT

In many species of protozoa, in the acrasiales (slime moulds) and in fungi, the nuclear membrane does not break down during mitosis. It elongates under the influence of a bundle of rectilinear MT which may or not be attached to specialized regions comparable to polar plaques or MTOC. In the micronucleus of the ciliate *Paracineta limbata,* at interphase, a paracrystalline structure probably representing some form of tubulin is visible. From this, intranuclear MT are formed at mitosis. One peculiarity of this species is that the first MT to appear have a diameter of 30 – 40 nm, while later they have the more orthodox size of 18 – 20 nm. These MT are parallel, and no definite attachments either to the nuclear membrane or to the poorly defined chromosomes is apparent [43, 93]. On the other hand, in the fungus *Pilobolus crystallinus* (zygomycete) the elongation of the nucleus by an intranuclear spindle shows definite attachment zones to the nuclear membrane, as multilaminated structures which resemble some kinetochores. No special attachment zones of the chromosomes (which are not individualized) are visible; the role of the spindle is mainly to elongate considerably the nucleus (Fig. 10.1; [18]). Similar intranuclear mitoses have been described in various species: in the slime-mould *Physarum polycephalum* [90] (where, however, during the amebic phase, typical centrioles are present, [1]), in *Trypanozoma rhodiense* [222], and in several species of haplosphoridia such as *Minchina nelsoni.* In this species, the MT radiate from an intranuclear "spindle pole body" and elongate until the pole bodies reach the nuclear membrane. The lengthening of the MT is such that the distance between the poles is multiplied by a factor of two or three [174]. This intranuclear spindle and its two polar bodies persist during interphase; no definite connections between MT and chromosomes are apparent.

10.3.2 Partially Intranuclear Spindle

As mentioned above, the role of the nuclear membrane varies from one species to another, and the behavior of the MT in yeasts and fungi shows that the relations between the MT and the nucleus may change during mitosis. These facts are interesting from an evolutionary point of view, for they could also indicate that the "primitive" MT were extranuclear and secondarily became intranuclear, contrary to the most widely accepted belief of a primitive intranuclear mitotic apparatus.

The mitosis of the heterobasidiomycetous yeasts such as *Leucosporidium (Candida) scotti* is remarkable. While in other yeasts such as *Saccharomyces cerevisiae* the spindle is entirely intranuclear, with a dense body of fusorial plate at each end [155, 177], in *Leucosporidium* and the related *Rhodosporidium* sp. the spindle arises outside

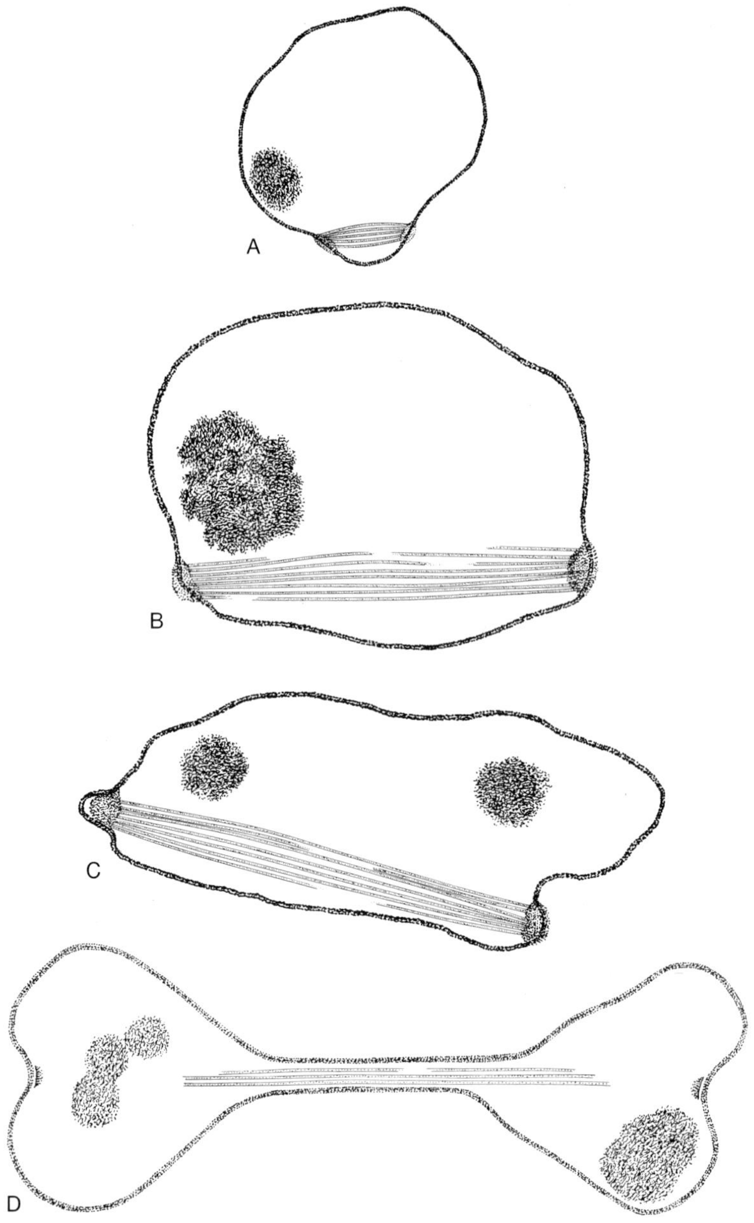

Fig. 10.1 A–D. Stages of intranuclear mitosis during sporangium formation in the zygomycete *Pilobolus crystallinus.* The intranuclear spindle is attached to the nuclear membrane by two distinct spindle bodies (SPB). The spindle is formed only of a bundle of parallel MT, and is entirely intranuclear. (A) Early stage of mitosis. Excentric spindle lying in a pocket on one side of the nucleus. (B) The excentrically located group of MT starts its elongation, separating the SPB. The nucleolus is undivided. (C) The nucleolus has divided into two masses. The MT have lengthened further. (D) Telophase: the nucleus has assumed a dumb-bell shape, the two halves being separated by a narrow zone where the continuous interzonal spindle is apparent (cf. the "telophasic bundle" as observed in vertebrate species, Fig. 10.13) (redrawn after Bland and Lunney [18])

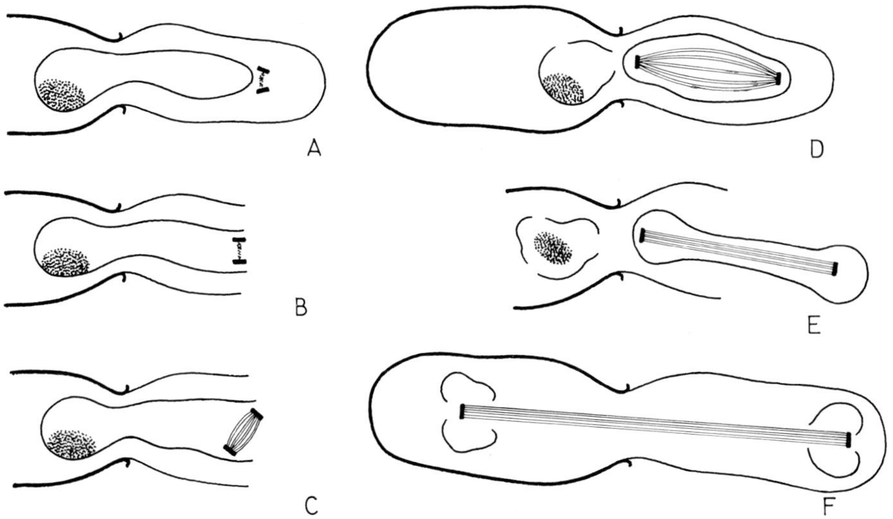

Fig. 10.2 A–F. Evolution of the spindle during the budding and mitosis of the heterobasidiomycetous yeast *Rhodosporidium* sp. (A) Cytoplasmic budding has started. At the nuclear pole, located in the bud, MT begin to assemble between two MTOC. (B) The nuclear membrane breaks down, and the early spindle becomes located inside the nuclear space. (C) A true spindle made of MT grows progressively. (D) The spindle is now completely encircled by a new nuclear membrane, the old one being left behind. (E) The spindle, made of parallel MT, elongates within the nucleus still located in the cytoplasmic bud. Remnant of nucleus degenerates in the mother cell. (F) The spindle and its parallel MT have considerably elongated, and now occupy the whole cell: the nuclear membrane is discontinued near the MT shaft and opposite the MTOC (telophase). This type of mitosis demonstrates the complex relations of the spindle with the nuclear membrane, and shows the role of MT in the elongation of the nucleus and the positioning of the daughter nuclei (redrawn from McCully and Robinow [142], simplified)

the nucleus and starts growing between two organizing centers after cytoplasmic budding. The two centers are separated by long MT, and the nuclear membrane opens, allowing the spindle to become intranuclear and to play a role in the elongation of the nucleus. At one moment, the whole nucleus and spindle is in the newly formed bud, but the considerable lengthening of the MT pushes back one daughter nucleus into the mother cell [141, 142]. This peculiar type of mitosis indicates that the relation of the nuclear membrane to the bundle of MT and to the polar bodies can vary considerably, while the MT activity is fundamentally similar to that seen in purely intranuclear spindles, i.e., a body indispensable for elongation.

10.3.3 Intranuclear Spindle with Extranuclear MT

The mitosis of the fungus *Fusarium oxysporium* provides another transition from intra- to extranuclear MT activity. Here also, there is an intranuclear mitosis, with a long bundle of MT attached at each end to a "polar body". However, this body—a plaque of dense material—is located on the outside of the nuclear membrane, and cytoplasmic MT are seen to radiate from it. Another important fact which is not apparent in yeasts, is that here the chromosomes are attached to the spindle by definite regions equivalent to kinetochores. The polar body, at the beginning of mitosis, divides, and a longitudinal spindle is formed as the two bodies migrate to occupy opposite ends of the elongated nucleus. After the chromosomes have been gathered at the poles of the nucleus, apparently by a shortening of the kinetochore MT, the continuous fibers elongate considerably.

A similar mitosis is found in the fungus *Uromyces phaseoli,* where intranuclear MT, attached to polar disks and to the kinetochores, move the chromosomes, and cytoplasmic MT are attached outside the nucleus to the same polar structures. Numerations of intranuclear MT indicate that some link one pole to another, while others are free in the spindle, without kinetochore attachment. The telophase elongation takes place by extension of the polar MT [95].

Other species show interesting connections between MT, chromosomes, and nuclear membrane. In the parasitic dinoflagellate *Oodinium,* the extranuclear MT (which are without any MTOC) channel through the nucleus, and are attached to the intact nuclear membrane at the same place as the chromosomes [162, 163].

In the slime-mould *Dyctiostelium,* spindle pole bodies are present, and the spindle is partly extranuclear; at anaphase, the MT attached to the chromosomes shorten, while the central spindle elongates, leading to the formation of a telophasic bundle [154].

The difference of behavior of continuous (polar) and chromosome (kinetochore) MT is evident; it is a property of MT in most mitoses of higher plants and animals, and is clearly demonstrated in the next example.

10.3.4 The Mitosis of *Syndinium* sp.

The description of the mitosis of this unicellular, classified with the dinoflagellates and found as a parasite of colonial radiolaria, gives a truly remarkable demonstration of the difference between kinetochore MT (K-MT) and continuous MT, while providing insight into the possible origin of the kinetochores. The chromosomes, which number four, have a V shape, and are attached at their middle to the nuclear membrane, suggesting a primitive mode of mitosis. From these attachment zones—which can be considered kinetochores—short MT link the chromosomes to the *two* centrioles which are located in a recess of the nuclear membrane. At mitosis, the kinetochore divides and each new one becomes attached to *one* of the two centrioles. These start moving apart, while long MT grow between them; at the same time, new centrioles are seen to bud laterally, so that before the end of mitosis, each pole will have two centrioles as in the mitosis of higher species. The long MT that separate the two centrioles, which come to occupy the two poles of the nucleus, are extranuclear, but they progressively tunnel through the invaginated nuclear envelope. Each daughter chromosome is linked to one centriole only. The MT linking the centrioles to the kinetochores during all these movements do not change length: as a result the chromosomes are separated, while maintaining their connection with the centrioles through the kinetochore [191].

This mitosis, which resembles that described in the dinoflagellate *Gyrodinium cohnii* [128], shows a primitive mode of chromosome attachment to the nuclear membrane, recalling what is seen in prokaryotes, and a complex set-up of MT: at the same time, in the same region of the cell, three groups of MT behave quite differ-

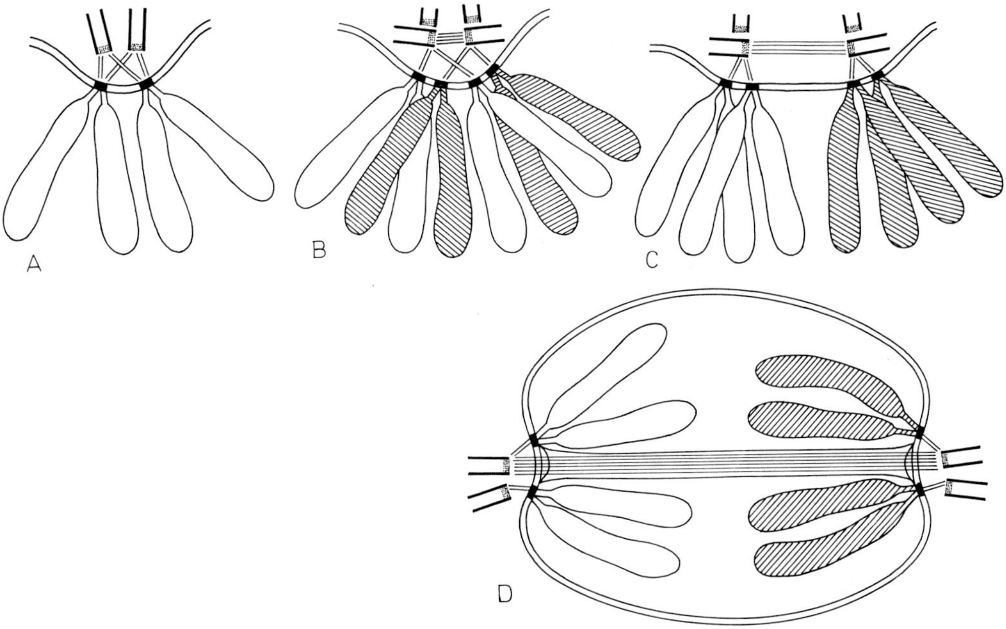

Fig. 10.3 A–D. Mitosis of the parasitic protozoan *Syndinium* sp. (A) Before the beginning of mitosis, the two still undivided mediocentric chromosomes are attached to two dense zones of the nuclear membrane, close to one another. From these, extranuclear MT extend to the two centrioles, in such a way that each chromosome is attached to both centrioles. (B) Beginning of mitosis: the chromosomes and their attachment points on the nuclear membrane have divided. Note that one chromosome of one pair (*hatched*) is now attached to one of the centrioles, the other to the second one. Between the centrioles, MT start growing. New centrioles also appear close to the old ones, at about right angles. (C) The mitosis proceeds by an elongation of the pole to pole MT extending between the centrioles, which are now in two pairs. The MT linking the centrioles to the chromosomes (chromosomal MT) do not elongate, and as a result the chromosomes are separated in two equal groups. The nuclear membrane remains intact, and in further stages the polar MT will tunnel through infoldings of this membrane. (D) Anaphase: the elongation of the bundle of pole to pole MT has pushed the two groups of centrioles to opposing poles of the nucleus. The chromosomes are segregated. The nuclear membrane will soon divide the nucleus in two. This mitosis illustrates the apparent independent but coordinated behavior of three types of MT: centriolar, pole to pole or "continuous" and chromosomal (redrawn from Ris and Kubai [191], simplified)

ently, and one may question whether all these MT are really identical (cf. Chap. 2). The K-MT do not change, the continuous MT elongate, and the centriole MT build two new centrioles with the typical ninefold symmetry of triplets. The myxamaebas of the slime-mould *Polysphondylium violaceum* display a somewhat similar type of division: ring-shaped extranuclear polar bodies migrate to both sides of the nucleus, while the nuclear membrane remains intact. There is an intranuclear spindle, comprizing kinetochore MT—one for each chromosome—and continuous MT. The first shorten and pull the chromosomes to the poles, while the others elongate and modify the nuclear shape. Cytoplasmic MT radiate from the polar bodies [194, 195].

While the mitosis of *Syndinium* has been particularly well studied at the ultrastructural level, many other protozoa show a similar behavior of polar MT and chromosome attachments to the poles through short MT (cf. [36, 103]).

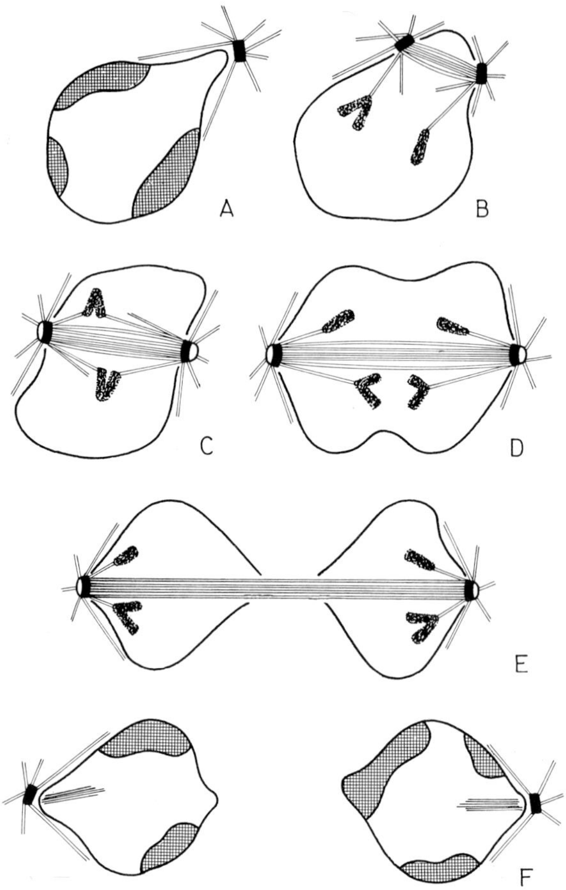

Fig. 10.4 A–F. Mitosis in the slime-mold *Polysphondylium violaceum.* (A) Interphase: the extranuclear MT radiate from a dense mass acting possibly as a MTOC, which is a precursor of the spindle pole body (SPB). (B) Prophase: the SPB has divided, and both parts are separated by a group of interpolar MT. Other MT extend from the SPB towards the chromosomes. (C) Metaphase: the SPB occupy the poles of the nucleus. They separate through the elongation of the polar MT, and are attached to the chromosomes by the equivalent of kinetochore-MT. (D) Anaphase: the chromosomes have divided; the "continuous" (polar) MT continue to increase in length. (E) Telophase: the two daughter nuclei are pushed apart by the lengthening of the polar MT, while the chromosomal MT show a slight shortening. (F) Late telophase: the intranuclear MT gradually fade away. This type of mitosis should be compared to that of yeasts (Fig 10.1) and to that of *Syndinium* (Fig. 10.3). It illustrates the differential behaviour of chromosomal and polar MT, and the role of the last in the nuclear elongation and chromosome separation (redrawn from Roos [195])

10.3.5 Extranuclear MT with Chromosomes Attached to the Nuclear Membrane

Chromosomes may be attached to the nuclear membrane at specialized (and often highly complex) regions which are equivalent to kinetochores, with a completely extranuclear spindle. This has been well described in the complex mitosis of the polymastigina *Barbulanympha* and *Trichonympha agilis* [87, 127]. In this last species, the chromosomes are fastened to the nuclear membrane at differentiated structures, to which the numerous intracytoplasmic spindle MT become attached. The nuclear membrane activity appears to mobilize the chromosomes, the extranuclear MT (which comprize K-MT and continuous MT) playing their role only in the late stages of the nuclear division. During the first stages the kinetochores, which are always in pairs, are attached to the intact nuclear membrane, which separates them from the cytoplasmic MT. Later, the nuclear envelope opens at the sites of the kinetochores, which move in relation to the spindle MT. The early movements of the

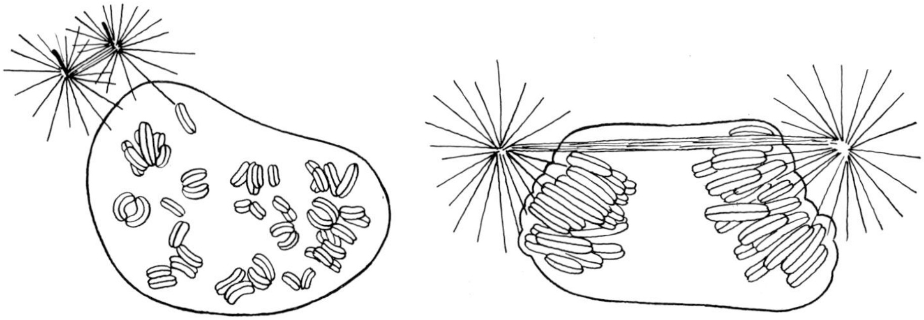

Fig. 10.5. Meiosis of *Trichonympha* (Grell). *Left:* Formation of tetrads. The centrioles have divided; they are linked by parallel MT. *Right:* Anaphase I. The centrioles are widely separated by the increase in length of the polar MT. Each centriole is connected to one group of chromosomes by relatively short MT (redrawn after Cleveland [36])

chromosomes are thus intranuclear and unrelated to MT [126]. This indicates that a "primitive" mode of chromatid separation can persist in organisms which have most complex cytoplasmic structures, comprizing many rows of basal bodies and cilia.

10.3.6 Mitosis in Higher Plants and Animals

These mitoses, if compared to those described above, are characterized by the fact that two types of spindle fibers, "continuous", from pole to pole, and kinetochoral, are intermingled. The presence or absence of a centriole at the pole is without importance, as all mitoses in higher plants are without centrioles and without any structure (polar body) resembling a MTOC. On the other hand, the role of the kinetochores becomes much more important in this type of mitosis where chromosomes, between prophase and metaphase, have to orient themselves on the spindle without any help from the nuclear membrane. This may result from an alignment of the K-MT with the continuous MT by lateral interactions, perhaps with formation of bridges [145]. During the transition from prophase to metaphase, when the polarity of the mitotic figure is not yet evident, the chromosomes undergo complex movements which are oriented in relation to the poles, although they take place in the nucleus: the forces which drive them at this moment remain poorly understood [160]. The movements of the centrioles, which, like in the mitosis of *Syndinium* described above, generate new daughter centrioles while in motion, are possibly related to

MT activity, as shown by the results of colchicine poisoning [47], which may lead to a "monopolar" mitosis (vide infra).

However, as clearly demonstrated in the newt *Taricha granulosa,* the prophasic migration of the centrioles is independent of the formation of the spindle polarity, which may be present some time before the centrioles reach the poles [156].

Another feature is the formation at telophase, from what is left of the "continuous" fibers, of a bundle uniting the two daughter cells, and showing at its center a denser zone where the overlapping MT are surrounded by dense extraneous material [145].

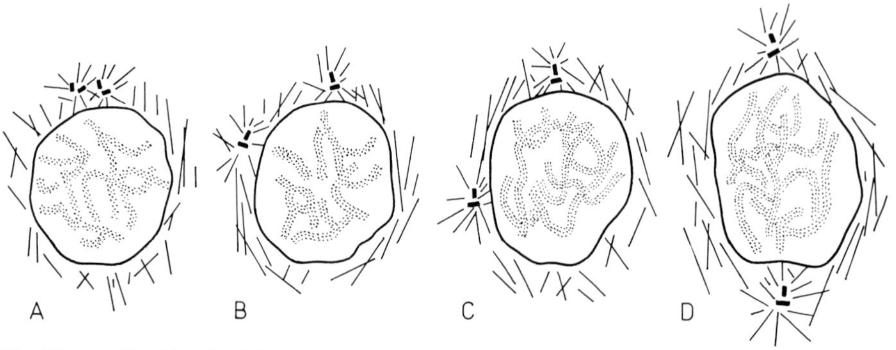

Fig. 10.6 A–D. Mitosis of the newt, *Taricha granulosa granulosa.* (A–D) Steps of migration of the centrioles before the breakdown of the nuclear envelope. The spindle polarity, as indicated by the direction of the MT (represented as *lines*), is already established before the centrioles occupy the poles of the achromatic figure (redrawn from Molè-Bajer [156], simplified)

It should also be recalled that the movement of chromosomes on the spindle is by no means synchronous, and that many facts indicate some chromosome autonomy: this may have some relation to the non-synchronous reduplication of their DNA. There are reasons to believe that chromosome movements are not only passively directed by the spindle MT and other proteins but also by a kinetochore-programmed activity.

10.4 Role of MT in Mitotic Movements

As it is clearly impossible to discuss all the types of mitoses which have been mentioned above, this survey of MT in mitotic movements will be limited to "typical" mitoses, that is to say to those of higher plants and animals. These may be defined as mitoses in which after prophase the nuclear membrane disappears, and the chromosomes, after becoming grouped at the equator of the cell, are in relation to a true spindle-shaped "mitotic apparatus" (MA) which is formed, inter alia, by two types of MT: those attached to the kinetochores (K-MT) and those extending between the poles, without any fixation on the chromosomes. These are often called "continuous" MT, although there is no definite evidence that they really run without interruption from one pole to another. It is better to call them polar MT or non-K-MT.

Plant and animal mitoses may be discussed together, the main difference being the absence of centrioles in the higher plants: this does not bring many changes in the course of division, and there are evidently other factors which determine the bipolar structure of the spindle.

The changes of MT during the steps of mitosis will be described, and further information given about the role of MT in the spindle structure or in the isolated MA. Experiments with physical or chemical agents which disrupt or inhibit the formation of MT will be mentioned. Special types of mitosis will only be considered if they bring important information about the role of MT: this is the case for meiosis and polar body formation in eggs. On the other hand, cells from various species of plants and animals will be described when good and recent studies on their MT are available [11].

10.4.1 Methods of Study

The analysis of mitotic movements has benefited from all the improvements in microscopical techniques. Light microscopy, with phase or interference contrast optics, and microcinematography, have permitted a detailed analysis of chromosome and spindle dynamics. Polarization optics, as improved by the group of Inoué [106], has provided remarkable cinematographical documents on the activity of metaphase and anaphase MT (cf [7]). Micromanipulation techniques have also helped to test the linkages between MT and kinetochores, and to study the movements of chromosomes artificially detached from the spindle. The more recent techniques of immunohistochemical staining of MT [24, 25, 166, 206, 223, 224, 225] have clearly indicated how and where MT are assembled, and their evolution through the steps of mitosis. Electron microscopy, of course, has brought considerable information on the behavior of centrioles and kinetochores, and has enabled students of mitosis to demonstrate the sites of attachments of MT. Several authors have published quantitative studies on the numbers of MT at various stages of mitosis, by careful analysis of serial sections. High voltage electron microscopy (1 MeV) has permitted the obtaining of good overall views of the whole metaphase spindle in relatively thick sections and has improved the comparison between light and electron microscopy [144, 145].

Another approach, inaugurated by the studies of Mazia [151] and his group, has been the isolation of the "mitotic apparatus" (MA) comprizing the spindle, the asters, and the chromosomes of large cells. This MA is a complex structure which has several properties of the mitotic spindle as seen in living cells, and which contains the MT and other proteins and organelles, and of course the chromosomes. Chemical studies of the MA, and experiments on its changes under the action of various chemicals, have helped to understand the quantitative importance of protein synthesis related to mitotic movements.

Isolated MA have also been studied in solutions of tubulin, which increase the number and the visibility of MT [110].

---

[11] Many important facts on spindle function were known before the discovery of MT (cf. [151, 209]). However, the technical advances in electron microscopy are such that many documents more than ten years old appear today to be of poor quality

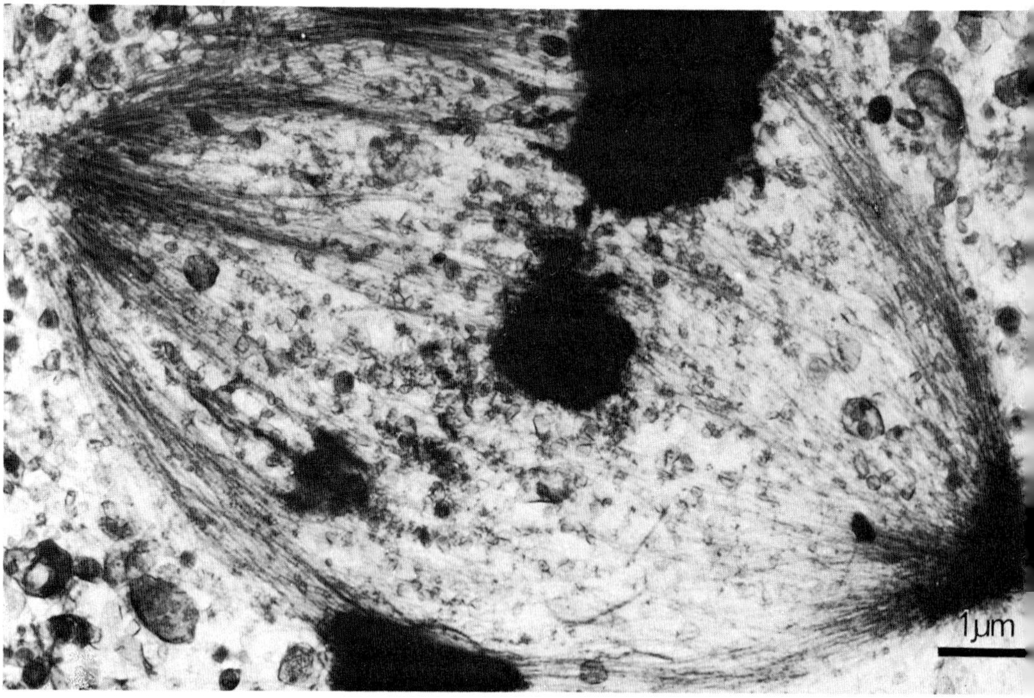

Fig. 10.7. Metaphase spindle of a PtK_1 cell as observed by high voltage electron microscopy (1 Mev). The curvature of the MT toward the poles is quite evident (from McIntosh et al. [144]) (Raven Press, 1975)

The various MT poisons, and physical agents such as cold, high hydrostatic pressure, and ultraviolet beams, have permitted a true "dissection" of mitosis, and demonstrated the interrelations between the different forms of intracellular motility related to cell division.

The integration of these various techniques, and the comparison of their results with the data obtained by the comparative study of mitosis, have considerably helped to further our understanding of mitotic movements and the role of MT. Two recent conferences have brought together most of the recent acquisitions in this field [83, 113].

10.4.2 Prophase

Most research work has been conducted on the dynamics of anaphase movements, and prophase and prometaphase appear to have been somewhat neglected. These last mitotic events are however of capital importance, for they involve the complex movements which will bring chromosome orientation into correspondence with the polarized structure of the spindle, and prepare the chromosome gathering at the metaphase plate. In cells with centrioles, it also involves the separation of these organelles (which may have started long before prophase) and their movement toward opposite sides of the nucleus, preparing the formation of the polar MT. In plant cells, other MT activities may be observed, such as the formation of an equatorial ring of MT located close to the cell membrane, indicating the plane of the future cell division: this ring is excentrically located in cells which divide unequally, like those leading to the formation of stomata in leaves [178, 180]. These MT disappear later when the spindle is formed and has assumed its correct orientation [171, 172].

A detailed study of prophase chromosome movement [190] has brought interesting information about the possible role of MT before the breakdown of the nuclear envelope. In spermatids of *Acheta domesticus,* the house cricket, these movements have been carefully recorded by microcinematography. Inside the nucleus, the chromosomes undergo "saltatory" movements. These are relatively rapid (up to 6 μm/min) and their amplitude may reach 9 μm and more. The velocity of these movements, which are arrested by colcemid (2×10^{-5} M for 30 min), decreases after the fragmentation of the nuclear envelope (to 2 μm/min) and is slower still at anaphase (0.2 – 0.5 μm/min). The remarkable fact is that the MT, which are related to the centrioles, are outside the nuclear envelope, and have no definite connections to the chromosomes. Micromanipulation studies do indicate however that some attachment between the chromosome ends (not the K) and the nuclear envelope may be present. After prophase, the chromosomes become attached to the spindle, this time by their kinetochores, and their movements become clearly polarized. The role of MT appears to be indirect, and the conclusion which is reached, after comparison with other movements such as those of melanin granules (cf. Chap. 7), is that a two-component model should be considered: "a non microtubule, linear force producer, together with microtubules with a skeletal, orientational role." The extranuclear MT would provide a "framework" which orients the force-producing fibers (actin?). These experiments provide a remarkable, although not unique, instance of colchicine poisoning of movements, while no definite link has been demonstrated at this stage of mitosis between MT and chromosomes. It should be remembered however that several facts (cf. Chap. 3) indicate topographical relations between extranuclear MT and the location of chromatin in interphasic nuclei. The possibility that colchicine may combine with elements of the nuclear envelope has been considered, but is unlikely, as the doses used are small [190].

It is worth noting that in the study of this dynamic condition the "skeletal" hypothesis of MT function, as mentioned in the preceding chapters, is considered the most likely.

The movements of centrioles and their relation with MT are some of the early changes in prophase. At this stage, each centriole has usually formed, at right angles, a smaller daughter centriole which will progressively grow to the same length. At metaphase four centrioles, two at each pole, are normally present (cf. Chap. 4). While centrioles are MT structures, the spindle MT are usually indirectly linked with these organelles—although in some of the simpler mitoses, as mentioned above, the continuous MT may arise from one extremity of the centrioles.

In vertebrates, the centriole may be considered a MTOC; more exactly, the dense pericentriolar bodies are the regions from which the MT grow [48, 192, 193]. These will mainly form the polar MT, the kinetochores appearing as the growth points of the K-MT. The movements of the centrioles at prophase may be linked with the elongation of the intercentriolar MT [21, 147], although this is not always evident [156, 185]. These movements, which are similar to those of the "polar plaques" in simpler organisms, deserve further study; they are inhibited by spindle poisons [20, 46]. The role of MT in controlling these movements is indicated by the fact that in multipolar mitoses, when more than two groups of centrioles are present, equal distances are observed between each group. While centrioles cannot be the cause of the symmetry of the mitotic figure, which may exist without them [156], they are indicators of the poorly understood factors which lead the cell to the bipolarity found at metaphase. The recent observations of the growth of MT as observed by immunofluorescence techniques [166] from a localized point which most probably is centriolar are a further indication of the role of these MT in the control of early stages of mitosis.

10.4.3 Metaphase

The transition from prophase to metaphase is probably one of the most complex steps of mitosis, and at metaphase all the structures needed for chromosome movements toward the future daughter cells are present. An attempt will be made to summarize the principal data from the vast literature on these movements: the MT at metaphase and their relations with kinetochores and centrioles (or poles) will be described first. Studies on the isolated MA and its chemistry will be summarized next, and a description of movements before, during, and shortly after metaphase will follow. The actions of various MT poisons will be considered later.

10.4.3.1 The Metaphase Spindle and its MT. Excellent descriptions will be found in the recent reviews by Bajer and Molè-Bajer [8, 9], Forer [65] and McIntosh et al. [145]. At this stage, the spindle shows mainly a variable number of MT which are attached at the vicinity of the centrioles and at the kinetochores. As already mentioned, although MT seem to extend from pole to pole, the existence of truly "continuous" MT is doubtful, and the polar spindle may be made of long MT extending only part of the spindle length [75]. Most of these MT are parallel but this is not a rule, and often MT making even large angles with the spindle axis may be seen (cf. [7, 9, 10]).

In many cells, the spindle at metaphase is barrel-shaped, the MT curving toward the poles—whether centrioles are present or not. From what is known about MT, such a shape indicates that lateral forces are exerted on them. One explanation of the barrel shape is that it results from the interaction of the K-MT and polar MT. It may also be related to another property of the metaphase spindle, that of excluding most cell organelles from the space it occupies. Exclusion of organelles is seen also in the asters, where it remains poorly understood [8].

The curved shape of metaphase spindles has been explained by the formation of interMT bridges maintaining the polar MT in a strained configuration [147]. The same shape is quite evident in spindles of cells lysed at metaphase in solutions of tubulin, under conditions favoring the assembly of tubulin into MT [145]. Curving MT are often seen in mitosis, and any theory of mitotic movement should explain the nature of the bending force.

The numbers of MT vary considerably from one type of cell to another. For instance, in the endosperm of *Haemanthus catherinae* Bak., which provides mitoses of exceptional visibility studied in great detail by Bajer et al. [5 – 10], the spindle contains from 5000 to 10,000 MT; from 70 to 150 MT are attached to each kinetochore at anaphase [6]. In the spindles of the crane-fly spermatocytes, *Pales ferruginea,* the number of MT could be measured in single cells as the spindle contains only MT, some ribosomes, and an amorphous or filamentous matrix. About 100 MT/μm^2 of spindle cross-section were counted, and more than 2000 MT were visible in a single micrograph [73]. The variations in MT number during mitosis and at different regions of the spindle have been carefully measured; in this species the autosomes have 20 – 25 MT at each kinetochore, and about 150 MT extend from the metaphase plate to each pole, a small fraction of the total number of MT [74]. From calculations of the total length of MT per spindle, and assuming that the MT are made of 13 protofilaments with monomers of about 4 nm in diameter, the total number of tubulin monomers has been estimated to be about 0.6×10^8, a figure which is of the same order of magnitude as that calculated from data from the sea-urchin MA [38]. In the rat kangaroo metaphase spindle, a similar number of MT has been counted—approximately 900 in the equatorial region [21] and 1500 between metaphase plate and pole [144].

The spindle MT have all the properties which have been described in Chapter 2, and several authors have observed bridges extending more or less regularly between them. The importance of these for any theory of mitotic movements is considerable. In HeLa cells, regularly spaced bridges were compared to similar structures seen in *Haemanthus* and to the bridges between actin and myosin fibers in muscle [98]. The regular spacing of spindle MT in the rat kangaroo and in hamster fibroblasts was attributed to bridges measuring 45 – 54 nm in length [21].

Lateral expansions of MT were also described in electron microscope studies of whole mount preparations of mouse oocytes MT: they were considered to be made either of matrix material or of tubulin molecules inserting laterally into the MT [31]. Similar structures have been reported in various mitoses: in amebae ([196]; *Chaos carolensis*), the green alga *Blastophysa rhizopus* Reinke [227], and in *Pales ferruginea* [74, 75]. In the spindle of *Barbulanympha,* translation of superposed micrographs has revealed a very clear periodicity, comparable to that seen in the dynein arms of cilia [111]. These bridges are probably a general feature of mitotic MT, and further data on their chemical nature are awaited.

The MT are attached to the chromosomes, in most cells, at specialized structures: the kinetochores (K-MT). These undergo during mitosis a progressive differentiation and display their full complexity at metaphase. In cells of the fetal rat they have the form of three-layered disks with a diameter of 200 – 245 nm [116]. Similar structures are found in various species, and also in higher plants and flagellates (cf. [139]). Kinetochores are also found in meiotic divisions; the problems resulting from their relation to the proper orientation of bivalents have been fully discussed by Luykx [139]. In the rat kangaroo, the kinetochore is also made of three layers, the MT terminating in the outer one [193]; the inner layer, curiously, is not visible in cells treated with colchicine. The outer layer appears between pro- and metaphase, and it is to this region that the K-MT become attached, the kinetochore apparently acting as a MTOC.

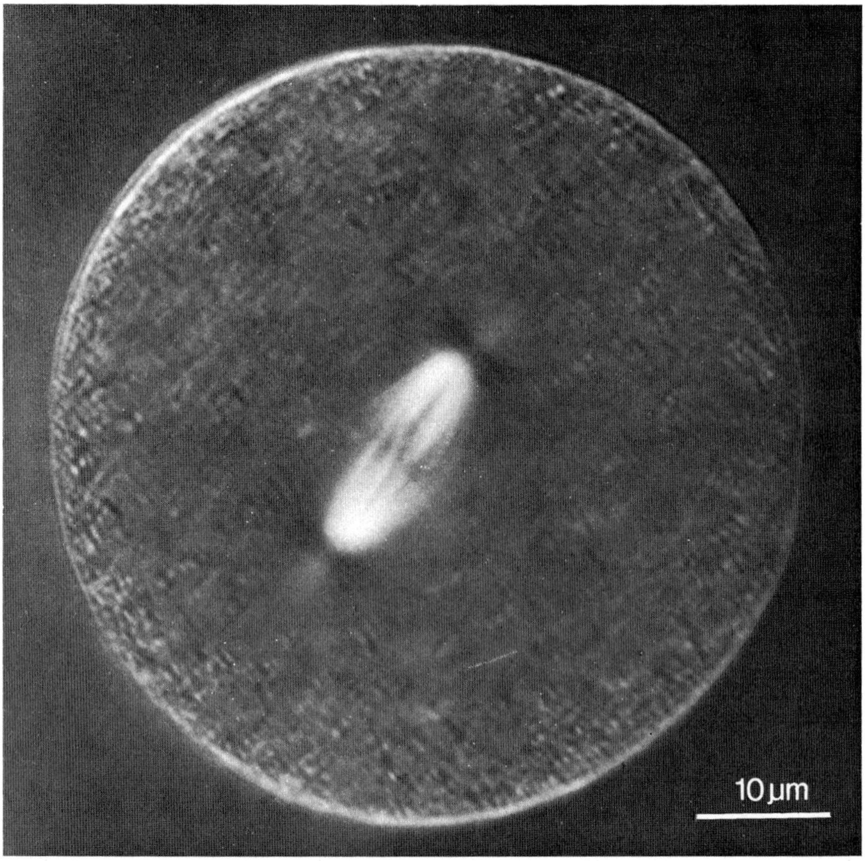

Fig. 10.8. Birefringence of the metaphase spindle in a mature oocyte of *Retinaria gouldi* (original document, by courtesy of H. Sato)

Before separation of the daughter chromosomes, the two kinetochores of each chromatid face in opposite direction: the proper orientation of the chromosomes demands that the K-MT extend before anaphase towards the two poles of the spindle. The mechanism of this orientation will be discussed below. It is not evident that the K exist at interphase: they are certainly not apparent.

There are many reasons to think that the kinetochores have important physiological functions: they may play a role in the asynchronous division and movements of chromosomes, and may be one of the main regulating structures of mitosis. Some of these problems have been discussed by Fuge [76, 77].

One of the properties of the spindle is its birefringence, already beautifully illustrated by Schmidt in 1936 [cf. 208] in the egg of the sea-urchin *Psammechinus miliaris* and considered by this author to be without doubt an indication of the fibrillar structure of the spindle and asters. As other structures made of MT, such as axonemes, are birefringent, it would appear likely that MT are the cause of this optical property. The demonstration of this is however difficult for, even in isolated MA, the MT are associated with other oriented cytoplasmic structures which may contribute to the optical properties [188]. There are several reasons to believe that the birefringent structures are the MT: in the crane-fly spermatocytes, a study by micromanipulation showed that the chromosomes display a stronger resistance toward movements when the birefringence is greatest and become loose when through the action of colchicine or VLB the birefringence disappears [14]. In the same cells a comparison of the distribution of MT in transverse sections of the spindle with the degree of optical activity also suggests that MT are the basis for birefringence: the MT form five bundles, as observed in transverse sections of the spindle, and microdensitometer readings of the polarized light image of the metaphase show a similar number of strongly birefringent zones [129]. Birefringence disappears with all treatments which destroy MT (colchicine, *Vinca* alkaloids, hydrostatic pressure, cold, ultraviolet irradiation) (but cf. [64, 65]).

However, recent studies of the sea-urchin egg spindle isolated in a mixture of glycerol and dimethylsulfoxide indicate that the pressure-induced loss of birefringence can destroy as much as 70% of the optical retardation without any appreciable change in the numbers of MT [70]. After cooling MT are still present when 55% of the birefringence is lost [71]. Isolated sea-urchin MA may be extracted with 0.5 M KCl, causing loss of 45% of the birefringence without any extraction of tubulin, as demonstrated by electrophoresis.

10.4.3.2 The Isolated Mitotic Apparatus. Great advances in the understanding of mitosis have resulted from the studies of Mazia and Dan [cf. 151] on the isolation from sea-urchin eggs of the MA defined as "the ensemble of structures constituting the 'chromatic' and 'achromatic' figures ... including spindles, asters, centrioles ... and chromosomal structures" [151]. Although this is evidently a complex structure, MT and tubulin are an important part of the MA and the study of this structure has thrown some light on the relation of MT to other constituents of the spindle. Most research has been carried on in invertebrate eggs (mainly sea-urchins); meiotic spindles of the eggs of *Chaetopterus* have also been studied after isolation [110].

Much work on the chemical composition, the methods of isolation and stabilization, and the ultrastructure of the MA has been published (cf. [67, 69]). It was rapidly understood that a large proportion of the cell proteins were present in the MA [226], indicating that the spindle and the asters resulted from the assembly of molecules already present before mitosis, as protein synthesis is considerably decreased when the chromosomes are in a condensed form. Already in 1966, the MT of the MA of the sea-urchin were recognized to have 13 subunits [121], and important observations in relation to the methods of preparation of the MA, which cannot be discussed here, indicated that -SH groups played a role in the stability of its fibrillar structures [198]. Two protein components were isolated, one with a sedimentation constant of 4 – 5S, the other of 22S [119]. While the first could later be identified as tubulin, the 22S protein, which represents about 8% of the total proteins of the unfertilized egg, has a molecular weight of about 880,000 daltons [217]. Ribosomes and fragments of cell membranes are also found in isolated MA [84].

A quantitative study of the amount of MT protein (tubulin) in the MA of *Arbacia punctulata,* based on counts of MT in sections of the spindle and asters, led to the following figures: the half-spindle contains about 2000 MT, and one aster 1300

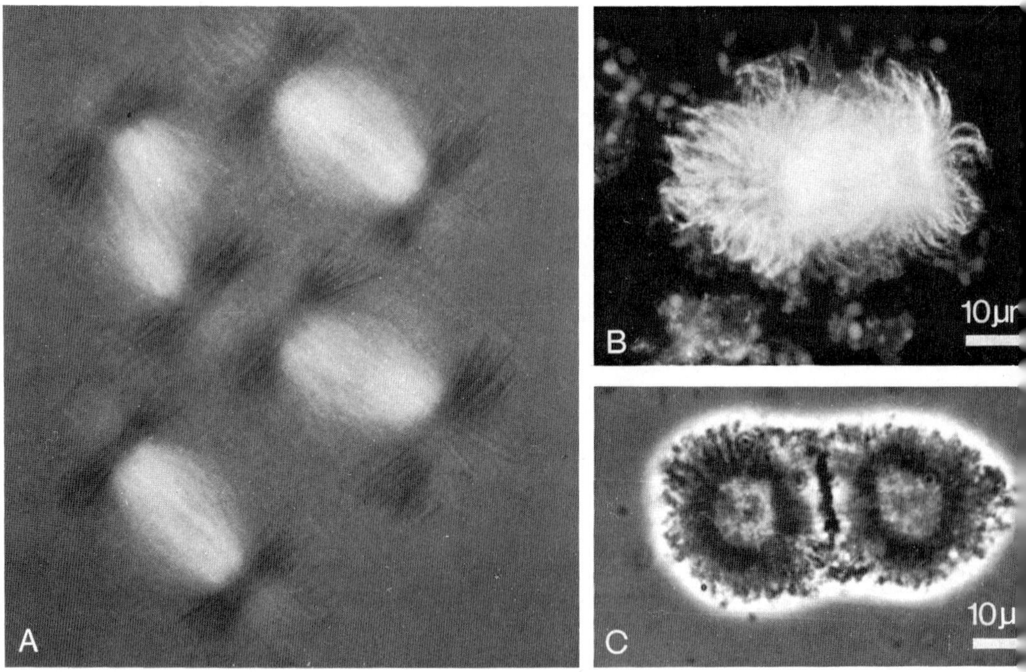

Fig. 10.9 A–C. (A) Isolated spindles from mature oocytes from *Pisaster ochraceus*. Polarized light. (B) Isolated spindle from mature oocyte of *Pisaster*. Indirect immunofluorescence. (C) Isolated metaphase spindle. First division of sea-urchin zygote. Phase contrast (original documents, by courtesy of H. Sato)

MT, the number approaching 7000 for the whole spindle. From a calculation based on MT 12.5 μm long, the weight of tubulin would be about $1.7-2.0 \times 10^{-11}$ g. This represents only one eighth of the total protein of the MA, and confirms that the 22S protein is not directly related to the MT [38]. The MT protein could be selectively extracted from the MA by an organic mercurial (indicating the role of -SH groups): this was confirmed by the properties of the extracted protein and by the selective removal of the MT as observed by electron microscopy [17].

The purification of MA from cytoplasmic material has been gradually improved, with the aim, not yet fulfilled, of obtaining a functional MA which could be made to contract or mobilize the chromosomes in vitro [66, 67]. The most recent techniques lead to preparations with a stable birefringence and stable MT over periods of two weeks, starting from the sea-urchin *Strongylocentrotus purpuratus* [69]. It would be interesting to obtain responses of isolated spindles to MT poisons, which has not yet been possible (cf. [110]).

Formalin-fixed MA stain strongly by immunofluorescence techniques as if the tubulin persisted although MT are no more visible. The fluorescence disappears after KCl extraction. These results have suggested that after formaldehyde fixation, tubulin may be present in a new condition; it appears however that linear arrays remain visible which could be the poorly preserved but still present MT [66].

10.4.3.3 Premetaphase and Metaphase Movements. Metaphase may appear to be a period of static equilibrium, the chromosomes being equally attracted to the two poles. In fact, considerable activity is going on at all moments in the spindle and in the cytoplasm located close to the spindle MT. Studies of changes in birefringence of the spindle by time-lapse cinematography show a pattern which has been compared quite rightly to the moving flashes of the "northern lights" [8, 109]. [12]

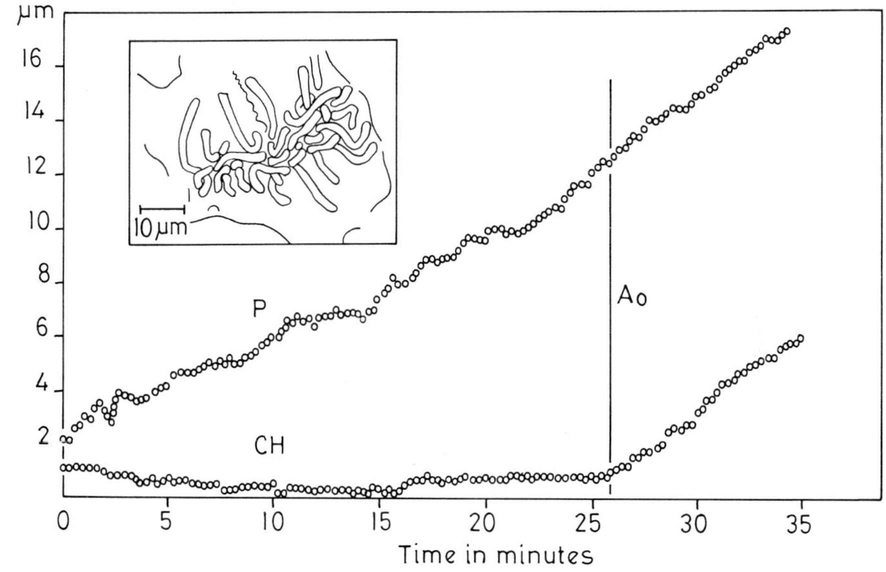

Fig. 10.10. Movement of particles and chromosomes at metaphase and anaphase in the endosperm of *Haemanthus katherinae* Baker. During the first 25 min of observation, the chromosomes (*CH*) remain at the equator, their centromeres not yet having divided. The movements of a particle (*P*) within one half-spindle have been plotted (cf. *wavy line* in *inset*). At A_0, anaphase begins, and the chromosomes move toward the poles: their movement is parallel to, and at the same speed as that of the particle, which continues to progress toward the pole, indicating the existence of some poleward flow in the spindle, even before any chromosome movement (cf. the results of X-irradiation of half-spindles, Figs. 10.15 and 10.16; from a microcinematographic study, Bajer[5])

The preprophase movements have been mentioned above; once the nuclear membrane disappears, the chromosomes progressively become grouped at the equator, their kinetochores oriented toward the poles. This often implies a rotation of the transverse K to K axis of the chromosomes which could result from an interaction between the K-MT and the polar MT [147]. This type of interaction also takes place later in mitosis and will be discussed at the end of this chapter.

An interesting movement which may be observed in the metaphase spindle is that of particles of small size which move toward the poles by rapid and relatively

[12] A note of caution about time-lapse cinematography of mitotic movements is necessary: these beautiful movies show what appear to be tremendous activity and rapid movements: in fact, the fastest movements of mitosis—for instance the anaphase migration of chromosomes—are two orders of magnitude slower than the movements of pigment granules in melanophores or of neurosecretory material in axons

large displacements, which have been compared to the saltatory movements reviewed by Rebhun [186]. In the endosperm of *Haemanthus catherinae* these movements, which differ from brownian agitation by their directionality, take place during meta- and anaphase: they have about the same speed as the anaphase chromosome movements, i.e., far slower than other types of "saltatory" displacements [6]. They are evidence of a general transport activity present in the half-spindles. In a discussion of the theories of mitosis, Bajer and Mole-Bajer [8] conclude that "it is an open question whether the mechanism of saltations is related to any mechanism in the spindle," and suggest the possible role of other components (microfilaments?).

The possibility that actin may be present in the spindle as the force-generating element, while MT would be mainly structural [139], will be discussed later.

10.4.4 Anaphase

This is the most spectacular phase of mitosis, when, as seen in time-lapse cinematography, the two groups of chromosomes—their centromere having divided—are pulled toward the poles, while the center of the cell shows a zone nearly empty of all organelles.

10.4.4.1 The Spindle MT at Anaphase. Many arguments about the mechanisms of mitosis result from the fact that two different motions take place at anaphase: the chromosomes move (or are pulled) towards the poles, and the poles separate from one another, the whole mitotic figure elongating. The study of various types of mitoses as summarized above should indicate that the evidence for two forces—a pulling and a pushing one—is clearer if one considers separately (although their MT are closely intermingled) the K-MT and the polar MT: evidently, the first shorten as the chromosomes come closer to the poles, while the second may remain stationary in some cells, and elongate considerably in others. This is made quite clear in *Barbulanympha* sp., a symbiotic protozoan found in the intestine of the wood roach, and studied before the advent of electron microscopy by Cleveland (cf. [113]). The mitosis is of the intranuclear type, with persistance of the nuclear envelope throughout. The nucleus becomes wrapped around the spindle, like in *Syndinium*. The first part of anaphase or anaphase A shows a movement of the chromosomes toward the centrosomes by a shortening of the K-MT, while the pole to pole distance does not vary. In anaphase B, which follows after a short pause, the central spindle starts to elongate and may reach a size five times greater than before. During this stage, the kinetochores are firmly attached to the centrosomes while the cell elongates before dividing into two [113].

While this type of mitosis is an interesting model, one question comes to mind at once: what is the mechanism of MT elongation and shortening? There are good reasons (and many observations) to think that the polar MT arise close to the centrioles or the polar structures, and grow towards the equator. Although counts of MT suggest that one half of the MT pass from one side to the other of the metaphase plate [148], it is doubtful that these MT extend from one pole to the other: it is more probable that the polar spindle is made of long MT lying side by side. This means that changes in length could take place anywhere in the spindle. Here is a problem which has already been met in previous chapters: do the spindle MT assemble and disassemble at their ends, or can they grow by lateral addition of subunits? The behavior of the K-MT indicates more clearly than that of the polar MT that growth is polarized: before metaphase, these MT are seen to grow from the kinetochores. Do they disassemble at the centriolar (polar) region—where their subunits could possibly be utilized for the elongation of the polar MT—or anywhere along their length? These questions, which are intimately linked with theories on the equilibrium assembly of MT [112], will be discussed later.

Several studies on the numbers and the length of MT during anaphase provide answers to some of these problems. The K-MT evidently, in most mitoses, shorten. It is quite difficult to count their numbers in the vicinity of the centrosphere, and it is not clear whether they are disassembled at this end, or closer to the kinetochores from which they grew in prometaphase. Some authors [51] suggest that disassembly may take place all along the K-MT. In *Haemanthus katherinae,* the number of MT fixed to the kinetochores was found to decrease by half between meta- and anaphase, and some fragmentation of the MT into shorter segments was noticed at anaphase [115]. This is related to a decrease in half-spindle birefringence, and most authors (cf. [8]) consider that the K-MT are disassembled near the pole. However, in the crane-fly, disassembly at the kinetochore end has recently been mentioned [68]. As the decrease in K-MT takes place (in some cells) when polar MT are growing, it is possible that the liberated tubulin subunits contribute to the elongation of the spindle [77].

One of the most precise approaches to anaphase dynamics is to study, by serial sections of transversally cut spindles, the numbers of MT at various steps of mitosis. This has been done by three groups of authors, with some differences which could be accounted for by the fact that different types of cells were studied.

1. McIntosh and Landis [148] studied the spindle of two types of human cells in tissue culture (WI-38 fibroblasts and HeLa epithelial). In HeLa cells the maximum number of MT, about 2400, was found between the metaphase plate and the pole. Counts made on cells at metaphase show a smaller number of MT at the equator: here the K-MT have ended, and the number should represent only the “continuous” MT. The ratio of “continuous” or polar MT to K-MT was 0.5 at metaphase and 0.6 at mid-anaphase. At the end of anaphase, when the so-called stem bodies (Flemming bodies) appear at the equator, leading later to the formation of the telophasic bundle, the number of MT in this region is (in WI-38 cells) about 1.3 – 1.5 times that counted elsewhere between the two anaphasic groups of chromosomes, indicating a partial overlap of MT. These results are important for the sliding filament theory proposed by the authors. They also indicate that at anaphase, while there is a shortening of the K-MT, the number of polar MT does not decrease, and may perhaps show some increase.

2. The same year, Brinkley and Cartwright [20] counted the MT during mitosis of rat kangaroo (strain PTK_1) and Chinese hamster (strain Don) cells. The rat kangaroo fibroblasts show an important lengthening of the spindle at anaphase, the pole to pole distance increasing from 10 μm at metaphase to 20 μm at telophase. About 25 MT are attached to each kinetochore. At the end of mitosis, the MT of each half-spindle become disoriented and reduced in number, while the parallel interzone MT remain constant and later form the telophasic body. About 50 – 60%

of the metaphasic spindle MT are polar ones (cf. [192]). The maximum number of MT was observed at the equator at anaphase, while the MT of the half-spindle decreased by 50 – 60%, resulting from the disassembly of the K-MT. In both cell types, the number of interzonal MT was found to increase at telophase. Some overlapping of interpolar MT takes place at the equator. The ratio between the numbers of MT at the equator at metaphase and anaphase never reaches the values found by McIntosh and Landis [148]. At anaphase, the length of interpolar MT increases, probably by an addition of subunits at their polar ends.

3. The MT counts of Fuge [73, 74, 75] are made on quite different cells, the meiotic divisions of crane-fly (*Pales ferruginea*) spermatocytes. The number of MT increases from prometaphase to metaphase (about 2000) and does not decrease until late anaphase. The maximum number of MT is found at metaphase, and at anaphase the greatest number of MT is found in front of the autosomes in each half-spindle. The study of serial sections demonstrates that the MT are far from being parallel, and at anaphase an interdigitation of K-MT and polar MT is evident. The author concludes that the presence of "continuous" MT passing from pole to pole "is highly improbable" [77]. Contrary to the findings in vertebrate cells, it is suggested that at anaphase a displacement of MT units could take place from the equator—where their number decreases—to the poles. This problem has been further discussed in a review article [76].

The interpretation of these results is made difficult by the fact that contradictory opinions are held about the way MT may assemble and disassemble, either subunit by subunit at one end of MT, or with interchange of tubulin molecules with MT all along their length. C-MT—as mentioned in Chapter 2—have been described in mitotic spindles: they may represent one form of MT (dis)assembly. They are visible in the interpolar spindle at anaphase in *Arbacia punctulata* isolated MA [37], in the anaphase of spermatocytes of *Pales ferruginea*, where they have been considered to be a step in assembly of MT [51], and in the spindle of *Haemanthus katherinae* [115], where they are interpreted as the "open" ends of MT, as in *Arbacia* and other cells [115]. In *Haemanthus*, C-MT are abundant in the equatorial region at metaphase; later, their number increases in the zone of the phragmoplast [8].

10.4.4.2 Anaphase Movement. While the movements of chromosomes towards the metaphase plate, before the separation of their kinetochores, are relatively easy to explain by balanced forces acting through the MT on the kinetochores, the movements of chromosomes at anaphase are the stumbling block of most theories of mitosis, as discussed at the end of this chapter.

One point is important to note: the movements of the chromosomes are not only very slow—about 1 μm/min (1.5 mm/day)—but the energy involved is minute. From data about the size of chromosomes and the viscosity of the surrounding cytoplasm, the force requirement for moving one chromosome from the equator to the pole has been figured at about 10^{-8} dynes [113, 159]. This is two orders of magnitude smaller than the force required for flagellar movements. It implies that a few molecules of ATP—or of dynein-like protein acting as ATPase—would be sufficient for moving the chromosomes. As recently calculated [160], if the spindle MT had as many dynein or dynein-like molecules along their length as flagella, the chromosomes could be expected to move at the speed of one *centimeter*/min. Either the energy-producing mechanism is quite different, or there must be some velocity-limiting process.

Let us first consider the movements of the chromosomes toward the poles. These have been compared to the displacements of many small particles—lipid droplets, fragments of nuclear envelope, pigment granules, yolk granules, mitochondria—toward the poles in the two half-spindles [107]. It has been mentioned already that such movements cannot be compared to other "saltatory" ones, as they are much slower. From the moment the kinetochores separate, the chromosomes move toward the poles at a rate which for some authors is constant throughout anaphase [6, 139], while others found that it decreased as the half-spindle shortened [105]. Luykx [139] summarizes the evidence indicating that the force acting on the kinetochore is proportional to its distance from the pole.

Most authors agree that the K-MT grow at prometaphase from the kinetochores toward the poles. At anaphase, it is evident that these MT shorten considerably and, as they are often poorly distinct in the polar region, it is believed that they undergo disassembly at the poles. This would imply that the K-MT move toward the poles, where they are changed into subunits. The number of the K-MT is known to decrease during anaphase [8]; as there is no known technique to ascertain that the K-MT move toward the poles, some believe that disassembly could take place all along their length [51].

Any explanation of anaphase movements must take into account the strange behavior of some chromosomes, and their non-random distribution. In the mitosis of *Pales ferruginea,* the sex chromosomes start to move only when all the autosomes are already grouped near the poles, X going to one side, Y to the other. This indicates that transport properties exist in the interzonal region [8]; it is also evidence of the autonomy of chromosomes, related perhaps to the functioning of their kinetochores and to genetic factors. In the spermatocytes of *Melanoplus differentialis,* analysis of anaphase movements has been made by displacing, with a micromanipulator, one chromosome: before metaphase and until shortly thereafter, the chromosome will move to the pole to which its kinetochore is turned. At anaphase, the displaced chromosome moves to the closest pole, even if its kinetochore is turned the other way: this can be interpreted as a decline in the activity and growth of K-MT at anaphase [161], a suggestion in agreement with the fact that K-MT decrease in numbers at anaphase. It should also be kept in mind that this is a period where the MT are in a very dynamic condition, as evidenced by the "northern lights" phenomenon described by Inoué [107]: this gives the impression that MT are continuously shifting places, and perhaps that assembly and disassembly of MT may be taking place in the half-spindles. A technique enabling a measure of the turnover of tubulin in anaphase would evidently be welcome.

Some hope of a better understanding of anaphase was that the movement of chromosomes could be observed in isolated MA. Several years ago, it was claimed that this was possible in glycerinated cells exposed to ATP [104]: however, the cytoplasmic contraction (under the action of actin) may push apart the two groups of chromosomes, independent of any spindle contraction [8]. One recent series of experiments on rat kangaroo cells lysed in the presence of tubulin may indicate however that anaphase movements can be restarted in vitro by the addition of dynein from spermatozoa [33, 145].

Movements of chromosomes toward the poles have also been observed in isolated MA of eggs of several species of sea-urchins, either after addition of ATP or in a medium containing brain tubulin. This last procedure increases the size and the birefringence of the MA; the elongation of the spindle would result from newly assembled MT. The movements observed were in both cases far slower than those observed in vivo [199].

As already mentioned, while the K-MT shorten, in many cells the poles move apart, and between the two anaphasic groups of chromosomes "continuous" fibers can be seen. Their lengthening can in some cases be considerable, and is the basis for the "Stemmkörper" theory of mitosis [15, 16]. While in several of the "primitive" intranuclear mitoses mentioned above it is apparent that the continuous fibers push the two sister nuclei apart, this is much less evident in higher organisms. Two possible mechanisms must be considered here. As it is improbable that true continuous fibers exist, the polar MT may grow at one extremity [8]; there may also be at anaphase a large number of MT with free ends, which could possibly (cf. Chap. 2) grow at both extremities. The numerous C-MT found at anaphase in *Arbacia punctulata* may be such free ends [37]. Another possibility is that the growth of the anaphase central spindle is mainly the apparent result of sliding of MT one in relation to another: if this hypothesis is correct, numerations of the MT should demonstrate some decrease at the extremities of the anaphase spindle, and a larger number of MT in the central region where they overlap. While the maximum number of MT is found at the equator in anaphase [21] and increases in the middle zone of the anaphase spindle [148], the interpretation of these changes is still under discussion. Some lengthening of the fibers by an assembly of subunits is possible [21], although it is not known where this assembly takes place.

10.4.5 The MT at Telophase and the Telophasic Bundle

At telophase, the daughter nuclei resume their interphasic pattern, the K-MT and the asters vanish progressively, while the continuous spindle persists as more and more tightly packed MT. At the end of anaphase, the MT located near the plane of cell division are seen to clump together in several more dense zones, the anaphasic bundles. As cytokinesis proceeds, these are progressively pulled together in one structure which persists for some time after the cytoplasms have separated, as a thin protoplasmic bridge containing densely packed MT with, midway between the two cells, an electron-dense matrix surrounding the MT, the mid-body. This structure was long known in animal cells and is often called the "Flemming body" [62]. In plant cells, the formation of the phragmoplast gives to telophase a quite different aspect, in which MT also play an important role.

The midbody is interesting for several reasons: its formation may be explained by the "sliding filament" theories of mitosis [145, 147], it demonstrates a curious association of MT with a dense matrix—which is somewhat similar to that which ensheathes the centriolar MT (cf. Chap. 4)—, and it is one of the "resistant" types of MT association.

The telophasic body has been described in most animal cells and may be readily observed in light microscopy: for instance, in smears of bone marrow, the erythroblasts are often seen to be linked, after telophase, by a narrow filament which may extend several μm between the cells: it is formed by continuous spindle fibers and a central density [29] and is probably shed by the cells when they become free. It has even been suggested that the proteins liberated in the blood by these shed bodies may contribute to the plasma proteins [117]. Whatever its fate, it is evident that relatively large amounts of tubulin are liberated daily in the bloodstream or the bone marrow, as about 3×10^6 mitoses of erythroblasts end every second in an adult man.

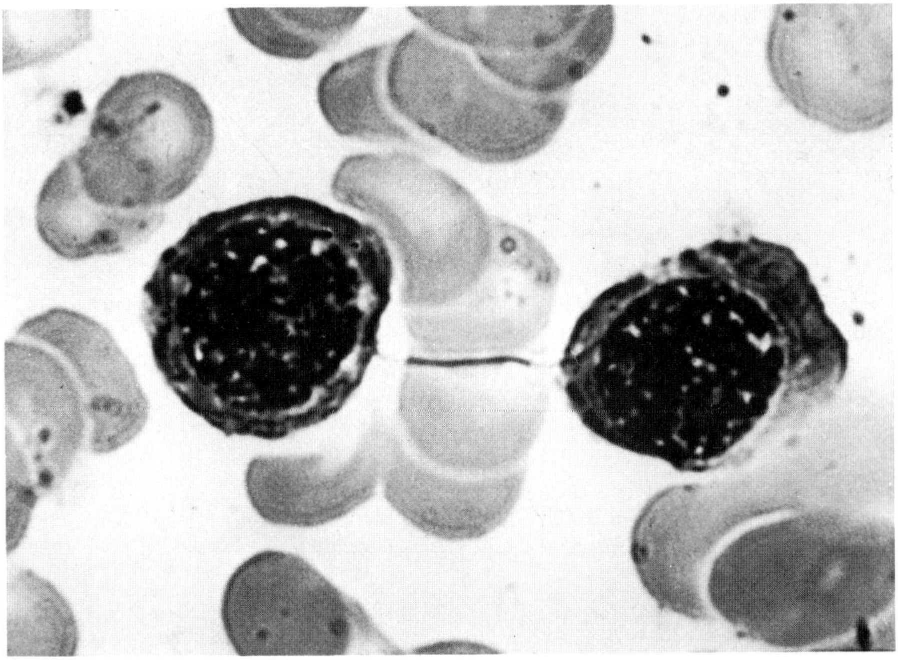

Fig. 10.11. Telophasic bundle of human erythroblasts: from a smear of bone marrow (May Grünwald Giemsa staining). The cells have been artificially separated from one another by the preparation of the smear. They remain linked by the telophasic bundle

In the chick embryo, evidence has also been presented that the midbody is discarded [2].

The sliding filament theories of mitosis imply that the polar MT slide one in relation to another until telophase, when they are arrested when only a small zone of overlapping is present, this zone becoming the central, dense part of the telophasic bundle. Counts of MT have substantiated this suggestion and close examination of midbodies indicates that the MT (or the majority of them) do not pass from one side to another and overlap [145].

This sliding implies some shearing force between MT, and certain observations on the movements of the cytoplasm surrounding the telophasic body demonstrate the dynamic condition of this thread of cytoplasm. In the cytokinesis of HeLa cells, when the cells have become rounded and are only separated by a thin cytoplasmic bridge, this appears to undergo an active elongation as indicated by "waves" of cytoplasmic swelling which move along its MT axis: these waves may play a role in the eventual rupture of the bridge. They appear as movements of cytoplasm around MT whose length does not change [32]. Time-lapse cinematography of another line of human cells (D-98S) has clearly confirmed this "peristaltic" activity of the bridge cytoplasm. These movements cease after cytochalasin treatment, which does not affect the MT. They are compared to the ruffling movements of cell membrane, and their purpose appears to be to thin out the remaining periMT cytoplasm [157].

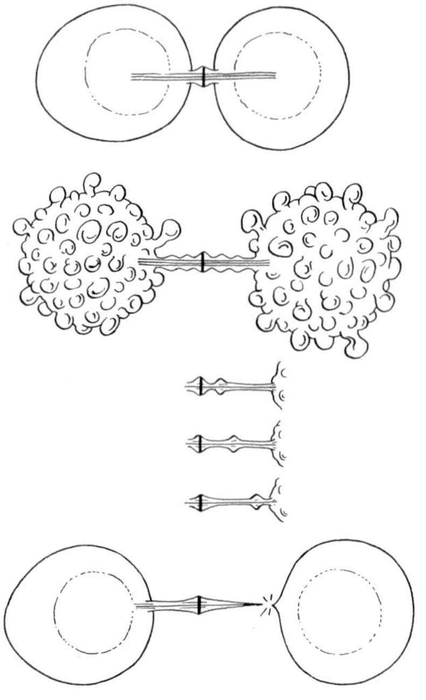

Fig. 10.12. Telophase of HeLa cells. The elongation of the telophasic bundle, with its central dense body, is evident. Waves of cytoplasmic motion are seen to move from the body to the cells, eventually breaking the intercellular bridge (redrawn from Byers and Abramson [32])

Like other MT, the telophasic bundle may be stained with fluorescent antitubulin [21, 25, 224, 225]: at the end of mitosis it may appear, in HeLa cells, whose cytoplasm contains less MT than non-neoplastic cells [224], as the only fluorescent remnant of the mitotic spindle. However, the central zone—the midbody—is not stained by this technique, as if the dense matrix embedding the MT prevented the binding of antitubulins. The nature of this matrix, which is found in all types of animal cells (cf. [192, 193]), is not known: it may play the role of a cementing substance, which would prevent the two groups of sliding MT from separating too early. It remains difficult to grasp why a structure which increases the duration of cytokinesis should be so constant, and what physiological role it may have.

Fig. 10.13 A–F. Indirect immunofluorescence of the mitotic spindle of rat kangaroo cells using antibodies to 6S bovine brain tubulin. (A) Prophase. (B) Prometaphase: the centrioles are still centrally located. (C) Metaphase (*right*) and beginning of anaphase (*left*). (D) Anaphase. (E) Tetrapolar spindle at metaphase. (F) Telophase showing the telophasic bundle (Flemming's body), with the central unstained zone corresponding to the zone where the MT overlap and are surrounded by a dense material (cf. Fig. 10.14) (from Brinkley and Chang [23])

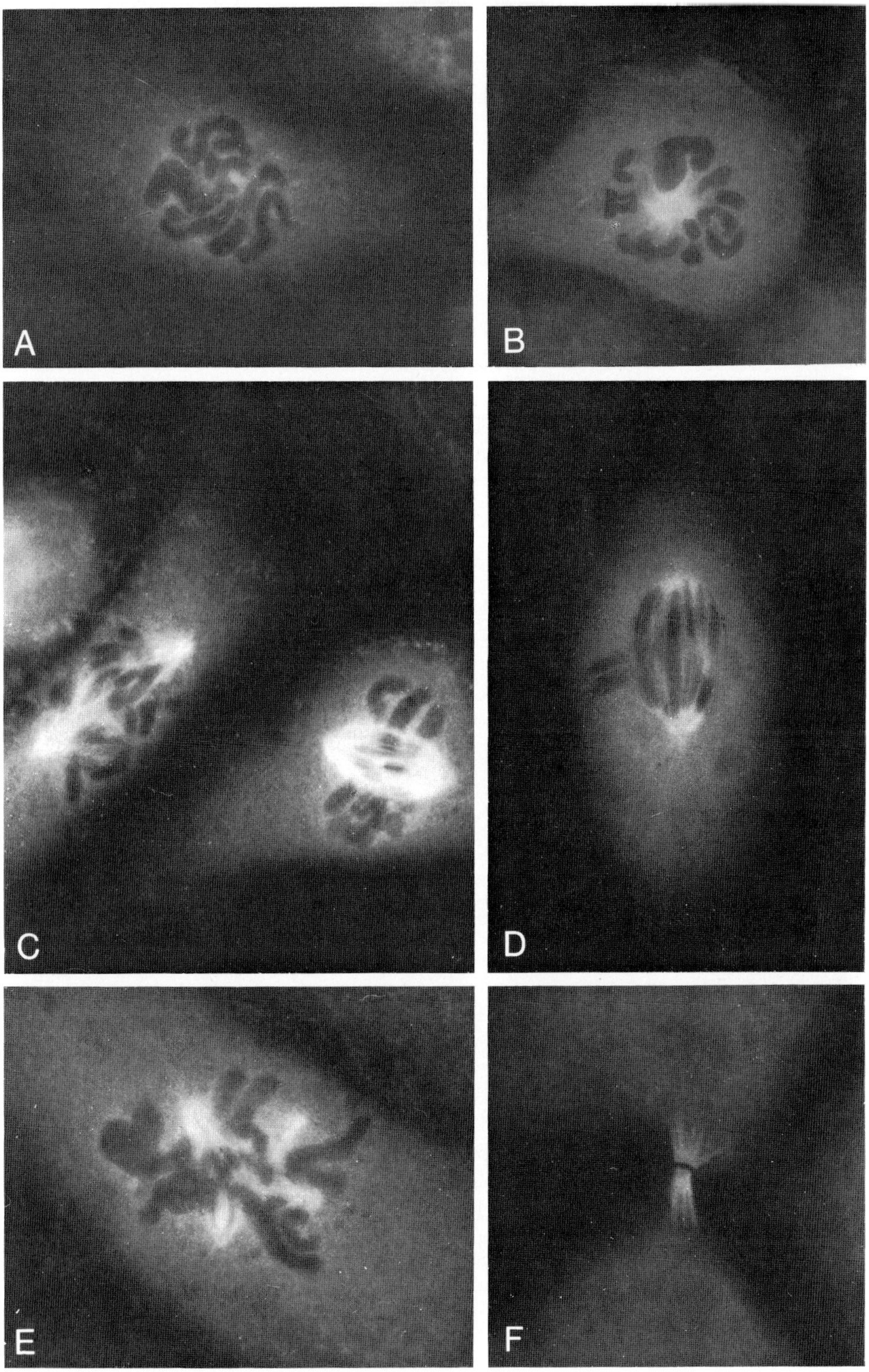
A
B
C
D
E
F

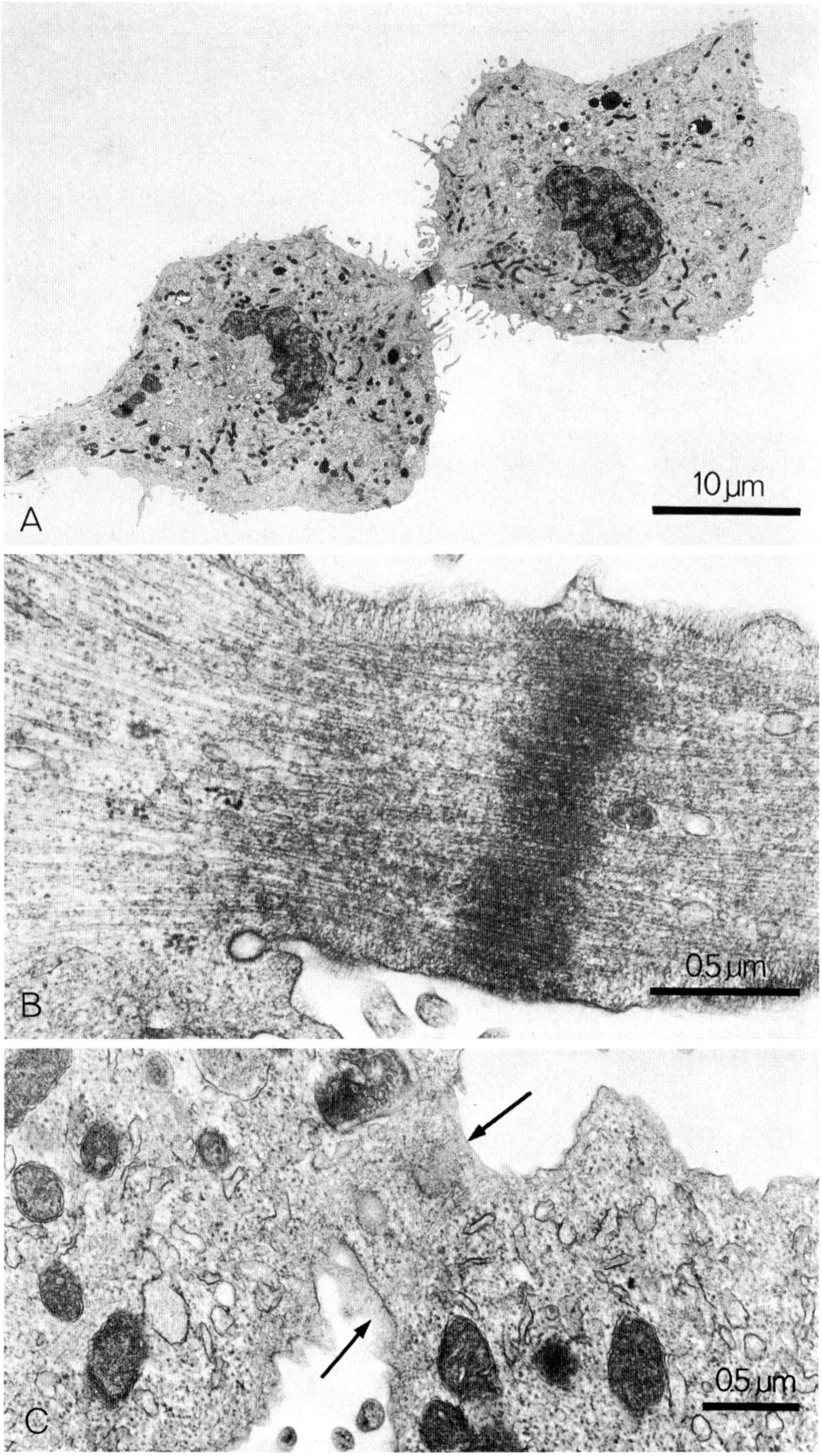
10 µm
A
0.5 µm
B
0.5 µm
C

The telophasic bundle of MT has another remarkable property: it is resistant to MT poisons, a fact which may be compared to the resistance of centrioles and basal bodies, and is related to the dense matrix of the midbody [24]. However, this resistance is not limited to this central, matrix-enveloped group of MT: in HeLa cells treated with dilute solutions of colchicine (5×10^{-8}) which block the formation of metaphase spindles, the anaphase MT and the terminal bundle are unaffected [165]. In the same type of cells, treatment by high hydrostatic pressures (680 atm) completely depolymerizes astral and interpolar MT, while the length and density of the telophasic bodies are not changed: not only the midbody, but also the MT extending in daughter cells at each side of this structure display a resistance comparable to that of centrioles or cilia [203].

In tricentric mitoses, the telophasic bundle may show MT oriented in three directions, with a star-shaped three-point midbody [123].

The movements described along the telophasic body are related to the cell contractility: this is indispensable for cytodieresis and depends on the acto-myosin microfilaments which are located in the superficial regions of the cytoplasm, close to the membrane [67]. Their activity may be, to some extent, guided by the MT apparatus, but their location and their mode of contraction do not imply any binding to MT, and will not be discussed here (cf. [210]). In HeLa cells the MF which form the "contractile ring" are located around the telophasic bundle [210].

Similar movements, leading to separation of two parts of cytoplasm, have been seen when MT are absent: they are clearly in relation to circular MF which pinch off the cytoplasmic bridge ([44], Fig. 10.15).

In plant cells, the phragmoplast formation is closely linked with MT. Although the rigid cellulose wall implies a quite different mechanism of telophase from that in animal cells, it is remarkable that the behavior of the MT shows many facts in common with those described above: here also, the MT assemble in discrete bundles which appear, in phase contrast, as fibers [96, 131, 132], and the central MT overlap and are surrounded by a dense zone similar to that of the telophasic body [96]. Vesicles formed from dictyosomes collect between the equatorial MT. At the beginning of cell plate formation this is crossed by the MT. The vesicles fuse later to form the new cell walls [169, 183]. The role of the MT is evidenced by the fact that their destruction by colchicine prevents the formation of the cell plate (179). No constant connections are however seen between the MT and the dyctiosomal vesicles (96); however, bridges have been described between the MT clustered in the region of the cell plate of *Haemanthus catherinae.* These bridges appear regularly spaced and could play a role in the mitotic movements, according to the sliding filament theory

Fig. 10.14 A–C. The telophasic bundle of MT in embryonic mouse cells in tissue culture. (A) Normal telophase. The two sister nuclei are visible. The cytoplasm shows many microvilli. The cells remain attached by a narrow cytoplasmic bridge, showing a dense line midway between the cells. (B) Higher magnification of the telophasic body, showing the two groups of parallel MT interdigitating at the middle of the body, where they are embedded in a dense matrix (cf. Fig. 10.13 showing that this dense zone does not stain by antitubulin fluorescent antibodies). (C) This cell has been treated with 1 μg/ml of the MT poison R 17934 (oncodazole). No more MT are visible. However, the general structure of the cytoplasmic bridge is not affected, and circular MF are visible (*arrows*). These are apparently the cause of the final separation of the cells (from De Brabander [44])

[98]. A thorough description of the complex movements of MT in *Haemanthus* during plate formation indicates that the MT do not remain parallel, as lateral movements take place while the vesicles which will form the cell plate move from the side of the cell to meet at the center of the telophasic spindle [132]. Many C-MT are found at this stage. These complex movements are fully described by Bajer and Molè-Bajer [8].

It can be concluded from this survey of telophase in cells as different as erythroblasts, HeLa cells, and plant meristems, that one common feature is observed, the interdigitation of interzone MT in relation to a specialized dense matrix. While animal cells end their mitosis with complex movements which are associated with the contraction of the cell surface, in plant cells the movements are intracellular and lead to the positioning of the vesicles which will fuse and separate the two daughter cells. In both types of mitosis, the MT provide a scaffolding for these movements, and indicate the location of the future plane of separation.

10.5 Action of Physical Agents and MT Poisons on Mitotic Movements

The action of "spindle poisons" is too well known for any detailed description here. The properties of these drugs have been summarized in Chapter 5, and in other parts of this book their manifold actions on tubulins and MT have been mentioned. As pointed out many years ago, mitotic poisons are not only useful drugs for medicine (cf. Chap. 11): they are "tools which enable the cytologist to dissect the interrelated phases of the mitotic cycle" [59]. The study of their action is most useful for an adequate understanding of mitosis mechanics, and the data which will be summarized now are intended as an introduction to the following discussion of the theories which attempt to explain the role of MT in mitosis.

10.5.1 Physical Factors

As explained earlier (Chap. 2), MT are destroyed by high hydrostatic pressures and by cold: the study of these agents in mitosis throws more light on the difference of sensitivity of various types of MT. Ultraviolet beams destroy the spindle birefringence: the study of the changes taking place after such localized lesions has brought surprizing facts to light. Spindle MT appear to be rapidly destroyed by many agents, this fragility resulting from some inherent properties or from their dynamic condition of equilibrium and exchange with tubulin molecules.

10.5.1.1 Cold. In 1943, the effects of cold were compared to those of colchicine in the newt, *Triturus vulgaris:* when animals were kept in water at 3° C, their mitoses were arrested with the chromosomes in a star configuration [12]. This was similar to colchicine mitosis, and later research showed that it resulted from a destruction of the polar MT and from the central location of the centrioles (cf. [47]). In an important work, the effect of low temperatures on the birefringence of the spindle of *Lilium longiflorum* pollen mother cells demonstrated a linear relation between temperature and optical properties, the thermodynamic analysis of which brought an indication that entropy of the system increased during MT assembly (cf. Chap. 2). When the cells are rewarmed, birefringence appears first at the continuous (polar) MT, a few minutes later at the K-MT. An interesting observation, which was to play a great role in one theory of mitosis, was that at anaphase the disassembly of mitotic MT by cold caused the chromosomes to "recoil", the spindle becoming shorter. Similar experiments on the egg of *Chaetopterus pergamentaceus* indicated that cooling and warming could be alternated several times in succession, the spindle reappearing each time. The size of the spindle decreased when the birefringence was destroyed. This important work [106, 107] several years before the discovery of MT suggested that an equilibirium existed between an anisotropic fibrillar structure and subunits. When the spindle reforms, chromosomes are seen to move away from the poles, indicating definite relations between MT length and chromosome movements [111].

The action of cold on MT was demonstrated later, in particular in the ameba *Chaos carolinensis:* all MT of the mitotic apparatus are destroyed at 2° C; they reappear when the cell is warmed [196]. However, all spindle MT do not react similarly to cold, as the observations on *Triturus* mentioned above suggested (the star metaphase configuration implies the persistance of K-MT): in rat kangaroo cells, cold destroyed only 30–40% of the chromosomal MT, while all polar MT disappeared [20]. The numbers of MT found at telophase by serial sectioning after cooling did not differ significantly from those of controls, while a considerable decrease of MT at metaphase and anaphase was found. In early anaphase, most of the interzonal MT were destroyed. It is difficult to assess whether these effects are the result of MT with different intrinsic properties, or of the environmental conditions [22, 24]. They are in agreement with those obtained with colchicine which also shows a greater resistance of the telophase MT.

The speed of reformation of MT after cooling in *Chaos carolinensis* has been measured: the assembly of MT begins at the kinetochore and proceeds at a rate of 1.5 μm/min. This figure is to be compared to the rate of movements of chromosomes and other particles at anaphase (vide supra; [85]). It was calculated that this could be explained by the diffusion of single subunits to one assembly point per MT—an explanation which is in agreement with some, but not all conceptions of assembly of the helical MT molecule (cf. Chap. 2).

10.5.1.2 Hydrostatic Pressure [13]. The spindle structures are readily destroyed by hydrostatic pressures, as shown already in 1941 in the eggs of *Urechis caupo:* the spindle fibrils are no longer visible at a pressure of 3000 psi, and the chromosome movements are retarded; they are completely blocked at 6000 psi. After release of the pressure, new cystasters and half-spindles are formed [173]. In *Arbacia punctulata,* higher pressures (10,000 psi) abolish all fibrous structures in the spindle. This effect is reversible in less than 10 min. The chromosome movements are retarded at 2000 psi, completely arrested at 4000 psi [229]. In HeLa cells, similar pressures

[13] Cf. note on pressure units, 1.1.5.

(10,000 psi) decrease the numbers of MT, although the midbody MT are not affected. This is reversible, and after 10 min at a normal pressure, the spindle is again visible although with shorter fibrils, sometimes abnormally oriented [86]. These results are in contradiction to those recently published on sea-urchin zygotes, where no loss of MT was found, even after 16,000 psi: this pressure destroys all spindle birefringence, which would then be caused by a non-MT constituent of the spindle [66].

There are other indications that spindle MT are modified by pressure: in the meiosis of *Chaetopterus pergamentaceus* no more birefringence of the spindle is visible after three minutes of 3500 psi applied pressure. The changes are biphasic, and an initial rapid decrease is followed by a phase of spindle shortening without change of retardation. It is suggested that birefringence is directly related to the number of MT, and that these may have different stabilities according to the attachment or non-attachment of their ends to other structures [200, 201, 204]. A thermodynamic analysis of these results confirms that MT are not homogeneous. The change in molar volume is large: $\Delta\bar{V} = 343$ ml/mol, and the changes of enthalpy and entropy of association (cf. Chap. 2) 28 Kcal/mol and 101.0 eu. respectively. These large values for positive entropy and molar volume change can be explained by the release of many moles of structural water during spindle assembly [202]. In *Chaetopterus*, spindle disassembly either by increased hydrostatic pressure, or by cold or MT poisons, results in a shortening of the spindle. The poles move toward the equator when the destruction of the spindle is not too rapid: this would imply that the pole to pole distance and the kinetochore to pole distance are controlled by the length of the MT. The conclusion of Salmon [203] is experimental evidence for the "dynamic equilibrium" model of mitosis movements (cf. [107]): "Polymerization of MT does produce pushing forces and, if controlled MT depolymerization does not actually produce pulling forces, at least it governs the velocity of chromosome-to-pole movement."

10.5.1.3 Deuterium Oxide. Early work—also on *Arbacia* and *Chaetopterus*—showed a reversible arrest of mitosis when eggs were placed in 50% solutions of D_2O: this was considered to result from "stabilization" of the spindle, and to indicate the importance of hydrogen bonds in this structure [88]. A study of birefringence changes in the oocytes of *Pectinaria gouldi* treated by D_2O indicated that the assembly of the spindle was endothermic with a high positive ΔH [35]. Experiments combining the effects of D_2O and colchicine on *Arbacia punctulata* eggs showed that concentrations of 25–35% of D_2O antagonized the effects of colchicine on mitosis, while larger concentrations (55%) intensified the arresting effects of colchicine. The antagonism of low concentrations of colchicine (less than 5.0×10^{-5} M) and D_2O remained unexplained [150]. Similar studies in wheat cells showed that D_2O increased the number of MT in the preprophasic equatorial band and in the spindle, while colchicine decreased considerably the number of MT. When the alkaloid was dissolved in a 70% solution of D_2O, no spindle was visible, and the preprophasic MT were scarce [30].

10.5.1.4 Ultraviolet Microbeams. The irradiation of spindle fibers by a small beam of UV locally destroys its birefringence [107]. The changes observed in various cells and at various times of the mitotic cycle have been described minutely by Forer [63,

64, 65]. Although the relation between birefringence and MT is not accepted by all authors—Forer, for one, considering that at least 30 – 50% of spindle birefringence is not of MT origin—and although no ultrastructural correlation has been presented indicating that UV irradiation destroys the MT in these experiments, these results are important for a proper understanding of movements within the mitotic spindle. On the other hand, several authors have demonstrated that spindle MT may be very rapidly (within seconds, [8]) disassembled by UV (cf. [6, 230]).

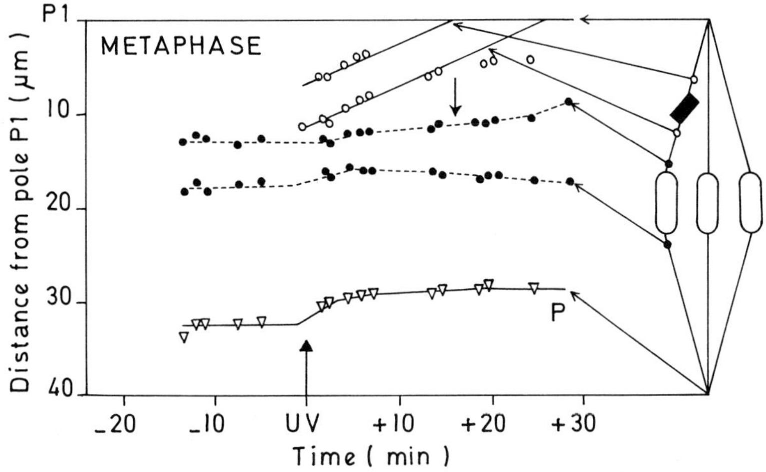

Fig. 10.15. Effects of UV localized irradiation of the metaphase spindle. The irradiated zone, which has lost its birefringence, is represented as a *black rectangle.* As indicated by the *line with open circles,* this zone moves gradually toward the pole, while the chromosomes, still undivided, remain at the equator (cf. Fig. 10.16; one pole is shown as fixed, the position of the other by the *triangles: P*). The movements of the chromosomes are indicated by the dashed lines (– – – ● – – –) (redrawn from Forer [63], slightly modified)

The experiments of Forer [63, 64] have been carried out on the spermatocytes of the crane-fly *Nephrotoma suturalis,* and must be related here in some detail. The results differ at metaphase and at anaphase, and should be considered separately. The technique was however identical: a limited small zone of the spindle was irradiated for a short time by a polychromatic beam of ultraviolet light, carefully focused on the spindle fibers, which were observed by polarized light and phase contrast.

At metaphase [63] the irradiated zone, which received about 10 ergs/μm^2 of radiant energy, lost its birefringence almost immediately (20 sec). The zone of decreased birefringence is shaped like the microbeam of UV and immediately begins to move toward the pole of the mitotic figure, without changing its form. The overall size of the spindle does not change, nor do the positions of the chromosomes. This poleward movement takes place at rates varying between less than 0.5 and 1 μm/min, i.e., at speeds comparable to those already mentioned for particle movements in the spindle. When it reaches the pole, the zone of decreased birefringence fades out, and the spindle is normal again (Fig. 10.16). This is indicated by resumed mitotic activity, with normal anaphase movements. This is only true for localized irradiations: if the whole half-spindle is irradiated, mitosis is arrested. Under some conditions, in particular with a 280 nm wavelength, chromosome movements may also be arrested without any change in the optical properties of the spindle. It is clear that what is observed is a poleward displacement of a darker zone: this may have several molecular explanations, and no ultrastructural data are available. Whatever the change of spindle organization (molecular ruptures, cross-linkings), the interesting point is the movement—at a speed comparable to that of the chromosomes at anaphase—of the altered zone of the spindle: this shows that changes are taking

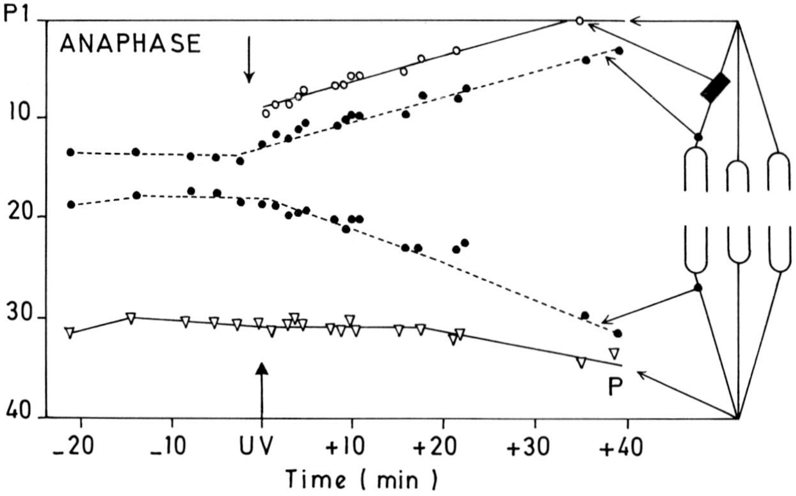

Fig. 10.16. Irradiation of the spindle at anaphase. The *black rectangle,* corresponding to the irradiated zone, which has lost its birefringence, moves toward the pole (*open circles*). The chromosome movements are not impeded and are as rapid on the side of the irradiated half-spindle as on the other half (*dotted lines*). The movements of the chromosomes are indicated by the dashed lines (– – – ● – – –) (redrawn from Forer [64], slightly modified)

place in this structure, and suggests that apart from the birefringent material (which would be the MT) some other structure is implicated in movement, unless one imagines that MT are disassembled at the distal side of the irradiated region and reassemble at the side closest to the kinetochore.

When similar experiments are conducted at anaphase, the same poleward movement of the irradiated zone is observed; however, the chromosome movements, which may be slightly slowed down, are not affected, a fact which would indicate that the traction apparatus is not affected and differs from the structure which confers on the spindle its optical properties [64].

Another interesting observation made with this technique is that the irradiation of the half-spindle attached to one dyad of chromosomes at anaphase may stop the movement of the symmetrical chromosome.

Before anaphase, the destruction of a localized zone of the birefringent spindle may arrest chromosome movements, so it appears that at that time of the cycle, the birefringent material is necessary for chromosome movements. At anaphase, on the other hand (Fig. 10.17), the chromosomes sometimes move even if their spindle fibers have lost their birefringence (cf. [65]). One explanation may be that before anaphase the chromosomes are still linked together and unable to separate or be influenced by the poleward flow taking place in the spindle, while at anaphase, they are carried along normally, but without being affected by the loss of birefringent fibers. Some other constituents of the spindle may play a part in movement: the possible role of actin has been repeatedly suggested (cf. [66, 68, 69]) and will be discussed when the various theories of mitotic movement are considered.

One more word about these results: if the birefringent material is made of MT, a fact which probably needs some qualification as other birefringent components may be present in the spindle, the movement of the UV lesions would be another indication of the dynamic condition of the meta- and anaphase MT, as previously mentioned in connection with the "northern lights" phenomenon. However, the movement of the irradiated zone precisely directed in space and time strongly suggests that the "motor" is different from the MT, a conclusion which is in agreement with many facts summarized in Chapters 6 and 9.

10.5.2 Chemical Agents: Colchicine

Although many chemicals have similar effects on the spindle and on mitotic movements, colchicine occupies a special place, not only because it was discovered first [52, 175], but mainly because it is one of the most specific poisons, acting on nearly all types of cells of the animal and vegetal kingdoms, combining electively with tubulin (cf. Chap. 5), and active at very low concentrations ($10^{-8} - 10^{-7}$ M) in some cells [28, 165]. While a few species appear to be far more resistant to colchicine, like *Mesocricetus auratus,* the golden hamster, this may result from the impermeability of some cell membranes to the alkaloid, as in *Amoeba sphaeronucleus* [39]. A more surprizing instance is the fungus *Saprolegnia ferax,* the mitoses of which resist colchicine and also VLB, griseofulvin, low temperatures, and 10,000 psi hydrostatic pressure [94].

The action of colchicine on mitoses in plants and animals has led to a wide use of this drug in applied biology: the possibility of arresting spindle activity without damaging the nuclear cycle has resulted in many applications in the field of artificial polyploid and amphidiploid plant production, and the increase in mitotic index which results, in animals, from the accumulation of "arrested" mitoses—or "stathmokineses"—, has been largely put to use for studies of cell growth and mitotic regulation. These aspects of applied colchicine research will not be treated here (cf. [60]).

In the first years of use of the drug, the spectacular increase of visible mitoses in germinative regions suggested that colchicine truly stimulated mitosis: this opinion is no longer tenable; some of the results which did show an increase of the mitotic index difficult to explain by spindle poisoning alone (cf. [54]) may result from the combined action of the drug and of other factors increasing, at the same time, the rate of new mitoses.

The action of colchicine in most cells is reversible, if the treatment does not last too long and if the doses are not too large. In mammals, however, the mitotic arrest, which may last from six to eight hours or more, kills a large number of cells, resulting in part from the fact that a cell maintained in metaphase is unable to form mRNA [58].

While studying the action of colchicine as the most typical "spindle" poison, the facts mentioned in Chapter 5 should be kept in mind, and also the many findings related to colchicine which have been gathered in the monograph: the alkaloid has more than one action, and if in this chapter the mitotic changes will be considered above all, side-effects on structures other than MT must not be forgotten.

Several reviews on the action of colchicine (and other spindle poisons) have been written, and detailed bibliographical information may be found in the book by Eigsti and Dustin [60] and the reviews by Gavaudan [81], Lettré [134], Deysson [49], Taylor [219], and Wilson and Bryan [228].

Colchicine mitosis has been described many times, and this survey will be limited to problems in relation to MT and the movements of mitosis. Early work, in particular that of Bucher [28] on tissue cultures of fibroblasts followed by cinematography, indicated clearly that the rate of mitoses was not changed, but that pathological divisions lasting for several hours accumulated progressively, explaining the increase of the mitotic index. These mitoses have often been called "arrested"—"prolonged" would be a better term—and the spindle effects described as a "destruction" of the fibrillar components. In fact, the changes are more complex than a total spindle (or MT) inactivation, and from recent work and electron microscopic analysis, the C-mitosis, as it is often called, displays various degrees of spindle inactivation.

Another early work [80] showed that according to the dose and the time of action of the drug on dividing cells, various changes of the mitotic figure could be ob-

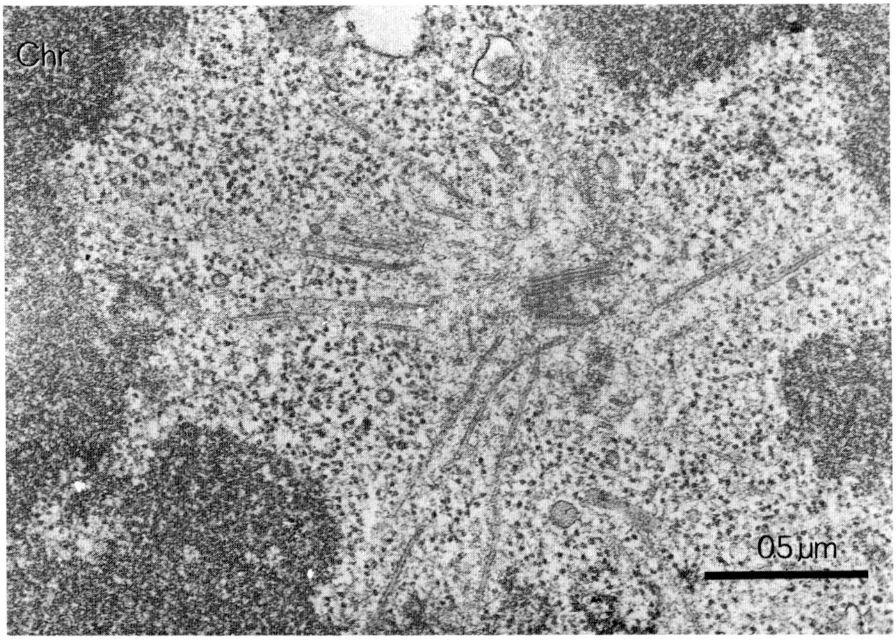

Fig. 10.17. Mouse intestinal mitosis, 4 h after injection of 1 mg/kg colchicine. Intact MT radiating from the centrally located centrioles towards the periphically located chromosomes. Chr: chromosome

served, from a complete scattering of chromosomes throughout the cytoplasm (3 h, 2.5×10^{-6} M) to more discrete changes, the most typical being the "star-metaphase" with the centromeres (kinetochores) pointing to the center of the cell (90 min, same concentration). The suggestion had already been made that centrioles may lie at the center of these stars, which are often seen during recovery from colchicine poisoning [176]. In the spermatogenesis of *Triturus helveticus*, similar star metaphases or "monopolar stathmokinesis" were correctly interpreted as resulting from a small central spindle [212].

When these mitotic abnormalities were studied with the electron microscope, the central location of the centrioles was confirmed [46]. In Chinese hamster cells cultured in vitro, the star metaphases are centered around two centrioles, the structure of which is not affected by colchicine (or colcemid); MT are seen to radiate from the centrioles toward the kinetochores (only one kinetochore per chromosome is connected by MT to the centrioles, the one facing outward being without MT). In favorable cases, the centrioles may have already divided and four were gathered together. When the cells are replaced in a normal medium, the two groups of centrioles move apart and continuous fibers are again observed between them, leading to a normal anaphase. The migration of centrioles may take place as soon as 5 min after removal of colcemid from the medium. It was concluded that C-mitosis is the result of an inhibition of the poleward movement of centrioles and of the smaller number of MT. The "continuous" fibers appear more sensitive than the K-MT. The migration of the centrioles may be the result of the assembly of continuous (polar) MT [26]; however, this opinion is not shared by all authors (cf. [8]). The C-mitosis is thus, as suggested earlier, a monopolar mitosis, and not a true "arrested metaphase". Identical mitotic abnormalities, with central centrioles, have been described in human leukemic cells 2 h after 0.02 μg/ml of colcemid [143].

In plant cells, the aspects of spindle inactivation also vary between star "metaphases" and complete spindle destruction with scattered chromosomes. Here, perhaps as a result of the absence of centrioles, the spindle material remains at the center of the cell (so-called "pseudo-spindle"; [49, 50, 179]). Plant cells differ from those of animals, in particular warm-blooded animals, by the fact that the chromosomal cycle may persist while the cells are under the influence of colchicine, as the alkaloid has no effect on DNA reduplication and chromosome division: this results in tetraploidy or higher degrees of polyploidy, and explains the success of colchicine as an agent for the creation of species with a doubled number of chromosomes. When plant cells are exposed for a short time to colchicine (0.1%, 30 min), many mitoses may be changed into multicentric anaphases; this is more marked at low temperature (15° C; [102]).

Once they have started, the later stages of mitosis are more resistant to MT poisons, and the resistance of the telophasic bundle has been mentioned above. The movements of cytokinesis are not affected: they are mainly related to MF. A study of the effects of low concentrations (5×10^{-8} M) of colchicine on the anaphase of HeLa cells has revealed a double action which may throw some light on the mechanics of the anaphase spindle. The chromosome and pole movements are normal, until the chromosomes are separated by a distance equal to the length of the mitotic spindle: this could take place by sliding of polar MT. Normally the chromosomes should separate further: in contrast, in presence of colchicine there is a slow tendency for the poles to move one toward the other. The possible explanation is that the further increase in spindle length at the end of anaphase implies an assembly of new MT subunits. Colchicine would in this experiment demonstrate its action—at the low doses used—on MT assembly, without interference with the sliding movements of MT [165].

The problems of colchicine resistance, and the few data on antagonists of colchicine have already been summarized in Chapter 5. Another question which is related to the antimitotic action of the drug, is whether colchicine destroys already formed MT or only prevents the assembly of MT from subunits by linking with the tubulin subunits. In mitosis, where there is much evidence that the MT are in a dynamic condition, the second alternative appears acceptable. It should however not be forgotten that physical agents—cold, pressure, UV—do destroy already formed MT and that this destruction can take place in a few minutes. On the other hand the rapid reversibility of colcemid poisoning of sea-urchin eggs, after irradiation with ultraviolet light has changed the molecule into the inactive lumicolchicine, indicates that spindle MT are readily reassembled [3] and apparently in a condition of "dynamic equilibrium".

Recent experiments have brought new evidence of this equilibrium. Sea-urchin eggs (*Lythechinus variegatus*) were treated, before mitosis, by 10^{-6} M colcemid for short periods—from 1 to 6 min. The spindle was reduced in length and its birefringence decreased in proportion to the duration of colchicine action. If these small spindles were irradiated by ultraviolet light, converting colchicine into lumicolchicine, the spindles increased in length and in birefringence faster than usual, a fact suggesting that the size of the spindle is controlled not only by the pool of available tubulin, but by other inhibitors preventing maximum assembly [215].

10.5.3 Other Chemical Agents

The list of substances which may affect the spindle MT is very long. It may even be claimed that this effect is devoid of all specificity, and that inert substances, provided they are present in high enough concentrations, may be active: this line of research was at the origin of the "narcosis" theory of C-mitosis as proposed by Östergren [167, 168]. However, while experiments with isolated cells or root-tips may bring interesting insights into cell biology, they do not throw much light on the action of the few drugs which modify specifically the tubulin assembly, even when present at low concentrations. The substances which will be discussed here belong mainly to this category, and their specificity of action is linked with the applications that several have found in medicine, in particular as inhibitors of neoplastic growth. This specificity, as shown in Chapter 2, indicates the presence of receptor sites on the tubulin molecules, and should one day help to understand better the mechanisms of tubulin assembly.

10.5.3.1 Podophyllin and Podophyllotoxins. As mentioned in Chapter 5, the active principles of the resin of *Podophyllum* combine with tubulin at the same site as colchicine (cf. [120]. They have similar actions on mitotic MT, leading to arrested mitoses. In rats, effects comparable to those of colchicine have been described, but the toxicity of podophyllin is greater and the effective dose close to the LD_{50} dose. A study of various derivatives indicated definite relations between structure and antimitotic activity [211]. Few detailed descriptions of podophyllin-arrested mitoses have been published: in a line of human lymphoblasts cultured in vitro, doses of 0.01 – 5 μg/ml acting for 24 h lead to an accumulation of up to 60% arrested mitoses. This effect was reversible. Other derivatives, such as the epipodophyllotoxins, showed a greater toxicity and arrested the growth at other stages of the cycle [125].

10.5.3.2 The Catharanthus (Vinca) Alkaloids. Apart from the precipitation of tubulin paracrystals (cf. Chap. 5) VLB and VCR are powerful mitotic inhibitors. comparable to colchicine [34, 42, 72]. The action on the spindle is specific, the other phases of the mitotic cycle not being affected, and concentrations as low as 10^{-7} g/ml are active, and lead to an accumulation of arrested mitoses which may last for 20 h, much longer than with colchicine [27]. In HeLa cells, mitotic arrest could already be demonstrated with doses of 10^{-9} g/ml. Larger doses (10^{-7} and 10^{-6} g/ml) are toxic and the effects become irreversible. The mitoses show aspects similar to those described for colchicine, with centrally located centrioles, the reduplication of which is not impeded [82, 118, 122]. This fact is in agreement with observations of centriolar multiplication (in intermitotic cells) in rats receiving repeated injections of VCR [61].

The reversibility of VLB action enables repeated destruction and reassembly of the spindle to be observed, for instance in the oocytes of *Pectinaria pectinaria gouldi,* a marine annelid (10^{-5} M; [140]).

Action of this drug upon the spindle is found also in plant cells; however, the mitotic index does not increase as after colchicine, and a depression of the mitotic cycle was noted [49, 50]. Multipolar anaphases are frequently observed in *Allium cepa* [101].

The low toxicity, the long duration of the arrested mitoses, and other effects on the mitotic cycle and on the cell surface (cf. [124]) are some explanations of the favorable results obtained with VLB and VCR in cancer chemotherapy, as opposed to colchicine (cf. Chap. 11).

10.5.3.3 Griseofulvin. The action of this fungistatic is different from that of colchicine and the *Catharanthus* alkaloids, and its mode of fixation on the mitotic spindle remains poorly understood. In rats, only large doses (from 100 to 200 mg/kg) lead to metaphase arrest in the germinative zones [170]. In plant cells, it is less active, although it demonstrates a synergic activity with colchicine in *Allium* [49]. In the oocytes of *Pectinaria,* on the other hand, griseofulvin decreased the size and the birefringence of the mitotic spindle in less than 6 min; this effect was found with small doses (10^{-5} M) and was rapidly reversible when the cells were placed in a normal medium: in about 10 min the spindle regained its normal properties. This could be done repeatedly, even with doses ten times larger. When the spindle shortens, the whole mitotic figure moves closer to the cortex, where one of the centrioles is anchored [140].

It is not surprizing that the fungistatic chemical arrest mitosis in fungi: in the myxomycete *Physarum polycephalum,* 20 μg/ml of griseofulvin inhibits growth and leads to the formation of polyploid cells. One remarkable action is the formation in the nuclei of paracrystalline arrays of densely packed MT. It is known that in this species mitosis is intranuclear, and the nucleus contains VLB-precipitable tubulin [89]. These results suggest that the action of griseofulvin is somewhat different from that of other MT poisons (cf. Chap. 5).

10.5.3.4 Isopropyl-N-Phenylcarbamate (IPC). Various carbamates have been studied as mitotic inhibitors. Some act mainly on the chromosome reduplication, such as ethyl-carbamate (urethane) which was used in the treatment of human leukemias [56]. Others are poisons of the achromatic apparatus. IPC is used as an herbicide. Its action on the mitosis of *Haemanthus katerinae* Baker endosperm is different from that of other spindle poisons. While the spindle loses its birefringence, electron microscopy demonstrates that this is the result of a disorientation, not a destruction, of MT, which become arranged in multiple radial arrays already visible 2 h after treatment with a 10 g/l solution [97]. Melatonin, which increases the spindle birefringence in these cells, partially antagonizes IPC [114]. Similar multipolar mitoses have been described in *Allium cepa* root-tips [100]. In the green alga *Oedogonium cardiacum*, IPC at the concentration of 5.5×10^{-4} M prevents the polimerization of MT. This effect is reversible, but not entirely, and leads to a disorientation of MT. The number

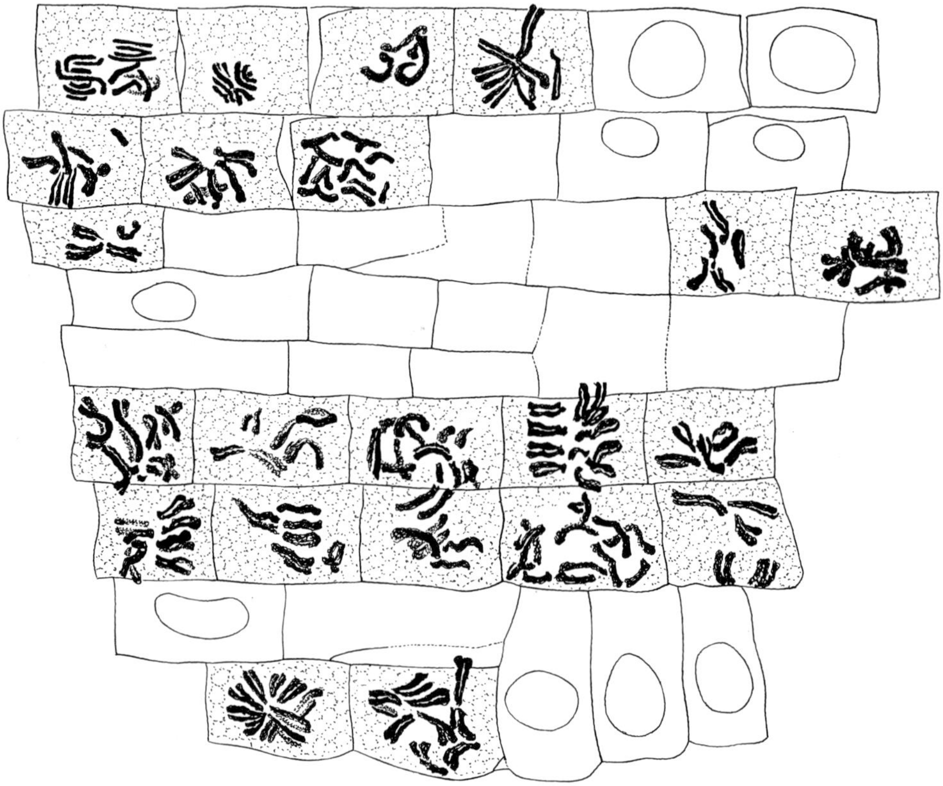

Fig. 10.18. Root-tips of *Vicia faba* treated for 3 h by a 1% solution of chloral hydrate and replaced for 24 h in water. Arrested mitoses (from Van Regemoorter [221])

of polar bodies (MTOC) increases, and it is suggested that IPC mainly affects the regulation of the MTOC [40, 41]. This is in agreement with the fact that the IPC does not prevent polymerization of tubulin in vitro, nor does it bind to tubulin [40].

10.5.3.5 Chloral Hydrate. In animal cells chloral hydrate is probably the first chemical to have been described as influencing mitosis, as early as 1886 [99].

In the spermatogenic cells of *Triturus alpestris,* after 8 h in a 1/500 solution of chloral hydrate, many star metaphases are found [212]. The mechanism of these is different from that of colchicine: in the fertilized eggs of *Pleurodeles waltii* Michah, the chromosomes are attracted to the center of the cell occupied by a granular mass which results from the destruction of MT [213]. Fragments of MT may be visible in this mass, to which the kinetochores are attached [214].

Chloral hydrate was also one of the earliest spindle poisons studied in the plant cells: after three hours of soaking root-tips of *Vicia faba* in a 10 g/l solution, many arrested, abnormal metaphases were observed [221]. It is however a weak poison, the threshold of action on mitosis being close to that of toxicity [49].

10.5.3.6 Other Spindle Poisons. Among the many substances which may affect the spindle and its MT, a few should be mentioned here, for their study may help to elucidate some of the problems of mitotic movements.

Sulfhydryl reagents should come first, as it was their study, as mentioned in Chapter 1, which led to the rediscovery of the action of colchicine on mitosis [53, 55, 56]. In mammals the effects of injections of sodium cacodylate or arsenious oxide are similar to those of colchicine, with a considerable increase in the mitotic index of germinative regions resulting from the accumulation of arrested metaphases. These are ultrastructurally identical to colchicine mitoses (personal data). There is evidence that sulfhydryl groups of tubulins play a role in MT assembly, as mentioned in Chapter 2. Several of the methods used for the isolation of the MA depend on the stabilization of S-S bonds of MT. It has also been demonstrated that an excess of free -SH groups may block metaphases [56, 57]. The action of arsenicals has been demonstrated in tissue cultures [138] and plant cells [149]. It would be important to know whether the arsenic and heavy metals which combine with -SH groups directly affect the tubulin molecule or the -SH enzymes necessary for mitosis.

Recent advances have confirmed the role of sulfhydryl groups in the dynamics of mitotic MT. Sea-urchin eggs treated by *mercaptoethanol,* a chemical which has been used for the isolation of the MA, show a mitotic arrest with centrally located centrioles. The spindle loses its birefringence, and decreases in size. However, the poleward movement of the centrioles is not inhibited, as if two pools of tubulin were present in the cell, one available for the formation of MT and the other blocked by mercaptoethanol [91]. The role of sulfhydryls, and in particular gluthatione, has been confirmed by observations on sea-urchin and clam eggs, where methylxanthine inhibits cell division, with a decrease in the levels of $NADP^+$and NADPH: this inhibition is reversed by glutathione. Moreover *diamide,* a thiol-oxidizing agent, completely inhibited cell division at a concentration of 10^{-3} M, the MA disappearing entirely, as if dissolved. Diamide could combine either with -SH groups of tubulin (cf. Chap. 2) or with glutathione: it is suggested that the second action is the most important, and that -SH$\rightleftarrows$S-S balance may regulate tubulin assembly [158].

The reaction between diamide and -SH takes place in two steps:

$$RS^- + (CH_3)_2NCON{=}NCON(CH_3)_2 \rightarrow (CH_3)_2NCO\underset{H^+}{N} - \overset{RS}{\underset{H^+}{N}}\,CON(CH_3)_2$$

$$+ \underset{H^+}{RS^-} \rightarrow RSSR + (CH_3)_2NCONHNOCON\,(CH_3)_2$$

In vitro assembly of brain tubulin is completely inhibited by diamide (3×10^{-4} M), and electron microscopy demonstrates that all the oligomeric rings are transformed into 6S dimers. These effects can be reversed by the addition of dithiothreitol, which reduces S-S links. As the free sulfhydryl titer of tubulin is decreased by 50% after diamide, the action may be directly on the tubulin molecule. However, in vivo the oxidation of glutathione could explain the rapid destruction of the spindle MT. A link between the concentration of Ca^{2+} and sulfhydryls, Ca^{2+} favoring the formation of disulfide bonds, is an interesting possibility [153].

Some poisons of cellular respiration also block the spindle action. Such is *ethylcarbylamine* ($C{=}N-C_2H_5$), which inhibits the Pasteur effect and arrests mitosis of mouse fibroblasts in vitro [220], and *rotenone,* a heterocyclic compound which specifically inhibits NAD-dependent mitochondrial respiration [19, 152]. Rotenone combines with tubulin and destroys the mitotic spindle, the chromosomes lying round centrally located centrioles. As it inhibits colchicine binding to tubulin, it is possible that rotenone becomes attached to the same receptor side [13]; its chemical structure shows, like colchicine and the *Catharanthus* alkaloids, an aromatic ring with two methoxy groups:

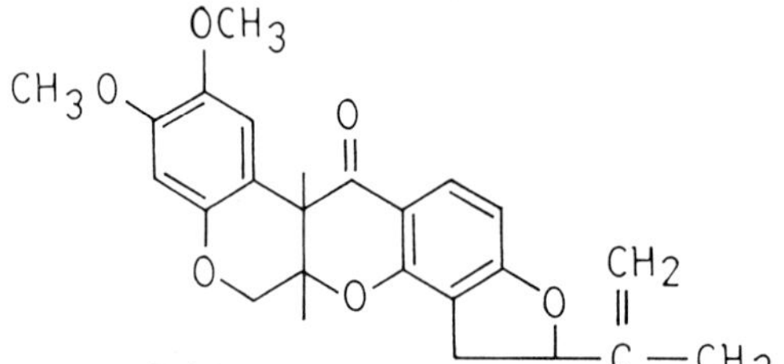

ROTENONE

Rotenone was known as a fish poison of plant origin, and belongs to the little studied group of natural mitotic poisons (cf. sanguinarine, cryptopleurine, protoanemonin) [60].

Some actions of *melatonin,* the pineal gland hormone, a tryptophane derivative, have already been mentioned. Although it may act in some circumstances as a colchicine antagonist, melatonin displays a colchicine-like inhibition of cilia regeneration in *Stentor* [184] and has been considered to be a true MT poison. A 8×10^{-4} M solution arrests mitoses and growth of *Allium cepa* cells. The CH_3O group attached to an aromatic ring has been suggested as related to its properties [11].

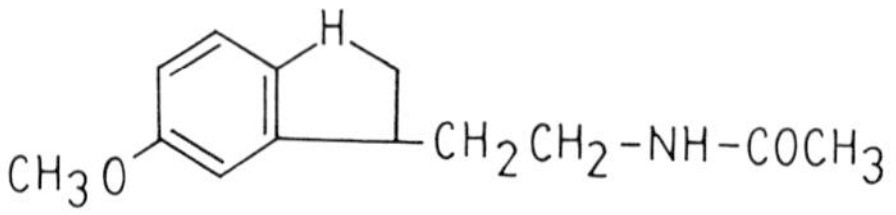

Melatonin

New aromatic molecules inhibiting tubulin assembly have recently been discovered, and may be of interest in cancer chemotherapy. While purine and pyrimidine antagonists are mainly inhibitors of the S phase of the mitotic cycle, unexpectedly *5-fluoropyrimidin-2-one* has shown metaphase-arresting properties on a line of hu-

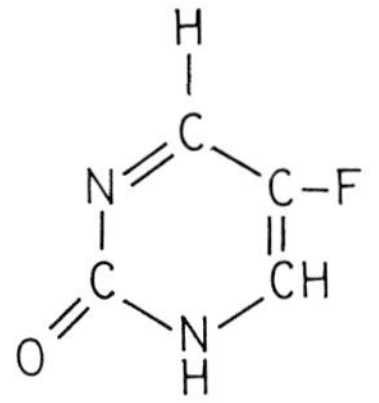

5-fluoropyrimidin-2-one

man liver cells in culture [164]. The increase of arrested mitoses is impressive, with doses of about 1.75×10^{-3} M. The action is reversible; its mechanism remains unknown.

A new series of benzimidazole derivatives are powerful spindle poisons. They have been recently tested on various cells, and on neoplastic growth: *oncodazole* (R 17934-NSC 238159) is one of the most active. It destroys MT in tissue cultures at the dose of 1 μg/ml, leading eventually to cell death [4, 44, 45]. Its chemical structure is quite different from that of other mitotic poisons (methyl [5-(2-thienylcarbonyl)-1H-benzimidazol-2-yl] carbamate) (cf. Chap. 5).

The rapid destruction of mitotic MT by the macrolide *maytansine*, which is one of the most powerful MT poisons in vivo, as studied on the divisions of sea-urchin eggs, has been mentioned in Chapter 5 (cf. [189]).

10.6 MT and the Theories of Mitosis

The fact that MT are present in all types of mitoses, from the most primitive to the most complex, and that agents acting on MT disrupt in one way or another the movements of chromosomes and other structures of mitosis is certainly evidence that any theory of mitosis which does not account for the MT would be invalid. On the other hand, it would be presumptuous to think that the complexities of mitotic movements may be explained by tubulin and MT alone: while the MT of spindles have yet to be purified, all that is known about MT in other structures, such as ciliary doublets, indicates that many other proteins, perhaps as important as tubulin, are intimately associated with MT. It would be most important to know whether or not "contractile" proteins of the actin-myosin type are present in the mitotic structures. As mentioned above, the movements of mitosis require only a fraction of the energy of ciliary (or muscle) movements, and quantitatively the contractile proteins could be present in quite small quantities, but representing the real dynamic factors, the MT being a convenient scaffolding.

Modern theories on the role of MT in mitosis result from two different lines of thought: the first implies that the assembly and disassembly of MT suffices to provide the energy for the main mitotic movements, from the gathering of chromosomes on the metaphase plate through to their anaphase separation. The other theories assume that movements take place by some interaction of MT with MT, the exact nature of the motor force being linked to the MT but of a different chemical nature: these theories are similar to those of ciliary bending, and involve sliding movements of MT, the assembly and disassembly of which is seen as secondary and not providing the energy for motion.

These theories will be discussed now, and it is understandable why mitosis was presented in this book after many other types of MT movements. It is believed that a clearer view of mitosis becomes possible when the general properties of MT and MT-associated proteins are considered. The aim will not be to oppose two main theories; on the contrary, it will become apparent, as so often in science, that both are useful and partially true, and that both help to understand the main problem: what makes chromosomes move?

10.6.1 The Assembly-Disassembly Theory

Inoué and his group proposed, following their studies on spindle birefringence and several years before the discovery of MT, that the spindle fibers are in equilibrium with a pool of subunits and that their growth and shortening could result from displacement of this equilibrium. The studies of pressure and cold disruption of spindles brought important arguments for this thesis, which was expressed in 1964 [107] in the following sentence "... we know there is an anisotropic distribution of material which could one way or another account for the development of mechanical forces required for pulling (or pushing) the chromosomes, most likely dependent on the shift of orientation equilibrium."

Shortly after the discovery of MT, these were considered to represent the "anisotropic material" in equilibrium with a pool of monomers (or dimers), an opinion which was in agreement with the following studies on tubulin assembly (cf. Chap. 2; [112]). The lengthening or shortening of the MT would result from the addition of tubulin molecules, and drugs such as colchicine were imagined to associate with tubulin subunits, which would thus be prevented from assembling into MT. This idea implied that a condition of rapid exchange, a "dynamic equilibrium" existed between tubulin and MT, for in eggs of *Pectinaria* for instance, the spindle birefringence disappears after 10 min of action of colchicine [112]. A comparable rapid dissolution has been described after many MT poisons; it is usually rapidly reversible. A similar "assembly hypothesis" of spindle activity was defended by Dietz [51] in his studies of *Pales ferruginea*. This theory has been repeatedly supported by Inoué (cf. [108, 111], although in his recent work the role played by interMT bridges in the elongation of the anaphase spindle has been mentioned.

New techniques have indicated that such an equilibrium does exist and can be modified if spindles are placed in a medium containing tubulin in conditions favorable for its polymerization (cf. Chap. 2). Isolated spindles of *Chaetopterus* oocytes [110], *Spisula* eggs [187] or mammalian cells [144, 146] become larger and their MT more apparent in such solutions, indicating that a transfer of subunits to spindle fibers is possible. This is of course in agreement with most facts which have been described in this chapter: spindle MT do grow, they do appear to be in a dynamic equilibrium in relation to the surrounding cytoplasm (the "northern lights" phenomenon), and elongations and shortenings (disassembly) are apparent in most stages of mitosis.

If MA of *Spisula solidissima* eggs are treated by cold and have lost their birefringence, this reappears after addition of tubulin from chick brain at 30° C and the spindles may reach three or four times their normal length. An interesting observation is that these restored MAs remain identical if placed in a solution of diluted tubulin, and are stable in 10^{-5} M Ca^{2+}, while they lose their birefringence in 10^{-3} M Ca^{2+} and by cold treatment. Apparently, the isolated MA can more easily incorporate new tubulin molecules than disassemble them [187].

Some objections can however be raised to the idea that assembly and disassembly are sufficient per se to explain all the complexities of spindle activities:

1. It is not certain that in all cells a single pool of tubulin is present, and that an equilibrium exists between MT and tubulin. In *Naegleria* evidence for two different pools has been presented ([79]; cf. Chap. 4), and mitoses such as that of *Syndinium* demonstrate that in the same cell, and nearly side by side, some MT may grow (the polar ones) while others are not changed (the K-MT). The equilibrium concept has often been mentioned as explaining the growth of the anaphasic polar MT from the tubulin subunits released by the shortening of the K-MT. This certainly cannot hold for all types of mitosis.

2. One of the principal objections is a "mechanical" one: while it is easy to understand that the assembly of tubulin molecules into MT may push aside chromosomes or other structures—and evidence for this has been presented, in particular in the "primitive" mitoses of yeasts—it is difficult to understand how the disassembly of MT could move anything. It is true that in some circumstances it seems as if disassembly—by hydrostatic pressure, by cold, or by griseofulvin (cf. [140])—does decrease the size of the spindle, and move the chromosomes toward one pole, but this has mainly been observed in *Chaetopterus* eggs, and the movements may well be explained either by the fact that the external spindle pole is attached close to the cell membrane, or by the pressure of the surrounding cytoplasm on the spindle which has shrunk by tubulin disassembly. Quantitative studies on these eggs indicate clearly that the rate of spindle disassembly under the action of high hydrostatic pressures parallels the rate of motion of the chromosomes, which, in this experiment, is much faster than in normal mitosis. These results indicate a coupling of the disassembly of mitotic MT and the action of pulling forces, but the author adds with caution that "the mechanism by which forces are generated ... is still open to question" [204]. The disassembly of the K-MT, which is evident, may be comparable to the pulling down of a scaffolding which is no longer needed, the positive force moving the chromosome not being the MT (cf. [66]).

3. Another problem requires more information: any equilibrium theory implies a balance between assembled MT and tubulin molecules. Information is needed about how the MT are assembled and disassembled: two opinions have been expressed, and are contradictory. Many facts, reported in other chapters, indicate that the growth of MT (and probably their disassembly) takes place by addition of subunits at one (eventually both) extremity. The helical structure of the polymer suggests this, although it is not impossible that MT may grow or disassemble as linear threads (the C-MT may for instance be evidence of this type of change). The experi-

ments related in this chapter indicate however that these changes may be quite rapid (a few minutes for disassembly under high hydrostatic pressure conditions), and some authors (cf. [51]) think that the MT may be disassembled all along their length, and fragment into small segments.

Whatever the answer, it is evident that mitotic MT (except of course those of centrioles and of the telophasic bundle) are in a most dynamic condition and that the turnover of their molecules may be rapid [cf. 108]. It has not, however, been demonstrated that all MT poisons, as the "dynamic equilibrium" theory suggests, combine only with tubulin mono- or dimers, and cannot influence whole MT: it seems probable—but clearly further metabolic data are required—that some agents may alter and destroy already assembled MT.

10.6.2 The Sliding Filament Theory

Several movements associated with MT can only be explained by the sliding, on the surface of the MT, of other structures, by a mechanism similar to that of muscle contraction. Sliding filament theories of mitosis do not answer the question of the true motor of chromosome movements, but analyze the changes of the spindle MT in terms of forces acting between MT sliding one in relation to another. They apply to mitosis ideas which have already been met in Chapters 7, 8 and 9.

McIntosh and collaborators have presented several times [144, 145, 147] such a model of mitosis, which may explain many aspects of the mitotic MT. For instance, the interrelations between polar and K-MT, as already stated, may provide an answer to the peculiar barrel shape of many MA. The counts of MT at various stages of metaphase and anaphase, notwithstanding some differences between the results of McIntosh and Landis [148], Brinkley and Cartwright [22], and Fuge [75], do strongly suggest, particularly for the polar fibers, that the two groups of MT slide one in relation to another, until they are arrested by the formation of the telophasic midbody. Such sliding of parallel MT which have probably grown from two separate MTOC could be explained by the fact that the helical structure of MT is associated with a definite polarity, and that MT of opposite polarity tend to move in opposite directions.

This theory, which helps to understand the formation of the metaphase plate (by the interactions of the K-MT with the polar MT and the ensuing proper orientation of the chromosomes) and that of the ana- and telophase movements, by sliding of the polar MT, cannot alone explain all mitotic movements:

a) If the relations between the polar MT and the K-MT are considered, the formation of the metaphase plate may be explained by a sliding of the K-MT in relation to the polar MT, bringing the chromosomes to a condition of equilibrium at the center of the cell. Later, the elongation of the continuous fibers would take place, as said above, by sliding of MT of opposite polarities. However, the anaphase movements of the K-MT, which slide toward the poles in close relation with the polar MT, imply that sliding takes place in the opposite direction from at prometaphase. This objection, recognized by McIntosh et al. [146], may however lose much importance when one remembers what happens in the nerves, where transport along MT is bidirectional. There are also facts about ciliary beating (one form of sliding filament mechanism) indicating the possibility of reversal movements [207], and sliding motions in mitosis could explain many facts if it was accepted that its relation to the polarity of MT could be reversed.

b) In sliding filament explanations, the elongation of the anaphase spindle would be possible while the MT maintain the same length. In vertebrate cells, this is far from certain, and in other types of mitosis, where a polar spindle is seen to increase considerably in length, it cannot be explained by sliding alone, and the assembly hypothesis is far more acceptable.

c) The sliding filament theories have difficulty in explaining the movements of isolated chromosomes. The experiments on the spermatocytes of *Melanoplus differentialis* [161] where, according to the phase of mitosis, detached chromosomes move either to the pole toward which their kinetochore is turned, or away from it, are perplexing, It is also interesting that in the same cells, the detached chromosomes move towards the spindle four times faster (2.6 μm/min, at 24° C) than the anaphasic movements, and move toward the equator first, not toward the pole.

d) Such theories imply some physical relation between the MT, and in several mitoses the spindle MT appear too distant for such action. One answer to this has been given by the "zipper hypothesis" of Bajer et al. (cf. [8, 9, 10]): in *Haemanthus katherinae,* while sliding movements were suggested in early work [6], further analysis has led to the discarding of this idea [115], and to more attention to the MT which show large angulations with the spindle MT. The lateral movements of the K-MT would lead their MT to "zip" with continuous MT, and this "zipping" would have a pulling action in the polar direction [9, 10, 132]. Whatever the value of this hypothesis for the case of *Haemanthus,* "zipping" is clearly unacceptable for the numerous types of spindles (in particular anaphase or intranuclear spindles) whose shape is cylindrical and where all MT are parallel.

10.6.3 The Role of Other Spindle Proteins

It should be apparent that mitosis may no more be explained only in terms of MT than axonal flow or the movements of melanin granules. MT are the most apparent structures and their relations with polar bodies (centrioles or MTOC) and with kinetochores clearly indicate their importance. Yet it appears that alone they could not move the chromosomes, first from prophase to the metaphase plate, later toward the anaphase, nor elongate the entire cell and separate the daughter cells. Some other factors would be required, and once more the MT appear as a temporary scaffolding, not the motor of mitosis.

The idea that some protein resembling actin, or actin itself, may be present in the spindle and explain the chromosome movements, has been attractive. This still remains without a proper solution, and the fact that during interphase anti-actin antibodies stain fibers with a quite different location in the cell from the MT and the mitotic spindle would suggest that actin is not involved. However, following the suggestion by Forer (vide supra) that two components are present in the spindle, this author, with several collaborators, has repeatedly indicated that actin is present in mitotic spindles. The technique used was to glycerinate cells and to study the "decoration" of cytoplasmic fibrils by heavy meromyosin. It has indicated the presence of such fibrils in the spindle and the cell cortex of the crane-fly (*Nephrotoma suturalis*, [66]) and more recently in the spindle of *Haemanthus katerinae* [68].

Immunofluorescence demonstration of actin has shown that apart from the bundles of filaments visible in the cytoplasm, the mitotic spindle and in particular the kinetochores and the centriolar regions fix anti-actin antibodies (in rat kangaroo cells; [205]). When these cells have been treated by colchicine, one sees only a diffuse peripheral stain, without staining of kinetochores and spindle. During disassembly, at anaphase, of the kinetochore MT, these actin filaments could provide the motive force for the chromosome movements, MT serving "as lengthening and cytoskeletal elements" [66].

The demonstration of myosin by similar techniques appears more difficult (perhaps because the number of molecules of myosin is much smaller than that of actin): fluorescent anti-myosin antibodies have however recently been shown to stain the mitotic spindle and the contractile ring in HeLa cells [78].

It is worth mentioning that in the intranuclear mitosis of *Paramecium bursaria*, MF ranging in size from 8 to 16 nm, and which could belong to the acto-myosin group, are present in large numbers inside the nucleus [136].

10.6.4 Conclusion

This rapid survey of some theories which have been much more thoroughly discussed by other authors in recent conferences (cf. [83, 216]) was intended to show that whatever the complexities of mitosis, some facts are evident:

- mitosis is polarized, either by MTOC, by centrioles, or by unknown polar forces (as in higher plants)
- MT grow mainly from two specialized structures: the kinetochores, towards the poles, and the poles, towards the opposite ones
- this growth takes place by assembly of tubulin molecules from a pool of tubulin; this is not necessarily the same pool as that which is used for other MT structures, such as cilia
- the prophase movements may start before the nuclear membrane disappears and are interesting evidence of some not yet understood relations between MT and nucleus
- the equilibrium attained at metaphase results from the assembly of two groups of intermingled MT, the K-MT and the polar ones. It is doubtful that polar MT extend from one pole to another
- all these MT are polarized in relation to their site of growth
- anaphase implies in most species (but not always) a shortening of K-MT. This can take place quite independent of other changes of MT, and clearly several regulatory mechanisms are at play, as there are perhaps several different tubulin compartments
- the most primitive type of mitosis results from the elongation of "continuous" MT, by the assembly of tubulin molecules
- disassembly of MT is evident at several steps of mitosis; it is doubtful that it alone can provide the motive force

— although chromosome movements are slow, and require a very small amount of energy, it is probable that this is transmitted by structures related to MT but different from tubulins
— the movements of cytokinesis and cell separation are in relation to the activity of cytoplasmic MF (probably actin)
— particles can move along the spindle, and UV lesions can be carried similarly to the poles: these observations indicate the role of components other than MT in spindle activity.

References

1. Aldrich, H. C.: The ultrastructure of mitosis in myxamoebae and plasmodia of *Physarum flavicomun.* Am. J. Bot. **56,** 290 – 299 (1969)
2. Allenspach, A. L., Roth, L. E.: Structural variations during mitosis in the chick embryo. J. Cell Biol. **33,** 179 – 196 (1967)
3. Aronson, J., Inoué, S.: Reversal by light of the action of N-methyl-N-desacetyl colchicine on mitosis. J. Cell Biol. **45,** 470 – 477 (1970)
4. Atassi, G., Tagnon, H. J.: R 17934-NSC 238159: a new antitumor drug. I. Effect on experimental tumors and factors influencing effectiveness. Eur. J. Cancer **11,** 599 – 607 (1975)
5. Bajer, A.: Notes on ultrastructure and some properties of transport within the living mitotic spindle. J. Cell Biol. **33,** 713 – 719 (1967)
6. Bajer, A.: Chromosomes movement and fine structure of the mitotic spindle. Symp. 22 Soc. Exp. Biol. Aspect of Cell Motility. pp. 285 – 310. Cambridge: Univ. Press 1968
7. Bajer, A. S.: Interaction of microtubules and the mechanism of chromosome movement (zipper hypothesis). 1. General principle. Cytobios **8,** 139 – 160 (1973)
8. Bajer, A. S., Molè-Bajer, J.: Spindle Dynamics and Chromosome Movements. International Review of Cytology, Suppl. 3. New York-London: Academic Press 1972
9. Bajer, A., Molè-Bajer, J.: Lateral movements in the spindle and the mechanism of mitosis. In: Molecules and Cell Movement (eds.: S. Inoué, R. E. Stephens), pp. 77 – 96. New York: Raven Press 1975
10. Bajer, A. S., Molè-Bajer, J., Lambert, A. M.: Lateral interaction of microtubules and chromosome movements. In: Microtubules and Microtubule Inhibitiors (eds.: M. Borgers, M. De Brabander), pp. 393 – 423. Amsterdam-Oxford: North-Holland, New York: Am. Elsevier 1975
11. Banerjee, S., Margulis, L.: Mitotic arrest by melatonin. Exp. Cell Res. **78,** 314 – 318 (1973)
12. Barber, H. N., Callan, H. G.: The effects of cold and colchicine on mitosis in the newt. Proc. Roy. Soc. London B **131,** 258 – 271 (1943)
13. Barham, S. S., Brinckley, B. R.: Action of rotenone and related respiratory inhibitors on mammalian cell division. 1. Cell kinetics and biochemical aspects. 2. Ultrastructural studies. Cytobios **15,** 85 – 96; 97 – 110 (1976)
14. Begg, D. A., Ellis, G. W.: The role of birefringent chromosome fibers in the mechanical attachment of chromosomes to the spindle. J. Cell Biol. **63,** 18 a (1974)
15. Belar, K.: Beiträge zur Kausalanalyse der Mitose. II. Untersuchungen an den Spermatocyten von *Chorthippus (Stenobothrus) lineatus* Panz. Roux' Arch. **118,** 359 – 484 (1929)
16. Belar, K.: Beiträge zur Kausalanalyse der Mitose. III. Untersuchungen an den Staubfadenhaarzellen und Blattmeristenzellen von *Tradescantia virginica.* Z. Zellforsch. **10,** 73 – 134 (1929)
17. Bibring, T., Baxandall, J.: Selective extraction of isolated mitotic apparatus. Evidence that typical microtubule protein is extracted by organic mercurial. J. Cell Biol. **48,** 324 – 339 (1971)
18. Bland, C. E., Lunney, C. Z.: Mitotic apparatus of *Pilobolus crystallinus.* Cytobiologie **11,** 382 – 391 (1975)
19. Brinkley, B. R., Barham, S. S., Barranco, S. C., Fuller, G. M.: Rotenone inhibition of spindle microtubule assembly in mammalian cells. Exp. Cell Res. **85,** 41 – 46 (1974)
20. Brinkley, B. R., Cartwright, J. Jr.: Organization of microtubules in the mitotic spindle: differential effects of cold shock on microtubule stability. J. Cell Biol. **57,** 25 a (1970)
21. Brinkley, B. R., Cartwright, J. Jr.: Ultrastructural analysis of mitotic spindle elongation in mammalian cells in vitro. Direct microtubule counts. J. Cell Biol. **50,** 416 – 431 (1971)

22. Brinkley, B. R., Cartwright, J. Jr.: Cold-labile and cold-stable microtubules in the mitotic spindle of mammalian cells. Ann. N. Y. Acad. Sci. **253,** 428 – 439 (1975)
23. Brinkley, B. R., Chang, J. P.: Mitosis in tumor cells: methods for light and electron microscopy. In: Methods in Cancer Research (ed.: H. Busch), Vol. 11, pp. 247 – 291. New York-San Francisco-London: Acad. Press 1975
24. Brinkley, B. R., Fuller, G. M., Highfield, D. P.: Studies of microtubules in dividing and non-dividing mammalian cells using antibody to 6-S bovin brain tubulin. In: Microtubules and Microtubule Inhibitors (eds.: M. Borgers, M. De Brabander), pp. 297 – 312. Amsterdam-Oxford: North-Holland; New York: Am. Elsevier 1975
25. Brinkley, B. R., Fuller, G. M., Highfield, D. P.: Tubulin antibodies as probes for microtubules in dividing and nondividing mammalian cells. In: Cell Motility (eds.: R. Goldman, T. Pollard, J. Rosenbaum), pp. 435 – 456. Cold Spring Harbor Lab. 1976
26. Brinkley, B. R., Stubblefield, E., Hsu, T. C.: The effects of colcemid inhibition and reversal on the fine structure of the mitotic apparatus of Chinese hamster cells in vitro. J. Ultrastr. Res. **19,** 1 – 18 (1967)
27. Bruchowsky, N., Owen, A. A., Becker, A. J., Till, J. E.: Effects of vinblastine on the proliferative capacity of L cells and their progress through the division cycle. Cancer Res. **25,** 1232 – 1237 (1965)
28. Bucher, O.: Zur Kenntnis der Mitose. VI. Der Einfluß von Colchicin und Trypaflavin auf den Wachstumsrhythmus und auf die Zellteilung in Fibrocyten-Kulturen. Z. Zellforsch. **29,** 283 – 322 (1939)
29. Buck, R. C., Tisdale, J. M.: The fine structure of the mid-body of the rat erythroblast. J. Cell Biol. **13,** 109 – 115 (1962)
30. Burgess, J., Northcote, D. H.: Action of colchicine and heavy water on the polymerization of microtubules in wheat root meristem. J. Cell Sci. **5,** 433 – 451 (1969)
31. Burkholder, G. D., Okada, T. A., Comings, D. E.: Whole mount electron microscopy of metaphase. I. Chromosomes and microtubules from mouse oocytes. Exp. Cell Res. **75,** 497 – 511 (1972)
32. Byers, B., Abramson, D. H.: Cytokinesis in HeLa: post-telophase delay and microtubule-associated motility. Protoplasma **66,** 413 – 436 (1968)
33. Cande, W. Z., Snyder, J., Smith, D., Summers, K., McIntosh, J. R.: A functional mitotic spindle prepared from mammalian cells in culture. Proc. Natl. Acad. Sci. USA **71,** 1559 – 1563 (1974)
34. Cardinali, G., Cardinali, G., Enein, M. A.: Studies on the antimitotic activity of leurocristine (vincristine). Blood **21,** 102 – 110 (1963)
35. Carolan, R. M., Sato, H., Inoué, S.: A thermodynamic analysis of the effect of D_2O and H_2O on the mitotic spindle. Biol. Bull. **129,** 402 (1965)
36. Cleveland, L. R.: Hormone-induced sexual cycles of flagellates. I. Gametogenesis, fertilization, and meiosis in *Trichonympha.* J. Morphol. **85,** 197 – 295 (1949)
37. Cohen, W. D., Gottlieb, T.: C-microtubules in isolated mitotic spindles. J. Cell Sci. **9,** 603 – 620 (1971)
38. Cohen, W. D., Rebhun, L. I.: An estimate of the amount of microtubule protein in the isolated mitotic apparatus. J Cell Sci. **6,** 159 – 176 (1970)
39. Comandon, J., DeFonbrune, P.: Action de la colchicine sur *Amoeba sphaeronucleus.* C. R. Soc. Biol. Paris **136,** 410 – 411; 423; 460 – 461; 746 – 747; 747 – 748 (1942)
40. Coss, R. A., Bloodgood, R. A., Brower, D. L., Pickett-Heaps, J. D., McIntosh, J. R.: Studies on the mechanism of action of isopropyl N-phenylcarbamate. Exp. Cell Res. **92,** 394 – 398 (1975)
41. Coss, R. A., Pickett-Heaps, J. D.: The effects of isopropyl-N-phenyl carbamate on the green alga *Oedogonium cardiacum.* I. Cell division. J. Cell Biol. **63,** 84 – 98 (1974)
42. Cutts, J. H.: The effect of Vincaleukoblastine on dividing cells in vivo. Cancer Res. **21,** 168 – 172 (1961)
43. Davidson, L., La Fountain, J. R. Jr.: Mitosis and early meiosis in *Tetrahymena pyriformis* and the evolution of mitosis in the phylum ciliophora. Biosystems **7,** 326 – 336 (1975)
44. De Brabander, M.: Onderzoek naar de rol van microtubuli in gekweekte cellen met behulp van een nieuwe synthetische inhibitor van tubulin-polymerisatie. Thesis, Brussels 1977
45. De Brabander, M., van de Veire, R., Aerts, F., Geuens, G., Borgers, M., Desplenter, L., de Cree, J.: Oncodazole (R 17934): a new anti-cancer drug interfering with microtubules. Effects on neoplastic cells cultured in vitro and in vivo. In: Microtubules and Microtubule Inhibitors (eds.: M. Borgers, M. De Brabander), pp. 509 – 521. Amsterdam-Oxford: North-Holland; New York: Am. Elsevier 1975

46. de Harven, E.: The centriole and the mitotic spindle. In: Ultrastructure in Biological Systems (eds.: A. J. Dalton, F. Haguenau), Vol. 3, pp. 197 – 227. New York-London: Acad. Press 1968
47. de Harven, E., Dustin, P. Jr.: Etude au microscope électronique de la stathmocinèse chez le rat. Coll. Intern. Centre Rech. Sci. Paris **88,** 189 – 197 (1960)
48. De Thé, G.: Cytoplasmic microtubules in different animal cells. J. Cell Biol. **23,** 265 – 275 (1964)
49. Deysson, G.: Antimitotic substances. Intern. Rev. Cytol. **24,** 99 – 148 (1968)
50. Deysson, G.: Microtubules and antimitotic substances. In: Microtubules and Microtubule Inhibitors (eds.: M. Borgers, M. De Brabander), pp. 427 – 451. Amsterdam-Oxford: North-Holland; New York: Am. Elsevier 1975
51. Dietz, R.: Die Assembly-Hypothese der Chromosomenbewegung und die Veränderungen der Spindellänge während der Anaphase I in Spermatocyten von *Pales ferruginea (Tipulidae, Diptera).* Chromosoma **38,** 11 – 76 (1972)
52. Dustin, A.P.: Contribution à l'étude des poisons caryoclasiques sur les tumeurs animales. II. Action de la colchicine sur le sarcome greffé, type Crocker, de la souris. Bull. Acad. Roy. Méd. Belg. **14,** 487 – 502 (1934)
53. Dustin, A. P.: L'action des arsenicaux et de la colchicine sur la mitose. La stathmocinèse. C. R. Ass. Anat. **33,** 204 – 212 (1938)
54. Dustin, A. P.: Recherches sur le mode d'action des poisons stathmocinétiques. Action de la colchicine sur l'utérus de Lapine impubère sensibilisé par l'injection préalable d'urine de femme enceinte. Arch. Biol. (Liège) **54,** 111 – 187 (1943)
55. Dustin, A., Havas, L., Lits, F.: Action de la colchicine sur les divisions cellulaires chez les végétaux. C. R. Ass. Anat. **32,** 170 – 176 (1937)
56. Dustin, P.: Some new aspects of mitotic poisoning. Nature (London) **159,** 794 (1947)
57. Dustin, P. Jr.: Mitotic poisoning at metaphase and -SH proteins. Exp. Cell Res. *Suppl. 1,* 153 – 155 (1949)
58. Dustin, P.: The quantitative estimation of mitotic growth in the bone marrow of the rat by the stathmokinetic (colchicinic) method. In: The Kinetics of Cellular Proliferation (ed.: F. Stohlman Jr.), pp. 50 – 56. New York-London: Grune and Stratton 1959
59. Dustin, P. Jr.: New aspects of the pharmacology of antimitotic agents. Pharmacol. Rev. **15,** 449 – 480 (1963)
60. Eigsti, O. J., Dustin, P. Jr.: Colchicine in Agriculture, Medicine, Biology, and Chemistry. Ames, Iowa: Iowa State Coll. Press 1955
61. Flament-Durant, J., Hubert, J. P., Dustin, P.: Centriolo- and ciliogenesis in the rat's pituicytes under the influence of microtubule poisons in vitro. Exp. Cell Res. **99,** 435 – 437 (1976)
62. Flemming, W.: Zur Kenntnis der Zelle und ihrer Lebenserscheinungen. Arch. Mikros. Anat. **16,** 302 – 436 (1879)
63. Forer, A.: Local reduction of spindle fiber birefringence in living *Nephrotoma suturalis* (Loew) spermatocytes induced by ultraviolet microbeam. J. Cell Biol. **25,** 95 – 117 (1965)
64. Forer, A.: Characterization of the mitotic traction system, and evidence that birefringent spindle fibers neither produce nor transmit force for chromosome movement. Chromosoma **19,** 44 – 98 (1966)
65. Forer, A.: Chromosome movements during cell division. In: Handbook of Molecular Cytology (ed.: A. Lima-de-Faria), pp. 553 – 604. Amsterdam-London: North-Holland; New York: Wiley Interscience Division 1969
66. Forer, A.: Actin filaments and birefringent spindle fibers during chromosome movements. In: Cell Motility (eds.: R. Goldman, T. Pollard, J. Rosenbaum), pp. 1273 – 1294. Cold Spring Harbor Lab. 1976
67. Forer, A., Goldman, R. D.: The concentrations of dry matter in mitotic apparatuses in vivo and after isolation from sea-urchin zygotes. J. Cell Sci. **10,** 387 – 410 (1972)
68. Forer, A., Jackson, W. T.: Actin filaments in the endosperm mitotic spindles in a higher plant, *Haemanthus katherinae* Baker. Cytobiologie **12,** 199 – 214 (1976)
69. Forer, A., Kalnins, V. I., Zimmerman, A. M.: Spindle birefringence of isolated mitotic apparatus: further evidence for two birefringent spindle components. J. Cell Sci. **22,** 115 – 132 (1976)
70. Forer, A., Zimmerman, A. M.: Spindle birefringence of isolated mitotic apparatus analysed by pressure treatment. J. Cell Sci. **20,** 309 – 328 (1976)
71. Forer, A., Zimmerman, A. M.: Spindle birefringence of isolated mitotic apparatus analysed by treatments with cold, pressure, and diluted isolation medium. J. Cell Sci. **20,** 329 – 340 (1976)
72. Frei, E. III, Whang, J., Scoggins, R. B., van Scott, E. J., Rall, D. P., Ben, M.: The stathmokinetic effect of vincristine. Cancer Res. **24,** 1918 – 1925 (1964)

73. Fuge, H.: Spindelbau, Mikrotubuliverteilung und Chromosomenstruktur während der I. meiotischen Teilung der Spermatocyten von *Pales ferruginea.* Eine elektronenmikroskopische Analyse. Z. Zellforsch. **120,** 579 – 599 (1971)
74. Fuge, H.: Verteilung der Mikrotubuli in Metaphase- und Anaphase-Spindeln der Spermatocyten von *Pales ferruginea.* Eine quantitative Analyse von Serienquerschnitten. Chromosoma **43,** 109 – 144 (1973)
75. Fuge, H.: An estimation of the microtubule content of crane fly spindles based on microtubule counts. Protoplasma **79,** 391 – 394 (1974)
76. Fuge, H.: Ultrastructure and function of spindle apparatus microtubules and chromosome during nuclear division. Protoplasma **82,** 289 – 320 (1974)
77. Fuge, H.: The arrangement of microtubules and the attachment of chromosomes to the spindle during anaphase in tipulid spermatocytes. Chromosoma **45,** 245 – 260 (1974)
78. Fujiwara, K., Pollard, T. D.: Fluorescent antibody localization of myosin in the cytoplasm, cleavage furrow, and mitotic spindle of human cells. J. Cell Biol. **71,** 848 – 875 (1976)
79. Fulton, C., Kowit, J. D.: Programmed synthesis of flagellar tubulin during cell differentiation in *Naegleria.* Ann. N. Y. Acad. Sci. **253,** 318 – 332 (1975)
80. Gaulden, M., Carlson, J.: Cytological effects of colchicine on the grass-hopper neuroblast in vitro, with special reference to the origin of the spindle. Exp. Cell Res. **2,** 416 – 433 (1951)
81. Gavaudan, P.: Les facteurs de la cytonarcose. In: Exposés Annuels de Biologie Cellulaire (ed.: J. A. Thomas), pp. 275 – 361. Paris: Masson et Cie. 1956
82. George, P., Journey, L. J., Goldstein, M. H.: Effect of vincristine on the fine structure of HeLa cells during mitosis. J. Natl. Cancer Inst. **35,** 355 – 376 (1965)
83. Goldman, R., Pollard, T., Rosenbaum, J. (eds.): Cell Motility. Cold Spring Harbor Lab. 1976
84. Goldman, R. D., Rebhun, L. I.: The structure and some properties of the isolated mitotic apparatus. J. Cell Sci. **4,** 179 – 209 (1969)
85. Goode, D.: Kinetics of microtubule formation after cold disaggregation of the mitotic apparatus. J. Mol. Biol. **80,** 531 – 538 (1973)
86. Goode, D., Salmon, E. D., Maugel, T. K., Bonar, D. B.: Microtubule disassembly and recovery from hydrostatic pressure treatment of mitotic HeLa cells. J. Cell Biol. **67,** 138 A (1975)
87. Grell, K.G.: Protozoology. Berlin-Heidelberg-New York: Springer 1973
88. Gross, P. R., Spindel, W.: The inhibition of mitosis by deuterium. Ann. N. Y. Acad. Sci. **84,** 745 – 754 (1960)
89. Gull, K., Trinci, A. P. J.: Ultrastructural effects of griseofulvin on the myxomycete *Physarum polycephalum.* Inhibition of mitosis and the production of microtubule crystals. Protoplasma **81,** 37 – 48 (1974)
90. Guttes, S., Guttes, E., Ellis, R. A.: Electron microscope study of mitosis in *Physarum polycephalum.* J. Ultrastruct. Res. **22,** 508 – 529 (1968)
91. Harris, P.: Structural effects of mercaptoethanol during mitotic block of sea urchin eggs. Exp. Cell Res. **97,** 63 – 73 (1976)
92. Hartmann, J. F., Zimmerman, A. M.: The mitotic apparatus. In: The Cell Nucleus (ed.: H. Busch). New York-London: Acad. Press 1974
93. Hauser, M.: The intranuclear mitosis of the ciliates *Paracineta limbata* and *Ichthyophtirius multifiliis.* I. Electron microscope observations on pre-metaphase stages. Chromosoma **36,** 159 – 175 (1972)
94. Heath, I. B.: Effect of antimicrotubule agents on growth and ultrastructure of Fungus *Saprolegnia ferax* and their ineffectiveness in disrupting hyphal microtubules. Protoplasma **85,** 147 – 176 (1975)
95. Heath, I. B., Heath, M. C.: Ultrastructure of mitosis in the cowpea rust fungus *Uromyces phaseoli* var. *vignae.* J. Cell Biol. **70,** 592 – 607 (1976)
96. Hepler, P. K., Jackson, W. T.: Microtubules and early stages of cell-plate formation in the endosperm of *Haemanthus katerinae* Baker. J. Cell Biol. **38,** 437 – 446 (1968)
97. Hepler, P. K., Jackson, W. T.: Isopropyl-*N*-phenylcarbamate affects spindle microtubules orientation in dividing endosperm cells of *Haemanthus katerinae* Baker. J. Cell Sci. **5,** 727 – 743 (1969)
98. Hepler, P. K., McIntosh, J. R., Cleland, S.: Intermicrotubule bridges in mitotic spindle apparatus. J. Cell Biol. **45,** 438 – 444 (1970)
99. Hertwig, R.: Über den Einfluss von Chloralhydrat auf die inneren Befruchtungserscheinungen. Anat. Anz. **1,** 11 – 16 (1886)

100. Hervas, J. P.: Sur l'activité antimitotique de l'isopropyl N-phényl-carbamate dans les cellules méristématiques: cinétique de la production d'anaphases multipolaires. Arch. Biol. **85,** 453 – 459 (1974)
101. Hervas, J. P.: Sur l'effet multipolarisant de la vinblastine dans la division cellulaire d'*Allium cepa* L. Experientia **31,** 170 – 171 (1975)
102. Hervas, J. P., Fernandez-Gomez, M. E., Gimenez-Martin, G.: Colchicine effect on the mitotic spindle: estimate of multipolar anaphase production. Caryologia **27,** 359 – 368 (1974)
103. Hollande, A.: Etude comparée de la mitose syndinienne et de celle des Péridiniens libres et des hypermastigines. Infrastructure et cycle évolutif des Syndinides parasites de Radiolaires. Protistologica **10,** 413 – 451 (1974)
104. Hoffmann-Berling, H.: Die Bedeutung des Adenosintriphosphat für die Zelle und Kernteilungsbewegungen in der Anaphase. Biochim. Biophys. Acta **15,** 226 – 236 (1945)
105. Hughes, A. F., Swann, M. M.: Anaphase movements in living cell. A study with phase contrast and polarized light on chick tissue cultures. J. Exp. Biol. **25,** 45 – 70 (1948)
106. Inoué, S.: Polarization optical studies of the mitotic spindle. I. The demonstration of spindle fibers in living cells. Chromosoma **5,** 487 – 500 (1953)
107. Inoué, S.: Organization and function of the mitotic spindle. In: Primitive Motile Systems in Cell Biology (eds.: R. D. Allen, N. Kamiya), pp. 549 – 598. New York: Acad. Press 1964
108. Inoué, S.: Chromosome movement by reversible assembly of microtubules. In: Cell Motility (eds.: R. Goldman, T. Pollard, J. Rosenbaum), pp. 1317 – 1328. Cold Spring Harbor Lab. 1976
109. Inoué, S., Bajer, A.: Birefringence in endosperm mitosis. Chromosoma **12,** 48 – 63 (1961)
110. Inoué, S., Borisy, G.G., Kiehart, D. P.: Growth and lability of *Chaetopterus* oocyte mitotic spindles isolated in the presence of porcine brain tubulin. J. Cell Biol. **62,** 175 – 184 (1974)
111. Inoué, S., Ritter, H.: Dynamics of mitotic spindle organization and function. In: Molecules and Cell Movement (eds.: S. Inoué, R. E. Stephens), pp. 3 – 30. New York: Raven Press 1975
112. Inoué, S., Sato, H.: Cell motility by labile association of molecules: the nature of mitotic spindle fibers and their role in chromosome movement. J. Gen. Physiol. **50,** 259 – 292 (1967)
113. Inoué, S., Stephens, R. E. (eds.): Molecules and Cell Movement. Society of General Physiologists Series, vol. 30. New York: Raven Press; Amsterdam: North-Holland 1975
114. Jackson, W. T.: Regulation of mitosis II. Interaction of isopropyl-N-phenyl carbamate and melatonin. J. Cell Sci. **5,** 745 – 755 (1969)
115. Jensen, C., Bajer, A.: Spindle dynamics and arrangement of microtubules. Chromosoma **44,** 73 – 90 (1973)
116. Jokelainen, P. T.: The ultrastructure and spatial organization of the metaphase kinetochore in mitotic rat cells. J. Ultrastruct. Res. **19,** 19 – 44 (1967)
117. Jones, O. P.: Elimination of midbodies from mitotic erythroblasts and their contribution to fetal blood plasma. J. Natl. Cancer Inst. **42,** 753 – 764 (1969)
118. Journey, L. J., Burdman, J., George, P.: Ultrastructural studies on tissue culture cells treated with vincristine. Cancer Chemother. Rep. **52,** 509 – 518 (1968)
119. Kane, R. E.: The mitotic apparatus. Identification of the major soluble component of the glycol-isolated mitotic apparatus. J. Cell Biol. **32,** 243 – 253 (1967)
120. Kelly, M. G., Hartwell, J. L.: The biological effects and the chemical composition of podophyllin. A review. J. Natl. Cancer Inst. **14,** 967 – 1010 (1954)
121. Kiefer, B., Sakai, H., Solari, A. J., Mazia, D.: The molecular unit of the microtubules of the mitotic apparatus. J. Mol. Biol. **20,** 75 – 80 (1966)
122. Krishan, A.: Time-lapse and ultrastructure studies on the reversal of mitotic arrest induced by vinblastine sulfate in Earle's L cells. J. Natl. Cancer Inst. **41,** 581 – 596 (1968)
123. Krishan, A., Buck, R. C.: Structure of the mitotic spindle in L strain fibroblasts. J. Cell Biol. **24,** 433 – 444 (1965)
124. Krishan, A., Frei, E. III: Morphological basis for the cytolytic effect of vinblastine and vincristine on cultured human leukemic lymphoblasts. Cancer Res. **35,** 497 – 501 (1975)
125. Krishan, A., Paika, K., Frei, E. III: Cytofluorometric studies on the action of podophyllotoxin and epipodophyllotoxins (VM-26, VP-16-213) on the cell cycle traverse of human lymphoblasts. J. Cell Biol. **66,** 521 – 530 (1975)
126. Kubai, D. F.: Unorthodox mitosis in *Trichonympha agilis:* kinetochore differentiation and chromosome movement. J. Cell Sci. **13,** 511 – 552 (1973)
127. Kubai, D. F.: Evolution of the mitotic spindle. Intern. Rev. Cytol. **43,** 167 – 228 (1975)

128. Kubai, D. F., Ris, H.: Division in the dinoflagellate *Gyrodinium cohnii* (Schiller). A new type of nuclear reproduction. J. Cell Biol. **40,** 508 – 528 (1969)
129. Lafountain, J. R. Jr.: Birefringence and fine structure of spindles in spermatocyte of *Nephrotoma suturalis* at metaphase of first meiotic division. J. Ultrastuct. Res. **46,** 268 – 278 (1974)
130. Lafountain, J. R. Jr.: Analysis of birefringence and ultrastructure of spindles in primary spermatocytes of *Nephrotoma suturalis* during anaphase. J. Ultrastruct. Res. **54,** 333 – 346 (1976)
131. Lambert, A. M.: Contribution à l'étude ultrastructurale du fuseau de caryocinèse: organisation et évolution. Doctoral Thesis (No. CNRS A.O. 6162) Univ. of Strasbourg 1971
132. Lambert, A. M., Bajer, A. S.: Dynamics of spindle fibers and microtubules during anaphase and phragmoplast formation. Chromosoma **39,** 101 – 144 (1972)
133. Lambert, A. M., Bajer, A.: Fine structure dynamics of the prometaphase spindle. J. Micros. Biol. Cell. **23,** 181 – 194 (1975)
134. Lettré, H.: Über Mitosegifte. Ergebn. Physiol. **46,** 379 – 452 (1950)
135. Levine, L. (ed.): The Cell in Mitosis. New York: Acad. Press 1963
136. Lewis, L. M., Witkus, E. R., Vernon, G. M.: The role of microtubules and microfilaments in micronucleus of *Paramecium bursaria* during mitosis. Protoplasma **89,** 203 – 220 (1976)
137. Lits, F. J.: Recherches sur les réactions et lésions cellulaires provoquées par la colchicine. Arch. Intern. Méd. Exp. **11,** 811 – 901 (1936)
138. Ludford, R. J.: The action of toxic substances upon the division of normal and malignant cells in vitro and in vivo. Arch. Exp. Zellf. Mikros. Anat. **18,** 411 – 441 (1936)
139. Luykx, P.: Cellular Mechanisms of Chromosome Distribution. International Review of Cytology, Suppl. 2. New York-London: Acad. Press 1970
140. Malawista, S. E., Sato, H., Bensch, K. G.: Vinblastine and griseofulvin reversibly disrupt the living mitotic spindle. Science **160,** 770 – 771 (1968)
141. McCully, E. K., Robinow, C. F.: Mitosis in heterobasidiomycetous yeasts. I. *Leucosporidium scottii (Candida scottii).* J. Cell Sci. **10,** 857 – 881 (1972)
142. McCully, E. K., Robinow, C. F.: Mitosis in heterobasidiomycetous yeasts. II. *Rhodosporidium* sp. *(Rhodotorula glutinis)* and *Aessosporon salmonicolor (Sporobolomyces salmonicolor).* J. Cell Sci. **11,** 1 – 31 (1972)
143 McGill, M., Brinkley, B. R.: Mitosis in human leukemic leukocytes during colcemid inhibition and recovery. Cancer Res. **32,** 746 – 755 (1972)
144. McIntosh, J. R., Cande, Z., Snyder, J. A.: Structure and physiology of the mammalian mitotic spindle. In: Molecules and Cell Movement (eds.: S. Inoué, R. E. Stephens), pp. 31 – 76. New York: Raven Press 1975
145. McIntosh, J. R., Cande, Z., Snyder, J., Vanderslice, K.: Studies on the mechanism of mitosis. Ann. N. Y. Acad. Sci. **253,** 407 – 427 (1975)
146. McIntosh, J. R., Cande, W. Z., Lazarides, E., McDonald, K., Snyder, J. A.: Fibrous elements of the mitotic spindle. In: Cell Motility (eds.: R. Goldman, T. Pollard, J. Rosenbaum), pp. 1261 – 1272. Cold Spring Harbor Lab. 1976
147. McIntosh, J. R., Hepler, R. K., Wie, D. G. van: Model for mitosis. Nature **224,** 659 – 663 (1969)
148. McIntosh, J. R., Landis, S. C.: The distribution of spindle microtubules during mitosis in cultured human cells. J. Cell Biol. **49,** 468 – 497 (1971)
149. Mangenot, G.: Effets cytotoxiques de l'arsenic pentavalent. C. R. Acad. Sci. (Paris) **210,** 412 – 415 (1940)
150. Marsland, D., Hecht, R.: Cell division: combined anti-mitotic effects of colchicine and heavy water on first cleavage in the eggs of *Arbacia punctulata.* Exp. Cell Res. **51,** 602 – 608 (1968)
151. Mazia, D.: Mitosis and the physiology of cell division. In: The Cell. Biochemistry, Physiology, Morphology (eds:. J. Brachet, A. E. Mirsky), Vol. III, pp. 77 – 412. New York-London: Acad. Press 1961
152. Meisner, H. M., Sorensen, L.: Metaphase arrest of Chinese Hamster cells with rotenone. Exp. Cell Res. **42,** 291 – 295 (1966)
153. Mellon, M. G., Rebhun, L. I.: Sulfhydryls and in vitro polymerization of tubulin. J. Cell Biol. **70,** 226 – 238 (1976)
154. Moens, P. B.: Spindle and kinetochore morphology of *Dictyostelium discoideum.* J. Cell Biol. **68,** 113 – 122 (1976)
155. Moens, P. B., Rapport, E.: Spindles, spindle plaques, and meisois in the yeast *Saccharomyces cerevisiae* (Hansen). J. Cell Biol. **50,** 344 – 361 (1971)
156. Molè-Bajer, J.: The role of centrioles in the development of the astral spindle (newt). Cytobios **13,** 117 – 140 (1975)

157. Mullins, J. M., Biesele, J. J.: Cytokinetic activities in a human cell line: the midbody and intercellular bridge. Tissue and Cell **5**, 47 – 62 (1973)
158. Nath, J., Rebhun, L. I.: Effects of caffeine and other methylxanthines on the development and metabolism of sea urchin eggs: involvement of $NADP^+$ and glutathione. J. Cell Biol. **68**, 440 – 450 (1976)
159. Nicklas, R. B.: Mitosis. In: Advances in Cell Biology (eds.: D. M. Prescott, L. Goldstein, E. H. McConkey), Vol. 2, pp. 1 – 116; 225 – 294. New York: Appleton-Century-Crofts 1971
160. Nicklas, R. B.: Chromosome movement: current models and experiments on living cells. In: Molecules and Cell Movement (eds.: S. Inoué, R. E. Stephens), pp. 97 – 118. New York: Raven Press 1975
161. Nicklas, R. B., Koch, C. A.: Chromosome micromanipulation. IV. Polarized motions within the spindle and models for mitosis. Chromosoma **39**, 1 – 26 (1972)
162. Oakley, B. R.: Mitosis in the Dinoflagellate in *Amphidinium carterae.* Biosystems **7**, 305 (1975)
163. Oakley, B. R., Dodge, J. D.: Kinetochores associated with the nuclear envelope in the mitosis of a dinoflagellate. J Cell Biol. **63**, 322 – 325 (1974)
164. Oftebro, R., Grimmer, Ø., Øyen, T. B., Laland, S. G.: 5-fluoropyrimidin-2-one, a new metaphase arresting agent. biochem. Pharmacol. **21**, 2451 – 2456 (1972)
165. Oppenheim, D. S., Hauschka, B. T., McIntosh, J. R.: Anaphase motions in dilute colchicine. Evidence for two phases in chromosome segregation. Exp. Cell Res. **79**, 95 – 105 (1973)
166. Osborn, M., Weber, K.: Cytoplasmic microtubules in tissue culture cells appear to grow from an organizing structure toward plasma membrane. Proc. Natl. Acad. Sci. USA **73**, 867 – 871 (1976)
167. Östergren, G.: Colchicine mitosis, chromosome contraction, narcosis and protein chain folding. Hereditas **30**, 429 – 467 (1944)
168. Östergren, G.: Narcotized mitosis and the precipitation hypothesis of narcosis. Coll. Intern. Centre Nat. Rech. Sci. **26**, 77 – 88 (1951)
169. Packard, M. J., Stack, S. M.: The preprophase band: possible involvement in the formation of the cell wall. J. Cell Sci. **22**, 403 – 411 (1976)
170. Paget, G. E., Walpole, A. L.: Some cytological effects of griseofulvin. Nature **182**, 1320 – 1321 (1958)
171. Palevitz, B. A., Hepler, P. K.: The control of the plane of division during stomatal differentiation in *Allium.* I. Spindle reorientation. Chromosoma **46**, 297 – 326 (1974)
172. Palevitz, B. A., Hepler, P. K.: The control of the plane of division during stomatal differentiation in *Allium.* II. Drug studies. Chromosoma **46**, 327 – 341 (1974)
173. Pease, D. C.: Hydrostatic pressure effects upon the spindle figure and chromosome movement. I. Experiments on the first mitotic division of *Urechis* eggs. J. Morphol. **69**, 405 – 442 (1941)
174. Perkins, F. O.: Fine structure of the haplosporian *Kernstab,* a persistent intranuclear mitotic apparatus. J. Cell Sci. **18**, 327 – 346 (1975)
175. Pernice, B.: Sulla cariocinesi delle cellule epiteliali e dell'endotelio dei vasi della mucosa dello stomaco e dell'intestino, nelle studio della gastroenterite sperimentale (nell'avvelenamento per colchico). Sicil. Med. **1**, 265 – 279 (1889)
176. Peters, J. J.: A cytological study of mitosis in the cornea of *Triturus viridescens* during recovery after colchicine treatment. J. Exp. Zool. **103**, 33 – 60 (1946)
177. Peterson, J. B., Ris, H.: Electron-microscopic study of the spindle and chromosome movement in the yeast *Saccharomyces cerevisiae.* J. Cell Sci. **22**, 219 – 242 (1976)
178. Pickett-Heaps, J. D.: Cell division in the formation of the stomatal complex in the young leaves of wheat. J. Cell Sci. **1**, 121 – 128 (1966)
179. Pickett-Heaps, J. D.: Effect of colchicine on the ultrastructure of dividing plant cells, xylem wall differentiation and distribution of cytoplasmic microtubules. Dev. Biol. **15**, 206 – 236 (1967)
180. Pickett-Heaps, J. D.: Preprophase microtubule bands in some abnormal mitotic cells of wheat. J. Cell Sci. **4**, 397 – 420 (1969)
181. Pickett-Heaps, J. D.: The evolution of mitosis and the eukaryotic condition. Biosystems **6**, 37 – 48 (1974)
182. Pickett-Heaps, J. D.: Aspects of spindle evolution. Ann. N. Y. Acad. Sci. **253**, 352 – 361 (1975)
183. Pickett-Heaps, J. D., Northcote, D. H.: Relationship of cellular organelles to the formation and development of the plant cell wall J. Exp. Bot. **17**, 20 – 26 (1966)

184. Probst, S., Banerjee, S., Kelleher, J. L., Margulis, L.: Inhibition of cilia regeneration by antineoplastic agents; delay of band migration by vinblastine (NSC-49842), griseofulvin (NSC-34533) and β-peltatin (NSC-24819). Cancer Chemother. Rep. **56,** 557 – 558 (1972)
185. Rattner, J. B., Berns, M. W.: Centriole behaviour in early mitosis of rat kangaroo cells (PTK_2). Chromosoma **54,** 387 – 395 (1976)
186. Rebhun, L. I.: Polarized intracellular particle transport: saltatory movements and cytoplasmic streaming. Intern. Rev. Cytol. **32,** 93 – 139 (1972)
187. Rebhun, L. I., Rosenbaum, J., Lefebvre, P., Smith, G.: Reversible restoration of the birefringence of cold-treated, isolated mitotic apparatus of surf clam eggs with chick brain tubulin. Nature (London) **249,** 113 – 115 (1974)
188. Rebhun, L. I., Sander, G.: Ultrastructure and birefringence of isolated mitotic apparatus of marine eggs. J. Cell Biol. **34,** 859 – 884 (1967)
189. Remillard, S., Rebhun, L. I., Howie, G. A., Kupchan, S. M.: Antimitotic activity of the potent tumor inhibitor Maytansine. Science **189,** 1002 – 1005 (1975)
190. Rickards, G. K.: Prophase chromosome movements in living house cricket spermatocytes and their relationship to prometaphase, anaphase, and granule movements. Chromosoma **49,** 407 – 455 (1975)
191. Ris, H., Kubai, D. F.: An unusual mitotic mechanism in the parasitic protozoan *Syndinium* sp. J. Cell Biol. **60,** 702 – 720 (1974)
192. Roos, U. P.: Light and electron microscopy of rat kangaroo cells in mitosis. I. Formation and breakdown of the mitotic apparatus. Chromosoma **40,** 43 – 82 (1973)
193. Roos, U. P.: Light and electron microscopy of rat kangaroo cells in mitosis. II. Kinetochore structure and function. Chromosoma **41,** 195 – 220 (1973)
194. Roos, U. P.: Fine structure of an organelle associated with the nucleus and cytoplasmic microtubules in the cellular slime mould *Polysphondylium violaceum.* J. Cell Sci. **18,** 315 – 326 (1975)
195. Roos, U. P.: Mitosis in the cellular slime mould *Polysphondylium violaceum.* J. Cell Biol. **64,** 480 – 492 (1975)
196. Roth, L. E.: Electron microscopy of mitosis in amebae. III. Cold and urea treatments: a basis for tests of direct effects of mitotic inhibitors on microtubule formation. J. Cell Biol. **34,** 47 – 59 (1967)
197. Ryter, A.: Association of the nucleus and the membrane of bacteria: a morphological study. Bact. Rev. **32,** 39 – 54 (1968)
198. Sakai, H.: Studies on sulfhydryl groups during cell division of sea-urchin eggs. VIII. Some properties of mitotic apparatus proteins. Biochim. Biophys. Acta **112,** 132 – 145 (1966)
199. Sakai, H., Hiramoto, Y., Kuriyama, R.: The glycerol isolated mitotic apparatus: a response to porcine brain tubulin and induction of chromosome motion. Dev. Growth Diff. **17,** 265 – 274 (1975)
200. Salmon, E. D.: Pressure-induced depolymerization of spindle microtubules. I. Changes in birefringence and spindle length. J. Cell Biol. **65,** 603 – 614 (1975)
201. Salmon, E. D.: Pressure-induced depolymerization of spindle microtubules. II. Thermodynamics of in vivo spindle assembly. J. Cell Biol. **66,** 114 – 127 (1975)
202. Salmon, E. D.: Spindle microtubules: thermodynamics of in vivo assembly and role in chromosome movement. Ann. N. Y. Acad. Sci. **253,** 383 – 406 (1975)
203. Salmon, E. D.: Pressure-induced depolymerization of spindle microtubules: production and regulation of chromosome movement. In: Cell Motility (eds.: R. Goldman, T. Pollard, J. Rosenbaum), pp. 1329 – 1342. Cold Spring Harbor Lab. 1976
204. Salmon, E. D., Goode, D., Maugel, T. K., Bonar, D. B.: Pressure-induced depolymerization of spindle microtubules. III. Differential stability in HeLa cells. J. Cell Biol. **69,** 443 – 454 (1976)
205. Sanger, J. W., Sanger, J. M.: Actin localization during cell division. In: Cell Motility (eds.: R. Goldman, T. Pollard, J. Rosenbaum), pp. 1295 – 1316. Cold Spring Harbor Lab. 1976
206. Sato, H., Ohnuki, Y., Fujiwara, K.: Immunofluorescent anti-tubulin staining of spindle microtubules and critique for the technique. In: Cell Motility (eds:. R. Goldman, T. Pollard, J. Rosenbaum), pp. 419 – 434. Cold Spring Harbor Lab. 1976
207. Satir, P.: The present status of the sliding microtubule model of ciliary motion. In: Cilia and Flagella (ed.: M. A. Sleigh), pp. 131 – 142. London-New York: Acad. Press 1974
208. Schmidt, W. J.: Die Doppelbrechung von Karyoplasma, Zytoplasma und Metaplasma. Protoplasma Monographien, Vol. 11. Berlin: Borntraeger 1937
209. Schrader, F.: Mitosis. The Movements of Chromosomes in Cell Division (2nd ed.). New York: Columbia Univ. Press 1953

210. Schroeder, T. E.: Dynamics of the contractile ring. In: Molecules and Cell Movement (eds.: S. Inoué, R. E. Stephens), pp. 305 – 334. New York: Raven Press 1975
211. Seidlova-Masinova, V., Malinsky, J., Santavy, F.: The biological effects of some podophyllin compounds and their dependence on chemical structure. J. Natl. Cancer Inst. **18,** 359 – 372 (1957)
212. Sentein, P.: L'action de la colchicine, de la podophylline, et de l'hydrate de chloral sur les mitoses spermatogénétiques chez quelques Urodèles. Arch. Anat. Microsc. Morph. Exp. **43,** 79 – 116 (1954)
213. Sentein, P., Ates, Y.: Ultrastructure des mitoses bloquées par l'hydrate de chloral dans les œufs en segmentation de *Pleurodeles waltlii* Michah. C. R. Acad. Sci. Paris D **277,** 793 – 795 (1973)
214. Sentein, P., Ates, Y.: Action de l'hydrate de chloral sur les mitoses de segmentation de l'œuf de Pleurodèle. Etude cytologique et ultrastructurale. Chromosoma **45,** 215 – 244 (1974)
215. Sluder, G.: Experimental manipulation of amount of tubulin available for assembly into spindle of dividing sea urchin eggs. J. Cell Biol. **70,** 75 – 85 (1976)
216. Soifer, D. (ed.): The biology of cytoplasmic microtubules. Ann. N. Y. Acad. Sci. **253** (1975)
217. Stephens, R. E.: The mitotic apparatus. Physical and chemical characterization of the 22S protein component and its subunits. J. Cell Biol. **32,** 255 – 276 (1967)
218. Symposium on the Evolution of Mitosis in Eukaryotic Microorganisms (Boston University, Boston, Mass., 1975). Biosystems **7,** 295 – 385 (1975)
219. Taylor, E. W.: Macromolecular assembly inhibitors and their action on the cell cycle. In: Drugs and the Cell Cycle (eds.: A. M. Zimmerman, G. M. Padilla, I. L. Cameron), pp. 11 – 24. New York-London: Acad. Press 1973
220. Tennant, R., Liebow, A. A.: The actions of colchicine and ethyl-carbylamine on tissue cultures. Yale J. Biol. Med. **13,** 39 – 49 (1940)
221. Van Regemoorter, D.: Les troubles cinétiques dans les racines chloralosées et leur portée pour l'interprétation des phénomènes normaux. Cellule **37,** 43 – 73 (1926)
222. Vickerman, K., Preston, T. M.: Spindle microtubules in the dividing nuclei of Trypanosomes. J. Cell Sci. **6,** 365 – 384 (1970)
223. Weber, K., Pollack, R., Bibring, T.: Antibody against tubulin: the specific visualization of cytoplasmic microtubules in tissue culture cells. Proc. Natl. Acad. Sci. USA **72,** 459 – 463 (1975)
224. Weber, K.: Specific visualization of tubulin containing structures by immunofluorescence microscopy: cytoplasmic microtubules, vinblastine induced paaracrystals and mitotic figures. In: Microtubules and Microtubule Inhibitors (eds.: M. Borgers, M. De Brabander), pp. 313 – 325. Amsterdam-Oxford: North-Holland; New York: Am. Elsevier 1975
225. Weber, K.: Visualization of tubulin-containing structures by immunofluorescence microscopy: cytoplasmic microtubules, mitotic figures and vinblastine-induced paracrystals. In: Cell Motility (eds.: R. Goldman, T. Pollard, J. Rosenbaum), pp. 403 – 418. Cold Spring Harbor Lab. 1976
226. Went, H. A.: The behaviour of centrioles and the structure and formation of the achromatic figure. Protoplasmatologia **6,** 1 – 109 (1966)
227. Wilson, H. J.: Arms and bridges on microtubules in the mitotic apparatus. J. Cell Biol. **40,** 854 – 859 (1969)
228. Wilson, L., Bryan, J.: Biochemical and pharmacological properties of microtubules. In: Adv. Cell Mol. Biol. Vol. 3, pp. 21 – 72 (ed.: E. Du Praw). New York: Acad. Press 1975
229. Zimmerman, A. M., Marsland, D.: Cell division: effects of pressure on the mitotic mechanisms of marine eggs *(Arbacia punctulata).* Exp. Cell Res. **35,** 293 – 302 (1964)
230. Zirkle, R. E.: Ultraviolet-microbeam irradiation of newt-cell cytoplasm: spindle destruction, false anaphase, and delay of true anaphase. Radiat. Res. **41,** 516 – 537 (1970)

Chapter 11 Pathology and Medicine

11.1 Introduction

Many threads link the study of MT to medicine and pathology. After all, the most specific of MT poisons, colchicine, has been used for centuries in the treatment of gout, and the discovery by Pernice in 1889 [90] of its action on mitosis was the result of studies on its toxicity. The other specific MT poisons, VLB and VCR, have become routinely used, with excellent results, in the treatment of many malignant conditions in man. Last, griseofulvin, used as a fungistatic, also belongs to the group of drugs modifying MT.

While the known facts about the pathology of MT or of MT structures such as cilia and flagella are still limited in number, it would be surprizing if in the coming years genetic modifications of tubulin were not discovered, even though this protein has apparently undergone very few changes during the long evolution of eukaryotes. Recent work on one disease of man and animals, the Chediak-Higashi anomaly of lysosomes and secretory granules, has demonstrated close links between the disturbances of leukocyte function, the abnormal formation of specific granules, and MT. The observation of several human cases with immotile cilia devoid of dynein arms [2] is an indication that anomalies of MT structures are not necessarily lethal.

The therapeutic uses of drugs such as colchicine have brought interesting results in diseases unrelated to neoplasia, and VLB has found some use in the treatment of thrombocytopenic purpura, while colchicine was found by several authors to affect favorably cases of periodic (Mediterranean) fever. On the other hand, all MT poisons, through their action on mitosis, are potentially dangerous and toxic.

The few descriptions of fatal human poisoning with colchicine published since the action of this drug on MT has been known have revealed some unexpected cellular changes. The toxicity of MT poisons, which results from the multiple functions of MT, may lead to curious clinical syndromes: the *Vinca* aklaloids, and in particular VCR, have demonstrated neurotoxicity in patients, which may be related to the role of MT in neurons and neuroplasmic flow. The syndrome of inappropriate secretion of antidiuretic hormone has been repeatedly described in the last few years, and raises interesting questions about the secretion of ADH and the role of neuroplasmic flow and pituicytes.

Returning to the starting point of these studies, the action of colchicine on gout deserves some discussion in the light of all which has been learnt about leukocytes, urate microcrystals, secretion, and phagocytosis. The same is true for the action of VCR and VLB in cancer chemotherapy: it is not always evident that the favorable results obtained can be explained as the consequence of the antimitotic activity.

This aspect of the use of MT poisons in medicine will not be studied in detail as these drugs have become routine tools in cancer centers, where their effects are most often combined, in complicated "chemotherapeutic cocktails," with other inhibitors of mitotic growth. The use of MT poisons in cancer chemotherapy has been discussed in several monographs [77, 107].

11.2 Colchicine and Gout

11.2.1 Pathogenesis of Acute Gouty Arthritis

Gout is a disease of uric acid metabolism, leading to the precipitation of sodium urate microcrystals in the vicinity of joints with an acute and painful inflammatory reaction. Urate may also be deposited in connective tissue and in organs such as the kidney, in a more chronic fashion, leading to large accumulations of crystalline material surrounded by a mononuclear cell reaction ("tophus"). Colchicine is mainly effective in the acute crisis and only this will be discussed here. It involves all the major signs of acute inflammation, with all its known complexity: liberation of histamine and other substances from connective tissue mastocytes, increased vascular permeability, diapedesis and chemotropic attraction of neutrophil polymorphonuclear leukocytes, phagocytosis of microcrystals of urate by these cells, activation of the complement system, formation of kinins from globulin precursors.

Colchicine—and other MT poisons, as mentioned below—remains today the most active drug alleviating the very painful acute gouty crisis. A dose of a few mg in an adult subject is effective, indicating that the local concentration may be quite small. The mechanisms of action of colchicine are probably complex, and it is far from certain that the drug has a single target: more probably several steps in the urate microcrystal arthritis, which may readily be reproduced experimentally in animals, may be modified through an action on MT or non-specific effects. The following steps have been mentioned as possible sites of colchicine action: the degranulation of polymorphonuclears and the intracellular movement of their granules (lysosomes) preceding their exocytosis; the movements of polymorphonuclears towards the site of deposition of urate crystals (chemotaxis); indirect effects on the liberation of the granules of mastocytes, or on other factors involved in acute inflammation, such as complement [18], prostaglandins, and the Hageman factor. Some of these actions have already been alluded to in Chapters 5 and 8. A recent review by Malawista [71], who was one of the first to point out the common factors between the antimitotic and the antiinflammatory properties of colchicine [70, 72], discusses these various sites of action. Whatever its mechanism, the treatment of acute gout by colchicine remains one of the best (cf. [125]). The experimental studies on urate microcrystal inflammation will be discussed first, before considering the more general antiinflammatory properties which have been attributed to colchicine and other MT poisons.

11.2.2 Experimental Acute Urate Microcrystal Inflammation

The action of crystals of urate, sufficiently small to be phagocytized by polymorphonuclear leukocytes (microcrystals), results from their needle-like shape: their penetration into the leukocyte lysosomes, after phagocytosis, ruptures the lysosomal membrane, liberating numerous hydrolases, first into the cell, leading to its death, and further into the connective tissue [113]. This would in turn attract to the site more polymorphonuclears which are known to have "necrotropic" properties. It is clear that several steps in this sequence involve movements which may be affected by MT: one is phagocytosis itself, another is fusion of primary lysosomes with the phagocytic vacuoles surrounding the microcrystals ("phagosomes"), the third, the movement of polynuclears toward the microcrystal depot (chemotaxis).

As mentioned in Chapter 6, cell movements may be slowed down by interference with MT (and probably associated MF), and chemotaxis was already found to be inhibited by colchicine in 1965 [19]. In the dog, intraarticular injection of urate crystals attracts polymorphonuclears, and this is prevented by colchicine [91, 92, 93]. This action appears however more complex than a simple disturbance of cell movement: urate crystals alone are not chemotropic, and the factor which attracts the white blood cells is liberated at the moment of phagocytosis. Colchicine would interfere with this last step of the reaction. Similar findings have been reported in comparative studies with urate and calcium pyrophosphate, a crystalline material which is responsible for similar tissular deposits ("pseudo-gout") [37]: a chemotropic factor is liberated upon rupture of the leukocyte granules, more with urate than with calcium crystals, a fact which is in agreement with the known more intense reaction to urate in man. An activation of the Hageman factor by urate microcrystals was reported in the same work, a factor which may further aggravate the inflammation through the formation of vasoactive kinins. On the other hand, leukocyte motility remains important and may be inhibited by low concentrations (10^{-8} M) of colchicine [69]. An important series of experiments on urate-induced inflammation in the rat have also shown that anti-inflammatory results could be obtained with various colchicine derivatives, and also with podophyllotoxin and VLB. The order of activity paralleled the antiMT activity: demecolcine, colchicamide, trimethylcolchicinic acid methyl ether, and trimethylcolchicinic ethyl ether were as effective as colchicine, as appreciated by the degree of local edema. Colchiceine was less active, and trimethylcolchicinic acid was inactive [138], indicating that MT are involved in the antiinflammatory actions of colchicine and derivatives. As mentioned above, pseudo-gout or "chondrocalcinosis" is hardly relieved by colchicine in man: experiments in chickens, by intraarticular injection of microcrystals of sodium urate and calcium pyrophosphate, also demonstrated that colchicine was more effective against urate inflammation. The doses of colchicine injected were relatively large: 0.5 – 1 mg/kg of weight. Colchicine was suggested to act mainly on the liberation of the chemotactic activity following urate phagocytosis [37]. Similar results with 1 mg/kg of colchicine have been reported in chickens and pigeons; it had a double action, decreasing the numbers of leukocytes at the site of injection and the liberation of lysosomal enzymes (β-glucuronidase) from polymorphonuclears [14, 15, 16, 17].

A factor which may be involved in this type of inflammation is prostaglandin PGE_1. Antagonistic effects of colchicine and PGE_1 were observed, suggesting that the main action of the alkaloid may be to antagonize PGE_1. However, colchicine effects on cell membranes may also play a role in its therapeutic action [26].

11.2.3 Other Antiinflammatory Effects of Colchicine

While the movement of leukocytes towards the inflammatory site is certainly important, colchicine may also affect the intracellular movements which take place in these cells during phagocytosis. Malawista suggested in 1968 several possible mechanisms: inhibition of lysosome fusion with phagocytic vacuoles in leukocytes, stabilization of lysosomes through actions on their membrane, inhibition of histamine release from mastocytes [70]. The role of leukocyte movements had been indicated in experiments on staphylococcal inflammation in guinea-pigs. Colchicine (0.4 mg/kg) increased the local lesions through a delay in the accumulation of polymorphonuclear leukocytes at the site of injection [71], confirming the inhibition of chemotaxis [19]; this could in turn be a consequence of the destruction of MT by the drug [73].

An important observation was made in man by injecting therapeutic doses of ^{14}C-labeled colchicine: its half-life in plasma was short (18.3 min) but concentrations in leukocytes were 3 – 17 times higher than in the plasma: this indicates that however small the doses administered in man, the intracellular concentration may be sufficient for MT changes [32]. After administration of 1.0 mg of ^{14}C-colchicine, the peak concentrations in the blood reached 0.32 ± 0.17 μg/100 ml. Two groups of populations were found in the ten volunteer subjects: in one the maximum was reached in ½ h, in the other after 2 h [127]. The rapid disappearance from the plasma indicates that colchicine crosses membranes rapidly; however, it may be fixed for some time to the MT of the gastrointestinal tract, and the rate of release from these may explain the differences observed in the two groups [127].

The intracellular movements of lysosomes which take place after phagocytosis may also be affected: studies on human leukocyte phagocytosis of heat-killed staphylococci in vitro confirmed that two drugs active in the treatment of gout, colchicine (2.5×10^{-5} M) and cortisol (5×10^{-4} M) inhibited both the intracellular movements of lysosomes and the liberation, in the extracellular space, of their enzymes (lysozyme, acid phosphatase). This is important, as ultrastructural data indicate that leakage of these lysosomal enzymes is one of the causes of gouty inflammation [113].

From all these data, it is apparent that colchicine may have several sites of action in the complex changes taking place in acute gout, and that it is doubtful whether a single effect can explain why it remains such an excellent medicament. Comparison with other MT poisons does indeed suggest that MT are involved, but this may affect in several different steps of urate inflammation.

11.2.4 Action on Gout of Other MT Poisons

Early observations [115, 128] have shown that griseofulvin may be effective in the treatment of acute gouty arthritis. It also inhibits the chemotaxis of human polymorphonuclears in vitro at concentration ranging from 0.1 to 1.0 μg/ml. This effect would be explained by changes of cytoplasmic MT [7]. Podophyllotoxin and VLB act like colchicine in the inflammation produced in the rat's paw by the injection of sodium urate microcrystals [138], while colchicine derivatives without antiMT action are inactive.

However, one author at least has claimed since 1961 that in the treatment of human gout, trimethylcolchicinic acid, which is definitely not a MT poison (cf. Chap. 5), is as effective as colchicine [124, 125, 126]. These results may indicate that colchicine has other actions unrelated to MT (cf. its antagonism to prostaglandin PGE_1; [26]). The possible transformation of trimethylcolchicinic acid to colchicine or a related molecule has been suggested [71]. There certainly remains a possibility that the complex interrelations of colchicine and acute urate inflammation involve other changes than those of polymorphonuclear MT. It is evident that many aspects of colchicine pharmacology (cf. Chap. 5), and the relations between MT, cell movement, and cell secretion, affect the therapeutic effects of this drug. It is not the only instance of a most efficient drug used for decades without a proper scientific explanation of its effects: aspirin (acetylsalicyclic acid), which recently has been suggested to act on prostaglandins, is another well-known case.

11.3 Other Therapeutic Uses of Colchicine

As colchicine was found to have remarkable antiinflammatory properties in gout, it is not surprizing that is should have been tested as an antiinflammatory drug in diseases of obscure etiology, where inflammation was one of the principal signs. Such is *periodic disease,* a syndrome individualized in 1948 [96] and characterized by the recurrence of fever, abdominal pain, arthralgia, myalgia, and signs of pleuritis, pericarditis, or peritonitis. This syndrome, known under diverse names (familial Mediterranean fever, familial paroxysmal polyserositis, periodic fever, recurrent polyserositis), is found frequently in Armenians, Jews, and Arabs. It may be complicated by amyloidosis (cf. [74]). In the last few years, many reports on the favorable effects of colchicine on this disease of unknown etiology have been published, following the observations of Mamou [74].

Several critical studies of relatively large series of cases have been reported. In a double-blind study of ten patients conducted over six months, a daily dose of three tablets of 0.6 mg of colchicine was compared to the action of a placebo. While 59 attacks were observed during placebo administration, only two patients treated by colchicine suffered five attacks of fever, a statistically quite significant ($P< 0.002$) result [46]. Another study, on 11 patients treated by the same dose of colchicine (three 0.6 mg tablets per day) or by placebo, showed a much larger number of attacks (38) under placebo than during colchicine administration (7): this was significant ($P<0.001$) and colchicine considered as "highly effective in preventing attacks of the disorder" [28]. A third study, published the same year, on 22 patients treated daily with two tablets of 0.5 mg colchicine or a placebo, also showed that the number of attacks decreased while the patients were on colchicine therapy [136]. The authors of these three studies give excellent summaries of the literature on this subject, which also contains some negative results. The mechanism of action of colchicine in this strange disease remains quite obscure, although actions on several steps of inflammation, as those mentioned above in gout, may be imagined. The beneficial results of this method, inaugurated by Goldfinger [45], have more recently been confirmed in 12 out of 14 patients who had for many years undergone unsuc-

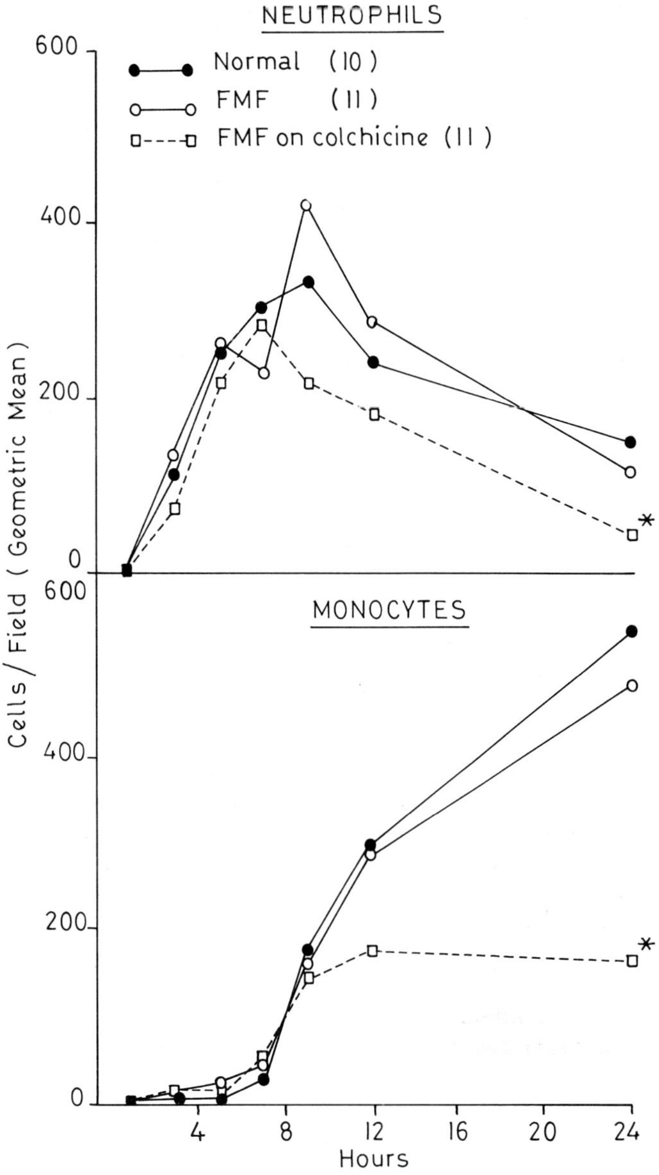

Fig. 11.1. Effect of daily colchicine on skin-window response in patients with familial Mediterranean fever (*FMF*). The numbers of individuals studied are in parentheses. Asterisked values are those which are significantly different (P<0.05) (redrawn from Dinarello et al. [29])

cessful treatment with various drugs, continuous administration of 0.65 mg of colchicine suppressing the episodes of recurrent peritonitis [97].

A detailed double-blind study of the functions of leukocytes in patients receiving colchicine for periodic fever showed only that the number of polymorphonuclears accumulating in skin-windows was decreased, although chemotaxis and random movements were normal. The authors think that the results of the "highly effective" treatment by colchicine could be explained by a decreased liberation, by the leukocytes, of secondary factors amplifying the inflammatory response [29].

One complication of this disease is *amyloidosis*, and the observations of Goldfinger [45] suggested experimental studies on casein-induced amyloidosis in mice. While daily injections of casein induce amyloidosis in all control mice within two weeks, in animals injected daily with 0.02 mg colchicine, no evidence of amyloidosis was observed. This experiment, on more than 60 mice, suggested that possibly colchicine interferes with the extracellular assembly of the amyloid fibrils. However, many other sites of action could be imagined [62]. These results have been confirmed: no mice receiving more than 0.015 mg/day developed amyloidosis after casein injections. Colchicine proved equally effective when given after the first nine injections of casein, in the "late pre-amyloid phase," or in animals in which amyloid formation was induced by the transfer of splenic tissue of mice in the late phase of amyloidosis [113 b].

Colchicine has also been tried quite empirically in diseases with a rheumatoid, articular component, for instance *sarcoid arthritis* [61], with unconvincing results [51]. As mentioned in Chapter 8, colchicine (and MT poisons) may interfere with the secretion of collagen by fibroblasts. In carbon tetrachloride-induced *liver cirrhosis* in rats, the formation of collagen has been demonstrated to be decreased in vitro under the action of colchicine. In experimentally cirrhotic rats, the importance of liver sclerosis was significantly lower than in controls [103] if colchicine was administrated at the dose of 50 μg/week for four weeks. Similar results have been mentioned in human liver cirrhosis [104], although the appreciation of variations of liver fibrosis in human patients is fraught with causes of error.

It was logical to attempt to use colchicine in the treatment of *sclerodermia*, an autoimmune disease in man with excessive collagen formation in subcutis and arterial walls. Following a first favorable report [5], studies on three patients led to negative results: no evidence of decreased collagen synthesis or increased collagen breakdown could be demonstrated [50].

It is evident from these scattered data that apart from periodic disease, in which the mechanism of action of colchicine certainly deserves further studies, the other medical applications lack a secure scientific basis. On the other hand, dangers of colchicine administration should be kept in mind.

11.4 Colchicine Poisoning in Man

In the nineteenth century, many reports of colchicine toxicity or overdosage were published: the main symptoms were diarrhea, loss of hair, and blood disorders. The first report of a fatal case of overdosage after the discovery of colchicine effects on mitosis was reported in 1941 [31]. Since then cases of acute colchicine poisoning have been published but few have been studied cytologically. While some of the findings could be expected, from the experimental data, other cellular changes have only been reported in man, and should be mentioned (cf. Table 11.1). A review of the clinical literature on this subject, with many other references, can be

Table 11.1. Cases of fatal colchicine poisoning with cytological study

Ref.	Sex	Age (yr)	Dose	Blood cells	Bone marrow	Intestinal tract and other organs	Time between poisoning and death
31	F	41	60 mg	leukopenia leukocytosis thrombocytopenia some immature cells Heinz bodies (8/1000 Rbc)	stimulation of granulopoiesis mitotic arrest of granular and erythroblastic cells	atrophy of intestinal mucosa with arrested mitoses liver: 40/1000 liver cells in mitosis pancreas: some arrested mitoses partial hemorrhagic destruction of anterior lobe of pituitary	8 days
22	F	16	30 mg	leukocytosis on 2nd day (24,800) followed by leukopenia (5400) thrombocytopenia	not studied	diarrhea, coma multiple small hemorrhagic foci in brain	3 days
118	M	60	?	leukopenia Döhle bodies in polymorphonuclear leukocytes	hypoplasia	multiple hemorrhages metaphase-arrested cells in gastric mucosa and oesophagus foci of liver necrosis	7 days
	M	50	?	severe leukopenia (1100) and thrombocytopenia		lung hemorrhage	18 h
	M	71	200 mg in 232 days			lung hemorrhagic infarct endocarditis renal abscesses villous atrophy of small intestine	24 h
52	F	14	35 mg	severe leukopenia on 4th day (1100) with leukocytosis on 14th day (40,400) thrombocytopenia Heinz bodies (3rd day)	hyperactive no abnormal mitoses	gastric hemorrhages bronchopneumonia with some hyaline membranes vacuolization of some muscle cells in myocardium and psoas muscle alopecia	13 days

found in the publications by Gaultier et al. [42, 43]: these authors observed 23 personal cases, but did not perform any autopsy.

Some comments on Table 11.1 can be made. In our personal case [31], a patient hospitalized three days after the absorption of 60 mg of colchicine in a suicide attempt, a bone marrow study was performed shortly before death on the 8th day, and autopsy took place less than 4 h post mortem, which probably explains the fact that more evident effects of colchicine on mitosis were visible than in other reports. The bone marrow, which appeared to be in a condition of active regeneration, showed a definite mitotic arrest of granulocytes and erythroblasts: this indicates that colchicine is slowly metabolized and that the MT poisoning effects may be of long duration, a fact which is not evident in animals, but is confirmed in one other report [116]. Another surprizing finding is that of Heinz bodies in the red blood cells. These structures are known to result from the denaturation of the globin molecule following the oxidation of hemoglobin. Their relation with colchicine is quite obscure, and would not have been mentioned here if another case of fatal colchicine poisoning had not quite recently also shown Heinz bodies [52]. They are probably a very indirect result of the poisoning, and their formation may perhaps be linked with the severe gastrointestinal disturbances shown by the patients.

Another finding which confirms the long-term action of colchicine is that of arrested mitoses in the liver and pancreas, eight days after poisoning [31]: these mitoses, which result from the metaphase arrest by colchicine of regenerating liver cells, are comparable to the so-called "late" mitoses found in animals injected with colchicine, several days after a single injection, as described in the princeps paper by Lits (cf. Chap. 1).

A recently developed radioimmunoassay for colchicine has shown in three patients that the alkaloid remained present in the body up to eight days after administration, a fact which would explain the findings of colchicine poisoning; the binding to tissue receptors (tubulin?) would explain the slow excretion of the drug [33].

Apart from the gastrointestinal changes resulting from the atrophy (and often ulceration) of the mucosa secondary to mitotic arrest (and similar to those found after many other antimitotic drugs and after irradiation), the main effects of colchicine poisoning are the leukopenia and thrombocytopenia, with a transient hypoplasia of the bone marrow: this has been well described in the non-fatal case of Hitzig [57] by a detailed study of bone marrow smears. The other cases reported often show hemorrhagic symptoms resulting from the thrombocytopenia. The exact doses are not known in the three cases reported by Stemmermann and Hayashi [116]: these were all three gouty patients who had been using colchicine extensively, often without proper medical control, and who apparently died of an overdose. Reports of similar toxicity, although non-fatal, have been mentioned for patients treated with colcemid (demecolcin) [30]. It was mentioned in Chapter 5 that *Gloriosa superba* is a plant which contains colchicine: two reports of poisoning by this plant have been described, one with severe depression of leukocytopoiesis [66], the other with alopecia, indicating the blocking of hair bulb mitoses ([48; cf. also 81]).

One remarkable effect of colchicine intoxication which is without relation to the antimitotic properties of the drug, but may be related to an action on MT, is the production of the syndrome of inappropriate secretion of antidiuretic hormone

(ADH). As will be described below, this syndrome is well-known after VCR treatment, with or without any overdosage, and its mechanism may be related to disturbances of the hypothalamo-pituitary function.

Two cases of this syndrome have been reported after accidental colchicine overdosage [41]. The first was a girl aged 15 who took 40 mg of colchicine. Apart from the usual signs of toxicity—diarrhea, hemorrhagic diathesis, bone marrow aplasia, alopecia—the blood sodium fell progressively, and on the 9th day reached only 113 mEq/l. This returned to normal figures on the 16th day. The second case was quite similar: a boy, aged 17, also swallowed 40 mg of colchicine. After a period of acute toxicity, with diarrhea, nervous symptoms appeared on the 4th day (headache, muscle pain and atrophy, absence of deep reflexes); these symptoms regressed and two months later the nervous system reacted normally. The blood sodium fell from the first to the 11th day, reaching a value of 115 mEq/l, and returning to normal values after the 17th day. It was corrected by restriction of fluid intake and administration of sodium. The authors point to the similarity of these observations and those described after VCR (cf. Table 11.2), suggesting that a common effect on MT (of the posterior lobe of the pituitary?) may explain the disturbance of ADH secretion.

At least one non-fatal case, although with severe toxicity, of poisoning after therapeutic use of podophyllin for the treatment of vaginal condylomas [80] has been described: it is interesting to note the severe blood changes which took place, apart from the intestinal toxicity. About 14 h after the application of the drug, the leukocytosis reached 50,200 with 27% of immature granulocytes. This rapid action on leukocytes should be compared to the well-known, although still unexplained, colchicine leukocytosis (cf. Chap. 5).

11.5 The Thrombocytopoietic Action of VCR

An interesting fact about the *Catharanthus* alkaloids is that minor changes of their structure modify their biological effects: while VLB and VCR are in many regards similar in their antiMT effects, the clinical tests have demonstrated that their indications were different (VLB for Hodgkin's disease, VCR for leukemias). While VCR has a broader spectrum of activity, it depresses less than VLB the number of platelets. This led to further studies which have shown that the use of VCR may be of some benefit in the treatment of various conditions of thrombocytopenia [98] and in particular in chronic idiopathic thrombocytopenic purpura. This condition is considered an autoimmune disease, resulting from the formation of antibodies against platelets. As early reports indicated a surprizing increase in the number of platelets in patients treated for various neoplasms by VCR (cf. [3]), clinical trials by several groups of authors confirmed that VLB also increased the number of platelets [58]. Favorable results in idiopathic thrombocytopenic purpura were reported in several cases [75]. In a study on a larger number of patients, some of whom were resistant to cortisone treatment and some to splenectomy and cortisone, the effects of a daily 2 mg dose of VCR for 7 – 10 days brought what was described as "prompt benefit" [3]. In one case, this treatment led to a remission lasting more than two years. It is not evident that immune depression may alone explain these results, and a direct action on bone marrow megakaryocytes is suggested. Some patients responded to VCR and not to VLB, but the difference between VCR and VLB in this respect may be minor [3]. VCR is also recommended in the treatment of other forms of thrombocytopenia with active bone marrow.

The mechanism of action of the *Vinca* alkaloids has led to several animal studies: in some regards, the increased number of blood platelets recalls the colchicine leukocytosis.

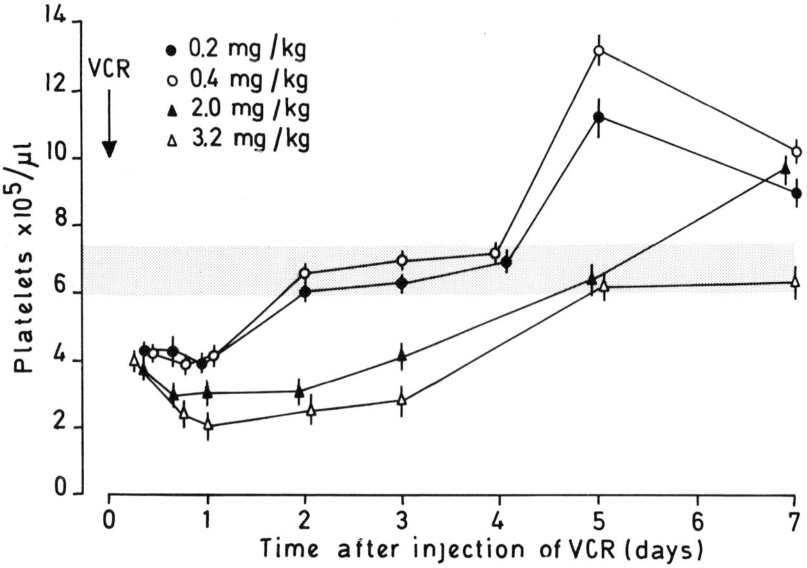

Fig. 11.2. Thrombocytosis after VCR in mice: action of different doses of VCR. Each group contained five mice or more. Results are shown as mean ± 1 SD. The *stippled area* shows the mean ± 1 SD for the circulating blood platelet count in normal mice (50) (redrawn from Rak [95], slightly modified)

In rats, the increased blood platelets after VCR [101] suggested that their formation was truly stimulated. Studies in mice showed that after a single injection of 0.2 – 3.2 mg/kg of VCR, the platelet number increased progressively for several days. This effect was preceded by an increase in the platelet incorporation of ^{75}Se-methionine, indicating an effect on thrombocytopoiesis [95]. Similar findings were reported in rats injected 0.1 – 0.5 mg/kg of VLB: the increase in blood platelets was preceded by an increase in the number of bone marrow megakaryocytes. There was no evidence that a phase of thrombocytopenia preceded the increase in blood platelets [63]. ^{75}Se-methionine experiments indicated a true increase of thrombocytopoiesis [102]. However, the possibility of a rapid toxic effect on platelets followed by an increased regeneration under the influence of thrombocytopoietin cannot be ruled out (cf. [135]). In mice, doses of 0.2 – 0.6 mg/kg of VCR produce a thrombocytopenia within 12 h, followed by a regeneration which reaches its peak five days later. The serum thrombopoietic activity was measured and found to be increased at the moment of platelet depression (12 h after the injection; [65]). Some results in rats do however contradict these findings: while a 30% increase in the platelet count was found five days after a single injection of 0.2 mg/kg of VCR, there was no significant increase in ^{75}Se-methionine incorporation, and as the protein content of the platelets decreased, it was concluded that megakaryocytes did not increase in number, but released a larger number of platelets from their cytoplasm [67]. It should be noted that although the destructive actions of MT poisons on platelets and megakaryocyte MT are well-known, these studies on apparent increased thrombopoiesis have been carried out without any ultrastructural studies of the bone marrow or the platelets. It is known that VLB and VCR bind to platelets, and this binding is enhanced by colchicine (cf. Chap. 5; [110]). VCR displaces ^{3}H-VLB much more slowly from the platelets than VLB; colchicine, which attaches to another site of tubulin, does not displacc ^{3}H-VLB [110].

A recent study of VCR action on rat bone marrow megakaryocytes with a careful study of size distribution of these cells reaches the conclusion that the dose of 0.5 mg/kg produces a rapid depression followed by an increased number of indifferentiated cells entering the megakaryocyte compartment, and an increased polyploidy of megakaryocytes. On the other hand, 0.2 mg/kg does not cause any thrombocytopenia and stimulated megakaryocytopoiesis [24]. Similar findings result from a study of rats recovering from a sublethal X-irradiation: VLB restores the platelet production, 17 – 20 days later, more rapidly than in controls, suggesting an increased formation of megakaryocyte precursors [64].

11.6 The Neurological Toxicity of *Catharanthus* Alkaloids

Most symptoms of "toxicity" as observed in patients treated with MT poisons result from the inhibition of mitotic growth in the bone marrow (aplasia), the intestine (diarrhea), or the hair bulbs (alopecia). The *Catharanthus* alkaloids used in human therapeutics (VLB and VCR) are known for their particular toxicity towards the nervous system. This is more apparent for VCR than for VLB: the reason is that VLB has a greater toxicity towards bone marrow, and that this limits its use, while VCR can be given for a longer time and in larger doses. These nervous disturbances deserve particular comment, for it is reasonable to imagine, although this is far from proven, that the neurotoxicity may have something to do with neurotubules and/or neuroplasmic flow (cf. Chap. 9). Colchicine, which is also very toxic for the nervous system, is never used in human therapeutics at doses which may affect the nerve cells: however, two cases of severe neurotoxicity with disturbances of antidiuretic hormone secretion, after accidental poisoning with large doses of colchicine, have been mentioned above [41].

11.6.1 Peripheral and Central Nervous System

Shortly after the first trials of VCR in cancer chemotherapy, disturbances related to lesions of the peripheral nerves were mentioned [27, 54]. Observations of serious central nervous troubles, leading eventually to a condition of coma or to convulsions, were mentioned later. The relatively large literature on the various aspects of VCR neurotoxicity has been reviewed several times, and a great number of cases described [21, 105, 106, 121]. Neurological toxicity has also been reported for other antineoplastic drugs, and this subject has been thoroughly reviewed [129].

The most frequent nerve disturbances can be summarized here. A depression of the musculotendinous reflexes (in particular the Achilles reflex) is found in a large number of patients, although it may remain asymptomatic [106]. Motor weakness and involvement of sensory nerves have also been observed. The cranial nerves may be affected. Mental depression, confusion, and hallucinations have been described after relatively long periods of VCR treatment, and several reports in the literature mention reversible comatose conditions [68, 76, 130]. Orthostatic hypotension has also been reported as a consequence of VCR therapy [4, 20].

Morphological studies have demonstrated that the most specific and apparent change is a demyelination of peripheral nerves followed by axonal degeneration [10]. These lesions resemble Wallerian degeneration. While some authors have called attention to the possibility that they may result from neurofibrillary changes as described experimentally [112], there is no clear evidence that VCR toxicity may be related to an action on neurotubules or neuroplasmic flow, although this hypothesis can by no means be discarded: in a boy aged 2½ years treated for lymphatic leukemia by intrathecal injection of 3 mg VCR, who died three days later, typical VCR tubulin crystals were found in the neurons of the anterior horn of the spinal cord, and the cytoplasm of these cells also contained an increased number of microfilaments [109].

11.6.2 The Syndrome of Inappropriate Secretion of Antidiuretic Hormone

This syndrome (SIADH) described by Bartter and Schwartz [8] may be found in various clinical conditions, one of the most familiar being ADH-secreting lung tumors. It is defined by the following features: (1) hyponatremia with hypoosmolality of serum; (2) normal renal excretion of sodium; (3) osmolality of the urine greater than appropriate for the concomitant blood osmolality; (4) normal renal and adrenal function; (5) improvement of hyponatremia and renal loss of sodium by fluid restriction [8, 117]. The retention of water leads to nervous disturbances, which disappear once the hyponatremia is corrected.

Table 11.2 gives a summary of the 14 cases from the literature. A few comments are justified, as a relation between this syndrome and MT poisoning is indicated by the fact that a similar syndrome was reported in two cases of accidental poisoning with colchicine [41]. The first report of hyponatremia was published in fact one year before the SIADH was described [34] in a child treated for rhabdomyosarcoma, and hyponatremia in a single case of leukemia treated with VCR was mentioned with no details in an early review of the therapeutic effects of VCR.

This complication of VCR therapy can occur at any age (from 11 months to 65 years), is not related to the dose of VCR administered (although it has been observed in accidental overdosage), and has been more frequently found in leukemic patients (it has never been reported in this disease without VCR therapy). No nervous lesions have been described in the cases which eventually (after the correction of their hyponatremia) came to autopsy, and no ultrastructural studies have been published.

In three cases, an increased secretion of ADH could be demonstrated. In the first (where a dose of VCR ten times too large had been injected by error), the serum level of ADH was about ten times above the normal values [119]. In the second, al-

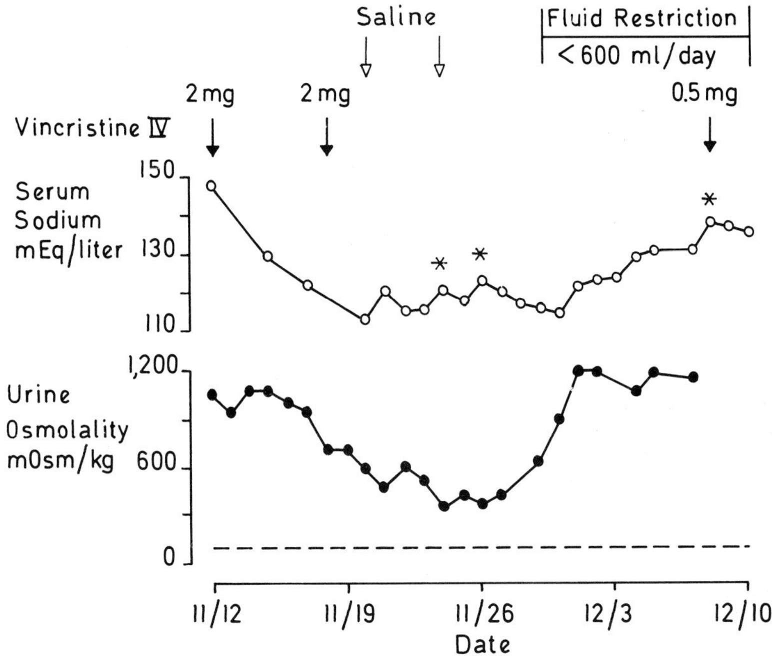

Fig. 11.3. Inappropriate secretion of ADH. Clinical course of a 50-year-old patient. The *horizontal broken line* indicates the normal level of urine dilution during water loading. The patient received 50 mg daily of prednisone and 150 mg daily of cyclophosphamide during the entire period depicted by the graph, and injections of 3% saline on 20th and 24th day. Plasma vasopressin was measured on the days indicated by the *asterisks* (redrawn from Robertson et al. [100], slightly modified) (Copyright 1973, American Medical Association)

though the patient, aged 50, had only received two injections of 2 mg of VCR, plasma arginine vasopressin was considerably above the values normally found in relation to the plasma osmolality [100]. In the third case, that of a child aged 4, the urinary excretion of ADH was found to be increased during the episode of SIADH. Later, several more doses of VCR were administered, without pathological or clinical manifestations, although each resulted in an increased urinary excretion of ADH [117]. This last observation, demonstrating an excessive secretion of ADH with no clinical complications, probably explains the reports of other patients who received VCR without trouble after experiencing a single episode of SIADH. In all the cases reported in Table 11.2 there was no indication that a cause other than VCR therapy could have modified the secretion of ADH.

Table 11.2. Inappropriate secretion of antidiuretic hormone after VCR therapy

Reference	Age of patient	Total dose (i.v., intravenously)	Natremia (lowest fig.)	Osmolality serum/urine (mOsm/kg)	Diagnosis	Other clinical and neurological findings	Evolution
Fine et al., 1966 [34]	11 mo	0.5 mg/wk, then 0.075 mg/wk for 8 wk	115 mEq/l	—	Rhabdomyosarcoma	Coma	
Haggard et al., 1968 [49]	child	—	—	—	Leukemia	—	
Loo and Zittoun, 1969 [68]	14 yr	7 mg/5 d (0.2 mg/kg)	Transitory and moderate hyponatremia	—		Diarrhea, coma (transient)	
Slater et al., 1969 [114]	7 yr	2×1.8 mg in 9 d	120 mEq/l	235/525	Leukemia	Coma	
Cutting, 1971 [25]	54 yr	4 mg (0.065 mg/kg) i.v.	106 mEq/l	240/780	Leukemia	Peripheral neuropathy ileus	No nervous system lesions at autopsy
Meriwether, 1971 [78]	34 yr	2×4.5 mg i.v. in 1 wk	125 mEq/l	—	Leukemia	Confusion, lethargy, parethesia, ileus	6 months remission
Nicholson and Feldman, 1972 [82]	2½ yr	2×2 mg/m²	118 mEq/l	—	Leukemia	Seizures. Increased ADH in urine. Normalization after fluid restriction	Remission. No disturbances during further periods of VCR treatment
Oldham and Pomeroy, 1972 [84]	52 yr	2×2.25 mg i.v. at 8 d interval	98 mEq/l	215/269	Sarcoma	Areflexia, dysarthria. Rapid improvement after fluid restriction	No recurrence
Suskind et al., 1972 [119]	3½ yr	0.5 mg/kg i.v.	118 mEq/l 138 mEq/l on 16th day	240/400	Wilms tumor	General toxicity (dose ten times too large). Serum ADH: 4.1 μU/ml (normal: 0.4±0.6)	
Robertson et al., 1973 [100]	50 yr	2×2 mg	113 mEq/l	—	Oat cell carcinoma	Increased plasma vasopressin (14.6 – 22.0 mg/ml)	No brain lesions at autopsy
Whittaker et al., 1973 [130]	65 yr	1.75 mg i.v.	120 mEq/l	—	Leukemia	Coma	No brain lesions at autopsy

Table 11.2. continued

Reference	Age of patient	Total dose (i.v., intravenously)	Natremia (lowest fig.)	Osmolality serum/urine (mOsm/kg)	Diagnosis	Other clinical and neurological findings	Evolution
Stuart et al., 1975 [117]	4 yr	2×2 mg/m²	118 mEq/l	—	Leukemia	Seizures. Increased urinary ADH	Remission. Further periods of VCR treatment with increased ADH secretion, without symptoms.
Wakem and Bennett, 1975 [123]	14 yr	15 mg/wk i.v. (by mistake)	188 mEq/l	246/407	Leukemia	Lethargy, confusion, ataxia. Leuko- and thrombocytopenia	
O'Callaghan and Ekert, 1976 [83]	21 mo	1.5 mg/m²	123 mEq/l	745 (urine)	Wilms tumor	Coma. Neuro- and thrombopenia.	

As no study of the pituitary and the hypothalamic neurons has been made during or shortly after SIADH, and as no similar syndrome can be reproduced experimentally, the mechanism of the excessive ADH secretion remains unknown, although its relation to MT poisoning is suggested. The studies on the action of MT poisons (colchicine and VCR) on posterior pituitary secretion have been summarized in Chapter 9. The morphological data suggest a decreased secretion of ADH, as the neuroplasmic flow of secretory granules towards the posterior lobe is blocked. However, the mechanisms of secretion itself, i.e., the liberation of the neurosecretory granules and their hormones in the posterior lobe, are poorly understood: the role of the pituicytes, which are closely associated with the nerve endings, and which undergo stimulation and mitotic growth when the secretion of ADH is increased, deserves further study.

11.7 Toxicity of MT Poisons Used in Cancer Chemotherapy

The application of MT poisons in cancer chemotherapy has grown considerably in the last twenty years, and in most diseases combinations of various drugs are used, a fact which makes appreciation of their toxicity difficult. However, the MT poisons are all known to affect mainly one phase of the cell cycle, mitosis, and more precisely, the movements of chromosomes related to spindle activity. They belong to what is now called the group of "phase-specific" chemotherapeutic drugs, as opposed to the "cycle-specific" ones [122]. These last affect cells whatever the phase of the mitotic cycle, and their effect is proportional to the dose, while the phase-specific drugs inhibit only one phase of growth (DNA reduplication, or spindle action), and their dose action curve reaches a plateau, as cells outside the specific phase are not affected, whatever the dose.

It would however be an error to think that this simplified view explains all the chemotherapeutic effects of MT poisons: if this were correct, all should have identical effects and find their place in cancer chemotherapy. Other factors, which are not necessarily related to the phase-specific effect, should be considered: the differences between VLB and VCR in regard to the bone marrow toxicity of the first, and the neurotoxicity of the second, have been mentioned above. Curiously, colchicine has found no place in cancer chemotherapy, although it was used locally many years ago for the treatment of benign skin tumors. Podophyllotoxin, and related molecules such as 4′-dimethyl-epidophyllotoxin-β-D-thenylidene and the similar -β-D-ethylidene have found some use in the treatment of acute monocytic leukemia, reticulosarcoma, renal tumors, and malignant gliomas, usually in combination with other drugs [77]. The indications of VLB are restricted to Hodgkin's disease (in which VLB alone gives up to 30% complete remissions), and chorio-epithelioma [77]. On the other hand, VCR is widely used, often in combination with other drugs, in many neoplasms (leukemia, histiocytosis X, melanoma, chorioepithelioma, neuroblastoma, Wilms' tumor, Ewing's tumor, rhabdomyosarcoma, Kaposi's sarcoma). Several conferences and reviews have been devoted to the applications of the *Catharanthus* alkaloids in therapeutics [6, 44, 60, 111, 118].

The toxicity of MT poisons results mainly from the inhibition of mitoses in the gastrointestinal tract and the bone marrow. Secondary effects also related to mitosis, such as transient alopecia and arrest of spermatogenesis, are of less importance [11]. Although MT are the main target of the drugs, unexplained differences are noted, such as the absence of thrombocytopenia after VLB treatment, as discussed above. Bone marrow aplasia, with all its consequences, and intestinal mucosal atrophy, with diarrhea, are the principal causes of the severe complications which may result from excessive use of all MT poisons. There are many reasons to think that neoplastic cells are more sensitive to MT poisons—one fact, recently discovered, being for instance the smaller number of cytoplasmic MT in malignant cells [12, 13]—, but the difference between fast growing normal tissues and neoplasms is small, and the wide use of combination therapy finds its origin in the necessity to harm normal mitoses minimally. It would also be an oversimplification to think that the MT poisons affect neoplastic cell growth only by acting on the spindle MT: there are many other possibilities, as indicated by the different results obtained in closely related tumors. One of these which should be kept in mind, and which may also have detrimental effects, is the destruction of lymphoid tissue and the resulting depression of immunological reactions. Cancer chemotherapy, whatever its advances, is far from having found the ideal agent, capable of affecting malignant growth without unpleasant side-effects, and it is doubtful whether an action on MT, which cannot be specific, can alone arrest malignant growths. In tumors where chemotherapy is spectacularly active, such as Burkitt's tumor and chorio-epithelioma, the immunological defenses of the host are known to play an important role in modifying the balance between the neoplastic growth and the normal tissues.

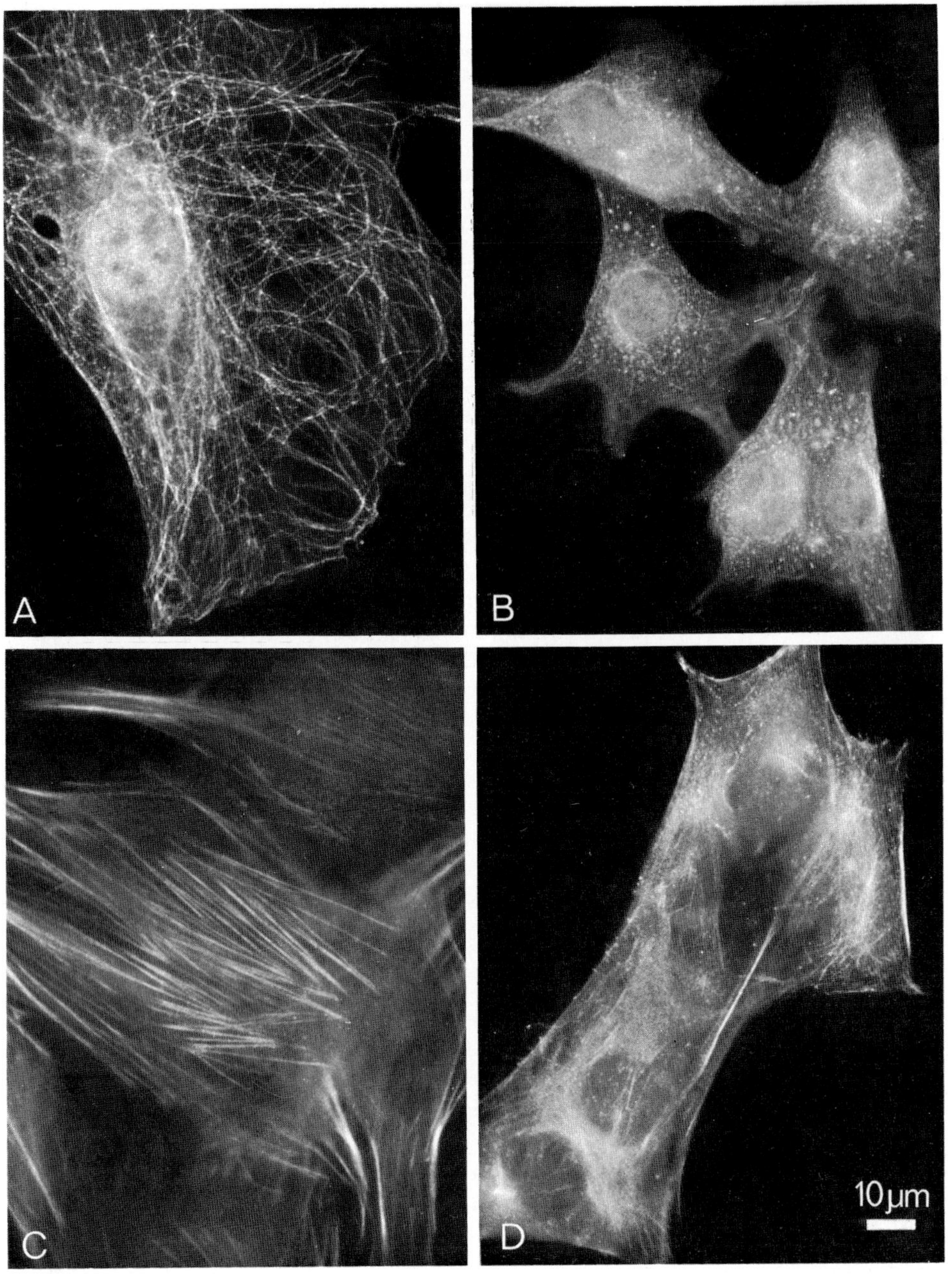

Fig. 11.4 A–D. Influence of malignant change on MT and MF in cultured mouse fibroblasts (line 3T3). (A) Normal fibroblasts: extensive network of MT (cytoplasmic microtubule complex: CMTC; cf. Chap. 2). (B) Transformed fibroblasts: greatly reduced CMTC. (C) Normal fibroblasts: large actin "cables" (cf. Chap. 7). (D) Transformed cells: decreased number of actin fibrils. (Tubulin and actin immunofluorescence staining with antibodies against bovine tubulin and actin) (from Brinkley et al. [13])

11.8 Tubulin and MT Pathology

As MT are indispensable for so many cell functions, it is not surprising that few hereditary abnormalities of MT have been described so far: integrity of MT is indispensable for the movement of the male gametes, for the mitoses of the egg, for the shaping of the embryo, and any important anomaly of tubulin would be lethal. However, as it is becoming more and more evident that all tubulins are not necessarily identical, the possibility must be kept in mind that in future years tubulins will be found with amino acid sequences modified as a consequence of genetic changes. On the other hand, disturbances of tubulin and MT assembly into more complex structures may be compatible with survival, and recent works on the relation between MT assembly and the abnormal lysosomes in the Chediak-Higashi syndrome are evidence for this. Abnormalities of complex structures such as cilia have been mentioned in Chapter 4, and the observations by Afzelius et al. [2] of normal though sterile males with immotile cilia indicate that serious anomalies may curiously be compatible with prolonged survival. Another aspect of MT pathology is the formation of fibrous structures comparable to those produced experimentally, and which may be found in some pathological conditions in man: while the relation of these fibrils to MT is by no means certain, they should be mentioned here. Another aspect of possible tubulin modifications concerns the selection of cells resistant to MT poisons: this resistance may result more from membrane permeability changes than from MT themselves, although the possibility that MT may become more resistant is compatible with what has been said of the variable resistance of MT to poisons in some cells.

11.8.1 MT Resistant to MT Poisons

The resistance to a substance such as colchicine is very variable in nature, and reference has been made to this in several chapters: most cases of resistance can probably be explained by a relative impermeability of cell membranes. The best known case of resistance is that of the golden hamster, *Cricetus auratus*, as mentioned in Chapter 5. This resistance is only relative and can be overcome by large doses.

Experimental research with unicellulars has led to the isolation of strains resistant to colchicine: in *Chlamydomonas reinhardii*, while the growth of normal cells is completely inhibited by 5×10^{-3} M colchicine, five strains have been isolated by cultivation on colchicine medium, showing a complete resistance to doses of $4-6 \times 10^{-3}$ M colchicine. These mutants differed from the wild type by a slowing of their logarithmic growth, and morphological changes (many cells held together in palmella envelopes; [1]). In a study of the inhibition of growth and of flagellar regeneration in the same species, only large doses of colchicine were found to be effective, a fact which was apparently related to a permeability barrier, as the addition of sodium deoxycholate increased colchicine effectiveness thirtyfold. Mutants were isolated after treatment with methyl methanesulfonate, a mutagen. Three colchicine-resistant mutants grew in concentrations of colchicine ten times higher than the wild type and could also regenerate flagella in higher concentration of the drug.

These strains were also resistant to a water-soluble derivative of podophyllotoxin. No true resistance of VLB could be induced [36]. In the same species, another interesting mutation affecting the formation of the basal body, has been isolated from cultures irradiated with UV light. This strain (*bald*-2) is incapable of forming normal basal bodies, and instead of nine triplets shows only a ring of nine singlet MT. They develop no flagella: a flagellar protuberance of the cytoplasm can be formed, but the singlet MT are unable to grow properly and form a functional cilium, with doublets. The mitotic MF of this strain are normal. A single gene would be responsible for this abnormality [47].

Another *Chlamydomonas* mutant, TS-60, showed resistance to 6×10^{-3} M colchicine, which did not affect either cell division or flagellar regeneration. This strain is thermolabile, and this property is inherited in a mendelian fashion, like colchicine resistance, from which it cannot be separated. The mutation may affect the colchicine binding of tubulin [108].

Resistant lines of haploid cells from *Rana pipiens* have been obtained by cultivation in vitro in the presence of toxic doses of various MT poisons. These cells had been previously treated with the mutagen acridine half-mustard ICR 191. A line with a threefold increased resistance to podophyllin was also resistant to colchicine, but not to VLB. A more surprizing fact, as the two poisons are supposed to become attached to the same site of the tubulin molecule, is that another strain, resistant to podophyllin, maintained a normal sensitivity to colchicine. Strains selected by exposure to VLB were specifically resistant to this MT poison [38]. No morphological study of the MT of these cells appears to have been carried out so far.

An interesting mutant has been described in the axolotl (*Amblystoma mexicanum*): the eggs contain a normal pool of unassembled tubulin, although they are unable to divide, as no spindle or asters are formed. This "assembly mutant" may be corrected partially by the injection of MT fragments or doublet MT, which permit a certain number of normal cell divisions [94].

Other colchicine-resistant strains of cells isolated from mice and Syrian hamsters have not demonstrated a genetic basis for this property. Resistance, which is instable, appears to be related to a decreased permeability and may affect other drugs such as actinomycin and puromycin [79].

11.8.2 Neuronal Fibrillary Changes and Alzheimer's Disease

In some cells, the destruction of MT by colchicine or other MT poisons leads to the formation of large bundles of filaments (cf. Chap. 5). The reasons for thinking that, contrary to the first hypotheses on this subject, these fibrils could not originate from a rearrangement of tubulin molecules, have already been given. However, as this change has a superficial resemblance to fibrillary alterations found in several nervous diseases, a few words of comment are justified.

In fact, fibrillar changes of neuronal cytoplasm had first been observed after injections of aluminium, which is by no means a MT poison [131, 132], and is used as an adjuvant in experiments of immunological stimulation. The same group of authors, observing fibrillar changes in rat neurons in tissue culture under the in-

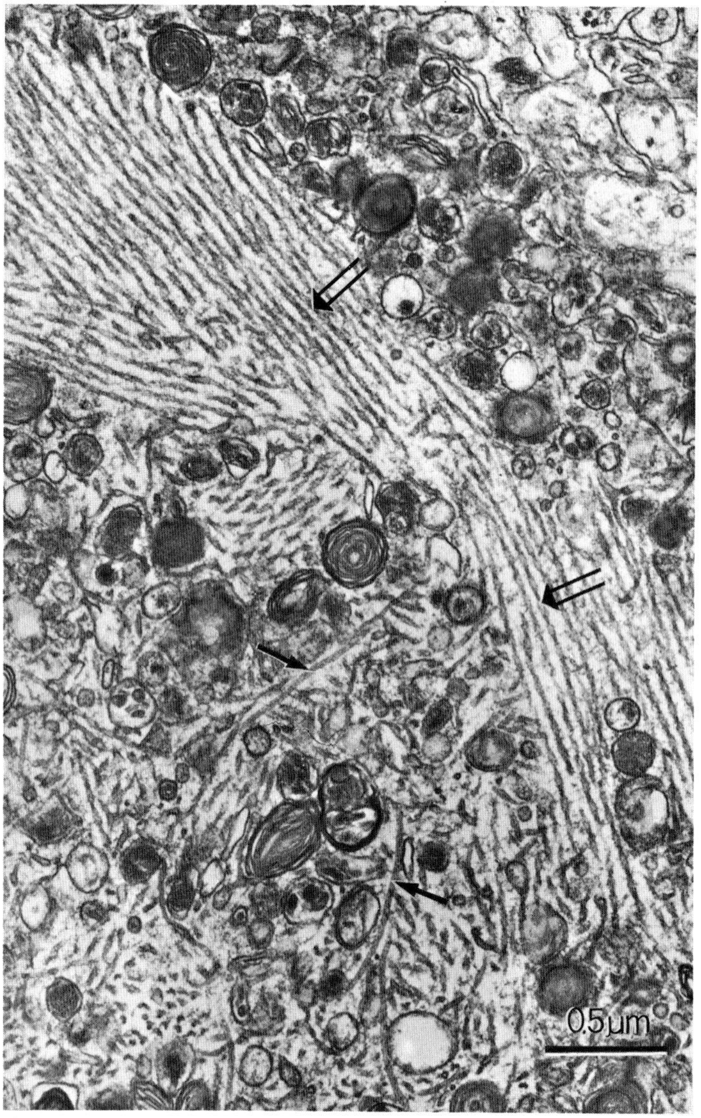

Fig. 11.5. Man. Cerebral biopsy. Alzheimer's disease. Normal MT (*arrows*) and twisted tubules (*double arrows*) in a neuron

fluence of colchicine [89], and similar changes in HeLa cells in tissue culture [99], injected solutions of colchicine in the cisterna magna of rabbit's brain. With doses between 50 and 1000 µg the animals died within six days, and striking accumulations of fibrils with a diameter of 10 – 14 nm were observed. The MT did not appear to be destroyed [132]. In further work, similar fibrils were found in neurons of the anterior horn in rabbits after injections of colchicine, VLB, and podophyllotoxin, by the subarachnoid route. These changes were reversible [132].

These fibrils were identical to those produced by aluminium injections, and appeared comparable to the fibrillar changes known to occur in human neurons in several diseases, in particular Alzheimer's presentile dementia, and some "slow virus" diseases such as kuru [55, 56, 120, 133, 134].

However, this resemblance was later shown to be only superficial, as the filaments found in Alzheimer's disease in fact contain a new protein, with a molecular weight of 50,000 daltons, which can be separated by electrophoresis both from tubulins and from the major neurofilament protein. This new protein is not found in normal brains; an interesting fact is that the aluminium concentration of Alzheimer's disease brains appears to be four times higher than in controls [59].

One observation does however indicate that similar twisted filaments can originate from MT: when their amino acids (mainly the tyrosine) are iodinated in vitro by lactoperoxidase, H_2O_2, and iodide, the MT with 0.1 iodotyrosines per dimer form filaments twisted around each other, which are stable at 4 °C. Iodinated tubulin cannot assemble in vitro: the twisted filaments are formed by MT and have been compared to similar filaments found in aged monkey's brain [39, 40].

The relations between the twisted tubules and the fibrillary material found after aluminium and colchicine and other MT poisons remains open to further study. The colchicine-induced fibrils are not twisted and their possible relation to other cytoplasmic fibrils, such as actin, deserves to be studied anew, particularly by immunofluorescence techniques.

11.8.3 Chediak-Higashi Disease and its Relation to MT

The Chediak-Higashi disease is a disturbance of lysosomal, secretory, and melanin granules which is known in man, mink, mice (the beige mouse), cattle, and cats. Anomalies of pigmentation, as in the aleoutian strain of mink, decreased resistance to infections, photophobia, and an increased frequency of malignant lymphoid tumors and leukemias are symptoms of this condition, which is transmitted as a recessive trait [23, 53].

Recent work has brought important information about the possible relation between the abnormal size of the leukocyte granules and MT. In Chapter 3, the formation of "caps" by leukocytes treated with the lectin Concanavalin A (Con A) has been mentioned. The Con A "capping" phenomenon is known to be affected by MT poisons: while in normal lymphocytes the surface antigens remain dispersed over the cell membrane, treatment with colchicine enables these receptors to move more freely, and this results in the agglomeration of the receptor sites at one side of the cell in the shape of a cap, which may later become internalized (cf. Chap. 3). Contrary to normal polymorphonuclears, those of the beige mouse cap spontaneously after treatment by Con A and a similar capping is found in polymorphonuclears from the blood of patients with the Chediak-Higashi syndrome [85, 87].

In comparison with what is known about the relations of cell surface receptors and MT, it was natural to seek an anomaly of cell MT and the possible means to correct it. In the mouse, the role of MT was indicated by the fact that the capping produced by Con A is inhibited by cGMP and substances such as the cholinergic drugs carbamylcholine and carbamyl-β-methylcholine, which increase the amount in intracellular cGMP. It was also observed that in cultured fibroblasts of beige mice, the giant lysosomes are not present after treatment with these drugs. cGMP and carbamylcholine were also found to reduce capping in human leukocytes from cases of Chediak-Higashi disease. Also, the large granules found in monocytes from these subjects when incubated in vitro decrease in the presence of carbamylcholine, the lysosomal granules assuming a normal size. Moreover, beige mice injected with carbamylcholine for several weeks showed a normalization of the granule morphology of leukocytes and a normal reaction to Con A, with no capping.

The effect of Con A on MT of Chediak-Higashi leukocytes was studied: while incubation with Con A of normal leukocytes increases markedly the number of cytoplasmic MT, this effect is completely missing in Chediak-Higashi leukocytes. On the other hand, it is known that cGMP enhances MT assembly in normal human polymorphonuclears ([137]; cf. Chap. 3).

Furthermore, while leukocytes from a Chediak-Higashi patient failed to react to Con A by an increase in the number of pericentriolar MT, this anomaly could be corrected by incubating the cells with cGMP or cholinergic agonists. Fibroblasts from this patient, which showed abnormally large lysosomes, became morphologically normal when cultured in the presence of the same drugs [88]. Similarly, the addition of cGMP or the cholinergic drug carbamyl methylcholine chloride (betanechol) to leukocytes of a child with Chediak-Higashi's disease affected favorably several leukocyte functions: the release of lysosomal β-glucuronidase was increased, and bactericidal activity as well as chemotaxis were normalized. These facts suggest that abnormal cyclic nucleotide levels are the cause of the cellular defects, and of the decreased MT activity [9].

These results could be explained by an anomaly of the tubulin molecules, which fail to assemble, but experiments have shown that the tubulin of beige mouse brain copolymerizes normally with rabbit brain tubulin [86]. It is also known that Chediak-Higashi cells divide normally, indicating that their spindle MT are not affected. It is suggested that there is a failure of the stimulation of MT assembly resulting from the binding of Con A to the membrane receptors, the anomaly residing in the mechanisms of assembly in the presence of some stimuli. This failure would also explain the formation of abnormally large secretion granules (cf. Chap. 8) and of large lysosomes. In some respects, the pathological cells resemble transformed neoplastic cells (cf. [14, 35]) which have a decreased amount of cytoplasmic MT and cap readily when treated with lectins such as Con A. The generation of cGMP may be the common factor in both conditions [86, 9].

Whatever the future findings in this interesting syndrome (several problems remain open for further research: why are the melanin granules too large? Is the axonal flow normal? Are ciliary structures affected?), it is an excellent demonstration of the significance of MT research in medicine.

References

1. Adams, M., Warr, J. R.: Colchicine-resistant mutants of *Chlamydomonas rheinhardii*. Exp. Cell Res. **71**, 473 – 475 (1972)
2. Afzelius, B. A., Eliasson, R., Johnsen, Ø., Lindholmer, C.: Lack of dynein arms in immotile human spermatozoa. J. Cell Biol. **66**, 225 – 232 (1975)

3. Ahn, Y. S., Harrington, W. J., Seelman, R. O., Eytel, C. S.: Vincristine therapy of idiopathic and secondary thrombocytosis. New Engl. J. Med. **291,** 376 – 380 (1974)
4. Aisner, J., Weiss, H. D., Chang, P., Wiernik, P. H.: Orthostatic hypotension during combination chemotherapy with vincristine (NSC-67574). Cancer Chemother. Rep. **58,** 927 – 930 (1974)
5. Alarcon-Segovia, D., Ibanez, G., Kerchenobich, D., Rojkind, M.: Treatment of scleroderma by modification of collagen metabolism. A double blind trial with colchicine-placebo. J. Rheumatol. **1** (Suppl.), 97 (1974)
6. Armstrong, J. G.: Vincristine (NSC-67574) symposium. Summary and prospectives. Cancer Chemother. Rep. **52,** 527 – 535 (1968)
7. Bandmann, U., Norberg, B., Simmingsköld, G.: Griseofulvin inhibition of polymorphonuclear leucocyte chemotaxis in Boyden chambers. Scand. J. Haemat. **15,** 81 – 87 (1975)
8. Bartter, F. C., Schwartz, W. B.: The syndrome of inappropriate secretion of antidiuretic hormone. Am. J. Med. **42,** 790 – 806 (1967)
9. Boxer, L. A., Rister, M., Allen, J. M., Baehner, R. L.: Improvement of Chediak-Higashi leukocyte function by cyclic guanosine monophosphate. Blood **49,** 9 – 17 (1977)
10. Bradley, W. G., Lassman, L. P., Pearce, G. W., Walton, J. N.: The neuromyopathy of vincristine in man. Clinical, electrophysiological and pathological studies. J. Neurol. Sci. **10,** 107 – 132 (1970)
11. Bremmer, W. J., Paulsen, C. A.: Colchicine and testicular function in Man. New Engl. J. Med. **294,** 1384 – 1385 (1976)
12. Brinkley, B. R., Fuller, G. M., Highfield, D. P.: Cytoplasmic microtubules in normal and transformed cells in culture. Analysis by tubulin antibody immunofluorescence. Proc. Natl. Acad. Sci. USA **72,** 4981 – 4985 (1975)
13. Brinkley, B. R., Miller, C. L., Fuseler, J. W., Pepper, D. A., Wible, L. J.: The cytoskeleton and cell transformation in malignancy. Microtubules, microfilaments, and growth properties in vitro. Cancer Bull. **29,** 13 – 15 (1977)
14. Brune, K., Bucher, K. E. Walz, D.: The avian microcrystal arthritis. II. Central versus peripheral effects of sodium salicylate, acetaminophen and colchicine. Agents and Actions **4,** 27 – 33 (1974)
15. Brune, K., Glatt, M.: The avian microcrystal arthritis. III. Invasion and enzyme-release from leukocytes at the site of inflammation. Agents and Actions **4,** 95 – 100 (1974)
16. Brune, K., Glatt, M.: The avian microcrystal arthritis. IV. The impact of sodium salicylate, acetaminophen and colchicine on leukocyte invasion and enzyme liberation in vivo. Agents and Actions **4,** 101 – 107 (1974)
17. Brune, K., Graf, P., Glatt, M.: Anti-inflammatory action of colchicine. In: Microtubules and Microtubule Inhibitors (eds.: M. Borgers, M. De Brabander), pp. 471 – 481. Amsterdam-Oxford: North-Holland 1975
18. Byers, P. H., Ward, P. A., Kellermeyer, R. W., Naff, G. B.: Complement as a mediator of inflammation in acute gouty arthritis. II. Biological activities generated from complement by the interaction of serum complement and sodium urate crystals. J. Lab. Clin. Med. **81,** 761 – 769 (1973)
19. Caner, J. E. Z.: Colchicine inhibition of chemotaxis. Arthritis Rheum. **8,** 757 – 764 (1965)
20. Carmichael, S. M., Eagleton, L., Ayers, C. R.: Orthostatic hypotension during vincristine therapy. Arch. Intern. Med. **126,** 290 – 293 (1970)
21. Casey, E. B., Jeliffe, A. M., Le Quesne, P. M., Millett, Y. L.: Vincristine neuropathy. Clinical and electrophysiological observations. Brain **96,** 69 – 86 (1973)
22. Castaing, R., Boget, J. C., Favarel-Garrigues, J. C., Vital, C., Cardinaud, J. P.: Intoxication aiguë mortelle par la colchicine. Bordeaux Méd. 2101 – 2104 (1968)
23. Chédiak, M.: Nouvelle anomalie leucocytaire de caractère constitutionel et familial. Rev. Hématol. (Paris) **7,** 352 – 357 (1952)
24. Choi, S., Simone, J. V., Edwards, C. C.: Effects of vincristine on platelet production. In: Platelets. Production, Function, Transfusion and Storage (eds.: M. G. Baldini, S. H. Ebbe), pp. 51 – 62. New York: Grune and Stratton 1974
25. Cutting, H. O.: Inappropriate secretion of antidiuretic hormone secondary to vincristine therapy. Am. J. Med. **51,** 269 – 271 (1971)
26. Denko, C. W.: Anti-prostaglandin action of colchicine. Pharmacology **13,** 219 – 227 (1975)
27. D'Agostino, A. N., Jarcho, L. W.: A new neuropathy associated with vincristine (leurocristine) therapy. Clin. Res. **12,** 106 (1964)
28. Dinarello, C. A., Wolff, S. M., Goldfinger, S. E., Dale, D. C., Alling, D. W.: Colchicine therapy for familial Mediterranean fever: a double-blind trial. New Engl. J. Med. **291,** 934 – 937 (1974)

29. Dinarello, C. A., Chusid, M. J., Fauci, A. S. Gallin, J. I., Dale, D. C., Wolff, S. M.: Effect of prophylactive colchicine therapy on leukocyte function in patients with familial Mediterranean fever. Arthritis Rheum. **19,** 618 – 622 (1976)
30. Dittman, W. A., Ward, J. R.: Demecolcine toxicity. A case report of severe hematopoietic toxicity and a review of the literature. Am. J. Med. **27,** 519 – 524 (1959)
31. Dustin, P.: Intoxication mortelle par la colchicine. Etude histologique et hématologique. Bull. Acad. Roy. Méd. Belg. VIe. série **6,** 505 – 529 (1941)
32. Ertel, N., Omokoku, B., Wallace, S. L.: Colchicine concentrations in leukocytes. Arthritis Rheum. **12,** 293 (1969)
33. Ertel, N. H., Mittler, J. C., Akgun, S., Wallace, S. L.: Radioimmunoassay for colchicine in plasma and urine. Science **193,** 233 – 234 (1976)
34. Fine, R. N., Clarke, R. R., Shore, N. A.: Hyponatremia and vincristine therapy. Am. J. Dis. Child. **112,** 256 – 259 (1966)
35. Fine, R. E., Taylor, L.: Decreased actin and tubulin synthesis in 3T3 cells after transformation by SV40 virus. Exp. Cell Res. **102,** 162 – 168 (1976)
36. Flavin, M., Slaughter, C.: Microtubule assembly and function in *Chlamydomonas:* inhibition of growth and flagellar regeneration by antitubulins and other drugs and isolation of resistant mutants. J. Bacteriol. **118,** 59 – 120 (1974)
37. Floersheim, G. L., Brune, K., Seiler, K.: Colchicine in avian sodium urate and calcium pyrophosphate microcrystal arthritis. Agents and Actions **3,** 20 – 23 (1973)
38. Freed, J. J., Ohlsson-Wilhelm, B. M.: Cultured haploid cells resistant to antitubulins. In: Microtubules and Microtubule Inhibitors (eds.: M. Borgers, M. De Brabander), pp. 367 – 378. Amsterdam-Oxford: North-Holland 1975
39. Gaskin, F., Gethner, J. S.: Characterization of the in vitro assembly of microtubules. In: Cell Motility (eds.: R. Goldman, T. Pollard, J. Rosenbaum), pp. 1109 – 1123. Cold Spring Harbor Lab. 1976
40. Gaskin, F., Litman, D. J., Cantor, C. R., Shelanski, M. L.: The formation of filamentous structures from iodinated neurotubules. J. Supramol. Struct. **3,** 39 – 50 (1975)
41. Gaultier, M., Bismuth, C., Autret, A., Pillon, M.: Antidiurèse inappropriée après intoxication aiguë par la colchicine. Deux cas. Nouv. Presse Méd. **4,** 31 – 32 (1975)
42. Gaultier, M., Bismuth, C., Conso, F.: L'aplasie médullaire de la colchicine. A propos de 20 cas. Sang and Toxiques **98,** 75 – 85 (1977)
43. Gaultier, M., Kanfer, A., Bismuth, C.: Données actuelles sur l'intoxication aiguë par la colchicine. A propos de 23 observations. Ann. Méd. Intern. (Paris) **120,** 605 – 618 (1969)
44. Gmachl, E. (ed.): Internationales Symposium über die Anwendung der *Vinca*-alkaloide Velbe und Vincristin. Berlin-Wien: Urban und Schwarzenberg 1969
45. Goldfinger, S. E.: Colchicine for familial Mediterranean fever. New Engl. J. Med. **287,** 1302 (1972)
46. Goldstein, R. C., Schwabe, A. D.: Prophylactic colchicine therapy in familial Mediterranean fever. Ann. Intern. Med. **81,** 792 – 794 (1974)
47. Goodenough, U. W., St. Clair, H. S.: *Bald-2*: a mutation affecting the formation of doublet and triplet sets of microtubules in *Chlamydomonas reinhardii.* J. Cell Biol. **66,** 480 – 491 (1975)
48. Gooneratne, B. W.: Massive generalized alopecia after poisoning by *Gloriosa superba.* Brit. Med. J. **5494,** 1023 – 1024 (1966)
49. Haggard, M. E., Fernback, D. J., Holcomb, T. M., Sutow, W. W., Vietti, T. J., Windmiller, J.: Vincristine in acute leukemia of childhood. Cancer **22,** 438 – 444 (1968)
50. Harris, E. D. Jr., Hoffman, G. S., Mc Guire, J. L., Strosberg, J. M.: Colchicine: effects upon urinary hydroxyproline excretion in patients with scleroderma. Metabolism **24,** 529 – 536 (1975)
51. Harris, E. D. Jr., Millis, M.: Treatment with colchicine of the periarticular inflammation associated with sarcoidosis: A need for continued appraisal. Arthritis Rheum. **14,** 130 – 133 (1971)
52. Heaney, D., Derghazarian, C. B., Pintteo, G. F., Ali, M. A. M.: Massive colchicine overdose: a report on the toxicity. Am. J. Med. Sci. **271,** 233 – 238 (1976)
53. Higashi, O.: Congenital gigantism of peroxidase granules. The first case ever reported of qualitative abnormity of peroxidase. Tohoku J. Exp. Med. **59,** 315 – 332 (1953)
54. Hildebrand, J., Coërs, S.: Etude clinique, histologique et électrophysiologique des neuropathies associées au traitement par la vincristine. Europ. J. Cancer **1,** 51 – 58 (1965)
55. Hirano, A.: Neurofibrillary changes in conditions related to Alzheimer's disease. In: Alzheimer's Disease and Related Conditions (eds.: G. E. W. Wolstenholme, M. O'Connor), pp. 185 – 201. London: Churchill 1970

56. Hirano, A., Zimmerman, H. M.: Some effects of vinblastine implantation in the cerebral white matter. Lab. Invest. **23,** 358 – 367 (1970)
57. Hitzig, W. H.: Colchicum-Vergiftung bei einem Kleinkind. 2. Beobachtungen über Blutbildungs- und Mitosestörungen. Acta Haematol. **21,** 170 – 186 (1959)
58. Hwang, Y. F., Hamilton, H. E., Sheets, R. F.: Vinblastine-induced thrombocytosis. Lancet **2,** 1075 – 1076 (1969)
59. Iqbal, K., Wisniewski, H. M., Grundke-Iqbal, I., Korthals, J. K., Terry, R. D.: Chemical pathology of neurofibrils. Neurofibrillary tangles of Alzheimer's presenile-senile dementia. J. Histochem. Cytochem. **23,** 563 – 569 (1975)
60. Johnson, I. S., Armstrong, J. G., Gorman, M., Burnett, J. P. Jr.: the *Vinca* alkaloids: a new class of oncolytic agents. Cancer Res. **23,** 1390 – 1427 (1963)
61. Kaplan, H.: Further experiences with colchicine in the treatment of sarcoid arthritis. New Engl. J. Med. **268,** 761 – 764 (1963)
62. Kedar, I., Ravid, M., Sohar, E., Gafni, J.: Colchicine inhibition of casein-induced amyloidosis in mice. Israel J. Med. Sci. **10,** 787 – 789 (1974)
63. Klener, P., Donner, L., Housková, J.: Thrombocytosis in rats induced by vinblastine. Haemostasis **1,** 73 – 78 (1972)
64. Klener, P., Donner, L., Hynčica, V., Šafránková, D.: Possible mechamism of vinblastine-induced thrombocytosis. Scand. J. Haematol. **12,** 179 – 184 (1974)
65. Krisza, F., Kovacs, Z., Dobay, E.: Effect of vincristine on the megacaryocyte system in mice. J. Med. **4,** 12 – 18 (1973)
66. Layani, F., Aschkenasy, A., Mouzon, M.: Intoxication aiguë par la colchicine; importantes altérations de la leucopoïèse. Bull. Soc. Méd. Hôp. Paris **63,** 10 – 16 (1947)
67. Legard, C., Eberlin, A., Caen, J. P.: Vincristine-induced thrombocytosis in the rat. Thrombopoiesis and platelet populations. Path. Biol. **21,** 41 – 45 (1973)
68. Loo, H., Zittoun, R.: Intoxication aiguë à forme comateuse par la vincristine. Gaz. Méd. Fr. **76,** 2693 – 2698 (1969)
69. McCarty, D. J.: Urate crystal phagocytosis by polymorphonuclear leukocytes and the effects of colchicine. In: Phagocytic Mechanisms in Health and Disease (eds.: R. C. Williams, H. H. Fudenberg), pp. 107 – 122. Stuttgart: Georg Thieme 1972
70. Malawista, S. E.: Colchicine: a common mechanism for its anti-inflammatory and antimitotic effects. Arthritis Rheum. **11,** 191 – 197 (1968)
71. Malawista, S. E.: The action of colchicine in acute gouty arthritis. Arthritis Rheum. **18** (Suppl.) 835 – 846 (1975)
72. Malawista, S. E., Andriole, V. T.: Colchicine: anti-inflammatory effect of low doses in a sensitive bacterial system. J. Lab. Clin. Med. **72,** 933 – 942 (1968)
73. Malawista, S. E., Bensch, K. G.: Human polymorphonuclear leukocytes: demonstration of microtubules and effect of colchicine. Science **156,** 521 – 522 (1967)
74. Mamou, H.: La Maladie Périodique. Paris: Expansion Sci. Fr. 1956
75. Marmont, A. M., Damasio, E. E.: Clinical experiences with cytotoxic immunosuppressive treatment of idiopathic thrombocytopenic purpura. Acta Haematol. **46,** 74 – 91 (1971)
76. Martin, J., Mainwaring, D.: Coma and convulsions associated with vincristine therapy. Brit. Med. J. **4,** 782 (1973)
77. Mathé, G., Kenis, Y.: La Chimiothérapie des Cancers (Leucémies, Hématosarcomes et Tumeurs Solides), 3rd ed. Paris: Expansion Sci. Fr. 1975
78. Meriwether, W. D.: Vincristine toxicity with hyponatremia and hypochloremia in an adult. Oncology **25,** 234 – 238 (1971)
79. Minor, P. D., Roscoe, D. H.: Colchicine resistance in mammalian cell line. J. Cell Sci. **17,** 381 – 396 (1975)
80. Montaldi, D. H., Giambrone, J. P., Courey, N. G., Taefi, P.: Podophyllin poisoning associated with the treatment of condyloma acuminatum: a case report. Am. J. Obst. Gyn. **119,** 1130 (1974)
81. Nagaratnam, N., de Silva, D. P. K. M., de Silva, N.: Colchicine poisoning following ingestion of *Gloriosa superba* tubers. Trop. Geograph. Med. **25,** 15 – 17 (1973)
82. Nicholson, R. G., Feldman, W.: Hyponatremia in association with vincristine therapy. Canad. Med. Ass. J. **106,** 356 – 357 (1972)
83. O'Callaghan, M. J., Ekert, H.: Vincristine toxicity unrelated to dose. Arch. Dis. Childh. **51,** 289 – 292 (1976)
84. Oldham, R. K., Pomeroy, T. C.: Vincristine-induced syndrome of inappropriate secretion of antidiuretic hormone. South. Med. J. **65,** 1010 – 1012 (1972)
85. Oliver, J. M.: Defects in cyclic GMP generation and microtubule assembly in Chediak-Higashi and malignant cells. In: Microtubules and Microtubule Inhibitors (eds.: M. Borgers, De Brabander), pp. 341 – 354. Oxford-Amsterdam: North-Holland 1975

86. Oliver, J. M.: Impaired microtubule function correctable by cyclic GMP and cholinergic agonists in the Chediak-Higashi syndrome. Am. J. Path. **85,** 395 – 418 (1976)
87. Oliver, J. M., Zurier, R. B., Berlin, R. D.: Concanavalin cap formation on polymorphonuclear leukocytes of normal and beige (Chediak-Higashi) mice. Nature (London) **253,** 471 – 473 (1975)
88. Oliver, J. M., Zurier, R. B.: Correlation of characteristic abnormalities of microtubule function and granule morphology in Chediak-Higashi syndrome with cholinergic agonists. Studies in vitro in Man and in vivo in beige mouse. J. Clin. Invest. **57,** 1239 – 1247 (1976)
89. Peterson, E., Murray, M.: Serial observations in tissue cultures on neurotoxic effects of colchicine. Anat. Rec. **154,** 401 (1966)
90. Pernice, B.: Sulla cariocinesi delle cellule epiteliali e dell'endotelio dei vasi della mucosa dello stomaco e dell'intestino, nelle studio della gastroenterite sperimentale (nell'avvelenamento per colchico). Sicil. Med. **1,** 265 – 279 (1889)
91. Phelps, P.: Polymorphonuclear leukocyte motility in vitro. III. Possible release of a chemotactic substance following phagocytosis of urate crystals by polymorphonuclear leukocytes. Arthritis Rheum. **12,** 197 – 211 (1969)
92. Phelps, P.: Polymorphonuclear leukocyte mobility in vitro. IV. Colchicine inhibition of the formation of chemotactic activity after phagocytosis of urate crystals. Arthritis Rheum. **13,** 1 – 9 (1970)
93. Phelps, P.: Appearance of chemotactic activity following intra-articular injection of monosodium urate crystals: effect of colchicine. J. Lab. Clin. Med. **76,** 622 – 631 (1970)
94. Raff, E. C., Brothers, A. J., Raff, R. A.: Microtubule assembly mutant. Nature (London) **260,** 615 – 616 (1976)
95. Rak, K.: Effect of vincristine on platelet production in mice. Brit. J. Haematol. **22,** 617 – 624 (1972)
96. Reimann, H. A.: Periodic disease. A probable syndrome including periodic fever, benign paroxysmal peritonitis, cyclic neutropenia and intermittent arthralgia. J. Am. Med. Ass. **136,** 239 – 244 (1948)
97. Reimann, H. A.: Colchicine for periodic peritonitis. J. Am. Med. Ass. **231,** 64 – 66 (1975)
98. Reiquam, C. W., Prosper, J. C.: Chronic idiopathic thrombocytopenia: treatment with prednisone, 6-mercaptopurine, vincristine, and fresh plasma transfusions. J. Pediatr. **68,** 885 – 889 (1966)
99. Robbins, E., Gonatas, N.: Histochemical and ultrastructural studies on HeLa cultures exposed to spindle inhibitors with special reference to the interphase cell. J. Histochem. Cytochem. **12,** 704 (1964)
100. Robertson, G. L., Bhoopalam, N., Zelkowitz, L. J.: Vincristine neurotoxicity and abnormal secretion of antidiuretic hormone. Arch. Intern. Med. **132,** 717 – 722 (1973)
101. Robertson, J. H., Crozier, E. H., Woodend, B. E.: The effect of vincristine on the platelet count in rats. Brit. J. Haematol. **19,** 331 – 337 (1970)
102. Robertson, J. H., Crozier, E. H., Woodend, B. E.: Vincristine-induced thrombocytosis studied with ^{75}Se selenomethionine. Acta Haematol. **47,** 356 – 360 (1972)
103. Rojkind, M., Kershenobich, D.: Effect of colchicine on collagen, albumin and transferrin synthesis by cirrhotic rat liver slices. Bioch. Biophys. Acta **378,** 415 – 423 (1975)
104. Rojkind, M., Uribe, M., Kershenobich, D.: Colchicine and the treatment of liver cirrhosis. Lancet **I,** 38 – 39 (1973)
105. Rosenthal, S., Kaufman, S.: Vincristine neurotoxicity. Ann. Intern. Med. **80,** 733 – 737 (1974)
106. Sandler, S. G., Tobin, W., Henderson, E. S.: Vincristine-induced neuropathy: a clinical study of fifty leukemic patients. Neurology **19,** 367 – 374 (1969)
107. Sartorelli, A. C., Johns, D. G. (eds.): Antineoplastic and immunosuppressive agents, Part. 2. In: Handbuch der experimentellen Pharmakologie (new series) 38/2 (eds.: O. Eichler, A. Farah, H. Herken, A. D. Welch). Berlin-Heidelberg-New York: Springer 1975
108. Sato, C.: A conditional cell division mutant of *Chlamydomonas reinhardii* having an increased level of colchicine resistance. Exp. Cell Res. **101,** 251 – 259 (1976)
109. Schochet, S. S. Jr., Lampert, P. W., Earle, K. M.: Neuronal changes induced by intrathecal vincristine sulfate. J. Neuropathol. Exp. Neurol. **27,** 645 – 658 (1968)
110. Secret, C. J., Hadfield, J. R., Beer, C. T.: Studies on the binding of (^{3}H) vincristine by rat blood platelets in vitro. Effects of colchicine and vincristine. Biochem. Pharmacol. **21,** 1609 – 1624 (1972)
111. Selawry, O. S., Holland, J. F., Wolman, I. J.: Effect of vincristine (NSC-67574) on malignant solid tumors in children. Cancer Chem. Rep. **52,** 497 – 500 (1968)
112. Shelanski, M. L., Wisniewski, H.: Neurofibrillary degeneration induced by vincristine therapy. Arch. Neurol. **20,** 199 –206 (1969)

113. Shirahama, T., Cohen, A. S.: Ultrastructural evidence for leakage of lysosomal contents after phagocytosis of monosodium urate crystals: a mechanism of gouty inflammation. Am. J. Path. **76,** 501 – 520 (1974)
113b. Shirahama, T., Cohen, A. S.: Blockage of amyloid induction by colchicine in an animal model. J. Exp. Med. **140,** 1102 – 1107 (1974)
114. Slater, L. M., Wainer, R. A., Serpick, A. A.: Vincristine neurotoxicity with hyponatremia. Cancer **23,** 122 – 125 (1969)
115. Slonim, R. R., Howell, D. S., Brown, H. E. Jr.: Influence of griseofulvin upon acute gouty arthritis. Arthritis Rheum. **5,** 397 – 404 (1962)
116. Stemmermann, G. N., Hayashi, T.: Colchicine intoxication. A reappraisal of its pathology based on a study of three fatal cases. Human Path. **2,** 321 – 332 (1971)
117. Stuart, M. J., Cuaso, C., Miller, M., Oski, F. A.: Syndrome of recurrent increased secretion of antidiuretic hormone following multiple doses of vincristine. Blood **45,** 315 – 320 (1975)
118. Sullivan, M. P.: Vincristine (NSC-67574) therapy for Wilms' tumor. Cancer Chem. Rep. **52,** 481 – 484 (1968)
119. Suskind, R. M., Brusilow, S. W., Zehr, J.: Syndrome of inappropriate secretion of antidiuretic hormone by vincristine toxicity (with bioassay of ADH level). J. Pediatr. **81,** 90 – 91 (1972)
120. Terry, R. D.: Neuronal fibrous protein in human pathology. J. Neuropath. Exp. Neurol. **30,** 8 – 19 (1971)
121. Tobin, W., Sandler, S. G.: Neurophysiologic alterations induced by vincristine (NSC-67574) Cancer Chem. Rep. **52,** 519 – 526 (1968)
122. Valeriote, F., van Putten, L.: Proliferation-independent cytotoxicity of anticancer agents: a review. Cancer Res. **35,** 2619 – 2630 (1975)
123. Wakem, C. J., Bennett, J. M.: Inappropriate ADH secretion associated with massive vincristine overdosage. Aust. N. Zeal. J. Med. **5,** 266 – 269 (1975)
124. Wallace, S. L.: Trimethylcolchicinic acid in the treatment of acute gout. Ann. Intern. Med. **54,** 274 – 279 (1961)
125. Wallace, S. L.: Colchicine and new antiinflammatory drugs in the treatment of gout. Arthritis Rheum. **18** (Suppl.) 847 – 852 (1975)
126. Wallace, S. L., Bernstein, D., Diamond, H.: Diagnostic value of the colchicine therapeutic trial. J. Am. Med. Ass. **199,** 525 – 528 (1967)
127. Wallace, S. L., Ertel, N. H.: Preliminary report: Plasma levels of colchicine after oral administration of a single dose. Metabolism **22,** 749 – 754 (1973)
128. Wallace, S. L., Nissen, A. W.: Griseofulvin in acute gout. New Engl. J. Med. **266,** 1099 to 1101 (1962)
129. Weiss, H. D., Walker, M. D., Wiernik, P. H.: Neurotoxicity of commonly used antineoplastic agents. New Engl. J. Med. **291,** 75 – 81; 127 – 133 (1974)
130. Whittaker, J. A., Parry, D. H., Bunch, C., Weatherall, D. J.: Coma associated with vincristine therapy. Brit. Med. J. **10,** 335 – 336 (1973)
131. Wisniewski, H., Karczewski, W., Wisniewska, K.: Neurofibrillary degeneration of nerve cells after intracerebral injection of aluminium cream. Acta Neuropathol. **6,** 211 – 219 (1966)
132. Wisniewski, H., Shelanski, M. L., Terry, R. D.: Experimental colchicine encephalopathy. I. Induction of neurofibrillary degeneration. Lab. Invest. **17,** 577 – 587 (1967)
133. Wisniewski, H., Terry, R. D.: An experimental approach to the morphogenesis of neurofibrillary degeneration and the argyrophilic plaque. In: Alzheimer's Disease and Related Conditions (eds.: G. E. W. Wolstenholme, M. O'Connor), pp. 223 – 240. London: Churchill 1970
134. Wisniewski, Terry, R. D.: Neurofibrillary pathology. J. Neuropath. Exp. Neurol. **19,** 163 – 176 (1970)
135. Zbinden, G., Hégélé, M., Grimm, L.: Toxicological effects of vincristine sulfate on blood platelets. Agents and Actions **2,** 241 – 245 (1972)
136. Zemer, D., Revach, M., Pras, M., Modan, B., Schor, S., Sohar, E., Gafni, J.: Colchicine to prevent attacks of familial Mediterranean fever. New Engl. J. Med. **291,** 932 – 933 (1974)
137. Zurier, R. B., Weissmann, G., Hoffstein, S.: Mechanisms of lysosomal enzyme release from human leukocytes. II. Effects of cAMP and cGMP, autonomic agonists, and agents which affect microtubule function. J. Clin. Invest. **53,** 297 – 309 (1974)
138. Zweig, M. H., Maling, H. M., Webster, M. E.: Inhibition of sodium urate-induced rat hindpaw edema by colchicine derivatives: correlation with antimitotic activity. J. Pharmacol. Exp. Ther. **182,** 344 – 350 (1972)

Chapter 12 Outlook

At the end of this survey of MT research, which is of necessity incomplete—more than one hundred interesting papers are published each year and only a fraction of the literature has been mentioned—, the task of covering the whole field with the purpose of indicating the well-established facts and the fields in which new research is needed, is by no means easy. MT were born at the time of the great change from pro- to eukaryotes, and for eons have been put to use by cells of all types, motile or immotile, undifferentiated or highly specialized, resting or dividing. It is not surprizing that under the apparent unity of their shape and chemistry a far greater complexity than was imagined ten years ago is becoming evident. The formation of more or less helical tubules is a quite general property of proteins, as exemplified by viruses, bacterical flagella, and some abnormal hemoglobins. The tubulins have maintained through evolution the property of becoming assembled into MT, which are in many conditions labile structures which can be disassembled when no longer required by the cell. These tubulins are also capable of assembling into other geometrical structures—rings, helices, sheets. Moreover, at a higher level, this polymorphism is maintained as indicated by the multiple patterns made of grouped MT. The typical cilia and flagella structure of 9+2 has played and is playing such an important role in cell motility that some authors have imagined that it may have preceded, as a symbiotic structure, that of single MT.

Confronted with the impressive mass of information gathered, in the last few years, about MT and tubulins, one senses clearly that this is a central problem of cell biology, comparable to the study of ribosomes, cell membranes, mitochondria, lysosomes, and other organelles. The purpose of an overall approach, as is attempted in this volume, is to bring together data covering many fields of research, and having MT as the common denominator. One danger of this method is the temptation, in the maze of ever-growing references, to simplify too much the MT concept. Although this has a heuristic value, it may overlook the incompleteness of many data and the hidden complexities of the various types of MT which have been observed in so many different cells. No attempt will be made in this last chapter toward a synthetic approach: the aim will be to point out some problems for future work and to discuss, this time without adding any more references to an already too long list, some of the aspects of MT research as they appear in 1977.

12.1 Unity or Diversity?

While all MT are clearly made of helically assembled proteins of about the same size, their biochemistry and their reactions are complex. Leaving aside at this moment the problem of the MT-associated proteins (MAPs), it is clear that two tubulins, α and β, exist in nearly all MT which have been analyzed. Two varieties of α tubulins have been described. The preparation of antibodies against tubulins and the techniques of visualizing MT by immunofluorescence or immunohistochemistry have been an important advance of the last few years. They have not confirmed that all tubulins are identical: in *Naegleria,* those which are required for ciliary growth differ from the cytoplasmic ones, and may perhaps also differ from the intranuclear MT involved in mitosis. Amino acid sequence analyses have indicated that α and β tubulins are closely related, and that tubulins from invertebrates do not differ essentially from those of man. It should however be pointed out that only a very limited number of tubulins has been investigated by these complex techniques and, from all that is known about the polymorphism of proteins, it would certainly be remarkable if some variation had not taken place in evolution. It appears premature to say that the tubulins have remained, like the histones, stable through millions of years. It is highly probable that when amino acid sequences are studied in more tubulins, some differences will become apparent, opening a whole new chapter of biochemical tubulin research.

The fact that specific antibodies can be prepared against tubulins is in itself interesting: it shows that there is no strong tolerance toward these ubiquitous proteins, a fact which is surprizing when one remembers that in mammals billions of telophasic bodies, made of MT, are discarded from erythroblasts each day. However, some tolerance is present, or tubulins are weakly antigenic, as adjuvants have always been required to obtain specific antibodies. The fact that in some pathological conditions antitubulins have been found in man is not surprizing, although it deserves further study, for this was found in diseases where various other auto-antibodies could be present, indicating a drop in natural tolerance.

The diversity of MT was first noted when "labile" and "stable" MT were observed in various cells. These great differences — for instance, between mitotic spindle MT and the telophasic bundle, or between axonemes of heliozoa and neurotubules — may be intrinsic or extrinsic. It appears probable that many stable MT are resistant to various chemicals or physical agents because they are embedded in a dense matrix made of other proteins: this is clearly the case with the telophasic bundle and probably with the centriole and basal bodies.

Opinions about stability and lability have often been obscured by the fact that MT were compared to more complex structures, such as ciliary doublets or centriolar triplets: the stability of these structures — quite relative, as indicated by the differential extraction and solubilization of MT from cilia — is most probably related to the association of tubulins with other proteins. But "simple" MT, with no specialized associated structures, show great differences in stability. One explanation is that the stable MT undergo only a slow (neurotubules) or no (erythrocytes) turnover, while the lability of spindle MT results from their dynamic condition, assembly and disassembly going on rapidly, the destruction of the MT — their fragility — resulting from a block in their renewal much more than from a direct action on assembled MT. However, this explanation cannot be generalized, for its is known that some MT — other than those of the mitotic spindle — can be rapidly disassembled by cold or high hydrostatic pressures. This is explained by the thermodynamics of tubulin assembly. However, many facts mentioned in previous chapters indicate that chemical influences may also disassemble MT without any drastic modification of the tubulin molecule. Very dilute solution of MT poisons (let us say, less than 10^{-5} M) may be effective, and recent work suggests that minute changes of Ca^{2+} concentration in the cell—in the micromolar range—could play a role in the equilibrium between tubulin and MT. It appears thus that the resistance of some MT, more than the fragility of others, deserves further consideration: of course, all stable MT may turn out to be maintained in this condition through associated proteins.

Morphological diversity is also apparent, for it is not demonstrated that all MT result from the helical assembly of 13 subunits. Without discussing the fact that the exact geometry of the MT helix with dimers of α and β tubulin requires studies of far more species before any generalization is possible, it has been clearly shown in recent years that MT with more or less than 13 subunits exist. All experimental work on tubulin assembly in vitro has also indicated the variety of possible associations of the tubulin molecules. The stability of MT will be better understood when the nature of the chemical links between the tubulin subunits is known. While many of the extraordinary patterns of MT in protozoa may be properly explained by geometrically defined links between MT with 13 subunits, it is far from certain that different numbers of subunits do not underlie some patterns.

The association of MT in complex arrays is related to the attachment, to the MT subunits, of various types of "arms", "bridges", and other extensions. The study of cilia has indicated that these have a definite localization in relation to the MT subunits: while this may be explained by factors exterior to the MT—for instance, in the case of the centrioles, by an "organizing" RNA similar in some senses to the ribosomal RNA—it may also suggest that the 13 subunits differ more one from another than is usually imagined. An explanation suggesting a MT as made of many different subunits assembled in a definite order appears to introduce an unnecessary complication; theories based on allosteric effects as shown in the axonemes of heliozoa may be more convincing.

The aim of tomorrow's research will be the unraveling of the detailed molecular structure of tubulins; and from that, the discovery of the nature and location of the several possible links between tubulin molecules. The role played by these proteins in so many biological functions and their ubiquity should justify an effort comparable to that which has led to the detailed structural analysis of other proteins. This will be the task of analytical chemistry: the morphological, ultrastructural data will then be easier to understand. However, such problems as the association of MT into complex patterns may well need a different approach, that of the new science of "supramolecular" biology.

12.2 Microtubule-Associated Proteins

MT are by definition tubular, rather rigid, non-contractile structures: when they are involved in various types of cell movement—ciliary beating, phagocytosis, movement of secretory granules, mitosis—it is through links with other cell structures. Even when MT play a prominent role in maintaining shapes, they must be attached to other cell constituents or fastened together to form more or less rigid structures, like the axonemes or the peripheral bundles of blood cells. These links are often clearly visible, as in cilia and centrioles, where at least two well-defined proteins, dynein and nexin, are present. The role of the ATPase dynein in ciliary movements and its connection with the sliding movements of MT in relation one to another is well understood. More and more information about cell movements is available now, and the presence of actin and myosin, and other muscular proteins (troponin, actinin) has been surmized from biochemical analysis and immunohistochemical techniques. The relations between these contractile proteins and the MT which may act as supports need however further studies. Some authors talk about a "MT-microfilament" system, but morphological studies often show that MF are located at a distance from MT, and in most cells no relation between actin and tubulin, comparable to that between dynein and ciliary doublets, has been demonstrated. The possible presence of actin in the mitotic spindle, where it would be closely associated with the MT, is still under discussion.

The nature of the other links between MT remains poorly understood in biochemical terms. These links are of various types. Nexin, already mentioned, plays a role in keeping together the triplets of centrioles and basal bodies. In cilia, two (at least) other types of links are well known: the radial spokes which extend from the doublets toward the central pair of MT, and the complex links (or sheath) which are found around this pair. All these links are located with the utmost precision, as clearly visible from longitudinal negatively stained pictures of cilia of flagella. They no doubt contribute to the mechanical properties of the cilium, the dynein arms providing the motive force. It may also be questioned whether the two dynein arms are identical, and whether the links observed between doublets 5 and 6 are not of a different nature.

Some authors have imagined that the interMT links—as seen for instance between the manchette MT of the spermatozoa—are identical with tubulin, and made of linearly assembled tubulin in molecules. The fact that these links may be modified by drugs such as colchicine was an indication of this similarity. However, more recent data on electron microscopy of MT assembled in vitro in the presence or absence of the high molecular weight (HMW) proteins which copurify with tubulin demonstrate that the fluffy material ordinarily seen around the MT is not present when tubulin is assembled without MAPs. This important observation indicates that the MAPs, which, it schould be remembered, purify stoichiometrically with tubulin, are closely associated with the MT. It does not however mean that all the bridges found between MT are of the same nature: a careful distinction should be made between lateral, irregularly shaped expansions of MT, and bridges firmly linking (as demonstrated by cell fractionation studies) MT one with another.

Once more, the danger of oversimplifying is evident. Merely because in cilia a special ATPase protein, dynein, is attached at regular intervals along the doublets, all similar extensions should not be imagined as contractile proteins. It would be presuming to think that the bridges between the axonemes of heliozoa, the radial links of cilia, and the interMT bridges of chick spermatocytes could be identical. They may all belong to the same family of proteins grouped now under the term of MAPs: the fact that more than twenty proteins may be associated with tubulin from some structures such as cilia is evidence that much remains to be learnt about these molecules.

Their role may be to link MT together, and also to govern their assembly, as suggested by the stoichiometric relations, and the role of MAPs in the formation of rings or other types of possible initiation sites for MT assembly.

Other proteins—perhaps also lipids or carbohydrates—may be closely related to MT. The nature of the clear exclusion zone which is visible in many different cells requires more information, and may help to understand the mechanism of some MT-associated movements. Other substances, very densely stained in electron microscopy, surround the triplets at the base of the centrioles and basal bodies, and the interlocked MT of the telophasic body: their nature in unknown. As the formation of MT—such as those of centrioles—often takes place in dense matrices, it could be imagined that these are nothing but unpolymerized tubulin. However, the absence of staining of the telophasic body by fluorescent antitubulins indicates that this is not correct.

MT have been shown to have definite links with many cytoplasmic structures: cell membranes, nuclear pores, Golgi apparatus, pericentriolar bodies, ribosomes. In some cases, this may be simply a continuity between MT and cell membranes (like in nervous cells); in others, a direct attachment of particles to the MT subunits, as in several types of viruses. But where the links either provide the force to bend MT into definite shapes (as in red blood cells, platelets, or insect eggs) or connect MT with particles of whatever nature which are carried along them—pigment granules, food particles, secretion droplets—it is most probable that this bridging is provided by special MT-associated proteins, which in some cases may be actin or closely related proteins, in other cases remain completely unidentified.

12.3 Assembly and Disassembly

One of the most extraordinary properties of MT is their rapid assembly—several μm in an hour—and disassembly—sometimes in a few seconds. Alone, this would not be so remarkable, for viral proteins are known to assemble at similar rates, and may build tubular structures. The strange fact about MT is that they form when and where the cells need them—implying some complex regulation of spatial localization—and that in many instances they may construct, always at the same speed, extraordinarily complex assemblies such as those described in the axonemes and the tentacles of many protozoa.

Let us consider first some problems related to assembly. As mentioned above, the exact mode of gathering of the subunits together to form tubules is not known, and it is quite possible that several mechanisms exist. As the MT are polarized helical structures, the fact that they grow with preference from one end is not surprizing; however, it is not an absolute rule, as experiments of assembly in vitro on preexisting MT have shown. Helical growth implies one or a quite limited number of places where new tubulin molecules become attached. This seems to be confirmed by experiments with poisons acting at very low dilutions, as if it would suffice to block a few growth points in order to arrest all assembly. However, it remains possible that in some, probably mostly artificial conditions, subunits may assemble linearly, the 13 rows of assembled subunits curving later to form the tubule. This model is more difficult to reconcile with the fact that in most cells the number of 13 subunits is found. Assembly of subunits along the MT has also been considered, and the simple fact that so many proposals have been presented indicates that several mechanisms may exist, although in living cells assembly from one extremity is evidenced by ciliary growth and regeneration, spindle fiber growth, and probably the growth of MT in non-dividing cells from the centriolar region towards the periphery.

This leads us to consider the role of the so-called MT organizing centers (MTOC): it is apparent in the formation of basal bodies, of asters in the fertilized egg, and in mitosis. However, it is also easy to think of many cells in which MT are formed independently of any preexisting structure, at least at the electron microscopical level. MTOC appear mainly to be necessary for the growth of complex structures such as cilia and the mitotic apparatus. In cilia, it is possible that some structures, difficult to observe, are present at the beginning of growth and explain the ninefold symmetry, which remains as mysterious as ever. In the formation of centrioles, better information should be gained about the dense bodies, deuterosomes, and other checker-board patterns from which these organelles with all their symmetry arise as if by magic. In mitosis, the pericentriolar bodies on one hand, the kinetochores on the other, behave as MTOC, and experimentally it can be shown that solutions of tubulin assemble in relation to these structures, which act as nucleating centers. Again, it would be unwise to compare the formation of single MT to that of complex assemblies of MT such as centrioles.

The main biological problem which requires more light is that of the location of "simple" MT: why, in stomatal plant cells does the equatorial group of MT appear asymmetrically when the future division is to be unequal; when and why do MT assemble in bundles, bent circularly, during platelet formation in megakaryocytes; why do the MT in spermatogenesis form at one time a spiral wrapping around the elongating nucleus, and later become parallel to the nuclear axis? These are problems of supramolecular coordination, closely akin to organogenesis and morphogenesis.

The problems linked with the assembly of MT are difficult and many remain without solution. First of all, it is evident that some kind of regulation must exist within the cell, as most studies indicate that the amount of non-polymerized tubulin is greater than that of the assembled MT: a mechanism must prevent the assembly of all tubulin and regulate the equilibrium between tubulin and MT. Similar controls are probable in the reverse direction, as shown by many observations that tubulin is not degraded or lost and is largely maintained in the cell in a reutilizable form.

A second problem is that of the links between tubulin molecules: experimental data indicate that tubulin is provided with many important sites, either for nucleotide fixation, or for attachment of molecules such as colchicine and vinblastine. Sulfhydryl groups are required for tubulin stability. The relation between these sites —four or more of them—and the sites of linkage of tubulin molecules one to another remains unknown. As the change of GTP to GDP on one site is necessary for tubulin assembly, it is suggested that the guanine nucleotides may play an important role in fastening the subunits one to another. However, this role remains obscure, and the arguments about the possible relation of tubulin to a cAMP-activated protein-kinase (which would phosphorylate either tubulin or some other substrate) do not throw much light on this problem. It remains doubtful whether tubulin itself has a kinase activity, and the protein-kinase may belong to the MAPs. The role of the guanine nucleotides may be of great importance in the regulation of tubulin assembly in the cell: data on the action of GDP on living cells, and on cAMP and cGMP, are an indication that the guanosine monophosphates may trigger the assembly of MT, and maybe even play a role in the synthesis of new molecules of tubulin. (This last problem will be left aside, as too few facts are known: the role of estrogens in ciliary growth, that of nerve growth factor, may be recalled here.)

Another factor which is more and more often mentioned as capable of controlling the formation of MT from tubulin is the concentration of calcium ions (magnesium is necessary for assembly, but does not appear to act as a controlling factor). The demonstration that MT could be assembled in vitro provided that the concentration of Ca^{2+} was sufficiently low was the first indication of the role of this ion. Further work has provided information on the possibility that a similar mechanism may be operating in the living cell. It had first been thought that the intracellular changes of Ca^{2+} concentration were too small to have an effect on tubulins. More recently, several authors have indicated that Ca^{2+} may modulate in vivo the assembly (and disassembly) of MT. The cytoplasmic level of Ca^{2+} could be controlled either by the mitochondria (which act as reservoirs for the ion) or by the smooth endoplasmic reticulum. When the role of Ca^{2+} in other forms of cell motility and its action on actomyosin is recalled, these facts acquire a larger dimension. Could it be imagined that Ca^{2+} controls at the same time the supportive structures (the MT) and the contractile proteins of the cell which are so often associated with the MT?

Another possibility of assembly control could be related to the MT poisons, which are substances of plant origin, and could be imagined to play some role in the parent plant. This may seem far-fetched, as the mitoses of *Colchicum* or *Catharanthus* are not affected by the alkaloids from these plants. It does indicate that molecules with a high specificity toward tubulins, attaching to definite sites, may be present in cells: the action of a hormone like melatonin on MT is a further suggestion that some control of MT assembly may be exerted in cells by small highly specific inhibitors of tubulin assembly.

12.4 Tubulins and MT Poisons

This brings us back to the beginning of MT research, the discovery of the specific and powerful action of drugs such as colchicine, vinblastine, and vincristine. One point should be stressed again at the start: too many research workers are nowadays using colchicine as a tool, and drawing conclusions from effects which may not at all be specific for MT. While the sensitivity of various cells to colchicine (and the *Catharanthus* alkaloids) varies considerably (mainly because of differences in cell permeability, as demonstrated recently by the study of "resistant" cells), any action which is obtained with concentrations greater than 10^{-5} M must be considered with caution, except if the authors have demonstrated a definite dose-action relationship. While colchicine and other poisons do become specifically attached to tubulin molecules, they may also affect other structures, in particular cell membranes: the action of *Catharanthus* alkaloids on nucleotide transport is one well-known effect which has no relation to tubulins.

Two main problems demand further research: the attachment sites of the drugs, and their respective action on tubulin molecules and on already assembled MT. Biochemical data with tritiated colchicine and VLB have made it clear that one dimer fixes one molecule of colchicine, and probably two of VLB. However, the most recent papers often mention that only 0.8 mol of colchicine is fixed per mol of dimer: this may be the result of technical difficulties. The *Catharanthus* alkaloids appear to have two sites of fixation, which differ in their properties. While studies on the fluorescence of colchicine-tubulin complexes, and some investigations on the changes of the tubulin molecule after fixation of colchicine promise to lead us to a better understanding of the reaction (role of hydrophobic bonds, importance of the methoxy groups of the alkaloid), the exact configuration of the colchicine to tubulin bonds must await the elucidation of the molecular structure of tubulin. One group of poisons—some of the earliest studied—, the –SH reagents, do modify tubulin in a chemically defined way, as several sulfhydryl groups are known to be present in the molecule, and some of these are apparently needed for a proper assembly of MT. It remains surprizing that the action of drugs such as arsenicals has not interested cytologists for so many years, with only a few exceptions. The recent work on diamide is a move in this direction.

It is generally thought that colchicine binds to tubulin molecules and prevents their assembly into MT. When assembled, MT only fix a hundredfold smaller amount of colchicine than tubulin subunits. It is often concluded from these findings—which are related to the action of colchicine on in vitro assembly of tubulin—that the apparent "destruction" of MT which is one of the striking features of the action of this drug can only be observed in cells where the MT are in a condition of "dynamic equilibrium" with tubulin—as is probably the case in mitosis—and that colchicine cannot affect assembled MT. It is true that the MT of centrioles and cilia are resistant to the drug; these are not simple MT but complex assemblies of MT which may be stabilized by other proteins. Neurotubules (the MT of nerve cells), and also the marginal bundle of erythrocytes, do show "single" MT which are only affected by relatively large doses of colchicine. However, it has been shown by many experimental examples that colchicine—like other MT poisons—can sometimes rapidly lead to the disappearance of MT, and at this moment it is difficult to affirm that this does not result from disassembly, and that colchicine may not influence assembled MT. This problem could find its solution if the turnover of the tubulin molecules were better known: colchicine would lead to the disappearance of MT only in cells where their turnover was rapid.

Another aspect of the action of these drugs is the relation between their chemical structure and their specificity. The importance of several methoxy groups is apparent from a comparison of the structural formulas of colchicine, *Catharanthus* alkaloids, and podophyllotoxin. Recent research has however introduced some much simpler molecules with powerful antitubulin effects.

Simpler derivatives of colchicine have also been recently studied, and the relation between the stereochemical shape of the drug and its specific attachment to tubulin should be explained in the coming years.

The MT poisons should also be considered from quite another point of view, that of drugs useful in cancer chemotherapy. It is surprizing that colchicine, which is active at such small concentrations, has not been used more in this field: a simple explanation of this may be the very slow metabolism of the drug, leading rapidly to toxicity because of slow excretion and destruction. The *Vinca* poisons also show how a small modification of the chemical structure can influence the medical uses: VLB is mainly interesting for its favorable action on platelet formation, and in the treatment of Hodgkin's disease, while VCR, although rather toxic, has gained an important place in the cancer armamentarium. Its surprizing effects, in some subjects, on the secretion of antidiuretic hormone, remain unexplained: one would like to link them with some modification of axonal flow in the hypothalamo-posterior pituitary axis, but then one would have expected, from the experimental data, a decreased secretion of the hormone.

12.5 MT and Membranes

The relations of MT with cell membranes should be considered from two different points of view: first, the physical connection between constituents of membrane and MT; second, the possibility that membranes contain proteins similar to tubulin and capable of fixing colchicine.

Many observations, some quite recent and obtained by novel methods of pretreatment of tissues before fixation, have shown that MT have close connections with cell membranes, and may end in intimate contact with the cell surface. Intracellular membranes are also linked with MT, as indicated by the changes in the Golgi apparatus and the endoplasmic reticulum observed in cells after destruction of the MT: in this respect, MT appear to behave like a skeleton keeping in place the various organelles. Close connections with the nuclear pores have also been mentioned.

The relations of MT with the nuclear membrane are of particular interest for those who have attempted to reconstitute the evolution of mitosis. As mentioned in Chapter 10, the connections of MT and nucleus may take place inside the nucleus, where MT play a role in the elongation which is one of the main changes in some primitive mitoses, or outside, when MT are seen to be attached to condensations of the nuclear membrane to which the chromosomes are fixed through what appear to be the precursors of true kinetochores. These connections have become further complicated by the presence of MTOC, like the polar plaques and eventually the centrioles and spindle. Later in mitosis, the role of MT in plants is well documented: preparing from elements of the endoplasmic reticulum the new cell plate separating the two daughter cells.

The translation of cell surface receptors, as observed in the capping phenomenon in leukocytes or other cells, is affected by MT poisons. Although the opinions about the action of colchicine are somewhat contradictory—and this probably depends on the type of cell studied, and the type of receptor—there are enough positive data to indicate that MT play some role in the capping of receptors, that is their gathering at one pole of the cell, the prelude to their later internalization. This implies that close connections must exist between MT and proteins located in the cell membrane. These observations should be compared to those of the authors who, after purifying membranes of various cells, in particular synaptosomal structures, found that this fraction fixed ^{3}H-colchicine. It is not impossible that tubulin molecules, or other proteins capable of fixing colchicine, were present in the membranes, although it appears more probable, as some ultrastructural data show a close contact between membranes and MT, that the procedures for purifying the membrane fraction entail the mixture of tubulin proteins with those of the membrane.

These observations on membranes should be studied in connection with the recent findings that in transformed neoplastic cells the amount of MT—as demonstrated by immunofluorescence techniques—is smaller than in the same cell lines before virus-induced neoplastic change. The importance of modifications of the cell surface in malignancy, as evidenced by the loss of contact inhibition of mitosis and invasive growth, is more and more emphasized. It is too early to draw conclusions from the published facts, which concern only a limited series of cell lines. It would however be quite interesting to find out if these changes of MT are truly linked with malignancy, and whether they may play a role in the chemotherapeutic effects of the MT poisons. Anyhow, the studies on capping have indicated that the cellular changes found in the hereditary Chediak-Higashi's disease—in animals as well as in man—are related to disturbances of MT assembly and function, maybe through an abnormal control by the cyclic nucleotides.

12.6 MT and Movement

Many facts indicate close relations between MT and movement: the beating of cilia or flagella, the movements of food particles along axonemes, the sliding of secretory droplets along intracellular MT, the rapid changes of place of melanin granules in pigment cells, the movements of chromosomes and of the spindle constituents during mitosis; all these facts, often beautifully illustrated by motion pictures, link MT with movement. Moreover, the wide use of MT poisons and of physical agents affecting MT—cold, high hydrostatic pressure—has brought evidence that the "destruction" of MT arrests many forms of movement. MT were also found to be capable of lengthening and shortening rapidly through assembly and disassembly, and although it was soon understood that MT had no contractile activity, these changes of length could be imagined as motors for various types of intracellular displacement. However, it was often found that movement could be arrested without any visible changes of the MT, and many types of cell movements were known to take place without any MT being present: the best known is cyclosis. The discovery that actomyosin, and other proteins similar to those of muscle, were constituents of nearly all cells—even of neurons and red blood cells—has marked an important turning-point in this field, showing that the microfilaments visible in many cells could be involved, much more than the MT, in contractility and other forms of movement. Of course, in cilia and flagella the discovery of dynein and the sliding filament theory of motion had already indicated that the MT bend under the influence of externally applied forces, and have no real motor function.

In the chapter on mitosis, the role of MT in the displacement of chromosomes at preprophase as well as at anaphase has been discussed. Although cytologists do not yet agree about the presence of proteins of the actin type in the spindle, the hypothesis that the assembly and disassembly of MT could alone provide the energy required for chromosome movement is debatable. Of course, it should be remembered that chromosomes move very slowly and need a minute amount of energy for their displacements, but the countinuous flow of matter, as exemplified by the motion of small particles in the metaphase spindle, and by the movement of UV-irradiated zones of the spindle, does not seem to be directly related to MT.

More and more, attention is turning towards the proteins associated with MT, to the side-arms and lateral expansions of these, and to hitherto poorly explained connections between MT and proteins of the actomyosin group. It is probable that in a short time many of the MT-associated movements may find an explanation as clear as that of ciliary bending under the influence of the dynein links. What then will be the role of MT? Are they really necessary for motion? The experiments on neurons are an indication that motion can be arrested withouth obvious change (at the electron microscopial level) of the MT; many other facts show that motion is related to cytoplasmic MF, and that its inhibition by MT poisons—often used at rather high concentrations—is only a secondary effect, linked with the changes of intracellular architecture resulting from the absence of MT. These appear more and more as supports for motion, not as motors. This does not decrease their importance: mitosis without the guidance of MT is, as the work on colchicine showed forty years ago, impossible.

12.7 Scaffolding by MT

The main properties of MT which explain most of their activities, and the role they have been playing in the eukaryotic cells for millions of years, are those of rapidly assembled, more or less rigid, tubular elements, made of a few standard proteins present in all cells, and rapidly taken apart (disassembled) when their role has ended. Such may have been the first MT, and only later did they assemble with other proteins to build complex structures such as the centrioles and cilia, organelles of the greatest utility for cell motion, and much more stable than MT. In unicellulars, MT associated to form also beautiful patterns which combined rigidity, necessary to maintain the long axopodia used in feeding, with a remarkable plasticity, demonstrated by their disassembly in a few minutes and their rapid assembly.

The tubular structure of these organelles—comparable to that of other tubular assemblies of proteins or lipids—gives them a certain strength and rigidity, about which more quantitative data are desirable. In some conditions MT are seen to break, while in several types of cells they are bent with sharp curves: the blood platelets, and the feeding tentacles of *Tokophyra* show this flexibility of MT. Their spiral wrapping around the nucleus in some types of spermatogenesis is another instance of their flexibility. As was indicated in Chapter 6, all these conditions where MT are found to be curved or bent imply some other factors exerting a force on them and maintaining a shape which is not that of MT as observed alone in vitro.

In these conditions, MT play a role in the general rigidity of the cell, but this differs from that of a skeleton by the fact that MT may vanish, changed into their subunits, in a short time. The term "scaffolding" seems best to describe their function, for they resemble in some ways the rapidly assembled tubular scaffoldings used by man while erecting complex structures. The comparison indicates also that the tubes alone could not fulfil their function: they must be assembled and fastened together by specialized links to create a structure, even quite ephemeral. But man's tubules remain as such when the scaffolding has been pulled apart, while those of the cell disperse in small, globular subunits which remain in the cytoplasm—for exactly how long?—ready to build new MT—under the effect of what controls?

This property of rapid disassembly has found its principal use in mitosis, where the "northern lights" phenomenon is an indication that throughout most of mitosis—at least in the higher types of cells—MT are continuously undergoing formation and destruction. This dynamic condition is another aspect of MT physiology. Here, a most complex type of scaffolding has to be assembled and slowly modified as the chromosomes move towards the poles. Slowly, if compared to other types of movement; quite rapidly, if the turnover of individual MT is considered, although here again precise quantitative data are lacking. Recent findings on the decrease of spindle size after very short treatments with colchicine are an indication that rapid exchanges are controlled with great precision, as only some of the tubulin molecules are involved in the spindle. It is this dynamic condition—made still more spectacular by cinematographic recording—which suggested that the MT themselves are the causes of motion, while it is more and more probable that other proteins are involved. However, in primitive types of mitosis, it is reasonable to think that the growth of an intranuclear bundle of MT may be the motor force pushing the nuclear extremities apart, and many telophasic spindles retain this property, indicating that the "Stemmkörper" theory was not ungrounded. But even here, it could be argued that the main function of the spindle is to guide the lengthening nucleus, to maintain its shape and its bipolarity, rather than to "push" apart its two ends.

In one of the first reviews on MT, written ten years ago, Porter insisted on the role of these organelles in maintaining the asymmetrical shapes of cells: while much has been learnt since, this conclusion remains valid today.

12.8 MT and Evolution

Very little is known about the transition from pro- to eukaryotes which took place about a billion years ago. This change was most probably gradual, but the transition links are missing today. Great changes took place in cytoplasmic structures: the protein synthesis machinery, while using the same code as before, and remaining compatible with that of prokaryotes, modified its ribonucleic acids and the proteins of the ribosomes; the genetic material became separated from the cytoplasm by a membrane, and the nucleus was born; new mechanisms of bipartition of the nuclear membrane had to be devised, and these became linked with new organelles, the MT. While one may find some tubular structures similar to MT in a prokaryote [14], the link between mitosis, even in its most primitive forms, and MT appears fundamental. Some authors have suggested that this is not so, and that MT, existing in prokaryotes, were taken up by eucaryotes as symbiotic structures, and that the 9+2 (or 9+0) structure of the cilium may have preceded the single MT found in eukaryotic cells: this hypothesis awaits confirmation, and it would be surprizing if the extremely complex structure of cilia had preceded the much simpler one of MT. Cilia require the association with MT of various other proteins, while simple MT, in their most primitive form, may result from the assembly of a single protein in tubular form. Of course, the cell would have needed a machinery not only to control the synthesis of tubulin, but also to assemble and disassemble it at will. This machinery could well have originated at the same time as the mitotic separation of chromosomes.

The question whether the first MT were intranuclear or extranuclear does not seem of great importance, as the nuclear membrane, even in the complex cells of today, is a transient structure made of cytoplasmic membranes. There is no reason to believe that in primitive eukaryotes the separation between the genetic material and the cytoplasm was as definite as in most cells now: the main difference with prokaryotes was that the genome was no longer attached to the cell wall, and its bipartition became more or less separated from that of the whole cell. In mitoses which appear "primitive" the chromosomes (or more exactly the genetic material) are attached to the nuclear membrane, and MT play their role in the elongation of the nucleus or in becoming attached, from outside the nucleus, to the sites of chromosome attachment.

It is most probable that these sites became specialized zones of the chromosomes, from which MT were assembled, and which could possibly control this assembly. This would be under genetic control, and one could suppose that the earliest MT were assembled in relation to these MTOC, while later, similar MTOC became more independent of the chromosomes in the form of polar plaques and, much later, pericentriolar structures. This hypothesis should await further experimental data on the presence of particular genes located in the kinetochore region, and possibly of nucleic acids present in other MTOC.

RNA has been demonstrated repeatedly in basal bodies and probably in centrioles. These structures, with their considerable complexity and their peculiar symmetry, need complex controls for their assembly and multiplication. While centrioles and basal bodies are essentially structures related to cell motility and cilia formation, and quite accessorily to mitosis, their symmetry remains a problem for further research. The role of the central bodies present in basal bodies—cart-wheel and other links—and of the RNA are most interesting objects for research.

During the billion years of MT evolution, many changes may have taken place: the differentiation of two types of tubulins is the best known. The variability of tubulins, which has been discussed above, and which is becoming more and more evident, would be natural for structures which were to assume functions as different as the skeleton of the rays of heliozoa, the force-transmitting organelles of arthropodia cuticules, the flagella of the spermatozoa with all their structural varie-

[14] MT 10 nm in diameter and 150 nm long have been described in one species of bacteria *(Anabaena)* by T. E. Jensen, R. P. Ayala, Arch. Microbiol. **111,** 1 – 5 (1976)

ties, the modified cilia of sensory cells, and the several forms of mitotic MT. Throughout this monograph, the concept of MT as organelles of similar or comparable structure and chemical build-up in many cells of all phyla of eukaryotes has been a leading theme which has enabled us to find our way through the multitude of publications of the last decade and to propose some general conclusions about these organelles. Simplicity is always attractive in science; future work usually shows that facts are more intricate than synthetic approaches suggest. Biology in particular has accustomed us to a degree of complexity which defies the imagination. Contemplating some of the most beautiful forms of MT assemblies—flagella, axonemes—one cannot refrain from thinking that the reality is still more fantastic than the long story which has been gathered in these pages.

List of Abbreviations

ATP	adenosine triphosphate
cAMP	cyclic adenosine monophosphoric acid
CMTC	cytoplasmic microtubular complex
C-MT	C-shaped microtubules
Con	concanavaline
dbcAMP	dibutyryl cyclic adenosine monophosphoric acid
EDTA	ethylene-diamine tetraacetic acid
EGTA	Ethylene glycol bis (β-aminoethylether)N,N,N′,N′,-tetraacetic acid
IPC	isopropyl phenylcarbamate
HVEM	high voltage electron microscope
HMW	high molecular weight
K-MT	kinetochore-microtubules
MA	mitotic apparatus
MAP	microtubule-associated proteins
MF	microfilaments
MT	microtubules
MTOC	microtubule organizing center
NF	neurofibrils
NGF	nerve growth factor
SEM	scanning electron microscope
TEM	transmission electron microscope
VCR	vincristine
VLB	vinblastine (vincaleukoblastine)
VLDL	very low density lipoproteins
SIADH	syndrome of inappropriate secretion of antidiuretic hormone

Subject Index

Italized numbers refer to the legends of figures. Numbers in **bold type** refer to headings of chapters or subchapters.
Italized names refer to Latin names of animals or plants.

Results and Problems in Cell Differentiation

A Series of Topical Volumes in Developmental Biology

Editors: W. Beermann, W. Gehring, J. B. Gurdon, F. C. Kafatos, J. Reinert

Volume 1

The Stability of the Differentiated State

Editor: H. Ursprung

1968. 56 figures. XI, 144 pages
ISBN 3-540-04315-2

This volume brings together authoritative reviews on the subject from the point of view of tissue culture (both in vitro and in vivo), regeneration, dedifferentiation, metaplasia, clonal analysis, and somatic hybridization. The ten contributing authors are actively working in the respective areas of interest and have written short, concise accounts of their own and related work. The book is intended for advanced students who want to keep up with recent developments in these areas of basic biology which are so important for the understanding of the development of higher organisms, both normal and malignant. The book as a whole uncovers similarities in experimental systems as widely different as plants and insects and provides new insights into fundamental processes.

Volume 2

Origin and Continuity of Cell Organelles

Editors: J. Reinert, H. Ursprung

1971. 135 figures. XIII, 342 pages
ISBN 3-540-05239-9

A great deal of precise information is available on structure and function of cell organelles. But much controversy exists on the question of their origin, and the mode of their replication. This book contains authoritative reviews relevant to this question on the following components of cells: mitochondria, vacuoles, desmosomes, centrioles, polar granules, Golgi apparatus, endosymbionts, plastids, microtubules, lysosomes, and membranes in general.

Volume 3

Nucleic Acid Hybridization in the Study of Cell Differentiation

Editor: H. Ursprung

1972. 29 figures. XI, 76 pages
ISBN 3-540-05742-0

The method of molecular hybridization of nucleic acids has made it possible to obtain more precise information on quantitative and – in some cases at least – qualitative composition of the genome. Both beginners and advanced students will find the book helpful for understanding the principles of the method, the various techniques that are used, and the biological implications of the results already obtained.

Volume 4

Developmental Studies on Giant Chromosomes

Editor: W. Beermann

1972. 110 figures. XV, 227 pages
ISBN 3-540-05748-X

This volume is devoted to the question of structural and functional organization of the eucaryotic genome in relation to cell differentiation as revealed by current work on the genetic fine structure and synthetic activities of giant polytene chromosomes and their subunits, the chromosomeres.

Results and Problems in Cell Differentiation

A Series of Topical Volumes in Developmental Biology

Editors: W. Beermann, W. Gehring, J. B. Gurdon, F. C. Kafatos, J. Reinert

Volume 5

The Biology of Imaginal Disks

Editors: H. Ursprung, R. Nöthiger

1972. 56 figures. XVII, 172 pages
ISBN 3-540-05785-4

The book summarizes the current state of knowledge concerning the biochemistry and ultrastructure of the imaginal disk and reviews experiments on the mechanisms of cell determination as studied in imaginal disks.

Volume 6

W. J. Dickinson, D. T. Sullivan

Gene-Enzyme Systems in Drosophila

1975. 32 figures, XI, 163 pages
ISBN 3-540-06977-1

This book brings together for the first time biochemical, genetic, and developmental work, on enzymes in Drosophila. It documents the usefulness of gene-enzyme systems as tools in the investigation of numerous problems, particularly those relative to the organization and regulation of the eukaryotic genome. Research methods are described as well as the impact of the findings upon other fields such as population genetics, evolution, biochemistry, and physiology.

Volume 7

Cell Cycle and Cell Differentiation

Editors: J. Reinert, H. Holtzer

1975. 92 figures. XI, 331 pages
ISBN 3-540-07069-9

The concept of the cell cycle is a relatively recent one but it plays an important part in modern cytology. In this book eighteen well-qualified authors, writing from a variety of points of view, seek to resolve the inconsistencies between the concept of the cell cycle and current ideas on cell differentiation.

Volume 8

Biochemical Differentiation in Insect Glands

Editor: W. Beermann

1977. 110 figures, 24 tables. XII, 215 pages
ISBN 3-540-08286-7

The spinning glands and other tissues with secretory functions in Lepidopterons and Dipterons are highly specialized in the production of one or only a few different proteins in huge quantities. Moreover, as these insects can easily by raised and handled in the laboratory they provide a unique possibility to perform cytogenetic and molecular investigations on genomes and to study the formation of giant cells with polyploid and polytene nuclei.

Springer-Verlag
Berlin
Heidelberg
New York